Systems Modeling and Computer Simulation

1. Rational Fault Analysis, *edited by Richard Saeks and S. R. Liberty*
2. Nonparametric Methods in Communications, *edited by P. Papantoni-Kazakos and Dimitri Kazakos*
3. Interactive Pattern Recognition, *Yi-tzuu Chien*
4. Solid-State Electronics, *Lawrence E. Murr*
5. Electronic, Magnetic, and Thermal Properties of Solid Materials, *Klaus Schröder*
6. Magnetic-Bubble Memory Technology, *Hsu Chang*
7. Transformer and Inductor Design Handbook, *Colonel Wm. T. McLyman*
8. Electromagnetics: Classical and Modern Theory and Applications, *Samuel Seely and Alexander D. Poularikas*
9. One-Dimensional Digital Signal Processing, *Chi-Tsong Chen*
10. Interconnected Dynamical Systems, *Raymond A. DeCarlo and Richard Saeks*
11. Modern Digital Control Systems, *Raymond G. Jacquot*
12. Hybrid Circuit Design and Manufacture, *Roydn D. Jones*
13. Magnetic Core Selection for Transformers and Inductors: A User's Guide to Practice and Specification, *Colonel Wm. T. McLyman*
14. Static and Rotating Electromagnetic Devices, *Richard H. Engelmann*
15. Energy-Efficient Electric Motors: Selection and Application, *John C. Andreas*
16. Electromagnetic Compossibility, *Heinz M. Schlicke*
17. Electronics: Models, Analysis, and Systems, *James G. Gottling*
18. Digital Filter Design Handbook, *Fred J. Taylor*
19. Multivariable Control: An Introduction, *P. K. Sinha*
20. Flexible Circuits: Design and Applications, *Steve Gurley, with contributions by Carl A. Edstrom, Jr., Ray D. Greenway, and William P. Kelly*
21. Circuit Interruption: Theory and Techniques, *Thomas E. Browne, Jr.*
22. Switch Mode Power Conversion: Basic Theory and Design, *K. Kit Sum*
23. Pattern Recognition: Applications to Large Data-Set Problems, *Sing-Tze Bow*
24. Custom-Specific Integrated Circuits: Design and Fabrication, *Stanley L. Hurst*
25. Digital Circuits: Logic and Design, *Ronald C. Emery*
26. Large-Scale Control Systems: Theories and Techniques, *Magdi S. Mahmoud, Mohamed F. Hassan, and Mohamed G. Darwish*
27. Microprocessor Software Project Management, *Eli T. Fathi and Cedric V. W. Armstrong (Sponsored by Ontario Centre for Microelectronics)*
28. Low Frequency Electromagnetic Design, *Michael P. Perry*
29. Multidimensional Systems: Techniques and Applications, *edited by Spyros G. Tzafestas*
30. AC Motors for High-Performance Applications: Analysis and Control, *Sakae Yamamura*
31. Ceramic Materials for Electronics: Processing, Properties, and Applications, *edited by Relva C. Buchanan*
32. Microcomputer Bus Structures and Bus Interface Design, *Arthur L. Dexter*
33. End User's Guide to Innovative Flexible Circuit Packaging, *Jay J. Miniet*
34. Reliability Engineering for Electronic Design, *Norman B. Fuqua*
35. Design Fundamentals for Low-Voltage Distribution and Control, *Frank W. Kussy and Jack L. Warren*

36. Encapsulation of Electronic Devices and Components, *Edward R. Salmon*
37. Protective Relaying: Principles and Applications, *J. Lewis Blackburn*
38. Testing Active and Passive Electronic Components, *Richard F. Powell*
39. Adaptive Control Systems: Techniques and Applications, *V. V. Chalam*
40. Computer-Aided Analysis of Power Electronic Systems, *Venkatachari Rajagopalan*
41. Integrated Circuit Quality and Reliability, *Eugene R. Hnatek*
42. Systolic Signal Processing Systems, *edited by Earl E. Swartzlander, Jr.*
43. Adaptive Digital Filters and Signal Analysis, *Maurice G. Bellanger*
44. Electronic Ceramics: Properties, Configuration, and Applications, *edited by Lionel M. Levinson*
45. Computer Systems Engineering Management, *Robert S. Alford*
46. Systems Modeling and Computer Simulation, *edited by Naim A. Kheir*

Additional Volumes in Preparation

Transformer and Inductor Design Handbook, Second Edition, Revised and Expanded, *Colonel Wm. T. McLyman*
Signal Processing Handbook, *edited by C. H. Chen*

Electrical Engineering-Electronics Software

1. Transformer and Inductor Design Software for the IBM PC, *Colonel Wm. T. McLyman*
2. Transformer and Inductor Design Software for the Macintosh, *Colonel Wm. T. McLyman*
3. Digital Filter Design Software for the IBM PC, *Fred J. Taylor and Thanos Stouraitis*

Systems Modeling and Computer Simulation

edited by

Naim A. Kheir

Oakland University
Rochester, Michigan

Marcel Dekker, Inc.

New York • Basel

Library of Congress Cataloging-in-Publication Data

Systems modeling and computer simulation.

    (Electrical engineering and electronics ; 46)
    Bibliography: p.
    Includes indexes.
    1. Computer simulation. 2. System design. I. Kheir,
Naim A.   II. Series.
QA76.9.C65S975   1988   001.4'34   87-27615
ISBN 0-8247-7812-X

MARCEL DEKKER, INC.
270 Madison Avenue, New York, New York 10016

Current printing (last digit):
10 9 8 7 6 5 4 3 2 1

PRINTED IN THE UNITED STATES OF AMERICA

To my grandparents
Who instilled faith in me

To my parents
Who taught me persistence and perseverance

To Ferial, Karen, Osama, and Trish
Who made my life more meaningful

Foreword

Anyone engaged in the writing or editing of a text dealing with modeling
and simulation is faced with the very difficult task of cutting the vast subject down to manageable proportions. The term "model" connotes a relatively
concise and abstract description of some aspects of a real-world object—the
system. All real-world systems have a myriad of physical, biological,
economic, aesthetic, etc., attributes. It is the essence of the art of modeling to assemble only that small subset, of all the characteristics or features
of the system, that is sufficient to serve the specific objectives of the
model as set by the modeler. In analyzing the dynamic characteristics of
an automobile, for example, the modeler might specify the magnitudes of all
masses, springs, and dampers as well as the manner in which these are
linked to each other. One would probably not include in the model the
electrical or chemical properties of the components, or their appearance, or
their cost. This kind of abstraction from reality is fundamental to all formal methods for system design or performance analysis and prediction. It
can be argued in fact that all intellectual activities involve modeling of sorts,
since we always make conceptual simplifications in order to comprehend a
real-world object. The construction of models with an acceptable validity
requires a profound understanding of the system being modeled, the
scientific area or discipline to which it belongs, and the techniques or
methodologies available for the detailed specification of the model.

Computer simulation entails the implementation of system models on a
computer so as to permit experimentation. In order to assure that the
computer generates a sufficiently accurate solution, i.e., that the computer
outputs are sufficiently close to the corresponding features of the system
being simulated, it is necessary to have a thorough understanding of computer technology and methodology.

System modeling and simulation therefore encompass a very broad
range of subjects, including among others:

Formulation of the modeling and simulation objectives
Construction of models in the many subareas and disciplines of the
    physical, technical, life, and social sciences

Validation of model(s) to assure that the system being modeled has
been adequately characterized

Adaptation of the models to the characteristics of the computer, using
numerical analysis, discretization, algorithms, etc.

Algorithm selection and evaluation

Programming, coding, and debugging of the (digital) computer
implementation

Verification and error analysis of the computed (simulation) results

Systematic experimentation using the simulator

Documentation

An exhaustive treatment of the subject would therefore include a survey of the formal representation of systems arising in all the physical, life,
and social sciences, a comprehensive study of a large number of mathematical techniques, as well as a major part of what has come to be known as
computer engineering/science. Since such an all-encompassing approach is
patently impractical, all editors and authors of textbooks dealing with
modeling and simulation must perforce focus their attention upon one or a
few of the above subjects to make judicious selections of topics from each
of these areas. The wisdom and skill exercised in delimiting the subject
matter very largely determine the ultimate usefulness of the resulting text.

In editing the present volume, Professor Kheir has chosen to address
readers with considerable insight into the systems that they are interested
in modeling and simulating but with relatively limited expertise in computers.
Relevant computer oriented topics are, therefore, developed in depth in
separate chapters dealing with data structures, programming languages,
operating systems, and the architecture of multiprocessor systems. Other
major chapters are devoted to the methodology of system simulation including the implementation of continuous system models, characterized by
ordinary differential equations or transfer functions and discrete-time
models. Special attention has also been given to the treatment of large-
scale systems as well as discrete-event systems. These subjects are
articulated in Part III, devoted to case studies taken from such diverse
areas as control systems, robotics, manufacturing, and economic systems.
This selection of subject matter makes the scope of this book unique among
available texts in the modeling and simulation field. Because of its depth
and breadth this text is suitable for formal university courses at the beginning graduate level as well as for practicing professionals in engineering
or the sciences interested in modeling and simulating complex systems.
Professor Kheir has rendered a most valuable service to the simulation community in assembling, unifying, and honing the chapters constituting this
important contribution to the technical literature.

Walter J. Karplus
Professor
University of California
Los Angeles, California

Preface

The popular and expanding use of computer models and simulation has been on the rise for the last three decades or so. The ever greater power of the ever smaller computers will continue to contribute significantly to the activities in systems design, analysis, development, modeling, simulation, and management. There is hardly a discipline or an industry that does not use simulation. More and more emphasis is being placed upon the technological advancement in accuracy, capacity, speed, and reliability, as well as convenience. Although accurate estimates of current worldwide expenditures on simulation activities are not available, it is not difficult to agree that it is probably in the billions of dollars annually.

Simulation is an almost indispensable tool when it comes to the study of complex and/or large systems. Among the applications where simulation has been widely used and accepted are those in civilian and military aerospace, manufacturing, computer-aided design, economics, health/medicine, power/energy, and education. The degree of accepting simulation results, however, depends on the extent of experience with simulation in the field of the specific application.

The objectives of using computer models and simulation range from design optimization to training (as with video flight simulators), safety, cost effectiveness, forecasting, even entertainment. The use of simulators (with or without man-in-the-loop) in a wide variety of complex applications has grown to the extent that it warranted the recent birth of an annual specialized conference. The cover of our book depicts an aerial refueling simulator.

A good number of publications on mathematical modeling and simulation (transactions, journals, and proceedings) exists. The literature provides designers, researchers, educators, and decision-makers with information about current efforts and development in: methodology, simulation languages, computer simulation techniques and hardware; verification, validation, and credibility of models; computer-aided engineering (CAE) and computer-aided design (CAD) as well as specific applications. However, when it comes to textbooks for mainly senior and/or first graduate level courses for engineers (regardless of their discipline), there seems to be a need for a text that gives an up-to-date comprehensive treatment of the important aspects of

systems modeling and simulation. This book addresses this need and
also should serve as a text for self-study or a reference for those already
in professional roles (engineers and scientists) from industry, government,
and the military.

Systems Modeling and Computer Simulation has two unique features:
the first is concerned with the overall understanding of the fundamentals
of modeling and simulation as well as those of software and hardware systems; the second is the scope and number of the carefully selected applications. The six applications chapters and the two appendices are expected
to bring to the classroom, as well as to the practicing simulationists, a
unique enriching awareness and experience. A detailed presentation of a
few case studies with simulation results and discussion are included.
References are expected to be most helpful for further readings. The
topics included here have never before been collected or given a unified
presentation in one volume. Because of the nature of the material and the
diversity of the applications, a number of outstanding contributors have
participated in the development of this interdisciplinary book. Emphasis
has been placed on continuity throughout the text.

The presentation is in a manner understandable to individuals who have
basic familiarity with differential and integral calculus, probability, elementary statistics, and computer programming. The book contains a large number of examples, figures, problems, and references.

Systems Modeling and Computer Simulation could serve as:

1. A textbook for senior and/or first-year graduate level courses in
   modeling and simulation for engineering students (regardless of
   their discipline) and possibly computer science majors (depending
   on graduate program requirements and students' background,
   which might be in mathematics, sciences, or engineering). To a
   lesser degree, selected chapters could be useful to those majoring
   in management, economics, or business administration.
2. A supplementary textbook in a general course on systems or
   CAE/CAD.

For a three-hour one-semester course, a carefully balanced syllabus can
be structured depending on the course prerequisites and objectives. Chapters 1, 2, 4, 5, part of 6, and 10 are suggested as a core; a couple of
applications may be chosen as a supplement. In a two-course sequence,
students are expected to be able to cover most of the material in Parts I
and II with as many applications chapters and/or appendices as time allows.
It is my hope that this book will provide enough material and opportunity
for such an educational experience, via self-study and/or classroom instruction. At the end of this preface is a list of available textbooks. In our
judgment none of these duplicates the thrust or has the unique features of
our text. Although some overlap exists, for the most part all these references deal with relatively narrow specialties (continuous systems, discrete-
event systems, methodology or statistics, etc.) or they are directed toward
specialized applications and/or specialized computational problems.

I am indebted to those individuals and organizations who have, in
various ways, contributed to my professional growth and to those who
made the completion of this book possible. In particular, I would like to
acknowledge the support and encouragement extended to me by Dean
Richard Wyskida, School of Engineering, Professor Alex Poularikas,

Chairman of the ECE Department, and Dr. Ned Audeh, Dean of Graduate Studies, all at the University of Alabama in Huntsville. The opportunities for almost a decade of funded research and consulting with the U.S. Army Missile Command and the Army Research Office are also gratefully acknowledged. Special thanks are due to Professors R. Husson and J. M. Kauffmann of the Research Center of Automatic Control of Nancy, France for my sabbatical opportunity, which made it possible to concentrate on this project. I am also grateful to the Society for Computer Simulation, with special thanks to Professor Walter Karplus of UCLA for his helpful and significant suggestions on the textbook project, as well as to Stewart Schlesinger of the Aerospace Corporation. Also, thanks are due to Professor M. Mansour, the Swiss Federal Institute of Technology (ETH), Zurich, Dean Hans R. van Nauta Lemke, Delft University, the Netherlands, and Professor Ralph Gonzalez, University of Tennessee, Knoxville, for their encouragement and useful discussions during the early phases of this project. I am also indebted to several anonymous reviewers for their constructive criticism, suggestions, and enthusiasm, and to my graduate students who read parts of the manuscript.

Special thanks are due to my coauthors for their contributions, cooperation, and understanding, and to Dr. Eileen Gardiner, Executive Editor of Marcel Dekker, Inc., for her encouragement and continued interest in this project. Also, my thanks are due to Brian Black, Production Editor of Marcel Dekker, Inc., for his help, coordination, and patience. I would also like to thank Mrs. Peggy Byrd for her typing and retyping several sections of this text.

An affectionate thanks to my wonderful wife, Ferial, for her encouragement, faith, counsel, patience, and companionship.

Naim A. Kheir

BIBLIOGRAPHY

Proceedings

MODELING and SIMULATION Proceedings of the Annual Pittsburgh Conference (currently publishing Vol. 16).
Proceedings of the Mathematics and Computers in Simulation, North-Holland (currently publishing Vol. 27).
Proceedings of the Summer Computer Simulation Conferences (currently publishing Vol. 17).
Proceedings of the Winter Computer Simulation Conferences (currently publishing Vol. 17).
SIMULATION Journal of the Society for Computer Simulation (currently publishing Vol. 45).
Simulation Series of the Society of Computer Simulation (currently publishing Vol. 15).
TRANSACTIONS of the Society for Computer Simulation (currently publishing Vol. 2).

Textbooks

Aris, R. (1979). *Mathematical Modeling Techniques*, Pitman Advanced Publishing Company.
Banks, J. and Carson II, J. S. (1984). *Discrete-Event System Simulation*, Prentice-Hall.

Bennett, A. W. (1974). *Introduction to Computer Simulation*, West Publishing Company.

Chu, Y. (1969). *Digital Simulation of Continuous Systems*, McGraw-Hill Book Co.

Close, C. M. and Frederick, D. F. (1978). *Modeling and Analysis of Dynamic Systems*, Houghton Mifflin Company.

Fishman, G. S. (1972). *Concepts and Methods in Discrete Event Digital Simulation*, Wiley-Interscience Publication.

Franta, W. R. (1977). *The Process View of Simulation*, North-Holland (operating and programming system series).

Giloi, W. K. (1975). *Principles of Continuous System Simulation*, B. G. Teubner, Stuttgart.

Gordon, G. (1978). *System Simulation*, 2nd ed., Prentice-Hall.

Kleijnen, J. P. C. (1985). *Statistical Tools for Simulation Practitioners*, Marcel Dekker, Inc., New York.

Law, A. M. and Kelton, W. D. (1982). *Simulation Modeling and Analysis*, McGraw-Hill Book Co.

Lewis, T. G. and Smith, B. J. (1979). *Computer Principles of Modeling and Simulation*, Houghton Mifflin Co.

MacFarlane, A. G. J. (1970). *Dynamical System Modeling*, George G. Harp and Co.

Maryanski, F. (1980). *Digital Computer Simulation*, Hayden Book Co., Inc.

McLeod, J. (1982). *Computer Modeling and Simulation: Principles of Good Practice*, Simulation Series Vol. 10, No. 2, The Society for Computer Simulation.

Nash, P. (ed. in chief) (1981). *Systems Modeling and Optimization*, Peter Peregrinus Ltd.

Ord-Smith, R. J. and Stephenson, J. (1975). *Computer Simulation of Continuous Systems*, Cambridge University Press.

Rosko, J. S. (1972). *Digital Simulation of Physical Systems*, Addison-Wesley Publishing Co.

Shannon, R. E. (1975). *Systems Simulation: The Art and Science*, Prentice-Hall.

Spriet, J. A. and Vansteenkiste, G. C. (1982). *Computer–Aided Modeling and Simulation*, Academic Press.

Zeigler, B. P. (1976). *Theory of Modeling and Simulation*, Wiley-Interscience New York.

Zeigler, B. P. (ed. in chief) (1979). *Methodology in Systems Modeling and Simulation*, North-Holland Publishing Co.

Zeigler, B. P. (1984). *Multifaceted Modeling and Discrete-Event Simulation*, Academic Press.

Contents

Foreword   v

Preface   vii

Contributors   xxi

Part I   FUNDAMENTALS OF SYSTEMS MODELING AND
         SIMULATION   1

Chapter 1   MOTIVATION, DEFINITIONS, AND AN OVERVIEW   3

   1.1   Background and Rationale   3
   1.2   Systems Modeling   5
         1.2.1   Definitions and Overview   5
         1.2.2   Concept of a Model and Model Building   7
         1.2.3   Model Classification and Mathematical
                 Representations   11
         1.2.4   Identification   13
         1.2.5   Nonlinearities and Discontinuities   14
   1.3   Computer Simulation: An Overview   16
         1.3.1   The Computer   17
         1.3.2   Evolution of Computers   18
         1.3.3   Simulation Software   19
   1.4   Outline of Text   21
         References   25

Chapter 2   CONTINUOUS-TIME AND DISCRETE-TIME SYSTEMS   29

   2.1   Introduction   29
   2.2   Continuous-Time Linear Systems   29
         2.2.1   Simple Electrical Circuits   30
         2.2.2   The Laplace Transform   33
         2.2.3   Transfer Functions   38
         2.2.4   State-Space Models   45
         2.2.5   Systems Representation   49
         2.2.6   Order of System   59

2.2.7 Reduced-Order Models 60
2.2.8 Model Overdescription 61
2.3 Discrete-Time Systems 62
2.3.1 The Z-Transform and Its Inverse ... 65
2.3.2 α-β Tracking System 71
2.4 Some Related Topics 74
2.4.1 Feedback Systems 74
2.4.2 Stability 79
2.4.3 Controllability and Observability ... 88
Problems 91
References 95

Chapter 3   DISCRETE-EVENT SYSTEMS 97

3.1 Introduction 97
3.2 Principles of Discrete-Event Systems Simulation ... 98
3.3 Discrete-Event Simulation Languages 100
3.4 Statistical Models in Simulation 101
3.4.1 Discrete Distribution 103
3.4.2 Continuous Distributions 105
3.4.3 Poisson Processes 112
3.4.4 Empirical Distributions 113
3.5 Queueing Models 114
3.5.1 Characteristics of Queueing Systems ... 114
3.5.2 Queueing Notation 115
3.5.3 Performance Measures for Queueing Systems ... 115
3.6 Inventory Systems 120
3.6.1 Deterministic Models 120
3.6.2 Probabilistic Models 123
3.7 Random Numbers 125
3.7.1 Properties of Random Numbers 125
3.7.2 Random Variate Generation 128
3.8 Analysis of Simulation Data 130
3.8.1 Input Data Analysis 130
3.8.2 Verification and Validation 132
3.8.3 Output Analysis 132
3.9 Summary and Conclusions 133
Problems 133
References 135

Chapter 4   COMPUTER SIMULATION 136

4.1 Introduction 136
4.2 Digital Representation of Signals 137
4.3 Numerical Integration 138
4.4 Errors in Numerical Integration 140
4.5 State Space Simulation Techniques 141
4.6 Simulation of Discrete-Time Systems 146
4.7 Digital Simulation of Transfer Functions 147
4.8 Digital Simulation Languages 151
4.9 Analog Simulation 155
4.9.1 Comparison of Analog and Digital Computers ... 155
4.9.2 Operational Amplifier 155
4.9.3 The Summing Amplifier 156

4.9.4	The Integrating Amplifier	157
4.9.5	Analog Simulation of Linear Systems	162
4.9.6	Magnitude Scaling	165
4.9.7	Time Scaling	167
4.9.8	Simulation of Nonlinear Systems	168
4.9.9	Simultaneous Equations	168
4.9.10	Transfer Function Simulator	169
4.9.11	State Variable Systems	170
4.10	Hybrid Simulation	171
4.11	Summary and Conclusions	173
	Problems	174
	References	177

Chapter 5 PRINCIPLES OF DESIGN AND ANALYSIS OF SIMULATION EXPERIMENTS 179

5.1	Introduction	179
5.2	Design of Simulation Experiments	180
5.2.1	Input Parameters	180
5.2.2	Estimation of Parameters	180
5.2.3	Main Effects and Interactions of Factors	181
5.2.4	Simulation Runs and Replications	182
5.2.5	Response Surface Methodology	185
5.2.6	Factorial and Fractional Factorial Designs	186
5.3	Analysis of Simulation Experiments	189
5.3.1	Static Simulation Output Analysis	189
5.3.2	Dynamic Simulation Output Analysis	190
5.3.3	Steady-State Simulation	190
5.3.4	Confidence Interval for Steady-State Simulation	190
5.3.5	Terminating Simulation	191
5.3.6	Confidence Interval for Terminating Simulation	192
5.3.7	Comparison of Alternative Models	193
5.3.8	Output Analysis for Two Models	193
5.3.9	Analysis of Variance	196
5.3.10	The F-Test	196
5.3.11	Multiple Comparisons	196
5.3.12	Multiple Rankings	196
5.3.13	Spectral Analysis	197
5.3.14	Sequential Sampling	197
5.4	Variance Reduction Techniques	197
5.4.1	Common Random Numbers	198
5.4.2	Antithetic Variates	199
5.4.3	Control Variates	204
5.4.4	Stratified Sampling	205
5.4.5	Importance Sampling	206
5.5	Summary and Conclusions	209
	Problems	209
	References	211

Chapter 6 INTRODUCTION TO LARGE-SCALE SYSTEMS 213

6.1	Introduction	213
6.2	Modeling and Model Reduction	214

	6.2.1	Aggregation	215
	6.2.2	Perturbation	232
6.3	Hierarchical Control		237
	6.3.1	Goal Coordination—Interaction Balance	239
	6.3.2	Interaction Prediction	242
6.4	Decentralized Control		247
	6.4.1	The Stabilization Problem	248
	6.4.2	Fixed Modes and Polynomials	249
	6.4.3	Dynamic Compensation	252
6.5	Structural Properties of Large-Scale Systems		256
	6.5.1	Stability	256
	6.5.2	Controllability and Observability	257
6.6	Summary and Conclusions		258
	Problems		258
	References		261

PART II   SOFTWARE/HARDWARE SYSTEMS   263

Chapter 7   DATA STRUCTURES AND DATABASES   265

7.1	Introduction		265
7.2	Basic Data Structures		267
	7.2.1	Elements	267
	7.2.2	Arrays	268
	7.2.3	Tables (Multidimensional Arrays)	269
	7.2.4	Records	271
	7.2.5	Sets	272
7.3	Advanced Data Structures		274
	7.3.1	Linked Lists	276
	7.3.2	Singly and Doubly Linked Lists	278
	7.3.3	Stacks and Queues	278
	7.3.4	Trees and Graphs	280
7.4	Databases		281
	7.4.1	Purpose of Database Systems	281
	7.4.2	Overview of Database Systems	283
	7.4.3	Database Applications	287
7.5	Conclusions and Trends		288
	References		289

Chapter 8   PROGRAMMING LANGUAGES   291

8.1	Introduction		291
8.2	Programming Language Evolution		292
8.3	Programming Language Data-Related Features		295
	8.3.1	Data Structures	296
	8.3.2	Data Referencing Environment	298
8.4	Programming Language Control Structures		300
	8.4.1	Control Structures for General Use	300
	8.4.2	Time Control in Simulation Languages	302
8.5	Support Environments for Programming Languages		306
8.6	Criteria for Programming Language Selection		307
8.7	Assessment of Some Languages for Use in Discrete-Event Simulation		310

8.8	Trends	311
	Problems	314
	References	314

Chapter 9 OPERATING SYSTEMS 319

9.1	Nature of an Operating System	319
	9.1.1 View as a Resource Manager	321
	9.1.2 View as an Interface Between the User and the Machine	323
	9.1.3 View as a Hierarchy	323
	9.1.4 Operating System Software and System Utility Software	326
9.2	Effect of the Operating System on Simulation	326
9.3	Concurrency, Process Synchronization, and Scheduling	327
	9.3.1 Effects of Scheduling on Simulation Software	327
	9.3.2 Desirability of Concurrency in Simulations	330
	9.3.3 Problems with Concurrent Processes Sharing Data	330
	9.3.4 Methods for Controlling the Access to Shared Data	333
	9.3.5 Methods for Communicating Between Processes	336
9.4	File Systems	339
	9.4.1 Organization of System's Files	340
	9.4.2 Access Methods and I/O Throughput	340
9.5	Input/Output	342
	9.5.1 Use of the I/O System by Simulation Software	342
	9.5.2 Handling of Simulation's I/O Requests by the Operating System	343
	9.5.3 Software Problems in Interfacing Special-Purpose Hardware to Real-Time Simulation Systems	343
9.6	Memory Management	344
	9.6.1 Concepts of Virtual Addressing and Virtual Memory	345
	9.6.2 Effects of Segmentation and Paging on Simulation Software	351
	Summary	351
	References	352

Chapter 10 HARDWARE AND IMPLEMENTATION 355

10.1	Introduction	355
	10.1.1 Analog Versus Digital Computers	355
10.2	Digital Computer Model	356
	10.2.1 Hardware	356
	10.2.2 Software	358
	10.2.3 Firmware	360
10.3	A Hypothetical Computer	361
	10.3.1 Organization	361
	10.3.2 Data Format	362

10.3.3	Instruction Set	363
10.3.4	Instruction Format	363
10.3.5	AHC Assembler	364
10.4	Types of Digital Computer Systems	367
10.4.1	Architecture Classification	369
10.5	A/D and D/A Converters	371
10.5.1	Analog-to-Digital Converters	371
10.5.2	Digital-to-Analog Converters	372
10.6	Sampling/Quantization Effects	374
10.6.1	Sampling Effects	374
10.6.2	Quantization Effects	375
10.7	Implementation Aspects	376
10.7.1	Analog Simulation	376
10.7.2	Analog Computer Programming	377
10.7.3	Amplitude and Time Scaling	382
10.7.4	Digital Simulation	384
10.8	Example of Digital Control System Using Microcomputer	388
10.8.1	Robot Arm Model	388
10.8.2	Analog/Digital Simulation	390
10.8.3	Programming 2920 Signal Processor	391
	Problems	394
	References	396

PART III   APPLICATIONS   399

Chapter 11   DIGITAL CONTROL SYSTEMS   401

11.1	Introduction	401
11.2	Basic Digital Control Structures	403
11.3	Design Approaches	405
11.3.1	Modeling	405
11.3.2	Design Methods	413
11.3.3	Cascade Compensation	415
11.3.4	Feedback Compensation	418
11.3.5	PID Controllers	419
11.3.6	State-Space Controllers	422
11.3.7	Quantitative Feedback Technique	423
11.4	Implementation	439
11.4.1	Effect of Finite Word Length	439
11.4.2	Computational Delays	443
11.4.3	Software	443
11.5	Summary	444
	Problems	445
	References	447

Chapter 12   ROBOTICS   449

12.1	Introduction	449
12.2	Background	450
12.2.1	Description and Functional Structure of a Robot	450
12.2.2	Programming of Industrial Robots	453

12.3	Modeling of Robots	455
	12.3.1 Mathematical Background	455
	12.3.2 Kinematic Modeling of Robots	462
	12.3.3 Kinematic Equations Solution	468
	12.3.4 Jacobian Model	471
	12.3.5 Dynamic Models of Robots	478
12.4	Control of Robots	486
	12.4.1 Joint Variable Determination	486
	12.4.2 Positional Control	486
	12.4.3 Resolved Motion Rate Control	487
	12.4.4 Resolved Acceleration Control	487
	12.4.5 The Computed Torque Technique	489
	12.4.6 Adaptive Control	490
12.5	Applications	492
	12.5.1 Robot Vision	492
	12.5.2 Welding by Robot	492
	12.5.3 Product Assembly Automation	492
	12.5.4 Other Applications	493
12.6	Current Directions of Research in Robotics	493
	12.6.1 Mobile Robots	493
	12.6.2 Vision and Scene Analysis in Conjunction with the Robot	493
	12.6.3 Artificial Intelligence	494
	12.6.4 Programming	494
	Problems	494
	References	496
Chapter 13	**WORLD MODELING: CONCEPTS AND APPLICATION**	**499**
13.1	Introduction	499
13.2	Birth of World Models	499
13.3	Approaches to World Modeling	501
	13.3.1 Regions and Sectors	502
	13.3.2 Variables in Model	502
	13.3.3 Demand, Supply, and Equilibrium	502
	13.3.4 Dynamic and Static Models	503
13.4	World Models	503
	13.4.1 World Economic Models	503
	13.4.2 Food and Agriculture Models MOIRA	521
	13.4.3 Environmental Models	523
	Long-Term Global Energy—CO_2 Model	525
13.5	Case Study Using the World Integrated Model	527
	13.5.1 AGB's Sustained-Growth Objective: The Problem	527
	13.5.2 Management Options	529
	13.5.3 Strategy Formulation	529
	13.5.4 Strategy Mix	530
	13.5.5 International Market Assessment	531
	13.5.6 Conclusion	538
	13.5.7 System Features Used in the Case Study	538
	References	539

Chapter 14　ECONOMIC SYSTEMS 541

14.1　Introduction 541
14.2　Econometric Model Development 543
　　14.2.1　Macroeconomic Model Structure 543
　　14.2.2　Model Parameter Estimation 547
　　14.2.3　Dynamic Characteristics 550
14.3　Economic Policy Application 551
14.4　Economic Forecasting 555
　　14.4.1　Forecasting Applications 557
　　14.4.2　"No-Change" Forecast 558
　　14.4.3　"Average" Forecast 558
　　14.4.4　"Exponential" Forecast 558
　　14.4.5　"Least Squares" Forecast 559
　　14.4.6　"Low-Order System" Forecast 560
　　14.4.7　Large Econometric Model Forecasts 562
14.5　Conclusions 563
　　Problems 564
　　References 565

Chapter 15　MANUFACTURING SYSTEMS SIMULATION USING SIMAN 567

15.1　Introduction 567
15.2　General-Purpose Modeling Features of SIMAN 568
15.3　Characteristics of Manufacturing Systems 574
15.4　Manufacturing Modeling Features of SIMAN 575
　　15.4.1　Modeling Workstations 575
　　15.4.2　Macro Submodels 575
　　15.4.3　Visitation Sequences 577
　　15.4.4　Resource Schedules 578
　　15.4.5　Modeling Materials-Handling Systems 579
15.5　Cinema: An Animation System 581
　　15.5.1　The Animation Layout 582
　　15.5.2　Runtime Features 590
15.6　An Example Application 590
　　15.6.1　Overview of the Problem 590
　　15.6.2　The SIMAN Model 591
　　15.6.3　The Cinema Layout 592
　　15.6.4　Results 592
15.7　Conclusions 596
　　References 596

Chapter 16　MULTIPROCESSOR SYSTEMS 597

16.1　Introduction 597
16.2　Enhancements to SISD 598
　　16.2.1　Control Unit 598
　　16.2.2　Arithmetic-Logic Unit 598
　　16.2.3　Memory Unit 599
　　16.2.4　Input/Output Unit 599
16.3　Multiple Processor Architectures 600
　　16.3.1　Single Instruction Multiple Data 600
　　16.3.2　Multiple Instruction, Single Data 602
　　16.3.3　Multiple Instruction, Multiple Data 604

      16.3.4 Multiprocessor Architecture Implications 609
16.4 Simulation-Oriented Commercial Systems 610
      16.4.1 Floating-Point Systems' AP-120B 610
      16.4.2 Applied Dynamic International's AD-10 612
      16.4.3 Intel's iPSC 616
16.5 Real-Time Multiprocessor Simulator 619
      16.5.1 Hardware 619
      16.5.2 Firmware 623
      16.5.3 Programming Language 624
      16.5.4 Operating Systems 626
      16.5.5 A Simulation Example 628
16.6 Summary 629
      References 629

APPENDIX A COMPUTER-AIDED CONTROL SYSTEMS:
TECHNIQUES AND TOOLS 631

A.1 Introduction 631
A.2 Development and Classifications of CACSD
     Techniques 637
A.3 Development and Classification of CACSD Tools 642
A.4 CACSD Tools—A Survey 648
A.5 Standardization Versus Diversification 660
A.6 Simulation and CACSD 663
A.7 Outlook 668
     References 674

APPENDIX B SIMULATION SOFTWARE: A SURVEY 681

Index 695

Contributors

A. Wayne Bennett, Ph.D. Professor and Head, Electrical and Computer Engineering Department, Clemson University, Clemson, South Carolina, Chapter 4

William Biles, Ph.D. Professor and Chairman, Industrial Engineering Department, Louisiana State University, Baton Rouge, Louisiana, Chapter 3

Francois E. Cellier, Ph.D. Associate Professor, Electrical and Computer Engineering Department, University of Arizona, Tucson, Arizona, Appendix A

Deborah A. Davis Graduate Student, Pennsylvania State University, University Park, Pennsylvania, Chapter 15

Ratan K. Guha, Ph.D. Professor, Computer Science Department, University of Central Florida, Orlando, Florida, Chapter 7

James W. Hooper, Ph.D. Associate Professor, Computer Science Department, The University of Alabama in Huntsville, Huntsville, Alabama, Chapter 8

Pei Hsia, Ph.D. Professor, Computer Science Engineering Department, The University of Texas at Arlington, Arlington, Texas, Chapter 7

Rene H. C. Husson, Ph.D. Professor and Director, Automatic Control Research Center, National Polytechnic Institute, Nancy, France, Chapter 12

Mohammad Jamshidi, Ph.D. Professor and Director, CAD Laboratory for Systems and Robotics, Electrical and Computer Engineering Department, The University of New Mexico, Albuquerque, New Mexico, Chapter 6

Naim A. Kheir, Ph.D.* Professor and Chairman, Electrical and Systems Engineering Department, Oakland University, Rochester, Michigan, Chapters 1 and 2

Previous Affiliation: Professor, Electrical and Computer Engineering Department, The University of Alabama in Huntsville, Huntsville, Alabama

Gary Lamont, Ph.D.* MSEE Visiting Professor, Computer Science and Computer Engineering, Wright State University, Dayton, Ohio, Chapter 11

M. Mesarovic, Ph.D. Professor, Systems Engineering Department, Case Western Reserve University at Cleveland, Cleveland, Ohio, Chapter 13

Mahmoud Mohadjer, Ph.D. Assistant Professor, Electrical and Computer Engineering Department, The University of Alabama in Huntsville, Huntsville, Alabama, Chapter 10

Bidhu B. Mohanty, Ph.D. Quantitative Business Analysis Department, Cleveland State University, Cleveland, Ohio, Chapter 13

Dennis Pegden, Ph.D. Associate Professor, Industrial Engineering Department, Pennsylvania State University, University Park, Pennsylvania, Chapter 15

Magnus Rimvall, Ph.D. Assistant Professor, Automatic Control Department, Swiss Federal Institute of Technology, Zurich, Switzerland, Appendix A

Yosef S. Sherif, Ph.D. Jet Propulsion Laboratory, California Institute of Technology, Pasadena, California, Chapter 5 and Appendix B

Sajjan G. Shiva, Ph.D. Professor, Computer Science Department, The University of Alabama in Huntsville, Huntsville, Alabama, Chapters 10 and 16

Lawrence W. Taylor, Jr. President, Investment Analysis Company, Williamsburgh, Virginia, Chapter 14

John J. Zenor, Ph.D. Professor, Computer Science and Computer Engineering Department, California State University at Chico, Chico, California, Chapter 9

Current Affiliation: Professor, Air Force Institute of Technology, School of Electrical Engineering and Computer Engineering Department, Wright-Patterson Air Force Base, Dayton, Ohio

Systems Modeling and Computer Simulation

Part I

Fundamentals of Systems Modeling and Simulation

1

Motivation, Definitions, and Overview

1.1 BACKGROUND AND RATIONALE

Basically simulation is the process by which understanding of the behavior
of an already existent (or to be constructed) physical system is obtained by
observing the behavior of a model representing the system. Thus, simulation is justly considered the art and science of experimenting with models.
A simulation study must have a purpose and there are many good reasons
why simulation is valuable. For example, simulation may be performed to
check and optimize the design of a system before its construction, thus
helping to avoid costly design errors and ensuring safe designs. Other
purposes include analysis, performance evaluation, tests of sensitivity, cost
effectiveness, forecasting, safety, man-in-the-loop training, teaching, and
decision making.

The diversity and interdisciplinary nature of modern complex technical
and nontechnical problems almost require some form of systems modeling and
simulation. Popularity and expansion of these activities have been on the
rise over the past three decades or so. The continuing advancement in
computer technology in terms of accuracy, speed, capacity, and reliability,
coupled with cost reduction, is expected to contribute to even more use of
simulation using mainframe, personal, and supercomputers. Also, the impact
of the tremendous growth in development of simulation languages and software packages should be recognized.

Examples of modern society's areas of applications include:

1. Technical/engineering: space flights, aircraft design, radar,
    communication, power/energy, chemical processes, manufacturing
    processes, controls, robotics and manipulators, computer-aided
    engineering (CAE), computer-aided design (CAD), battlefield envi-
    ronment, aviation safety traffic control, etc.
2. Science-oriented: human eye visual system, brain cell growth,
    nervous system, anticancer drugs, disease control, physics of the
    atmosphere, planetary systems, surface water studies, etc.
3. Humanities/social sciences: economics, labor force, food and agri-
    culture, natural resources, banking, politics, population growth,

learning processes, human behavior, health industry, migration in urban and regional systems, water quality in streams, etc.

Characterization of the reliability of simulation studies by the hardness or softness of the underlying discipline has been discussed by Karplus (1976). Studies in the engineering and physical sciences (hard systems) are expected to yield numerical results with reliable accuracy. In the humanities and social studies (soft areas), however, studies undertaken help identify trends and critical areas rather than quantifying social or behavioral phenomena. Karplus also arranges the various fields of applications, which utilize mathematical models, according to the grayness of the box representing them. In this regard, electric circuits define the "white box" end of the spectrum and economic and social problems are on the "black box" end.

It is interesting at this point to note a basic distinction between design and modeling. The major goal in "design" is to build a real system to meet a set of performance specifications. Thus, many configurations are tried in order to avoid overlooking promising ideas. This includes allowing basic changes in the way system elements (components) are connected; additional components may be added and/or system values (parameters) may be subjected to change. Zeigler (1979) mentioned that while design is the manipulation of elements, and thus represents real possibilities, it relies on "credible models" for its efficacy. In practice, only when simulation results of a specific design are acceptable may a "prototype" of the system be produced.

On the other hand, in "modeling" there exists, in general, a real system whose behavior has been partially observed, and the goal is to reproduce, through simulation, this known behavior and be able to predict future behavior. A more concrete discussion of systems and their modeling is presented in the next section.

With this in mind, let us now focus on the work environment especially in modern industrialized societies. It is not difficult to note that the evolution over the past three decades or so not only has affected the hardware and software available for simulation, but also has resulted in changes in the types of people practicing the art of simulation. We now see engineers, scientists, economists, and others involved in modeling, analysis, management, and/or simulation activities of complex projects. Adequately specialized training for its newly hired engineers. Complicating the situation is the fact that most existing textbooks (see the bibliography at the end of ized training for its newly hired engineers.) Complicating the situation is the fact that most existing textbooks (see the bibliography at the end of the Preface) and courses on simulation seem to be too specialized. Recognition of this situation provided the motivation for this text.

Systems Modeling and Computer Simulation is developed with students as well as those already in professional roles from industry, government and the military in mind. This book has two unique features: The first is concerned with the presentation of systems modeling and computer simulation techniques as well as the needed fundamentals of software and hardware systems. We deal with these topics in Parts I (six chapters) and Part II (four chapters). The second is the scope of the six carefully selected applications of Part III. The two appendices present languages and simulation packages for computer-aided design of control systems, and a survey of simulation software. More on the text will be discussed in Section 1.4.

1.2 SYSTEMS MODELING

1.2.1 Definitions and Overview

A "system" (reality) is usually defined as any ordered set of interrelated physical (or abstract) objects. Our interest in a specific system may be in one or more of these activities: analysis, design, control, or improved understanding/performance. By "modeling" we mean the study of the mechanisms inside a system, and through using basic physical (biological, economic, etc.) laws and relationships, a model is inferred. Models are therefore not reality, and a model, no matter how complex, is only a representation of reality and should never be confused with it (Bekey, 1977). As we discuss below, observations representing the behavior of a system (if they exist) are essential in assessing the quality of a model and the success of a simulation.

So, what is being called a system's model in Figure 1.1 is actually a reflection of the modeler's understanding of the reality, its components, and their interrelations. The "computerized model" is an operational computer program that implements a system's model. A record of predicted behavior of the system is obtained from computer run(s). Measurements, on the other hand, make it possible to obtain a record (table, graph, etc.) of observations of physical system behavior. Before one gets involved in the quality of the correspondence between observed and predicted simulation data, an effort related to model verification is necessary. "Model verification" is defined as the "substantiation that a computerized model represents the system's model within specified limits of accuracy" (see Society for Computer Simulation's report by its Technical Committee on Model Credibility, 1979). Now, the level of agreement between observed and predicted behavior of the physical system is the essence of "model validation." Validating a model requires, as shown in Figure 1.1, comparing its behavior (simulation results) with that of the real system (measured or observed data). This presupposes, as implied earlier, that the verification step has been performed to avoid confusing faults of the program with faults in the model. It is obvious that the evolution of an acceptable (valid) model is an iterative procedure (Kwatny and Mablekos, 1975); in other words, a model is modified so as to reduce the differences between model and system behaviors.

One should keep in mind that the nature of the system (technical, nontechnical, etc.) is helpful in defining the "range of accuracy" of the computerized model for a certain domain of applicability. For instance, trying to get a quick feeling for the predominant effects out of a certain study is a very different situation from requiring the most accurate predictions of future behavior, and the ranges of accuracy are therefore expected to be different.

Differences between model behavior and corresponding real-world behavior are well illustrated by an example of a helicopter. The problem here is a result of "the model being removed further and further from the observations." Sources of this difficulty (Bekey, 1977) are mainly due to:

1. Observations and measurements of airframe and rotor behavior contain errors due to imperfect instrumentation and the presence of measurement noise.
2. Model simplifications may be due to omitting those aspects of helicopter behavior which are not well understood or for which data are difficult or impossible to obtain.

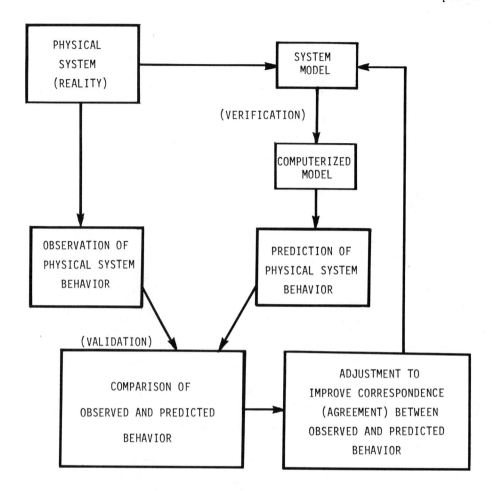

Figure 1.1   Building a credible model.

3. Further model simplification may be needed to make solution computationally feasible.
4. Linearization of model equations.

In summary, always remember that simulations are, in general, partial
in the sense that they never represent the entire system. One, therefore,
should keep in mind what has been omitted and what assumptions have been
made in modeling.

Zeigler (1976) has presented "degrees" of strength of validity if the
system under consideration does exist. A "replicatively" valid model is one
that generates data that matches data acquired from the real system. "Predictively" valid model applies when simulation data accurately match data
to be acquired from the system. A model is "structurally" valid if, in
addition to its ability to reproduce observed real-system behavior, it reflects
the way in which the system operates to produce such behavior. Additional
discussion on model credibility and validation can be found in Kheir et al.
(1985), Kheir (1978), and Sargent (1979).

In practice, documentation to communicate information concerning a model's credibility should not be taken lightly. Such information may include a statement of purpose for which the model has been built, a description of the model and its computerized version, specification of the domain of applicability, and a description of tests conducted for model verification and validation. Once a "model user" accepts the above-outlined documentation as adequate evidence that a computerized model can be effectively used for a specific application, "model acceptance" is finally at hand (SCS report by its Technical Committee on Model Credibility, 1979).

1.2.2 Concept of a Model and Model Building

"System" is a key word in this book and can be viewed as the collection of interacting elements for which there are cause-and-effect relationships among the variables. These elements (parts, components, or subsystems) are included inside a specified boundary (often arbitrary). "Dynamic" systems are those where one or more aspects of the system are time-dependent (vary with time) and whose behavior can be described by differential equations. Figure 1.2 illustrates the concept of a dynamic system. The boundary of such a system is usually chosen for a convenient conceptual separation of the system from its environment. The influences (variable) that act on the system from the outside, and are not affected by what happens inside it, are termed "inputs" to the system. Two types of input variables exist: those which can be controlled (or changed as desired) and those we have no control over (wind direction and speed, for example). The latter type is undesirable and is called a "disturbance input." Now, the property of the system that results from the influence of the inputs on the system is the "output." Output variables are observable quantities and are measurable. In a dynamic system, variables change over time. "Static" situations result when the system is in equilibrium (balance).

As an illustration, consider the example of a moving automobile. The pedal position and the slope of the road are the system's inputs; the speed is the output. Of course, wind (specially if it is strong) is expected to affect the speed as well. As a second example, consider a simple electromagnetic device consisting of a ball rolling on the rim of a hoop (Wellstead, 1983). Although the device is seemingly simple in construction, it is actually

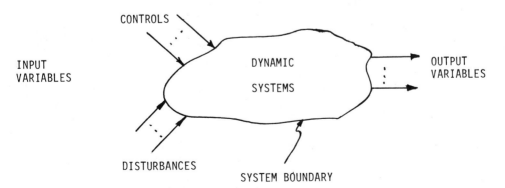

Figure 1.2 Concept of a dynamic system.

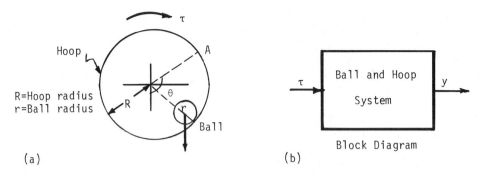

Figure 1.3   An electromagnetic device.   (a) ball/hoop illustration (b) block diagram.

rich in dynamics.   One may be interested here in seeking a model that has the essential dynamic characteristics of an oscillating load.   The angular position θ of the hoop is controlled via torque τ acting on the hoop from a torque motor applied to the hoop drive shaft (see Figure 1.3); τ is therefore an input.   The output y is the position of the ball on the inner periphery of the hoop measured with respect to a datum point A.

An example of a more sophisticated dynamic system is one of a steam boiler-generator plant.   The design objective here is to produce electricity that meets the customer's demand reliably and without interruption.   This includes guaranteeing the supply quality in terms of voltage and frequency. The input variables to this system are the fuel (coal, gas, or oil), water, and air.   Output variables are the generated voltage and frequency, in addition to temperature and pressure of generated steam by the boiler.   A computer-controlled operation is commonly in use where desired levels of output variables are compared with actual measurements, and adjustments (regulations) of the valves controlling the fuel, water, or air follow accordingly.   This is a "multivariable" example, in contrast to the "single-input single-output" ball-and-hoop system discussed earlier.

Now, for a specific system and with an established purpose of building a model, the modeler has a few options, depending on many factors including whether the system is in existence (or to be constructed), its complexity, and cost and time alloted for this important task, in addition to whether it is possible to experiment with the physical system while it is in operation.   Next, we briefly discuss the distinction between mathematical and experimental modeling. Also presented are prototype, pilot, and scale models.   In the remainder of this book, mathematical and experimental models will be highlighted and used.

The process of "mathematical modeling" involves three separate tasks: (a) identification and idealization of a system's individual elements (subsystems), (b) identification and idealization of their interaction, and (c) systematic application of basic (physical, biological, economic, etc.) laws (Cannon, 1967; Shearer et al., 1967; Luenberger, 1979); this process involves "deduction." "Experimental modeling," on the other hand, is the selection of mathematical relationships (through "induction") of an already existing system by fitting its observed input-output data.   Later in this chapter, more is presented on experimental models and their identification.

Another interesting technique used in describing the interrelationship between system variables is use of diagrams.   Representing the various

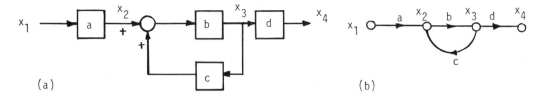

Figure 1.4 (a) Block diagram. (b) Signal flow graph.

components by blocks and connecting them by connecting lines is straight-forward and useful in expressing relations between components. In many complex situations, a block diagram (or, similarly, an equivalent signal flow graph) is not only very useful in visualizing the structure of the model, but also helpful in the communication between builders and uses of models. Simplifications of block diagrams and signal flow graphs are possible; how-ever, some intermediate variables may disappear in the simplification process. Figure 1.4 depicts a block diagram and its corresponding signal flow graph. Equations representing the relations between input variable x_1 and output variable x_4 are: $x_2 = ax_1$, $x_3 = b(x_2 + cx_3)$, and $x_4 = dx_3$. One can find an expression relating x_4 and x_1 where x_2 and x_3 are eliminated.

Next, other types of models are presented for the sake of completeness. These are prototype, pilot, and scale models. Our brief presentation here follows (Colella et al., 1974). A "prototype model" is a physical one that possesses nearly a one-to-one correspondence with the system under con-sideration. A prototype model is usually costly and time consuming in ad-dition to presenting a modest level of flexibility in terms of modification. One should contrast this with the great flexibility for modification and the relative savings in cost and time associated with mathematical models.

A "pilot model," on the other hand, may be a scaled-down physical rep-resentation of a physical system. Chemical processes are usually modeled in this fashion within a laboratory environment to gain knowledge about general operational characteristics. Prototype models, however, can be used to study macroscopic and microscopic characteristics of physical processes. Now, if the purpose of modeling is to get information concerning a specific characteristic (variable) of the system, then a "scale model" would be an appropriate choice. Such a model will offer almost no flexibility with respect to the behavior of other variables. Wind-tunnel facilities are examples of this type. An important element in this presentation is the possibility of utilizing one model at one time or another, or a combination of several types of models.

As an extension to this approach of coupling of models, "hardware-in-the-loop (HWIL)" facilities have been developed and are in use (Grider and Dublin, 1978). In such cases, part(s) of an overall complex physical system, say a missile, is included as physical hardware subsystem(s), and the re-maining components are represented by their mathematical models simulated on a computer. This real-time approach leads to significant cost savings and has proven to be invaluable in systems evaluation. Subsystems that have been used in the example of missile HWIL include missile guidance electronics, autopilot, or the control system. Grider (1984) has discussed the complex radiofrequency simulator with its real-time dynamic environment and HWIL capability (see Figure 1.5). This facility also offers the option

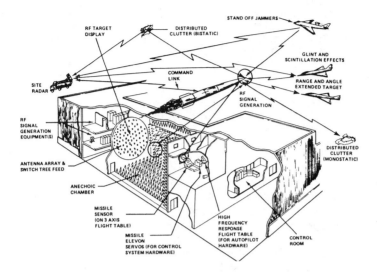

Figure 1.5   Radiofrequency simulation system. [From Grider (1984) with permission from AGARD/North Atlantic Treaty Organization.]

of using it for a "man-in-the-loop" environment. A piloted flight simulator has been presented by Baarspul and Vande Moesdijk (1976).

A few summary remarks on "model building" are in order:

The success of the model-building process largely depends on the validity of the assumptions made by the modeler in arriving at the model.

The model's relative simplicity should be an important feature of model construction.

No real physical system should be expected to dynamically behave exactly according to the equations of its math model which describe its "ideal" elements and interactions.

Models with successive degrees of complexity might be necessary to have all important aspects of a system's behavior included.

The need to develop models with successive degrees of complexity (hierarchies) is illustrated next by two examples. First, an automobile and suspension system is considered (Shearer et al., 1967). The simplest case is one that considers only a lumped mass model; next in complexity is the mass, spring, and damper model; third is the model including tire stiffness and wheel in addition to the axle mass. All these models provide some information about the system's behavior. An even more elaborate version results from considering the presence of two sets of tires, wheels, springs, and shock absorbers; this is a two-dimensional model. This model becomes more complex when the nonlinearity of the system elements is included.

The second example presents one of today's major environmental problems—the increase of pollution levels in streams. (Even though we have cited this example here, it will probably require the reader a good level of preparation to fully appreciate this problem.) Efforts have been directed toward maintaining pollution levels within reasonable bounds to better meet ecological requirements and water quality standards. The modeling of water quality in streams generates nonlinear models with high dimensionality. The basic element in such a model is number of reaches, where a reach is defined as a water stretch of some convenient length that receives one major controlled effluent discharge from, for example, a sewage treatment facility. A transportation delay between adjacent reaches is therefore expected. In a recent study, three models with various degrees of complexity have been summarized and discussed. For further reading consult Mahmoud, Hassan, and Darwish (1985).

1.2.3 Model Classification and Mathematical Representations

In this section, system models will be classified according to the type of equations used to describe them. First we distinguish between continuous, discrete, and hybrid systems. A "continuous" system is one for which variables change continuously with respect to time. However, in a "discrete-time" system, variables change only at distinct (finite) instants of time (such as hourly, daily, quarterly, etc.). In some physical applications, both discrete-time and continuous variables may coexist; thus, they are termed "hybrid" systems. A digitally controlled missile-launching system, for example, belongs to this hybrid classification (Saucedo and Schiring, 1968).

Physical applications may arise where a quantity may vary as a function of position in one or more directions and also as a function of time. In such cases, any dependent variable is likely to be a function of more than one variable and therefore will possess not ordinary derivative, with respect to a single variable, but partial derivatives with respect to several variables.

Models can be divided also into distributed-parameter and lumped-parameter models. In a "distributed-parameter" model, the dynamic behavior is described in terms of "partial differential equations," for example as in the expressions for voltages and currents at all points along a transmission line. An often desirable and less complicated approximation is the equivalent "lumped-parameter model." This is characterized only by a finite number of elements. For a physical inductor, a lumped-model representation includes only a resistance R (in ohms) connected in series with an equivalent single inductance L (in henries). In a lumped-parameter model, "ordinary differential equations" are used. When dealing with a lumped-parameter model, one may discretize in time, resulting in a discrete-time model that is described by a "difference equation." In such representation, for example,

$$x(k + 1) = ax(k)$$

relates the magnitude of the variable at step $(k + 1)$ to its value at step k and the value of constant a; $x(0)$ is the value of initial step.

The "superposition property" is helpful in distinguishing between linear and nonlinear systems. "Linear systems" are those satisfying the following conditions: (a) multiplying system input by a constant results in multiplication of its output by the same constant, and (b) the system response to a number of inputs applied together (simultaneously) is the sum of individual responses when each input is applied individually. "Nonlinear systems" are the ones for which these two conditions do not hold.

A differential equation (DE) of order n has the form

$$a_n(t) \frac{d^n x(t)}{dt^n} + a_{n-1}(t) \frac{d^{n-1} x(t)}{dt^{n-1}} + \cdots + a_o(t)x(t) = f(t)$$

where $f(t)$ is the input (forcing function) and x is the desired output (response). This equation is linear if the coefficients $a_i(t)$ do not depend on x but are functions of t, and the equation contains x and its derivatives as linear terms (terms of first degree). If $f(t)$ is zero, the DE is said to be "homogeneous" and system behavior is determined by the initial conditions. In many instances, $f(t)$ may consist of two or more individual functions of time. The following are examples of nonlinear DE:

$$2 \frac{d^2 x}{dt^2} + x^2 = 3$$

$$\left(\frac{dx}{dt}\right)^2 - 2x = 0$$

If all the coefficients of the equation are constant, the system is termed "stationary" or time-invariant. A "time-varying" system is one for which one or more coefficients may vary as a function of time. For example, the mass of a missile changes as a function of time as the fuel it carries is consumed during flight.

Another classification is into deterministic and stochastic models. In a "stochastic model," the relationships between system variables are described in a probabilistic fashion, whereas in a "deterministic" case the probabilities do not exist. A dynamic system that is stochastic and discrete is usually represented by a "discrete-event" model (e.g., the number of customers in a bank, and queueing and sequencing applications in general).

One further distinction is related to the complexity of the system and the number of variables; thus we have "large-scale" versus "smaller-scale" systems. These two classes differ significantly in the degree of difficulty in handling them.

Classification based on the type of system variables is considered next. In a deterministic lumped-parameter system, each element is described by a relationship between two physical variables: a "through-variable," which has the same value at both terminals of the element, and an "across-variable," which is specified in terms of a relative value (or difference) between element terminals. Examples of through-variables are force, torque, electric current, fluid flow, and heat flow. Velocity difference, voltage difference, pressure difference, and temperature difference are examples of across-variables. In linear networks, Kirchoff's voltage and current laws are often utilized to obtain mathematical models based on loop (mesh) or nodal methods. Loop equations involve all unknown currents, whereas nodal equations are in terms of a network's unknown nodal voltages. In a given situation, the number of unknowns (loop currents or nodal voltages) helps select the method requiring less effort.

Another representation is based on the "state variable" approach in which a new set of independent equations is formulated. These "state equations" differ from the loop and nodal sets described above. The state variables of a system are selected to present the "state" of the system; there is no unique method; however, in making this selection. Some of the state variables, as in the example of a network, are voltages and some are currents. In general, the state variables of an nth order system $x_1(t)$, $x_2(t)$, $\cdots$, $x_n(t)$ with initial values $x_i(t_0)$ and the system inputs are sufficient to describe completely the future behavior (response) of the system for $t > t_0$ (DeRusso et al., 1965).

1.2.4 Identification

Our concern here is with identification of the parameters of the system being modeled in the form of a differential equation, difference equation, transfer function, etc. This is usually one of the early steps dealt with in practice. Various identification techniques exist and employ a number of different concepts concerning the form of the identification model. Moreover, each of these techniques has its own range(s) of applicability.

Fundamental to the identification problem is the distinction between situations that call for different treatments: linear and nonlinear, continuous and discrete-time, single-input and multi-input, and deterministic and stochastic. Perhaps the most important is the classification according to the "degree of a priori knowledge" available regarding the system (Graupe, 1972). Obviously, a knowledge of the dimension of the system is important, as the knowledge of any nonlinearities that might exist, in addition to component (subsystem) interaction, i.e., structure of the model. One should note that identification techniques that assume less a priori knowledge are not only less accurate, but are also more complex in terms of their mathematical representation and require more computation time than those with more prior knowledge.

Now, if the model structure is known in advance, "parametric" identification methods can be used; otherwise, "nonparametric" procedures have to be employed. Algebraic equations, differential equations, and transfer functions are examples of parametric models. Nonparametric models result from using experimental modeling, discussed previously, based on a history of input-output measurements; it is expected that identification improves with an increased number of measurements (Box and Jenkins, 1970). The input signals used can be either operating signals of the system or artificial, periodic or nonperiodic, signals. Regardless of the type of signal employed, measurements should be taken when the system is in a transient state; i.e., the dynamic parameters cannot be identified when the system reaches its steady-state. A parametric model can be obtained from a nonparametric one; also, identification may be performed "off-line" or "on-line."

The identification problem may be employed in "deductive models," which are to be distinguished from the "black-box" approach. In the former case, identification is carried out at the component level, while in the latter, it is applied at the system level. Moreover, in deductive models, database is not generally obtained by actual experimentation of the system, but rather from earlier (available) design models, from component experimentation, or from the manufacturer's data, if available.

Identification methods based on experimental data and which are suitable for implementation on a digital computer are: Fourier analysis, correlation analysis, spectral analysis, model fitting, and parameter estimation. Perhaps the most widely used parameter estimation methods are those based on the principles of least-squares and the maximum-likelihood methods (Fasol and Jorgl, 1980). The availability and the accuracy of the data may

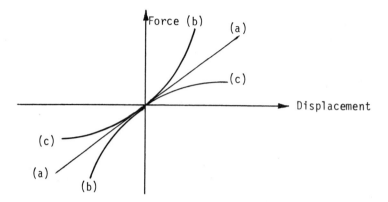

Figure 1.6   Spring characteristics.   (a) Linear spring.   (b) Hard spring.
(c) Soft spring.

present significant limitations that should be recognized.   For further
readings consult the works of Astrom and Eykhoff (1971), Eykhoff (1974),
and Isermann (1980).

1.2.5   Nonlinearities and Discontinuities

Strictly speaking, linear systems do not exist in practice, since all physical
systems are nonlinear to some extent.   The modeling and analysis of systems with nonlinear components is not an easy task.   Consider the oscil-
lation of a mass hung from a spring where elongation is possible.   Figure
1.6 (curve a) depicts the linear case where spring rigidity (force/displace-
ment) is constant.   The characteristics of a hard spring are shown in
curve b, where rigidity increases as a function of displacement; the characteristics of a soft spring are displayed in curve c (Blaquiere, 1966).

Figure 1.7 depicts important typical input-output relationships.   Unsymmetrical characteristics are shown in a, and symmetrical relationships
describing the dependency of the output on the absolute value of the input,
or the sign of the input, are shown in b and c, respectively.   An amplifier,
for instance, often exhibits saturation nonlinearities when input signals are
large.   This same property is characteristic of actuators (valves) often used
in process control applications, where nonlinearity corresponds to a fully
open or closed valve.   Often, nonlinearities are introduced to improve
systems performance.   For example, the ideal relay (also termed on-off or
bang-band) is used in many spacecraft or missile control systems to
achieve minimum time control.

Now, the instantaneous transition from one state, of a dynamic discontinuous system, into another is well illustrated by the example of a tracked
vehicle (A) and a buffer (B) in Figure 1.8.   In this vehicle-buffer system,
differential equations are used to describe the dynamics of the vehicle and
those of the buffer when they are apart.   This independence in dynamics
becomes invalid whenever the vehicle is in contact with the buffer.   The
dynamics then are linked (Hay and Griffin, 1980).   The transition conditions
may be expressed in terms of discontinuity function ϕ defined as: $\phi = (x - y)$.
Obviously, negative ϕ implies that A and B are not in contact.

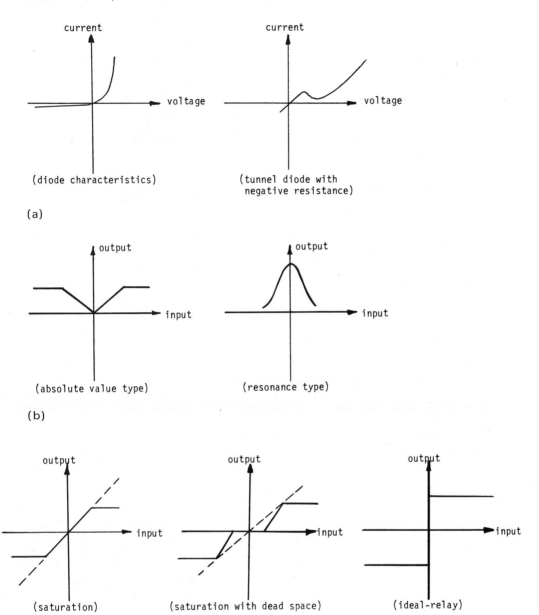

Figure 1.7 Typical nonlinearities. (a) Unsymmetrical characteristics. (b) Symmetrical characteristics: output depends on the absolute value of the input. (c) Symmetrical characteristics: output depends on the sign of the input.

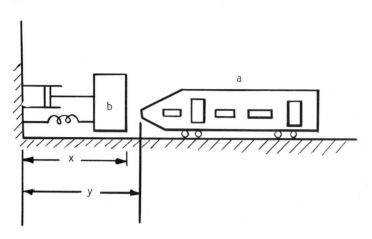

Figure 1.8   A vehicle-buffer system.

    Treatment of this class of models is complex; the superposition property
of linear systems cannot be applied.   Various methods, however, have been
suggested.   Graphic methods based on the assumption of a small nonlinearity
allow differential equations to reasonably describe physical phenomena.
Also, linearization of nonlinear equations is another possibility when high-
order terms are omitted from a Taylor series expansion.   The justification
of such linearization is that it yields reasonably accurate results when compared with those obtained analytically or experimentally from the given nonlinear system.   In nonlinear control systems applications, describing function
and phase plane methods have been used (a phase plane is a plot of the derivative of a function versus the function itself).

1.3   COMPUTER SIMULATION: AN OVERVIEW

Having discussed the aspects of system modeling in the previous section, we
now state the working definition of simulation for the purpose of this text.
Simulation is the process of building and experimenting with (manipulating)
a computerized system model such that a specific purpose of the study is
achieved through observing model's behavior under assumptions defined by
the experimenter.   The experimenter is meant to be modeler, user, and/or
decision-maker.   Therefore, the simulation process, as defined above, encompasses the activities included in model building as well as the analytical
use of it.
    Earlier, a list of reasons was given for which simulation has been, and
is being, widely used.   But we may raise the question about any constraints
one may encounter in performing system simulation.   There are, as a matter
of fact, a number of constraints; first is the cost of model definition, software development (programming), running the model on a computer, data
collection, refining the model, etc. (Gray, 1976).   Also one should not
underestimate the importance and value of the human element.   Experience
and knowledge, not only in the system under study and its environment
but also in programming language(s), specifying input data, instrumentation

and data collection, hardware implementation, design of experiments, documentation, analysis of results, and decision-making, are required. These become more critical when one deals with mankind's "large and soft" modern systems. Other constraints include: time allotment for simulation activity, hard versus soft system type, severity of nonlinearities, availability of adequate hardware, etc.

In the sequel, we introduce an overview of computers and the software aspects of simulation. As mentioned earlier, these topics are important to successful simulation studies and are, therefore, further explored in other chapters of this text. An account of the evolution of computers and languages is also included.

1.3.1   The Computer

The basic computing tools for simulating dynamic systems, represented by their mathematical models, are analog, digital, and hybrid (combined analog and digital) computers. Of these, the digital is by far the most widely used. Simulation techniques and other related issues are the subject of Chapter 4; here we discuss analog and digital computers, their functional requirements, advantages, and disadvantages.

Today's electronic analog computer consists of basic elements that are capable of performing summation, integration, multiplication, nonlinear function generation, and logical operations (Jackson, 1960). In simulating linear and nonlinear systems, the analog computer performs all the required operations necessary, with dc voltages (on the computer) representing the system variables (pressure, temperature, distance, etc.). The variables are continuous and time is the independent variable. Simulation results are generally accessible to the user on a voltmeter, oscilloscope, or x-y plotter (with the x-axis being time). Equations (the math model) are programmed and carefully patched (by interconnecting the computer elements), and as a result, the analog computer program (setup) has a unique relationship to the system being simulated. In developing a solution, the analog computer performs parallel operations (occurring simultaneously), thus making computational speed a salient feature of analog computation. This parallel operation performance and the operational amplifiers' capability of integration make "real-time" (or even faster) solutions feasible.

Two types of scaling are often required: these are magnitude and time scaling. Scaling is usually a time-consuming task and is considered a disadvantage of the analog computer. Difficulty of storing information is a second disadvantage. On the positive side, the fact that the user can change the value of a coefficient just by adjusting a potentiometer setting on the computer allows repeated (relatively easy) real-time simulation runs, and thus insight into the system being simulated is gained at a relatively low cost.

Rather than imitating a physical phenomenon as on an analog computer, a digital computer operates in a logical manner in solving the mathematical model of a system numerically. On a digital computer the variables are discretely defined (in contrast with the continuous nature in analog simulation); variables are defined only at specified intervals of time (the independent variable). Digital signals are binary in nature and represented by a two-state signal, with the higher voltage level called the "one" state and the lower voltage called the "zero" state level. The basic digital computer can be divided into logic elements and storage elements, thus allowing the performance of a variety of arithmetic and logical functions as well as storage of information. The physical subsystems of a general-purpose digital computer are the input, control, storage (memory), arithmetic, and output units. Basic arithmetic operations possible are counting, addition, subtraction, multiplication, and division. Floating-point arithmetic and digital computer large memory are among its important features. Also, the need for magnitude scaling is eliminated owing to floating-point arithmetic operations and the capability of representing a wide range of variables (Wegner, 1968). However, the need to use analog/digital (A/D) and digital/analog (D/A) converters results in "quantization errors" which may be significant.

For an analog computer, computing time is invariant, but it is proportional to problem complexity on a digital computer. Also, an analog computer hardware configuration (number of amplifiers, integrators, etc.) grows in proportion to problem size, whereas a digital computer hardware configuration is not usually dependent on problem size.

Even though accuracy is considered one of the most important attributes of a digital computer, its sequential nature of operating on problem variables may make it unable to perform real-time simulation of complex systems. However, increased component speeds as technology continues to advance rapidly, in addition to the methodology of using "parallel computation," are offering promising opportunities (Karplus, 1977) (here simultaneous execution of two or more operations is possible by a computer). Another point is the inability of the digital computer to perform integration and differentiation except through numerical approximations; these require consideration of accuracy requirements and solution-time constraints. Special-purpose digital computers have been widely used in several applications, including aerospace and process control systems. Hybrid computers are discussed in the next section.

1.3.2   Evolution of Computers

Following is a summary of the evolution and capabilities of analog and digital computers. Also discussed are the features and development of hybrid computers. In 1927, Vannevar Bush at MIT started to work on the first general-purpose mechanical analog differential analyzer. In the 20 years that followed, improvement was introduced through the addition of servomechanism devices. The mechanical differential analyzer was capable of performing mathematical operations and solving differential equations with relatively good accuracy, but very low operational speed. Interestingly, special-purpose electromechanical analog computers were widely used during World War II as fire control computers, aircraft autopilots, and navigational computers (Gioli, 1975). Many special-purpose electronic analog computers were also built which led to the construction of the general-purpose analog computer known as project TYPHOON. The early analog computers were, however, fairly sizable and not very reliable. Because they were limited, simplification of mathematical models, of relatively large systems, was necessary so that problems could be accommodated on the then available computers.

The 1950s machines included the digital differential analyzer (DDA) and the magnetic drum; by 1958, TRICE/DDA was built with solid-state intergrators. Of course, general purpose DDAs are outperformed by general-purpose digital computers. Early digital computers offered their users the required accuracy and precision, but not speed. The use of the digital computer for simulation of physical systems in real-time started in the

mid-1960s when faster digital computers were developed. The historic replacement of vacuum tubes with transistors and the development of magnetic
tape, drum, and core memories were instrumental in reducing computer size
and significantly increased speed and reliability. The dramatic development
of high-speed large-scale integrated (LSI) circuits in the 1970s led to smaller,
but more powerful, reliable, and less expensive digital computers. The 1980s
have also witnessed the extensive use of minicomputers along with mainframes.
So the user today has a spectrum of digital computers to choose from, ranging from microprocessors on a single LSI chip to very large systems with
great speed. Also possible is the use of digital computers in "time-sharing"
capacity with many users each on a remote terminal. Terminals allowed the
addition of "graphics" capability and therefore added a new dimension to
man-machine communication (interactive capability). The user is able to
interact (on-line) with the computer and decides on the next step based on
the available results obtained so far.

Among the recent advancements in digital computer hardware technology
is the development of peripheral array processors which greatly enhance the
capability of a host computer to process numerical data at a very high speed.
Also noted is the new generation of large general-purpose "supercomputers";
examples are CRAY-1, STAR-100, and ILLIAC-IV (Karplus, 1984), Fujitsu
VP200, and Hitachi Ltd S810 (Wallich, 1985), with a reported capability in
the neighborhood of 400 million operations per second. For extensive reporting on recent advancement in computer technology, the interested reader
may consult with IEEE *Spectrum* issues of January 1984, 1985, and 1986.

The development of the hybrid computer resulted from realizing the
potential, advantages, and disadvantages of the analog and digital computers
of the 1950s (Howe, 1976). The main advantages of the digital, as mentioned
earlier, were high precision and large storage capabilities, but low speed
was the principal disadvantage. The analog, on the other hand, offered
high speed and good man-machine interaction. The scaling requirements
and limited precision were, and still are, among the analog computer disadvantages. The hybrid computer emerged as a combination of analog-dig-
ital computers, mainly to solve aerospace problems, based on optimum utilization of analog and digital computing elements. An interface converting
analog to digital (A/D) and digital to analog (D/A) allowed successful communication between the two types of computers and real-time simulation.
Subdivision of the tasks performed on the analog and digital subsystems
allowed using the analog for the dynamic part of simulation (including integration), while the digital mainly performed computations that are difficult
or impossible on the analog. The hybrid computer today is in use for time-
critical applications such as hardware-in-the-loop simulation (Grider, 1984).
The SIMSTAR system. recently developed by Electronics Associates Inc.,
offers the user the option to select its operation from a parallel, serial
(sequential), or paralles/serial combination.

1.3.3 Simulation Software

In this section we focus on the software aspects of digital computer simulation only. This is done because analog computer simulation does not
require a language to computerize a system model. Programming on the
analog is done by connecting the appropriate computer elements in a manner
that often exhibits a near one-to-one correspondence to the system being
simulated. Also, the rather limited usage of hybrid computers does not

warrant covering their software requirements here. There will be no attempt
to present the full details of any of the languages.

A digital computer program is a defined sequence of necessary operations
that a programmer uses to solve the equations that describe a system's model.
Because of the early extensive use of the analog computer, digital programming languages based on simulating an analog computer, rather than solving
arithmetic and logic equations, were developed during the late 1950s and
1960s. These were called "digital-analog simulators"; a summary of such
languages has been reported (see Linebarger and Brennan, 1964).

Available to today's simulationists is a wide spectrum of digital computer
software which mainly consists of (a) programming languages, (b) simulation
languages, and (c) simulation packages. These will be focused on in more
details in future chapters and in Appendices A and B.

In selecting a computer language, one would consider the following
three areas to assess its efficiency and suitability: program execution speed,
computer memory utilization, and language availability.

1. If one surveys the spectrum of "programming languages," machine
language and assembly language are at the spectrum's lowest level. Instructions in "machine language" are written in binary notation which corresponds
directly to machine functions; this is recognized by most as a tedious and
impractical task. On the other hand, an "assembly language" program is
a string of mnemonic symbols that correspond to machine language functions
and are translated to a basic machine language program by an "assembler."

Higher-level programming languages (also called general-purpose or
compiler languages) allow the programmer to be removed from concerns related to machine operations (machine independent). Examples of these general-purpose languages are: BASIC, FORTRAN, COBOL, and ALGOL. An
efficiently written program in such languages may require, in general, less
execution time than when the problem is programmed using a simulation language. BASIC and FORTRAN are available on almost every computer and
are probably the most popular.

2. High-level "simulation languages" are also compiler oriented, similar
to high-level programming languages, but are specifically used for simulation
applications. Most simulation languages require less programming time; moreover, it is simpler to change a model after being written. It is also easier
to debug such programs. A unique feature of simulation languages is their
basic building blocks. Among the earlier simulation languages are MIDAS,
DYSAC, DSL, GASP, MIMIC, DYNAMO, GPSS, SIMULA, CSSL (Continuous
System Simulation Language), and CSMP. The standard functional blocks
in CSMP, for example, include first- and second-order transfer functions
in addition to nonlinear functions (dead space, hysteresis, saturation, etc.)
(see Speckhart and Green, 1976). More recent simulation languages include
ACSL (Advanced Continuous Simulation Language), DARE-P and DARE-Inter-
active, C-SIMSCRIPT and SIMSCRIPT II-5, Ada, GASP IV, SDL, SIMNON,
SLAM, and SIMAN. In later chapters of this book , these languages, and
others, will be highlighted, classified in terms of their applicability to
continuous systems, discrete, or combined (continuous/discrete) systems,
and their features presented. For a recent summary of the features of 15
(widely used) simulation languages, the reader may consult Cellier (1983);
for a recent simulation software listing, see the SCS *Catalog* (1984, 1985).
Sargent suggests a list of factors to be considered in selecting languages
for a specific problem or for an organization (Sargent, 1978).

3. In its simplest form, a digital "simulation package" is a collection of
routines [programs to be possibly compiled separately and then included as

part of other program(s)]. Today's simulation packages have not only matured in their development, but, coupled with the available hardware, provide one of the most powerful tools for modeling and simulation activities in an interactive fashion.

The "interactive" (conversational) mode of simulation means that the simulation process on the computer be interrupted for the purpose of asking, or reporting to, the user (modeler, programmer, or decision-maker). When the computer asks questions and the user selects from predefined answers, it is called "menu-driven interaction." Based on the information available from the computer (output), the human partner decides on what is to be modified, executed, etc., next. This capability is very helpful in the iterative process of model building as well as in validation studies.

Over the past dozen years or so, attention has focused on developing simulation packages that are useful in many areas of applications. These packages have impacted (and are expected to continue to do so) on the educational processes and on activities in computer-aided design (CAD), computer-aided manufacturing (CAM), and, in general, in computer-aided engineering (CAE). There will be no attempt here to list all available simulation packages; however, on should note the tremendous growth in the number of workshops and symposia on the subject (examples are: IFAC symposia in 1979 and 1985, introducing packages for control systems and engineering systems design). Jamshidi and Herget (1985) have introduced a summary of 15 packages for control systems engineering applications; Appendix A of this book summarizes simulation packages in controls education and research. Other applications with active development of simulation packages include transportation systems, steam boilers, electric power systems, manufacturing systems, and robot control education.

The last item in this section on software is concerned with expert systems. An "expert system" is considered an intelligent computer program which uses knowledge and procedures to solve problems. This artificial-intelligence software is often capable of giving expert advice within a particular domain. Applications of such knowledge-based systems include: medicine, neurology, geology, and chemistry (Heuristic Programming Project, 1980). As promising and powerful simulation tools, expert systems have recently been introduced in air traffic control (Cross, 1985), in the automation of space stations (Georgeff and Firschein, 1985), and in the next-generation integrated avionics systems to assist pilots in the 1990s (Frankovich et al., 1986). In these applications, one capitalizes on the shared knowledge between man and the machine. Another recent study presents similarities between simulation and expert systems (O'Keefe, 1986).

1.4 OUTLINE OF TEXT

The remaining chapters of Part I continue to deal broadly with fundamentals of systems modeling and simulation. Chapter 2 treats continuous-time and discrete-time systems; emphasis is placed on Laplace transform, transfer function models and state-space models, and estimation of system's order. Also presented are simplification methods using block diagram algebra and signal-flow graphs. The z -transform and its properties are presented and used in an α-β tracking system example. This chapter ends with a discussion of the notions of stability, controllability and observability of feedback systems; Routh-Hurwitz criterion and the root locus method for systems stability are presented.

Discrete-event systems is the subject of Chapter 3. The chapter begins with a review of important terminology followed by a summary of languages that are widely used in discrete-event simulations. Statistical models described in terms of discrete and continuous distributions, Poisson processes and empirical distributions are developed. Characteristics of queueing and inventory systems are examined next as typical discrete-event applications. Methods of analysis of input and output data are also presented with model verification and validation in mind.

Computer simulation is the subject of Chapter 4. Topics covered include analog, digital and hybrid simulations with more emphasis on digital simulation. Numerical integration methods and associated errors are examined; also presented are techniques used in simulating state-space models, transfer function models and discrete-time systems. Although most simulations are performed digitally, the reader is expected to gain from the presentation of the analog approach to systems simulation; magnitude and time-scaling requirements are also discussed.

Successful simulations require a good understanding of principles of design and analysis of simulation experiments. Sections 5.2 and 5.3 treat the statistical design and analysis aspects, respectively. In using statistical design of experiments, the modeler is often concerned with determining the effects of various factors on system response variable at the lowest experimental error possible. On the other hand, one attempts in the analysis of experiments, to define some acceptable measure of confidence on response variables. Variance reduction techniques are briefly discussed in section 5.4.

Chapter 6, the last one in Part I, introduces the modeling and control aspects of large-scale systems. Aggregation and perturbation schemes are presented as modeling approaches; while the concern, in the aggregation scheme, is to be able to retain key qualitative properties of the system, the modeler ignores in the perturbation approach certain interactions of the dynamic or structural nature in the system. Hierarchical (multi-level) control is presented in section 6.3 where subsystems are positioned on levels with different degrees of hierarchy; here subsystems' solutions are coordinated by a higher-level coordinator (supervisory control). In section 6.4 on decentralized control, the system's output measurement is shared among a finite number of controllers which collectively help control the system. In many large-scale applications, some degree of restriction is assumed to prevail on the transfer of information. The stability, controllability and observability concepts are discussed in section 6.5. This chapter uniquely presents several CAD examples.

The four chapters (7 through 10) of Part II focus on software/hardware systems. Data structures and databases are the subject of Chapter 7; the presentation starts with basic data structures (e.g., elements, arrays, tables, records, sets) and proceeds to cover advanced data structures including linked lists, stacks and queues, and trees and graphs. An overview of database systems is given followed by a brief survey of database applications. Although the basic and advanced data structures covered are by no means exhaustive, they provide the simulationist with some basic understanding required in more complex situations.

Programming languages are the subject of Chapter 8 with their evolution and development reviewed in section 8.2. The focus in section 8.3 is on the support provided by programming languages for representing data and data relationships for simulation applications. The modern features for expressing flow-of-control within a program with consideration to time control in simulation languages are presented in section 8.4. Section 8.5 discusses current programming language support environments including processors (i.e., compilers, interpreters and translators). Not only does the remainder of Chapter 8 define and apply criteria for programming languages' assessment, but it also gives a perspective on current trends in the evolution of simulation languages and support systems.

Chapter 9 deals with Operating Systems and their effect on simulation. An understanding of the basic operating system concepts such as segmentation, paging, I/O and scheduling is necessary to develop successful simulation under the control of an operating system. Section 9.3 discusses concurrency, process synchronization and scheduling; problems encountered with concurrent processes sharing data as well as methods for controlling the access to shared data are included. While section 9.4 presents file systems, section 9.5 treats in details the Input/Output systems with emphasis on software problems arising in interfacing special-purpose hardware to real-time simulation systems. Memory management is the topic of section 9.6.

Chapter 10 on hardware implementation begins by introducing a typical digital computer model and gives the details of the organization and programming of a simple hypothetical computer. Various digital computer architectures are classified next in section 10.4 which also discusses computer use as a component of a simulation environment. The hardware components designed to perform analog-to-digital and digital-to-analog conversion are the subject of section 10.5. The quality of representation of a continuous signal by a sampled and digitized one makes it important to define and discuss the sampling and quantization effects. Implementation aspects of analog and digital algorithms are given in section 10.7 for continuous-time and discrete-time systems. Analog programming is revisited using direct programming, parallel, cascade or the M-programming method. An example of a digital control of a single robot arm using a microcomputer is given in section 10.8.

Part III includes six applications chapters. Chapter 11 is devoted to the basic concepts of modeling and designing of digital control systems. Several design approaches are presented in section 11.3; these include cascade and feedback compensation, proportional- plus- integral- plus- derivative (PID) controllers and the quantitative feedback technique (QFT). Special attention has been given to meeting acceptable responses despite uncertainty in the plant and disturbance inputs; time-domain specifications are translated into bounds in the frequency-domain. The effects of finite word length and computational delays on overall digital controller performance are examined in section 11.4.

Chapter 12 on robotics starts off with a description and functional structure of a robot. In the modeling section 12.2, robots' kinematic modeling is developed followed by kinematic equations solution; in this context, solvability and computational aspects are discussed as well as computer-aided setting of the kinematic model. Also presented are the Jacobian and robot dynamic models. Section 12.3 is on control of robots to achieve a prescribed motion for the robot hand along a desired trajectory with a specified orientation and velocity. Positional control, resolved motion rate control and resolved acceleration control are discussed along with the computed torque and adaptive control techniques. Robot vision, welding and assembly automation are among the applications of robots included in section 12.4. Section 12.5 presents an up-to-date summary of research directions in robots.

World models, of Chapter 13, are computerized mathematical simulations to study the world's economic, physical and political systems on a relatively long-term basis. Objectives of such models influence their designs and structures. In section 13.2 approaches to world modeling are presented including regions and sectors; demand, supply and equilibrium; and dynamic and static models. A review of the most widely used world models follows in section 13.3. These are classified as economic models, food and agriculture models and environmental models. More attention has been given to the first category of economic models where World 3, project Link, world integrated model (WIN) and the United Nations world models are presented. A detailed case study using WIN is the subject of 13.4 where ARISTOTLE (a software system) is used in this simulation.

The treatment of economic systems is covered in Chapter 14. It is well known that the economic principles governing the dynamics of such systems are not as precise as for physical systems; however, it is possible and useful to model and simulate national economies with reasonable accuracy. It is desirable to study, for example, the problems of controlling inflation and unemployment using an econometric model of the United States. Section 14.2 is concerned with the development of econometric models; attention has been given to their dynamic characteristics and to model parameter estimation. Section 14.4 deals with economic forecasting techniques where the simpler techniques require only the autocorrelation of historical economic data for formulation and error estimation of the time-series being forecasted. Among the forecasting applications presented are the average forecast, exponential forecast, least-squares forecast and the low-order system forecast.

Chapter 15 presents manufacturing systems simulation using SIMAN. The overall characteristics of manufacturing systems are reviewed, so are the main features of the general-purpose simulation language, SIMAN. The areas of manufacturing systems modeled in section 15.4 are workstations, macro submodels, visitation sequences, resource schedules and material handling systems. The animation system Cinema is presented in adequate detail in section 15.5 with emphasis on its layout and runtime features. Cinema provides immediate visual feedback on system changes. The example application in section 15.6 is based on a manufacturing system installed at a large Japanese machine tool manufacturer; the simulation study was performed to determine whether the automatic guided vehicles (AGV's) track layout would be feasible to handle the expected production lead, and how many AGVs would be required.

Multiprocessor systems of Chapter 16 are concerned with parallel processing which allows parallelism of simulation algorithms through a simulation process. In today's technology, parallel processing capability allows algorithm parallelism to be retained through the coding, compiling and program execution. Three multiple processor architectures have been examined in section 16.3 while section 16.4 sheds some light on three simulation-oriented commercial systems; these are the peripheral array processor, the applied dynamics international (AD-10) and microprocessor networks (Intel iPSC). In section 16.5, the hardware, firmware, programming language and operating system of a real-time multiprocessor simulator (RTMPS) are discussed along with a simulation example. Here, a small turboshaft engine in the 20,000 pound thrust class was modeled for simulation on RTMPS; the simulation equations, to use all the processors concurrently, have been examined with the minimization of interprocessor data transfers and the increase of simulation speed in mind.

Appendix A focuses on Computer-Aided Control System Design (CACSD) techniques and tools. This appendix should be most useful to students and professionals with interest in control systems. The classification of CACSD techniques (algorithms) is treated in section A.2 from the point of view of: frequency domain, state-space, structure of the selected controller, robust controller design, and sensitivity analysis. Section A.3 on development and classification of CACSD tools compares and defines interactive interface, command-driven, batch-operated, code-driven or data-driven, menu-driven, graphics-driven, question-and-answer-driven, or window-driven. A survey of twenty currently available and used CACSD packages are compared in section A.4; classification of these packages is based mainly on user interface, data/program structures, built-in algorithms including transformation routines, identification, analysis and design routines. Also real-time capabilities as well as portability and software reliability have been included in the classification criteria. A discussion of tools standardization versus diversification is included in section A.5.

Appendix B presents an extensive survey of available simulation software; discrete, continuous and combined discrete-continuous languages are surveyed with major features highlighted.

It is our hope that this book will provide enough material and opportunity for such educational experience, via self-study and/or classroom instruction. The above outline is expected to be of help in the selection process of material to be covered in one course or a two-course sequence

REFERENCES

Astrom, K. J., and Eykhoff, P. (1971). System identification—A survey, *Automatica 7*, 123–162.

Baarspul, M., and Vande Moesdijk, G. A. J. (1976). The moving-base visual flight simulator of the Department of Aerospace Engineering, in *Simulation of Systems* (L. Dekker, ed.), North Holland, Amsterdam, pp. 1061–1066.

Bekey, G. A. (1977). Models and reality: Some reflections on the art and science of simulation, *SIMULATION 29*(5), 161–164.

Blaquiere, A. (1966). *Nonlinear System Analysis*, Academic Press, New York.

Box, G. E. P., and Jenkins, G. M. (1970). *Time Series Analysis, Forecasting and Control*, Holden Day, San Francisco.

Cannon, R. H., Jr. (1967). *Dynamics of Physical Systems*, McGraw-Hill, New York.

Cellier, F. E. (1983). Simulation software features, in *Proceedings of IMACS*, Nantes, France.

Colella, A. M., O'Sullivan, M. J., and Calino, D. J. (1974). *Systems Simulation: Methods and Applications*, Lexington.

Cross, S. E. (1985). Model-based reasoning in expert systems: An application to enroute air traffic control, *IEEE/AESS Magazine*, pp. 6–12.

DeRusso, P. M., Roy, R. J., and Close, C. M. (1965). *State Variables for Engineers*, Wiley, New York.

Eykhoff, P. (1974). *System Identification—Parameter and State Estimation*, Wiley, New York.

Fasol, K. H., and Jorgl, H. P. (1980). Principles of model building and identification, *Automatica 16*, 505–518.

Frankovich, K., Pedersen, K., and Bernsteen, S. (1986). Expert system applications to the cockpit of the 90's, *IEEE/AESS Magazine*, pp. 13–19.

Georgeff, M. P., and Firschein, O. (1985). Expert systems for space station automation, *IEEE/Control Systems Magazine*, pp.3–8.

Gioli, W. K. (1975). *Principles of Continuous System Simulation*, Teubner, Stuttgart.

Graupe, D. (1972). *Identification of Systems*, Van Nostrand Reinhold.

Gray, P. (1976). The economics of simulation, in *Proceedings of the Winter Simulation Conference* (Dec. 6–8), pp. 17–25.

Grider, K. V. (1984). Seeker systems simulation, present capability and future technology, *NATO/AGARD, Lecture Series No. 135*, paper #6.

Grider, K. V., and Dublin, D. H. (1978). Simulation in missiles development—Yesterday, today and tomorrow, in *Proceedings of the AIAA Annual Meeting and Technical Display*, Washington, D.C. (paper #78–336).

Hay, J. L., and Griffin, A. W. J. (1980). Simulation of discontinuous dynamical systems, in *Simulation of Systems '79*, (L. Dekker, ed.), North Holland, New York.

Heuristic Programming Project (1980). Published by Heuristic Programming Project, Department of Computer Science, Stanford University, Stanford, CA.

Howe, R. M. (1976). Tools for continuous system simulation: Hardware and software, *Simulation of Systems* (L. Dekker, ed.), North Holland, Amsterdam, pp. 21–29.

IFAC Symposium (1979) on Computer Aided Design of Control System, Zurich, Switzerland (M. A. Cuenod, ed.).

IFAC Symposium (1985) on Computer Aided Design in Control and Systems Engineering, Copenhagen, Denmark (N. Hansen and M. Larsen, eds.).

Isermann, R. (1980). Practical aspect of process identification, *Automatica 16*, 575–587.

Jackson, A. S. (1960). *Analog Computation*, McGraw-Hill, New York.

Jamshidi, M., and Herget, C. J. (1985). *Computer-Aided Control Systems Engineering*, North Holland, Amsterdam.

Karplus, W. J. (1976). The spectrum of mathematical modeling and systems simulation, *Simulation of Systems* (L. Dekker, ed.), North Holland, Amsterdam, pp. 5–13.

Karplus, W. J. (1977). Peripheral processors for high speed simulation, *SIMULATION 29*(5), 143–153.

Karplus, W. J. (1984). Selection criteria and performance evaluation methods for peripheral array processors, *SIMULATION 43*, 125–131.

Kheir, N. A. (1978). On validating simulation models of missile systems, *SIMULATION 30*, 117–128.

Kheir, N. A., Damborg, M., Zucker, P., and Lalwani, C. S. (1985). Credibility of models, *SIMULATION* (technical note), 87–89.

Kwatny, H. G., and Mablekos, V. E. (1975). The modeling of dynamical processes, in *Proceedings of the Energy Research and Development Administration (ERDA) Conference*, Washington, D.C., 368–393.

Linebarger, R. N., and Brennan, R. D. (1964). A survey of digital simulation: Digital analog simulator programs, *SIMULATION 3(6)*.

Luenberger, D. G. (1979). *Introduction to System Dynamics*, Wiley, New York.

Mahmoud, M. S., Hassan, M. F., and Darwish, M. G. (1985). *Large-Scale Control Systems: Theories and Techniques*, Dekker, New York.

O'Keefe, R. (1986). Simulation and expert systems: A taxonomy and some examples, *SIMULATION 46*, 10–16.

Sargent, R. G. (1978). Introduction to simulation languages, in *Proceedings of the Winter Computer Simulation Conference*, Miami Beach, Fla, pp.15–17.

Sargent, R. G. (1979). Validation of simulation models, in *Proceedings of the Winter Simulation Conference*, San Diego, CA, pp. 497–503.

Saucedo, R., and Schiring, F. F. (1968). *Introduction to Continuous and Digital Control Systems*, Macmillan, New York.

Shearer, J. L., Murphy, A. T., and Richardson, H. H. (1967). *Introduction to System Dynamics*, Addison Wesley, Readings, MA.

Society for Computer Simulation (SCS) Technical Committee on Model Credibility (1979). Terminology for Model Credibility, *SIMULATION 31*, 103–104.

Society for Computer Simulation (SCS) (1984). Catalog of Simulation Languages, *SIMULATION 43*, 180–192.

Society for Computer Simulation (SCS) (1985). Additions to the catalog of simulation software, *SIMULATION 44*, 106–108.

Speckhart, F. H., and Green, W. L. (1976). *A Guide to Using CSMP: The Continuous System Modeling Program*, Prentice-Hall, Englewood Cliffs, NJ.

Wallich, P. (1985). Minis and mainframes, *IEEE Spectrum* (January), pp. 42–44.

Wegner, P. (1968). *Programming Languages Information Structure and Machine Organization*, McGraw-Hill, New York.

Wellstead, P. E. (1983). The ball and hoop system, *Automatica 19*(4), 401–406.

Zeigler, B. P. (1976). *Theory of Modeling and Simulation*, Wiley, New York.

Zeigler, B. P. (ed. in chief) (1979). *Methodology in System Modeling and and Simulation*, North Holland, Amsterdam.

2

Continuous-Time and Discrete-Time Systems

2.1 INTRODUCTION

This chapter provides an extensive discussion of the important fundamentals of continuous-time and discrete-time systems. Section 2.2 is concerned with continuous-time systems and their mathematical models in terms of differential equations, transfer functions, and state-space models. Systems representation by block diagrams and signal flow graphs is given with simplification and reduction techniques leading to overall transfer functions. This section also introduces a method for system order estimation followed by a discussion on system order reduction, as well as problems encountered in model overdescription.

Discrete-time systems is the subject of Section 2.3, where the Z-transform, its inverse, and properties are introduced. Attention has been given to a detailed treatment, using the Z-transform, of a radar tracking system.

Section 2.4 introduces a few important topics. First we focus on the complete response of a system with a discussion of its transient and steady-state components. The concepts of stability (in the s- and z-planes), controllability, and observability are introduced next. The two stability methods covered are the Routh-Hurwitz and the root locus.

2.2 CONTINUOUS-TIME LINEAR SYSTEMS

In this section, mathematical models in terms of differential equations representing linear continuous-time physical systems are derived. The Laplace transform (and its inverse) are introduced; thus, the concept of transfer functions is developed. State-space models, based on state variables, are then introduced as an alternative. Systems representation by block diagrams or signal flow graphs is given, with simplification and reduction techniques leading to overall system transfer functions. It is useful to conclude this section with a method for system order estimation of physical systems; this is followed by a brief discussion of system order reduction as well as problems of model overdescription.

The mathematical model for a broad class of linear, or linearized, physical systems can be represented by an nth-order differential equation:

$$a_n \frac{d^n x}{dt^n} + a_{n-1} \frac{d^{n-1} x}{dt^{n-1}} + \cdots + a_o x = b_m \frac{d^m f}{dt^m} + \cdots + b_1 \frac{df}{dt} + b_o f \tag{2.1}$$

where the variations of output $x(t)$ and input $f(t)$ are functions of time t, the a's and b's are real constants, and ($n \geqslant m$).

2.2.1 Simple Electrical Circuits

The passive elements in electrical circuits are the resistance R (ohms), inductance L (henries), and capacitance C (farads). The relationship between the voltage $v(t)$ across each element and the current $i(t)$ through it is illustrated in Figure 2.1.

Circuits equations are obtained through the applicability of Kirchoff's voltage law (KVL) and current law (KCL). This is illustrated by the following two examples.

Example 2.2.1. $v_s(t)$ is an independent voltage source (input) which produces current i through the series R, L, and C circuit (Figure 2.2). Applying KVL around the circuit:

$$v_s = R\,i + L \frac{di}{dt} + \frac{1}{C} \int i \; dt \tag{2.2a}$$

Differentiating this equation with respect to time and rearranging yields:

$$\frac{dv_s}{dt} = L \frac{d^2 i}{dt^2} + R \frac{di}{dt} + \frac{1}{C} i \tag{2.2b}$$

which has the same form as Eq. (2.1) with $n = 2$, $a_2 = L$, $a_1 = R$, $a_o = 1/C$, and $b_o = 1$. In this circuit, one may view the current i or the voltage across any of the elements as an output. The voltage v_o across the capacitor is:

$$v_o = \frac{1}{C} \int i \; dt = v_C$$

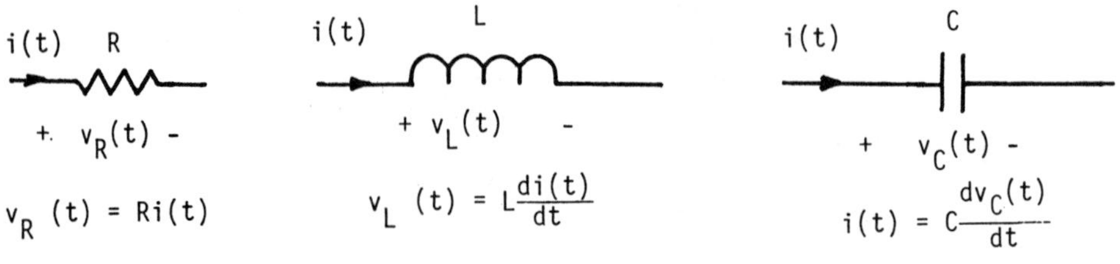

i(t)   R

+. $v_R(t)$ -

$v_R(t) = Ri(t)$

i(t)   L

+ $v_L(t)$ -

$v_L(t) = L\frac{di(t)}{dt}$

i(t)   C

+ $v_C(t)$ -

$i(t) = C\frac{dv_C(t)}{dt}$

Figure 2.1   Voltage-current relationships.

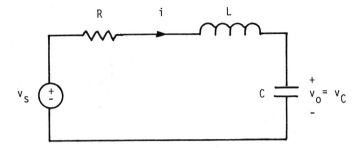

Figure 2.2   Series RLC circuit.

Example 2.2.2.   To illustrate the applicability of KCL at a node, examine the parallel circuit of Figure 2.3 excited by an independent current source $i_s(t)$. Summing all the currents at node A gives

$$i_s = \frac{v}{R} + \frac{1}{L} \int v \; dt + C \frac{dv}{dt} \tag{2.3}$$

Note that Ψ, the magnetic flux linkage, is described by

$$\frac{d\Psi}{dt} = v \tag{2.4}$$

Substituting in Eq. (2.3) and rearranging, one gets

$$i_s = C \frac{d^2\Psi}{dt^2} + \frac{1}{R} \frac{d\Psi}{dt} + \frac{1}{L} \Psi \tag{2.5}$$

which is of the same form as Eq. (2.1).

In the above examples, the situation may arise where initial conditions are expressed numerically in terms of initial current through L and/or initial voltage across C. Such conditions must be taken into consideration for a complete solution of the system differential equation (Kreyszig, 1972).

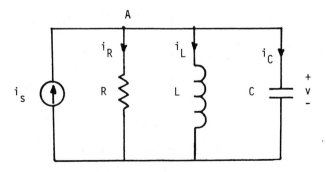

Figure 2.3   Parallel RLC circuit.

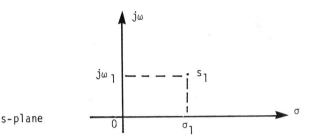

Figure 2.4   The complex s-plane.

Linear differential equations can be transformed into algebraic equations in a complex variable. The Laplace transform, based on complex variable theory, allows a simpler solution of a given differential equation; the obtained solution contains both the "transient" (source-free) component and the "steady-state" (source-dependent) component, simultaneously.

Before defining the Laplace transform of a function of time f(t), we review briefly the concept of a complex variable s. A complex variable s has two components: a real component (σ) and an imaginary component (ω). Thus, in the complex s-plane of Figure 2.4, any arbitrary value $s = \sigma + j\omega$ is defined by its corrdinates.

Now, if G(s) is a function of s, the complex variable, then for each value of s there is a corresponding value (or values) of G(s). To illustrate this dependency, G(s) is usually represented by the complex G-plane with the horizontal axis representing Re G(s), the real part, and the vertical axis, the imaginary part, labeled Im G(s). For a "single-valued" function, the correspondence (or mapping) between the s-plane and the G(s) plane is as shown in Figure 2.5.

$G(s_1)$ is uniquely determined for a selected s_1 for a given G(s). A complex function G(s) is said to be "analytic" in a region if G(s) and all its derivatives exist in that region. The important points in the s-plane at which G(s) is not analytic are called "singular" points. For example, let

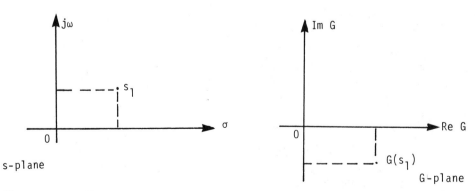

Figure 2.5   Single-valued mapping.

$$G(s) = \frac{K(s + z)}{s(s + p_1)(s + p_2)} \tag{2.6}$$

Here $G(s)$ or its derivatives approach infinity at the "poles" where $s = 0$, $s = -p_1$, and $s = -p_2$. Points in the s-plane at which $G(s)$ equals zero are called "zeros." The above function has three zeros, a finite zero at $s = z$ and two zeros at infinity [with the points at infinity included, $G(s)$ has an equal number of poles and zeros]. Now if $p_1 = p_2$, the function is said to have a pole with multiplicity two.

2.2.2   The Laplace Transform

A function $f(t)$ is transformable if

$$\int_0^\infty |f(t)| e^{-\sigma_1 t} dt < \infty \tag{2.7}$$

for some real, positive σ_1, known as the abscissa of absolute convergence. $F(s)$, the Laplace transform of $f(t)$, is

$$F(s) = \int_0^\infty f(t) e^{-st} dt = \mathcal{L}[f(t)] \tag{2.8}$$

Likewise, given $F(s)$, one expresses the inverse Laplace transform, $f(t)$, as

$$f(t) = \mathcal{L}^{-1}[F(s)] \tag{2.9}$$

which is defined as

$$f(t) = \frac{1}{2\pi j} \int_{c-j\infty}^{c+j\infty} F(s) e^{st} ds \tag{2.10}$$

where c is a real constant that is greater than the real part of all the singularities of $F(s)$. For many applications, $\mathcal{L}^{-1}$ is obtained by using the Laplace transform table.

Example 2.2.3.   Obtain the Laplace transform of $f(t) = e^{-at}$ for $t \geq 0$ where "a" is a constant.

$$F(s) = \int_0^\infty e^{-at} e^{-st} dt = \left. \frac{e^{-(s + a)t}}{-(s + a)} \right|_0^\infty = \frac{1}{(s + a)} \tag{2.11}$$

Example 2.2.4.   Evaluate $F(s)$ for $f(t)$, a ramp function, defined as

$$f(t) = 0 \qquad \text{for } t < 0$$

$$= Kt \qquad \text{for } t \geq 0 \qquad K \text{ constant}$$

$$\mathcal{L}[f(t)] = K \int_0^\infty t e^{-st} dt$$

$$= K\left[t\frac{e^{-st}}{-s}\bigg|_0^\infty - \int_0^\infty \frac{e^{-st}}{-s} dt\right]$$

$$= \frac{K}{s}\int_0^\infty e^{-st} dt = \frac{K}{s^2} \tag{2.12}$$

Laplace Transform Theorems. The following are useful Laplace transform theorems that are presented without proof.

1. If K is a constant, then

    $$\mathcal{L}[K f(t)] = KF(s) \tag{2.13}$$

2. If c_1 and c_2 are constants, then

    $$\mathcal{L}[c_1 f_1(t) \pm c_2 f_2(t)] = c_1 F_1(s) \pm c_2 F_2(s) \tag{2.14}$$

3. The Laplace transform of the first derivative of f(t) is found to be

    $$\mathcal{L}\frac{df(t)}{dt} = sF(s) - f(0) \tag{2.15}$$

    and for higher-order derivatives

    $$\mathcal{L}\left[\frac{d^n f(t)}{dt^n}\right] = s^n F(s) - s^{n-1}f(0) - s^{n-2}\dot{f}(0) - \cdots - f^{n-1}(0) \tag{2.16}$$

    If all the initial conditions of f(t) and its derivatives $\dot{f}(t)$, $\ddot{f}(t)$, . . ., $f^{n-1}(t)$ are zero, then

    $$\mathcal{L}\left[\frac{d^n f(t)}{dt^n}\right] = s^n F(s) \tag{2.17}$$

4. If $g(t) = \int f(t) dt$, then

    $$\mathcal{L}[g(t)] = \frac{F(s)}{s} + \frac{f^{-1}(0)}{s} \tag{2.18}$$

    where $f^{-1}(0) = \int f(t) dt$ evaluated at t = 0. Again, for zero initial conditions, the transform of the nth-order integral is $F(s)/s^n$.

5. A function f(t) delayed by time T is represented by f(t − T); see Figure 2.6. The Laplace transform of the delayed function is

    $$\mathcal{L}[f(t - T) u_{-1}(t - T)] = e^{-Ts}F(s) \tag{2.19}$$

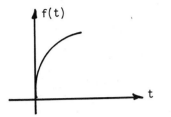 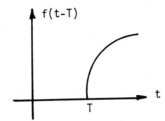

Figure 2.6   Delaying f(t) by T.

where $u_{-1}(t - T)$ is a unit step function shifted to the right (delayed) by time T. This is often termed "real-translation."

6. $\mathcal{L} [e^{-\alpha t} f(t)] = F(s + \alpha)$

This theorem is called "complex-translation."

Example 2.2.5. If the $\mathcal{L}[\sin\omega t] = \omega/(s^2 + \omega^2)$, then $\mathcal{L}[e^{-\alpha t}\sin\omega t] = \omega/[(s + \alpha)^2 + \omega^2]$.

7. The "final-value" theorem:

If $F(s) = \mathcal{L}[f(t)]$, then

$$\lim_{t\to\infty} f(t) = \lim_{s\to 0} sF(s) \qquad (2.20)$$

8. The "initial-value" theorem is

$$\lim_{t\to 0} f(t) = \lim_{s\to\infty} sF(s) \qquad (2.21)$$

Table 2.1 lists some Laplace transform pairs; for a more complete listing see, for example, Ogata (1970).

The Inverse Laplace Transform by Partial Fraction Expansion. As discussed later, a system transfer function F(s) may be represented in the form:

$$F(s) = \frac{N(s)}{D(s)} \qquad (2.22a)$$

where N(s) and D(s) are polynomials in s and the degree of D(s) is equal to or higher than that of N(s). Assuming distinct poles, the partial fraction expansion of this nth-order expression requires its rewriting as

$$F(s) = \frac{K(s + z_1) (s + z_2) \cdots (s + z_m)}{(s + p_1) (s + p_2) \cdots (s + p_n)} \qquad (2.22b)$$

Table 2.1   Some Laplace Transform Pairs

f(t)	F(s)
Unit impulse $u_0(t)$	1
Unit step $u_{-1}(t)$	$\dfrac{1}{s}$
Ramp t	$\dfrac{1}{s^2}$
$e^{-\alpha t}$	$\dfrac{1}{s + \alpha}$
t^n	$\dfrac{n!}{s^{n+1}}$
$t^n e^{-\alpha t}$	$\dfrac{n!}{(s + \alpha)^{n+1}}$
$\sin \omega t$	$\dfrac{\omega}{s^2 + \omega^2}$
$\cos \omega t$	$\dfrac{s}{s^2 + \omega^2}$
$e^{-\alpha t} \sin \omega t$	$\dfrac{\omega}{(s + \alpha)^2 + \omega^2}$
$e^{-\alpha t} \cos \omega t$	$\dfrac{s + \alpha}{(s + \alpha)^2 + \omega^2}$

where the zeros and poles are either real or complex quantities, and is
expanded in this form:

$$F(s) = \frac{K_1}{s + p_1} + \cdots + \frac{K_n}{s + p_n} \tag{2.23}$$

The K_i 's are constants and called the residues at the corresponding poles.
Here n is assumed to be greater than m [note that if this is not the case,
one must divide $N(s)$ by $D(s)$]. The evaluation of the residues is illustrated
by the following examples.

Example 2.2.6. Given

$$F(s) = \frac{(s + 5)}{(s + 1)(s + 3)}$$

find f(t) using the partial fraction expansion

$$F(s) = \frac{K_1}{s + 1} + \frac{K_2}{s + 3} \tag{2.24}$$

It is desired to obtain constants K_1 and K_2; K_1 is obtained by multiplying $F(s)$ by $(s + 1)$ and letting $s = -1$:

$$\therefore K_1 = (s + 1)F(s)\Big|_{s = -1} = \frac{s + 5}{s + 3}\Big|_{s = -1} = \frac{4}{2} = 2$$

$$K_2 = (s + 3)F(s)\Big|_{s = -3} = \frac{2}{-2} = -1$$

This procedure is often termed the "coverup" technique. The partial fraction expansion is therefore

$$F(s) = \frac{2}{s + 1} + \frac{-1}{s + 3} = F_1(s) + F_2(s) \tag{2.25}$$

The components $f_1(t)$ and $f_2(t)$ of solution $f(t)$ (using Table 2.1) result in

$$f(t) = 2e^{-t} - e^{-3t} \tag{2.26}$$

Now, if $D(s)$, the denominator of Eq. (2.22), is such that

$$D(s) = 0 = (s + p_1)^r (s + p_2) \cdots (s + p_n)$$ then, in this case the partial fraction expansion takes the form

$$F(s) = \frac{K_{11}}{(s + p_1)^r} + \frac{K_{12}}{(s + p_1)^{r-1}} + \cdots + \frac{K_{1j}}{(s + p_1)^{r-j+1}} + \cdots$$

$$+ \frac{K_{1r}}{(s + p_1)} + \frac{K_2}{(s + p_2)} + \cdots + \frac{K_n}{(s + p_n)} \tag{2.27}$$

For the distinct (simple) poles at $-p_2, \ldots, -p_n$, the corresponding residues $K_2, \ldots, K_n$ are evaluated as before. The residues, however, for a repeated pole, say at $-p_k$ with multiplicity r, are $K_{k1}, K_{k2}, \ldots, K_{kj}, \ldots, K_{kr}$ [in Eq. (2.27) $k = 1$] and are evaluated as follows:

$$K_{kj} = \frac{1}{(j - 1)!} \lim_{s \to -p_k} \frac{d^{j-1}}{ds^{j-1}} [(s + p_k)^r F(s)]$$

$$j = 1, 2, \ldots, r \tag{2.28}$$

Example 2.2.7. Find the inverse Laplace transform if

$$F(s) = \frac{s^3 + 3s^2 + 2s + 2}{(s + 1)^2(s + 2)} \tag{2.29}$$

It is evident that we must express F(s) as

$$F(s) = 1 - \frac{(s^2 + 3s)}{(s + 1)^2(s + 2)} = 1 + \frac{K_{11}}{(s + 1)^2} + \frac{K_{12}}{(s + 1)} + \frac{K_2}{(s + 2)} \qquad (2.30)$$

Using Eq. (2.28) for j = 1 and 2 yields

$$K_{11} = (s + 1)^2 F(s) \Big|_{s = -1} = 2$$

$$K_{12} = \frac{d}{ds} (s + 1)^2 F(s) \Big|_{s = -1} = \frac{d}{ds} \left[\frac{s^3 + 3s^2 + 2s + 2}{(s + 2)} \right]_{s = -1} = -3$$

The residue K_2 is

$$K_2 = (s + 2) F(s) \Big|_{s = -2} = 2$$

Rewriting Eq. (2.30) as

$$F(s) = 1 + \frac{2}{(s + 1)^2} + \frac{-3}{(s + 1)} + \frac{2}{(s + 2)}$$

and using Table 2.1 gives

$$f(t) = u_0(t) + 2te^{-t} - 3e^{-t} + 2e^{-2t} \qquad (2.31)$$

2.2.3   Transfer Functions

A transfer function of a linear time-invariant system is defined as the ratio of the Laplace transform of the output to the Laplace transform of the input. All initial conditions are assumed to be zero. This input-output relationship is useful; however, it does not provide any information concerning the physical structure of the system. This is due to the fact that a transfer function may represent a number of systems; for instance, a mechanical system and an electrical system may both end up having identical transfer functions (such systems are called analogous systems). A few examples are selected for illustration.

   Example 2.2.8.   To find the transfer of the simple R-C circuit shown in Figure 2.7(a), we redraw the circuit using frequency-dependent variables, as in part b, with

$$I(s)   = \mathcal{L}[i(t)]$$

$$V_s(s) = \mathcal{L}[v_s(t)]$$

$$V_o(s) = \mathcal{L}[v_o(t)]$$

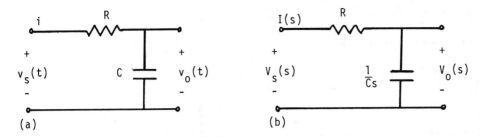

Figure 2.7   First-order RC circuit.

Recalling the voltage-current relationships in Figure 2.1, we write

$$V_o(s) = (1/Cs)I(s)$$

Also,

$$V_i(s) = (R + 1/Cs)I(s)$$

thus

$$F(s) = \frac{V_o(s)}{V_i(s)} = \frac{1/RC}{s + 1/RC} \qquad (2.32)$$

Example 2.2.9.   A second-order mechanical system is considered next (Figure 2.8).   In this spring-mass–dashpot system, M is the mass, K is spring constant, B is the damping coefficient of the damper; p(t) and x(t) are the input (force) and output (displacement), respectively.   The differential equation of this system is

$$M\ddot{x} + B\dot{x} + Kx = p \qquad (2.33)$$

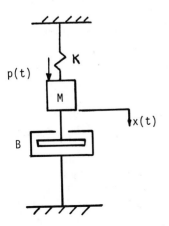

Figure 2.8   Second-order mechanical system.

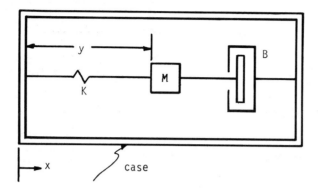

Figure 2.9   An accelerometer.

The analogy between the circuit of Figure 2.2 and this system is evidenced when their differential equations, Eq. (2.5) and Eq. (2.33), are compared. Using Eq. (2.33), the transfer function can be written as follows:

$$F(s) = \frac{X(s)}{P(s)} = \frac{1/M}{s^2 + Bs/M + K/M} \tag{2.34}$$

Example 2.2.10.   Figure 2.9 shows a simple mechanical accelerometer where K, M, and B are as defined earlier. The case position is x, while y is the position of the mass with respect to the case and is proportional to the acceleration of the case. It is desired to obtain the transfer function relating the accceleration (input) to the mass position y (output).

Considering the forces acting on M, one obtains

$$M \frac{d^2}{dt^2} (y - x) = -B \frac{dy}{dt} - Ky \tag{2.35}$$

$$M \frac{d^2 y}{dt^2} + B \frac{dy}{dt} + Ky = M \frac{d^2 x}{dt^2} \tag{2.36}$$

Letting $a(t) = d^2 x/dt^2$ be input acceleration and A(s) be its transform, we rewrite Eq. (2.36) as follows:

$$(Ms^2 + Bs + K) Y(s) = M.A(s)$$

thus

$$F(s) = \frac{Y(s)}{A(s)} = \frac{M}{Ms^2 + Bs + K} = \frac{1}{s^2 + Bs/M + K/M} \tag{2.37}$$

Example 2.2.11.   DC motors are either armature controlled (with fixed field current) or field controlled (with fixed armature current). Here we consider an armature-controlled motor with an input signal v_a applied to the

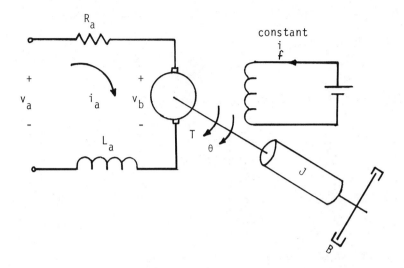

Figure 2.10   Armature-controlled motor.

armature terminals (see Figure 2.10).   The motor variables and parameters are

R_a; L_a :   armature resistance and inductance
  i_a: armature current
  ϕ :   air-gap flux
  v_b :   back electromotive force (emf) voltage
  i_f: field current
K_b; K_i :   back emf constant; torque constant
  T: motor-developed torque
  J: motor and load moment of inertia
  B: motor and load viscous friction coefficient
  θ: angular displacement of motor shaft

Since i_f is constant, the torque T and the armature current i_a are related as follows:

$$T = K_i i_a \qquad\qquad (2.38)$$

The back emf v_b is directly proportional to the motor speed:

$$v_b = K_b \frac{d\theta}{dt} = K_b \omega \qquad\qquad (2.39)$$

Applying KVL around the armature circuit one gets

$$v_a = R_a i_a + v_b + L_a \frac{di_a}{dt} \qquad\qquad (2.40)$$

The torque produced is applied to the inertia and the friction, thus:

$$T = J \frac{d^2\theta}{dt^2} + B \frac{d\theta}{dt} = K_i \, i_a \tag{2.41}$$

The Laplace transform of Eq. (2.39), (2.40), and (2.41) yields

$$V_b(s) = K_b \, s\theta(s) \tag{2.42}$$

$$V_a(s) = V_b(s) + (R_a + L_a s) \, I_a(s) \tag{2.43}$$

$$T(s) = K_i \, I_a(s) = (Js^2 + Bs) \, \Theta(s) \tag{2.44}$$

The transfer function is obtained as

$$\frac{\theta(s)}{V_a(s)} = \frac{K_i}{s[L_a Js^2 + (R_a J + BL_a)s + (K_b K_i + BR_a)]} \tag{2.45}$$

Example 2.2.12. Consider a simple thermal system made out of a wall, of surface area A, separating two regions with temperature difference $(T_1 - T_2)$. The heat flow rate is defined as

$$q = hA(T_1 - T_2) \tag{2.46}$$

where h is the heat transfer coefficient. If R_t, thermal resistance, is defined as

$$R_t = 1/hA, \text{ then} \tag{2.47}$$

$$q = (T_1 - T_2)/R_t \tag{2.48}$$

Another relationship exists between the heat flow rate and the temperature gradient $(T_1 - T_2)/d$ where d is the wall thickness. This takes the form

$$q = kA \frac{(T_1 - T_2)}{d} \tag{2.49}$$

here k is the thermal conductivity. From Eq. (2.46) and Eq. (2.49), $h = k/d$.

The analogy between this situation and an electric resistance becomes obvious when $(T_1 - T_2)$ is compared with v, q with the current i, and R_t with resistance R [see Eq. (2.48)].

Next, we present the concept of thermal capacitance of a material with volume V, mass density ρ, and specific heat c. C_t, thermal capacitance, which is analogous to an electrical capacitance (recall $i = Cdv/dt$), is related to the rate of change in temperature as follows:

$$q = C_t \frac{dT}{dt}$$

$$C_t = \rho Vc; \qquad \text{thus } q = (\rho Vc) \frac{dT}{dt} \qquad (2.50)$$

In the above, (ρV) is the mass of the material and (ρVCT) is an expression of the heat stored at a temperature T.

Example 2.2.13. Consider the problem of space heating. Let us define q_i as the heat inflow to the space to be heated (say from an electrical heater) where T is the difference with a constant ambient temperature; R_t and C_t are the thermal resistance and capacitance, respectively; and q_o is the heat loss. Thus from Eq. (2.48) we get $q_o = T/R_t$; also

$$(q_i - q_o) = C_t \frac{dT}{dt} \qquad (2.51)$$

Thus, the transfer function of this thermal system can be easily found to be:

$$\frac{T(s)}{Q_i(s)} = \frac{R_t}{1 + R_t C_t s} \qquad (2.52)$$

Example 2.2.14. This example follows the treatment by Lewis et al. (1969) and deals with a two-input mechanical system as in Figure 2.11, where p_1 and p_2 are external driving forces. The three blocks with masses M_1, M_2, and M_3 are allowed to slide longitudinally with respect to the stationary machine bed. The three unknown velocities are $\dot{x}_1$, $\dot{x}_2$, and $\dot{x}_3$, and K is stiffness constant of the spring which restrains the middle block. The symbolic diagram of Figure 2.12 makes the derivation of the mathematical model somewhat easier. The elements are connected to the three vertical lines which have the velocities as shown. Following the relationships previously used in Ex. (2.2.9) and (2.2.10), the equations expressing the summation of forces are

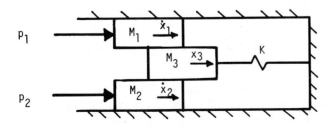

Figure 2.11   System of Ex. (2.2.14).

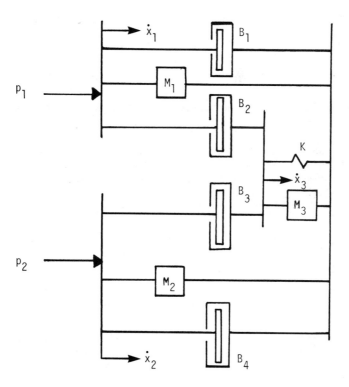

Figure 2.12   Symbolic diagram [Ex. (2.2.14)].

$$p_1 = B_1 \dot{x}_1 + B_2(\dot{x}_1 - \dot{x}_3) + M_1 \ddot{x}_1$$

$$p_2 = B_3(\dot{x}_2 - \dot{x}_3) + M_2 \ddot{x}_2 + B_4 \dot{x}_2$$

$$0 = B_2(\dot{x}_3 - \dot{x}_1) + B_3(\dot{x}_3 - \dot{x}_2) + M_3 \ddot{x}_3 + K x_3 \qquad (2.53)$$

Rearranging these equations results in

$$p_1 = \left(B_1 + B_2 + M_1 \frac{d}{dt}\right) \dot{x}_1 - B_2 \dot{x}_3$$

$$p_2 = \left(B_3 + B_4 + M_2 \frac{d}{dt}\right) \dot{x}_2 - B_3 \dot{x}_3$$

$$0 = -B_2 \dot{x}_1 - B_3 \dot{x}_2 + \left(B_2 + B_3 + M_3 \frac{d}{dt} + K \int dt\right) \dot{x}_3 \qquad (2.54)$$

A set of equations similar to Eq. (2.54) represents the electric circuit of Figure 2.13 driven by two current sources and resulting in the nodal voltages v_1, v_2, and v_3. This is a force-current and a velocity-voltage

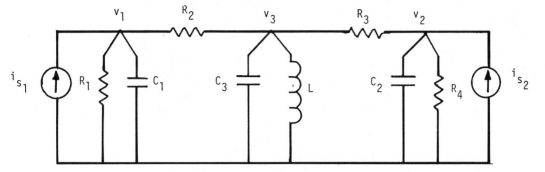

Figure 2.13 An analogous electric circuit [Ex. (2.2.14)].

correspondence. The circuit of Figure 2.13 has the following mathematical model when KCL is applied at the three non-reference nodes.

$$i_{s_1} = \left(\frac{1}{R_1} + \frac{1}{R_2} + C_1 \frac{d}{dt} \right) v_1 - \frac{1}{R_2} v_3$$

$$i_{s_2} = \left(\frac{1}{R_3} + \frac{1}{R_4} + C_2 \frac{d}{dt} \right) v_2 - \frac{1}{R_3} v_3$$

$$0 = -\frac{1}{R_2} v_1 - \frac{1}{R_3} v_2 + \left(\frac{1}{R_2} + \frac{1}{R_3} + C_3 \frac{d}{dt} + \frac{1}{L} \int dt \right) v_3 \qquad (2.55)$$

Of course one should perform the Laplace transform on Eq. (2.54) and/or Eq. (2.55) to obtain the required transfer functions of these analogous systems.

2.2.4   State-Space Models

The derivation of state-space models is similar to that of transfer functions, described above, in the sense that differential equations describing the system are derived first. The key advantage of transfer functions is in their compactness, which makes them suitable for frequency-domain analysis and stability studies. However, the transfer function approach suffers from neglecting the initial conditions. Not only does state-space representation serve as an alternative to transfer functions, but also it is not limited to linear or time-invariant systems.

Basically, the equations in the state model are arranged as a set of first-order differential equations in terms of the selected "state variables." All other variables, however, are expressed in terms of their state variables. The state variables $x_1(t)$, . . . , $x_n(t)$ of an nth-order system describe the dynamics of the system under study and are the minimal amount of information required at time t_1 to completely determine future behavior for all $t > t_1$. The values at a selected initial time t_0 define the initial status of the system. Knowledge of $x_1(t_0)$, ..., $x_n(t_0)$ is sufficient to determine the behavior of the system for any $t > t_0$ for a given input.

Let $\underline{x}$ be the n-state vector and $\underline{u}$ the m-input vector; then the n-state equations can be written as

$$\dot{\underline{x}} = \underline{A}\underline{x} + \underline{B}\underline{u} \tag{2.56}$$

where A is an (n × n) square matrix, B is an (n × m) matrix known as the distribution matrix, and $\dot{\underline{x}} = d\underline{x}/dt$. The output equation describes output vector $\underline{y}$ in terms of state vector $\underline{x}$ and the input vector $\underline{u}$ as follows:

$$\underline{y} = \underline{C}\underline{x} + \underline{D}\underline{u} \tag{2.57}$$

where $\underline{C}$ and $\underline{D}$ are matrices with appropriate dimensions. The next few examples illustrate the derivation of state models.

Example 2.2.15. Let us reconsider the second-order mechanical system of Ex. (2.2.9) with its differential Eq. (2.33) repeated here as

$$M\ddot{x} + B\dot{x} + Kx = p$$

Let the state variables be $x_1 = x$ and $x_2 = \dot{x}$, the input u = p, and output $y = x_1$. Thus, the state model is

$$\dot{x}_1 = x_2$$

$$\dot{x}_2 = -\frac{B}{M}x_2 - \frac{K}{M}x_1 + \frac{u}{M}$$

$$\dot{\underline{x}} = \begin{bmatrix} 1 & 1 \\ -\dfrac{K}{M} & -\dfrac{B}{M} \end{bmatrix} \begin{bmatrix} x_1 \\ x_2 \end{bmatrix} + \begin{bmatrix} 0 \\ \dfrac{1}{M} \end{bmatrix} u$$

$$\underline{y} = [1 \qquad 0] \ \underline{x}; \qquad \underline{D} = 0 \tag{2.58}$$

Example 2.2.16. The RLC series circuit of Ex. (2.2.1) shown in Figure 2.2 is revisited. The differential Eq. (2.2a) of this circuit was found to be

$$v_s = L\frac{di}{dt} + Ri + \frac{1}{C} \int i \, dt \tag{2.59}$$

It is convenient in electrical circuits to choose voltages across capacitors and currents through inductors as state variables. Thus, the state variables are defined as

$$x_1 = i \qquad \text{and} \qquad x_2 = v_c \tag{2.60}$$

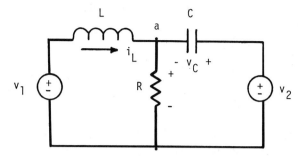

Figure 2.14   Circuit of Ex. (2.2.17).

Using Eq. (2.59) and Eq. (2.60), we get the two state equations as

$$\dot{x}_1 = -\frac{R}{L}x_1 - \frac{1}{L}x_2 + \frac{1}{L}v_s$$

$$\dot{x}_2 = \frac{1}{C}x_1 \tag{2.61}$$

It is easy to get $\underline{A}$ and $\underline{B}$ for this second-order example from Eq. (2.61).

Example 2.2.17. As another example, let us obtain the state equations for the RLC network of Figure 2.14. The state variables are $x_1 = i_L$ and $x_2 = v_C$, and the state equations are derived through, say, applying KVL around the outer loop and KCL at node a. First,

$$-v_1 + L\frac{d}{dt}i_L + (v_2 - v_c) = 0$$

yields

$$\frac{dx_1}{dt} = \frac{1}{L}x_2 + \frac{1}{L}(v_1 - v_2) \tag{2.62}$$

The second equation is

$$i_L - \frac{1}{R}(v_2 - v_c) + C\frac{dv_C}{dt} = 0 \tag{2.63}$$

Rewriting Eq. (2.63) in terms of x_1, x_2, and the input voltages we get

$$\frac{dx_2}{dt} = -\frac{1}{C}x_1 - \frac{1}{RC}x_2 + \frac{1}{RC}v_2 \tag{2.64}$$

Rewriting Eq. (2.63) and Eq. (2.64) in the general form of Eq. (2.56), we have

$$\underline{\dot{x}} = \begin{bmatrix} 0 & \dfrac{1}{L} \\ -\dfrac{1}{C} & -\dfrac{1}{RC} \end{bmatrix} \begin{bmatrix} x_1 \\ x_2 \end{bmatrix} + \begin{bmatrix} \dfrac{1}{L} & -\dfrac{1}{L} \\ 0 & \dfrac{1}{RC} \end{bmatrix} \begin{bmatrix} v_1 \\ v_2 \end{bmatrix} \qquad (2.65)$$

Here

$$\underline{x} = \begin{bmatrix} i_L \\ v_c \end{bmatrix} \qquad \text{and} \qquad \underline{u} = \begin{bmatrix} v_1 \\ v_2 \end{bmatrix}$$

Example 2.2.18. We have developed the third-order transfer function for the armature-controlled motor in Ex. (2.2.11), and it is desired that we obtain its state model. Refer to Figure 2.10 and let

$$x_1 = i_a \qquad \text{(the armature current)},$$

$$x_2 = \omega = \frac{d\Theta}{dt}$$

$$x_3 = \Theta$$

Now, for Eq. (2.40) we obtain

$$\frac{di_a}{dt} = -\frac{R_a}{L_a} i_a - \frac{K_b}{L_a} \omega + \frac{v_a}{L_a}$$

Rearranging Eq. (2.39) and using $\omega = \dfrac{d\Theta}{dt}$, the second-state equation is obtained:

$$\frac{d\omega}{dt} = \frac{K_i}{J} i_a - \frac{B}{J} \omega$$

Thus the state model is

$$\begin{bmatrix} \dot{x}_1 \\ \dot{x}_2 \\ \dot{x}_3 \end{bmatrix} = \begin{bmatrix} -\dfrac{R_a}{L_a} & -\dfrac{K_b}{L_a} & 0 \\ \dfrac{K_i}{J} & \dfrac{B}{J} & 0 \\ 0 & 1 & 0 \end{bmatrix} \begin{bmatrix} i_a \\ \omega \\ \Theta \end{bmatrix} + \begin{bmatrix} \dfrac{1}{L_a} \\ 0 \\ 0 \end{bmatrix} v_a \qquad (2.66)$$

Example 2.2.19. Consider the system defined by this transfer function:

$$\frac{W(s)}{P(s)} = \frac{6}{s^3 + 6s^2 + 11s + 6}$$

The differential equation for this system is

$$\dddot{w} + 6\ddot{w} + 11\dot{w} + 6w = 6p \tag{2.67}$$

where w and p are the output and input, respectively. It is desired to obtain the corresponding state model. Let us choose the state variables as

$$x_1 = w$$

$$x_2 = \dot{w}$$

$$x_3 = \ddot{w}$$

From these and Eq. (2.67) we obtain

$$\dot{\underline{x}} = \begin{bmatrix} 0 & 1 & 0 \\ 0 & 0 & 1 \\ -6 & -11 & -6 \end{bmatrix} \begin{bmatrix} x_1 \\ x_2 \\ x_3 \end{bmatrix} + \begin{bmatrix} 0 \\ 0 \\ 6 \end{bmatrix} p$$

Also, the output w is expressed as follows:

$$w = \begin{bmatrix} 1 & 0 & 0 \end{bmatrix} \begin{bmatrix} x_1 \\ x_2 \\ x_3 \end{bmatrix} \tag{2.68}$$

2.2.5 Systems Representation

"Block diagrams" have proven to be an effective tool in portraying systems. A block diagram shows how a system's components are interconnected and also represents cause-and-effect relationships between variables. A block may represent a linear or a nonlinear element; in a linear system, however, the overall system transfer function can be obtained through block-diagram algebra. In the next section, signal flow graphs are presented as a simplified notation for block diagrams.

Figure 2.15 illustrates the usefulness of block diagrams in representing systems; part a depicts an engine-speed control system where a set of flyweights is used to measure the speed; part b illustrates the economic law of supply and demand where a non-zero error (difference between supply and demand) allows an initiation of price change (by pricer) such that eventually supply equals demand. While the engine system has one feedback signal (the actual speed), the economic system has two.

The relationships between the variables at a summing node and at a pickoff point are illustrated in Figure 2.16 (a and b, respectively).

The reduction of a system, the components of which are connected in cascade (tandem), is shown in Figure 2.17 where $F_3(s)/F_1(s) = G_1(s)G_2(s)$ in both cases. This is easily illustrated when F_3/F_1 is expressed as

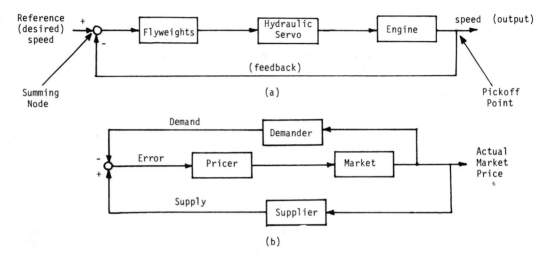

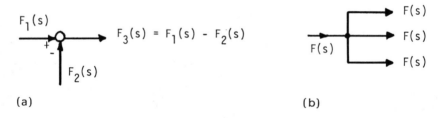

Figure 2.15   Block-diagram representations.

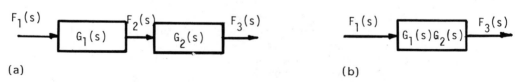

(a)   (b)

Figure 2.16   (a) A summing node.   (b) A pickoff point.

Figure 2.17   (a) Cascaded subsystems.   (b) Equivalent system.

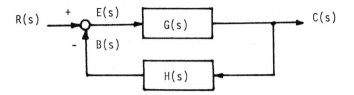

Figure 2.18   Feedback system.

$$\frac{F_3(s)}{F_1(s)} = \frac{F_3(s)}{F_2(s)} \qquad \frac{F_2(s)}{F_1(s)} = G_2(s)G_1(s) = G_1(s).G_2(s)$$

The single-feedback-loop system is a fundamental configuration and may, or may not, be with unity feedback. Consider the system of Figure 2.18 where $R(s)$ is input, $C(s)$ output, and $E(s)$ error; $G(s)$ and $H(s)$ are the feed-forward and feedback blocks, respectively. It is desired to obtain the $C(s)/R(s)$, the closed-loop transfer function.

Since $C(s) = G(s)E(s) = G(s)[R(s) - B(s)]$, and $B(s) = H(s)C(s)$, rewriting we obtain

$$\frac{C(s)}{R(s)} = \frac{G(s)}{1 + G(s)H(s)} \qquad (2.69)$$

This is a negative nonunity feedback system.

Table 2.2 presents a summary of block-diagram-reduction (transformation) aiding tools. The original and the reduced (or equivalent) configurations are shown on the left and right sides of the table, respectively.

Example 2.2.20. It is desired to construct the block diagram of the armature-controlled motor of Ex. 2.2.11 and Figure 2.10. Using the system Eq. (2.42), Eq. (2.43), and Eq. (2.44) and choosing $V_a(s)$, the armature voltage, to be system input and $\Theta(s)$, armature displacement of motor shaft, to be its output, the closed-loop system of Figure 2.19 is obtained. To get the transfer function $\theta(s)/V_a(s)$ of Figure 2.19, one may use the proposed simplifications of Figures 2.17 and 2.18; the result should be the same as Eq. (2.45).

Example 2.2.21. Draw the block diagram of a two-input, two-output system where the transfer relations are described by

$$\begin{bmatrix} C_1(s) \\ C_2(s) \end{bmatrix} = \begin{bmatrix} G_{11}(s) & G_{12}(s) \\ G_{21}(s) & G_{22}(s) \end{bmatrix} \begin{bmatrix} R_1(s) \\ R_2(s) \end{bmatrix} \qquad (2.70)$$

Figure 2.20 depicts the block diagram for this multivariable system.

Example 2.2.22.   The overall transfer function of the system depicted in Figure 2.21(a) is obtained as shown in parts b, c, d, e.

"Signal flow graph (SFG)" models represent an alternative method for determining the relationship between system variables. This representation

Table 2.2   Block-Diagram Algebra

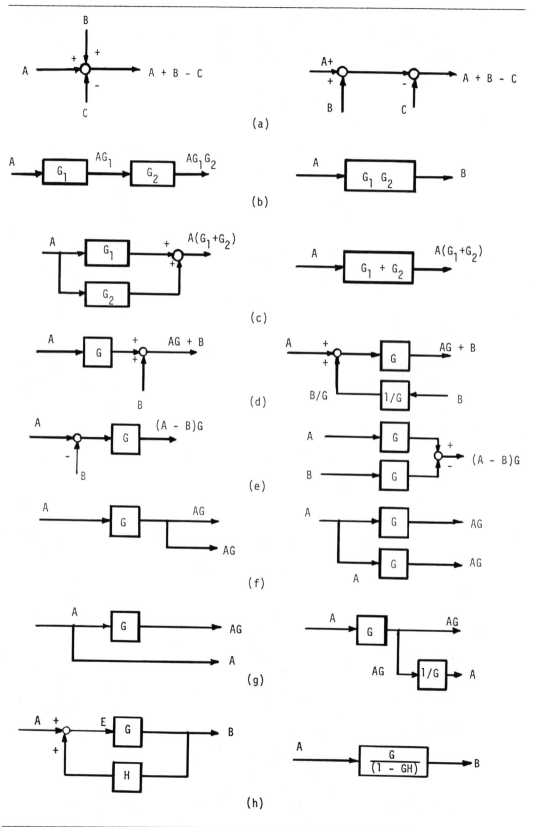

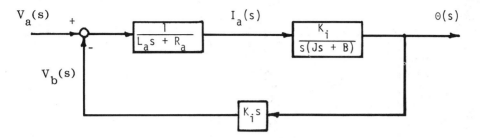

Figure 2.19   System of Ex. (2.2.20).

is attractive since block-diagram-reduction procedure may become difficult for complex systems. A SFG is a diagram consisting of nodes that are connected by line segments called branches. Each branch is a unidirectional path of signals and is labeled with the gain (transmittance) describing the relation between the input and output variables. In Figure 2.22, similar to block-diagram representation, $X_2(s) = G(s)X_1(s)$. Mason's gain formula for SFG provides the necessary relations between system variables without any required manipulation or reduction.

Consider the multivariable system depicted in Figure 2.20; the equivalent SFG is shown in Figure 2.23 where $R_1(s)$ and $R_2(s)$ represent the input nodes, while $C_1(s)$ and $C_2(s)$ represent the output nodes.

As another example, we show in Figure 2.24 the SFG corresponding to the system of Figure 2.21(c).

It is important to note that a node adds the signals of all incoming branches and transmits this sum (the value of the variable) to all outgoing branches. Thus, in Figure 2.24 we have

$$E = R - GX, \qquad X = KE, \qquad \text{and } C = \left(\frac{1}{s + 2}\right)X$$

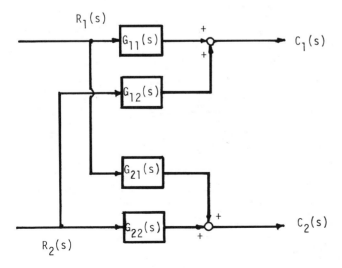

Figure 2.20   A multivariable system.

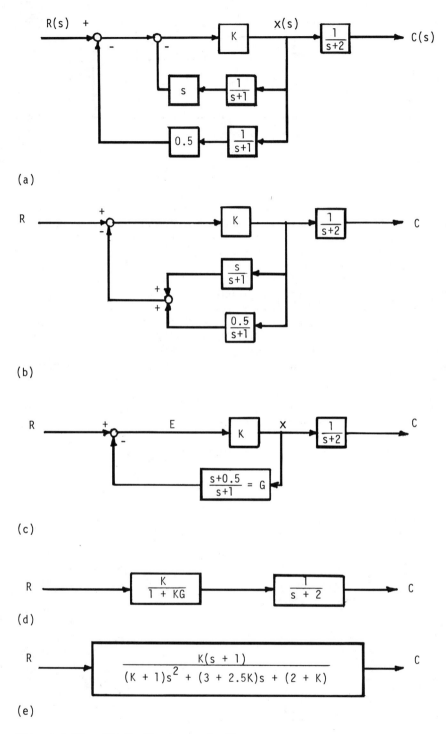

(a)

(b)

(c)

(d)

(e)

Figure 2.21   Block-diagram reduction.

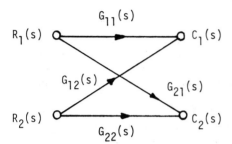

Figure 2.22 Definition of a branch.

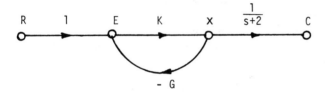

Figure 2.23 Multivariable example (See Fig. 2.20).

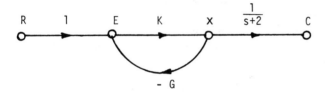

Figure 2.24 SFG for Figure 2.21(c).

Mason's gain formula (Mason, 1956) is very useful in obtaining the closed-loop transfer function for a SFG model. The gain

$$P = \frac{1}{\Delta} \sum_k P_k \Delta_k \tag{2.71}$$

where

P_k is the path gain of the kth forward path from the input node to the output node (any such path must not cross any node more than once)

Δ is the determinant of the graph defined as: $= 1 -$ (sum of all different loop gains) $+$ (sum of gain products of all possible combinations of two nontouching loops)
$-$ (sum of gain products of all possible combinations of three nontouching loops) $+ \cdot \cdot \cdot$

A loop is a closed path again with no node crossed more than once, except the node when the loop begins and ends.

Δ_k is the cofactor of the forward path P_k (it is obtained from Δ by removing the loops that touch the kth forward path).

In general, a path is considered touching a loop (or loop touching a loop) if they have a common node.

We illustrate the use of Mason's formula with two examples.

Example 2.2.23. Draw the SFG for the system shown in Figure 2.21(a) and obtain its C(s)/R(s) using Mason's gain formula; see Figure 2.25.

$$P_1 = (1)(1)(k)\left(\frac{1}{s+2}\right) = K/(s+2)$$

$$\Delta = 1 - \left[\frac{-Ks}{s+1} - \frac{0.5K}{s+1}\right]$$

$$= 1 + \frac{Ks}{s+1} + \frac{0.5K}{s+1}$$

$$\Delta_1 = 1$$

thus,

$$\frac{C(s)}{R(s)} = \frac{P_1 \Delta_1}{\Delta}$$

$$= \frac{K/(s+2)}{1 + \frac{Ks}{s+1} + \frac{0.5K}{s+1}}$$

$$= \frac{K(s+1)}{(K+1)s^2 + (3+2.5K)s + (2+K)} \tag{2.72}$$

Compare with the result in Figure 2.21(e).

Example 2.2.24. Construct the SFG for the system given in Figure 2.26(a) and obtain C(s)/D(s) where D(s) is considered a disturbance input. The required SFG is depicted in Figure 2.26(b); with inspection, there is only one path between D and C; the graph has four loops with only two pairs of two nontouching loops.

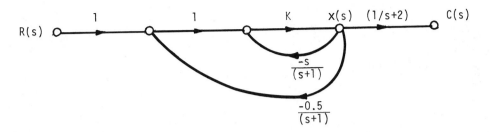

Figure 2.25   SFG of Ex. (2.2.23).

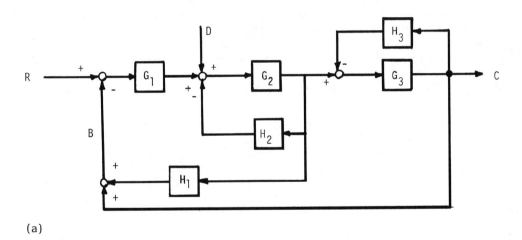

(a)

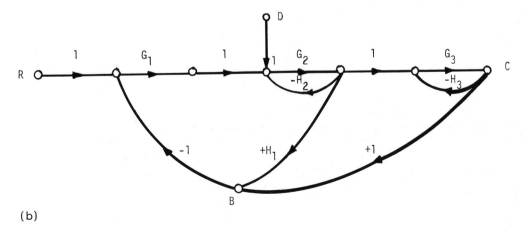

(b)

Figure 2.26   (a) System of Ex. (2.2.24).   (b) The required SFG.

$$P_1 = G_2 G_3$$

$$\Delta = 1 - (-G_1 G_2 G_3 - G_1 G_2 H_1 - G_2 H_2 - G_3 H_3)$$

$$+ \left[(-G_2 H_2)(-G_3 H_3) + (-G_3 H_3)(-G_1 G_2 H_1) \right]$$

$$\Delta_1 = 1$$

Therefore,

$$\frac{C(s)}{D(s)} = \frac{G_2 G_3}{1 + G_1 G_2 G_3 + G_1 G_2 H_1 + G_2 H_2 + G_3 H_3 + G_2 G_3 H_2 H_3 + G_1 G_2 G_3 H_1 H_3}$$

In a similar fashion, one may develop C/R, and if our system is linear, the response to both R and D becomes readily available.

In the example that follows, we illustrate the use of SFG in representing "state diagrams." Let us recall the armature-controlled motor presented earlier where the three states were: i_a, ω, θ.

Example 2.2.25. Establish the state diagram for the armature-controlled motor of Ex. (2.2.18).

Taking the Laplace transform of Eq. (2.66), the state model, we obtain:

$$s I_a(s) = -\frac{R_a}{L_a} I_a(s) - \frac{K_b}{L_a} \omega(s) + \frac{1}{L_a} V_a(s)$$

$$s \omega(s) = \frac{K_i}{J} I_a(s) - \frac{B}{J} \omega(s)$$

$$s \theta(s) = \omega(s) \tag{2.73}$$

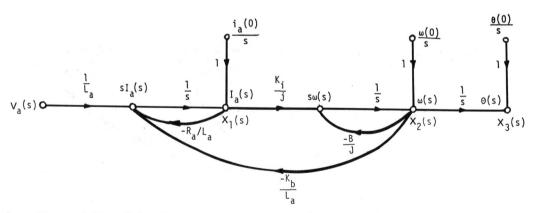

Figure 2.27   State diagram.

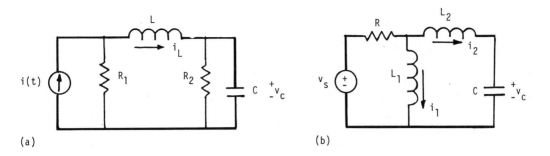

Figure 2.28 (a) Second-order RLC circuit. (b) Third-order RLC circuit.

Figure 2.27 shows the state diagram using the above relations assuming non-zero initial-state values: $i_a(0)$, $\omega(0)$, and $\Theta(o)$. In this third-order system model, three integrators are needed, and the Laplace of the state variables $I_a(s)$, $\omega(s)$, and $\Theta(s)$ appear as the output nodes, $X_1(s)$, $X_2(s)$, and $X_3(s)$, of these integrators.

2.2.6 Order of System

It is apparent from the earlier sections of this chapter that the order of the system output terms is the order of the system. This is most obvious with differential equation models. The state-space models, as you may also recall [as in Ex. (2.2.16) and Ex. (2.2.17)], showed that the order of the system is equal to the number of energy storage elements (capacitors and inductors in electrical circuits). It becomes not only interesting but also valuable to be able to predict the order of the system before actually deriving the mathematical model. Figure 2.28 depicts two circuits where the selected state variables and system order are as shown.

It is possible, however, for some systems to have an order that is lower than the number of energy storage elements. Consider, for example, a system with two parallel springs; these obviously can be replaced by one element only, and this system order is expected to be lower than the total number of energy storage elements. This same situation arises also in electrical circuits if an all-capacitor circuit or an all-inductor cutset (often called degenerate circuit and degenerate cutset, respectively) exists (Chan et al., 1972). In such cases, the order of the circuits (number of state variables) is expected to be equal to the number of capacitors "plus" the number of inductors in the circuit "minus" the number of any existing all-capacitor and/or all-inductor sections. This is demonstrated by the two circuits given in Figure 2.29. By examining the RC circuit in part a, it is easily seen that v_{c1} and v_{c2} are sufficient in this case since the voltage across C_3 is dependent on v_{c1} and v_{c2}; thus the circuit is a second-order one. Likewise, i_1, i_2, and v_c are the selected state variables for the RLC circuit of part b; the current through L_3 again is dependent on i_1 and i_2 and is no longer arbitrary. In these degenerate circuits, we can state that the order is actually equal to the number of "independent" energy storage elements. Let us examine Figure 2.30, the last example in this section. The voltage across C_1 is not an unknown (it is specified by the voltage source); now,

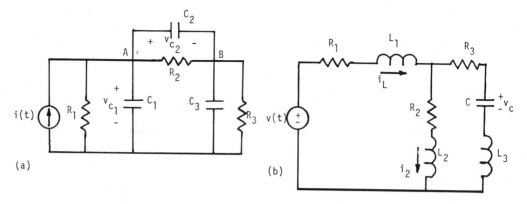

Figure 2.29   (a) Second-order RC circuit.   (b) Third-order RLC circuit.

if the voltage v_{c2} is selected as the state variable, the voltage across C_3 will be specified; thus this is a circuit of order one only.

In the sequel we briefly present two items that are related to system order; first "reduced-order models" are presented followed by what is termed "overmodeling" or "state overdescription." These become important especially with large and complex systems.

2.2.7   Reduced-Order Models

It is desirable in many practical situations, especially when dealing with complex systems, to use "simplification(s)" that lead to reduced-order models. This is in contrast to what we have presented so far, full-order models. This process of simplification and order reduction has the merit of cost reduction (related to programming activities and computing-time); moreover, the simpler-to-interpret results usually improve the vehicle of communication between the model builder, user, and decision maker.

In a typical study of, for example, a large power system, the number of power-generating units considered may be in the hundreds. With each unit represented by a ninth-order differential equation, the total number of

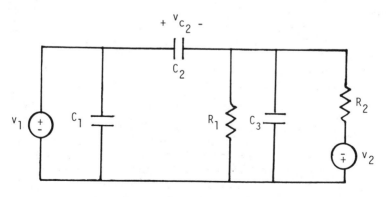

Figure 2.30   A circuit of order one.

differential equations will obviously be in the thousands. Complicating the problem even further is the consideration of the algebraic equations for the energy transmission network; these must be solved simultaneously with the differential equations. Certainly, reduction of the model order would be very desirable.

Our second example is drawn from the analysis of electric machinery, namely, the induction motor; where both the stator and rotor circuits are usually considered in a complete model. Two reduced-order models are often employed in calculating the electromagnetic torque during large transient excursions. One model results from neglecting transients in both stator and rotor circuits, while the other results from ignoring transients in the stator voltage equations only. Krause (1986) compares the motor's behavior predicted by the complete (full-order) model with those obtained from the reduced-order models. For synchronous machines as well, a large number of reduced-order models are used to predict large-excursion dynamic behavior.

A third example originates from applications where the assumption that all state variables can be measured may not be valid. That is, certain components of the state vector may correspond to inaccessible (internal) variables, which may not be available for measurement. The theory of "observers" (Luenberger, 1979) has been proposed to construct an approximation to the state vector based only on available measurements. Being a dynamic system in itself, the observer is designed to track the n state variables of the original system. However, reduced-order observers may be used to construct all $(n - p)$ nonaccessible state variables where p independent output variables are known precisely.

Several methods have been suggested for the reduction of the order of dynamic models (Van Ness, 1977); these are listed here without further presentation at this point: the "classical" method, the "modal" method, and the "topological" method. Other methods include aggregation principle, singular perturbation, error minimization, and the Pade-type approximation (Kwatny and Mablekos, 1975).

Regardless of the method used, the modeler faces the question of how close the results obtained from a study by a reduced-order model correspond to (agree with) those observed from the actual system and/or its original (full-order) model. Although there is no easy or unique answer, one may choose, say, in a nonlinear application to compare behaviors from the original model and a reduced-order model when both are linearized around one (or a set of) selected operating point(s). The results must agree in some sense if the reduced-order model is to be an adequate approximation, and of practical use in place of the original model.

2.2.8   Model Overdescription

When the system/process being modeled is complex, it is easy to err by overmodeling the state of the system. The terms "state overdescription" and "overmodeling" (Johnson, 1974) are often used. The problem basically arises from the fact that the modeler is often faced, in complex systems, with a large number of system variables, and it may not be at all clear which can be chosen as independent and which should be selected as dependent. This is a realistic assumption, especially in situations that are not fully understood, as in chemical processes, economic studies, etc. What complicates the situation is that the modeler does not know precisely the order of the system.

As discussed earlier, $\underline{x}$, the state vector of a system, is defined as the minimal amount of information required at time t_1 to completely determine future behavior for all $t > t_1$. Thus, x_1, x_2, ..., x_n are the n states when the system is appropriately modeled. This means that the system is neither undermodeled (with a few state variables) nor overmodeled (with more-than-enough variables). So, if the modeler chooses x_1, x_2, ..., x_N, where $N > n$, the value $(N - n)$ is the degree of overmodeling. Johnson has reported that the overmodeling of a dynamical system/process induces a "distinctive pattern of behavior" for the states of the system's mathematical model; it was also discussed that a model can be tested for detection of the presence of overmodeling.

2.3   DISCRETE-TIME SYSTEMS

In contrast with continuous-time systems studied so far, systems whose essential signals are all discrete-type (known only at specific instants of time) are termed "discrete-time systems." Applications range from a simple savings account to rather complicated examples such as automatic aircraft landing or a study of population. A discrete-time signal $r(t_k)$ is identified only at the instants . . . , t_{-2}, t_{-1}, t_0, t_1, t_2, . . . ; thus, the sequence of numbers . . . , $r(t_{-2})$, $r(t_{-1})$, $r(t_0)$, $r(t_1)$, $r(t_2)$, . . . specifies the signal. Often this sequence $r(t_k)$ is written as $r(k)$ for convenience, implying that $r(t_1) = r(1)$, etc. The elements of the sequence $r(t_k)$ are usually displayed as in Figure 2.31 with vertical lines, equidistant from one another, and the amplitudes of which correspond to the sequence. Here, one may specify this sequence as

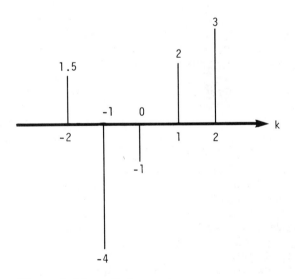

Figure 2.31 $r(t_k)$ sequence.

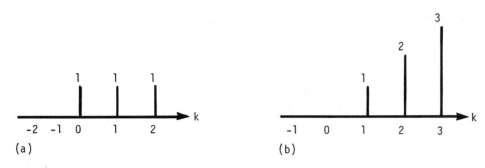

Figure 2.32 (a) Unit-step sequence. (b) Unit-ramp sequence.

$$r(-2) = 1.5$$
$$r(-1) = -4$$
$$r(0) = -1$$
$$r(1) = 2$$
$$r(2) = 3$$

Using this illustration, it is therefore simple to specify signals such as unit step and unit ramp in the following manner:

1. Unit-step sequence: $r(k) = 0$, for $k = -1, -2, \ldots$
                            $r(k) = 1$, for $k = 0, 1, 2, \ldots$
2. Unit-ramp sequence: $r(k) = 0$, for $k = -1, -2, \ldots$
                            $r(k) = k$, for $k = 0, 1, 2, \ldots$

These sequences are illustrated in Figure 2.32(a) and (b), respectively.
In a linear discrete system, the following operations are straightforward:

1. If the sequences $r_1(k)$ and $r_2(k)$ are added, then $r(t)$, their sum, is $r(k) = r_1(k) + r_2(k)$ for $k = 0, \pm 1, \pm 2, \ldots$
   The difference between $r_1(k)$ and $r_2(k)$ can be specified in a like manner.
2. If C is a constant, and $r_1(k)$ is a specified sequence, then $Cr_1(k)$ represents the multiplication of each element of r_1(at $k = 0, \pm 1, +2, \ldots$) with constant C.

"Difference equations" are used to describe discrete-time systems. For example, an equation specifying the relation between the value $r(k)$ and the values at other points may take the form

$$r(k + 2) = b \; r(k) \qquad k = 0, 1, 2, \ldots \qquad (2.74)$$

Thus, it is expected that an equation will result for every value of k considered in Eq. (2.74); in other words:

$$r(2) = b \; r(0) \qquad \text{for } k = 0$$
$$r(3) = b \; r(1) \qquad \text{for } k = 1$$

The "order" of a difference equation is the difference between the highest and the lowest indices appearing in the equation. Thus, the difference

Eq. (2.74) is second-order.   Equation (2.75) is an example of a first-order difference equation:

$$r(k + 1) = a\, r(k) + b \tag{2.75}$$

where a and b are constants.

The general form of an nth-order linear difference equation is

$$a_n(k)\, y(k + n) + a_{n-1}(k)\, y(k + n - 1) + \ldots + a_1(k)\, y(k + 1)$$

$$+ a_0(k)\, y(k) = x(k) \tag{2.76}$$

This linear difference equation is said to be "homogeneous" if $x(k)$, the forcing (driving) function, is equal to zero for all k.   For the first-order difference Eq. (2.75), one notes that the knowledge of the "initial condition" $r(0)$ is essential for evaluation of subsequent elements.   Similarly, two initial conditions, $r(0)$ and $r(1)$, are required for the second-order case of Eq. (2.74).   It is necessary, therefore, to specify the first n values of $y(k)$, that is, $y(0)$, $y(1)$, . . . , $y(n - 1)$, in the nth-order problem represented by Eq. (2.76).   Next, we present two examples.

Example 2.3.1.   Develop the response of the first-order system characterized by

$$y(k + 1) + \alpha y(k) = \beta x(k)$$

to a unit-step input $x(k)$, assuming that $y(0)$, the initial condition, is zero; α and β are constants.

The unit-step sequence, defined above, is used, and the difference equation is rewritten as follows:

$$y(k + 1) = \beta x(k) - \alpha y(k) \tag{2.77}$$

Therefore, at $k = 0$, we have

$$y(1) = \beta x(0) - \alpha y(0)$$

Since $x(0) = 1$ for the step function and $y(0) = 0$ (given), thus:

$$y(1) = \beta \tag{2.78}$$

Likewise, one may obtain future values:

$$y(2) = \beta x(1) - \alpha y(1)$$

$$= \beta - \alpha\beta = (1 - \alpha)\beta \tag{2.79}$$

and

$$y(3) = \beta x(2) - \alpha y(2)$$

$$= \beta - \alpha(1 - \alpha)\beta$$

$$= (1 - \alpha + \alpha^2)\beta \tag{2.80}$$

Example 2.3.2. It is desired to generate a model for national income based on the properties developed by Samuelson (1939). In any specific accounting period (say, quarterly), the national income may be assumed to be equal to the sum of consumer expenditures (to buy consumer goods), private investments (for purchasing capital equipment), and expenditures by the government. Samuelson postulated that in any specific accounting period (k)

1. Consumer expenditure $C(k)$ is proportional to the national income of the preceding accounting period.
2. Private investment $i(k)$ is proportional to the increase in consumer expenditure of that period over the one preceding it.
3. When it comes to government spending $g(k)$, Samuelson assumed that it remains the same for all periods.

Denoting the national income, at period k, by $y(k)$, thus:

$$y(k) = C(k) + i(k) + g(k) \qquad (2.81)$$

From property 1 above,

$$C(k) = a\, y(k - 1) \qquad (2.82)$$

where a is a positive constant.
Using property 2, $i(k)$ is expressed as

$$i(k) = b[C(k) - C(K - 1)] \qquad (2.83)$$

b is assumed as a positive constant.
From Eq. (2.82), it is seen that $C(k - 1) = a\, y(k - 2)$. Using this relation and Eq. (2.82) in Eq. (2.83) yields

$$i(k) = a\, b\ [y(k - 1) - y(k - 2)] \qquad (2.84)$$

Now, the national income $y(k)$, based on Eq. (2.81), becomes

$$y(k) = a(1 + b)y(k - 1) - aby(k - 2) + g(k) \qquad (2.85)$$

From Eq. (2.82) relating $C(k)$ to $y(k - 1)$, it is seen that the discrete signal $C(t_k)$ is dependent on another signal, namely, $y(t_{k-1})$, which is a "delayed signal" by one unit of discrete time. This helps the interpretation of $y(k)$, in Eq. (2.85), and its dependency on $y(k - 1)$ and $y(k - 2)$ as well as $g(k)$; obviously this is a second-order model.

2.3.1 The Z–Transform and Its Inverse

In Section 2.2 we dealt with continuous systems; the Laplace transform of differential equations was introduced and used as a tool for the development of transfer functions. In this section we introduce the Z-transform, which is a rule by which a sequence of numbers (a discrete-time sequence) is transformed into a function of the complex variable z. The treatment of difference equations is rather simplified through this transform, as will be illustrated through examples.

Given a sequence of numbers r(k), which is zero for all negative discrete time, the Z-transform of this sequence r(0), r(1), r(2), . . . is defined by

$$Z[r(k)] = R(z) = \sum_{k=0}^{\infty} r(k)z^{-k} = r(0) + \frac{r(1)}{z} + \frac{r(2)}{z^2} + \cdots \qquad (2.86)$$

R(z) is the sum of complex numbers and is, therefore, a complex number itself. The relation between the complex variables s and z is $z = e^{sT}$, where T is the "sampling period," i.e., the time between any two consequent numbers in the discrete-time sequence.

Example 2.3.3. Develop R(z) if $r(k) = a^k$.

Using $R(z) = \sum_{k=0}^{\infty} r(k)z^{-k}$

thus

$$R(z) = 1 + \frac{a}{z} + \frac{a^2}{z^2} + \frac{a^3}{z^3} + \cdots$$

$$= 1 + \frac{a}{z}\left(1 + \frac{a}{z} + \frac{a^2}{z^2} + \cdots\right) \qquad (2.87)$$

It could easily be verified that the term between parentheses is a geometrical sequence and is reduced to the closed form $1/[1 - (a/z)]$; through long division one may obtain the geometrical sequence from the closed form. Therefore, Eq. (2.87) becomes

$$R(z) = 1 + \frac{a/z}{1 - (a/z)} = \frac{z}{z - a} \qquad (2.88)$$

Example 2.3.4. Obtain the Z-transform of r(t) if

$$r(t) = 0 \qquad \text{for } t < 0$$

$$= e^{-at} \qquad \text{for } t \geq 0$$

Using Eq. (2.86) $Z[e^{-at}] = \sum_{k=0}^{\infty} e^{-at} z^{-k} \qquad (2.89)$

Now, since our interest is in the value of r(t) at specific discrete instants of time (i.e., to generate the desired sequence), it is convenient to replace t with kT, where T is the sampling period. Rewriting Eq. (2.89) accordingly yields

Table 2.3 Z-Transforms of Common Functions

R(k) for k > 0	
1	$\dfrac{z}{z-1}$
a^k	$\dfrac{z}{z-a}$
k	$\dfrac{z}{(z-1)^2}$
k^2	$\dfrac{z(z+1)}{(z-1)^3}$
sin kωT	$\dfrac{z(\sin \omega T)}{z^2 - 2z(\cos \omega T) + 1}$
cos kωT	$\dfrac{z(z - \cos \omega T)}{z^2 - 2z(\cos \omega T) + 1}$
t = kT	$\dfrac{Tz}{(z-1)^2}$
$e^{-at} = e^{-akT}$	$\dfrac{z}{z - e^{-aT}}$
$te^{-at} = kTe^{-akT}$	$\dfrac{zTe^{-aT}}{(z - e^{-aT})^2}$

$$Z[e^{-at}] = \sum_{k=0}^{\infty} e^{-akT} z^{-k}$$

$$= 1 + \frac{e^{-aT}}{z} + \frac{e^{-2aT}}{z^2} + \cdots$$

$$= \frac{1}{1 - \dfrac{e^{-aT}}{z}} = \frac{z}{z - e^{-aT}} \tag{2.90}$$

A brief table of Z-transforms of common functions is given in Table 2.3. For a more complete listing of pairs see Cadzow (1973).

Important properties of the Z-transform:

<u>Linearity</u>. Given $r(k) = a\, r_1(k) + b\, r_2(k)$ for $k = 0, 1, 2, \ldots$ then

$$Z[r(k)] = aZ[r_1(k)] + bZ[r_2(k)]$$

or

$$R(z) = a\, R_1(z) + b\, R_2(z) \tag{2.91}$$

Initial-Value Theorem.   The initial value of r(k) or r(t) is

$$r(0) = \lim_{z \to \infty} R(z) \tag{2.92}$$

Final-Value Theorem.   As k approaches infinity,

$$r(\infty) = \lim_{z \to 1} [(z - 1)R(z)] \tag{2.93}$$

Right-Shifting Property.   If a function y(k) is identical for k = 0, 1, 2, . . . to a function f(k) that is shifted (delayed) m units to the right [here f(k) is assumed identical to zero for negative time] then it can be shown that

$$Z[y(k)] = Z[f(k - m)]$$

or

$$Y(z) = z^{-m} F(z) \tag{2.94}$$

For a linear discrete-time system, the Z-transform of the output signal, say Y(z), is related to the Z-transform of the input signal, U(z), through the discrete transfer function H(z), that is:

$$H(z) = \frac{Y(z)}{U(z)} \tag{2.95}$$

If F(z) is the Z-transform of a sequence f(k), then

$$Z[kf(k)] = \frac{-z \; dF(z)}{dz}$$

This property is a helpful tool in generating new Z-transform pairs.   For proofs of the above properties, the reader may consult Ogata (1970) or Cadzow (1973).

Example 2.3.5.   Develop H(z) for the system characterized by

$$y(k) = a \, u(k) + b \, y(k - 1) \qquad \text{for } k = 0, 1, 2, \ldots$$

Let

$$Y(z) = Z[y(k)] \qquad \text{and} \qquad U(Z) = Z[u(k)]$$

thus

$$Z[y(k)] = aZ[u(k)] + bZ[y(k - 1)]$$

Applying the right-shifting property of Eq. (2.94), we get

$$Y(z) = a \, U(z) + b \, z^{-1} \, Y(z)$$

Rearranging,

$$Y(z) = \left[\frac{az}{(z - b)} \right] U(z) \tag{2.96}$$

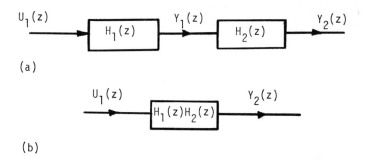

(a)

(b)

Figure 2.33   (a) Cascaded subsystems. (b) Equivalent system.

From Table 2.3, $U(z) = z/(z - 1)$ when $u(k)$ is a unit-step function; thus,

$$Y(z) = \frac{az^2}{[(z - 1)(z - b)]} \qquad (2.97)$$

It is interesting to note that for linear discrete systems, an equivalent (overall) transfer function is obtainable in a manner similar to that used earlier in the continuous case. For example, in Figure 2.33(a), two subsystems are cascaded with $H_1(z) = Y_1(z)/U_1(z)$ and $H_2(z) = Y_2(z)/Y_1(z)$. The equivalent system of Figure 2.33(b) has the overall discrete transfer function $H(z) = H_1(z) H_2(z) = Y_2(z)/U_1(z)$. It can easily be shown that for the negative feedback system shown in Figure 2.34(a), the overall transfer function is as given in part b.

The inverse Z-transform is considered next; it is desired to obtain the sequence $f(k)$ given $F(z)$; thus:

$$f(k) = Z^{-1}F(z) \qquad (2.98)$$

Again, as in the case of continuous systems, we may use the partial fraction-expansion method where poles may be distinct or multiple; it is often convenient to expand $F(z)/z$ rather than $F(z)$. A rather simpler method is the direct-division method; here we obtain the inverse Z-transform by

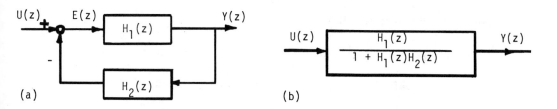

(a)

(b)

Figure 2.34   (a) Feedback system. (b) Equivalent system.

expanding F(z) into an infinite-power series in the variable z^{-1}. Two examples follow to illustrate these two methods.

Example 2.3.6. Find f(k) if F(z) is given by

$$F(z) = \frac{5z}{(z - 1)(z - 3)}$$ (2.99)

Expanding F(z)/z into partial fraction expansion yields:

$$\frac{F(z)}{z} = \frac{5}{(z - 1)(z - 3)}$$

$$= -\frac{5}{2} \left(\frac{1}{z - 1} - \frac{1}{z - 3} \right)$$

Therefore,

$$F(z) = -\frac{5}{2} \left(\frac{z}{z - 1} - \frac{z}{z - 3} \right)$$ (2.100)

Using Table 2.3, we obtain

$$Z^{-1} \left[\frac{z}{z - 1} \right] = 1 \qquad \text{and} \qquad Z^{-1} \left[\frac{z}{z - 3} \right] = 3^k$$

Therefore,

$$f(k) = -\frac{5}{2} [1 - 3^k] \qquad k = 0, 1, 2, 3, \ldots$$

The number sequence is easily obtained with

$$f(0) = 0 \qquad f(1) = 5 \qquad f(2) = 20 \qquad f(3) = 65$$

Example 2.3.7. Consider the F(z) we used in the previous example and develop f(k) using the infinite-power series expansion.

$$F(z) = \frac{5z}{(z - 1)(z - 3)}$$

Rewrite F(z) as

$$F(z) = \frac{5z^{-1}}{1 - 4z^{-1} + 3z^{-2}}$$ (2.101)

Using long division, we find that

$$F(z) = 5z^{-1} + 20z^{-2} + 65z^{-3} + \cdots$$

Recalling the definition of the Z-transform in Eq. (2.86), it is seen that

$$f(0) = 0 \qquad f(1) = 5 \qquad f(2) = 20 \qquad f(3) = 65$$

This, as expected, is the same result obtained through the partial fraction expansion method.

We next present the state space model for linear discrete-time systems. Recall that Eq. (2.56), repeated here for convenience

$$\dot{\underline{x}} = \underline{A}\underline{x} + \underline{B}\underline{u} \tag{2.102}$$

is the state space model for linear continuous systems; likewise, for the discrete case, the system takes this form:

$$\underline{x}(k+1) = \underline{A}(k)\,\underline{x}(k) + \underline{B}(k)\,\underline{u}(k) \tag{2.103}$$

where $\underline{x}(k)$ is an n-state vector, $\underline{A}(k)$ is a square (n × n) matrix, $\underline{B}(k)$ is an (n × m) distribution matrix, and $\underline{u}(k)$ is an m-input vector. If no forcing input exists, then the homogenous system is represented by

$$\underline{x}(k+1) = \underline{A}(k)\,\underline{x}(k) \tag{2.104}$$

2.3.2 α–β Tracking System

We conclude this section by presenting an example of practical importance on radar tracking; this simplified presentation follows the treatments by Cadzow (1973) and Benedict and Bordner (1962).

Example 2.3.8. The objective in a radar-tracking problem is to be able to estimate an object's (plane, missile, etc.) present range and its velocity (rate of change of range) and also to have a future good estimate of its range. In such a system, a signal, transmitted by the radar every T seconds, propagates through space, at speed of light c, and hits the object being tracked. A portion of this energy is reflected back and received by the receiving unit of the radar. Figure 2.35(a and b) shows an ideal transmitted pulse and an ideal received signal, respectively. If Δt seconds denote the time needed for the pulse to travel to the object and return to the tracking system, then the object is at a distance $u = c(\Delta t/2)$ meters from the radar ($c = 3 \times 10^8$ meters/sec). Obviously, Δt plays an important role in the evaluation of u. But owing to the environment,

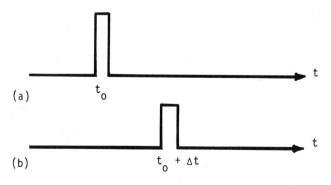

Figure 2.35 (a) Ideal transmitted. (b) Received pulses.

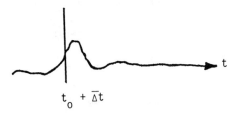

t$_0$ + $\overline{\Delta t}$

Figure 2.36   A realistic received pulse.

including the target's (object) shape and size, other objects that might be present, noise, etc., the received signal may take the form shown in Figure 2.36, where Δt, in general, is not equal to $\overline{\Delta t}$. Since Δt is the only available information to work with, the range equation becomes

$$\overline{u} = c \left(\frac{\overline{\Delta t}}{2}\right) \tag{2.105}$$

The successive transmission of radar signals, every T second, makes it possible to obtain periodic measurements of the travel time, thus improving the system's tracking accuracy. The error in the range is due to the fact that $\Delta t \neq \overline{\Delta t}$ and is given by

$$\Delta u = c \, \frac{\overline{\Delta t} - \Delta t}{2} \tag{2.106}$$

If $u(0)$, $u(1)$, $u(2)$, . . . represents the succession of range measurements, with error based on Eq. (2.105), then, in general, $u(k)$ is obtained from the kth radar pulse return. Data processing uses this sequence in arriving at the above-stated objectives of this tracking system. Let us define three new variables, namely, $y(k)$, $\dot{y}(k)$, and $y_p(k)$, "after" data processing as

$\quad y(k)$: the smoothed range, i.e., an estimate of the object's range at
$\qquad\quad$ the kth radar pulse
$\quad \dot{y}(k)$: the smoothed range velocity, i.e., an estimate of the object's
$\qquad\quad$ range velocity at the kth radar pulse
$\quad y_p(k)$: predicted object's range at the kth radar pulse obtained at the
$\qquad\quad$ time of previous pulse $(k - 1)$

It has been shown that the $\alpha-\beta$ tracking system is represented by the following set of difference equations where α and β are positive numbers:

$$y_p(k) = y(k - 1) + T\dot{y}(k - 1)$$

$$y(k) = y_p(k) + \alpha\,[u(k) - y_p(k)]$$

$$\dot{y}(k) = \dot{y}(k - 1) + \frac{\beta}{T}[u(k) - y_p(k)] \tag{2.107}$$

The term $[u(k) - y_p(k)]$ is essential here, and its sign will indicate whether the predicted range $y_p(k)$ is actually an underestimate or over-estimate of $u(k)$, the kth measurement of the range.

As stated earlier, u(k) is the only input signal to our system, while the three output signals are $y_p(k)$, $y(k)$, and $\dot{y}(k)$; their Z-transforms are $Y_p(z)$, $Y(z)$, and $\dot{Y}(z)$. Three transfer functions are therefore desired to relate each output to $U(z)$, the Z-transform of $u(k)$. These are

$$H_1(z) = \frac{Y(z)}{U(z)}$$

$$H_2(z) = \frac{\dot{Y}(z)}{U(z)}$$

$$H_3(z) = \frac{Y_p(z)}{U(z)} \tag{2.108}$$

Performing the Z-transform on the set of Eq. (2.107) yields

$$Y_p(z) = z^{-1} Y(z) + z^{-1} T\dot{Y}(z)$$

$$Y(z) = Y_p(z) + \alpha[U(z) - Y_p(z)]$$

$$\dot{Y}(z) = z^{-1} \dot{Y}(z) + \frac{\beta}{T}[U(z) - Y_p(z)] \tag{2.109}$$

Dividing both sides of each of these equations by $U(z)$ and using the relations in Eq. (2.108) results in:

$$\frac{Y_p(z)}{U(z)} = z^{-1}H_1(z) + z^{-1}TH_2(z) = H_3(z)$$

$$\frac{Y(z)}{U(z)} = H_3(z) + \alpha[1 - H_3(z)] = H_1(z)$$

$$\frac{\dot{Y}(z)}{U(z)} = z^{-1} H_2(z) + \frac{\beta}{T}[1 - H_3(z)] = H_2(z) \tag{2.110}$$

Combining similar terms and rearranging, the set of Eq. (2.110) becomes

$$z^{-1}H_1(z) + z^{-1}TH_2(z) - H_3(z) = 0$$

$$H_1(z) - (1 - \alpha)H_3(z) = \alpha$$

$$(1 - z^{-1})H_2(z) + \frac{\beta}{T}H_3(z) = \frac{\beta}{T} \tag{2.111}$$

The solution of this set of equations in the three unknowns $H_1(z)$, $H_2(z)$, and $H_3(z)$ gives

$$H_1(z) = \frac{z[\alpha z + \beta - \alpha]}{\Delta}$$

$$H_2(z) = \frac{\beta}{T} \; \frac{z(z-1)}{\Delta}$$

$$H_3(z) = \frac{[z(\alpha + \beta) - \alpha]}{\Delta} \tag{2.112}$$

Δ, the denominator, is the system's characteristic equation and in our case
is

$$\Delta = z^2 + (\alpha + \beta - 2)z + (1 - \alpha) \tag{2.113}$$

For a given u(k) and U(z), the expressions of Eq. (2.112), Eq. (2.113),
and Eq. (2.108) give $Y(z)$, $\dot{Y}(z)$, and $Y_p(z)$; taking the inverse Z-transform,
one obtains the desired sequences $y(k)$, $\dot{y}(k)$, and $y_p(k)$ for any assumed
(α, β) pair. For a useful discussion of a criterion to select the β parameter, not only to obtain a fast-responding system's dynamical behavior (to
be able to follow a target that might undergo sudden flight changes), but
also to have an acceptable noise-suppression capability, the reader may
consult the work of Cadzow (1973).

2.4   SOME RELATED TOPICS

In this section we focus our attention on a few related important topics,
including components of the complete response of a system (transients and
steady state) and the notions of stability, controllability, and observability of
systems. These are significant qualitative properties of linear systems.

2.4.1   Feedback Systems

Typical test signals are often used in connection with the testing and performance evaluation of systems. A mathematical model might be driven by
any of these signals: a step function, an impulse, a ramp, or a sinusoidal
function. When system output (behavior) is obtained as a function of time,
the "transient" component is the one concerned with the response from the
initial state to the final state, while the "steady state" represents the output
at a relatively large value of time (as $t \to \infty$). For practical systems, behavior often exhibits oscillations before reaching steady condition. The
difference between a system's input and steady-state response is defined as
"steady-state error."

Example 2.4.1.   Develop the time response for a unity negative feedback system due to a unit-step input if G(s), the feedforward function, is
1/Ts, as shown in Figure 2.37.
    Since $C(s)/R(s) = 1/(1 + Ts)$, and $R(s) = 1/s$, then the expanded C(s)
yields:

$$C(s) = \frac{1}{s} - \frac{T}{Ts + 1}$$

Using the $\mathcal{L}^{-1}$, we find that

$$c(t) = 1 - e^{-t/T} \qquad \text{for } t \geq 0 \tag{2.114}$$

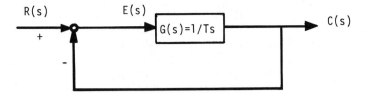

Figure 2.37   A unity-feedback system.

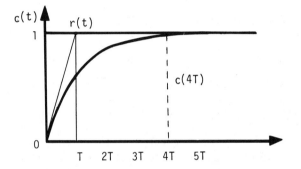

Figure 2.38   System's response of Ex. (2.4.1).

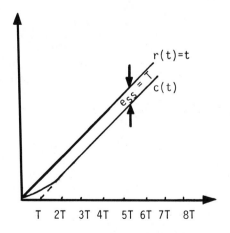

Figure 2.39   Steady-state error.

T is termed the "time constant" of this first-order system. The plot of c(t) versus t is depicted in Figure 2.38. As is seen, the slope of c(t) at the origin is $1/T$; also noted is the system's initial value of zero and steady-state value of 1 (zero steady-state error). For a small time constant T, the system is a quick-responding one, while large T identifies a slow-responding system. It can be verified that the response after four time constants, i.e., c(4T), is within 1.8% of the steady-state (final) value; 4T is usually termed the "settling time" in this case.

Example 2.4.2. For the system in Figure 2.37 and using a unit-ramp input signal, develop c(t) and evaluate the steady-state error. Following the procedure used in the previous example

$$C(s) = \left(\frac{1}{s^2}\right) \left(\frac{1}{1 + Ts}\right)$$

Using partial fractions and inverse Laplace transform, one gets

$$c(t) = t - T + T\, e^{-t/T} \qquad t \geqslant 0 \qquad\qquad (2.115)$$

The error signal $e(t) = r(t) - c(t) = t - c(t) = T(1 - e^{-t/T})$. As t approaches ∞, the steady-state error e_{ss} is equal to T; see Figure 2.39. The smaller T is, the smaller the steady-state error in following the ramp input.

Example 2.4.3. In this example we examine the step response of the second-order system shown in Figure 2.40. Here, the plant is G(s) and

$$\frac{C(s)}{R(s)} = \frac{\omega_n^2}{s^2 + 2\zeta\omega_n s + \omega_n^2} \qquad\qquad (2.116)$$

where ζ and ω_n are termed the "damping factor" and the "undamped natural frequency," respectively. The closed-loop transfer function, Eq. (2.116), has two poles located at

$$s = -\zeta\omega_n \pm j\omega_n \sqrt{1 - \zeta^2} \qquad\qquad (2.117)$$

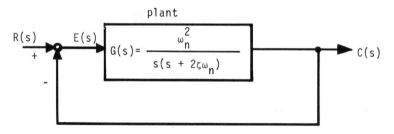

Figure 2.40   System of Ex. (2.4.3).

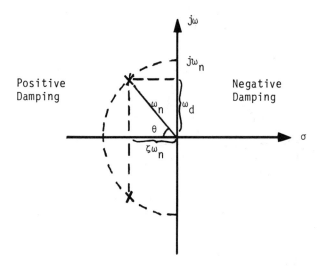

Figure 2.41   Possible pole locations in s-plane.

Figure 2.41 shows a possible poles location in the left half of the s-plane where $\omega_d = \omega_n \sqrt{1 - \zeta^2}$ is the "damped natural frequency." Note that $\zeta = \cos \theta$ and ω_n is the distance of either pole from the origin. ζ plays an important role in the dynamic behavior of the system; for example, for

1. $\zeta = 1$: the system has repeated (two equal) real poles at $-\omega_n$ (critical damping)
2. $\zeta = 0$: the system has two equal poles on the $j\omega$ axis at $\pm j\omega_n$ (undamped case)
3. $0 < \zeta < 1$: the system has two complex conjugate poles (underdamped case)
4. $\zeta > 1$: the system has two real negative poles (overdamped case).

The step response of this system for various values of ζ is shown in Figure 2.42. For $0 < \zeta < 1$, it can be shown that the step response for $t \geq 0$ is

$$c(t) = 1 - \left(\frac{1}{\sqrt{1 - \zeta^2}}\right) e^{-\zeta \omega_n t} \sin\left(\omega_d t + \tan^{-1}\frac{\sqrt{1 - \zeta^2}}{\zeta}\right) \qquad (2.118)$$

This second-order system has a time constant value of $1/\zeta\omega_n$ and a settling time of $4(1/\zeta\omega_n)$. Another characteristic of this response is the time and magnitude of the first peak value. It is obvious from Figure 2.42 that the amount of "overshoot" at the first peak increases as ζ becomes smaller. For practical systems, the range of ζ is usually chosen between 0.3 and 0.7. It can be shown (Ogata, 1970) that t_p, the time for the first peak, is given by

$$t_p = \frac{\pi}{\omega_d}$$

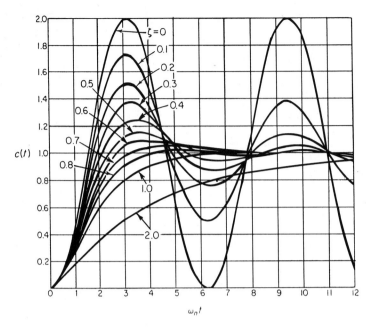

Figure 2.42   Unit-step response.

and the peak overshoot, in a unit-step response, is specified by

$$e^{-\left(\dfrac{\zeta}{\sqrt{1-\zeta^2}}\right)\pi}$$

The transient response of a feedback system can be modified depending on the type of control one selects. In practical cases, systems have to demonstrate acceptable levels of steady-state error as well as overshoot. Four arrangements are often used in modifying closed-loop-system performance. In Figure 2.43, feedback systems with proportional, derivative, integral, and rate-feedback controls are shown. Here, the plant function G(s) is the same as used earlier in Figure 2.40. It can be shown that it is impossible to improve both transient behavior and steady-state performance by using any one of these controllers. The main features of these controllers are listed below:

K_p, gain of the "proportional controller," is usually an adjustable gain
The "derivative control" is an anticipatory type of control which allows an early proper correcting action, before error becomes too large, based on measurements of rate of change of error, i.e., de/dt; this type also adds damping to the system
The "integral controller" tends to eliminate, or reduce, steady-state error by introducing a signal that is proportional to the time integral of the error. This integral action, when added to the proportional controller, will result in a third-order system in place of the original second-order system, and as a result the system may become unstable
As in the case of derivative controller, the "rate feedback" improves system's damping by feeding back the derivative of the output signal. For more information on controllers, consult any of these textbooks: Ogata (1970), Kuo (1975), Van de Vegte (1986).

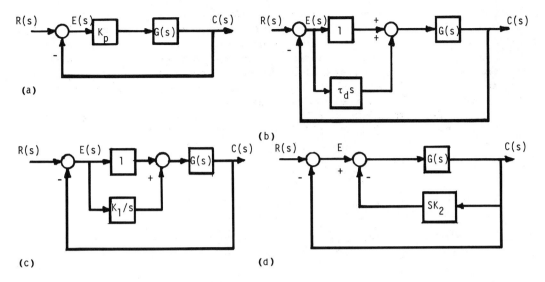

Figure 2.43   (a) Proportional controller.   (b) Derivative controller.
(c) Integral controller.   (d) Rate-feedback controller.

2.4.2   Stability

To be useful, a system must be stable. Stability at all times is viewed as
the most important characteristic of a system. A "stable system" is defined
as one that demonstrates a bounded response if the input is also bounded
in magnitude. Several methods have been developed to investigate the
stability of a linear system, but the most direct one is based on determining
the location of the poles of a given system's transfer function in the s-plane.
Recalling Eq. (2.118) for c(t), the step response, of a second-order system
for $0 < \zeta < 1$, it is seen that the real part of the poles positions is $-\zeta \omega_n$
(as in Figure 2.41). If this is positive, the poles belong to the right half
of the s-plane, and the term $e^{-\zeta \omega_n t}$ will continue to grow instead of decay
over time. Thus, for stability all the poles must be in the left half of
the s-plane; i.e., all have negative real parts.

Several methods for determining stability of linear systems have been
developed. Often used are: Routh-Hurwitz criterion, Nyquist criterion,
the root locus method, Bode diagrams, and Lyapunov's method. In this
book, we shall discuss only two methods, the Routh-Hurwitz and the root
locus (references on control systems listed at the end of this chapter are
excellent for a more complete presentation).

Routh-Hurwitz Criterion. This criterion was developed to help determine
the number of roots of a polynomial D(s) in the right half of the s-plane
without evaluating the actual location of all roots. For practical systems,
it is sufficient to determine whether one or more roots lie on or to the
right of the imaginary axis in the s-plane. Let the nth-order "characteristic
polynomial" D(s), which is the denominator of a transfer function, be

$$D(s) = a_0 s^n + a_1 s^{n-1} + \cdots + a_{n-1} s + a_n = 0 \qquad (2.119)$$

One should, before proceeding with the criterion and the Routh table, inspect the "characteristic equation" $D(s) = 0$ for (a) any missing coefficients, and (b) any change in sign of its coefficients. If the equation passes these two tests, i.e., all coefficients are present and are all positive (or negative), then there is no guarantee that the system is stable and one should proceed to apply the criterion. Thus, these two conditions are necessary but insufficient. However, if either test fails, the system is unstable.

The Routh table has $(n + 1)$ rows, with its first two rows obtained directly from the coefficients of $D(s)$ (regardless of n being even or odd) as follows:

$$
\begin{array}{c|cccc}
s^n & a_0 & a_2 & a_4 & \cdots \\[2mm]
s^{n-1} & a_1 & a_3 & a_5 & \cdots
\end{array}
\qquad (2.120)
$$

The coefficients of the third row are obtained from the elements of first and second rows; let the third row be

$$
\begin{array}{c|ccc}
s^{n-2} & b_1 & b_3 & b_5 \quad \cdots
\end{array}
$$

Using Eq. (2.120) and through a process of cross-multiplication, we obtain

$$
b_1 = \frac{1}{a_1}(a_1 a_2 - a_3 a_0), \quad b_3 = \frac{1}{a_1}(a_1 a_4 - a_5 a_0), \quad \cdots
$$

Likewise, elements of the fourth row c_1, c_3, c_5, $\ldots$ are obtained from second and third rows. This process continues until the table is complete.

Now, if all the elements of the first column of the array (i.e., a_0, a_1, b_1, c_1, $\ldots$) exist *and* are all of the same sign, then there are no roots of the characteristic equation on or to the right of the imaginary axis; thus, the system is stable. If all the elements in the first column exist but are not of the same sign, then the number of roots with a positive real part is equal to the number of changes in sign among the elements of the first column.

Example 2.4.4. Determine whether this characteristic equation

$$
D(s) = s^4 + 6s^3 + 23s^2 + 40s + 50 = 0
$$

represents a stable system or not.

Since all coefficients are present and have the same sign, we develop the Routh table:

$$
\begin{array}{c|ccc}
s^4 & 1 & 23 & 50 \\[1mm]
s^3 & 6 & 40 & \\[1mm]
s^2 & 16.33 & 50 & \\[1mm]
s^1 & 21.63 & & \\[1mm]
s^0 & 50 & &
\end{array}
$$

No sign change in the first column indicates no roots in the right half of the s-plane, and hence the system is stable.

Example 2.4.5. Discuss the location of the roots of the characteristic equation

$$D(s) = s^4 + s^3 - 5s^2 - s + 6 = 0$$

By inspection of the signs, the system is unstable. The Routh table for this equation is

s^4	1	-5	6
s^3	1	-1	
s^2	-4	6	
s^1	1/2		
s^0	6		

Two changes in the first column indicate two roots in the right half of the s-plane (unstable system) and the other two roots in the left half of the s-plane.

Two difficulties may arise in Routh tabulation: (a) an element in the first column may turn out to be zero, or (b) an entire row of the table may be zeros. In the former case, one should replace the zero element by a small positive number ε and proceed to complete the array. In case b, on the other hand, an auxiliary equation A(s) is formed from the row preceding (above) the zero row. The order of the auxiliary equation is always even, and this order indicates the number of existing pairs of roots, of the characteristic equation, that are "equal in magnitude and opposite in sign." To complete the table, one should replace the row of zeros with the coefficients of the equation resulting from differentiating the auxiliary equation with respect to s. Two examples illustrate these special cases.

Example 2.4.6. Examine the stability of the system whose characteristic equation is given by

$$D(s) = 3s^4 + 5s^3 + 3s^2 + 5s + 8 = 0$$

The Routh array is as follows:

s^4	3	3	8
s^3	5	5	
s^2	0	8	

Replace the zero element by ε; thus the third and the following rows become:

s^2	ε	8
s^1	$\dfrac{5\varepsilon - 40}{\varepsilon}$	
s^0	8	

With ε, a small positive number, it is seen that two changes in sign in the first column exist; therefore, the system is unstable with two roots in the right half of the s-plane.

Example 2.4.7. Discuss the roots of this characteristic equation:

$$s^3 + 3s^2 + s + 3 = 0$$

Four rows in the Routh table are expected.

$$
\begin{array}{c|cc}
s^3 & 1 & 1 \\
s^2 & 3 & 3 \\
s^1 & 0 & 0
\end{array}
$$

As discussed above, A(s) (the auxiliary equation) is $3s^2 + 3 = 0$, and its derivative $dA(s)/ds = 6s + 0$; thus, $(3s^2 + 3)$ is a factor of the characteristic equation. The complete array is given by

$$
\begin{array}{c|cc}
s^3 & 1 & 1 \\
s^2 & 3 & 3 \\
s^1 & 6 & 0 \\
s^0 & 3 &
\end{array}
$$

By inspection, there is no change is sign in the first column, indicating no roots in the right half of the s-plane. Moreover, one pair of equal and opposite roots lies on the jω axis at ±j1; therefore, the system is unstable, with one root in the left half of the s-plane.

Another application of the Routh criterion is in feedback systems design, where it is often used to determine the range of the open-loop gain for which a feedback system is stable. This is done by determining the gain values that will guarantee no changes in the signs of the elements of the first column in the Routh table.

Root Locus Method. The root locus, developed by W. R. Evans in 1948, is a graphical method for sketching, in the s-plane, the movement of the roots of the system's characteristic equation (which are the poles of the closed-loop transfer function) as a result of varying a parameter from zero to infinity. For example, consider that the open-loop transfer function G(s) of a unity-feedback system is given by

$$G(s) = \frac{K}{[s(s + a)]} \tag{2.121}$$

where K and a are constants. The system's closed-loop transfer function is given by

$$\frac{C(s)}{R(s)} = \frac{K}{(s^2 + a s + K)}$$

Letting a = 2, the characteristic equation is

$$s^2 + 2s + K = 0 \tag{2.122}$$

It is a simple exercise to find out that as gain K varies from 0 to ∞, the movements of the roots of Eq. (2.122) in the s-plane ($s_{1,2}$) are as shown in Figure 2.44. The root locus starts at the poles of $G(s)$, namely, at s = 0 and s = -2 for K = 0; the two roots become equal to s = -1 when K = 1; and for larger values of K, one root tends to move upward towards A and the other downward toward B. Since the roots are confined, for $0 < K < \infty$, to the left half of the s-plane, this second-order system is always stable.

To generalize, let us consider a negative-feedback system (see Figure 2.18) whose closed-loop transfer function is

$$\frac{C(s)}{R(s)} = \frac{G(s)}{1 + G(s)\ H(s)}$$

The roots of the characteristic equation are obtained by having

$$1 + G(s)\ H(s) = 0$$

or

$$G(s)\ H(s) = -1 \tag{2.123}$$

Equation (2.123) is satisfied if the magnitude condition and the phase condition are met, i.e.:

$$\left|\ G(s)\ H(s)\ \right| = 1$$

$$\underline{/\ G(s)\ H(s)} = (2k + 1)\ \pi \qquad k = 0, \pm 1, \pm 2, \ldots \tag{2.124}$$

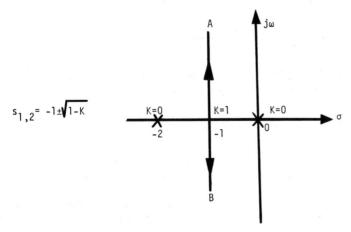

Figure 2.44   A root locus.

Now if G(s) H(s) is expressed as

$$G(s) \ H(s) = \frac{K(s + z_1)(s + z_2) \cdots (s + z_m)}{(s + p_1)(s + p_2) \cdots (s + p_n)}$$

then the conditions of Eq. (2.124) yield

$$| \ G(s) \ H(s) | \ = \frac{\lfloor K \rfloor \ \prod\limits_{i=1}^{m} | \ s + z_i |}{\prod\limits_{j=1}^{n} | \ s + p_j |} = 1$$

and

$$\underline{/G(s) \ H(s)} = \underline{/K} + \sum_{i=1}^{m} \underline{/s + z_i} - \sum_{j=1}^{n} \underline{/s + p_j} = (2k + 1)\pi \qquad (2.125)$$

For any given value of $0 < K < \infty$ and based on Eq. (2.125), any point in the s-plane that satisfies these two conditions is a point on the root locus.

In analysis as well as design situations, the construction of the root locus is simplified if one follows this set of rules (Kuo, 1975):

1. The points on the root locus corresponding to K = 0 are at the location of the poles of G(s) H(s)
2. The terminating points on the root locus corresponding to K = ∞ are at the location of the zeros of G(s) H(s), including the finite ones and those at infinity
3. The root locus is made out of a number of separate segments equal to the greater of n and m (the number of poles and zeros, respectively)
4. The root locus is symmetrical with respect to the real axis of the s-plane
5. The segments of the root locus terminating at ∞ are asymptotic to straight lines intersecting with the real axis at the point (centroid)

$$\sigma = \frac{\text{sum of all poles} - \text{sum of all zeros}}{n - m}$$

   Remember that all complex poles and zeros exist as conjugate pairs. The asymptotes' angles with the real axis are

$$\Theta_q = \frac{(2q + 1)\pi}{n - m} \qquad \text{where } q = 0, 1, \ldots, (n - m - 1)$$

6. Segments of the real axis are part of the locus only if the total number of real poles and real zeros of G(s) H(s) to the right of the section is "odd"
7. The value of the frequency and gain where the root locus intersects with the imaginary axis of the s-plane may be obtained by the Routh-Hurwitz criterion

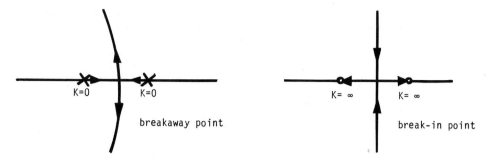

Figure 2.45   Breakaway and break-in points.

8. A breakaway (or break-in) point where two or more separate
    segments of the locus meet (this does not have to be on the real
    axis of the s-plane) is determined by finding the roots of d[G(s)
    H(s)]/ds = 0; this is a necessary condition only. If these points
    are not on the real axis, they occur in complex-conjugate pairs.
    Figure 2.45 shows a breakaway and break-in points.

Example 2.4.8.   Construct the root locus given

$$G(s) = \frac{K}{s(s + 1)(s + 2)} \qquad \text{and} \qquad H(s) = 1$$

By inspection there are no finite zeros (m = 0) for G(s) H(s); however,
there are three zeros at infinity; the G(s) H(s) poles are at s = 0, s = -1,
s = -2.
  Following the set of rules, one gets
  1.   The root locus starts at the three poles
  2.   The root locus terminates at infinity in three directions
  3.   Number of separate segments is three (n = 3)
  4.   Root locus is symmetrical
  5.   The centroid where the asymptotes intersect with the real axis is
at σ = (0 - 1 - 2)/3 = -1.
    The asymptotes go off toward infinity and have these angles with
the real axis (n - m = 3 here); thus:

$$\theta_1 = \frac{\pi}{3} = 60° \qquad \text{for } q = 0$$

$$\theta_2 = \frac{3 \cdot \pi}{3} = 180° \qquad \text{for } q = 1$$

$$\theta_3 = \frac{5 \cdot \pi}{3} = 300° \qquad \text{for } q = 2$$

  6.   The root locus exists on the real axis for -1 < s ≤ 0 and for -∞
< s ≤ -2.
    At this point, it is appropriate to put the information obtained so
far together; the resulting root locus appears in Figure 2.4.6.

7. It is of interest to find out the value of ω and K where the locus intersects with the $j\omega$ axis. The characteristic equation is required to be able to apply the Routh-Hurwitz criterion:

$$1 + G(s)\,H(s) = \frac{s^3 + 3s^2 + 2s + K}{s(s + 1)\,(s + 2)} = 0$$

The Routh table is:

$$
\begin{array}{c|cc}
s^3 & 1 & 2 \\
s^2 & 3 & K \\
s^1 & \dfrac{(6 - K)}{3} & \\
s^0 & K &
\end{array}
$$

The s^1 row becomes a row of zeros if K = 6, and this becomes the critical value of gain at which the locus crosses the $j\omega$ axis; the auxiliary equation A(s) is

$$3s^2 + K = 3s^2 + 6 = 0$$

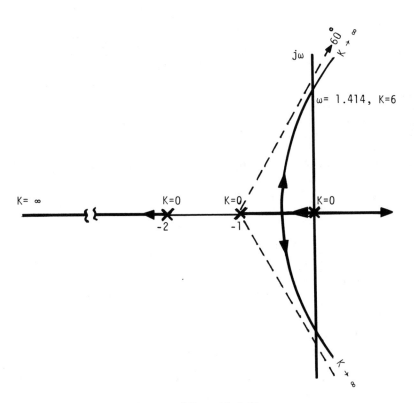

Figure 2.46   Root locus of Ex. (2.4.8).

from which $s^2 = -2$ or $s = \pm j\sqrt{2}$ at the crossing. It is obvious now that our third-order closed-loop system is stable for values of gain K < 6 and will have two poles in the right half of the s-plane (unstable) for K > 6.

8. There is one breakaway point on the real axis between 0 and -1; performing $d[G(s) H(s)]/ds = 0$ results in two roots at $s = -0.423$ and -1.577, and our choice should be at -0.423 (note that there is no locus where -1.577 is).

Example 2.4.9. Sketch the root locus if

$$G(s) H(s) = \frac{K}{s(s + 4)(s^2 + 4s + 20)}$$

Following the rules we applied in the previous example, and realizing that n = 4 and m = 0 in this case, we obtain:

$$\sigma = -8/4 = -2$$

$$\theta_1 = 45° \qquad \theta_2 = 135° \qquad \theta_3 = 225° \qquad \theta_4 = 315°$$

The characteristic equation is given by

$$1 + G(s) H(s) = s^4 + 8s^3 + 36s^2 + 80s + k = 0$$

and the corresponding Routh table is as follows:

s^4	1	36	K
s^3	8	80	
s^2	26	K	
s^1	(2080 − 8K)/26		
s^0	K		

The gain value obtained from the s^1 row is K = 260 and corresponds to the root locus crossing with the imaginary axis at $s = \pm j\sqrt{10}$. From rule 8, three breakaway points are at -2 and $-2 \pm j2.45$. The complete root locus is shown in Figure 2.47.

Stability in the z-plane. The discussion so far has focused on stability considerations for continuous-time systems, and as stated earlier a linear system is stable if all poles of its transfer function lie in the left half of the s-plane. The stability of discrete-time systems is discussed next in the z-plane. Through mapping of the left half s-plane into the z-plane, one would arrive at stability requirements for discrete-time systems. Recalling that

$$z = e^{sT} \qquad \text{and} \quad s = \sigma + j\omega$$

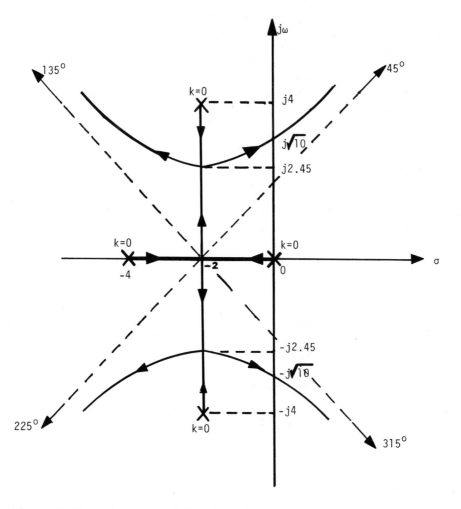

Figure 2.47   Root locus of Ex. (2.4.9).

we obtain

$$|z| = e^{\sigma T} \qquad \underline{/z} = \omega T \tag{2.126}$$

Thus, any point, say at $s = -a + jb$, could be mapped into the z-plane. Since the value of $\sigma = 0$ (the imaginary axis) is the boundary of the stable region in the s-plane, it follows from Eq. (2.126) that this maps into a circle with unit radius; that is to say, the inside of the unit circle in the z-plane corresponds to the left half of the s-plane. A modification to the Routh stability criterion has been developed to determine whether a polynomial in z contains one or more roots on, or outside, the unit circle (Ogata, 1970).

2.4.3   Controllability and Observability

Controllability and observability are important concepts that were introduced by Kalman. In the sequel we present their definitions as well as conditions for their existence. Let us recall the state space model given earlier and repeated here for convenience:

$$\dot{\underline{x}} = \underline{A}\underline{x} + \underline{B}\underline{u} \tag{2.127}$$

$$\underline{y} = \underline{C}\underline{x} \tag{2.128}$$

State controllability is concerned with the desire to transfer the system from its initial state $\underline{x}(t_0)$ to any other state, in a finite interval of time, by means of an unconstrained input vector $\underline{u}$. This simply implies that if any one of the state variables is not affected by the control effort, it will be impossible to drive this particular state to a desired state in a finite time. A linear time-invariant continuous system described by Eq. (2.127) and Eq. (2.128) is said to be completely controllable if, and only if, the rank of the controllability matrix defined as

$$\underline{S} = [\underline{B} \ \underline{A}\underline{B} \ \underline{A}^2\underline{B} \ . \ . \ . \ \underline{A}^{n-1}\underline{B}] \tag{2.129}$$

is equal to n (the dimension of the state vector). This necessary and sufficient condition is also valid for discrete-time systems described by Eq. (2.103).

Observability is analogous to controllability, and they are often referred to as dual concepts. A system is said to be completely observable if its state can be determined from the knowledge (measurement) of the output vector $\underline{y}$ over a specified interval of time. In other words, through this knowledge, the value of the initial state $\underline{x}(t_0)$ can be inferred if the system is completely observable. The importance of this concept becomes apparent in systems control problems where control inputs are determined on the basis of available outputs which are in turn expected to have full information about the state vector. Obviously, if one of the states cannot be observed from the available outputs, this particular state is said to be unobservable and the system, accordingly, is not completely observable (unobservable). Now, the system of Eq. (2.127) and Eq. (2.128) is completely observable if, and only if, rank of the observability matrix defined as

$$\underline{V} = \begin{bmatrix} \underline{C} \\ \underline{C}\underline{A} \\ \underline{C}\underline{A}^2 \\ . \\ . \\ . \\ \underline{C}\underline{A}^{n-1} \end{bmatrix} \tag{2.130}$$

is equal to n.
Likewise a discrete-time system

$$\underline{x}(k+1) = \underline{A}(k) \ \underline{x}(k) \tag{2.131}$$

$$\underline{y}(k) = C(k) \ \underline{x}(k) \tag{2.132}$$

is completely observable if the knowledge of the output values $\underline{y}(0)$, $\underline{y}(1)$, ... , $\underline{y}(N-1)$ is sufficient to determine the initial state $\underline{x}(0)$. The condition for observability of such a discrete-time system is, as expected, equivalent to that for the continuous case based on the rank of matrix $\underline{V}$ (Eq. 2.130).

Example 2.4.10. Obtain $\underline{S}$, and discuss the controllability of the system described by

$$\dot{\underline{x}} = \begin{bmatrix} 3 & -2/3 \\ 4 & -1/3 \end{bmatrix} \underline{x} + \begin{bmatrix} 1 \\ 3 \end{bmatrix} \underline{u}$$

Here, we have

$$\underline{B} = \begin{bmatrix} 1 \\ 3 \end{bmatrix} \qquad \text{and} \qquad \underline{A}\underline{B} = \begin{bmatrix} 1 \\ 3 \end{bmatrix}$$

Thus,

$$\underline{S} = \begin{bmatrix} 1 & 1 \\ 3 & 3 \end{bmatrix}$$

Since this matrix has two columns that are actually the same (linearly dependent), its rank is 1 and the system is not, therefore, a controllable one.

Example 2.4.11. Examine the observability of the system represented by

$$\underline{A} = \begin{bmatrix} 0 & 1 \\ -1 & -2 \end{bmatrix} \qquad \underline{C} = [1 \quad 0]$$

In this case

$$\underline{V} = \begin{bmatrix} \underline{C} \\ \underline{C}\,\underline{A} \end{bmatrix} = \begin{bmatrix} 1 & 0 \\ 0 & 1 \end{bmatrix}$$

Thus, the system is completely observable.

Example 2.4.12. Consider the system described by

$$\dot{\underline{x}} = \begin{bmatrix} a_1 & a_2 \\ a_3 & a_4 \end{bmatrix} \underline{x} + \begin{bmatrix} 1 \\ 1 \end{bmatrix} \underline{u}$$

$$y = [1 \quad 0] \underline{x}$$

Find the conditions on the a's for complete controllability, and complete observability.

Let us first examine the controllability matrix:

$$\underline{S} = [\underline{B} \quad \underline{A}\underline{B}] = \begin{bmatrix} 1 & a_1 + a_2 \\ 1 & a_3 + a_4 \end{bmatrix}$$

For complete controllability, the rank of $\underline{S}$ must be two, implying that $(a_1 + a_2) \neq (a_3 + a_4)$. Similarly, for complete observability

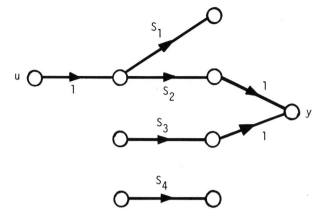

Figure 2.48 Four possible subsystems.

$$\underline{V} = \begin{bmatrix} \underline{C} \\ \underline{CA} \end{bmatrix} = \begin{bmatrix} 1 & 0 \\ a_1 & a_2 \end{bmatrix}$$

should have a rank of two. This would require that $a_2 \neq 0$ (otherwise the two rows of $\underline{V}$ become dependent).

Gilbert (1963) has shown that a system may be partitioned into four possible subsystems. A partitioned system is made of

1. A controllable and not observable subsystem S_1
2. A controllable and observable subsystem S_2
3. An uncontrollable but observable subsystem S_3
4. An uncontrollable and unobservable subsystem S_4

The subsystems are shown in Figure 2.4.8, with u and y being the system's input and output.

PROBLEMS

2.1. Classify the following differential equations according to whether they are linear or nonlinear.

   a. $t \dfrac{dx(t)}{dt} + x(t) = 0$

   b. $x(t) \dfrac{dx(t)}{dt} + x(t) = 0$

   c. $\dfrac{dx(t)}{dt} + x^2(t) = 0$

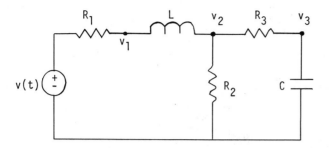

Figure 2.49   RLC circuit (Problem 2.2).

d. $(\sin t) \dfrac{d^2 x(t)}{dt^2} + (\sin 2t)\, x(t) = 0$

e. $\sin x(t)\, \dfrac{d^2 x(t)}{dt^2} + \cos 2x(t) = 0$

2.2.   Use the nodal analysis method to develop the mathematical model (integrodifferential equations) for the circuit given in Figure 2.49.

2.3.   Develop the differential equations representing the dynamic absorber shown in Figure 2.50 where M_1 is assumed to be a relatively small mass.

2.4.   Evaluate the Laplace transform of

a.   cosh at

b.   sinh bt

c.   at

d. at^2

e. t^n

f.   cos ωt

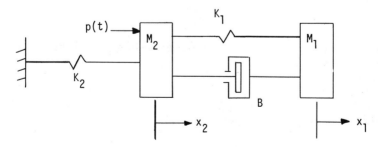

Figure 2.50   Mechanical absorber (Problem 2.3).

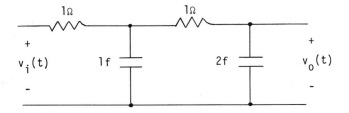

Figure 2.51 RC circuit (Problem 2.6).

2.5. Find the inverse Laplace transform of each of the following

   a. $F_a(s) = \dfrac{(s + 1)(s + 6)(s + 10)}{(s + 5)(s + 8)}$

   b. $F_b(s) = \dfrac{(s + 1)(s + 6)(s + 10)}{(s + 5)(s + 8)^2}$

   c. $F_c(s) = \dfrac{1}{s(s + 3)^2}$

   d. $F_d(s) = \dfrac{2(s + 3)}{(s + 1)(s + 6)}$

2.6. Develop the transfer function $V_0(s)/V_i(s)$ for the circuit shown in Figure 2.51.

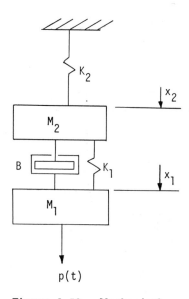

Figure 2.52 Mechanical system (Problem 2.8).

2.7.   Find the transfer functions $X_1(s)/P(s)$ and $X_2(s)/P(s)$ for the dynamic absorber defined in Problem 2.3 and depicted in Figure 2.50.

2.8.   Figure 2.52 shows a mechanical system that is described by the following differential equations:

$$M_1\ddot{x}_1 + B\dot{x}_1 + K_1 x_1 - B\dot{x}_2 - K_1 x_2 = p(t)$$

$$-B\dot{x}_1 - K_1 x_1 + M_2 \ddot{x}_2 + B\dot{x}_2 + (K_1 + K_2)\, x_2 = 0$$

a.   Assume zero initial conditions and obtain the transfer function $X_2(s)/P(s)$.

b.   For this fourth-order system, let the state variables be

$$z_1 = x_1 \qquad z_2 = \dot{x}_1 \qquad z_3 = x_2 \qquad z_4 = \dot{x}_2$$

and obtain the state-variable representation for this system.

2.9.   Obtain the closed-loop transfer function $C(s)/R(s)$ for the multiloop feedback system shown in Figure 2.53 by

a.   Using block-diagram-reduction technique

b.   Sketching the system's signal flow graph and applying Mason's formula

2.10. Find the z-transform for

a. $f(t) = 2\,e^{-3t} \cos 4t$

b. $G(s) = \dfrac{a}{s^2(s + a)}$

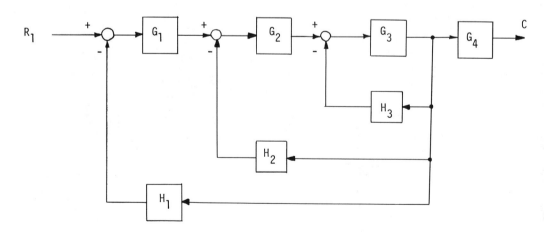

Figure 2.53   Multiloop feedback system (Problem 2.9).

2.11. Find the inverse z-transform for

a. $\dfrac{10z(z + 0.4)}{(z - 1)(z - 0.2)(z - 0.5)}$

b. $\dfrac{3z(z + 0.3)}{(z - 0.5)^2(z^2 + 0.8z + 0.25)}$

2.12. Apply the Routh criterion and determine the number of roots in the right half of the s-plane for each of the following equations:

a. $s^3 + 13s^2 + 33s + 30 = 0$

b. $s^4 + 12s^3 + 54s^2 + 108s + 80 = 0$

c. $s^5 + s^4 + 2s^3 + 2s^2 + 4s + 1 = 0$

2.13. Draw the root loci for the following open-loop transfer functions:

a. $K(s + 1)/s^2$

b. $Ks/(s^2 + 4s + 8)$

c. $\dfrac{K(s + 10)(s + 15)}{(s^3 + 3s^2 + 3s + 1)}$

2.14. Determine whether this system is controllable and observable.

$$\dot{\underline{x}} = \begin{bmatrix} -1 & 0 & 3 \\ 2 & -1 & -1 \\ -3 & 1 & -2 \end{bmatrix} \underline{x} + \begin{bmatrix} 1 \\ 0 \\ 0 \end{bmatrix} u$$

$$y = \begin{bmatrix} 1 & 2 & 1 \end{bmatrix} x$$

REFERENCES

Benedict T. R., and Bordner, G. W. (1962). Synthesis of an optimal set of radar track-while-scan smoothing equations, *IRE Trans. Automatic Control*, *AC-7(4)*, 27–32.

Cadzow, J. A. (1973). *Discrete-Time Systems: An Introduction with Interdisciplinary Applications*, Prentice-Hall, Englewood Cliffs, NJ.

Chan, S. P., Chan, S. Y., and Chan, S. G. (1972). *Analysis of Linear Networks and Systems*, Addison Wesley, Reading, MA.

Gilbert, E. G. (1963). Controllability and observability in multivariable control systems, *J. SIAM Control 1*, 128–151.

Johnson, C. D. (1974). Overmodeling of dynamical processes, in *Proceedings of the Annual Pittsburgh Conference on Modeling and Simulation*, pp. 1021–1026.

Krause, P. C. (1986). *Analysis of Electric Machinery*, McGraw-Hill, New York.

Kreyszig, E. (1972). *Advanced Engineering Mathematics*, Wiley, New York.

Kuo, B. C. (1975). *Automatic Control Systems*, 3rd ed., Prentice-Hall, Englewood Cliffs, NJ.

Kwatny, H. G., and Mablekos, V. E. (1975). The modeling of dynamical processes, in *Proceedings of the IEEE Decision and Control Conference.*

Lewis, L. J., Reynolds, D. K., et al. (1969). *Linear System Analysis,* McGraw-Hill, New York.

Luenberger, D. G. (1979). *Introduction to Dynamic Systems: Theory, Models and Applications,* Wiley, New York.

Mason, S. J. (1956). Feedback theory: Further properties of signal flow graphs, *Proc. IRE 44*(7), 920–926.

Ogata, K. (1970). *Modern Control Engineering,* Prentice-Hall, Englewood Cliffs, NJ.

Samuelson, P. A. (1939). Interaction between the multiplier analysis and the principle of acceleration, *Rev. Econ. Statistics 21,* 75–78.

Van de Vegte, J. (1986). *Feedback Control Systems,* Prentice-Hall, Englewood Cliffs, NJ.

Van Ness, J. E. (1977). Methods of reducing the order of power systems models in dynamic studies, *Proceedings of the IEEE International Symposium on Circuits and Systems.*

3
Discrete-Event Systems

3.1 INTRODUCTION

Simulation, as defined in Section 1.3, is the development of a mathematical-logical model of a system and the experimental manipulation of the model on a computer. Two basic steps in simulation are cited in this definition: (a) model development, and (b) experimentation. *Model development* involves the construction of a mathematical-logical representation of the system and the preparation of a computer program that allows the model to mimic the behavior of the system. Once we have a valid model of the system, the second phase of simulation study takes place—*experimentation* with the model to determine how the system responds to changes in the levels of the several input variables.

The terms "model" and "system" are also very important in the definition of simulation given above. A *system* is a collection of items from a circum-scribed sector of reality that is the focus of study. A system is a relative thing, and we define the boundaries of the system to include those items that are deemed most important to our objectives and to exclude items of lesser importance. For instance, if our focus is the operation of an out-patient clinic in a hospital, we will define the boundaries of the system to include the physicians, nurses, staff, facilities, and services of the out-patient clinic, while excluding all other areas of the system.

A *model* is the means we choose to capture the salient features of the system under study. The model must possess some representation of the entities or objects in the system, and reflect the activities in which these entities engage.

The steps in a simulation study are as follows:

1. *Problem formulation*—a statement of the problem that is to be solved. This includes a general description of the system to be studied and a preliminary definition of the boundaries of that system.
2. *Setting objectives*—a delineation of the questions that are to be answered by the simulation study. This step allows further defini-tion of the system and its boundaries.

3. *Model building*—the process of capturing the essential features of a
    system in terms of its entities (objects), the attributes or charac-
    teristics of each entity, the activities in which these entities engage,
    and the set of possible states in which the system can be found.
4. *Data collection*—gathering data and information that will allow the
    modeler to develop the essential description of each of the system
    entities and developing probability distributions for the important
    system parameters.
5. *Coding*—the process of translating the system model into a computer
    program which can be executed on an available processor.
6. *Verification*—the process of ascertaining that the computer program
    performs properly.
7. *Validation*—the process of ascertaining that the model mimics the
    real system, by comparing the behavior of the model to that of the
    real system where the system can be observed and altering the
    model to improve its ability to represent the real system. The
    combined steps of verification and validation are crucial to establish-
    ing the credibility of the model, so that decisions reached about the
    system on the basis of the simulation study can be supported with
    confidence.
8. *Experimental design*—determining the alternatives that can be
    evaluated through simulation, choosing the important input variable(s)
    and their appropriate level(s), selecting the length of the simulation
    run and the number of replications.
9. *Production runs and analysis*—assessing the effects of the chosen
    input variables on the selected measures of system performance and
    determining whether additional runs are needed.
10. *Simulation report*—documenting the simulation program, reporting
    the results of the simulation study, and making recommendations
    about the real system on the basis of the simulation study. The
    implementation of these recommendations is usually the result of a
    decision by the appropriate manager in the organization.

3.2   PRINCIPLES OF DISCRETE-EVENT
SYSTEMS SIMULATION

The concepts of a system and a model of a system have already been
broadly discussed. This section expands on those concepts and establishes
a general framework for discrete-event simulation. The major concepts are
briefly defined and then illustrated with examples:

System   A collection of entities that interact together over time to
     accomplish a set of goals or objectives.
Model   A mathematical-logical representation of the system in terms
     of its entities and their attributes, sets, events, activities, and
     delays.
System state   A collection of variables the values of which define the
     state of the system at a given point in time.
Entity   An object, item, or component of the system that requires
     explicit representation in the model.
Attributes   The properties or characteristics of a given entity.
Set   A collection of associated entities.

> *Event* An instantaneous occurrence in time that alters the state of the system.
>
> *Activity* A duration of time in which one or more entities are engaged in the performance of some function that is germane to the system under study, the length of which is known at the outset.
>
> *Delay* A duration of time of unspecified length, the length of which is not usually known until it ends.

To illustrate these concepts, consider the example of an outpatient clinic in a hospital. The entities in this system include patients, physicians, nurses, and examining rooms. The attributes of a patient include, for example, the nature of the disorder, the time the patient arrives at the clinic, and the type of insurance coverage available. In fact, the patient's entire medical history could be attributes of the patient. The attributes of a physician might include type of specialty, number of patients in-process, and number of nurses assigned. A set might include the patients waiting for service, ordered by severity of disorder and first-come-first-serve within disorder priority. An activity might be typified by the examination of a patient by a doctor, or the time required to perform an x-ray procedure. A delay might be the time a patient spends waiting to see a doctor. An event might be the arrival of another patient into the clinic, the completion of the examination of a patient by a physician, or the completion of an x-ray by a laboratory technician.

The scheduling of events is accomplished by a *next-event* approach. Time is advanced from the time of a current event, t_j, to the time of the next scheduled event, t_{j+1}. The *calendar* of events consists of a file containing the time and type of each of n events, arranged in chronological order in the memory of the computer. The simulation program must have a mechanism for fetching the next event that is scheduled to occur, automatically advancing simulation time to the scheduled time of occurrence of that event, and transferring control to the appropriate event program. Figure 3.1 illustrates the time-advance procedure in next-event simulation.

The method by which events are placed in the event calendar is extremely important. One or more events must be initialized at the outset of the simulation. Two techniques are then applied to generate "future" events in the course of the simulation. One such technique is *bootstrapping*, which refers to the process by which the occurrence of an event is used to generate the next occurrence of the same type of event. This technique is perhaps best illustrated by the arrival process of patients in the outpatient clinic. The first arrival is initialized to occur just as the clinic opens or

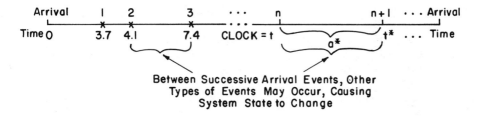

Figure 3.1 Time-advance procedure in next-event simulation.

shortly thereafter. When that arrival event occurs, the time of the next arrival is immediately scheduled and placed in the event calendar. Thus, arrivals generate new arrivals as the simulation proceeds. The second method for generating events is the *next-logical-event* approach. For example, when a physician has examined a patient, he/she can either (a) discharge the patient, (b) schedule one or more laboratory tests, (c) refer the patient to another specialist on the current visit, or (d) schedule another visit. Having done one or more of these, the physician can then take another patient from a queue, which is another example of the bootstrapping approach to event scheduling; that is, service completion events generate subsequent service completion events.

There are two basic approaches to discrete-systems simulation—event scheduling and process interaction. In the event-scheduling approach, we concentrate on events and their effect on *system state*. In the outpatient clinic example, system state is reflected by such variables as the number of patients in the system (clinic), the number of busy doctors, the number of busy nurses, the number of busy examining rooms, etc. For instance, when a patient arrival event occurs, it increases the number of patients in the system by one, possibly increases the number of busy nurses by one, and possibly increases the number of busy physicians by one. The word "possibly" is important, since the newly arrived patient could only activate a nurse or physician if the nurse or physician is idle. A "patient discharge" event reduces the number of patients in the system by one and possibly deactivates a nurse or physician.

Thus, in the event-scheduling approach to discrete-systems simulation, the occurrence of any of the events that make up the model will bring about a resultant change in system state. The model must provide for recording these state changes. This is usually managed by collecting one or more types of *statistics* at the occurrence of each event. For instance, when a patient enters or leaves the system, we would update a *time-dependent statistic* to reflect the change in the number of patients, whereas we would collect *sample statistics* on the duration of time the patient spends in the system. One might also update a *counter* as the patient leaves an examining room to keep track of the number of times that facility is utilized.

A different outlook on discrete-event simulation is provided by the *process-interaction* approach. A *process* is a time-ordered collection of events, activities, and delays that are somehow related to an entity. For example, the sequence of events, activities, and delays encountered by a patient in the outpatient clinic constitutes a process. Process-oriented simulations usually have many processes ongoing simultaneously and involve extremely complex interactions among these many processes.

3.3   DISCRETE-EVENT SIMULATION LANGUAGES

Discrete-event simulation models can be developed using one of three general classes of languages: (a) high-level programming languages such as FORTRAN, Pascal, Ada, or C; (b) general-purpose simulation languages such as GASP-IV, Simscript, SLAM-II, or SIMAN; or (c) special-purpose simulation languages such as GPSS, GEMS, GERTS, XCELL, or PASAMS.

FORTRAN, Pascal, Ada, and C are high-level languages that were developed for a wide range of computing applications. FORTRAN is the oldest of these and has been widely applied as the base language in simulation modeling, owing mainly to its being so well known and so widely available on almost any computer of sufficient size and speed to be able to accommodate computer simulation. Almost all simulation models developed directly from FORTRAN have utilized the event-scheduling approach. The recent emergence of smaller microcomputers has led to greater use of Pascal as a base language for simulation, and the growing popularity of the Unix operating system has led to greater use of the C language. These languages have not yet been applied in commercially available simulation languages, but soon will be.

GPSS (Schriber, 1974), one of the first process-oriented simulation languages, is a special-purpose simulation language that was designed for use primarily with queueing systems. The later, general-purpose simulation languages, notably GASP-IV (Pritsker, 1974) and Simscript, were largely event-oriented, but afforded more general constructs for model building. These languages initially found favor among simulation modelers who had previously relied on FORTRAN as the base language. SLAM-II (Pritsker, 1985) and Simscript II.5 (Kiviat et al., 1973) evolved from GASP-IV and Simscript, respectively, and offer the analyst a choice of either event or process orientations. SIMAN (Pegden, 1986) also provides a choice of orientations, but differs from the other general-purpose simulation languages in that it enables the analyst to develop separate model and experiment frames, thus permitting greater ease in experimenting with the simulation model once it has been developed. A detailed treatment of SIMAN modeling is given in Chapter 16.

The special-purpose simulation languages were designed to allow easy modeling in a highly specialized area of application. For instance, GPSS focuses on queueing environments. GEMS (Phillips, 1978) concentrates on the material flows in a manufacturing environment. PASAMS (Biles and Bathina, 1986) was developed to provide microcomputer users a Pascal-based modeling package for manufacturing applications. XCELL (Conway et al., 1986) is less a simulation tool than a factory-modeling system, but has many of the features of a specialized simulation modeling technique.

Other than GPSS, the special-purpose simulation languages have not found widespread application simply because analysts prefer to acquire simulation tools that afford them greater flexibility in a broader range of application environments.

3.4   STATISTICAL MODELS IN SIMULATION

In simulation the modeler sees a probabilistic world. The time it takes a machine to fail is a random variable, as is the time it takes a maintenance mechanic to repair it. Modeling requires skill in recognizing the random behavior of the various phenomena that must be incorporated into the model, analyzing data to determine the nature of these random processes, and providing appropriate mechanisms in the model to mimic these random processes. This section discusses the basic concepts in probability and statistics as they relate to discrete-event simulation modeling. These basic terms are as follows:

Random variable   X is one that can assume any of several possible values over a range of such possible values.

Discrete random variable   X is one in which the range of possible values is finite or countably infinite. For x_1, x_2, . . ., the probability mass function of X is

$$p(x_i) = P(X = x_i) \tag{3.1a}$$

$$p(x_i) \geqslant 0 \text{ for all } i \tag{3.1b}$$

$$\sum_i p(x_i) = 1 \tag{3.1c}$$

Continuous random variable   X is one in which the range of possible values is the set of reals $-\infty < x < -\infty$. If $f(x)$ is the probability density function of X, then

$$P(a \leqslant X \leqslant b) = \int_a^b f(x) \, dx \tag{3.2a}$$

$$f(x) \geqslant 0 \qquad \text{for all } x \text{ in } R_x \tag{3.2b}$$

$$\int_{R_x} f(x) = 1 \tag{3.2c}$$

Cumulative distribution function   Denoted $F(x)$, measures the probability that a random variable X has a value less than or equal to the value x, that is,

$$F(x) = P(X \leqslant x) \tag{3.3a}$$

If X is discrete, then

$$F(x) = \sum_{x_i \leqslant x} p(x_i) \tag{3.3b}$$

If X is continuous, then

$$F(x) = \int_{-\infty}^x f(x) \, dx \tag{3.3c}$$

Expectation   The expected value of the random variable X is given by

$$E(X) = \sum_i x_i p(x_i) \qquad \text{if X is discrete} \tag{3.4a}$$

and by

$$E(X) = \int_{-\infty}^\infty x \, f(x) \, dx \qquad \text{if X is continuous} \tag{3.4b}$$

The expected value is also called the *mean* and denoted by μ. It is also the first moment of X. We can also define the nth moment of X as

$$E(X^n) = \sum_i x_i^n \, p(x_i) \qquad \text{if X is discrete} \qquad (3.5a)$$

and as

$$E(X^n) = \int_{-\infty}^{\infty} x^n f(x) \, dx \qquad \text{if X is continuous} \qquad (3.5b)$$

Based on these, we next define the *variance* of the random variable X as

$$V(X) = E\{[X - E(X)]^2\} = E[(X - \mu)^2] \qquad (3.6)$$

Useful statistical models in discrete-event simulation include *queueing systems, inventory systems,* and *reliability and maintainability systems.* We shall examine some of these systems in subsequent sections. Underlying these systems, however, are several very important discrete and continuous probability distributions that are examined next.

3.4.1 Discrete Distribution

Discrete random variables are used to describe random phenomena in which only integer values of the random variable X can occur. The probability distributions of four such random variables which are fundamentally important in discrete-event simulation follow.

1. *Bernoulli trial and the Bernoulli distribution.* Consider a random experiment consisting of n trials in which each trial can produce one of only two outcomes, *success* and *failure.* Let $X_j = 1$ if the jth trial results in a success and $X_j = 0$ if the jth trial produces a failure. For example, let the random experiment consist of the inspection of a manufactured assembly, and let "success" be defined as finding a defective assembly. Thus, a defective assembly yields the value $X_j = 1$, while an acceptable assembly yields the value $X_j = 0$. (This viewpoint can, of course, be reversed without altering the model.) The n Bernoulli trials are called a *Bernoulli process* if the successive trials are independent.

The probability distribution for the Bernoulli trial is

$$p_j(x_j) = p(x_j) = \begin{cases} p & x_j = 1 \quad j = 1, 2, \ldots, n \\ q = 1 - p & x_j = 0 \quad j = 1, 2, \ldots, n \\ o & \text{otherwise} \end{cases} \qquad (3.7a)$$

For one trial, this equation is called the *Bernoulli distribution.* The mean and variance of the Bernoulli distribution are given by

$$E(X_j) = p \qquad (3.7b)$$

$$V(X_j) = p(1 - p) = pq \qquad (3.7c)$$

2. *Binomial distribution.* If we define the random variable X as the number of successes in n independent Bernoulli trials, then X has the binomial distribution given by

$$p(x) = \binom{n}{x} p^x q^{n-x} \qquad x = 0, 1, 2, \ldots, n \qquad (3.8a)$$

with mean and variance

$$E(X) = np \qquad\qquad\qquad (3.8b)$$

$$V(X) = npq \qquad\qquad\qquad (3.8c)$$

Example 3.4.1. For instance, if we are inspecting manufactured assemblies and our sample size is n assemblies, the number X of defective assemblies in the sample is binomially distributed. Suppose n = 30 and the probability of a defective assembly is p = 0.02. The probability of fewer than three defective assemblies is

$$P(X \leqslant 2) = \sum_{x=0}^{2} \binom{30}{x} (0.02)^x (0.98)^{30-x}$$

$$= \frac{30!}{0!\,30!} (.02)^0 (.98)^{30} + \frac{30!}{1!\,29!} (.02)^1 (.98)^{29}$$

$$+ \frac{30!}{2!\,28!} (.02)^2 (.98)^{28}$$

$$= 0.5455 + 0.3340 + 0.0988 = 0.9783$$

The mean number of defectives in the sample is

$$E(X) = np = 30(0.02) = 0.6$$

and the variance is

$$V(X) = npq = 30(0.02)(0.98) = 0.588$$

3. *Geometrical distribution.* The geometrical distribution applies to the random variable X, which is defined as the number of Bernoulli trials until the *first* success and is given by

$$p(x) = \begin{cases} q^{x-1} p & x = 1, 2, \ldots \\ 0 & \text{otherwise} \end{cases} \qquad (3.9a)$$

with mean and variance

$$E(X) = \frac{1}{p} \qquad\qquad\qquad (3.9b)$$

$$V(X) = \frac{q}{p^2} \qquad\qquad\qquad (3.9c)$$

Example 3.4.2. In the above example, the mean number of assemblies inspected until the first defective unit is found is

$$E(X) = \frac{1}{p} = \frac{1}{0.02} = 50$$

4. *Poisson distribution*. The Poisson distribution describes so-called "rate processes." For instance, the rate of occurrence of arrivals of patients at the outpatient clinic in our earlier example might well follow a Poisson distribution. The rate of occurrence of bubbles per square meter of plate glass, of pits per linear meter of extruded copper wire, and of failures of a milling machine might also follow Poisson distributions. The Poisson probability distribution is given by

$$p(x) = \frac{e^{-\alpha}\alpha^x}{x!} \qquad x = 0, 1, 2, \ldots \tag{3.10a}$$

with mean and variance

$$E(X) = \alpha \tag{3.10b}$$

$$V(X) = \alpha \tag{3.10c}$$

3.4.2 Continuous Distributions

Continuous random variables are used to describe random phenomena in which the random variable X can take on any value in some real interval. Six continuous distributions that are fundamentally important in discrete-event simulation are the uniform, exponential, gamma, Erlang, normal, and Weibull distributions.

1. *Uniform distribution*. A random variable X is uniformly distributed in the interval [a,b] if its *probability density function*, or pdf, is given by

$$f(x) = \begin{cases} \dfrac{1}{b-a} & a < x \leqslant b \\[3mm] 0 & \text{otherwise} \end{cases} \tag{3.11a}$$

The *cumulative distribution function*, or cdf, is given by

$$F(X) = \begin{cases} 0 & x < a \\[2mm] \dfrac{x-a}{b-a} & a \leqslant x \leqslant b \\[2mm] 1 & x > b \end{cases} \tag{3.11b}$$

The pdf and cdf for the uniform distribution are shown in Figure 3.2. The mean and variance of the uniform distribution are given by

$$E(X) = \frac{a+b}{2} \tag{3.11c}$$

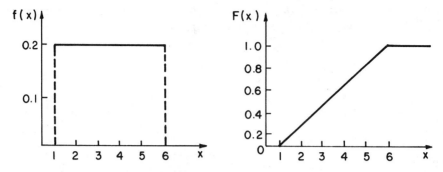

Figure 3.2   Uniform pdf and cdf.

and

$$V(X) = \frac{(b - a)^2}{12}$$

(3.11d)

The uniform distribution plays a vital role in discrete-event simulation. *Random numbers*, uniformly distributed in the interval [0,1], are used to generate the *random variates* X which give rise to random events. Random-number generation is discussed in a later section.

2. *Exponential distribution.* The random variable X is said to be exponentially distributed with parameter $\lambda > 0$ if its pfd is given by

$$f(x) = \lambda e^{-\lambda x} \qquad x \geqslant 0$$

(3.12a)

The exponential distribution is used to model interarrival times in a random process in which the rate of arrivals is Poisson distributed. In this case, λ is the mean rate of arrivals and $1/\lambda$ is the mean time between arrivals. The mean and variance of the exponential distribution are

$$E(X) = \frac{1}{\lambda}$$

(3.12b)

and

$$V(X) = \frac{1}{\lambda^2}$$

(3.12c)

The cumulative distribution function is given by

$$F(X) = 1 - e^{-\lambda x} \qquad x \geqslant 0$$

(3.12d)

To illustrate the exponential distribution, suppose that the life of a cathode ray tube, in thousands of hours, is exponentially distributed with $\lambda = 1/4$; that is, the mean life is 4000 hr. Then the probability that a tube will last 3000 hr is

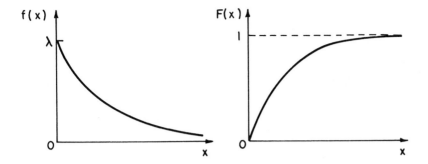

Figure 3.3   Exponential pdf and cdf.

$$P(X > 3) = 1 - (1 - e)^{-3/4} = e^{-3/4} = 0.472$$

Figure 3.3 shows the exponential pdf and cdf.

3. *Gamma distribution.* A function used in defining the gamma distribution is the *gamma function*, which is defined for all $\beta > 0$

$$\Gamma(\beta) = \int_0^\infty x^{\beta-1} e^{-x} \, dx \qquad (3.13a)$$

Integrating by parts we obtain

$$\Gamma(\beta) = (\beta - 1)\Gamma(\beta - 1) \qquad (3.13b)$$

If β is an integer, then $\Gamma(1) = 1$ and

$$\Gamma(\beta) = (\beta - 1)! \qquad (3.13c)$$

The random variable X has a gamma distribution with parameters β and Θ if its pdf is given by

$$f(x) = \frac{\beta\Theta}{\Gamma(\beta)} \; (\beta\Theta x)^{\beta-1} e^{-\beta\Theta x} \qquad x > 0 \qquad (3.13d)$$

The parameter β is called the *shape parameter*, and Θ is called the *scale parameter*. Figure 3.4 illustrates several gamma distributions for $\Theta = 1$ and various values of β. Observe that when $\beta = 1$, we have the exponential distribution. The mean and variance of the gamma distribution are given by

$$E(X) = \frac{1}{\Theta} \qquad (3.13e)$$

$$V(X) = \frac{1}{\beta\Theta^2} \qquad (3.13f)$$

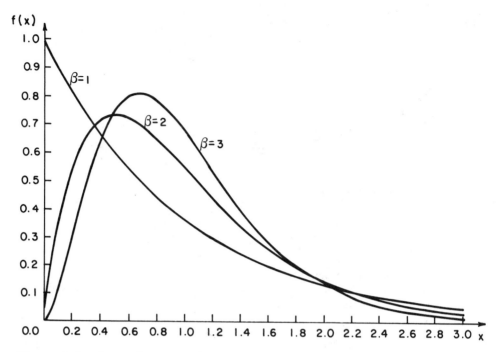

Figure 3.4   Several gamma distributions for $\Theta = 1$.

The cdf is given by

$$F(x) = 1 - \int_{x}^{\infty} \frac{\beta\Theta}{\Gamma(\beta)} \, (\beta\Theta x)^{\beta-1} e^{-\beta\Theta x} \, dx \qquad (3.13g)$$

4. *Erlang distribution.* If the parameter β in the gamma distribution is an integer k, the gamma distribution has a special relationship to the exponential distribution. If the random variable X is the sum of k-independent, exponentially distributed random variables X_j, each with parameter $k\Theta$, then X has an Erlang distribution with pdf

$$f(x) = \frac{k\Theta}{\Gamma(k)} \, (k\Theta x)^{k-1} e^{-k\Theta x} \qquad x > 0 \qquad (3.14a)$$

and cdf

$$F(x) = 1 - \sum_{i=0}^{k-1} \frac{e^{-k\Theta x}(k\Theta x)^{i}}{i!} \qquad x > 0 \qquad (3.14b)$$

The mean and variance of the Erlang distribution are given by

$$E(X) = \frac{1}{\Theta} \qquad (3.14c)$$

and

$$V(X) = \frac{1}{k\theta^2} \qquad (3.14d)$$

Example 3.4.3.   As an example of the Erlang random variable, a manufactured part is tested in an automatic inspection machine in which there are three test stations, each with an exponentially distributed service time with mean 8.5 sec.   The part is sequenced through the three tests and is released from the inspection machine only after all three tests are completed. Clearly, inspection time is Erlang-distributed with k = 3 and $\theta = 1/(8.5)(3)$. The mean and variance of inspection time are therefore

$$E(X) = 3(8.5) = 25.5$$

and

$$V(X) = 8.5$$

5.   *Normal distribution*.   A random variable X with mean μ and variance σ^2 has a normal distribution if its pdf is given by

$$f(x) = \frac{1}{\sigma\sqrt{2\pi}}\, e^{-[1/2[(x-\mu)/\sigma]^2]} \qquad -\infty < x < \infty \qquad (3.15a)$$

The normal distribution is illustrated in Figure 3.5.

The cdf for the normal distribution $N(\mu, \sigma^2)$ is given by

$$F(x) = \int_{-\infty}^{x} \frac{1}{\sigma\sqrt{2\pi}}\, e^{-[1/2[(x-\mu)/\sigma]^2]}\, dx \qquad (3.15b)$$

The evaluation of this cdf for given values for μ and σ^2 is extremely difficult, so that the preferred procedure is to employ the transformation $z = (x - \mu)/\sigma$.   The random variable Z has the so-called *standard normal distribution* N(0,1), with probability density function $\phi(z)$, as shown in Figure 3.6.   Tables of standard normal distribution are common to every text on probability and statistics.

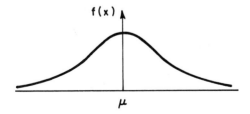

Figure 3.5   Normal distribution.

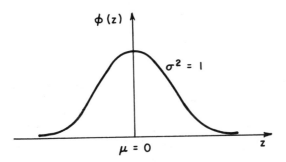

Figure 3.6   Standard normal distribution.

Example 3.4.4.   As an example of the normal distribution, consider the case in which project completion time for a construction follows a normal distribution with mean $\mu = 50$ days and variance $\sigma^2 = 9$ days^2. Let us determine the probability that a given project will require longer than 56 days, or $P(X > 56)$. Clearly, $P(X > 56) = 1 - P(X \leqslant 56)$. Using the transformation to the standard normal distribution, $z = (56 - 50)/3 = 2.0$, so that $P(Z \leqslant 2.0) = 0.977$. Therefore, $P(Z > 2.0) = P(X > 56) = 0.023$. This concept is illustrated in Figure 3.7.

6. *Weibull distribution.*   The random variable X has the Weibull distribution if its pdf has the form

$$f(x) = \begin{cases} \frac{\beta}{\alpha}\left(\frac{x - \nu}{\alpha}\right)^{\beta-1} e^{\left[-\left(\frac{x-\nu}{\alpha}\right)^{\beta}\right]} & x \geqslant \nu \\ 0 & \text{otherwise} \end{cases} \qquad (3.16a)$$

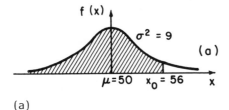

(a)

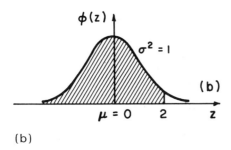

(b)

Figure 3.7   Transformation to the standard normal.

where $\nu(-\infty < \nu < \infty)$ is the *location parameter,* $\alpha(\alpha > 0)$ is the *scale parameter,* and $\beta(\beta > 0)$ is the *shape parameter.* Figure 3.8 shows various Weibull pdf's for $\nu = 0$, and $\alpha = 1$.

The cdf of the Weibull distribution has the form

$$F(x) = 1 - e^{\left[-\left(\frac{x-\nu}{\alpha}\right)^{\beta} \right]} \qquad x \geqslant \nu \qquad (3.16b)$$

The mean and variance of the Weibull distribution are given by the following expressions:

$$E(X) = \nu + \alpha \Gamma \left(\frac{1}{\beta} + 1 \right) \qquad (3.16c)$$

$$V(X) = \alpha^2 \ \Gamma \left(\frac{2}{\beta} + 1 \right) - \left[\Gamma \left(\frac{1}{\beta} + 1 \right) \right]^2 \qquad (3.16d)$$

The Weibull distribution is important in reliability and maintainability systems.

Example 3.4.5. As an example of the Weibull distribution, suppose a class of vacuum pumps has a time to failure that follows a Weibull distribution with $\nu = 0$, $\beta = 1/3$, and $\alpha = 200$ hr. The mean time to failure is therefore

$$E(X) = 200 \Gamma (3 + 1) = 200(3!) = 1200 \text{ hr}$$

The probability that a vacuum pump fails before 2000 hr is determined from the cdf as

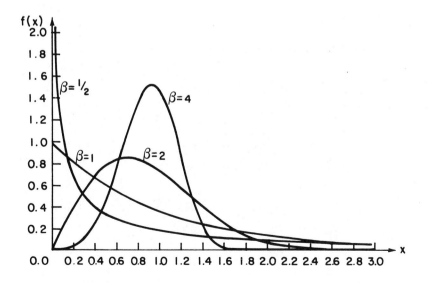

Figure 3.8   Weibull distributions for $\nu = 0$ and $\alpha = 1$.

$$F(2000) = 1 - e^{\left[- \left(\frac{2000}{200} \right)^{1/3} \right]} = 1 - e^{-2.15} = 0.884$$

3.4.3 Poisson Processes

The arrival processes described earlier in which the number of occurrences of an event per unit time is a Poisson-distributed random variable are *Poisson processes* if they conform to the following definition. Let a counting function $\{N(t), t \geqslant 0\}$ represent the number of occurrences of an event in $[0,t]$. Thus $N(t)$ is a discrete random variable with possible values 0, 1, 2, ... The counting process $\{N(t), t \geqslant 0\}$ is said to be a Poisson process with mean rate λt if the following assumptions are met:

1. Arrivals occur one at a time.
2. $\{N(t), t \geqslant 0\}$ has *stationary increments,* which means that the number of arrivals between t and t + s depends only on the length of the interval s and not on the starting point t.
3. $\{N(t), t \geqslant 0\}$ has *independent increments,* which means that the numbers of arrivals in nonoverlapping intervals are independent random variables.

If arrivals occur according to a Poisson process, meeting the assumptions stated above, then the probability that $N(t) = n$ is given by the expression

$$P[N(t) = n] = \frac{e^{-\lambda t}(\lambda t)^n}{n!} \qquad \text{for } t \geqslant 0 \quad \text{and} \quad n = 0, 1, 2, \ldots$$

$$(3.17a)$$

so that $N(t)$ has a Poisson distribution with parameter $\alpha = \lambda t$. Thus, its mean and variance are

$$E[N(t)] = V[N(t)] = \alpha = \lambda t \qquad\qquad (3.17b)$$

Table 3.1   Lot Sizes in Production Orders

Lot size per order	Frequency	Relative frequency	Cumulative relative frequency
1	30	0.10	0.10
2	110	0.37	0.47
3	45	0.15	0.62
4	71	0.24	0.86
5	12	0.04	0.90
6	13	0.04	0.94
7	7	0.02	0.96
8	12	0.04	1.00

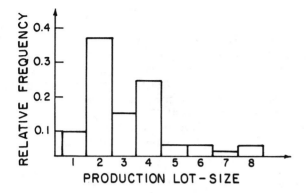

Figure 3.9   Histogram for a sample distribution.

3.4.4   Empirical Distributions

Sometimes it is impossible to match a distribution of data to a known distribution. In such instances we resort to fitting an *empirical distribution* to the sample distribution. Suppose that the lot sizes in production orders followed the distribution shown in Table 3.1. By plotting a histogram of the sample distribution we obtain Figure 3.9, which shows the relative distribution, usually called the *probability mass function* when dealing with discrete random variable. A similar procedure is followed for continuous random variables. Figure 3.10 shows an empirical cdf for a sample of 100 observations of the time to repair, say, an overhead trolley conveyor.

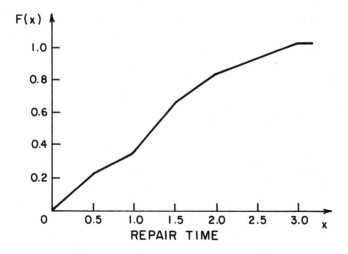

Figure 3.10   Empirical cdf.

3.5   QUEUEING MODELS

One of the most important types of random processes involved in simulation studies is the *queueing system*.   Shown in Figure 3.11, a queueing system is composed of a calling population of potential customers, a waiting line for customers, and a service process.   For instance, the population of potential customers might be a group of machines in a factory.   As a machine fails, it becomes a "customer" for the repair service.   If the repair mechanic is busy servicing another failed machine, the newly failed machine must join a queue.   After a machine has been repaired, it rejoins the calling population of operating machines.

3.5.1   Characteristics of Queueing Systems

The key elements of a queueing system are the customers and the servers. The term "customers" can refer to people, parts, machines, aircraft, computer jobs, etc.   "Servers" are bank tellers, machine operators, maintenance mechanics, air traffic controllers, computer operators (or the computer itself), etc.   This section examines some of the more important characteristics of the queueing system.   The "arrival process" describes the manner in which customers seek service.   The "departure process" describes the manner in which customers leave the service system.

> *Calling population*   The population of potential customers; it may be finite or countably infinite.   The arrival rate into the system depends on the size of the calling population.
> *System capacity*   A limit on the number of customers the system can accommodate at any one time.
> *Arrival process*   Arrivals may occur at scheduled times, or at random times, the latter usually by an assumed probability distribution. The Poisson distribution is the most common.
> *Queue discipline*   The behavior of the queue in reaction to its current state, or the manner in which the service facility organizes the queue.
> *Service mechanism*   Service times may be constant or of some random duration.   The exponential, Weibull, gamma, and normal distributions are commonly used in scheduling services.   Service can be in single or multiple channels.

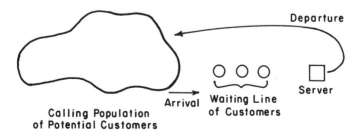

Figure 3.11   Single-channel queueing system.

3.5.2 Queueing Notation

The standard notation for queueing systems is one involving a five-character code A |B |c |N |k where:

A represents the interarrival-time distribution.
B represents the service-time distribution.
c represents the number of parallel servers.
N represents system capacity.
k represents the size of the population.

For example, M |M |1 |∞ |∞ indicates a single-server system with unlimited queue capacity and infinite calling population. Interarrival times and service times are exponentially distributed. This system is typically abbreviated as M |M |1, with the absence of the last two characters indicating default to values of infinity. Table 3.2 lists queueing notation for parallel server systems.

3.5.3 Performance Measures for Queueing Systems

One of the principal uses of *analytic* queueing models in simulation is in the validation of discrete-event simulation models of queueing systems that are too complex to be modeled analytically. The results of the simulation model, executed under the same set of assumptions employed for the closest

Table 3.2 Queueing Notation for Parallel Server Systems

P_n	Steady-state probability of having n customers in system
$P_n(t)$	Probability of n customers in system at time t
λ	Arrival rate
λ_e	Effective arrival rate
μ	Service rate of one server
μ_e	Effective service rate of one server
ρ	Server utilization
A_n	Interarrival time between customers n − 1 and n
S_n	Service time of the nth arriving customer
W_n	Total time spent in system by the nth arriving customer
W_n^Q	Total time spent in the waiting line by customer n
$L(t)$	The number of customers in system at time t
$L_Q(t)$	The number of customers in queue at time t
L	Long-run time-average number of customers in system
L_Q	Long-run time-average number of customers in queue
w	Long-run average time spent in system per customer
w_Q	Long-run average time spent in queue per customer

corresponding analytic model, are compared with those of the analytic model to ascertain the validity of the simulation model.

A queueing system typically exhibits two stages of behavior—short-term or *transient* behavior, followed by long-term or *steady-state* behavior. If the system is started from a so-called *empty and idle* condition, it must operate for a period of time before reaching steady-state conditions. A simulation model must be run for a sufficiently long period of time to exceed the transient period before measures of steady-state performance can be determined.

The long-run measures of queueing system performance are as follows:

Server utilization   The fraction of available time that the server is busy.

Customers in system   The steady-state time-average number of customers in the system.

Customers in queue   The steady-state time-average number of customers in the queue.

Time in system   The average time a customer spends in the system, including time in queue and time in service.

Time in queue   The average time a customer spends in the queue waiting for service.

Probability of n customers in system   The steady-state probability that there are n customers in the system.

Using the notation and definitions of Table 3.2, the analytical models for various classes of queueing systems are summarized in Tables 3.3–3.8.

$\underline{M}\,|\underline{M}\,|\underline{1}$ Queues, $\underline{M}\,|\underline{G}\,|\underline{1}$ Queues, and $\underline{M}\,|\underline{E}\,|\underline{1}$ Queues.   Service times are exponentially distributed (M), generally distributed (G), or Erlang-distributed (E).

$\underline{M}\,|\underline{D}\,|\underline{1}$ Queues.   Service times are constant.

$\underline{M}\,|\underline{M}\,|\underline{1}\,|\underline{N}$ Queues.   System capacity is limited to N, as might be the case in an outpatient clinic or an airport.

Table 3.3   Steady-State Parameters for $M\,|M\,|1$ Queues

L	$\dfrac{\lambda}{\mu - \lambda} = \dfrac{\rho}{1 - \rho}$
w	$\dfrac{1}{\mu - \lambda} = \dfrac{1}{\mu(1 - \rho)}$
w_Q	$\dfrac{\lambda}{\mu(\mu - \lambda)} = \dfrac{\rho}{\mu(1 - \rho)}$
L_Q	$\dfrac{\lambda^2}{\mu(\mu - \lambda)} = \dfrac{\rho^2}{1 - \rho}$
P_n	$\left(1 - \dfrac{\lambda}{\mu}\right)\left(\dfrac{\lambda}{\mu}\right)^n = (1 - \rho)\rho^n$

Table 3.4   Steady-State Parameters for $M|G|1$ Queues

ρ	$\dfrac{\lambda}{\mu}$
L	$\rho + \dfrac{\lambda^2(\mu^{-2} + \sigma^2)}{2(1 - \rho)} = \rho + \dfrac{\rho^2(1 + \sigma^2\mu^2)}{2(1 - \rho)}$
w	$\mu^{-1} + \dfrac{\lambda(\mu^{-2} + \sigma^2)}{2(1 - \rho)}$
w_Q	$\dfrac{\lambda(\mu^{-2} + \sigma^2)}{2(1 - \rho)}$
L_Q	$\dfrac{\lambda^2(\mu^{-2} + \sigma^2)}{2(1 - \rho)} = \dfrac{\rho^2(1 + \sigma^2\mu^2)}{2(1 - \rho)}$
P_0	$1 - \rho$

Table 3.5   Steady-State Parameters for $M|E_k|1$ Queues*

L	$\dfrac{\lambda}{\mu} + \dfrac{1 + k}{2k}\,\dfrac{\lambda^2}{\mu(\mu - \lambda)} = \rho + \dfrac{1 + k}{2k}\,\dfrac{\rho^2}{1 - \rho}$
w	$\dfrac{1}{\mu} + \dfrac{1 + k}{2k}\,\dfrac{\lambda}{\mu(\mu - \lambda)} = \mu^{-1} + \dfrac{1 + k}{2k}\,\dfrac{\rho\mu^{-1}}{1 - \rho}$
w_Q	$\dfrac{1 + k}{2k}\,\dfrac{\lambda}{\mu(\mu - \lambda)} = \dfrac{1 + k}{2k}\,\dfrac{\rho\mu^{-1}}{1 - \rho}$
L_Q	$\dfrac{1 + k}{2k}\,\dfrac{\lambda^2}{\mu(\mu - \lambda)} = \dfrac{1 + k}{2k}\,\dfrac{\rho^2}{1 - \rho}$

*Service times are Erlang distributed.

Table 3.6   Steady-State Parameters for $M|D|1$ Queues[†]

L	$\dfrac{\lambda}{\mu} + \dfrac{1}{2}\,\dfrac{\lambda^2}{\mu(\mu - \lambda)} = \rho + \dfrac{1}{2}\,\dfrac{\rho^2}{1 - \rho}$
w	$\dfrac{1}{\mu} + \dfrac{1}{2}\,\dfrac{\lambda}{\mu(\mu - \lambda)} = \mu^{-1} + \dfrac{1}{2}\,\dfrac{\rho\mu^{-1}}{1 - \rho}$
w_Q	$\dfrac{1}{2}\,\dfrac{\lambda}{\mu(\mu - \lambda)} = \dfrac{1}{2}\,\dfrac{\rho\mu^{-1}}{1 - \rho}$
L_Q	$\dfrac{1}{2}\,\dfrac{\lambda^2}{\mu(\mu - \lambda)} = \dfrac{1}{2}\,\dfrac{\rho^2}{1 - \rho}$

[†]Service times are constant.

Table 3.7   Steady-State Parameters for M|M|1|N Queues*

L	$\begin{cases} \dfrac{a[1 - (N + 1)a^N + Na^{N+1}]}{(1 - a^{N+1})(1 - a)} & \lambda \neq \mu \\[4mm] \dfrac{N}{2} & \lambda = \mu \end{cases}$	
$1 - P_N$	$\begin{cases} \dfrac{1 - a^N}{1 - a^{N+1}} & \lambda \neq \mu \\[4mm] \dfrac{N}{N + 1} & \lambda = \mu \end{cases}$	
λ_e	$\lambda(1 - P_N) = \mu(1 - P_0) = \mu_e$	
ρ	$\dfrac{\lambda_e}{\mu} = 1 - P_0$	
w	$\dfrac{L}{\lambda_e}$	
w_Q	$w - \dfrac{1}{\mu}$	
L_Q	$\lambda_e w_Q = L - (1 - P_0)$	
P_n	$\begin{cases} \dfrac{(1 - a)a^n}{1 - a^{N+1}} & \lambda \neq \mu \\[4mm] \dfrac{1}{N + 1} & \lambda = \mu \end{cases}$	$n = 0, 1, 2, \ldots, N$

*(N = system capacity; $a = \lambda/\mu$.) Here system capacity is limited to N, as might be the case in an outpatient clinic or an airport.

Table 3.8   Steady-State Parameters for $M|M|c$ Queueing Systems

ρ

$$\frac{\lambda}{c\mu}$$

P_0

$$\left\{\left[\sum_{n=0}^{c-1} \frac{(\lambda/\mu)^n}{n!}\right] + \left[\left(\frac{\lambda}{\mu}\right)^c \left(\frac{1}{c!}\right) \left(\frac{c\mu}{(c\mu - \lambda)}\right)\right]\right\}^{-1}$$

$$= \left\{\left[\sum_{n=0}^{c-1} \frac{(c\rho)^n}{n!}\right] + \left[(c\rho)^c \left(\frac{1}{c!}\right) \frac{1}{1-\rho}\right]\right\}^{-1}$$

$P[L(t) \geqslant c]$

$$\frac{(\lambda/\mu)^c P_0}{c!(1 - \lambda/c\mu)} = \frac{(c\rho)^c P_0}{c!(1-\rho)}$$

L

$$c\rho + \frac{(c\rho)^{c+1} P_0}{c(c!)(1-\rho)^2} = c\rho + \frac{\rho P[L(t) \geqslant c]}{1-\rho}$$

w

$$\frac{L}{\lambda}$$

w_Q

$$w - \frac{1}{\mu}$$

L_Q

$$\lambda w_Q = \frac{(c\rho)^{c+1} P_0}{c(c!)(1-\rho)^2} = \frac{\rho P[L(t) \geqslant c)}{1-\rho}$$

$L - L_Q$

$$\frac{\lambda}{\mu} = c\rho$$

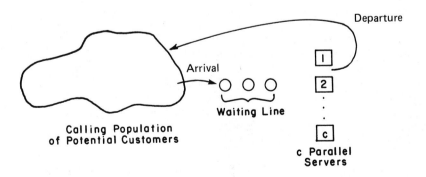

Figure 3.12   Parallel Server Queueing System.

M |M̲|c Queues. Suppose c channels operate in parallel, as illustrated
in Figure 3.12. The traffic intensity, λ/μ, cannot exceed the number of
channels c, or $\lambda < c\mu$. Table 3.8 lists the steady-state parameters of the
M |M|c queueing system.

Banks and Carson (1984) present a more complete discussion of queueing systems.

3.6   INVENTORY SYSTEMS

Inventory systems are encountered in a variety of applications, and simulation is often the only means of solving complex inventory problems. Hadley
and Whitin (1963) is a classic text on analytical approaches to inventory
systems modeling. As with queueing models, the principal use of analytic
models of inventory systems is validation of simulation models under baseline conditions.

The primary measure of effectiveness of an inventory system is *total
system cost*. Contributing to total inventory cost are the following

Item cost   The actual cost of the Q items acquired.
Order cost   The cost of initiating a purchase or setup.
Holding cost   The cost of maintaining items in inventory, including
    interest, insurance, deterioration, etc.
Shortage costs   The costs of failing to satisfy demand: either cost of
    lost sales or backorder cost.

Inventory policies are generally of two types: periodic-review policies
and continuous-review policies. *Periodic review* involves a process in
which inventory level is checked at regular intervals to determine whether
an order for additional stock is needed. If additional stock is needed,
either an amount Q or an amount Q − x is ordered, where x is the current
level at the time the stock level is checked. Q is either the order quantity
or the maximum inventory level, depending on the order policy. *Continuous
review* involves checking the stock level each time a sale is made and ordering up to Q if the stock level drops below a specified level L, so that a
quantity Q − x is ordered where x is the stock level at the instant the
order is placed.

3.6.1   Deterministic Models

Although most real-world situations involve elements of uncertainty, such
as demand or lead time for receipt of a shipment of goods, deterministic
models are much simpler and give a good understanding of inventory situations. *Deterministic* models are those in which all parameters of the model
are known with certainty.

Several deterministic inventory models are in common use in inventory
management systems. Although more complex systems require the development and simulation of probabilistic models, deterministic models afford
means for assessing the validity of such complex models under simplifying
assumptions. This section examines four deterministic inventory models.

Lot-Size Model with No Stockouts and Zero Lead Time. In this model,
shortages are not allowed and replenishment is assumed to occur instantaneously. Two factors contribute to the total cost model, namely, purchase
cost and holding cost, giving the equation

$$C_T = C_p + C_H = \frac{AD}{Q} + \frac{QiC}{2} \qquad (3.18a)$$

where

C_T = total cost per inventory cycle, \$

C_p = purchase cost per inventory cycle, \$

C_H = holding cost per inventory cycle, \$

A = ordering cost per order

N = number of periods per order cycle

C = item cost, \$ per unit

i = interest rate, expressed as a proportion

D = demand per period, units

Q = economic order quantity

Figure 3.13 shows the behavior of this model. Clearly, by setting the derivative of C_T, total cost, with respect to Q, the order quantity, equal to zero and solving for Q, we obtain the so-called *economic order quantity* (EOQ)

$$Q^* = \sqrt{\frac{2AD}{iC}} \qquad (3.18b)$$

The minimum total cost per period, C_T^*, may be found by substituting Q^* into the total cost equation. Thus,

$$C_T^* = \sqrt{2AiCD} \qquad (3.18c)$$

<u>EOQ Model with Deterministic Lead Time</u>. Instead of assuming zero lead time, suppose lead time is a known, positive quantity T. An order must therefore be placed T time periods ahead of the point at which inventory is exhausted, or when the inventory level is

$$L^* = DT \qquad (3.19)$$

This is verified by examining Figure 3.14.

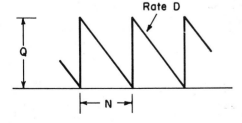

Figure 3.13   EOQ inventory model.

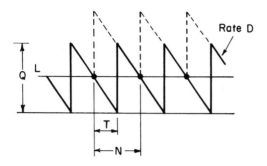

Figure 3.14   EOQ model with deterministic lead time.

EOQ Model with Backorders.   The backorders model is illustrated in
Figure 3.15.   Define the number of periods in which inventory is held as
H, and the number of shortage periods as S.   The total variable cost for
the inventory period is given by

$$C_T = C_P + C_H + C_s \tag{3.20a}$$

which can be shown to yield

$$C_T = \frac{AD}{Q} + \frac{iC(Q + L - DT)^2}{2Q} + \frac{\pi'(DT - L)^2}{2Q} \tag{3.20b}$$

Therefore, the economic order quantity is

$$Q^* = \sqrt{\frac{2AD}{iC} + \frac{2AD}{\pi'}} \tag{3.20c}$$

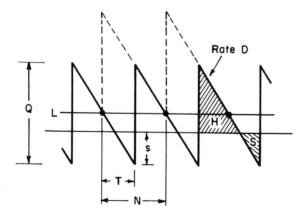

Figure 3.15   EOQ model with backorders.

The reorder point is

$$L^* = DT - \sqrt{\frac{2AiC\ D}{\pi'(iC + \pi')}} \qquad (3.20d)$$

and the minimum total cost is

$$C_T^* = \sqrt{\frac{2AiC\pi'D}{iC + \pi'}} \qquad (3.20e)$$

Manufacturing Lot-Size Model. The EOQ models discussed above assumed that stock was purchased from a vendor, and that replenishment was instantaneous when the shipment arrived. Another situation occurs when the needed stock is manufactured in-house. Stock is produced at a finite rate R while satisfying demand at rate D. Therefore, the replenishment rate is (R − D) during the production cycle of n_1 periods, and demand is satisfied at rate D during n_2 periods. The length of the total inventory cycle is $N = n_1 + n_2$. This model is illustrated in Figure 3.16. As in the case of the purchase model, the total inventory cost is made up of an "order" cost, which we now call the *setup* cost, and the holding cost. The relevant equations are as follows:

$$C_T = C_P + C_H \qquad (3.21a)$$

$$= \frac{AD}{Q} + \frac{Q(1 - D/R)iC}{2} \qquad (3.21b)$$

$$C_T^* = \sqrt{2A(1 - D/R)iCD} \qquad (3.21c)$$

$$Q^* = \sqrt{\frac{2AD}{iC(1 - D/R)}} \qquad (3.21d)$$

3.6.2   Probabilistic Models

The deterministic models discussed above assist us in building complex simulation models of inventory systems; however, demand and lead time in

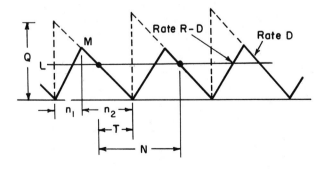

Figure 3.16   Manufacturing lot-size model.

most real-world inventory systems are random variables. In this section we examine how to account for uncertainty in inventory modeling.

Single-Period Model. Consider an inventory process of fixed duration. A single order is placed. Demand is probabilistic with the probability mass function p(x) for a discrete random variable X, and the pdf f(x) for a continuous random variable X. Only holding cost and shortage cost are relevant. If the cost of an item is C, the selling price is P, and an unsold item is sold for scrap at price B, the expected holding cost is

$$E(C_H) = (C - B) \int_{x=0}^{M} (M - x) f(x) dx \qquad (3.22a)$$

where M is the inventory purchased at the beginning of a period. The expected shortage cost is

$$E(C_S) = (P - C) \int_{x=M}^{\infty} (x - M)f(x) dx \qquad (3.22b)$$

Therefore, the expected total cost is

$$E(C_T) = (C - B)\int_{x=0}^{M} (M - x) f(x) dx + (P - C)\int_{x=M}^{\infty} (x - M)f(x) dx$$

$$(3.22c)$$

Taking the derivatives of this total cost equation with respect to M, setting the result equal to zero, and solving for M gives

$$\int_{x=0}^{M^*} f(x) dx = \frac{P - C}{P - B} \qquad (3.22d)$$

where M* is the optimum value of M.

Example 3.6.1. To illustrate this model, suppose a bookstore is purchasing a popular preseason football magazine. The store must place its order 3 months in advance of the shipment. Magazines cost $1.50 and sell for $2.50. The store owner does not know what the demand for the magazine will be, but believes it is between 100 and 200. Magazines that do not sell are sold as scrap for $0.10 per copy.
Treating demand as a uniformly distributed random variable,

$$f(x) = 1/100 \qquad\qquad 100 \leqslant x \leqslant 200$$

$$P = \$2.50 \qquad\qquad C = \$1.50 \qquad\qquad B = \$0.10$$

$$\frac{X}{100} \bigg|_{100}^{M^*} = \$\frac{2.50 - 1.50}{2.50 - 0.10} = \$0.417 \qquad \therefore \qquad M^* = 142$$

Banks and Carson (1984) also discuss models for periodic review with backorders and continuous review with backorders. The principal use of these models is to provide a baseline against which more complex simulation models can be tested.

3.7 RANDOM NUMBERS

Random numbers are an essential component of discrete-event simulation models. Most computer languages have a function or subroutine that will generate random numbers that can in turn be used to generate random events or other random variables. In this section, we shall examine how to generate (a) random numbers, and (b) random variates from several common probability distributions.

3.7.1 Properties of Random Numbers

A sequence of random numbers R_1, R_2, . . ., must possess two important properties to be useful in discrete-event simulation: uniformity and independence. *Uniformity* requires that each random number R_i is an independent sample drawn from a continuous uniform distribution in the interval $[0,1]$. That is, its pdf is given by

$$f(x) = \begin{cases} 1 & 0 \leqslant x \leqslant 1 \\ 0 & \text{otherwise} \end{cases} \qquad (3.23)$$

as shown in Figure 3.17. The mean and variance of this distribution are $1/2$ and $1/12$, respectively.

The only way to generate truly random numbers on a digital computer is to store a table of such size to provide all numbers needed during the course of a simulation. This technique would require either very large storage in main memory or many accesses to off-line memory, neither of which is desirable. The alternative is to use a *pseudorandom* number generator. Desirable properties of such algorithms include:

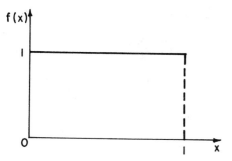

Figure 3.17   Uniform pdf for random numbers.

The routine should be fast.

The routine should not require large storage in main memory.

The routine should have a sufficiently long *cycle*, which is the length of the sequence until the series begins to repeat itself in the previous order.

The routine should avoid *degeneration*, which is the condition of continuously repeating the same number.

The random numbers should be *replicable*, given a particular starting seed.

The generated random numbers should be uniform and independent.

Five techniques for generating random numbers are introduced next:

Midsquare Technique. This technique starts with an initial number, or *seed*, consisting of n digits. This number is squared, and the middle n digits taken as the next random number. The random number R_i is found by simply placing a decimal before the first digit in the n-digit set.

Example 3.7.1. Let

$$X_0 = 3187$$

Then

$$X_0^2 = (3187)^2 = 10,156,969 \rightarrow X_1 = 1569 \quad \text{and} \quad R_1 = .1569$$

$$X_1^2 = (1569)^2 = 2,461,761 \rightarrow X_2 = 4617 \quad \text{and} \quad R_2 = .4617$$

$$X_2^2 = (4617)^2 = 21,316,689 \rightarrow X_3 = 3166 \quad \text{and} \quad R_3 = .3166$$
$$\vdots$$

Midproduct Technique. Two initial seeds of n digits each are used to start the sequence. These numbers are multiplied, and the middle n digits taken as the next number in the sequence. This process is repeated as often as needed.

Example 3.7.2. Let

$$X_0' = 7143 \qquad X_0 = 1689$$

Then

$$U_1 = X_0'X_0 = (7143)(1689) = 12,064,527 \rightarrow X_1 = 0645 \quad \text{and} \quad R_1 = 0.0645$$

$$U_2 = X_0X_1 = (1689)(0645) = 1,089,405 \rightarrow X_2 = 0894 \quad \text{and} \quad R_2 = 0.0894$$

$$U_3 = X_1X_2 = (0645)(0894) = 576,630 \rightarrow X_3 = 5766 \quad \text{and} \quad R_3 = 0.5766$$
$$\vdots$$

Constant Multiplier Technique. A constant K is used as a multiplier of a seed X. The middle n digits are taken as the next number in the series, and the process is repeated.

Example 3.7.3. Let

$$K = 4552 \qquad X_0 = 6129$$

Then

$$V_1 = KX_0 = (4552)(6129) = 27,899,208 \rightarrow X_1 = 8992 \quad \text{and} \quad R_1 = 0.8992$$

$$V_2 = KX_1 = (4522)(8992) = 40,931,584 \rightarrow X_2 = 9315 \quad \text{and} \quad R_2 = 0.9315$$

$$V_3 = KX_2 = (4522)(9315) = 42,401,880 \rightarrow X_3 = 4018 \quad \text{and} \quad R_3 = 0.4018$$

Additive Congruential Method. The method by which values are generated is

$$X_i = (X_{i-1} + X_{i-n}) \bmod m$$

where, by definition, a = b mod m if (a − b) is divisible by m with zero remainder.

Example 3.7.4. Let

$$X_1 = 57 \qquad X_2 = 34 \qquad X_3 = 89 \qquad X_4 = 92 \qquad X_5 = 16 \qquad n = 5 \qquad m = 100$$

Then

$$X_6 = (X_5 + X_1) \bmod 100 = (16 + 57) \bmod 100 = 73 \bmod 100 = 73$$

$$X_7 = (X_6 + X_2) \bmod 100 = (73 + 34) \bmod 100 = 107 \bmod 100 = 7$$

$$X_8 = (X_7 + X_3) \bmod 100 = (\ 7 + 89) \bmod 100 = 96 \bmod 100 = 96$$

Linear Congruential Method. Using an initial seed X_0, constant multiplier a, constant increment c, and a modulus m, random numbers are generated using the following recursive relationship:

$$X = (aX + c) \bmod m \qquad i = 0, 1, 2, \ldots$$

Example 3.7.5. Using X = 27, a = 17, c = 43, and m = 100, we obtain the series

$$X_0 = 27$$

$$X_1 = (17 \times 27 + 43) \bmod 100 = 502 \bmod 100 = 2 \quad \text{and} \quad R_1 = 0.02$$

$$X_2 = (17 \times 2\ + 43) \bmod 100 = \ 77 \bmod 100 = 77 \quad \text{and} \quad R_2 = 0.77$$

$$\vdots$$

It is necessary to test random numbers for validity as follows:

Frequency test—using either the Kolmogorov-Smirnov or the Chi-square test; tests whether the distribution of the generated set is U[0,1].

Runs test—tests run up or down, or above or below the mean, using
a **Chi-square statistic.**

Autocorrelation test—tests the correlation between numbers in the
series.

A *gap test* and a *poker test* should also be used. Banks and Carson
(1984) describe how to apply these tests to uniformly distributed random
numbers. Once a reliable random number generator has been ascertained,
the analyst can go on to using these numbers to generate random variables
in the simulation.

3.7.2   Random Variate Generation

Assuming the availability of a source of random numbers from U[0,1] distribution, random variates from specified probability distributions can be
generated from one of the following three methods: the inverse-transform
technique, the convolution method, and the acceptance-rejection technique.

<u>Inverse-Transform Technique</u>. A four-step procedure is described below
and illustrated with the exponential distribution:

Step 1.   Compute the cdf of the desired random variable X. For the
exponential distribution

$$F(x) = 1 - e^{-\lambda x} \qquad x \geqslant 0 \qquad\qquad (3.24a)$$

Step 2.   Set $F(X) = R$.

$$1 - e^{-\lambda x} = R \qquad\qquad (3.24b)$$

Step 3.   Solve for X in terms of R.

$$1 - e^{-\lambda x} = R$$

$$e^{-\lambda x} = 1 - R$$

$$-\lambda x = \ln(1 - R)$$

$$X = \frac{-1}{\lambda} \ln(1 - R) \qquad\qquad (3.24c)$$

Step 4.   Generate random numbers R_1, R_2, R_3, . . . and compute
the random variates X_1, X_2, X_3, ...

<u>Convolution Method</u>. The probability distribution of the *sum* of two or
more independent random variables is called a *convolution* of the distributions of the original variables. For instance, the Erlang random variable X
with parameters (K,Θ) can be shown to be the sum of K independent exponential random variables X_1, X_2, . . ., X_K. An Erlang variate can be
generated by summing K exponential variates.

Normal variates are generated by exploiting a result from the central-
limit theorem, which asserts that the sum of n independent and identically
distributed random variables X_1, X_2, normally distributed with mean μ_x

and variance σ_X^2. Applying this result to uniform variates $U[0,1]$, which have mean 0.5 and variance $1/12$, it follows that

$$Z = \frac{\sum_{i=1}^{n} R_i - 0.5n}{\left(\frac{n}{12}\right)^{1/2}} \tag{3.25a}$$

is approximately normally distributed with mean zero and variance one. One could sum various numbers of $U[0,1]$ variates, but $n = 12$ is convenient and efficient. To generate a variate $N[\mu_\gamma, \sigma_\gamma^2]$, simply apply the transformation

$$Y = \mu_\gamma + \sigma_\gamma Z \tag{3.25b}$$

Example 3.7.5. Suppose machine repair times are normally distributed with mean $\mu = 15.5$ min and variance $\sigma^2 = 16.5$. Generating 12 random numbers from $U[0,1]$, we get

| 0.0960 | 0.3912 | 0.4646 | 0.0977 | 0.1312 | 0.4397 |
| 0.7080 | 0.6478 | 0.9003 | 0.5081 | 0.0989 | 0.2801 |

Then

$$Z = \sum_{i=1}^{12} R_i - 6 = 4.7636 - 6 = -1.2364$$

Therefore

$$Y = 15.5 + \sqrt{16.5}(-1.2364) = 10.48 \text{ min}$$

Acceptance-Rejection Method. In this method, the analyst generates a random number or random variate, subjects this quantity to a test to determine whether it meets preestablished criteria, *accepts* the quantity if it passes the test, or *rejects* the quantity if it fails the test. The procedure continues to generate random numbers or variates and test them until a sufficient number of them pass the test. For example, suppose it is desired to generate variates from the uniform distribution $U[0.25,0.75]$. The procedure would be as follows:

Step 1. Generate a random number R.
Step 2a. If $R \geqslant 0.25$ and $R \leqslant 0.75$, accept R and go to step 3.
Step 2b. If $R < 0.25$ or $R > 0.75$, reject R and return to step 1.
Step 3. Repeat the procedure until all needed variates are generated.

Banks and Carson (1984) describe procedures for using the acceptance-rejection method to generate Poisson and gamma random variables. Fishman (1973) and Law and Kelton (1982) give extensive treatments of the subject of random variate generation.

3.8 ANALYSIS OF SIMULATION DATA

Simulation activities require careful analysis of input data as well as output results. As stated in the discussion on identification in Section 1.2.4, input data form the base on which realistic models are derived. The simulationist should therefore exercise extreme caution in treating input data.

3.8.1 Input Data Analysis

There are four steps in developing a valid simulation model: data collection; identifying the underlying probability distributions; estimating the parameters of those distributions; and using the estimated parameters and the assumed distribution, test goodness-of-fit. This section discusses procedures for input data analysis.

Data Collection. Plan the data collection activity. Be alert to unusual circumstances that might warrant discarding certain data elements. Perform data analysis as the collection activity is underway. Employ scatter diagrams to visually examine the data for correlation among variables. Look for so-called *autocorrelation* in time-series data.

Identifying the Distribution. Construct frequency *histograms* to give an indication of the shape of the probability distribution.

Estimating Parameters. If the observations in a sample of size n are $X_1, X_2, \ldots, X_n$, compute the sample mean and variance using

$$\overline{X} = \frac{1}{n} \sum_{i=1}^{n} x_i \tag{3.26a}$$

and

$$S^2 = \frac{1}{n-1} \left[\sum_{i=1}^{n} x_i^2 - n\overline{x}^2 \right] \tag{3.26b}$$

If the data are grouped into a frequency distribution, it is usually more convenient to compute the sample mean and variance from the relation

$$\overline{X} = \frac{1}{n} \sum_{j=1}^{k} f_j x_j \tag{3.26c}$$

and

$$S^2 = \frac{1}{n-1} \left[\sum_{j=1}^{k} f_j x_j^2 - n\overline{x}^2 \right] \tag{3.26d}$$

Suggested estimators for the common distributions employed in discrete-event simulation are shown in Table 3.9.

Table 3.9 Estimators for Common Distributions

Distribution	Parameter(s)	Suggested estimator(s)
Poisson	α	$\hat{\alpha} = \overline{X}$
Exponential	λ	$\hat{\lambda} = \dfrac{1}{\overline{X}}$
Gamma	β, θ	$\hat{\beta}$ (see Banks and Carson, 1984, p. 346) $\hat{\theta} = \dfrac{1}{\overline{X}}$
Uniform on $(0, b)$	b	$\hat{b} = \dfrac{n+1}{n}[\max(X)]$ (unbiased)
Normal	μ, σ^2	$\hat{\mu} = \overline{X}$ $\hat{\sigma}^2 = S^2$ (unbiased)
Weibull with $v = 0$	α, β	$\hat{\beta}_0 = \dfrac{\overline{X}}{S}$ $\hat{\beta}_j = \hat{\beta}_{j-1} - \dfrac{f(\hat{\beta}_{j-1})}{f'(\hat{\beta}_{j-1})}$

$$f(\beta) = \frac{n}{\beta} + \sum_{i=1}^{n} \ln x_i - \frac{n \sum_{i=1}^{n} x_i^{\beta} \ln x_i}{\sum_{i=1}^{n} x_i^{\beta}}$$

Iterate until convergence

$$f'(\beta) = -\frac{n}{\beta^2} - \frac{n \sum_{i=1}^{n} x_i^{\beta} (\ln x_i)^2}{\sum_{i=1}^{n} x_i^{\beta}} + \frac{n \left(\sum_{i=1}^{n} x_i^{\beta} \ln x_i \right)^2}{\left(\sum_{i=1}^{n} x_i^{\beta} \right)^2}$$

$$\hat{\alpha} = \left(\frac{1}{n} \sum_{i=1}^{n} x_i^{\beta} \right)^{1/\beta}$$

Goodness-of-Fit Tests. The two most important tests of goodness-to-fit of an assumed distribution to a theoretical distribution are the Kolmogorov-Smirnov (K-S) test and the Chi-square (χ^2) test. The K-S test is typically applied to samples with fewer than 50 observations. Banks and Carson (1984) give a detailed treatment of these techniques.

Other techniques that are often useful in analyzing input data are *linear and multiple regression, correlation analysis,* and *time-series* analysis. When a thorough analysis of input data has been completed, the modeler is then prepared to construct the simulation model.

3.8.2   Verification and Validation

During the process of modeling building, the analyst must employ procedure to ascertain the *credibility* of the simulation model. The first of these steps is called *verification*. The purpose of model verification is to assure that the conceptual model is accurately reflected in the computer code. The first step in this procedure is to develop a *flowchart* of the model. This helps in organizing the computer code as it is being developed. The unfinished computer code should provide for printing numerous traces and intermediate results so that one can find flawed results easily should they occur.

Model *validation* is the process of comparing the behavior of the model to that of the system it is intended to represent. The first goal of the simulation modeler is to construct a model that appears reasonable on its face to prospective model users. This should be followed by examining the structural assumptions and the data assumptions of the model. *Structural* assumptions are those which relate to how the system operates. *Data* assumptions are those related to the form of distributions, parameter estimates, goodness-of-fit tests, etc. We have already presented methods by which to perform this analysis. The ultimate test of the model is its ability to predict future behavior of a real system, including perhaps under conditions that may be only proposed. To accomplish this, it is necessary to compare the behavior of the model to that of the system under conditions at which the system can be observed. An important means of model validation is to compare the performance of the model to that of selected, applicable *analytic* models.

3.8.3   Output Analysis

Once a simulation model is ascertained to be valid, the next step in the simulation study is to use the model to predict the performance of the system. If the performance is measured by a parameter θ, the result of a simulation experiment will be an estimate $\hat{\theta}$ of θ. The accuracy of $\hat{\theta}$ can be measured by its variance. These estimates are called *point estimates* and are computed from the time series of n observations $\{Y_1, Y_2, \ldots, Y_n\}$ by the relation

$$\hat{\theta}_r = \sum_{i=1}^{n_r} \frac{Y_{ri}}{n_r} \qquad r = 1, \ldots, R \qquad (3.27)$$

A second type of estimate produced by a simulation run is the *interval estimate*. For instance, the $100(1 - \alpha)\%$ confidence interval for the parameter θ is given by

$$\hat{\theta} - t_{\alpha/2,f}\hat{\sigma}(\hat{\theta}) \; \leqslant \; \theta \; \leqslant \; \hat{\theta} + t_{\alpha/2,f}\hat{\sigma}(\hat{\theta}) \tag{3.28}$$

where $t_{\alpha/2,f}$ is student's t statistic at $\alpha/2$ and f is the number of degrees of freedom, which is a function of the estimator $\hat{\sigma}(\hat{\theta})$. If $\{Y_1, Y_2, \ldots, Y_n\}$ are statistically independent observations, then the sample variance is computed from

$$s^2 = \frac{1}{n-1} \sum_{i=1}^{n} (Y_i - \hat{\theta})^2 \tag{3.29}$$

If they are not independent, other means must be employed. Fishman (1973) is an excellent reference for this purpose.

An important part of the output-analysis phase of a simulation study is the design of simulation experiments. This involves identifying a set of *dependent* or *response* variables Y_j, j = 1, . . ., m which are assumed to be functions of the set of *controllable* or *independent* variables X_i, i = 1, . . ., n. Some of the techniques used for this purpose include (a) designed experiments, (b) regression analysis, and (c) statistical tests of hypothesis. Kleijnen (1975) has presented a thorough treatment of the statistical techniques used in simulation output analysis.

3.9   SUMMARY AND CONCLUSIONS

This chapter has reviewed the concepts, principles, and techniques of discrete-event simulation. This discussion should be viewed as a guide to the approaches one should follow in conducting a simulation study of discrete-event systems, but should not be construed as an exhaustive treatment of the subject. This text as well as its references enable the simulationist to select a modeling language and perform the input and output analyses requisite to a sound simulation study.

PROBLEMS

3.1. Using an available random number generator, generate 30 random numbers. Test for uniformity and independents.

3.2. Failure times for a microcomputer floppy disk drive are exponentially distributed with mean 2500 hr. Generate 10 failures for this random variable.

3.3. If a microcomputer has two floppy disk drives, each with the failure time distribution described in Problem 3.2, what distribution describes the time until the microcomputer has no available disk drives? Generate five microcomputer failures.

3.4. Repair time for the microcomputer disk drives in Problems 3.2 and 3.3 is normally distributed with mean 6.5 hr and variance 2.25 hr^2. A repairman is paid for a minimum of 4 hr if a service call is made. Generate repair charge times for 20 floppy disk failures.

3.5.   If a service facility has maintenance contracts on 500 microcomputers of the type described in Problems 3.2, 3.3, and 3.4, discuss the queueing system for machines with failed floppy disk drives.

3.6.   A mainframe computer has jobs arriving according to a Poisson distribution with a mean arrival rate of 2.4 jobs/sec. Service times are exponentially distributed with a mean time of .33 sec. Compute the steady-state mean queue length, mean number of units in the system, mean time in queue, mean time in the system, and traffic intensity. Compute the probability distribution of n jobs in the system, $n = 0, 1, 2, \ldots$.

3.7.   An item costing $2000 has a demand rate of 88 units/month. The ordering cost per order is $135.00. The inventory carrying cost fraction is 0.24 on an annual basis. The lead time for the unit is 1.5 months. The cost of a backorder is $4.50 per unit.

       a.   Determine Q^*, L^*, and C_T^*.

       b.   Construct the inventory geometry for this system.

3.8.   A geological survey is being planned. Special-purpose 12-volt batteries are needed for this survey. Demand for these batteries over the duration of the survey is Poisson distributed with mean 3 units. Each battery costs $400. They can be sold for scrap for $80 each. The cost of a battery that is unavailable when needed is $1500. How many batteries should be taken on the survey? If you were simulating this system, outline several alternative policies you might wish to consider.

3.9.   A large hotel serves banquets and several restaurants from a central kitchen in which labor is shifted among various jobs. Salad consumption (demand) is virtually constant at a rate of 40,000 salads/year. Salads can be produced at a rate of 60,000 salads/year. Salads cost $0.55 each, and it costs $12.00 to set up a salad preparation line. Carrying costs of salads, high due to spoilage is 90% of the cost of a salad. No stockouts are allowed. The hotel would like to establish an operations plan for salad preparation. What are Q^* and C_T^*?

3.10. Four workers are evenly spaced along a conveyor belt. Items needing processing arrive according to a Poisson process at rate 2/min. Processing time is exponentially distributed with a mean of 1.6 min. If a worker becomes idle, then he or she takes the first item to come by on the conveyor. If a worker is busy when an item comes by, that item moves down the conveyor to the next worker, taking 20 sec between two successive workers. When a worker finishes processing an item, the item leaves the system. If an item passes by the last worker, it is recirculated on a loop conveyor and will return to the first worker after 5 min.
       Management is interested in having a balanced workload; that is management would like worker utilization to be equal. Let p_i be the long-run utilization of worker i, and let p be the average utilization of all workers. Thus, $p = (p_1 + p_2 + p_3 + p_4)/4$. Using queueing theory, p can be estimated by $p = \lambda/c\mu$ where $\lambda = 2$ arrivals/min,

c = 4 servers, and $1/\mu$ = 1.6 min is the mean service time. Thus, $p = \lambda/c\mu$ = (2/4)1.6 = 0.8; so on the average, a worker will be busy 80% of the time.

a. Make five independent replications, each of run length 40 hr preceded by a 1-hr initialization period. Compute 95% confidence intervals for p_1 and p_4. Draw conclusions concerning workload balance.

b. Based on the same five replications, test the hypothesis H_0: p_1 = 0.8 at a level of significance α = 0.05. If a difference of +0.05 is important to detect, determine the probability that such a deviation is detected. In addition, if it is desired to detect such a deviation with probability at least 0.9, determine the sample size needed to do so.

c. Repeat b for H_0: p_4 = 0.8.

d. Based on the results from a–c, draw conclusions for management about the balancing of workloads.

REFERENCES

Banks, J. and Carson, J. S. (1984). *Discrete-Event System Simulation,* Prentice-Hall, Englewood Cliffs, N.J.

Biles, W. E. and Bathina, V. R. (1986). PASAMS: Pascal simulation and analysis of manufacturing systems, *Material Flow 3,* 99–112.

CACI, Inc. (1976). *SIMSCRIPT II.5 Reference Handbook,* Los Angeles.

Conway, R., Maxwell, W. L., and Worons, S. L. (1986). *User's guide to XCELL Factory Modeling System,* The Scientific Press, Palo Alto, CA.

Fishman, G. S. (1973). *Concepts and Methods in Discrete-Event Digital Simulation,* Wiley, New York.

Hadley, G. and Whitin, T. M. (1963). *Analysis of Inventory Systems,* Prentice-Hall, Englewood Cliffs, NJ.

Kiviat, P. J., Villanueva, R., and Markowitz, H. M. (1973). *SIMSCRIPT II.5 Programming Language,* CACI, Los Angeles, CA.

Kleijnen, J. P. C. (1975). *Statistical Techniques in Simulation,* Parts I and II, Dekker, New York.

Law, A. M. and Kelton, W. D. (1982). *Simulation Modeling and Analysis,* McGraw-Hill, New York.

Pegden, C. C. (1986). *Introduction to SIMAN,* Systems Modeling Corporation, State College, PA.

Phillips, D. H. (1978). *GEMS: Generalized Manufacturing Simulator,* Industrial Engineering Department, Texas A&M University, College Station, TX.

Pritsker, A. A. B. (1974). *The GASP-IV Simulation Language,* Wiley, New York.

Pritsker, A. A. B. (1985). *Introduction to Simulation and SLAM-II,* Wiley, New York.

Schriber, T. J. (1974). *Simulation Using GPSS,* Wiley, New York.

4

Computer Simulation

4.1 INTRODUCTION

In the discussion of the fundamental concepts of modeling and simulation in Chapter 1, it was noted that the first phase of the process focuses on understanding the system to be simulated and on developing mathematical equations to describe the operation of the system. Once an accurate mathematical model has been developed, attention can be focused on implementing the resulting equations on a computer. This is the second phase of modeling and simulation, and it is the primary emphasis of this chapter. Once the computer model is operational, the third phase, model credibility, can begin. A thorough validation study of the model should be carried out before moving to the fourth phase, which is data collection. In large-scale systems, the lack of knowledge and the complexity of the system may dictate a need to work on the modeling, computer implementation, and model validation phases simultaneously. Thus, modeling and simulation is often an evolutionary process.

In this chapter, it is assumed that the modeling process is essentially complete (see Chapter 2 for details) and that the continuous systems to be simulated are represented in terms of differential equations, transfer functions, and/or state space models. This chapter also presents computer simulation of discrete-time systems represented by difference equations or their Z-transform equivalents.

This chapter presents both digital and analog techniques with emphasis on digital simulation because of its widespread use. Hybrid simulation techniques are also discussed briefly. Digital simulation techniques are presented first, and the discussion includes user-written and general-purpose, "canned" simulation programs (languages). Individuals who use general-purpose languages will find the material on user-written programs helpful in understanding how simulation languages represent a system; they should gain considerable insight into the digital simulation process. For this reason, the material on user-written simulation programs is presented first. However, it is assumed that most individuals will utilize a simulation language, and for that reason, only elementary digital simulation algorithms are presented in detail.

The first portion of this chapter, Sections 4.2, 4.3, and 4.4, is devoted to digital representation of continuous signals and to numerical integration. The application of state space techniques in simulation is covered in Section 4.5, followed by material on simulation of discrete-time systems in Section 4.6. The simulation of systems represented in transfer function form is covered in Section 4.7. Digital simulation languages are discussed in Section 4.8, which includes an example application of CSMP. Section 4.9 is devoted to analog computing elements, examples of linear system simulation, and presentation of magnitude and time scaling. Section 4.9 contains a few comments on simulation of nonlinear systems, simultaneous equations, transfer functions, and state variable equations. A brief overview of hybrid computer simulation is presented in Section 4.10. The final section is a chapter summary.

4.2   DIGITAL REPRESENTATION OF SIGNALS

As background for the discussion of digital simulation of continuous systems, a few comments about digital representation of continuous signals are in order. As noted in Chapter 1, the variables for a continuous system have values for every point in time and change continuously with time. On the other hand, a digital computer calculates values for the continuous variables of a system at distinct points in time in a sequential, one-after-the-other manner. Thus, a digital computer simulation of a continuous system is in fact a discrete-time system, in which variables change only at distinct (finite) instants of time.

The digital simulation of a continuous system is further complicated by the discretization process. A digital computer represents variables with a finite number of binary variables (bits), and accuracy is limited by the value of the least significant bit. In general, this means that the digital equivalent of a continuous signal at a given point in time will be ±1/2 the value of the least significant bit, which is referred to as quantization error. The number of bits used, and hence the accuracy of the results, is determined by the word length of the digital computer being used for simulation.

Digital simulation is also complicated by the fact that the time between adjacent values (simulation results) usually represents the time required to calculate new values. Thus, the time interval will not be equal to the time elapsed in the system being simulated, unless special effort is made to synchronize time by making simulation time (computer time) proceed at the same rate as the real system operates (real time).

In most instances, it is desirable to make computer time proceed as rapidly as possible so that the digital computer can be used for other tasks or, more likely, for repeated simulations of the system being studied. In some applications, such as man-in-the-loop simulators or process control, the computer model *must* operate "in step" with real time. On the other hand, there are many real-world systems that operate faster than they can be simulated on currently available digital computers, and in these applications, real-time simulation is impossible.

In these instances, the parallel nature of analog and hybrid computers has utility. For this reason, the latter part of the chapter will provide brief coverage of analog simulation techniques and an introduction to hybrid simulation. It should be noted, however, that the speed and power of digital computers are increasing steadily, and parallel processing digital computers are likely to replace analog computers.

In order to use a digital computer to solve the equation describing a continuous system, numerical integration must be performed. Since it is fundamental to digital simulation, the next two sections will be devoted to numerical integration techniques.

4.3   NUMERICAL INTEGRATION

As noted in Section 4.2, the digital computer determines values for the continuous signals of the system being simulated by producing a series of discrete values. For example, the continuous function $x(t)$ becomes a sequence of discrete values $x(t_0)$, $x(t_1)$, $x(t_2)$, ..., $x(t_k)$, $x(t_{k+1})$, ..., $x(t_n)$, as noted in Section 2.3 (see Figure 2.31). Usually, the time interval between adjacent values is constant and represented by $T = t_{k+1} - t_k$.

Ideally, $x(t_k)$, the discrete value of the function at a particular point in time, should be identical to its continuous equivalent at $t = t_k$; but, as noted earlier, the accuracy is limited by the number of bits the digital computer uses to represent a value and by the accuracy of the simulation equations (the model). Typically, the word length (number of bits) of the digital computer determines the discretization accuracy. This is the reason for the "double precision" available in some computer programming languages (i.e., using two computer words to represent a variable more accurately).

Although the discretization error is often critical, the primary source of error in representing a simulation variable $x(t)$ at $t = t_k$ is found in the method used to calculate derivatives, commonly referred to as numerical integration. The root of the problem can be seen by a careful look at Taylor's series expansion x_{i+1} in terms of x_i, which can be written in the form

$$x_{i+1} = x_i + \frac{T\,dx_i}{dt} + (T^2/2!)\frac{d^2 x_i}{dt^2} + \cdots + (T^n/n!)\frac{d^n x_i}{dt^n} \qquad (4.1)$$

where T is the time interval and dx_i/dt is the derivative of x_i at $t = t_i$. Basically, the series gives the value of x at $t = t_{i+1}$ in terms of x_i and its derivatives. Taylor's series can be used to derive several numerical integration formulas, but more important, it is the criterion used for evaluating almost all numeral integration techniques.

As an example of a numerical integration method, consider the approximation by using only the first two terms of Taylor's series:

$$x_{i+1} = x_i + \frac{T\,dx_i}{dt} \qquad (4.2)$$

Solving for the derivative gives

$$\frac{dx_i}{dt} = \frac{x_{i+1} - x_i}{T} = \frac{x_{i+1} - x_i}{t_{i+1} - t_i} \qquad (4.3)$$

which is commonly referred to Euler's method, or the rectangular rule.

In order to illustrate the use of Euler's method to develop a digital computer simulation, consider the first-order differential equation

$$\frac{dx}{dt} + ax(t) = r(t) \tag{4.4}$$

with $x(t = 0) = 0$ and $r(t) = 1$, a unit step input. To develop a discrete equivalent suitable for programming on a digital computer, use Eq. (4.3) for dx/dt in Eq. (4.4). This yields

$$\frac{x_{i+1} - x_i}{T} + ax_i = r_i \tag{4.5}$$

which can be solved for x_{i+1} in terms of x_i, a, and the input r_i to give

$$x_{i+1} = x_i - aTx_i + Tr_i \tag{4.6}$$

Note that T is the time interval between adjacent discrete values and a is a parameter in the original equation. Equation (4.6) is a difference equation of the form discussed in Section 2.3 and is quite similar to Ex. (2.3.1). Almost any programming language can be used to write a simple program to calculate successive values of x_{i+1} from the proceeding value x_i, the time interval T, the parameter a, and the input r_i. This equation describes the system used in Ex. (2.2.8) in Figure 2.7.

Considerable insight into numerical integration and the impact of the time interval T on simulation accuracy can be gained by calculating the response to a unit step input [i.e., $r(t) = 1$ for $t \geqslant 0$]. First, note that

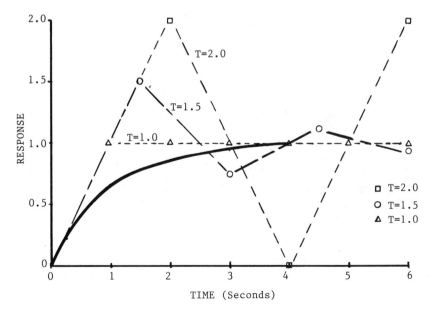

Figure 4.1   Solutions to Eq. (4.6).

the continuous system response for a = 1 and r(t) = 1 is an exponential rise, as shown by the solid curve in Figure 4.1 with x_0 = 0 (assumed). If a time interval of 2 sec (T = 2) is selected, then x_1 = (x_0 − aTx_0 + Tr_0) = 2. Likewise, the value for x_2 is x_2 = (x_1 − aTx_1 + Tr_1) = 0, and in a similar manner, x_3 = 2, x_4 = 0, ... The resulting output is as shown (labeled T = 2) in Figure 4.1 and is obviously incorrect. Smaller values of T will cause the values of x_i to gradually approach the correct value, x = 1.0. Note that T = 1 results in output values of 1.0 at all intervals of time. As T becomes smaller, the discrete solution approaches the correct exponential rise shown in Figure 4.1; it is obvious that the time interval for the digital simulation must be much shorter than the time constant of the system being simulated.

The example problem in this section illustrates that a simple method exists for developing an equivalent difference equation for a differential equation. The example also illustrates the sensitivity of the solution to the size of the time interval T and clearly indicates a need to discuss the relationships between accuracy, computation time, and time interval.

4.4   ERRORS IN NUMERICAL INTEGRATION

In the example in the preceding section, a little forethought would have indicated problems with T ⩾ 2. The time-constant for the system described by the differential equation in Eq. (4.4) is assumed to be 1 sec. Trying to calculate values for the output for integration time intervals larger than the system time constant is likely to create problems. Also, problems should have been anticipated from the use of only two terms of Taylor's series expansion because the contributions of the second and all higher-order derivatives are missing. Thus, Euler's method is referred to as a "first-order" numerical integrated technique. Techniques that include the effect of second-order derivatives are referred to as "second-order" methods and, in general, are more accurate than first-order techniques. Note that the inclusion of higher-order derivatives requires additional calculations for each value and will also need more computer memory.

Over the years, a number of numerical integration formulas have been developed that have the accuracy of including up to four or five terms of Taylor's series without the corresponding burden of calculating and storing values for the extra derivatives at every integration time interval. It is important to note that for T < 1, the T^n terms of the series become quite small for large values of n, and the contributions from these terms become insignificant, compared to other sources of error.

Since a variety of numerical integration methods are available, it might be useful to note that they have been classified into single-pass and multiple-pass techniques and are further distinguished by whether or not extrapolation is used (Kochenburger, 1972). The multiple-pass methods use one cycle to calculate a predicted value of the variable and then use that value to calculate a more accurate, "corrected" value. Thus, the term predictor-corrector integration is used. The improved integration method most commonly encountered is the trapezoidal rule, or the modified Euler method, in which the derivative is replaced by

$$\frac{dx_i}{dt} = \frac{x_{i+1} - x_{i-1}}{2T} \tag{4.7}$$

Note that at t = 0, the x_{i-1} value is unknown and special provisions are required for the initial cycle of calculations. (Usually, Taylor's series is employed to get started.) Other methods are more involved and are usually identified by their originator (e.g., Simpson, Adams, Adams-Moulton, Adams-Bashforth, Tustin, Runge-Kutta, or Milne), and the interested reader is referred to Rosko (1972), Benyon (1968), and Martens (1969) for a comparison of numerical integration techniques. Howe (1985) provides considerable insight into error analysis of integration algorithms and Kochenburger (1972) presents an excellent summary of the various methods.

The previous discussion might lead one to believe that smaller and smaller integration (time) intervals will result in improved accuracy. This is not always the case. Note that the derivative is approximated by a difference $(x_{i+1} - x_i)$ and that smaller values of T cause the values of x_i and x_{i+1} to approach each other. Since the digital computer represents variables with a finite binary sequence, each value will have a finite truncation error. If T is small enough so that x_{i+1} and x_i differ only by their truncation error, then $(x_{i+1} - x_i)$ becomes zero and the difference equation is no longer correct. Thus, the observation that accuracy improves directly with smaller and smaller time intervals is not correct.

Also, the oscillatory response (0, 2, 0, 2, . . .) for T = 2 in the example in the preceding section (Figure 4.1) is an indicator of the stability problems associated with numerical integration. (Note that the response grows without bound for T > 2 in Figure 4.1.) As a result of the problems identified thus far, a number of numerical integration formulas have been developed and tested. Several criteria have been used to evaluate numerical integration techniques; among the more important are sensitivity to truncation error, stability, per-step solution time (computational complexity), and ease of programming.

Before we end the discussion on errors in numerical integration, it should be noted that an error in calculating a variable at a given sample period is propogated to the next sample period if the erroneous value is used in calculating the next value. Since the algebraic sign of the error may vary, it is difficult to anticipate the net effect of error propagation on the stability and accuracy of a particular integration technique. A technique that works well for a given system may not be appropriate for another system. In addition, changing the parameters of a given system or varying the interval of integration may produce radical changes in performance. A knowledge of the factors affecting the performance of integration routines is essential. Programming and implementation aspects of simulation are also discussed in Section 10.7.

A concern for the ease of programming leads to further consideration of the techniques developed in Section 2.2 as means of digital simulation of continuous systems.

4.5 STATE SPACE SIMULATION TECHNIQUES

The most obvious application of Euler's method is to systems expressed in the state space form of Eq. (2.56) and Eq. (2.57) in Chapter 2. In order to reinforce the material of Chapter 2 and illustrate the technique, reconsider the mechanical system of Ex. (2.2.9) which is shown in Figure 4.2. The system differential equation [Eq. (2.33)] is

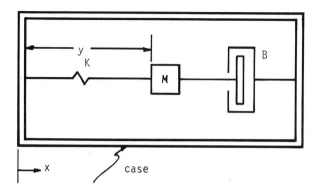

Figure 4.2   An accelerometer.

$$M\ddot{x} + B\dot{x} + Kx = p \tag{4.8}$$

where the state variables $x_1 = x$ and $x_2 = \dot{x}$ and the input $u = p$ and output $y = x_1$.   Equation (2.58) gives the state model as

$$\dot{x}_1 = x_2 \tag{4.9}$$

$$\dot{x}_2 = -\left[\frac{B}{M}\right]x_2 - \left[\frac{K}{M}\right]x_1 + \left[\frac{1}{M}\right]u \tag{4.10}$$

Applying Euler's method on these two equations yields

$$\frac{x_1(i+1) - x_1(i)}{T} = x_2(i) \tag{4.11}$$

and

$$\frac{x_2(i+1) - x_2(i)}{T} = -\left[\frac{B}{M}\right]x_2(i) - \left[\frac{K}{M}\right]x_1(i) + \left[\frac{1}{M}\right]u(i) \tag{4.12}$$

Note the need to put the time subscripts in parenthesis because the subscripts are now used to distinguish state variables.   Solving Eq. (4.11) for $x_1(i+1)$ yields

$$x_1(i+1) = Tx_2(i) + x_1(i) \tag{4.13}$$

and Eq. (4.12) becomes

$$x_2(i+1) = -\left[\frac{BT}{M}\right]x_2(i) - \left[\frac{KT}{M}\right]x_1(i) + \left[\frac{T}{M}\right]u(i) + x_2(i) \tag{4.14}$$

Programming language expressions for Eq. (4.13) and Eq. (4.14) can be written in a straightforward manner.

The initial conditions and the first few iterations of the calculations are important considerations and should be handled before going further. In this example, the displacement x is the first state variable x_1, and its initial value $x_1(0)$ must be known. Also, the second state variable x_2 is the velocity of the mass, and its initial value $x_2(0)$ must also be known. Thus, the first iteration cycle calculates $x_1(1)$ and $x_2(1)$ from the parameters M, B, and K; the initial conditions $x_1(0)$ and $x_2(0)$; and the input $u(0)$. Once this is accomplished, succeeding values of x_1 and x_2 can be calculated from the values determined in the previous iteration. Unless it is desired to save and store all values of x_1 and x_2, the value of x_1 and x_2 at a given iteration can be calculated by a single program loop. It should be noted that the predictor-corrector integration routines mentioned earlier would require several additional programming steps, but their improved accuracy would likely permit larger time intervals and hence provide more efficient use of computer time.

As noted in Chapter 2, the state space approach provides a compact notation for expressing the system differential equations. It is instructive to see the discrete equivalents (based on Euler's approximations) in the same form. Equations (4.13) and (4.14) become

$$\begin{bmatrix} x_1(i+1) \\ x_2(i+1) \end{bmatrix} = \begin{bmatrix} 1 & T \\ \dfrac{-KT}{M} & 1 - \dfrac{BT}{M} \end{bmatrix} \begin{bmatrix} x_1(i) \\ x_2(i) \end{bmatrix} + \begin{bmatrix} 0 \\ \dfrac{T}{M} \end{bmatrix} u(i) \qquad (4.15)$$

and can be written in a form similar to Eq. (2.103), such as:

$$\underline{x}(i+1) = \underline{A}\underline{x}(i) + \underline{B}\underline{u} \qquad (4.16)$$

This choice of state variables is referred to as phase variables because of the displacement-velocity relationship between the state variables. Other choices of state variables could be used and in some instances might be advantageous.

This is also a good place to note that the state space approach lends itself nicely to nonlinear systems and time-varying parameters. Note that known changes in system parameters could easily be handled by using the values of $K(i)$, $M(i)$, and $B(i)$ as the simulation progresses from iteration to iteration.

If the input is known and can be expressed analytically, the state space model can be expanded to include the input as a state variable, as shown later. This would permit the state space model to be expressed in the form

$$\dot{\underline{x}}(t) = \underline{A}\underline{x}(t) \qquad (4.17)$$

which has a solution of the form

$$\underline{x}(t) = e^{\underline{A}t}\underline{x}(0) \qquad (4.18)$$

If the input u in the previous example is a unit step, then it is the solution to the differential equation $\dot{u} = 0$, with the initial condition $u(0)$ equal to the value of the step input. If the input u is incorporated as the third state variable, then $x_3 = u$ and the equation for the third state variable can be adjoined to the original model of Eq. (4.9) and Eq. (4.10). The state equations are now

$$\dot{x}_1 = x_2 \tag{4.19}$$

$$\dot{x}_2 = -\left[\frac{B}{M}\right] x_2 - \left[\frac{K}{M}\right] x_1 + \left[\frac{1}{M}\right] x_3 \tag{4.20}$$

$$\dot{x}_3 = 0 \tag{4.21}$$

where $x_3(0) = u$.

The matrix-vector form for Eq. (4.19), Eq. (4.20), and Eq. (4.21) is

$$\underline{\dot{x}}(t) = \underline{A}\underline{x}(t) \tag{4.22}$$

and its solution is

$$\underline{x}(t) = e^{\underline{A}t}\underline{x}(0) \tag{4.23}$$

The solution can also be written in the form

$$\underline{x}(t_{i+1}) = e^{\underline{A}(t_{i+1}-t_i)}\underline{x}(t_i) \tag{4.24}$$

or

$$\underline{x}(i + 1) = e^{\underline{A}t}\underline{x}(i) \tag{4.25}$$

where $T = t_{i+1} - t_i$.

The term $e^{\underline{A}T}$ is referred to as the transition matrix. Since most high-level programming languages contain matrix-vector operators, this method is easily implemented on a digital computer. It should be noted that this method is not well suited for problems with time-varying parameters, since the transition matrix would have to be recalculated at every time interval instead of only once.

This is a good point to discuss changing the time interval T, while simulation is in progress. This is usually done to conserve computer time, since it is possible to move rapidly through relatively quiescent periods by using large time intervals and, conversely, improve the accuracy during periods of rapid transients by using small values of T. Since integration routines use past values of the system variables and assume constant T, any change in T will require the calculation of intermediate values of system variables to use in calculating values at the next interval. Usually, one or two interpolations are required. As a result, any attempt to change the period of integration or time step T should be carefully evaluated.

As an additional example of digital simulation of systems in state space form, consider the RLC circuit of Ex. 2.2.17 (Figure 2.14). The circuit diagram is repeated in Figure 4.3 and the state space model of Eq. (2.65) is restated for convenience

$$\underline{\dot{x}} = \begin{bmatrix} 0 & \dfrac{1}{L} \\[2ex] \dfrac{-1}{C} & \dfrac{-1}{RC} \end{bmatrix} \underline{x} + \begin{bmatrix} \dfrac{1}{L} & \dfrac{-1}{L} \\[2ex] 0 & \dfrac{1}{RC} \end{bmatrix} \underline{v} \tag{4.26}$$

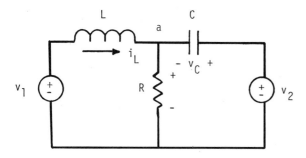

Figure 4.3   Circuit of Ex. (2.2.17).

To develop a difference equation suitable for programming on a digital computer, write the state equations separately as

$$\dot{x}_1 = \left[\frac{1}{L}\right] x_2 + \left[\frac{1}{L}\right] v_1 - \left[\frac{1}{L}\right] v_2 \tag{4.27}$$

and

$$\dot{x}_2 = -\left[\frac{1}{C}\right] x_1 - \left[\frac{1}{RC}\right] x_2 + \left[\frac{1}{RC}\right] v_2 \tag{4.28}$$

Now, using Euler's method as before yields

$$\frac{x_1(i + 1) - x_1(i)}{T} = \left[\frac{1}{L}\right] x_2(i) + \left[\frac{1}{L}\right] v_1(i) - \left[\frac{1}{L}\right] v_2(i) \tag{4.29}$$

and

$$\frac{x_2(i + 1) - x_2(i)}{T} = -\left[\frac{1}{C}\right] x_1(i) - \left[\frac{1}{RC}\right] x_2(i) + \left[\frac{1}{RC}\right] v_2(i) \tag{4.30}$$

Solving for $x_1(i + 1)$ and $x_2(i + 1)$ gives

$$x_1(i + 1) = x_1(i) + \left[\frac{T}{L}\right] x_2(i) + \left[\frac{T}{L}\right] v_1(i) - \left[\frac{T}{L}\right] v_2(i) \tag{4.31}$$

and

$$x_2(i + 1) = x_2(i) - \left[\frac{T}{C}\right] x_1(i) - \left[\frac{T}{RC}\right] x_2(i) + \left[\frac{T}{RC}\right] v_2(i) \tag{4.32}$$

which can be easily programmed in a variety of high-level programming languages.

As in previous examples, the initial conditions must be carefully handled to get the simulation started. In this example, the first state variable x_1 corresponds to the current in the inductor i_L; therefore, $x_1(0)$ is the initial current in the inductor. The second state variable x_2 represents the voltage across the capacitor v_c; thus, $x_2(0) = v_c(0)$. As noted previously, this approach permits the consideration of time-varying parameters and inputs.

4.6   SIMULATION OF DISCRETE-TIME SYSTEMS

The preceding sections of this chapter were devoted to techniques that permit digital simulation of continuous-time systems in which the major task is to convert the system differential equations into difference equations suitable for programming in a high-level language. In the case of discrete-time systems, the describing equations are difference equations; as a result, they are already in a form suitable for digital simulation. Therefore, it might appear that there is little to be gained from a closer investigation of simulation techniques for discrete-time systems. However, a number of important concepts are involved, and the material in Section 2.3 on discrete-time systems is very important to digital simulation of both continuous- and discrete-time systems.

Perhaps the most important concept presented in Section 2.3.1 is the Z-transform. The ability to take the Z-transform of a difference equation permits the use of the initial value theorem [Eq. (2.92)], the final value theorem [Eq. (2.93)], and discrete transfer functions [Eq. (2.95)]. The Z-transform also facilitates stability studies, which are particularly useful in evaluating numerical integration methods; interested readers should see Bekey and Karplus (1968).

As an example of the application of the Z-transform, consider the first-order differential Eq. (4.4). Euler's method was used to develop the difference Eq. (4.6), which is repeated here for convenience:

$$x_{i+1} = x_i - aTx_i + Tr_i \tag{4.33}$$

The shifting properties of Z-transforms [see Eq. (2.94)] can be used to take the transform of Eq. (4.33):

$$zX(z) = X(z) - aTX(z) + TR(z) \tag{4.34}$$

The discrete transfer function for this system can be determined by solving for $X(z)/R(z)$, which is

$$\frac{X(z)}{R(z)} = \frac{T}{(z - 1 + aT)} \tag{4.35}$$

The stability properties of the system can be studied using the material from Chapter 2. Also, for a variety of inputs, the time domain response can be determined by using the appropriate Z-transform; see Table 2.3 to obtain $R(z)$.

In this instance, it is instructive to compare the denominator of Eq. (4.35) with the denominator of the Z-transform for the exponential function e^{-at}, which is the natural response of the first-order system. From Table 2.3, note that the denominator of the Z-transform for e^{-aT} is $(z - e^{-aT})$. Now, if the first two terms of the series expansion of e^{-aT} are used, the term becomes $(z - 1 + aT)$, which is identical to the denominator of Eq. (4.35). This verifies that the Euler approximation used to develop Eq. (4.35) is a first-order technique and emphasizes the importance of keeping the interval of integration T small compared to the system time constant $1/a$. This result also underlies efforts to develop numerical integration formulas that are more efficient than Euler's method.

To provide additional insight into the simulation of discrete-time systems expressed in Z-transform notation, a difference equation, suitable for programming, will be developed for the transfer function

$$\frac{Y(z)}{U(z)} = \frac{5}{z - 3} \tag{4.36}$$

Cross multiplication gives

$$Y(z)(z - 3) = 5U(z) \tag{4.37}$$

Inverse transformation yields

$$y(i + 1) - 3y(i) = 5u(i) \tag{4.38}$$

which can be written in the form

$$y(i + 1) = 3y(i) + 5u(i) \tag{4.39}$$

If the input u is a unit step, the output $y(i + 1)$ becomes

$$y(i + 1) = 3y(i) + 5 \tag{4.40}$$

For $y(0) = 0$, the first few values can be calculated to give

$$y(1) = 5, \ y(2) = 20, \ y(3) = 65, \ \ldots \tag{4.41}$$

which are identical to those obtained for Ex. (2.3.6) and Ex. (2.3.7). The utility of converting between difference equations and the Z-transform expressions for discrete-time systems is very important to digital simulation.

Another important use of the Z-transform is to develop algorithms for simulating systems expressed in Laplace transform notation and block diagram form. This approach will be presented in the next section.

4.7 DIGITAL SIMULATION OF TRANSFER FUNCTIONS

The Laplace transform and transfer function techniques presented in Chapter 2 are fundamental tools of continuous-systems analysis and design; they are integral to the block diagram and signal flow graph methods for representing a system. Therefore, it is important to be able to develop digital simulation algorithms for systems expressed in these forms. As noted in the previous section, the Z-transform provides a convenient method for relating continuous- and discrete-time systems. This relationship will be utilized in this section.

Transfer function and block-diagram techniques permit a complex system to be divided into functional units that correspond to subsystems. The subsystems are represented by transfer functions that can be studied and simulated individually. An important use of simulation is the "hardware-in-the-loop" situation in which portions of the system are simulated and interconnected to the physical components in a manner that permits study of the entire system.

The subsystem components, or blocks, range in complexity from a simple linear gain with a transfer function K to a system of several poles and zeros. Most physical systems will exhibit some form of nonlinearity which must also be incorporated into the simulation process. The entire process is best illustrated by several examples.

Consider a continuous system as shown in Figure 4.4(a) with a transfer function

$$G(s) = \frac{Y(s)}{U(s)} = \frac{a}{s + a} \tag{4.42}$$

The discrete equivalent system is shown in Figure 4.4(b) where the sample-and-hold is assumed to be a zero-order hold, which has a transfer function given by the equation

$$G_H(s) = \frac{1 - e^{-Ts}}{s} \tag{4.43}$$

The implementation of this device is depicted in Figure 10.7. The zero-order hold is recommended for step and discontinuous inputs, and a more complex triangular hold is recommended for more regular functions; see Rosko (1972). The next step is to obtain the transfer function of the sample-and-hold and continuous system. In this instance, the combined transfer function is

$$G(z) = \frac{Y(z)}{U(z)} = Z[G_H(s)G(s)] = Z\left[\frac{1 - e^{-Ts}}{s}\right]\{G(s)\} \tag{4.44}$$

Recall that the relationship between the z- and s-domains is $z = e^{sT}$; which can be used to simplify the above expression to give

$$\frac{Y(z)}{U(z)} = \left(\frac{z - 1}{z}\right) Z\left[\frac{G(s)}{s}\right] \tag{4.45}$$

and substituting $G(s)$ results in the expression

$$\frac{Y(z)}{U(z)} = \left[\frac{z - 1}{z}\right] Z\left\{\frac{a}{[s(s + a)]}\right\} \tag{4.46}$$

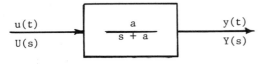

(a)

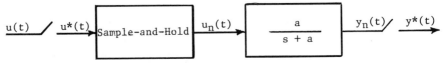

(b)

Figure 4.4   Discrete-time equivalent system.

The Z-transform table gives

$$\frac{Y(z)}{U(z)} = \left[\frac{z-1}{z}\right] \left\{\frac{(1-e^{-aT})z}{[(z-1)(z-e^{-aT})]}\right\} \qquad (4.47)$$

which can be simplified to yield

$$\frac{Y(z)}{U(z)} = \frac{1-e^{-aT}}{z-e^{-aT}} \qquad (4.48)$$

Cross-multiplying gives

$$zY(z) - e^{-aT}Y(z) = (1-e^{-aT})U(z) \qquad (4.49)$$

The difference equation becomes

$$y(i+1) = e^{-aT}y(i) + (1-e^{-aT})u(i) \qquad (4.50)$$

The difference Eq. (4.50) can be programmed for digital simulation once the values of a and T have been specified.

Similar results, which are not as accurate, could be obtained by using the methods of Boxer and Thaler (1956), Madwed, and Tustin (Rosko, 1972). Difference equations could be developed by taking the inverse Laplace transform of $G(s)$ and applying Euler's or other numerical integration methods on the resulting differential equations.

The method of Eq. (4.50) applies to linear, continuous systems of higher order, but the transforms and algebraic manipulations become quite involved. The number of prior terms that must be used in each calculation corresponds to the order of the system, with a second-order system requiring two terms, and so on. Transfer function simulation can be applied to multielement feedback systems by using the procedure outlined above on each system element. For example, the system of Figure 4.5(a) could be simulated by developing a discrete equivalent expression for each block. Block 2 is identical to the previous example, and its difference equation is given in Eq. (4.50). The notation is changed to conform to the variable selected for the system in Figure 4.5:

$$c(i+1) = e^{-aT}c(i) + (1-e^{-aT})e_2(i) \qquad (4.51)$$

The Z-transform of block 1 can be derived by a similar process and is

$$\frac{E_2(z)}{E_1(z)} = \frac{kT}{(z-1)} \qquad (4.52)$$

which gives the difference equation

$$e_2(i+1) = e_2(i) + KTe_1(i) \qquad (4.53)$$

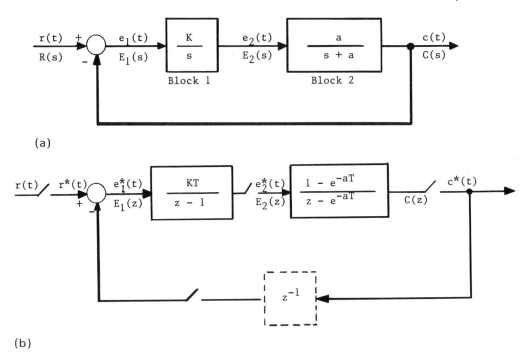

(a)

(b)

Figure 4.5   Multielement feedback system.

The digital solution can be accomplished by writing and programming the difference equation for the complete system including the summing junction

$$e_1(i) = r(i) - c(i) \tag{4.54}$$

The procedure begins by calculating the error $e_1(0)$ from the initial values of r and c, $r(0)$ and $c(0)$. Next, $e_2(1)$ is calculated from $e_1(0)$ and $e_2(0)$, using Eq. (4.53). Then, $c(1)$ is calculated from $c(0)$ and $e_2(0)$, using Eq. (4.51). If the error signal $e_1(i)$ is to be recorded, it may be necessary to insert a delay z^{-1} in the feedback loop, shown with dashed lines in Figure 4.5(b). This is the result of the absence of a delay in Eq. (4.54).

The above process can be used to calculate the response of systems containing nonlinear elements by using the input/output relationships of the nonlinearity to determine a gain change for each sample period; see Rosko (1972).

It should be noted that transfer functions assume zero initial conditions and therefore must be treated carefully when implementing a digital simulation. Also, it should be emphasized that digital simulation of a continuous system is subject to the errors and stability problems discussed in Section 4.4; see also Howe (1985). The approximations and errors associated with the process of obtaining discrete transfer functions (i.e., using a zero-order hold) can be cumulative in a feedback loop and may introduce significant error into the simulation process. If internal variables are not required as simulation results, accuracy can be improved by reducing the closed-loop system to a single transfer function before developing difference equations. However, the resulting equations are much more complex.

4.8 DIGITAL SIMULATION LANGUAGES

For a number of years, considerable effort has been devoted to the development of digital simulation languages for continuous- and discrete-event systems (see Appendix B for a summary of simulation languages). Chapter 3 discusses discrete-event systems, and attention is again called to the distinction between discrete-event and discrete-time systems.

A variety of continuous-system simulation languages have been developed since Ralph Selfridge wrote the first paper on the subject in 1955; see Selfridge (1955). A brief summary of the first 20 years of activity in this area is contained in a book by Bennett (1974). An excellent source of information on recent progress is presented in the *Proceedings of the Conference on Continuous System Simulation Languages*, edited by Cellier (1986). More than 30 simulation languages have been proposed, but continuous system modeling program (CSMP) (see Brennan and Silberberg, 1968), and continuous systems simulation language (CSSL) (see SCi Software Committee, 1967) appear to be the most widely used. Each language has salient features, and familiarity with one is easily transferred to the other. Both have been implemented on small computers and are widely used in education and industry.

Regardless of the simulation language selected, there are several requirements for digital simulation that must be fulfilled by the user. The first task is to describe the system to be simulated. This involves specifying the type of elements, or functional blocks, the system consists of and describing how the elements are interconnected. This task is accomplished by means of "structure statements," which include all the standard functional blocks of feedback control systems and analog computation, as well as provisions for writing equations to specify functional relationships. The output of each block in terms of its inputs and parameters, must be specified. In most cases, all that is necessary is to select the appropriate function from a menu and specify the inputs and parameters. The user can specify meaningful names for system variables and parameters to make the simulation outputs correspond to the quantities of the system being studied.

A second set of statements is required to provide values for constants, parameters, and initial conditions and to specify arbitrary functions. These statements are called "data" or "input" statements.

The third requirement for using a digital simulation language is the specification of start and run times, integration interval, integration method, and the allowable integration error. Conditions for halting simulation, changing parameters, and restarting a simulation run can also be included. These statements are called "control statements;" they provide considerable flexibility, even in batch mode operation on a time-shared, mainframe computer.

The final task is specification of the variable(s) to be outputted and their format. The user can select numerical or graphical outputs and provide titles, labels, and other descriptive information by means of "output statements."

Modern digital simulation languages are user-oriented and incorporate default conditions which enable a novice to get meaningful results. The more sophisticated user can choose from seven or eight numerical integration methods or, if desirable, define a unique integration routine. An appreciation of the available variety can be gained by reviewing Table A.1 of Appendix A, which contains data comparing CAD programs.

The process is best illustrated with several examples which include the various forms for describing the model of a system. As an example, using CSMP (Bennett, 1974), consider the generalized second-order system [Eq. (2.116)], which is repeated here in the form

$$\ddot{x} + 2\zeta\omega_n\dot{x} + \omega_n^2 x = \omega_n^2 f \tag{4.55}$$

with $\dot{x}(0) = x(0) = 0$. Solving for $\ddot{x}$ yields

$$\ddot{x} = -2\zeta\omega_n\dot{x} - \omega_n^2 x + \omega_n^2 f \tag{4.56}$$

which is the basis for the first structure statement

$$X2D = -2.0*ZETA*OMEGA*X1D - X*OMEGA**2 + FORCE*OMEGA**2 \tag{4.57}$$

where * indicates multiplication and ** is used for exponentiation. The statement

$$X1D = INTGRL (0.0, X2D) \tag{4.58}$$

specifies that $\dot{x}_1$ is the integral of $\dot{x}_2$, with an initial condition of 0. The final structure statement is

$$X = INTGRL (0.0, X1D) \tag{4.59}$$

which indicates that x is the integral of $\dot{x}_1$ with an initial condition of zero.

An input statement is used to specify constants: $\omega_n = 2.0$, f = 1.0, and $\zeta = 0.5$. This statement is

$$CONSTANT\ OMEGA = 2.0 \qquad FORCE = 1.0 \qquad ZETA = 0.5 \tag{4.60}$$

The simulation statements shown in Figure 4.6 provide the output shown in Figure 4.7. Note that statement 12 identifies variables to be printed and statement 13 specifies that x will be print-plotted, with statement 14 providing a label.

If desired, parameters for successive (modified) simulation runs can be specified by means of a parameter statement such as

$$PARAM\ ZETA = (0.2, 0.4, 0.6, 0.8) \tag{4.61}$$

which will produce four simulation runs with the corresponding values of ζ.

If a system is described in a transfer function format, the appropriate structure statements can be used to specify the system for simulation. For

```
1    TITLE SECOND ORDER SYSTEM
2    *
3       CONSTANT   OMEGA=2.0,FORCE=1.0,ZETA=0.5
4    *
5    *
6           X2D=-2*ZETA*OMEGA*X1D-X*OMEGA**2+FORCE*OMEGA**2
7           X1D=INTGRL(0.0,X2D)
8             X=INTGRL(0.0,X1D)
9    *
10    TIMER DELT=0.005,FINTIM=6.0,PRDEL=0.20,OUTDEL=0.20
11   *
12   PRINT X,X1D
13   PRTPLT X
14   LABEL X VERSUS TIME
15   END
16   STOP
```

Figure 4.6   CSMP program for second-order generalized system.

```
                    MINIMUM      X          VERSUS TIME        MAXIMUM
                    0.0                                        1.1630E 00

TIME          X            I                                      I
0.0           0.0          +
2.0000E-01    6.9413E-02   --+
4.0000E-01    2.3704E-01   ----------+
6.0000E-01    4.4868E-01   -------------------+
8.0000E-01    6.6229E-01   ----------------------------+
1.0000E 00    8.4943E-01   --------------------------------------+
1.2000E 00    9.9446E-01   --------------------------------------------+
1.4000E 00    1.0924E 00   ------------------------------------------------+
1.6000E 00    1.1460E 00   ---------------------------------------------------+
1.8000E 00    1.1630E 00   ----------------------------------------------------+
2.0000E 00    1.1531E 00   ---------------------------------------------------+
2.2000E 00    1.1266E 00   --------------------------------------------------+
2.4000E 00    1.0923E 00   ------------------------------------------------+
2.6000E 00    1.0574E 00   ----------------------------------------------+
2.8000E 00    1.0264E 00   --------------------------------------------+
3.0000E 00    1.0023E 00   -------------------------------------------+
3.2000E 00    9.8579E-01   ------------------------------------------+
3.4000E 00    9.7658E-01   ------------------------------------------+
3.6000E 00    9.7346E-01   -----------------------------------------+
3.8000E 00    9.7482E-01   -----------------------------------------+
4.0000E 00    9.7901E-01   ------------------------------------------+
4.2000E 00    9.8455E-01   ------------------------------------------+
4.4000E 00    9.9027E-01   ------------------------------------------+
4.6000E 00    9.9537E-01   -------------------------------------------+
4.8000E 00    9.9940E-01   -------------------------------------------+
5.0000E 00    1.0022E 00   -------------------------------------------+
5.2000E 00    1.0037E 00   -------------------------------------------+
5.4000E 00    1.0043E 00   -------------------------------------------+
5.6000E 00    1.0041E 00   -------------------------------------------+
5.8000E 00    1.0035E 00   -------------------------------------------+
6.0000E 00    1.0026E 00   -------------------------------------------+
```

Figure 4.7   Print/plot output of CSMP program.

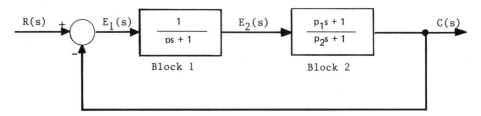

Figure 4.8   Transfer function block diagram.

example, the system of Figure 4.8 is described as follows: block 1, a real pole with the transfer function

$$\frac{E_2(s)}{E_1(s)} = \frac{1}{ps + 1} \tag{4.62}$$

is specified by the statement

$$E2 = REALPL (IC, P, E1) \tag{4.63}$$

where E2 is the output, E1 is the input, and IC is the initial value of E2. The transfer function of block 2 is

$$\frac{C(s)}{E_2(s)} = \frac{p_1 s + 1}{p_2 s + 1} \tag{4.64}$$

and the equivalent CSMP statement is

$$C = LEDLAG(IC, P1, P2, E2) \tag{4.65}$$

In this example, the remaining structure statement is for the summing junction:

$$E1 = R - C \tag{4.66}$$

As before, the parameters and input must be specified with a constant statement and the variables to be recorded, as well as the form of the output, must be selected by means of an output statement. The numerical integration routine can be selected by a method statement such as

$$METHOD \ RECT \tag{4.67}$$

which specifies the rectangular rule.

Most simulation languages include logical functions such as AND, OR, NOR, EOR, NOT, and so forth, as well as analog comparators and switches. A typical expression is of the form

$$Y = AND \ (X1, X2) \tag{4.68}$$

Obviously, digital simulation languages are much easier to use than any of the digital simulation techniques presented in the earlier sections of this chapter. The tradeoff for ease of use is efficiency. The simulation of a given system by means of a simulation language will require a great deal more computer time than the direct-programming techniques. However, with the cost of computer time decreasing rapidly in recent years, simulation languages are more popular than ever. It should be noted that special applications requiring real-time simulation may have computation time restraints that require the use of direct-programming techniques. Also, the knowledge of direct methods is useful in the application of microprocessors in real-time systems.

4.9   ANALOG SIMULATION

As noted in the introduction, analog computer simulation developed well before digital simulation. The first reference to a general-purpose analog computer was by Ragazzini, Randall, and Russell (1947). The field of analog simulation grew rapidly, and by the mid-1950s, analog computers were routinely used in the design and simulation of complex systems. The advent of high-level programming languages and user-friendly simulation software for digital computers led to a dramatic reduction in the use of analog computers. However, analog simulation has distinct advantages that will be explored in this section.

4.9.1   Comparison of Analog and Digital Computers

Before discussing analog programming techniques, it is worthwhile to compare analog and digital computers. First, it should be noted that the analog computer is named for the manner in which it operates; the voltages of its amplifiers are "analogous" to the variables of the system being simulated. The most dramatic difference between the two types of computers is the method of representing variables. The digital computer utilizes a combination of binary signals, whereas the analog uses a continuous signal. Also, the digital computer produces calculated values at distinct instants of time, in contrast to the continuous-time operation of an analog computer. Another major difference is the sequential nature in which a digital computer calculates, one after the other, the values of the system being simulated. On the other hand, the analog computer operates in parallel, producing all system variables simultaneously. There are other differences between the two types of simulation that will emerge as analog programming techniques are discussed. Before beginning a discussion of analog simulation, it is interesting to note that the first reference to digital simulation was a paper on programming a digital computer to emulate an analog computer (Selfridge, 1955).

4.9.2   Operational Amplifier

The basic element of the analog computer is the operational amplifier, which enables the summation of several variables, the integration of a variable (or sum of variables), and multiplication of variables by a constant. The operational amplifier on current analog computers is an inverting, direct-coupled device with a gain of at least 1×10^7, a high input impedance, a low output impedance, and an excellent (broad) bandwidth.

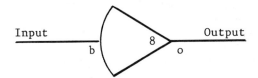

Figure 4.9   The operational amplifier.

The operating range is usually ±10 volts or ±100 volts. For most applications, analog computing amplifiers are assumed to have infinite gain, infinite input impedance, zero output impedance, zero drift or offset, and zero phase shift over the frequencies of interest. The symbol for the operational amplifier is shown in Figure 4.9, where the number in the apex of the symbol identifies which amplifier on the computer is being represented. The symbol o indicates the output and b identifies the input (referring to the base of the input transistor).

4.9.3   The Summing Amplifier

In order to understand how an operational amplifier can produce the sum of two or more variables, consider the diagram of Figure 4.10. The currents in the resistors are

$$i_1 = (e_1 - e_b)/R_1 \tag{4.69}$$

$$i_2 = (e_2 - e_b)/R_2 \tag{4.70}$$

$$i_3 = (e_3 - e_b)/R_3 \tag{4.71}$$

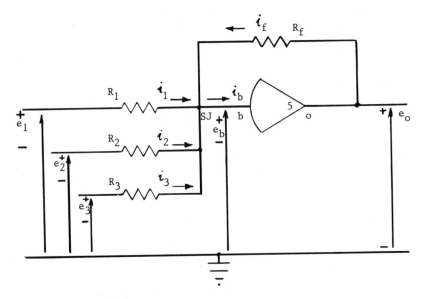

Figure 4.10   Summing amplifier circuit.

and

$$i_f = (e_o - e_b)/R_f \qquad (4.72)$$

where e_b refers to the voltage at the amplifier input or summing junction (sj). The sum of the currents at the summing junction is

$$i_1 + i_2 + i_3 + i_f - i_b = 0 \qquad (4.73)$$

Since the output voltage is less than ±100 or ±10 and the amplifier gain is at least 10^7, the input voltage e_b must be less than $100/10^7$ or less than 10^{-5}. In other words, the input voltage e_b is negligible and the amplifier input can be considered a "virtual" ground ($e_b \cong 0$). As a result of the large impedance, the input current can also be neglected ($i_b \cong 0$). By utilizing the assumptions that both e_b and i_b are zero, Eq. (4.69) through (4.72) can be simplified and substituted into Eq. (4.73) to give

$$\frac{e_1}{R_1} + \frac{e_2}{R_2} + \frac{e_3}{R_3} + \frac{e_o}{R_f} = 0 \qquad (4.74)$$

which can be solved for the output voltage

$$e_o = -\left[\left(\frac{R_f}{R_1}\right)(e_1) + \left(\frac{R_f}{R_2}\right)(e_2) + \left(\frac{R_f}{R_3}\right)(e_3) \right] \qquad (4.75)$$

Thus, the output voltage is the negative of the weighted sum of the input voltages, where the weighting factors are R_f/R_1, R_f/R_2, and R_f/R_3.

It is important to note the inversion that occurs due to the minus sign in Eq. (4.75).

Most computers provide several resistors on each operational amplifier that can be used as input resistors and as a feedback resistor with values that result in weighting factors (gains) of 0.1, 1.0, and 10. The resistor values on 10-volt analog computers are 10 K-ohms and 100 K-ohms (the ratios are 10/100, 100/100, and 100/10), while 100-volt computers use 100 K-ohms and 1 M-ohm. The resistors are usually not shown on a simulation diagram, and a "shorthand" symbol of the type shown in Figure 4.11(a) is used. Also note that the electrical ground is not shown, but all voltages are relative to the ground on the computer.

If only one input is used and the input resistor is equal to the feedback resistor, the output is the negative of the input. This configuration is referred to as an inverter, $e_o = -e_{in}$; a typical symbol is shown in Figure 4.11(b).

4.9.4   The Integrating Amplifier

The operational amplifier can also be connected in a manner to provide integration. Consider the circuit of Figure 4.12. The input currents are the same as in the previous example, but the feedback current i_f in the capacitor is

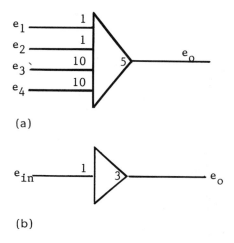

(a)

(b)

Figure 4.11   (a) Summing and (b) inverting amplifier symbols

$$i_f = C \left[\frac{d(e_b - e_1)}{dt} \right] \tag{4.76}$$

Since the sum of currents at the input is zero and e_b and i_b are approximately zero, the resulting currents are as follows:

$$\frac{e_1}{R_1} + \frac{e_2}{R_2} + \frac{e_3}{R_3} + \frac{Cde_o}{dt} = 0 \tag{4.77}$$

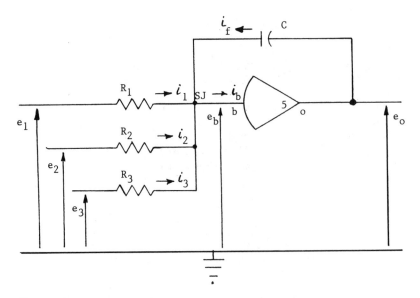

Figure 4.12   Integrating amplifier circuit.

which can be solved to give

$$\frac{de_o}{dt} = -\left[\frac{e_1}{R_1 C} + \frac{e_2}{R_2 C} + \frac{e_3}{R_3 C}\right] \qquad (4.78)$$

Multiplying by dt and integrating both sides gives

$$\int_{e(t_0)}^{e(t)} de_o = -\int_{t_0}^{t}\left[\frac{e_1}{R_1 C} + \frac{e_2}{R_2 C} + \frac{e_3}{R_3 C}\right] dt \qquad (4.79)$$

which results in

$$e_o(t) - e_o(t_0) = -\int_{t_0}^{t}\left[\frac{e_1}{R_1 C} + \frac{e_2}{R_2 C} + \frac{e_3}{R_3 C}\right] dt \qquad (4.80)$$

The initial condition $e_o(t_0)$ can be moved to the right-hand side to given an expression for $e_o(t)$

$$e_o(t) = -\int_{t_0}^{t}\left[\frac{e_1}{R_1 C} + \frac{e_2}{R_2 C} + \frac{e_3}{R_3 C}\right] dt + e_o(t_0) \qquad (4.81)$$

Thus, the output is the *negative* of the integral of the weighted sum of the inputs, plus the initial value of the input. One should note that the initial condition is inverted through the integrator. Thus, the negative of the initial value must be connected to the amplifier at the special initial condition (IC) input.

As with resistors, integrating capacitors are provided on most analog computers, and their values are selected to give gains of 0.1, 1, and 10. The resulting "shorthand" symbol is shown in Figure 4.13; attention is called to the input for the initial value to emphasize that the initial value of the output is of the same sign as the initial condition. In order to provide additional understanding of analog integrators, an example with input and output waveforms is provided in Figure 4.14. The initial

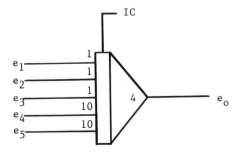

Figure 4.13   Integrating amplifier symbol.

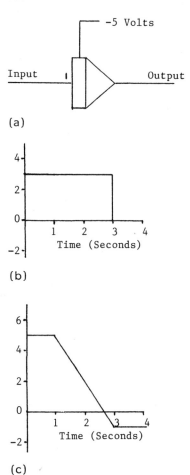

(a)

(b)

(c)

Figure 4.14   Pulse response of an integrator.   (a) Integrator.   (b) Input. (c) Output.

condition input is −5v and the reset-hold-operate switch is placed in operate at time = 1.0 sec.   The positive input causes the integrator output to go negative until the input goes to zero.

The limitation of fixed weighting factors or "gains" of the previous examples can be overcome by use of potentiometers, as shown in Figure 4.15.   If the output current is limited to very small values, the output e_0 is equal to the input e_1 multiplied by the relative position of the potentiometer, as shown in Figure 4.15(a).   If both terminals of the potentiometers have input values instead of one side connected to ground, the output is $e_0 = K(e_1 - e_2)$, as shown in Figure 4.15(b).   Since the output current will change when the potentiometer (more commonly referred to as "pot") is connected to different resistive loads, the potentiometer should always be set to the desired ratio between input and output, *after pot is connected to the circuit.*

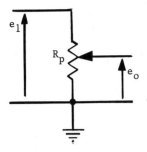

(a)

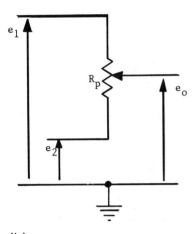

(b)

Figure 4.15   Potentiometer circuits.   (a) Grounded potentiometer.   (b) Ungrounded potentiometer.

This is a good point to provide an illustrative example to help understand operational amplifiers connected as summing and integrating devices. Note that the output in Figure 4.16(a) is

$$e_o = 0.2x - 3y - 4z \tag{4.82}$$

and the output of the integrator in Figure 4.16(b) is

$$e_o = \int_{t_0}^{t} (4x + y - 5.0z)dt + 1.0 \tag{4.83}$$

The characteristics of analog computers vary, and the user's manual should be reviewed before operating the machine. There are several factors worth special notice. First, the balance (i.e., offset) of each amplifier should be checked before operating the machine, and a special switch is usually provided, for this purpose. If the amplifiers are out of balance, the basic assumptions of $e_b \cong 0$ and $i_b \cong 0$ are not true, and error is introduced into the simulation. Another important consideration is

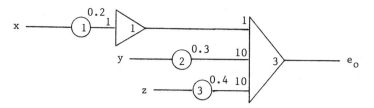

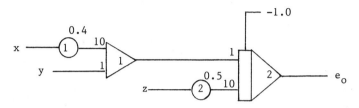

(a)

(b)

Figure 4.16   Examples of summing and integrating amplifiers. (a) Diagram for $e_o = 0.2x - 3.0y - 4.0z$. (b) Diagram for $e_o = \int_o^t (4.0x + y - 5.0z)dt + 1.0$.

"overload." If the input to an amplifier is overdriven or if the output draws too much current, the assumptions $e_b \cong 0$ and $i_b \cong 0$ will also be violated. Most analog computers provide an "overload" warning light for each amplifier.

As noted previously, the manner in which potentiometers are set is very important. The pot should be set after connecting *the load or input resistors*. Therefore, pot settings should always be checked when the interconnections are changed.

A final practical note on integrating amplifiers calls attention to the fact that most computers have a switching network which puts the integrator into the "reset" mode. This causes all integrators to assume the value of the corresponding initial condition (the initial-condition voltage is across the integrating capacitor). Integration occurs when the control switch is put into the "operate" mode. This essentially removes the initial-condition input and allows the integrator output to change according to the integral of the sum of the inputs, starting at the value of the initial-condition voltage. The simulation can be stopped and held at any point by switching to the "hold" mode. During the hold period, the output voltage is across the integrator capacitor. Thus, the operator has complete control of the simulation, and the process can be started, stopped, and restarted as often as desired.

4.9.5 Analog Simulation of Linear Systems

The simulation of a system described by a linear ordinary differential equation is very straightforward. The four-step process is as follows:

1. Solve for the highest-order derivative.
2. Assume the highest-order derivative is known and available.

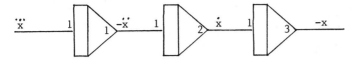

Figure 4.17   Repeated integration.

3. Use repeated integration on each derivative until the dependent variable is available.
4. Use a summing amplifier to sum the weighted values of the derivatives to produce the highest-order derivative that was assumed to be known.

This process is best illustrated by means of an example. Consider the differential equation

$$\dddot{x} + a\ddot{x} + b\dot{x} + cx = dy \tag{4.84}$$

According to the first step, the highest-order derivative is

$$\dddot{x} = -a\ddot{x} - b\dot{x} - cx + dy \tag{4.85}$$

where x is the dependent variable and y is the input or independent variable. If $\dddot{x}$ is known and available (step 2), it can be integrated (step 3), as shown in Figure 4.17, to give $-\ddot{x}$, which can be integrated twice to give $\dot{x}$ and $-x$.

A summing amplifier can be used to sum the weighted values of $\ddot{x}$, $\dot{x}$, and x to get $\dddot{x}$, as shown in Eq. (4.85), which is step 4 of the process. The inversion of integrating and summing amplifiers makes it necessary to use caution in interconnecting variables. The summing amplifier to implement Eq. (4.85) will invert the signals to produce $-\dddot{x}$, if the inputs are $-a\ddot{x}$, $-b\dot{x}$, $-cx$, and dy. Since it was assumed that $\dddot{x}$ was known and available, the summing amplifier inputs should be $+a\ddot{x}$, $+b\dot{x}$, $+cx$, and $-dy$, as shown in Figure 4.18.

Since the integrators of Figure 4.17 produce $-\ddot{x}$, $\dot{x}$, and $-x$, inverters will be required for $-\ddot{x}$ and $-x$ to produce $\ddot{x}$ and x, respectively. Also, the input or independent variable y must be supplied with a negative sign (i.e., inverted). The coefficients or weighting factors a, b, c, and d are simulated by means of the potentiometers mentioned previously. The

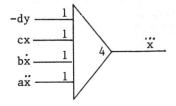

Figure 4.18   Summing amplifier.

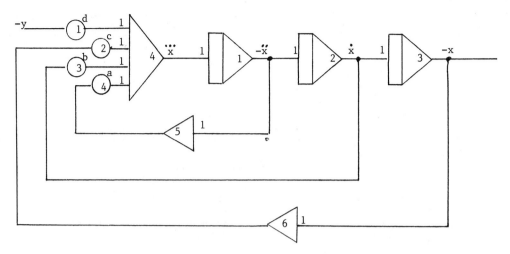

Figure 4.19   Simulation diagram for $\dddot{x} + a\ddot{x} + b\dot{x} + cx = dy$.

combined result of all steps in the process is shown in Figure 4.19. Note the need to invert the initial condition on each integrator.

A great deal can be learned by a careful analysis of the simulation diagram of Figure 4.19. First, note that assuming a value of $-\dddot{x}$, instead of $+\dddot{x}$, would have produced $\ddot{x}$, $-\dot{x}$, and x and resulted in a sign change in all the variables of the simulation diagram. Thus, the input is $+y$, instead of $-y$, which may avoid the need for an additional inverter. This is analogous to a sign change on all the variables of an equation, a process frequently used in mathematical manipulation.

It is also worthwhile to examine the number of amplifiers in each loop of the simulation diagram of Figure 4.19. The loop consisting of amplifiers 4, 1, and 5 has three inversions, as does the loop made up of amplifiers 4, 1, and 2. The outside loop includes five inversions (amplifiers 4, 1, 2, 3, and 6), which verifies that all closed loops in this example contain an odd number of amplifiers. This is reasonable to expect, since an even number of amplifiers would result in positive feedback and probably produce unstable operation (see Chapter 2 for a discussion of stability). Only in rare instances will simulation diagrams contain closed loops with an even number of inversions. Thus, it is a good idea to check each loop of a diagram and be very suspicious of loops containing an even number of amplifiers.

A final alteration of the simulation diagram in Figure 4.19 provides additional insight into the simulation process. Most analog computers contain resistive inputs for integrators which permit algebraic summation prior to integration. This feature can be used to combine the functions of the summing process (amplifier 4) and the integration taking place in amplifier 1. Instead of implementing Eq. (4.85) by assuming that the highest-order derivative $\dddot{x}$ is known and available, integrate the equation to give

$$\int \dddot{x}\,dt = \ddot{x} = -\int [a\ddot{x} + b\dot{x} + cx - dy]\,dt \tag{4.86}$$

which states that $\ddot{x}$ is equal to the negative of the integral of the sum $a\ddot{x} + b\dot{x} + cx - dy$. This is implemented in hardware in Figure 4.20. Additional

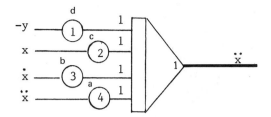

Figure 4.20   Summing integrator.

integrators and inverters can be used to complete the simulation diagram, as shown in Figure 4.21. This simulation diagram requires only four amplifiers, two fewer than the previous implementation. Again, note the odd number of inversions in each closed loop. Also, note that the initial conditions are also inverted.

In the example currently being discussed, numerical values have not been assigned to the coefficients a, b, c, and d. It was previously noted that potentiometers are used to simulate coefficients or weighting factors, and the range of practical values is 0.05 to 0.99. This range can be extended by using different values of gain on the amplifier inputs. Most analog computers provide standard gains ranging from 0.1 to 10. Special provisions easily extend the low value to 0.01 and the high value to 100. This permits a wide range of values for coefficients, since the pot setting is multiplied by the gain of the corresponding amplifier input.

4.9.6   Magnitude Scaling

As noted in the introduction to this section, the output voltage of an operational amplifier is "analogous" to the variables of the system being simulated. Since the operating range of commercial analog computers is ±10 volts or ±100 volts, using 1 volt to represent one unit of the problem variable

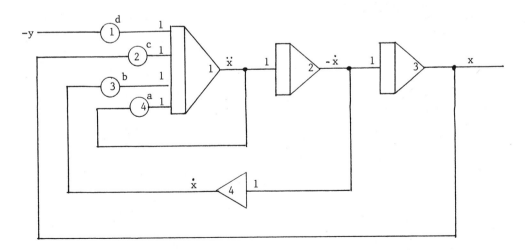

Figure 4.21   Simulation diagram.

would limit the upper range of values and result in poor accuracy for extremely small values. Therefore, some method of scaling the problem variables to a suitable voltage range is necessary. Although several techniques for scaling have been developed (Rekof, 1967; Jackson, 1960), only the "per unit" method will be presented here.

The first task of scaling is to determine the range of problem variables, and several methods have been developed for this process (Bennett, 1974). For example, in a simulation of a physical system, it is necessary to determine the approximate maximum values of the anticipated displacements, velocities, and accelerations. These values are used as a basis for choosing expected maximum values that are convenient for recording, reporting, and using the simulation results. This method results in per-unit computer variables that are the ratio of the problem variable to the expected maximum value. Computer variables have a per-unit value of 1 when the problem variable is equal to the expected maximum value. Also, when the problem variable is equal to the maximum expected value, the amplifier output voltage is a maximum (e.g., ±10 volts or ±100 volts).

As an example, if it is anticipated that the largest displacement will be 41.5 meters, the expected maximum value is "rounded up" to 50 and the corresponding computer variable will be [x/50]. (Note that square brackets are used to indicate computer variables.) If the amplifier producing the displacement variable has an output of 7.5 volts on a ±10-volt computer, the per-unit value of 7.5/10 or 0.75 and the displacement is [x/50] = 0.75, or x = (0.75)(50) = 37.50 meters. Figure 4.22 provides a list of several typical values and the selected computer variables for each case. Programs scaled by the per-unit method will run on a ±10-volt or ±100-volt computer without change.

In order to gain a better understanding of the process, consider a mechanical process that can be modeled by a second-order differential equation

$$\ddot{x} + 3\dot{x} + 9x = 9y \tag{4.87}$$

where the initial values of $\dot{x}$ and x are assumed to be zero and the forcing function is a unit step of 2.0 meters. A review of Ex. (2.4.3) will result in determining an undamped natural frequency ω_n of 3 radians and a damping ratio ζ of 0.5, which gives an overshoot of approximately 17% to a step input. The steady-state solution is 2.0 meters and the maximum value

Problem variable	Anticipated maximum	Scaling maximum	Computer variable
x (feet)	13.6	20	$\dfrac{x}{20}$
x (degrees k)	276	300	$\dfrac{x}{300}$
$\dot{x}$ (meters/sec)	1.15	1.25	$\dfrac{\dot{x}}{1.25}$
$\ddot{x}$ (kg/sec/sec)	0.021	0.025	$\dfrac{\ddot{x}}{0.025}$

Figure 4.22   Example of scaling values.

Problem variable	Anticipated maximum	Scaling maximum	Computer variable
x(meters)	2.34	2.5	[x/2.5]
$\dot{x}$(meters/sec)	7.02	10.0	[$\dot{x}$/10]
$\ddot{x}$(meters/sec/sec)	21.06	25.0	[$\ddot{x}$/25]
y(meters)	2.00	2.0	[y/2.0]

Figure 4.23 Example of scaling values.

is approximately 2.34. Therefore, choose x_{max} = 2.5. A conservative estimate for $\dot{x}_{max}$ is $\omega_n x_{max}$ (Bennett, 1974), which is $\dot{x}_{max} \cong$ (3)(2.34) = 7.02; for $\ddot{x}_{max}$, use $\ddot{x}_{max} \cong \omega_n \dot{x}_{max}$ = 3(7.02) = 21.06. These values are used to arrive at the computer variables of Figure 4.23.

Once computer variables have been selected, the scaling process can be completed by substituting computer variables into Eq. (4.87), which yields

$$25 \left[\frac{\ddot{x}}{25}\right] + (3)(10) \left[\frac{\dot{x}}{10}\right] + (9)(2.5) \left[\frac{x}{2.5}\right] = (9)(2) \left[\frac{y}{2}\right] \tag{4.88}$$

Note that the computer variable is multiplied by the scaling maximum to keep the equation properly balanced. Multiplication of coefficients of Eq. (4.88) and division by 25 gives

$$\left[\frac{\ddot{x}}{25}\right] + 1.2 \left[\frac{\dot{x}}{10}\right] + 0.9 \left[\frac{x}{2.5}\right] = 0.72 \left[\frac{y}{2}\right] \tag{4.89}$$

which has coefficients easily obtainable directly by potentiometer settings with amplifier gains of 1.0 or 10. The scaled computer diagram is shown in Figure 4.24. Note that one amplifier, and the need to provide [−y/2], could be eliminated by using a summing integrator, as discussed in Section 4.9.5. A good point to note is that the gains of the integrators, 2.5 and 4.0, are close to the undamped natural frequency ω_m = 3.0.

4.9.7 Time Scaling

The response times of systems of interest to engineers and scientists vary over a wide range. In some instances, the transients occur in a few nanoseconds, while other systems may take days or even years to respond to disturbances. The operational amplifiers and recording devices limit the lower bound on response time, while the drift characteristics (and patience of the operator) limit the upper bound on response time. One major advantage of computer simulation is the ability to study many cycles of system operation in a short period of time instead of waiting days or weeks for the actual system to respond. The opposite also holds, since it is often instructive to slow down system response to permit detailed analysis of system variables.

The need for time scaling in analog simulation is indicated by integrator gains that are out of the range 0.1 < gain < 10, where low gains indicate

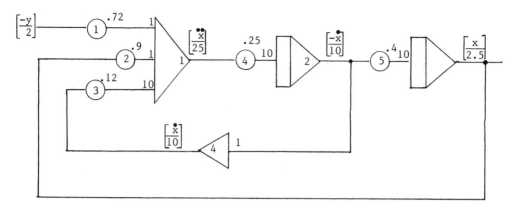

Figure 4.24   Scaled computer diagram.

a very slow system response and high gains indicate a rapid response. In order to time-scale, all that is necessary is to change the gains on all *integrators* by the same ratio. This has the effect of changing closed-loop gains. This may require that pot settings, as well as integrating amplifiers, be changed in the simulation diagram, and time scaling should be very carefully carried out. To repeat: change the gain of all integrators (and *only* integrators) by the same ratio. Summing amplifier gains are not changed in time scaling.

To illustrate, the previous example could be "speeded up" by a factor of two by doubling the gains of both integrators. This could be accomplished by increasing pot 4 to 0.5 and pot 5 to 0.8. The simulation of this equation could be made to run slower by reducing integrator gains to 1.0. Again, note that summing amplifier and inverter gains are *not* effected by time scaling.

Time scaling can also be accomplished by changing the original differential equation, but this is a much more involved process (Bennett, 1974). Magnitude and time scaling are further discussed in Section 10.7.3.

4.9.8   Simulation of Nonlinear Systems

The more commonly encountered nonlinearities, such as saturation, dead space, or hysteresis, can be simulated by incorporating diodes into the simulation diagram. Many analog computers include nonlinear function generators and special modules which permit multiplication, division, square root, and so forth (Bennett, 1974). The use of these elements requires considerable care, and the operating manual for the computer being used should be carefully reviewed.

4.9.9   Simultaneous Equations

Simulation of simultaneous equations can be accomplished by carrying out the process of the previous discussion for the equations and by making the appropriate interconnections between each equation. In this respect, analog simulation is very instructive, since there is a direct correlation (analogy) between the simulation diagram and the system being simulated.

4.9.10   Transfer Function Simulator

The ability to simulate systems expressed in transfer function form is very important to control system design and analysis. Transfer functions and block diagrams can be implemented directly on an analog computer, and this approach has been widely used for simulating systems with one or more of the actual system components in the loop. The simulation diagrams for several of the more frequently used transfer functions are shown in Figure 4.25. Note that the summing junction, Figure 4.25(a); the integrator, Figure 4.25(b); and the first-order diagram, Figure 4.25(c), can be combined to simulate higher-order, closed-loop systems. The simulation diagram in Figure 4.25(d) utilizes the fact that the gain of an operational amplifier is the ratio of the impedance of the feedback path to the impedance of the input path. This property can be utilized to develop a complex transfer function with a single operational amplifier and the appropriate

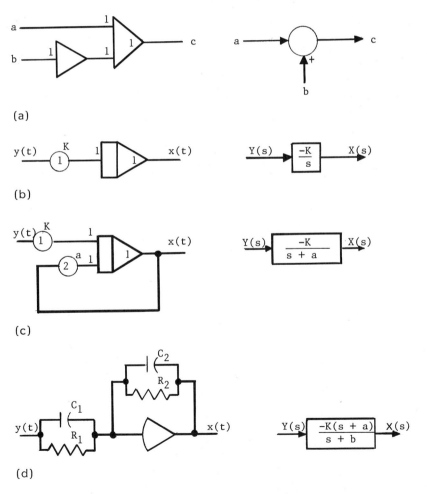

Figure 4.25   Transfer function simulation.   (a) Summing junction.   (b) Integrator.   (c) First-order term.   (d) Lead-lag term.

combination of resistors, capacitors, and inductors. However, the circuit analysis must be carried out carefully. This approach should be utilized only when amplifier economy is essential. Also, if the feedback path and/or input path has points connected to electric ground, the gain must be determined by detailed circuit analysis; the simple ratio $-Z_{FB}/Z_{IN}$ does not apply. A comprehensive list of transfer function simulation diagrams is included in Jackson (1960).

4.9.11   State Variable Systems

Systems expressed in state variable form can easily be programmed for analog simulation, since each state equation is a first-order differential equation. An integrator is used to produce each state variable, and the integrator inputs represent each term in A, which distributes state variables, and B, which distributes the control. For example, the state equation

$$\dot{\underline{x}} = \underline{A}\underline{x} + \underline{B}u \qquad (4.90)$$

is programmed in the simulation diagram of Figure 4.26. Note that $\underline{x}$ is a two-dimensional state vector, $\underline{A}$ is a 2×2 matrix, $\underline{B}$ is a 2×1 matrix, and u is a scalar.

There are direct relationships between state variable and transfer function simulation techniques that are instructive to explore. First, note that the number of integrators corresponds to the order of the system, while the output of each integrator corresponds to each state variable. Also, the

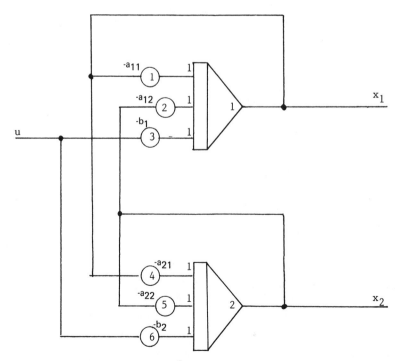

Figure 4.26   Program for $\dot{\underline{x}} = \underline{A}\underline{x} + \underline{B}u$.

use of phase variables (i.e., each state variable is the derivative of another state variable) results in an $\underline{A}$ matrix of the form of Eq. (2.68).

Magnitude and time scaling for systems expressed in state variable form can be accomplished by a series of matrix-vector operations; see Cannon (1973). If the $\underline{A}$ matrix is diagonal, the n first-order equations are uncoupled, and scaling can be handled individually. However, the same time scale factor must be used with every one of the n first-order equations.

4.10 HYBRID SIMULATION

The combination of analog and digital computing elements to take advantage of the parallelism, speed, and direct-system analogy of analog simulation plus the ease of programming, memory, and control capabilities of digital simulation began in the late 1950s and has resulted in a variety of "hybrid" computers. The simplest of these "hybrid" systems is predominantly an analog computer with a few digital elements to assist in controlling the analog simulation and to provide logic operations. These units are frequently referred to as "parallel hybrid" computers. This approach to simulation will be discussed briefly as an introduction to hybrid computation.

The basic element of the analog computer is the operational amplifier, which can be connected to provide integration and summation, and its inputs and output are analog signals. The basic element of the digital computer is the logic gate, which can be configured to provide logic operations on digital (binary) signals such as AND, OR, and INVERSION. Logic gates can be interconnected to make FLIP-FLOPS, which provide memory. In contrast to analog computing elements, the input and output signals of digital elements are binary. There are several hybrid computing elements which are very useful in simulation, and their input and output signals are both analog and digital in nature.

Perhaps the most useful hybrid computing element is the comparator shown in Figure 4.27. Its digital output will be a logic "1" if the sum of the analog inputs is positive. Its output will be a logic "0" if the sum of the analog inputs is less than zero. If a single analog input is used, the output will be a logic "1" when the signal is positive and a logic "0" otherwise. The latch causes the output to be in the "1" or "latched" state when the latch input is logic "1." Obviously, comparators can be used to monitor the levels of analog signals. A comparator can also be used to

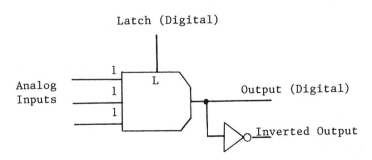

Figure 4.27   Comparator.

start or stop analog simulation on the basis of one or more analog signals. Some parallel hybrid computers include digitally controlled analog switches which can be used with the comparators to control simulation and, if necessary, alter the analog program connections to simulate changes in system dynamics. For additional information, see Bekey and Karplus (1968), Hausner (1972), and Bennett (1974).

Another important hybrid computing element is the track-store unit, which has analog inputs and outputs and digital controls as shown in Figure 4.28. When the track-store unit is in the track mode, the analog output tracks the sum of the analog inputs (i.e., the output equals the sum of the inputs). When the track-store is switched to the hold mode, the unit stores (holds) the value of the analog output that was present at the time. This device can be used to track and store the value of one or more analog signals under the control of digital signals. This is a convenient way to store (hold for later examination) the values of important analog signals at key instants of time during simulation. Track-store units can also be used to provide analog signals that make step increases in value on each cycle of simulation to assist in optimization studies requiring multiple simulation runs for a range of values. Examples of these techniques are found in Bekey and Karplus (1968), Hausner (1972) and Bennett (1974).

By adding digital gates to the analog computer, the power of logic decisions can be brought to system simulation. A two-input AND gate can be used to sense when condition A *and* condition B have occurred, as indicated by the output of two comparators. An OR gate would detect when *either* condition has occurred. The incorporation of logic gates into simulation can greatly enhance the power of the programmer to carry out optimization studies or to simulate systems under a series of changing conditions.

The real power of hybrid computing can be achieved by linking a full-scale digital computer to an analog computer incorporating hybrid elements such as track-stores, comparators, digital logic, and digitally controlled analog switches. The linkage (or interface) between the two computers requires a means of converting analog signals to digital form (A-to-D conversion) and a similar capability in the opposite direction, digital-to-analog

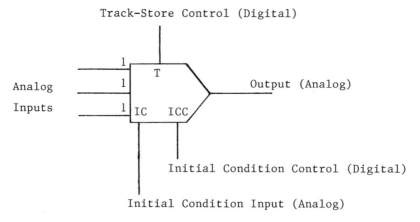

Figure 4.28   The track-store unit.

conversion (D-to-A) (see Chapter 10). A full-scale hybrid computer interface also needs connections which permit the digital computer to control the analog computer (i.e., put the analog computer into the operate, hold, or reset mode). In addition, the interface should include circuitry that enables the digital computer to sense the operating state of the analog computer and determine the outputs of the hybrid computing elements on the analog computer patchpanel.

A fully implemented hybrid computer can simplify many of the difficulties associated with analog computing. The digital computer can be programmed to completely check out the analog computer, automatically set motor-controlled potentiometers, and interconnect the analog computing elements by means of digitally controlled switches.

Programming the analog and digital portions of a hybrid computer is a very complex task. However, the computational speed of the parallel analog elements, coupled with the memory, convenience, and flexibility of the digital computer, is indeed a powerful simulation device. Hybrid simulation has been instrumental in real-time applications in the aerospace and nuclear fields, and hybrid computation is especially well suited to optimization applications requiring millions of iterations. However, the continued improvement in speed and power of digital computers, coupled with more accurate and efficient digital simulation algorithms, is making it increasingly difficult to justify the investment in analog and hybrid computers. Furthermore, the full advantage of parallel digital computation is yet to be realized. However, the one-to-one correspondence between analog computing elements and the components of the system being simulated, along with the direct, hands-on programming of analog computation, continues to provide excellent insight into the operation of a system under study.

4.11 SUMMARY AND CONCLUSIONS

This chapter presents an overview of techniques for simulating continuous systems represented in terms of differential equations, transfer functions, and state space models. Methods for simulating discrete-time systems represented by difference equations and Z-transforms are also reviewed. Although primary emphasis is placed on digital simulation techniques, analog simulation is included, along with a brief discussion of hybrid computation.

The digital representation of continuous signals is presented first to provide background for an overview of numerical integration. In the discussion of numerical integration, only the simplest routine, Euler's, is covered in detail, and the emphasis is on providing an appreciation for the process and the errors likely to be encountered. Some of the more advanced integration methods are mentioned, along with references to more detailed treatments of the subject. The intent is to provide a simple, straightforward technique that is appropriate for small-scale applications and, at the same time, enable readers to understand and better utilize general-purpose simulation software programs. This presentation is not intended to fulfill the needs of sophisticated users, who are advised to consult the references.

The chapter includes examples of the use of state space representation for system simulation. The only numerical integration technique illustrated is Euler's, but the utility of state space representation for nonlinear and time-varying systems is noted. The problems associated with initial conditions and variations in the interval of integration (sample time) are acknowledged.

Discrete-time systems are discussed briefly, and the Z-transform is used to illustrate some of the relationships between difference equations and numerical integration techniques. The utilization of Z-transform techniques for stability analysis of numerical integration methods is noted, and references for additional information are provided. The Z-transform is also used to develop methods for digital simulation of systems represented in transfer function form. However, only zero-order holds are illustrated, and readers are directed to other sources of more complete treatments.

Digital simulation languages for continuous systems are discussed briefly, and an example of CSMP is provided. Similarities among languages are noted, and references for additional details are provided.

A survey of analog simulation is included along with examples of simple applications. The discussion includes magnitude and time scaling, but non-linear systems, simultaneous equations, and transfer function simulation are only mentioned briefly. Hybrid simulation is also discussed briefly, but references for more extensive coverage are provided.

The references included with each section of this chapter indicate that the material covered could easily be the subject of several lenghty presentations. However, the intent of this chapter is to present an overview of continuous-system simulation techniques and to provide references for additional information if more detailed coverage is required. It is hoped that this chapter will provide the readers with needed background and direction for more advanced work.

PROBLEMS

4.1. Assume that an eight-bit binary word on a digital computer is used to represent a continuous signal. Further assume that the rightmost (least significant) bit is assigned a value of 0.1 volt when it is a binary one and 0.0 volt when it is a binary zero. Let the assigned value of the next six bits each be double that of the bit to its right (e.g., 0.2, 0.4, 0.8. 1.6, 3.2, 6.4 volts) and use the leftmost (most significant) bit to indicate the algebraic sign (e.g., 0 if positive, 1 if negative). (Note: This is the sign-magnitude format.) Determine the maximum positive and negative values that can be represented.

4.2. Refer to Problem 4.1 and determine the closest representation that can be made to an analog voltage of +5.26 volts. What is the pattern of binary ones and zeros for this case?

4.3. Repeat Problem 4.2 for −3.82 volts.

4.4. Refer to Problem 4.1 and determine the analog voltage represented by the following binary values: (a) 10110101, (b) 00011110.

4.5. If the circuitry requires 0.001 seconds to determine the value for each of the eight bits in each digital word: (a) How much time is required to complete each analog value? (b) How would this affect simulation?

4.6. Repeat Problem 4.1 for a 16-bit computer word. (Maintain the assignment of 0.1 volt for the least significant bit.)

4.7.   Select the analog values to be assigned to an eight-bit word if the desired range of coverage is ±10 volts.

4.8.   Repeat Problem 4.7 for a 16-bit computer word.

4.9.   Use Euler's method to develop a difference equation to solve the equation

$$\frac{dx}{dt} + 2x(t) = 1.0 \qquad \text{with } x(0) = 0$$

   a.   Calculate the first 20 values for T = 0.1 sec and also T = 0.05 sec.

   b.   Determine the exact solution, and compare the values at corresponding points in time.

   c.   Comment on the impact of integration time interval on the magnitude of the error.

4.10.   Repeat Problem 4.9 using Euler's improved method, Eq. (4.7). Note that it will be necessary to use Taylor's series to calculate the value of x at the first interval of time.

4.11.   Repeat Problem 4.9 with x(0) = −0.5.

4.12.   Use Euler's integration to develop difference equations to solve the equation

$$\frac{d^2x}{dt^2} + 2\frac{dx}{dt} + 4x = 4$$

with both initial conditions equal to zero.

4.13.   Discuss the criteria for selecting an appropriate interval for the integrating process for Problem 4.12.

4.14.   Solve the equations developed for Problem 4.12 for the first 10 values of x for T = 0.5, and compare the results with the exact solution.

4.15.   Solve Eq. (4.13) and Eq. (4.14) for u = a step of 2 units, M = 0.25, B = 0.1, K = 0.9, and T = 0.1 sec. Comment on the impact of the size of the integration interval on simulation accuracy.

4.16.   Repeat Problem 4.15 by adjoining another state equation as in Eq. (4.19), (4.20), (4.21), and (4.25). Discuss the influence of the integration interval T on accuracy.

4.17.   Repeat Problem 4.15 assuming that the mass M decreases from 0.25 to 0.15 linearly over the first 10 integration intervals. Comment on the impact of the relative size of the period of integration T.

4.18. Repeat Problem 4.16 assuming that the mass M decreases linearly from 0.25 to 0.15 over the first 10 integration intervals. Comment on the appropriateness of the integration interval and the computation required to change a problem variable.

4.19. Assume a = 0.5 and r = 1.0, and determine the Z-transform of Eq. (4.6). Use the stability analysis techniques in Chapter 2 to determine bounds on the interval of integration T.

4.20. Repeat Problem 4.19 for the difference equation developed in Problem 4.9.

4.21. Repeat Problem 4.19 for the difference equation developed in Problem 4.10.

4.22. Write a computer program to implement the simulation of Eq. (4.42). Assume that the parameter a = 0.5 and a unit-step input, and compare the results with Problem 4.19 and the exact solution for T = 0.1 sec.

4.23. Repeat Problem 4.22 assuming that the parameter a changes linearly from 0.5 to 1.0 over the first five time intervals.

4.24. Write a computer program to solve the difference equations for the system in Figure 4.5. Assume a = 0.5, K = 1.0, and a step input, and use T = 0.05 sec.

4.25. Reduce the system of Figure 4.5(a) to a single block and develop a simulation program for the resulting transfer function using the Z-transform method of Section 4.7. Assume T = 0.05 sec, a = 0.5, K = 1.0, and a step input, and write a computer program to generate the system response.

4.26. Repeat Problem 4.25 using state space techniques and Euler's integration.

4.27. Use an available continuous-simulation language to generate responses for the systems in the following:

a. Problem 4.9

b. Problem 4.12, with the assumptions of Problem 4.14

c. Problem 4.15

d. Problem 4.17

e. Problem 4.22

f. Problem 4.23

g. Problem 4.24

h. Problem 4.25

i. Problem 4.26

4.28. Use an available continuous simulation language to simulate the system of Figure 4.8 if P = 1.0, P_1 = 0.2, and P_3 = 0.5.

4.29. Repeat Problem 4.28 and select a different integration algorithm from the set available with the simulation program used for the first solution.

4.30. Given the differential equation

$$\dot{x} + 8x = 3$$

with x(0) = 0, determine the expected maximum value and the steady-state solution.

4.31. Repeat Problem 4.30 with x(0) = −6.0.

4.32. Given the differential equation

$$3\ddot{x} + 24\dot{x} + 1200x = 1600$$

with x(0) = $\dot{x}$(0) = 0, determine expected maximum values and choose appropriate computer variables.

4.33. Repeat Problem 4.32 with x(0) = 0 and $\dot{x}$(0) = 5.0.

4.34. Repeat Problem 4.32 with x(0) = 10 and $\dot{x}$(0) = 0.0.

4.35. Use an analog computer to generate input/output recordings for the systems described in the following:

a. Problem 4.9

b. Problem 4.12

c. Problem 4.15

d. Problem 4.22

e. Problem 4.28

f. Problems 4.30 and 4.31

g. Problems 4.32, 4.33, and 4.34

REFERENCES

Bekey, G. A. and Karplus, W. J. (1968). *Hybrid Computation*, Wiley, New York.

Bennett, A. W. (1974). *Introduction to Computer Simulation*, West, St. Paul, MN.

Benyon, P. R., (1968). A review of numerical methods for digital simulators, *Simulation 11*(5), 219–238.

Boxer, R. and Thaler, S. (1956). A simplified method of solving linear and nonlinear systems, *Proc. IRE 44*, 89–101.

Brennan, R. D. and Silberberg, M. Y. (1968). The system/360 continuous system modeling program, *Simulation 11*(6), 301–308.

Cannon, M. R. (1973). Magnitude and time scaling of state-variable equations for analog/hybrid computation, *Simulation 21*(2), 23–28.

Hausner, A. (1972). *Analog and Analog/Hybrid Computer Programming*, Prentice-Hall, Englewood Cliffs, NJ.

Howe, R. M. (1985). Transfer function and characteristic root error for fixed-step integration algorithms, *Trans. Soc. Computer Simulation 2*(4), 293–317.

Jackson, A. S. (1960). *Analog Computation*, McGraw-Hill, New York.

Kochenburger, R. J. (1972). *Computer Simulation of Dynamic Systems*, Prentice-Hall, Englewood Cliffs, NJ.

Martens, H. R. (1969). A comparative study of digital integration methods, *Simulation 12*(2), 87–95.

Ragazzini, J. R., Randall, R. H., and Russell, F. A. (1947). Analysis of problems in dynamics by electronic circuits, *Proc. IRE*, pp. 444–452.

Rekof, M. J., Jr. (1967). *Analog Computer Programming*, Merrill, Columbus, OH.

Rosko, J. S. (1972). *Digital Simulation of Physical Systems*, Addison-Wesley, Reading, MA.

SCi Software Committee (1967). The continuous system simulation language (CSSL), *Simulation 9*(6), 281–303.

Selfridge, R. G. (1955). Coding a general purpose digital computer to operate as a differential analyzer, in *Proceedings 1955 Western Joint Computer Conference, IRE*.

5

Principles of Design and Analysis of Simulation Experiments

5.1   INTRODUCTION

Simulation experiments are experiments performed with a mathematical model over time. As explained in Section 3.1, the steps in a simulation study include problem formulation, setting objectives, model building, data collection, coding, verification, validation, experimental design, production runs and analysis, and simulation report. Simulation experiments permit: (a) evaluation of estimates of one or more of the performance measures of the system under study, (b) evaluation of operating performance of a system prior to its implementation, (c) comparison of various operational alternatives of a real system without perturbing the system, (d) comparison of alternative system designs on the basis of some performance measures, and (e) time compression or expansion so that timely policy decisions and/or detailed system evolution can be made without ambiguity.

The statistical design of simulation experiments comprises rules and principles for designing and evaluating experiments to determine the effects of various factors set by the modeler on the response variables at the lowest experimental error possible.

These factors are the input (independent) variables, which include (a) the decision (policy) variables, (b) the structural assumptions about the system, and (c) the parameters of the random variables. Factors may be: continuous or discrete, controllable or uncontrollable, observed or unobserved, and interdependent (nested or interactions). Examples of continuous applications are flow of oil in pipelines, temperature, time, etc. Some discrete examples are detergent brands, varieties of vitamins, etc. Decision variables, such as the ordering policy in an inventory system or the number of tellers in a bank, are controllable variables, whereas the demand rate or the number of checks cashed is not controllable. Nevertheless, the simulationist may want to vary some of these uncontrollable variables so as to evaluate their effects on response variables. Other factors, such as nuclear fuel meltdown, may be uncontrollable but can be observed and taken into account in the analysis. Factors may also be nested and interrelated; for example, nested variables may be data files in a database or the output from multiparallel processors.

The statistical analysis of simulation experiments attempts to put some acceptable measure of confidence on the response variables or the estimates of performance measures of the system under study. This is due to the fact that simulation output data contain some amount of random variability, and without assessment of its magnitude, the estimates cannot be used with any degree of reliability, as discussed by Banks and Carson (1985).

5.2 DESIGN OF SIMULATION EXPERIMENTS

The design of simulation experiments involves choosing the important factors (input parameters) and their appropriate levels; selecting the length of the simulation run and the number of replications; and determining the alternatives that can be evaluated through the simulation.

5.2.1 Input Parameters

Choosing the important input parameters will strongly influence the scientific quality of the simulation experiment. Here, the concern is in the selection of probability models that appropriately describe the input to a given simulation and also the assignment of numerical values to the parameters of these probability models. When simulation models are designed to represent systems with a known environment, the characteristics of the input variables are known and expected not to vary greatly; thus the selection of input specifications may be based on theoretical considerations. On the other hand, when simulation models are designed to represent complex systems that are anticipated for the future, the characteristics of the input may not be known or may be partially known. One approach to improve this situation is to infer these characteristics from inputs of systems that are similar and available. In general, the specifications of input parameters depend on the comprehensiveness of the theory regarding the input, on the volume and quality of available data, and on the body of statistical theory available to make useful statistical inferences from the data, as discussed by Fishman (1978).

5.2.2 Estimation of Parameters

Estimation of the parameters of a distribution in generally accomplished by obtaining some types of point estimates or confidence bounds. Point estimation is concerned with inference about the unknown parameter(s) of a distribution from a sample. Here, some widely used estimation methods are discussed briefly (Sherif, 1981).

The Method of Moments. This method consists of equating the first few moments of a population with the corresponding moments of a sample, thus getting as many equations as needed to solve for the unknown parameters of the population distribution. Its advantages are: (a) the estimators obtained by this method are squared-error consistent and their asymptotic distribution is normal; (b) the method can be used when the parameters to be estimated happen to be themselves the moments of the population distribution; it can also be used even without knowledge of the exact functional form of the population; and (c) the method's simplicity and ease of application. Its disadvantages are that estimators are not uniquely defined; estimators are not generally efficient and may not be functions of sufficient or complete statistics. This method cannot be applied when the population moments do not exist, or when events are zero or sparse.

The Method of Maximum Likelihood. This method consists of selecting that value of the parameter under consideration for which the probability or the value of the joint density of obtaining the sample value is a maximum. Its advantages are: (a) estimators are consistent as well as squared-error consistent, and their asymptotic distribution is normal; (b) estimators are efficient and sufficient whenever they exist; and (c) this method also yields asymptotically minimum variance unbiased estimators which have the invariance property. Its disadvantages are: (a) estimators may not be unique, consistent, and (b) the solution of the likelihood functions may be difficult.

The Standard Bayesian Method. This method is in fact an extension of Bayes' rule to the continuous case. Its advantages are: (a) it is very useful when events are sparse; (b) estimation will improve as new data are acquired; and (c) it utilizes judgment or subjective probability which, if properly exercised, will contribute to solve problems that were thought to be impossible to solve. Its disadvantages are: (a) selection and fitting the prior distribution usually induce a source of error, and (b) subjective probability may be a mixed blessing.

The Method of Least Squares. This is a method of curve fitting or establishing relationships between random variables when their joint density is not given. Its advantages are: (a) estimators possess all the desirable asymptotic properties of the maximum likelihood estimators; (b) estimators are unbiased, efficient, and consistent; and (c) this method is easy to use and to extend to more complex models. Its disadvantage is that it cannot lend itself to extrapolation and thus no inference should be considered beyond the observed range of the independent variables.

Empirical Bayesian. Whereas the standard Bayesian method for parameter estimation requires a completely known and specified prior distribution, empirical Bayesian method acknowledges the existence of a prior distribution without requiring its explicit estimation. Another class of empirical Bayesian method considers an approximation to the prior distribution function. This method has the same disadvantages as the standard Bayesian method presented earlier.

5.2.3   Main Effects and Interactions of Factors

In simulation experiments, the modeler may specify several design factors (at several levels) that influence the understanding of the simulation output. If a complete investigation of all the factors involved (at all their respective levels) is required, then a full factorial design can be used. In this design we require $\prod_{i}^{n} l_i$ experiments where l_i is the number of levels for factor i. So studying K factors at each of two levels requires 2^K experiments, and for T levels we require T^K experiments.

In a full factorial design we solve for the main effects and the interactions of all orders, and this obviously requires a large number of experiments. To reduce this problem of size, especially when the experiment does not call for a complete investigation of all the factors' interactions, fractional factorial designs can be used.

5.2.4 Simulation Runs and Replications

A simulation run is an uninterrupted recording of the system's performance under a specified combination of controllable variables. A replication of a run is a recording of the system's performance still under a specified combination of controllable variables, but with different random variations.

For simulation experiments that are developed to study the steady-state characteristics of a system, a larger statistical sample can often be obtained efficiently by making the computer simulation run longer. The longer run will help improve confidence in the estimate of system performance. For example, if we assume that the observations within a run are "independent," then we can estimate the average performance of the system as

$$\hat{\mu} = \sum_{i=1}^{n} \left(\frac{x_i}{n} \right) \tag{5.1}$$

where x_i's are the individual observations taken at each event or time unit, and n is the number of observations in the sample. Now, the confidence in the estimate of the mean can be measured by

$$\hat{\sigma}_\mu^2 = \frac{\sigma^2}{n} \tag{5.2}$$

where σ^2 is the variance of an individual observation over the test run. So if the simulation is run for a long time, the number of observations will become very large and $\hat{\sigma}_\mu^2$ will be reduced, and consequently our confidence in the simulation performance is increased.

However, if the observations within a run are "correlated," then the estimate of the output is

$$\hat{\mu} = \sum_{i=1}^{n} \left(\frac{x_i}{n} \right) \tag{5.3}$$

where

$$\hat{\sigma}_\mu^2 = \frac{\sigma^2}{n} \left[1 + 2 \sum_{s=1}^{n-1} \left(1 - \frac{s}{n} \right) \rho_s \right] \tag{5.4}$$

$$\rho_s = \frac{E\{[x_t - \mu][x_{t+s} + s - \mu]\}}{\sigma^2} \tag{5.5}$$

and x_t, x_{t+s} are individual observations of output at simulated times t and t + s, respectively; μ is the average output of the simulation run; σ^2 is the variance; and ρ_s is the correlation coefficient between an observation at any simulation time t and an observation at time t + s.

To remove the effects of autocorrelated observations, we group the time series output data into blocks of consecutive observations such that each block represents an independent observation. Emshoff and Sisson (1970) describe the steps of the blocking method as follows:

1. Group the n-autocorrelated observations, x_1, x_2, . . ., x_n into k consecutive blocks with the following properties:
   (a) An observation is redefined to be the average of the m observations within a block. Let these new observations be

Y_1, Y_2, . . . , Y_k where

$$Y_1 = \frac{x_1 + x_2 \cdots + x_m}{m} \tag{5.6}$$

$$\vdots$$

$$Y_k = \frac{x_{m(k-1)+1} + x_{m(k-1)+2} + \cdots + x_n}{m} \tag{5.7}$$

where m = n/k.
(b) Choose the block size large enough so that the Y's are independent of each other; i.e., $\rho_s \cong 0$ for all s.
2. The mean and the variance of the blocked output can now be estimated as follows:

$$\hat{\mu} = \sum_{i=1}^{k} \frac{Y_i}{k} = \sum_{j=1}^{n} \left(\frac{x_j}{n}\right) \tag{5.8}$$

and

$$\hat{\sigma}_\mu^2 = \frac{\sigma_y^2}{k} \tag{5.9}$$

It can be seen from the above that as the simulation run length is increased, the block size can be increased and thus the confidence in the output estimates is improved.

Mechanic and McKay (1966) suggest a method for determining the proper block size or duration of a subrun that would make the new observations Ys independent of each other thus make the correlations among subrun averages negligible. The method is explained as follows:

1. Obtain a sample of N data, Y_1, Y_2, . . ., Y_N generated from a single simulation run.
2. Decompose the single run into subruns of size a, b = 4a, c = 4b, d = 4c, . . ., t, with N/t $\geq$ 25.
3. Compute the mean of each subrun, i.e.,

$$Y_{ai} = \frac{1}{a} \sum_{j=1}^{a} Y_{(i-1)a+j} \qquad i = 1, 2, \ldots \left\lceil \frac{N}{a} \right\rceil \tag{5.10}$$

where [N/a] is the integral part of N/a.
4. Compute the estimates of the variance of the overall average $\bar{Y}$ as follows:

$$S^2_{aN} = \frac{\sigma^2}{(N/a)} \tag{5.11}$$

$$\sigma^2_a = \frac{1}{([N/a] - 1)} \cdot \left[\sum_{i=1}^{[N/a]} (Y_{ai} - \bar{Y})^2 \right] \tag{5.12}$$

$$\bar{Y} = \frac{1}{N} \cdot \sum_{i=1}^{n} Y_i \tag{5.13}$$

5. Compute estimates of autocorrelation between subrun means $\hat{\rho}_{ab}$, $\hat{\rho}_{bc}, \ldots, \hat{\rho}_{rs}$ where $r = s/4 = t/16$.
6. Determine whether the first $\hat{\rho} = \rho_{m,4m}$, which is either
   (a) in the range $0.05 \leqslant \hat{\rho} \leqslant 0.5$ and is not the first $\hat{\rho}$ in the sequence;
   (b) it is less than its predecessor;
   (c) the following $\hat{\rho}$ values, if any, form a monotonically decreasing sequence with the exception of a last unbroken chain of $\hat{\rho}$ values which may oscillate, but they are all $\leqslant 0.05$, or $\hat{\rho} \leqslant 0.005$.
7. Now, we can calculate the confidence interval that will contain the true mean μ, i.e.,

$$\bar{Y} \pm Z_{\frac{\alpha}{2}} \cdot \hat{S}_{16m} \tag{5.14}$$

where $\hat{S}_{16m}$ is the estimate of the standard error from the next largest subrun of size 16m, and α is the significance level. According to Kleijnen (1975a) and Bobillier et al. (1976), the method of Mechanic and McKay seems to have several advantages since it permits calculation of the confidence interval during a single simulation run without storing all the data. The method is also attractive and simple to program, and the data do not require large storage areas.

Another way of obtaining a sample is to replicate the simulation run several times. The mean from each replication is then treated as one observation, and the set of observations (or the sample) is used to estimate the system's desired characteristics. For example, suppose we make two replications of a simulation run, A and B, where both have the same observational variance σ^2. Suppose also that each replication has an equal number of independent observations, $n/2$, and assume that the observations within a replication are independent. The estimate of the mean is therefore

$$\hat{\mu} = \sum_{i=1}^{n/2} \frac{\left(\frac{A_i + B_i}{2} \right)}{n/2} \tag{5.15}$$

and the confidence in the estimate of the mean is

$$\hat{\sigma}^2_\mu = \frac{\sigma^2}{n} \qquad\qquad (5.16)$$

i.e., the variance estimate is exactly identical to that of one replication of twice the run length.

However, if the two replications are correlated (with correlation coefficient ρ), then the variance estimate is given by

$$\hat{\sigma}^2_\mu = \frac{\sigma^2}{n}(1 + \rho) \qquad\qquad (5.17)$$

It can be seen from Eq. (5.17) that if we introduce negative correlation between pairs of observations in the two replications, the resulting variance will be less than the variance of one continuous run of n observations.

In summary, long simulation runs are commonly used to remove the effects of transients, since the data from the transient period are insignificant relative to the data in the steady state. Long runs also give large numbers of observations, and thus the value of $\hat{\sigma}^2_\mu$ will be reduced and our confidence level in the estimate of the mean is increased. However, this increase in confidence level is achieved at the cost of additional computation. One should note that replications are useful for simulation experiments that are developed specifically to study the transient rather than the steady-state characteristics of a system. Here the experiment run is replicated so as to obtain a distribution of results, and a large number of replications will ensure a high probability of including the unlikely, or extreme, events in the sample. Replication of runs can also be utilized in reducing the variance for the within-run estimate when negative correlation between replications of a run is introduced, or in reducing the variance for the difference-between-runs estimate when positive correlation is introduced between runs under different operating conditions. This is advantageous when the transient phase does not exist or if it is relatively short. If this is the case, then the variance can be reduced at no increase in computation time.

5.2.5   Response Surface Methodology

Response surface methodology offers a useful approach to the problem of designing computer simulation experiments. A response surface can be described as a relationship between some quantitative factors and a quantitative or numerical response. The basic feature of response surface methodology is the experimental investigation of a response surface using observations of the response, at various factor levels, as data. Burdick and Naylor (1966), and Myers (1971) explain that the experimental investigation of a response surface is divided into two phases, the design phase and the analysis phase. In the design phase decisions are made to set the factors at certain values and levels. This is equivalent to a selection of design points in the factor space at which the experimental runs are to be performed. The selection of design points constitutes an experimental design, and this design may be simultaneous or sequential. In the simultaneous design, results from earlier runs have no bearing on the selection of design points for later runs. However, in a purely sequential design, each design point is selected using information from previous runs. With this in mind, the objective of simultaneous designs is therefore to explore the response surface, while the objective of sequential designs is to find the location of an optimum response.

5.2.6   Factorial and Fractional Factorial Designs

The most widely used types of simultaneous designs are the factorial and fractional factorial designs. Factorial designs attempt to cover the relevant range of a factor by a series of uniformly spaced values. The total number of design points in the full factorial design is equal to the product of the numbers of levels for each factor. Fractional factorial designs, on the other hand, require only a fraction of the design points required by the full factorial design. These designs are recommended if we do not require a complete investigation of the factors in the experiment. Latin square and Greco-Latin square designs are special cases of fractional factorial designs.

   Example 5.2.1.   The accuracy of images received from a geostationary satellite is to be investigated. The factors under consideration are random displacements in the X, Y, and Z directions. It is assumed that each movement has an upper and lower level, beyond which action by the control center should be taken.
   The design for this experiment is then a 2^3 factorial design which requires eight experiments (16 if replicated). The factors and their levels are as follows:

Factor	Low level	High level
X	90	168
Y	18	38
Z	0.003	0.007

The levels of the factors are then coded and the experiment is run with replications according to the matrix in Table 5.1. The experimental results, which give an index of image accuracy are given in Table 5.2.

Table 5.1   Coded Design Matrix

Run	# of replications	X	Y	Z
1	2	−1	−1	−1
2	2	+1	−1	−1
3	2	−1	+1	−1
4	2	+1	+1	−1
5	2	−1	−1	+1
6	2	+1	−1	+1
7	2	−1	+1	+1
8	2	+1	+1	+1

Table 5.2   Calculation Matrix and Experimental Results

Run	X	Y	Z	XY	XZ	YZ	XYZ	Replicate 1	Replicate 2	Mean replicate $\overline{R}$
1	−1	−1	−1	+1	+1	+1	−1	61.6	55.7	58.65
2	+1	−1	−1	−1	−1	+1	+1	48.9	58.9	53.90
3	−1	+1	−1	−1	+1	−1	+1	70.5	77.1	73.80
4	+1	+1	−1	+1	−1	−1	−1	45.9	40.1	42.95
5	−1	−1	+1	+1	−1	−1	+1	64.4	70.9	67.65
6	+1	−1	+1	−1	+1	−1	−1	52.3	61.9	57.10
7	−1	+1	+1	−1	−1	+1	−1	84.9	78.8	81.85
8	+1	+1	+1	+1	+1	+1	+1	52.2	42.2	47.20

The main effects E(X), E(Y), E(Z) and interactions E(XY), E(XZ), E(YZ), and E(XYZ) are calculated as follows.

$$E(X) \;=\; \sum \frac{X\bar{R}}{\text{number of plus signs in column X}}$$

$$=\; \left(\frac{-58.65 + 53.90 - \cdots + 47.20}{4}\right) \;=\; -20.20$$

$$E(XY) \;=\; \sum \frac{XY\bar{R}}{\text{number of plus signs in column XY}}$$

$$=\; \left(\frac{58.65 - 53.90 - \cdots + 47.20}{4}\right) \;=\; -12.55$$

$$E(XYZ) \;=\; \sum \frac{XYZ\bar{R}}{\text{number of plus signs in column XYZ}}$$

$$=\; \left(\frac{-58.65 + 53.90 + \cdots + 47.20}{4}\right) \;=\; 0.50$$

The main effect and interactions are found to be

$$E(X) = -20.20$$
$$E(Y) = 2.13$$
$$E(Z) = 6.13$$
$$E(XY) = -12.55$$
$$E(XZ) = -2.40$$
$$E(YZ) = -1.23$$
$$E(XYZ) = 0.50$$

A 95% confidence interval (CI) on the main effects and interactions yields the results given in Table 5.3, where 95% CI on E(X) = E(X) $\pm\, t_{0.025,f} \cdot$ $\sigma(E_1)$ with

$$\sigma(E_1) \;=\; (1/2)\sigma = 1/2\, S_p$$

and

$$S_p^2 \;=\; \frac{\displaystyle\sum_{i=1}^{8} f_i S_i^{\,2}}{\displaystyle\sum_{i=1}^{8} f_i}$$

Table 5.3   95% C.I. for Main Effects
and Interactions

Effect of interaction	95% CI
E(X)	−26.50 to −13.80
E(Y)	−4.20 to 8.40
E(Z)	−0.20 to 12.50
E(XY)	−18.90 to −6.20
E(XZ)	−8.70 to 3.90
E(YZ)	−7.50 to 5.10
E(XYZ)	−5.80 to 6.80

Here, S_i is the estimate of the variance of the ith run, with

$$S_1^2 = \frac{(61.6 - 58.65)^2 + (55.7 - 58.65)^2}{(2 - 1)}$$

and f_i is the degree of freedom for the variance of the ith run (f = 2 − 1 = 1).

It can be seen from the above that only the main effect E(X) and inter-action E(XY) have a significant effect on the accuracy of the image, while the rest of the effects and interactions are not important.

5.3   ANALYSIS OF SIMULATION EXPERIMENTS

The analysis of simulation experiments refers to the analysis of the simulation-generated output data. The generated data give estimates of one or more of the system's performance measures that are of interest to the modeler. The purpose of the simulation analysis is to obtain some confidence that these estimates are sufficiently accurate, or to determine the number of observations required to achieve a desired accuracy.

5.3.1   Static Simulation Output Analysis

Static simulation involves sampling experiments that are designed to characterize estimators whose properties are not available analytically. The output in most static simulations consists of statistically independent events that follow the same probability model. For example, given n independent observations $x_1, \ldots, x_n$ generated in a particular run, one can use these data to infer the characteristics of the underlying probability model in three methods, as described by Fishman (1978). These methods are: (a) estimate one or several population parameters that are most common to all probability models, such as the mean, median, mode, variance, skewness, quantile, histogram, etc., (b) fit the data to a known probability distribution function, and then test the fit to check the adequacy of the choice, and (c) replicate the simulation experiment for the same sample size n and generate information about the distribution of the estimate, or replicate the experiment for different sample sizes and generate estimates of the 0.05, 0.1, 0.90, and 0.95 points of estimate as a function of n. The above-mentioned alternatives of inferring the true output behavior will only give us sample statistics that converge to their corresponding population parameters as n increases. To see how closely our numerical-valued estimate of a parameter approaches its unknown value, confidence intervals should be calculated. In replicating experiments, it is recommended to have large sample sizes (long run) to decrease the bias in the estimator and also to run an acceptable number of replications to provide a distributional theory for confidence intervals.

5.3.2   Dynamic Simulation Output Analysis

Whereas time evolution does not play any role in static simulation, dynamic simulation involves experiments where events evolve over time. Population parameters such as mean, median, variance, etc., are still of interest here; however, we also seek to know the relationship between events over time and their dependence or independence. In dynamic simulation, a distinction should be made between steady-state simulation and terminating or transient simulation.

5.3.3   Steady-State Simulation

The objective of steady-state simulation is to study the long-run behavior of a system, in the absence of the influence of the initial conditions of the model at time 0. For example, if we observe a process Y(t) that has been evolving over a long period of time, and we choose an arbitrarily selected point T in time for our observation, then E[Y(T)] will be the expected value of the process at time T, i.e., its steady-state mean. However, if the simulation run length is not long enough, the point estimators will be subject to initialization bias as discussed by Schruben (1982). To remove this problem several procedures are proposed, such as increasing the simulation run length, deletion of data of the first phase of simulation run (to be decided by the modeler, and given that the second phase is representative of the system's steady-state behavior), independent replications, blocks or batch means, time series regression techniques, and the regenerative method as discussed by Crane and Lemoine (1977) and Kelton and Law (1984).

5.3.4   Confidence Interval for Steady-State Simulation

Consider the estimation of a parameter $\overline{Y}$ of a simulated system, where $\overline{Y}$ may be regarded as the mean measure of the system's performance. Let the simulation output data be of the form $\{y_1, y_2, y_3, \ldots, y_n\}$. Then the point estimator of $\overline{Y}$ based on the output data is

$$\hat{\overline{Y}} = \frac{1}{n}\left(\sum_{i=1}^{n} y_i\right) \tag{5.18}$$

where $\hat{\overline{Y}}$ is a sample mean based on a sample of size n. Generally the length of the interval estimate is a measure of the accuracy of the point estimate. Therefore, the two-sided $100(1 - \alpha)\%$ CI for $\overline{Y}$ is

$$\hat{\overline{Y}} - t_{\frac{\alpha}{2},f} \cdot \hat{\sigma}(\hat{\overline{Y}}) \leqslant \overline{Y} \leqslant \hat{\overline{Y}} + t_{\frac{\alpha}{2},f} \cdot \hat{\sigma}(\hat{\overline{Y}}) \tag{5.19}$$

where $t_{\frac{\alpha}{2},f}$ is the $100(1 - \alpha)\%$ point of a t distribution with f degrees of freedom, and

$$\hat{\sigma}^2(\hat{\overline{Y}}) = \frac{S^2}{n} \tag{5.20}$$

is an unbiased estimator of the variance of $\hat{\overline{Y}}$. Also,

$$S^2 = \sum_{i=1}^{n} \left(\frac{(y_i - \hat{\overline{Y}})^2}{n - 1} \right) \tag{5.21}$$

is the sample variance, which is an unbiased estimator of the population variance Var (y_i), when the y_i's are independent and identically distributed.

If the y_i's are not statistically independent, then the observations are autocorrelated, and $\hat{\sigma}^2(\hat{\overline{Y}})$ given by Eq. (5.20) is a biased estimator of the true variance $\sigma^2(\hat{\overline{Y}})$ of the point estimator. In this case, standard statistical methods may not be employed; however, to remedy this situation, the methods of independent replications or blocks or batch means can be used as explained in Section 5.2.

Example 5.3.1. A process is observed for a long period of time, and a sample of 14 observations is: 6.1, 2.6, 3.5, 4.3, 3.1, 5.2, 3.6, 3.5, 5.4, 4.2, 3.2, 2.8, 4.0, 3.7. Construct a 90% confidence interval for the mean of the above observations.
From the data we find that

$$\hat{\overline{Y}} = 3.9429 \qquad s = 1.0188 \qquad n = 14$$

The resulting 90% CI for $\overline{Y}$ is therefore

$$3.9429 \pm t_{0.05,13} \cdot \left(\frac{1.0188}{\sqrt{14}} \right) = 3.461 \text{ to } 4.425$$

5.3.5 Terminating Simulation

The objective of the simulation here is to estimate

$$\overline{Y} = E\left[\sum_{i=1}^{n} \left(\frac{y_i}{n} \right) \right]$$

where $\overline{Y}$ is a parameter of the system, and $\{y_1, y_2, \ldots, y_n\}$ are observations from a terminating simulation that is run over a simulated time interval $(0, t_k)$. For terminating simulation, we use the method of independent replications, where each replication uses a different random-number stream and independently chosen initial conditions. Here, the sample means of all the replications $\hat{\overline{Y}}_1$, $\hat{\overline{Y}}_2$, . . ., $\hat{\overline{Y}}_N$ are statistically independent and identically distributed and are unbiased estimators of $\overline{Y}$. Therefore, the overall point estimate $\hat{\overline{Y}}$ is

$$\hat{\overline{Y}} = \frac{1}{N} \sum_{i=1}^{N} \hat{\overline{Y}}_i \qquad (5.22)$$

5.3.6   Confidence Interval for Terminating Simulation

For the terminating simulation case, classical methods of confidence interval estimation can be applied. A two-sided $100(1 - \alpha)\%$ CI for $\overline{Y}$ is given by

$$\hat{\overline{Y}} - t_{\frac{\alpha}{2},f} \cdot \hat{\sigma}(\hat{\overline{Y}}) \leqslant \overline{Y} \leqslant \hat{\overline{Y}} + t_{\frac{\alpha}{2},f} \cdot \hat{\sigma}(\hat{\overline{Y}}) \qquad (5.23)$$

where

$$\hat{\overline{Y}} = \frac{1}{N} \sum_{i=1}^{N} \hat{\overline{Y}}_i \qquad (5.24)$$

is the overall point estimate, and

$$\hat{\sigma}^2(\hat{\overline{Y}}) = \frac{1}{N(n - 1)} \sum_{i=1}^{N} (\hat{\overline{Y}}_i - \hat{\overline{Y}})^2 \qquad (5.25)$$

is the estimate of the variance of $\hat{\overline{Y}}$ where N is the number of replications, n is the number of observations per replication and $f = N - 1$ is the number of degrees of freedom. It can be seen from Eq. (5.25) that as the number of replications increases, the standard error $\hat{\sigma}(\hat{\overline{Y}})$ of the point estimate $\hat{\overline{Y}}$ tends to become smaller and approaches zero.

Example 5.3.2. For a terminating simulation, the observations for seven replications yield:

{7.9, 1.3, 2.5, 5.7, 7.8, 0.3, 0.6}
{0.1, 0.0, 7.6, 5.1, 4.0, 3.0, 6.5}
{3.0, 4.0. 1.0, 5.0, 0.1, 0.3, 6.0}
{7.5, 7.0, 2.0, 1.0, 3.0, 0.1, 1.0}
{6.0, 3.0, 4.0, 5.6, 0.1, 3.3, 6.0}
{5.0, 6.0, 7.0, 3.0, 1.0, 0.2, 0.1}
{0.1, 1.0, 7.5, 3.0, 6.0, 5.0, 3.0}

Construct a 95% CI for the mean of the system.

From the data:

$$\hat{\overline{Y}}_1 = 3.72 \qquad \hat{\overline{Y}}_2 = 3.76 \qquad \hat{\overline{Y}}_3 = 2.77 \qquad \hat{\overline{Y}}_4 = 3.08$$

$$\hat{\overline{Y}}_5 = 4.0 \qquad \hat{\overline{Y}}_6 = 3.18 \qquad \hat{\overline{Y}}_7 = 3.65$$

Using Eq. (5.24), the overall point estimate $\hat{\overline{\overline{Y}}} = 3.40$, and from Eq. (5.25) the estimate of

$$\hat{\sigma}^2(\hat{\overline{\overline{Y}}}) = \frac{1}{7(7-1)} [(3.72 - 3.40)^2 + \cdots + (3.65 - 3.40)^2]$$

$$= 0.0286$$

and

$$\hat{\sigma}(\hat{\overline{\overline{Y}}}) = 0.1691$$

The resulting 95% CI for Y is therefore

$$3.40 \pm t_{0.025,6} \cdot (0.1691) = 2.98 \text{ to } 3.81$$

5.3.7 Comparison of Alternative Models

The objective of the simulation experiment here is to obtain point as well as interval estimates of the difference in mean performance of alternative models. For example, suppose that a simulationist wants to compare two possible system designs, such as two queue disciplines or two possible inventory ordering policies. It is required then to obtain point estimates and interval estimates of $(Y_1 - Y_2)$ where Y_i (i = 1, 2) is the mean performance measure of system i.

5.3.8 Output Analysis for Two Models

Consider that the method of replication is used to analyze the output data; therefore a two-sided $100(1 - \alpha)$% CI for $(Y_1 - Y_2)$ will be of the form

$$(\overline{Y}_{.1} - \overline{Y}_{.2}) \pm (t_{\frac{\alpha}{2},f} \cdot \hat{\sigma}(\overline{Y}_{.1} - \overline{Y}_{.2}) \tag{5.26}$$

where

$$\overline{Y}_{.i} = \frac{1}{K_i} \sum_{r=1}^{K_i} Y_{ri} \tag{5.27}$$

is the sample mean performance measure for system i over all replications K, and f is the degrees of freedom associated with the variance estimator. Y_{ri} is a sample mean of observations from replication r. It is assumed that the data Y_{ri} are approximately normally distributed.

Now, if independent sampling is used to simulate the two designs and assuming that they are homogeneous systems, then the standard error of the point estimate is

$$\hat{\sigma}(\overline{Y}_{\cdot 1} - \overline{Y}_{\cdot 2}) = S_p \sqrt{\frac{1}{K_1} + \frac{1}{K_2}} \tag{5.28}$$

where

$$S_p^2 = \frac{S_1^2(K_1 - 1) + S_2^2(K_2 - 1)}{(K_1 + K_2 - 2)} \tag{5.29}$$

is a pooled estimate of the variance σ^2, given $\sigma_1^2 \cong \sigma_2^2 = \sigma^2$, and S_i^2 is an unbiased estimator of the variance σ_i^2.

$$S_i^2 = \frac{1}{(K_i - 1)} \sum_{r=1}^{K_i} (Y_{ri} - \overline{Y}_{\cdot 1})^2 \tag{5.30}$$

Therefore Eq. (5.26) becomes

$$(\overline{Y}_{\cdot 1} - \overline{Y}_{\cdot 2}) \pm t_{\frac{\alpha}{2},f} \cdot \left[\frac{S_1^2(K_1 - 1) + S_2^2(K_2 - 1)}{(K_1 + K_2 - 2)} \cdot \left(\frac{1}{K_1} + \frac{1}{K_2} \right) \right]^{1/2} \tag{5.31}$$

where $f = K_1 + K_2 - 2$ degrees of freedom.

If the homogeneity assumption of the two designs cannot safely be made, i.e., if $\sigma_1^2 \neq \sigma_2^2$, then a two-sided $100(1 - \alpha)\%$ CI for $(Y_1 - Y_2)$ will be of the form

$$(\overline{Y}_{\cdot 1} - \overline{Y}_{\cdot 2}) \pm t_{\frac{\alpha}{2},f} \cdot \left[\frac{S_1^2}{K_1} + \frac{S_2^2}{K_2} \right]^{1/2} \tag{5.32}$$

where s_i^2 is as given in Eq. (5.30), and degrees of freedom

$$f = \left(\frac{S_1^2}{K_1} + \frac{S_2^2}{K_2} \right)^2 \Bigg/ \left[\frac{(S_1^2/K_1)^2}{K_1 - 1} + \frac{(S_2^2/K_2)^2}{K_2 - 1} \right] \tag{5.33}$$

If, for each replication, the same random numbers are used to simulate both systems, i.e., we introduce positive correlation between estimates Y_{r1} and Y_{r2} (for each replication r), then a two-sided $100(1 - \alpha)\%$ CI for $(Y_1 - Y_2)$ will be of the form

$$(\overline{Y}_{\cdot 1} - \overline{Y}_{\cdot 2}) \pm t_{\frac{\alpha}{2},f} \cdot \left[\frac{1}{K(K - 1)} \sum (d_r - \overline{d})^2 \right]^{1/2} \tag{5.34}$$

where

$$d_r = Y_{r1} - Y_{r2}$$

$$\bar{d} = \frac{1}{K} \sum_{r=1}^{K} d_r$$

$K_1 = K_2 = K$ = number of replications

$f = K - 1$ = degrees of freedom

This case is also known as "correlated sampling," or common random numbers, as explained further in Section 5.4.

Example 5.3.3. In the simulation for comparing two alternative models, three replications were obtained for each model. The observations from the first model are

{7.9, 1.3, 2.5, 5.7, 7.8, 0.3, 0.6}
{0.1, 0.0, 7.6, 5.1, 4.0, 3.0, 6.5}
{3.0, 4.0, 1.0, 5.0, 0.1, 0.3, 6.0}

and the observations from the second model are

{7.5, 7.0, 2.0, 1.0, 3.0, 0.1, 1.0}
{6.0, 3.0, 4.0, 5.6, 0.1, 3.3, 6.0}
{5.0, 6.0, 7.0, 3.0, 1.0, 0.2, 0.1}

Construct a 95% CI on the difference between the mean performance measures of the two models, given that the models are homogeneous and the data are normally distributed.

From the data of the three replications per model, we obtain

For model 1: $Y_{11} = 3.72$ $Y_{21} = 3.76$ $Y_{31} = 2.77$

For model 2: $Y_{12} = 3.08$ $Y_{22} = 4.0$ $Y_{32} = 3.18$

Using Eq. (5.27), the corresponding sample mean performance measures are

$$\bar{Y}_{.1} = 3.42 \qquad \bar{Y}_{.2} = 3.42$$

$$S_1 = 0.56 \qquad S_2 = 0.50$$

$$S_p^2 = 0.28 \qquad S_p = 0.53$$

The resulting 95% CI on $(Y_1 - Y_2)$ is

$$(3.42 - 3.42) \pm t_{0.025, 4} \cdot (0.28)^{1/2} \cdot \left(\frac{1}{3} + \frac{1}{3}\right)^{1/2}$$

or

   -1.20 to 1.20

i.e., the two models are not significantly different (at $\alpha = 0.05$).

5.3.9   Analysis of Variance

The analysis of variance is a collection of techniques for data analysis that are appropriate when qualitative factors are present, although quantitative factors are not excluded, as discussed by Naylor (1969).   Qualitative factors typically represent structural assumptions in a model such as the queue discipline, and quantitative factors are those that assume numerical values, such as the number of servers in a service station.

   Many techniques for data analysis exist today; however, some of these techniques may not be of interest to us in analyzing simulation experiments. This is due to the fact that these techniques were formulated for some agricultural or industrial experiments where the experimenter does not have complete control over the experimental factors.   For the researcher performing simulation experiments, several special cases of the analysis of variance will be considered and presented next.   These techniques include the F-test, multiple comparisons, multiple rankings, spectral analysis, and sequential sampling.

5.3.10   The F-Test

The F-test rests on three important assumptions: (a) normality of observation, (b) homogeneity of variances, and (c) statistical independence. For data generated by simulation experiments, the F-test is a straightforward procedure for testing the null hypothesis that the expected values of the response variable for each of the observations (combinations of factors) are equal.   If the null hypothesis is accepted, then we can conclude that the sample differences between treatments are attributable to random fluctuations rather than to actual differences in population values, as discussed by Naylor (1969).

5.3.11   Multiple Comparisons

Multiple-comparison methods emphasize the use of confidence intervals rather than the testing of hypothesis.   These methods are designed to show how means of many treatments differ by constructing CIs for the differences between population means of interest.   This procedure also assumes normality of observations, equality of variances, and statistical independence, as explained by Kleijnen (1977).

5.3.12   Multiple Rankings

Multiple-ranking methods attempt to rank treatments' effects on the response variable of a simulation experiment.   Since ranking a set of treatments is simply the ranking of their respective sample means, and since it is desirable to isolate the differences of treatments, the methods of multiple ranking are useful because they put a measure of confidence on the ranking of sample means as representative of the ranking of the population means. These methods also assume normality, common variances, and statistical independence.

5.3.13 Spectral Analysis

Spectral analysis is a statistical technique used to analyze the behavior of time series. It is a nonparametric approach that attempts to decompose a time series into basic components that can be represented as sine and cosine functions. Once these periodic effects are exposed, the process becomes easier to interpret and analyze. Emshoff and Sisson (1970), and Duket and Pritsker (1978) explain that spectral analysis provides a way to convert an apparent non-steady-state time series to one in which steady-state conditions can be assumed. By removing periodic effects, observations can be defined as time invariant, and thus autocorrelation is removed. Naylor, Wertz, and Wonnacott (1969) conclude that spectral analysis is a good technique for analyzing data generated by simulation, and they cite the following reasons.

1. Since output data generated by simulation experiments are usually highly correlated; the use of classical statistical estimating techniques will lead to underestimates of sampling variance and inefficient predictions. Spectral analysis offers a sampling theory that avoids this problem.
2. Spectral analysis offers more information in the analysis of data than other techniques. So, in addition to information about the mean and variance, spectral analysis provides information about the length of time that deviations from the average response will last once they occur.
3. Spectral analysis provides a means of comparing time series generated with a computer model.
4. Spectral analysis can be used for validation of simulation models.

5.3.14 Sequential Sampling

The objective of sequential sampling is to minimize the number of observations for obtaining the information that is required from the experiment. For simulation experiments, and as explained by Naylor (1969), the sequential sampling method sets a procedure for deciding at the Kth observation whether to accept a given hypothesis, reject it, or continue sampling by taking the (K + 1)th observation. The modeler must specify for the Kth observation a division of the K-dimensional space of all possible observations into three mutually exclusive and exhaustive sets. These sets are: (a) an area of preference for accepting the hypothesis, (b) an area of preference for rejecting the hypothesis, and (c) an area of indifference where further observations are necessary. Kiefer and Sacks (1963) discuss optimization procedures for sequential sampling that can be used in analyzing output data generated by simulation experiments.

5.4 VARIANCE REDUCTION TECHNIQUES

Simulation experiments may generally be regarded as statistical experiments driven by random inputs. Therefore, the results of any simulation model represents estimates (random outputs) characterized by experimental errors. To obtain improved estimates and to increase the statistical efficiency of the simulation (as measured by the variances of the output random variables), various types of variance-reduction techniques are utilized. These techniques include common random numbers, antithetic variates, control variates, stratified sampling, and importance sampling.

The basic tenet of common random numbers, antithetic variates, and control variates is to introduce correlation among observations so as to reduce variance. Stratified sampling and importance sampling utilize weighted sampling based on a priori qualitative or quantitative information for variance reduction.

5.4.1   Common Random Numbers

This method, known also as correlated sampling, tries to minimize or, in effect, block the differences between alternative system designs that are due to experimental conditions (such as generated random variables). In so doing, any observed performance differences between alternatives will be attributed to differences in system design, rather than to fluctuations of experimental conditions, and thus a more precise experimental design is achieved.

For example, consider the case of two alternative systems, where X_{1j} and X_{2j} are observations from the first and second system, respectively on the jth independent replication. Let

$$\mu_i = E(X_{ij}) \tag{5.35}$$

be the mean response of interest, and suppose we want to estimate

$$T = \mu_1 - \mu_2 \tag{5.36}$$

If we take n runs of each system and let

$$Y_j = X_{1j} - X_{2j} \tag{5.37}$$

for j = 1, 2, . . ., n, then

$$E(Y_j) = T \tag{5.38}$$

and

$$\overline{Y}(n) = \sum_{j=1}^{n} Y_j/n \tag{5.39}$$

also

$$Var[\overline{Y}(n)] = 1/n \ Var \ (Y_j)$$

$$= 1/n \ [Var(X_{1j}) + Var(X_{2j}) - 2 \ Cov \ (X_{1j}, \ X_{2j})] \tag{5.40}$$

It can be seen from Eq. (5.40) that if the simulations of the alternative systems are performed with different random number streams, then X_{1j} and X_{2j} are independent and

$$\text{Cov}(X_{1j}, X_{2j}) = 0 \tag{5.41}$$

On the other hand, if X_{1j} and X_{2j} are positively correlated, then

$$\text{Cov}(X_{1j}, X_{2j}) > 0 \tag{5.42}$$

and thus the variance of the estimator $Y(n)$ is reduced.

To facilitate the utilization of the common-random-numbers technique for variance reduction, we must use the same random number stream for all the alternative models (designs) under study. This matching or synchronization of the same random numbers may be achieved by many methods, the easiest of which would be the generation of the required random variables ahead of time and storing them for use in each of the alternative systems simulations.

Drawbacks for using common random numbers include backfiring, or causing the covariance to become negative, i.e.,

$$\text{Cov}(X_{1j}, X_{2j}) < 0 \tag{5.43}$$

and also the difficulties of comparing multisystem designs with induced correlation, as discussed by Wright and Ramsay (1979), Heikes, Montgomery, and Rardin (1976), Kleijnen (1975b), and Law and Kelton (1982).

Example 5.4.1. A simulation study for two alternative system designs A and B of a service facility is performed. The first study used the same random variables for both designs. The estimated average waiting time in seconds per hour in a 10-hr-shift day is depicted as T_A and T_B in Table 5.4. The second study used different random variables for each design, as discussed by Payne (1982). The estimated average waiting time is depicted in Table 5.5. It can be seen that the estimated variance of $(T_A - T_B)$ under conditions of random numbers is 18.68, compared to 66.93 under conditions of different random numbers. In effect, the estimated variance is reduced by about 70% when common random variates are utilized.

5.4.2 Antithetic Variates

This method, developed by Hammersley and Morton (1956), tries to induce negative correlation between separate runs of a single system by using complementary random numbers to drive the two runs in a pair. In so doing, a small observation on the first run will be offset by a large observation on the second run, or vice versa. This means that the two runs will produce results on opposite sides of the population mean, and when taken together, their average will tend to be closer to the expectation of an observation than would be likely otherwise.

Let $X_{1i} = (X_{11}, X_{12}, \ldots, X_{1n})$, $X_{2i} = (X_{21}, X_{22}, \ldots, X_{2n})$ represent the outcomes of two runs. Therefore, an estimate of the mean μ is given by

$$\overline{X} = \frac{1}{n} \sum_{i=1}^{n} \frac{X_{1i} + X_{2i}}{2} \tag{5.44}$$

Table 5.4   Simulation with Common Random Numbers

	Alternative systems		Difference
Hour	T_A	T_B	$(T_A - T_B)$
1	84	52	32
2	98	63	35
3	98	65	33
4	88	56	32
5	77	47	30
6	73	47	26
7	86	49	37
8	86	51	35
9	93	66	27
10	87	47	40
$\overline{X}$	87.00	54.30	32.70
S^2	65.11	59.34	18.68

Table 5.5   Simulation with Different Random Numbers

	Alternative systems		Difference
Hour	T_A	T_B	$(T_A - T_B)$
1	84	50	34
2	98	58	40
3	98	62	36
4	88	50	38
5	77	56	21
6	73	60	13
7	86	57	29
8	86	53	33
9	93	61	32
10	87	59	28
$\overline{X}$	87.00	56.60	30.40
S^2	65.11	18.71	66.93

and

$$\text{Var}(\overline{X}) = \left(\frac{1}{4n}\right) [\text{Var}(X_{1i}) + \text{Var}(X_{2i}) + 2 \text{ Cov}(X_{1i}, X_{2i})] \qquad (5.45)$$

Now, if the two runs are independent, then

$$\text{Cov}(X_{1i}, X_{2i}) = 0 \qquad (5.46)$$

However, if we can induce negative correlation between X_{1i} and X_{2i}, then the covariance will be negative, i.e.,

$$\text{Cov}(X_{1i}, X_{2i}) < 0 \qquad (5.47)$$

This will reduce $\text{Var}(\overline{X})$ expressed in Eq. (5.45). Law and Kelton (1982) argue that there is no guarantee for the success of the method of antithetic variables. Furthermore, it is not known beforehand how great a reduction in variance might be achieved. Franta (1975) gives examples of the failure of this method. However, Fishman (1972a) and Arvidsen and Johnson (1982) give strategies that are successful in inducing negative correlation and thus favor this method.

Example 5.4.2. This example will demonstrate the characteristics of the antithetic variate technique in reducing the sampling error. Suppose we want to estimate the mean value of the exponential distribution

$$f(x) = 0.1 \exp(-0.1x) \qquad x > 0 \qquad (5.48)$$

whose exact mean value is $1/\lambda = 10$.

We sample the exponential distribution by using a sequence of random numbers $[R_i]$ and its complementary sequence $[1 - R_i]$. If $[t_{1i}]$ and $[t_{2i}]$ are exponential random variables given $[R_i]$ and $[1 - R_i]$, respectively, then

$$t_{1i} = -10 \ln R_i \qquad (5.49)$$

$$t_{2i} = -10 \ln (1 - R_i) \qquad (5.50)$$

and the antithetic technique is applied by computing

$$\overline{Z}_i = \frac{1}{2}(t_{1i} + t_{2i}) \qquad (5.51)$$

Table 5.6 shows the necessary computations for n = 10 random numbers with the antithetic computations based on 5 and 10 pairs of $\overline{Z}$. Table 5.7 shows the computations for n = 50 random numbers with the antithetic computations based on 50 pairs of $\overline{Z}$. It can be seen that the antithetic technique yields the smallest variance. Note that in actual simulation t_{1i} and t_{2i} may represent a measure of performance of the system under study when two replications are used to gather the above-mentioned observations. Here, the two observations should be based on equal time intervals such that negative correlation between them can be maintained.

Table 5.6   Antithetic Variates Technique

i	R_i [a]	t_{1i}	$1 - R_i$	t_{2i}	$\overline{Z}_i$	
1	0.4617	7.7283	0.5383	6.1933	6.9608	
2	0.1192	21.2695	0.8808	1.2692	11.2693	$\overline{Z} = 10.45$
3	0.1289	20.4871	0.8711	1.3799	10.9355	$S^2_{\overline{Z}} = 4.24$
4	0.9079	0.9662	0.0921	23.8488	12.4075	
5	0.1376	19.8340	0.8624	1.4803	10.6571	
6	0.2802	12.7225	0.7198	3.2878	8.0051	
7	0.6212	4.7610	0.3788	9.7074	7.2342	
8	0.2701	13.0896	0.7299	3.1484	8.1190	
9	0.6144	4.8710	0.3856	9.5295	7.2002	
10	0.1918	16.5130	0.8082	2.1294	9.3212	
$\hat{\mu}$		12.2242		6.1974	9.2107	
S^2		53.7908		48.7835	3.9095	

[a]Random numbers R_i are given in Table 5.10, column 10.

Table 5.7   Antithetic Variates Technique

i	R_i	t_{1i}	$1 - R_i$	t_{2i}	$\overline{Z}_i$
1	0.4617	7.7283	0.5383	6.1933	6.9608
2	0.1192	21.2695	0.8808	1.2692	11.2693
3	0.1289	20.4871	0.8711	1.3799	10.9335
4	0.9079	0.9662	0.0921	23.8488	12.4075
5	0.1376	19.8340	0.8624	1.4803	10.6571
⋮	⋮	⋮	⋮	⋮	⋮
46	0.0833	24.8530	0.9167	0.8697	12.8613
47	0.6154	4.8548	0.3846	9.5555	7.2051
48	0.7788	2.5000	0.2212	15.0868	8.7934
49	0.8844	1.2284	0.1156	21.5761	11.4022
50	0.7252	3.2130	0.2748	12.9171	8.0650
$\hat{\mu}$		10.8799		7.4736	9.1722
S^2		75.1591		39.0627	6.6094

Example 5.4.3. Suppose we want to sample a Poisson distribution with mean λt from the exponential distribution with mean $1/\lambda$. It is known that in the Poisson distribution, the outcome is expressed as the number of events n occurring in a specified time period. So, all that needs to be done is to sample the exponential distribution as many times as necessary until the sum of the generated exponential random variables exceeds the specified period Z for the first time. Here the sampled Poisson value n will be taken equal to the number of times we sampled the exponential distribution less one. Let λ = 3 events/hr during a period of 1 hr. Table 5.8 shows the necessary computations for n by using a sequence of random numbers $[R_i]$ and its complementary sequence $[1 - R_i]$, where t_n is the interarrival time.

Table 5.8   Sampling the Poisson Distribution

	n:	1	2	3	4	5	6	7	8
1.	R_n	0.2444	0.9039	0.4893	0.8568	0.1590			
	t_n	0.4696	0.0336	0.2444	0.0515	0.6129			
	Σt_n	0.4696	0.5032	0.7476	0.7991	1.4120			
	$1 - R_n$	0.7556	0.0961	0.5197					
	t_n	0.0934	0.7807	0.2181					
	Σt_n	0.0934	0.8741	1.0922					
2.	R_n	0.5748	0.7767	0.2800	0.6289	0.2814			
	t_n	0.1845	0.0842	0.4243	0.1545	0.4226			
	Σt_n	0.1845	0.2687	0.6930	0.8475	1.2701			
	$1 - R_n$	0.4252	0.2233	0.7200	0.3711				
	t_n	0.2850	0.4997	0.1095	0.3304				
	Σt_n	0.2850	0.7847	0.8942	1.2246				
	$\vdots$	$\vdots$	$\vdots$	$\vdots$	$\vdots$	$\vdots$			
20.	R_n	0.8587	0.0764	0.6687					
	t_n	0.0507	0.8572	0.1341					
	Σt_n	0.0507	0.9079	1.0420					
	$1 - R_n$	0.1413	0.9236	0.3313					
	t_n	0.6522	0.0264	0.3682					
	Σt_n	0.6522	0.6786	1.0486					

Table 5.9   Summary of Results

	n_{R_i}	n_{1-R_i}	$Z = 0.5(n_{R_i} + n_{1-R_i})$	
1	4	2	3.0	
2	4	3	3.5	$\bar{Z} = 2.7$
3	3	1	2.0	
4	3	1	2.0	$S^2 = 0.45$
5	3	3	3.0	
6	1	4	2.5	
7	4	0	2.0	$\bar{Z} = 3.0$
8	5	1	3.0	
9	2	4	3.0	$S^2 = 0.875$
10	2	7	4.5	
11	3	3	3.0	
12	2	7	4.5	$\bar{Z} = 3.2$
13	4	1	2.5	
14	4	2	3.0	$S^2 = 0.575$
15	5	1	3.0	
16	4	2	3.0	
17	4	3	3.5	$\bar{Z} = 3.0$
18	3	2	2.5	
19	6	2	4.0	$S^2 = 0.625$
20	2	2	2.0	
	$\bar{n} = 3.40$	$\bar{n} = 2.55$	$\bar{Z} = 2.975$	
	$S^2 = 1.51$	$S^2 = 3.41$	$S^2 = 0.565$	

Table 5.9 gives a comparison of the computations for n_{R_i}, n_{1-R_i} and $Z = (1/2)[n_{R_i} + n_{1-R_i}]$. It can be seen that the antithetic technique yields the smallest variance ($S^2 = 0.565$) as compared to ($S^2 = 1.51$) for n_{R_i} and ($S^2 = 3.41$) for n_{1-R_i}.

5.4.3   Control Variates

This method is used to compare the (complex) model to be simulated with a similar, but simplified model, or a pilot model, whose analytic solution is known. The sampling error of simulation is then estimated by simulating the simplified model (using identical sampling numbers) and comparing its results with the known theoretical results. Now the output of the simplified model simulation provides control variates or a statistical control on the (complex) model simulation as discussed by Cheng and Feast (1980). To illustrate, let X be an output random variable, and suppose we want to estimate $\mu = E(X)$. Let Y be another random variable provided by a similar, but simplified model (correlated positively with X since common random numbers are used), which has known expectation since the model's analytic solution is known.

Now, the random variable

$$Z = X - \alpha[Y - E(Y)] \tag{5.52}$$

is also an unbiased estimator of

$$\mu = E(X) \tag{5.53}$$

For any real number α,

$$Var(Z) = Var(X) + \alpha^2 Var(Y) - 2\alpha Cov(X,Y) \tag{5.54}$$

So as long as X and Y are positively correlated and the value of $2\alpha Cov(X,Y) > \alpha^2 Cov(Y)$, Z will have a reduced variance, as described by Bratley, Fox, and Schrage (1983), and Law and Kelton (1982).

5.4.4 Stratified Sampling

Our interest here is in estimating the mean value of a specified distribution whose cumulative distribution function is as follows:

X	1	2	3	4	5
F(X)	0.1	0.3	0.7	0.9	1.0

Assume that we use a uniform random number generator to generate 50 random variates x_1, x_2, x_3, ..., x_{50} from the above-specified distribution. Now, the expected value of the sample x_i is

$$\text{Mean} = E(X) = \int_0^5 xf(x)dx = 2.5 \tag{5.55}$$

and the variance is

$$Var(X) = E[(X - \mu)^2]$$

$$= \int_0^5 (X - \mu)^2 \cdot f(x) \, dx = 1.28 \tag{5.56}$$

The variance of the mean value of the 50 observations is therefore

$$Var(\overline{X}) = \frac{1.28}{50} = 0.0256 \tag{5.57}$$

The method of stratified sampling approaches the above problem in a different way. For example, first divide the range of the variable x into subintervals or strata, say 5 (0 to 1, 1 to 2, . . ., 4 to 5). Then generate 50 random variates, 10 values for X that lie in each of the five strata. Now estimates of the mean value of the original distribution and its variance are

$$\text{Mean (x stratified)} = \overline{X}_s = \sum_{j=1}^{5} P_j \overline{X}_j \tag{5.58}$$

$$= 0.1\overline{X}_1 + 0.2\overline{X}_2 + 0.4\overline{X}_3 + 0.2\overline{X}_4 + 0.1\overline{X}_5$$

$$\text{Var}(\overline{X}_s) = \sum_{j=1}^{5} P_j^2 \text{ Var}(\overline{X}_j) = 0.0021 \tag{5.59}$$

where $\overline{X}_j$ is the average value for the 10 samples x_1, x_2, . . ., x_{10} in strata j. Thus, the variance of the mean value evaluated by stratified sampling, as given in Eq. (5.59), is about 8% of the variance of the mean obtained, without stratification, using Eq. (5.57). In other words, stratified sampling, in this case, reduced the variance by about 90%. This reduction in the variance is due to the fact that stratification actually replaces a random variable (the number of samples taken in a stratum) by a constant (its expectation), as discussed by Payne (1982). In a simulation program, stratified sampling is used in the same manner, as explained above. For example, suppose we are to take 200 samples of the uniform distribution in the subinterval 0 to 1, then we can stratify the range into 10 strata, each of duration 0.1, and allow only 20 samples per stratum.

5.4.5 Importance Sampling

This method is somewhat analogous to stratified sampling except that the proportions of samples taken per stratum are not adhered to; as a matter of fact, samples are taken at the important parts of the range more frequently than otherwise. To illustrate this technique, we describe an application considered by Carter and Ignall (1975), to an inventory system.

Consider a period model of an inventory system with

D_t = demand during period t

I_t = inventory level at the end of period t

Φ_t = quantity ordered at beginning of period t

Demand is generated from a known distribution, and the order quantity is determined according to the following rule

$$\Phi_t = \begin{cases} 0 & I_{t-1} > s \\ S - I_{t-1} & I_{t-1} \leq s \end{cases} \quad s < S \tag{5.60}$$

Orders are assumed to be delivered instantaneously so that

$$I_t = I_{t-1} + \Phi_t - D_t \tag{5.61}$$

Let h = unit holding cost, b = unit backorder cost, K = setup cost for ordering. The generating cost in period t is therefore

$$c_t = h \, I_t^+ + b(-I_t)^+ + K\delta_t \tag{5.62}$$

where

$$x^+ = \max(x, 0)$$

and

$$\delta_t = \begin{cases} 1 & \Phi_t > 0 \\ 0 & \Phi_t \leqslant 0 \end{cases} \tag{5.63}$$

For an n period run, the sample mean operating cost period is

$$\bar{c}_n = \frac{1}{n} \sum_{t=1}^{n} c_t \tag{5.64}$$

From Eq. (5.62) and for a given I_{t-1}, C_t is seen to be a function of D_t only. Now since demands are assumed to be independent, it follows that $[c_t | I_{t-1}]$ is a sequence of independent observations. Let $I_{t-1} = j$, generate K_j independent demands $D_{1j}, D_{2j}, \ldots, D_{kj}$; then the cost of replication i is

$$c_{ij} = \begin{cases} h(S - D_{ij})^+ + b(D_{ij} - S)^+ + K & j \leqslant s \\ h(j - D_{ij})^+ + b(D_{ij} - j)^+ & j > s \end{cases} \tag{5.65}$$

If c_{ij} has variance σ_j^2, then

$$\bar{c}_{Kj} = \frac{1}{K_j} \sum_{i=1}^{K_j} C_{ij}$$

is an estimate of the conditional cost, given $I_{t-1} = j$, and has variance σ_j^2 / K_j. Then

$$C_n = \sum_{J=-\infty}^{\infty} \bar{c}_{Kj} \, \text{Prob} \, (I_{t-1} = j) \tag{5.66}$$

Table 5.10   Random Digits

2444	9039	4803	8568	1590	2420	2547	2470	8179	4617
5748	7767	2800	6289	2814	8281	1549	9519	3341	1192
7761	8583	0852	5619	6864	8506	9643	7763	9611	1289
6838	9280	2654	0812	3988	2146	5095	0150	8043	9079
6440	2631	3033	9167	4998	7036	0133	7428	9702	1376
8829	0094	2887	3802	5497	0318	5168	6377	9216	2802
9845	4796	2951	4449	1999	2691	5328	7674	7004	6212
5072	9000	3887	5739	7920	6074	4715	3681	2721	2701
9035	0553	1272	2600	3828	8197	8852	9092	8027	6144
5562	1080	2222	0336	1411	0303	7424	3713	9278	1918
2757	2650	8727	3953	9579	2442	8041	9869	2887	3933
6397	1848	1476	0787	4990	4666	1208	2769	3922	1158
9208	7641	3575	4279	1282	1840	5999	1806	7809	5885
2418	9289	6120	8141	3908	5577	3590	2317	8975	4593
7300	9006	5659	8258	3662	0332	5369	3640	0563	7939
6870	2535	8916	3245	2256	4350	6064	2438	2002	1272
2914	7309	4045	7513	3195	4166	0878	5184	6680	2655
0868	8657	8118	6340	9452	7460	3291	5778	1167	0312
7994	6579	6461	2292	9554	8309	5036	0974	9517	8293
8587	0764	6687	9150	1642	2050	4934	0027	1376	5040
8016	8345	2257	5084	8004	7949	3205	3972	7640	3478
5581	5775	7517	9076	4699	8313	8401	7147	9416	7184
2015	3364	6688	2631	2152	2220	1637	8333	4838	5699
7327	8987	5741	0102	1173	7350	7080	7420	1847	0741
3589	1991	1764	8355	9684	9423	7101	1063	4151	4875
2188	6454	7319	1215	0473	6589	2355	9579	7004	6209
2924	0472	9878	7966	2491	5662	5635	2789	2564	1249
1961	1669	2219	1113	9175	0260	4046	8142	4432	2664
2393	9637	0410	7536	0972	5153	0708	1935	1143	1704
7585	4424	2648	6728	2233	3518	7267	1732	1926	3833
0197	4021	9207	7327	9212	7017	8060	6216	1942	6817
9719	5336	5532	8537	2980	8252	4971	0110	6209	1556
8866	4785	6007	8006	9043	4109	5570	9249	9905	2152
5744	3957	8786	9023	1472	7275	1014	1104	0832	7680
7149	5721	1389	6581	7196	7072	6360	3084	7009	0239
7710	8479	9345	7773	9086	1202	8845	3163	7937	6163
5246	5651	0432	8644	6341	9661	2361	8377	8673	6098
3576	0013	7381	0124	8559	9813	9080	6984	0926	2169
3026	1464	2671	4691	0353	5289	8754	2442	7799	8983
6591	4365	8717	2365	5686	8377	8675	9798	7745	6360
0402	3257	0480	5038	1998	2935	1306	1190	2406	2596
7105	7654	4745	4482	8471	1424	2031	7803	4367	6816
7181	4140	1046	0885	1264	7755	1653	8924	5822	4401
3655	3282	2178	8134	3291	7262	8229	2866	7065	4806
5121	6717	3117	1901	5184	6467	8954	3884	0279	8635
3618	3098	9208	7429	1578	1917	7927	2696	3704	0833
0166	3638	4947	1414	4799	9189	2459	5056	5982	6154
6187	9653	3658	4730	1652	8096	8288	9368	5531	7788
1234	1448	0276	7290	1667	2823	3755	5642	4854	8844
8949	8731	4875	5724	2962	1182	2930	7539	4526	7252

Source: Owen (1962).

is an unbiased estimate of the mean cost and has variance

$$\text{Var}(C_n) = \sum_{J=-\infty}^{\infty} \frac{\sigma_j^2}{K_j} [\text{Prob } (I_{t-1} = j)]^2 \qquad (5.67)$$

Since our principal concern is to obtain accurate cost data for $I_{t-1} < s$, we should take more replications for $j < s$, thereby reducing the variances of the c_{Kj} (for $j < s$) to a greater relative extent. This approach is called importance sampling because it concentrates the sampling effort on the j's that are most likely to contribute to backorder cost, as discussed by Fishman (1978).

5.5   SUMMARY AND CONCLUSIONS

In any statistical experiment careful design and analysis are desirable. The design aspects should ensure that the experiment contains as much relevant information as possible, whereas analysis helps in extracting information of interest to the modeler and should reveal the limitations of conclusions that are based on the sampling data.

Before one attempts to design a simulation experiment, the following steps should be considered: (a) identification of the objective of the experiment and whether it is intended to show the general relationship between the response variable and the factors or to show that combination-of-factors level at which the response variable is optimized, (b) identification of the relevant factors to be included in the experiment so that the size of experiments can be manageable, (c) identification of techniques to be used for variance reduction since increasing sample size or number of runs is costly.

For the analysis of simulation experiments, it is appropriate to distinguish between terminating simulation analysis and steady-state simulation analysis. Terminating simulation analysis should be performed when interest lies in studying the transient behavior of a system, or when the system under consideration is physically terminating, or when the input models for the system change over time. However, steady-state simulation analysis is performed when the system under study satisfies steady-state conditions and when interest is in the study of long-run behavior of the system.

PROBLEMS

5.1.   A sample of 25 observations from a steady-state simulation experiment yields the following:

> {22.8, 29.3, 27.2, 30.2, 24.0, 23.2, 22.9, 30.3, 27.1, 31.2,
> 27.0, 32.0, 28.6, 24.1, 28.9. 26.8, 26.6, 23.4, 25.1, 26.6,
> 25.7, 28.1, 31.5, 24.8, 25.2}

Construct a 90% CI for the mean of the data.

5.2.   A sample of 18 observations from a steady-state simulation experiment yields the following:

{9.4, 8.6, 11,1, 7.5, 9.1, 10.4, 8.8, 10.8, 11.4,
6.8, 7.8, 11.8, 12.1, 10.7, 8.4, 9.5, 9.3, 10.1}

Construct a 95% CI for the mean of the data.

5.3. For a terminating simulation, the observations for eight replications
yield

{3.1, 4.2, 5.3, 3.0, 5.0, 4.0}
{2.1, 3.2, 6.3, 2.0, 6.0, 3.0}
{3.2, 4.3, 5.0, 2.5, 5.5, 4.5}
{2.5, 3.5, 4.5, 5.5, 6.5, 1.5}
{3.0, 2.0, 1.0, 5.0, 7.0, 6.0}
{2.9, 3.9, 4.1, 4.5, 5.1, 2.7}
{3.1, 2.2, 5.2, 3.0, 6.0, 2.9}
{4.0, 2.0, 2.9, 1.9, 5.9, 7.0}

Construct a 95% CI for the mean of the system.

5.4. In the simulation for comparing two alternative models, three replications were made for each model. The observations for the first
model are

{3.1, 4.2, 5.3, 3.0, 5.0, 4.0}
{2.1, 3.2, 6.3, 2.0, 6.0, 3.0}
{3.2, 4.3, 5.0, 2.5, 5.5, 4.5}

and the observations for the second model are

{5.1, 6.2, 7.3, 5.0, 7.0, 6.0}
{4.1, 5.2, 8.3, 4.0, 8.0, 5.0}
{5.2, 6.3, 7.0, 4.5, 7.5, 6.5}

Construct a 95% CI on the difference between the mean performance
measures of the two models, given that the models are homogeneous
and the data are normally distributed.

5.5. A simulation study for two alternative system designs A and B is
performed. The first study used the same random variables for both
designs. The second study used different random variables for each
design. The estimated average waiting time in seconds per hour of
an 8-hr-shift day is given as T_A and T_B. Compare the estimated
variance of $(T_A - T_B)$ for the two studies and discuss your results.

	First study		Second study	
Hour	T_A	T_B	T_A	T_B
1	94	62	94	60
2	108	73	108	68
3	108	75	108	72
4	98	66	98	60
5	87	57	87	66
6	83	57	83	70
7	96	59	96	67
8	96	61	96	63

5.6. Suppose you want to estimate the mean value of the exponential distribution $f(x) = 0.05 \exp(-0.05x)$, $x > 0$. Sample the exponential distribution by using a sequence of random numbers (R_i), and its complementary sequence $(1 - R_i)$ and utilize the antithetic technique to show that it yields the smallest variance. Use a sample size of $n = 10$.

5.7. If X_i, X_j are output data from a simulation experiment,

a. Show that $Cov(X_i, X_j)$ is dimensionless.

b. Show that $Cov(X_i, X_j) = 0$ if X_i, X_j are independent.

c. Discuss positive correlation.

5.8. List and compare three methods for variance reduction in simulation experiments.

5.9. In simulation experiments, when do we use the same stream of random numbers, and when do we use different streams? Why?

5.10. In a simulation study we make R runs, so our estimator is $X = (X_1 + X_2 + \cdots X_R)/R$. Find $Var(X)$.

5.11. For the factorial design simulation experiment of the geostationary satellite (2^3 with replications), the experimental results are as follows:

Run	Replicate 1	Replicate 2
1	60	55
2	40	45
3	70	65
4	50	55
5	60	50
6	50	45
7	80	85
8	50	60

Construct a 95% CI on main effects and interactions, and conclude which effects are significant.

REFERENCES

Arvidsen, N. I. and Johnson, T. (1982). Variance reduction through negative correlation, a simulation study, *J. Stat. Comp. Simulation 15*, 119–127.

Banks, J. and Carson II, J. S. (1985). *Discrete-Event System Simulation*, Prentice-Hall, Englewood Cliffs, NJ.

Bobillier, P. A., Kahan, B. C., and Probst, A. R. (1976). *Simulation with GPSS and GPSS V.*, Prentice-Hall, Englewood Cliffs, NJ.

Bratley, P., Fox, B. L., and Schrage, L. E. (1983). *A Guide to Simulation*, Springer-Verlag, New York.

Burdick, D. and Naylor, T. (1966). Design of computer simulation experiments for industrial systems, *Comm. ACM* 9(5), 329–338.

Carter, G. and Ignall, E. J. (1975). Virtual measures: A variance reduction technique for simulation, *Management Sci.* 21(6), 607–616.

Cheng, R. C. H. and Feast, G. M. (1980). Control variables with known mean and variance, *J. Opl. Res. Soc.* 31, 51–56.

Crane, M. A. and Lemoine, A. J. (1977). *An Introduction to the Regenerative Method for Simulation Analysis*, Lecture Notes in Control and Information Sciences, Vol. 4, Springer-Verlag, New York.

Duket, S. D. and Pritsker, A. A. (1978). Examination of simulation output using spectral methods, *Math. Comput. Simulation* 20, 53–60.

Fishman, G. S. (1972a). Bias consideration in simulation experiments, *Operations Res.* 20, 785–790.

Fishman, G. S. (1972b). Variance reduction in simulation studies, *J. Stat. Comp. Simulation* 1, 173–182.

Fishman, G. S. (1978). *Principles of Discrete Event Simulation*, Wiley, New York.

Franta, W. R. (1975). A note on random variate generators and antithetic sampling, *INFOR* 13, 112–117.

Hammersley, J. M. and Morton, K. W. (1956). A new Monte Carlo technique: Antithetic variates, *Proc. Camb. Phil. Soc.* 52, 449–475.

Heikes, R. G., Montgomery, D. C., and Rardin, R. L. (1976). Using common random numbers in simulation experiments—an approach to statistical analysis, *Simulation* 25, 81–85.

Kelton, W. D. and Law, A. M. (1984). An analytical evaluation of alternative strategies in steady-state simulation, *Operations Res.* 32, 169–184.

Kiefer, J. and Sacks, J. (1963). Asymptotically optimum sequential inference and design, *Ann. Math. Stat.* 34, pp. 705–750.

Kleijnen, J. P. C. (1975a). Antithetic variates, common random numbers, and optimal computer time allocation in simulation, *Management Sci.* 21(10), 1176–1185.

Kleijnen, J. P. C. (1975b). *Statistical Techniques in Simulation*, pt. II, Dekker, New York.

Kleijnen, J. P. C. (1977). Design and analysis of simulations: Practical statistical techniques, *Simulation* 28, 81–90.

Law, A. M. and Kelton, W. D. (1982). *Simulation Modeling and Analysis*, McGraw-Hill, New York.

Mechanic, H. and McKay, W. (1966). *Confidence Intervals for Averages of Dependent Data in Simulations II*, Technical Report 17-202, Advanced Systems Development Division, IBM Corp., Yorktown Heights, NY.

Myers, R. H. (1971). *Response Surface Methodology*, Allyn and Bacon, Boston.

Naylor, T. H. (1969). *The Design of Computer Simulation Experiments*, University Press, Durham, NC.

Naylor, T. H., Wertz, K., and Wonnacott, T. H. (1969). Spectral analysis of data generated by simulation results with econometric models, *Econometrica* 37(2), 333–352.

Payne, J. A. (1982). *Introduction to Simulation*, McGraw-Hill, New York.

Schruben, L. W. (1982). Detecting initialization bias in simulation output, *Operations Research*, 30, pp. 569–590.

Sherif, Y. S. (1981). Reliability, risks, resources, rewards, Int. J. Reliability Eng. 2(3), 167–178.

Wright, R. D. and Ramsay, Jr., J. E. (1979). On the effectiveness of common random numbers, *Management Sci.* 25, 649–656.

6
Introduction to Large-Scale Systems

6.1 INTRODUCTION

Over the past several years systems theory has evolved from a semi-heuristic discipline directed toward design and analysis of electronic and/or aerospace systems consisting of a handful of components into a very sophisticated theory capable of treating complex and large systems arising in a myriad of applications. The theory must be able to cope not only with the electronic/aerospace systems whose complexities have increased by several orders of magnitudes, but also with a vast number of real-life systems in society, economy, industry, government, etc.

As a first attempt, the system engineers tried to cope with this increasing system complexity through the development of sophisticated numerical techniques to allow the application of classical systems theory for large systems. This approach, however, soon reached a point of diminishing return and it became apparent that new theoretical techniques are needed to handle large and complex systems. Although many such techniques are still being developed and need a great deal of fine tuning, it is generally accepted that a key to successful treatment of large-scale systems is to fully exploit their structure interconnection. This exploitation has traditionally taken place in several ways. It can be a full use of, say, sparse matrix techniques or it may be through the "decomposition" of a larger system into a finite number of smaller systems.

The ever-changing high technology of today continues to create new processes which are complex, "*large*," and stochastic by nature. The notion of "large-scale" is a very subjective one in that one may ask: How large is *large*? A universal definition of a large-scale system has not been agreed upon in the literature. There are, however, three possible definitions for a large-scale system.

One common definition has been that a system is considered large-scale if it can be partitioned, decoupled, or decomposed into a number of subsystems, i.e., small-scale systems. Another viewpoint has been that a system is termed large-scale if its dimensions are so high that conventional techniques of modeling, analysis, control, design, estimation, computational procedures, etc., fail to give reasonable solutions with reasonable efforts. A third definition is based on the concept of "*centrality*." Until the development of large-scale systems, almost all control systems analysis and design procedures were limited to having the system's components and information flow from one point to another localized or centralized in one geographical location or center, such as a laboratory. Thus, a third possible definition is simply that a system is said to be large-scale if the concept of centrality fails in it. This may be due to either lack of centralized computing capability or a centralized information structure. These definitions, although seemingly far apart, do have at least one point in common. That is, as a given system's structure or information flow becomes complex, new methodologies for system modeling, analysis, and design are needed. Examples of large-scale systems in real life are too numerous to detail here. For now it suffices to note that large-scale systems appear in such diversified fields as society, business, management, the economy, the environment, energy, data networks, computer networks, power systems, flexible space structures, transportation, aerospace, and navigational systems, to name a few.

The first and third definitions of a large-scale system play major parts in two new control methodologies—hierarchical (multilevel) and decentralized. These concepts will be discussed in detail later.

This chapter presents an introduction to large-scale systems. In the next section modeling and model reduction of large-scale systems are discussed. Concepts of aggregation and perturbation are presented. Hierarchical control and some of its key issues and optimizing algorithms are briefly discussed in Section 6.3. The celebrated control of large-scale systems based on the decentralized system's output information, i.e., decentralized control, is given in Section 6.4. Structural properties such as stability, controllability, and observability are briefly discussed in Section 6.5. Finally, in Section 6.6 some initial conclusions are drawn and future trends of large-scale systems theory are discussed.

6.2 MODELING AND MODEL REDUCTION

As already noted in Chapters 1 and 2, the first step in any scientific or technological study of a system is to develop a "mathematical model" to substitute for the real problem. In any modeling task, two often conflicting factors prevail—"simplicity" and "accuracy." On one hand, if a system model is oversimplified, presumably for computational effectiveness, incorrect conclusions may be drawn from it in representing an actual system. On the other hand, a highly detailed model would lead to many unnecessary complications, and should a feasible solution be attainable, the extent of resulting details may become so vast that further investigations on the system behavior would become impossible with questionable practical value. Clearly, a mechanism by which a compromise can be made between a complex, more accurate model and a simple, less accurate model is needed. Finding such a mechanism is not a simple undertaking.

In the area of large-scale systems there have been two general approaches for modeling: "aggregation" and "perturbation" schemes. An aggregated model of a system is described by a "coarser" set of variables. The underlying reason for aggregating a system model is to be able to retain the key qualitative properties of the system, such as stability, which is viewed, by Siljak (1978), as a natural process through the

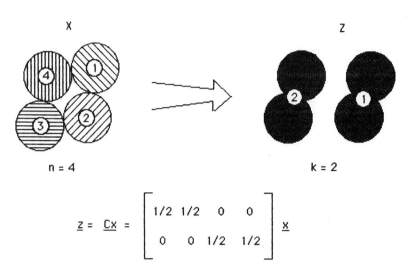

Figure 6.1 A pictorial presentation of the aggregation process.

second method of Lyapunov. In other words, the stability of a system described by several state variables is entirely represented by a single variable—the Lyapunov function. Figure 6.1 is a pictorial presentation of the aggregation process. The system on the left is described by four variables (circles), and the system on the right represents an aggregated model for it, where two variables now describe the system. Variable No. 1, called z_1 is an average of the full model's first two variables, while the second aggregated variable z_2 is an average of the third and fourth variables.

The other scheme for large-scale systems modeling has been perturbation, which is based on ignoring certain interactions of the dynamic or structural nature in a system. Here, again, the benefits received from reduced computations must not be at the expense of key system properties. Although both perturbation and aggregation schemes tend to provide reduction in computations and perhaps simplification in structure, there has been no hard evidence that they are the most desirable for large-scale systems. Figure 6.2 provides a pictorial presentation of perturbation of a singular type. The fast variable of the system can be approximated by an auxiliary or quasi steady-state variable, and through a concept called "boundary layer" (see Section 6.2.2) the missing initial information of this variable can be accounted for. This would lead to a reduced model based on the slow variable.

6.2.1 Aggregation

Aggregation has long been a technique for analyzing static economic models. The current treatment of aggregation is probably due to Malinvaud (Jamshidi, 1983), whose formulation is shown in Figure 6.3. In this

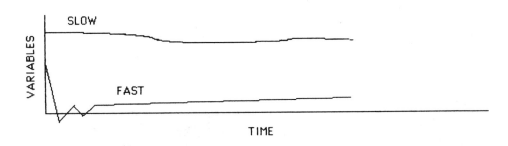

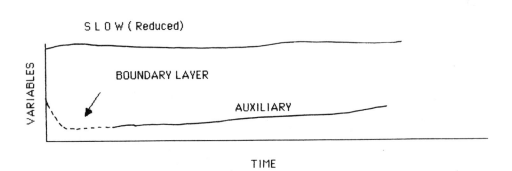

Figure 6.2   A pictorial presentation of the perturbation process.

diagram, x, y, z, and v are topological (or vector) spaces, f represents a linear continuous map between the exogenous variable $x \varepsilon x$ and endogenous variables $y \varepsilon y$.   The aggregation procedures h: $x \rightarrow z$ and g: $y \rightarrow v$ lead to aggregated variables $z \varepsilon Z$ and $v \varepsilon V$.   The map k: $z \rightarrow v$ is to represent a simplified or an aggregated model.   The aggregation is said to be "perfect" when k is chosen such that the relation

$$gf(\underline{x}) = kh(\underline{x}) \tag{6.1}$$

holds for all $\underline{x} \varepsilon x$.   The notion of perfect aggregation is an idealization at best, and in practice it is approximated through two alternative procedures, according to econometricians (Jamshidi, 1883).   These are (a) to impose some restrictions on f, g, and h while leaving x unrestricted and (b) to require Eq. (6.1) to hold on $\underline{x}_0$ some subset of $\underline{x}$.

   <u>Regular Aggregation</u>.   Consider a linear controllable system

$$\dot{\underline{x}}(t) = \underline{A}\underline{x}(t) + \underline{B}\underline{u}(t), \qquad \underline{x}(t_o) = \underline{x}_o \tag{6.2}$$

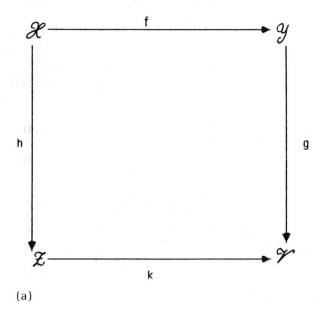

(a)

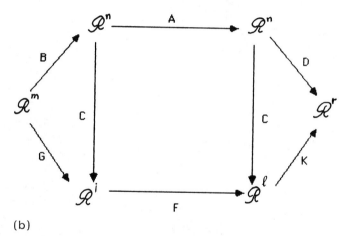

(b)

Figure 6.3 A pictorial representation of aggregation (a) a static system, and (b) a dynamic linear system.

$$\underline{y}(t) \;=\; \underline{D}\underline{x}(t) \tag{6.3}$$

where $\underline{x}(t)$, $\underline{u}(t)$, and $\underline{y}(t)$ and $n \times 1$, $m \times 1$, and $r \times 1$ state, control, and output vectors, respectively, $\underline{A}$, $\underline{B}$, and $\underline{D}$ are $n \times n$, $n \times m$, and $r \times n$ matrices. Assume that n is large and hence called unaggregated for our discussion. Moreover, consider an aggregated linear time-invariant model for the same large-scale system

$$\underline{\dot{z}}(t) \;=\; \underline{F}\underline{z}(t) + \underline{G}\underline{u}(t), \qquad \underline{z}(0) = \underline{z}_0 \tag{6.4}$$

$$\underline{\hat{y}}(t) \;\stackrel{\sim}{=}\; \underline{K}\underline{z}(t) \tag{6.5}$$

where $\underline{z}(t)$ is k-dimensional aggregated state vector, $\underline{F}$ and $\underline{G}$ are $k \times k$ and $k \times m$ dimensional constant matrices, respectively. The vector $\underline{\hat{y}}(t)$ is an $r \times 1$ approximate output. Assume that $\underline{x}$ is observed directly and Eq. (6.2) is controllable (Jamshidi, 1983). The aggregation condition of the system model (Eq. (6.4)] to aggregate model [Eq. (6.2)] is

$$\underline{z}(t) \;=\; \underline{C}\underline{x}(t), \qquad \underline{z}_0 = \underline{C}\underline{x}_0 \tag{6.6}$$

where $\underline{C}$ is a $k \times n$ constant aggregation matrix, $k \leqslant n$, and rank $\underline{C}$ is assumed to be k. A closer look at Eq. (6.2)–(6.5) indicates that "dynamic exactness" (perfect aggregation or aggregability) is guaranteed if, and only if, the following matrix identities hold

$$\underline{F}\underline{C} \;=\; \underline{C}\underline{A} \tag{6.7}$$

$$\underline{G} \;=\; \underline{C}\underline{B} \tag{6.8}$$

$$\underline{K}\underline{C} \;\stackrel{\sim}{=}\; \underline{D} \tag{6.9}$$

Making use of the generalized (pseudo-) inverse of $\underline{C}$ in Eq. (6.7),

$$\underline{C}\underline{A} \;=\; \underline{C}\underline{A}\underline{C}^{+}\underline{C} = \underline{C}\underline{A}\underline{C}^{T}(\underline{C}\underline{C}^{T})^{-1}\underline{C} \;=\; \underline{F}\underline{C} \tag{6.10}$$

where $\underline{C}^{+} \stackrel{\Delta}{=} \underline{C}^{T}(\underline{C}\underline{C}^{T})^{-1}$ is the generalized inverse of $\underline{C}$ and $(\ )^{T}$ denotes transpose. A comparison of both sides of the latter part of identity [Eq. (6.10)] reveals that

$$\underline{F} \;=\; \underline{C}\underline{A}\underline{C}^{T}(\underline{C}\underline{C}^{T})^{-1} \tag{6.11}$$

which indicates that once the aggregation matrix $\underline{C}$ is known, the aggregated matrix $\underline{F}$ is obtained by Eq. (6.11) and the aggregated control matrix $\underline{G}$ is determined from Eq. (6.8). Analysis of these four identities gives some insight in the choice of the "aggregation matrix" $\underline{C}$. Aoki (1971) has shown that analysis of Eq. (6.7) will lead to a description of the aggregated state vector $\underline{z}(t)$, which is a linear combination of certain modes of $\underline{x}(t)$. It is noted that aggregated matrix $\underline{F}$ is obtained from Eq. (6.11) only if the conditions Eq. (6.7)–(6.9) are satisfied. Under these circumstances, the eigenvalues of F constitute a subset of eigenvalues of $\underline{A}$. As mentioned earlier for the static models of econometrics, the dynamic exactness (perfect aggregation) is an idealized situation. The use of Eq. (6.11) is an approximation which, in fact, minimizes the square of the norm $\|\underline{F}\,\underline{C} - \underline{C}\,\underline{A}\|$ unless a consistency relation

$$\underline{C}\,\underline{A}\,\underline{C}^{+}\underline{C} = \underline{C}\,\underline{A} \tag{6.12}$$

is satisfied.

If an error vector is defined as $\underline{e}(t) = \underline{z}(t) - \underline{C}\underline{x}(t)$, then its dynamic behavior is given by $\underline{\dot{e}}(t) = \underline{F}\underline{e}(t) + (\underline{F}\underline{C} - \underline{C}\underline{A})\underline{x}(t) + (\underline{G} - \underline{C}\underline{B})\underline{u}(t)$, which reduces to $\underline{\dot{e}}(t) = \underline{F}\underline{e}(t)$ if conditions Eq. (6.7)–(6.8) hold. Hence, if $\underline{e}(0) = 0$, then $\underline{e}(t) = 0$ for all $t \geqslant 0$. Should $\underline{e}(0) \neq 0$ but F be a stable matrix, then $\lim_{t\to\infty}\underline{e}(t) = 0$; i.e., the dynamic exactness condition Eq. (6.7)–(6.8) is asymptotically satisfied.

An alternative approach that does not require knowledge of the eigenvalues and eigenvectors of $\underline{A}$ has also been proposed by Aoki (1971). Recalling Section 2.4.3 and considering the controllability matrix of Eq. (6.2)

$$\underline{W}_A \stackrel{\Delta}{=} [\underline{B}, \underline{A}\,\underline{B}, \ldots, \underline{A}^{n-1}\underline{B}] \tag{6.13}$$

and a modified controllability matrix for Eq. (6.4),

$$\underline{W}_F \stackrel{\Delta}{=} [\underline{G}, \underline{F}\,\underline{G}, \ldots, \underline{F}^{n-1}\underline{G}] \tag{6.14}$$

it can be seen from Eq. (6.7) and Eq. (6.8) that these matrices are related by

$$\underline{W}_F = \underline{C}\,\underline{W}_A \tag{6.15}$$

Thus, using the generalized inverse, C can be obtained by

$$\underline{C} = \underline{W}_F\underline{W}_A^{+} = \underline{W}_F\underline{W}_A^{T}(\underline{W}_A\underline{W}_A^{T})^{-1} \tag{6.16}$$

since the initial controllability assumption rank $\underline{W}_A = n$. Therefore, if $\underline{F}$ is specified, $\underline{F} = \text{diag}(\lambda_1, \lambda_2, \ldots, \lambda_k)$, for example, and choosing $\underline{G}$ to make aggregated system Eq. (6.4) completely controllable, i.e., rank $\underline{W}_F = k$, then C is obtained by Eq. (6.16).

Example 6.2.1. For a third-order unaggregated system described by

$$\underline{\dot{x}} = \begin{bmatrix} -0.1 & 1 & 2 \\ 0 & -0.2 & 1 \\ 0 & 0 & -2 \end{bmatrix} \underline{x} + \begin{bmatrix} 1 & 1 \\ 0 & 1 \\ 2 & 1 \end{bmatrix} \underline{u} \tag{6.17}$$

find an aggregated system.

This example is solved by the two methods described above.

1. The first solution is obtained by the use of eigenvalues of $\underline{A}$ of Eq. (6.17), which are, clearly, $\lambda_1 = -0.1$, $\lambda_2 = -0.2$, and $\lambda_3 = -2$. From the relative magnitudes of λ_i's it is clear that the slowest mode of the

system is the first while the third mode is much faster. Thus, a choice for aggregation matrix $\underline{C}$ can be

$$\underline{C} = \begin{bmatrix} 1 & 0 & 0 \\ 0 & 1 & 0 \end{bmatrix} \tag{6.18}$$

which implies that the first two aggregated states $z_1(t)$ and $z_2(t)$ are chosen to be the slowest modes of the full model. From Eq. (6.11) and Eq. (6.8) the aggregated matrices $\underline{F}$ and $\underline{G}$ are obtained and the aggregated model is

$$\dot{\underline{z}}(t) = \underline{F}\underline{z}(t) + \underline{G}\underline{u}(t) = \begin{bmatrix} -0.1 & 1 \\ 0 & -0.2 \end{bmatrix} \underline{z}(t) + \begin{bmatrix} 1 & 1 \\ 0 & 1 \end{bmatrix} \underline{u}(t) \tag{6.19}$$

2. This solution will be obtained by evaluating controllability matrices. Following the discussion above, let $\underline{F}$ = diag $(-0.1, -0.2)$. Then, by trial and error a $G = \begin{pmatrix} 1 & 1 \\ 0 & 1 \end{pmatrix}$ matrix can be found so that the pair $(\underline{F},\underline{G})$ is controllable. The controllability matrices $\underline{W}_A$ and $\underline{W}_F$ are given by

$$\underline{W}_A = \begin{pmatrix} 1 & 1 & 3.9 & 2.9 & -6.39 & -3.49 \\ 0 & 1 & 2 & 0.8 & -4.4 & -2.16 \\ 2 & 1 & -4 & -4 & 8 & 4 \end{pmatrix}$$

$$\underline{W}_F = \begin{pmatrix} 1 & 1 & -0.1 & -0.1 & 0.01 & 0.01 \\ 0 & 1 & 0 & -0.2 & 0 & 0.04 \end{pmatrix}$$

and hence a possible aggregation matrix is obtained from Eq. (6.16):

$$\underline{C} = \begin{pmatrix} 0.1 & 0.57 & 0.4 \\ -0.15 & 0.66 & 0.22 \end{pmatrix}$$

Modal Aggregation. We next consider an aggregation scheme of Davison (1966, 1968), which is a special case of regular aggregation discussed in the last section.

Consider the system Eq. (6.2)–(6.3) and let the aggregated (reduced) model be Eq. (6.4)–(6.5). Then the aggregated matrix pair (F,G) is given by

$$\underline{F} = \underline{M}_k \underline{P} \underline{A} \underline{P}^T \underline{M}_k^{-1} \tag{6.20}$$

$$\underline{G} = \underline{M}_k \underline{P} \underline{M}^{-1} \underline{B} \tag{6.21}$$

where $\underline{M}$ is the modal matrix of Eq. (6.4) consisting of the eigenvectors of $\underline{A}$ arranged in ascending (or descending) order of the $\mathrm{Re}[\lambda_i\{\underline{A}\}]$. $\underline{M}_k$ is a $k \times k$ matrix which includes the k-dominant eigenvectors of $\underline{A}$ corresponding to the retained modes of the original system, and

$$\underline{P} = \{\underline{I}_k | 0\} \tag{6.22}$$

is a $k \times n$ transformation matrix. Modal matrix M can be represented by

$$
\begin{array}{c}
k \\
\underline{M} \ = \ \begin{array}{c} k \ \{ \\ \\ \end{array}
\overbrace{\left[\begin{array}{c|c} \underline{M}_k & \underline{M}_{12} \\ \hline \underline{M}_{21} & \underline{M}_{22} \end{array}\right]}\left.\vphantom{\begin{array}{c}\\\\\end{array}}\right\} \ n \\
\underbrace{} \\
n
\end{array}
\tag{6.23}
$$

The aggregation matrix $\underline{C}$ in $\underline{z} = \underline{C}\underline{x}$ is given by

$$\underline{C} = \underline{M}_k \underline{P} \underline{M}^{-1} \tag{6.24}$$

It is noted that this scheme works for the case where $\underline{A}$ has complex or repeated eigenvalues as well. Under those conditions the columns of $\underline{M}$ can be real and imaginary parts of the complex eigenvectors or generalized eigenvectors in addition to regular eigenvectors.

Example 6.2.2. Consider the third-order system of Ex. (6.2.1) and find a modal aggregated model for it.

The eigenvalues of $\underline{A}$ are -0.1, -0.2, and -2, which indicates that the first two modes are dominant. The modal matrix $\underline{M}$ arranged in the order of dominancy of $\lambda\{\underline{A}\}$ is given by

$$
\underline{M} \ = \ \begin{bmatrix} 1 & 1 & 1 \\ 0 & -0.1 & 0.73 \\ 0 & 0 & -1.31 \end{bmatrix}
$$

while the diagonal matrix $\underline{A} = \mathrm{diag}\,(-0.1,\ -0.2,\ -2.0)$ and the input modal matrix is

$$
\underline{\Gamma} \ = \ \begin{pmatrix} 13.67 & 17.34 \\ -11.15 & -15.6 \\ -1.53 & -0.76 \end{pmatrix}
$$

The aggregated system is given by

$$
\underline{\dot{z}} \ = \ \begin{pmatrix} -0.1 & 1 \\ 0 & -0.2 \end{pmatrix} \underline{z} + \begin{pmatrix} 2.52 & 1.76 \\ 1.11 & 1.55 \end{pmatrix} \underline{u}
$$

with aggregation matrix

$$\underline{C} = \begin{bmatrix} 1 & 0 & 0.76 \\ 0 & 1 & 0.55 \end{bmatrix}$$

Chained Aggregation. One of the more recent approaches in aggregating large-scale linear time-invariant systems is "chained aggregation," developed by Perkins and colleagues (Tse et al., 1978). Based on the unaggregated large-scale system's information structure, the system is described through a "chain" of "aggregations," by a so-called "generalized Hessenberg representation" (GHR). Consider a large-scale linear time-invariant system

$$\dot{\underline{x}} = \underline{A}\,\underline{x} + \underline{B}\,\underline{u} \tag{6.25}$$

$$\underline{y} = \underline{C}_1\underline{x} \tag{6.26}$$

where dimensions of $\underline{x}$ and $\underline{y}$ are n and r_1, respectively, and without any loss of generality it is assumed that rank $(\underline{C}_1) = r_1$, and furthermore suppose that through an ordering of the states, $\underline{C}_1$ can be represented by

$$\underline{C}_1 = [\underline{C}_{11}, \underline{C}_{12}], \qquad \det \underline{C}_{11} \neq 0 \tag{6.27}$$

Let us define a nonsingular transformation matrix

$$\underline{T}_1 = \begin{bmatrix} \underline{C}_{11} & | & \underline{C}_{12} \\ \hline \underline{0} & | & \underline{I}_{n-r_1} \end{bmatrix} \tag{6.28}$$

which would transform Eq. (6.25) to

$$\dot{\underline{z}} = \begin{bmatrix} \dot{\underline{z}}_1 \\ \dot{\underline{z}}_2 \end{bmatrix} = \begin{bmatrix} \underline{F}_{11} & | & \tilde{\underline{F}}_{12} \\ \tilde{\underline{F}}_{21} & | & \tilde{\underline{F}}_{22} \end{bmatrix} \begin{bmatrix} \underline{z}_1 \\ \underline{z}_2 \end{bmatrix} + \begin{bmatrix} \underline{G}_1 \\ \tilde{\underline{G}}_2 \end{bmatrix} \underline{u} \tag{6.29}$$

If the system of Eq. (6.25) is "completely aggregable" (Tse et al., 1978), i.e., the "dynamic exactness" conditions of Aoki (1971) expressed in Eq. (6.7)-(6.9) are satisfied, then $F_{12} = 0$. The upper partition of Eq. (6.29), i.e.,

$$\dot{\underline{z}}_1 = \tilde{\underline{F}}_{11}\underline{z}_1 + \tilde{\underline{F}}_{12}\underline{z}_2 + \tilde{\underline{G}}_1\underline{u} \tag{6.30}$$

is termed the "aggregated subsystem" of Eq. (6.29), while the second partition

$$\dot{\underline{z}}_2 = \tilde{\underline{F}}_{21}\underline{z}_1 + \tilde{\underline{F}}_{22}\underline{z}_2 + \tilde{\underline{G}}_2\underline{u} \tag{6.31}$$

is called the "residual subsystem" of Eq. (6.29). If $\tilde{\underline{F}}_{12} = 0$, the "strict aggregation" condition

$$\underline{C}_1\underline{A} = \underline{F}_{11}\underline{C}_1 \tag{6.32}$$

holds and the large-scale system can be presented by two tandem sub-systems shown in Figure 6.4.

However, if $\underline{F}_{12} \neq 0$, as it can in general, it may be possible to find an aggregate model by enlarging the output vector. This process would begin by finding a matrix $\tilde{C}_1$ of maximum rank $\rho < n - r_1$ so that Eq. (6.25) is completely aggregable with respect to

$$\underline{C} = \begin{bmatrix} \underline{C}_1 \\ \tilde{\underline{C}}_1 \end{bmatrix} \tag{6.33}$$

In order to find the aggregable subsystem with $\tilde{C}_1$, the procedure follows by considering $v_2 = \tilde{F}_{12}z_2$ as the output of the residual subsystem in Eq. (6.31) and obtaining an aggregation of this subsystem with respect to $\underline{v}_2$. It is noted that since $\tilde{F}_{12}$ has a dimension of $r_1 \times n_2$, $n_2 = n - r_1$, then it does not have in general a full rank (Tse et al., 1978). However, there exists a nonsingular matrix $\underline{E}_{12}$ such that

$$\underline{E}_2\underline{v}_2 = \underline{E}_2\tilde{F}_{12}\underline{z}_2 = \begin{bmatrix} \underline{C}_2 \\ 0 \end{bmatrix} \underline{z} \tag{6.34}$$

which extracts linearly independent components of $\underline{v}_2$. In Eq. (6.34) matrix $\underline{C}_2$ has a dimension of $r_2 \times n_2$; $r_2 \leqslant r_1$. Tse et al. (1978) note that $\underline{E}_2$ is obtained by the product of matrices of the type found in Gaussian elimination.

Now let us define $\underline{y}_2 = \underline{C}_2\underline{z}_2$, which can be used to find an expression for $\underline{v}_2$ from Eq. (6.34), i.e.,

$$\underline{v}_2 = \underline{E}_2^{-1} \begin{bmatrix} \underline{C}_2 \\ 0 \end{bmatrix} \underline{z}_2 = \underline{E}_2^{-1} \begin{bmatrix} I \\ 0 \end{bmatrix} \underline{y}_2 \triangleq \underline{F}_{12}\underline{y}_2 \tag{6.35}$$

Denoting $\underline{z}_1 = \underline{x}^1$, then the residual subsystem [Eq. (6.31)] can be rewritten as

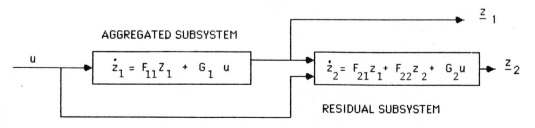

Figure 6.4   A block diagram of the chained aggregation.

$$\dot{\underline{z}}_2 = \tilde{\underline{F}}_{21}\underline{x}^1 + \tilde{\underline{F}}_{22}\underline{z}_2 + \tilde{\underline{G}}_2\underline{u} \tag{6.36}$$

$$\underline{y}_2 = \underline{C}_2\underline{z}_2 \tag{6.37}$$

Recalling the initial ordering of states of the original system as indicated by Eq. (6.27), one can now proceed with chained aggregation for the residual subsystem [Eq. (6.36)–(6.37)] and partition $\underline{C}_2$:

$$\underline{C}_2 = [\underline{C}_{22} \mid \underline{C}_{23}], \qquad \det \underline{C}_{22} \neq \underline{0} \tag{6.38}$$

and determine an $n_2 \times n_2$ nonsingular transformation matrix D_2:

$$\underline{D}_2 = \left[\begin{array}{c|c} \underline{C}_{22} & \underline{C}_{23} \\ \hline \underline{0} & I_{-n_2-r_2} \end{array} \right] \tag{6.39}$$

Once again, applying Eq. (6.39) to Eq. (6.36) and repeating the two aggregation steps similar to Eq. (6.28)–(6.31), i.e., by letting

$$\underline{D}_2\underline{x}^2 = \left[\begin{array}{c} \underline{x}_2 \\ \underline{z}_3 \end{array} \right] \tag{6.40}$$

then the new aggregated set of subsystems becomes

$$\dot{\underline{x}}^1 = \underline{F}_{11}\underline{x}^1 + \underline{F}_{12}\underline{x}^2 + \underline{G}_1\underline{u} \tag{6.41}$$

$$\dot{\underline{x}}^2 = \underline{F}_{21}\underline{x}^1 + \underline{F}_{22}\underline{x}^2 + \tilde{\underline{F}}_{23}\underline{z}_3 + \underline{G}_2\underline{u} \tag{6.42}$$

$$\dot{\underline{x}}_3 = \tilde{\underline{F}}_{31}\underline{x}^1 + \tilde{\underline{F}}_{32}\underline{x}^2 + \tilde{\underline{F}}_{33}\underline{z}_3 + \tilde{\underline{G}}_3\underline{u} \tag{6.43}$$

At this point if $\tilde{\underline{F}}_{23} = 0$, then the chained aggregation process halts, and if not, then the new residual subsystem [Eq. (6.43)] is aggregated with respect to a new output $\underline{v}_3 = \tilde{\underline{F}}_{23}\underline{z}_3$. This chain of aggregation would terminate in $1 \leqslant k \leqslant n_2 = (n - r_1)$ iterations—a finite number. The final format of the transformed system state matrix after a chain of k aggregations will be of the form

$$\underline{F} = \left[\begin{array}{cccccccc} \underline{F}_{11} & \underline{F}_{12} & 0 & \cdot & \cdot & \cdot & \cdot & 0 \\ \underline{F}_{21} & \underline{F}_{22} & \underline{F}_{23} & 0 & \cdot & \cdot & \cdot & 0 \\ \cdot & \cdot & \cdot & & & & & \\ \cdot & \cdot & \cdot & & & & & \\ \underline{F}_{j1} & \underline{F}_{j2} & & \cdot & \underline{F}_{jj} & \underline{F}_{j,j+1} & 0 & 0 \\ \cdot & \cdot & \cdot & & & & & \\ \underline{F}_{k-1,1} & \underline{F}_{k-1,2} & & & & & & \underline{F}_{k-1,k} \\ \underline{F}_{k,1} & \underline{F}_{k,2} & \cdot & \cdot & \cdot & \cdot & \cdot & \underline{F}_{k,k} \end{array} \right] \tag{6.44}$$

control and output matrices

$$\underline{G} = \begin{bmatrix} \underline{G}_1 \\ \underline{G}_2 \\ \cdot \\ \cdot \\ \underline{G}_j \\ \cdot \\ \cdot \\ \underline{G}_k \end{bmatrix} \qquad \underline{D} = [\underline{I}_{r_1} \quad 0 \quad \cdot \quad \cdot \quad \cdot \quad \cdot \quad 0] \qquad (6.45)$$

where the submatrices F_{ii} have $r_i \times r_i$ dimensions, $r_i \geqslant r_{i+1}$, $n = r_1 + r_2 + \cdots + r_k$, and I_{r_1} is an identity matrix of order r.

To summarize this discussion, one can state the main result of chained aggregation as follows: Any time-invariant system [Eq. (6.25)–(6.26)] is transformable to

$$\dot{\underline{z}} = \underline{F}\underline{z} + \underline{G}\underline{u} \qquad (6.46)$$

$$\underline{y} = \underline{D}\underline{z} \qquad (6.47)$$

where matrices $(\underline{F}, \underline{G}, \underline{D})$ are the GHR form [Eq. (6.44)–(6.45)].

Before the chained aggregation is illustrated by a numerical example, it is worth making a few more remarks on possible interpretation of the aggregation procedure. The F matrix in Eq. (6.44) of GHR is a generalization of block matrices in the lower Hessenberg forms which are commonly used in numerical transformations of matrices such as QR-algorithm (Tse et al., 1978). A possible interpretation of chained aggregation is a string of similarity transformations (Tse et al., 1978).

$$\underline{T}_j = \begin{bmatrix} \underline{I}_1 & & & & O \\ & \underline{I}_2 & & & \\ & & \ddots & & \\ & & & \underline{I}_{j-1} & \\ O & & & & \underline{D}_j \end{bmatrix} \qquad (6.48)$$

on the original system for $j = 2, 3, \ldots, k - 1$, and D_j is defined by

$$\underline{D}_j = \begin{bmatrix} \underline{C}_{jj} & \underline{C}_{j,j+1} \\ 0 & \underline{I}_{nj} \end{bmatrix} \qquad n_j = n - \sum_{i=1}^{j} r_i \qquad (6.49)$$

$$\underline{P} = \underline{T}_{k-1}\,\underline{T}_{k-2}\,\underline{T}_{k-3} \cdots \underline{T}_2\underline{T}_1 \qquad (6.50)$$

Matrix $\underline{P}$ would be of lower block triangular form. However, should it be necessary to permute states of the residual subsystems at intermediate steps to make sure matrices $\underline{C}_{ii}$ are nonsingular, $\underline{P}$ would no longer be a lower-block triangular and would be given by

$$\underline{P} = \underline{T}_{k-1} \, \underline{H}_{k-1} \, \underline{T}_{k-2} \, \underline{H}_{k-2} \cdots \underline{T}_1 \underline{H}_1 \tag{6.51}$$

where $\underline{H}_i$, $i = 1, 2, \ldots, k-1$ are the state permutation matrices

Below a CAD example illustrates chained aggregation using a CAD package called *LSSPAK/PC* (Jamshidi and Banning, 1985).

CAD Example 6.2.3. Using program "*CHAINAGG*," a fourth-order system is reduced below to a number of reduced-order models.

<<<program CHAINED aggregation>>> For further information, see Jamshidi (1983, pp. 23–30).

Number of states = 4

Number of inputs = 1

Number of outputs = 2

Matrix $\underline{A}$
```
 -.200E+01   -.100E+01   -.500E+00   -.100E+00
0.500E+00   -.100E+01   0.200E+00   -.400E+01
0.100E+01   -.100E+00   -.250E+00   0.000E+00
0.000E+00   0.100E+01   -.100E+00   -.100E+00
```
Matrix $\underline{B}$
```
0.100E+01
-.250E+01
0.100E-01
0.100E+02
```
Matrix $\underline{C}$
```
0.100E+01   0.000E+00   0.500E+00   0.250E+00
0.000E+00   0.100E+01   0.000E+00   0.500E-01
```
STEP 1 BEGINS HERE...

Enter the value of aggregated system #1, max. = 4

Aggregation step number 1: aggregated order = 3

Matrix $\underline{T1}$
```
 -.800E+00   0.100E+00   0.290E+00
 -.950E+00   -.550E-01   -.408E+01
0.000E+00   0.000E+00   0.100E+01
```
Matrix $\underline{F}$
```
 -.150E+01   0.100E+01   0.000E+00
0.500E+00   -.822E+00   -.122E+01
0.100E+01   0.469E+01   0.408E+01
```
Matrix $\underline{G}$
```
0.351E+01
0.450E+01
-.389E+02
```
Matrix $\underline{D}$
```
0.100E+01   0.100E+01   0.298E-06   0.000E+00
-.298E-07   0.100E+01   0.100E+01   0.000E+00
```
STEP 2 BEGINS HERE...

Enter the value of aggregated system #3, max. = 2

Aggregation step number 2: aggregated order = 2

Matrix $\underline{T1}$
```
0.100E+00   0.290E+00
-.550E-01   -.408E+01
```

Matrix F
    $-.150E+01$ $-.800E+00$
    $0.500E+00$ $-.950E+00$
Matrix G
    $0.351E+01$
    $-.200E+01$
Matrix D
    $0.100E+01$ $-.745E-08$ $0.000E+00$ $0.000E+00$
    $0.596E-07$ $0.100E+01$ $0.000E+00$ $0.000E+00$
STEP 3 BEGINS HERE...
    Enter the value of aggregated system #2, max. = 1
    Aggregation step number 3: aggregated order = 1
Matrix T1
    $0.139E+01$
Matrix F
    $-.150E+01$
Matrix G
    $0.351E+01$
Matrix D
    $0.100E+01$ $0.100E+01$ $0.298E-06$ $0.000E+00$
    $0.546E-01$ $0.100E+01$ $0.100E+01$ $0.000E+00$

This program provides a number of possible chains of aggregations for any linear time-invariant system. For this fourth-order full-order model, three reduced-order models are suggested by "CHAINAGG," i.e., a third-order, a second-order, and a first-order reduced model.

    <u>Balanced Aggregation</u>. In this section a method based on balancing the controllability and observability Grammian matrices and retaining the most controllable and observable modes of the system is given. Consider a linear time-invariant system $S(\underline{A}, \underline{B}, \underline{C})$ defined by

$$\dot{\underline{x}} = \underline{A}\underline{x} + \underline{B}\underline{u}$$

$$\underline{y} = \underline{C}\underline{x}$$

(6.52)

where $\underline{x}$, $\underline{u}$, $\underline{y}$, $\underline{A}$, $\underline{B}$, and $\underline{C}$ are as defined earlier. This system is assumed to be asymptotically stable, controllable, and observable. The associated observability and controllability Grammians are, respectively, given by

$$\underline{W}_o = \int_0^\infty e^{\underline{A}^T t} \underline{C}^T \underline{C} e^{\underline{A} t} dt$$

(6.53)

$$\underline{W}_c = \int_0^\infty e^{\underline{A} t} \underline{B}\underline{B}^T e^{\underline{A}^T t} dt$$

If the system is controllable and observable, the above Grammian matrices are positive, definite, and satisfy the following Lyapunov equations:

$$\underline{W}_o\underline{A} + \underline{A}^T\underline{W}_o + \underline{C}^T\underline{C} = 0$$

(6.54)

$$\underline{W}_c \underline{A}^T + \underline{A}\underline{W}_c + \underline{B}\underline{B}^T = 0 \tag{6.55}$$

More (1981) found a similarity transformation matrix $\underline{T}$, such that the matrices $\underline{W}_c$ and $\underline{W}_0$ are both equal and diagonal (i.e., balanced). A more efficient method to compute the balancing transformation is given by (Laub, 1980)

$$\underline{T} = \underline{L}_c \underline{U} \underline{D}^{-1/2} \tag{6.56}$$

where $\underline{L}_c$ is lower triangular of the Cholesky decomposition of W_c, U is an orthogonal modal matrix, and D is the diagonal matrix to the symmetrical eigenvalue/eigenvector problem of

$$\underline{U}^T(\underline{L}_c^T \underline{W}_0 \underline{L}_c)\underline{U} = \underline{D}^2 \tag{6.57}$$

It can easily be shown that $\hat{\underline{W}}_0 = \underline{T}^T \underline{W}_0 \underline{T}$ and $\hat{\underline{W}}_c = \underline{T}^{-1}\underline{W}_c\underline{T}$ are diagonalized and are equal to $\underline{D}$.

The elements along the diagonal matrix $\underline{D}$ (Moore's second-order modes) are ordered such that $\underline{D} = \text{diag}(\sigma_1, \sigma_2, \ldots, \sigma_n)$ and $\sigma_1 > \sigma_2 > \ldots > \sigma_n > 0$. If $\sigma_k > \sigma_{k+1}$, then $\underline{A}$, $\underline{B}$, and $\underline{C}$ are partitioned appropriately with k as the order of the reduced order model. The second-order modes provide a measure of error of how the reduced-order model approximates the original full-order model. The partitioned matrices are denoted as follows:

$$\hat{\underline{A}} = \begin{bmatrix} \underline{F} & | & \hat{\underline{A}}_{12} \\ \hline \hat{\underline{A}}_{21} & | & \hat{\underline{A}}_{22} \end{bmatrix} \quad \hat{\underline{B}} = \begin{bmatrix} \underline{C} \\ \hline \hat{\underline{B}}_2 \end{bmatrix} \qquad \hat{\underline{C}} = [\underline{H} \mid \hat{\underline{C}}_2] \tag{6.58}$$

where

$$\hat{\underline{A}} = \underline{T}^{-1}\underline{A}\underline{T} \qquad \hat{\underline{B}} = \underline{T}^{-1}\underline{B} \qquad \hat{\underline{C}} = \underline{C}\underline{T}$$

Matrices $\underline{F}$, $\underline{G}$, $\underline{H}$ describe Moore's reduced-order model, which is referred to as a direct-elimination technique (residual subsystem is directly eliminated). Thus, the reduced-order system is

$$\dot{\underline{z}}(t) = \underline{F}\underline{z}(t) + \underline{G}\underline{u}(t) \tag{6.59}$$

$$\tilde{\underline{y}}(t) = \underline{H}\underline{z}(t) \tag{6.60}$$

while the residual system is given by

$$\dot{\underline{w}}(t) = \hat{\underline{A}}_{22}\underline{W}(t) + \hat{\underline{B}}_2\underline{u}(t) \tag{6.61}$$

$$\hat{\underline{y}}(t) = \hat{\underline{C}}_2\underline{W}(t) \tag{6.62}$$

Example 6.2.4. Consider a fourth-order system $S(\underline{A},\underline{B},\underline{C})$:

$$\underline{A} = \begin{bmatrix} 0 & 0 & 0 & -150 \\ 1 & 0 & 0 & -245 \\ 0 & 1 & 0 & -113 \\ 0 & 0 & 1 & -19 \end{bmatrix} \qquad \underline{B} = \begin{bmatrix} 4 \\ 1 \\ 0 \\ 0 \end{bmatrix} \qquad \underline{C} = [0 \ 0 \ 0 \ 1]$$

Following the balancing procedure [Eq. (6.58)], one gets

$$\hat{\underline{A}} = \begin{bmatrix} -0.4378 & -1.1168 & -0.4143 & 0.05098 \\ 1.1680 & -3.1350 & -2.8350 & 0.32880 \\ -0.4143 & 2.8350 & -12.480 & 3.24900 \\ -0.05098 & 0.3288 & -3.2490 & -2.95200 \end{bmatrix}$$

$$\hat{\underline{B}} = [-0.1181 \quad 0.1307 \quad -0.05634 \quad -0.006875]^T$$

$$\hat{\underline{C}} = [-0.1181 \quad -0.1307 \quad -0.05634 \quad 0.006875]$$

with Moore's second-order modes in Eq. (6.56) yielding

$$\underline{D} = \text{diag} \ (0.0159 \ , \ 0.00272, \ 0.000127, \ 0.000008)$$

Consider a two-dimensional reduced-order model using Moore's direct-elimination technique and describe as

$$\underline{F} = \begin{bmatrix} -0.4378 & -1.1168 \\ 1.168 & -3.135 \end{bmatrix}$$

$$\underline{G} = [-0.1181 \quad 0.1307]$$

$$\underline{H} = [-0.1181 \quad -0.1307]$$

The eigenvalues of $\underline{A}$ are -1, -3, -5, and -10, the eigenvalues of subsystem $S(\underline{F},\underline{G},\underline{H})$ are -1.113 and -2.460, and the eigenvalues of subsystem $S(\hat{\underline{A}}_{22},\hat{\underline{B}}_2,\hat{\underline{C}}_2)$ are -4.232 and -11.200.

Unstable Systems. Santiago and Jamshidi (1986) have extended the balanced approach to handle unstable, uncontrollable, and unobservable systems. Here, we test only the unstable case. The concept of system stability was introduced briefly in Chapter 2.

As before, consider a linear stable time-invariant system given as

$$\begin{aligned} \dot{\underline{x}} &= \underline{A}\underline{x} + \underline{B}\underline{u} \\ \underline{y} &= \underline{C}\underline{x} \end{aligned} \qquad (6.63)$$

and transfer matrix

$$\underline{G}(s) = \underline{C}(s\underline{I} - \underline{A})^{-1}\underline{B} \tag{6.64}$$

where $x \epsilon R^n$ is the state, $\underline{u} \epsilon R^r$ is the input, and $\underline{y} \epsilon R^m$ is the output. A, B, and C are appropriately dimensioned constant matrices. The above system is assumed to be asymptotically stable, controllable, and observable.

Enns (1984) has discussed the use of the balancing technique for unstable systems. His method is to separate the unstable modes from the stable ones by fraction expansion of the full-order transfer matrix described as

$$\underline{G}(s) = \underline{G}_{us}(s) + \underline{G}_s(s)$$

where $\underline{G}_{us}(s)$ represents a transfer matrix with unstable modes and $\underline{G}_s(s)$ represents a transfer matrix with stable modes. After this separation is performed, the balancing technique is applied to the stable part of the system. It will now be shown that this may not be necessary by noting the following theorem of Kalman (1963), also found in Jamshidi (1983):

Theorem 6.1 (Kalman, 1963). The real parts of the eigenvalues of A are less than k, a real number, if and only if given any symmetrical positive definite matrix $\underline{Q}$, there exists a symmetrical positive definite matrix $\underline{P}$ which is the unique solution to

$$\underline{P}(\underline{A} - \underline{kI}) + (\underline{A} - \underline{kI})^T\underline{P} + \underline{Q} = 0$$

Essentially, all the real parts of the eigenvalues of A are shifted by k. Therefore, for an unstable system, we would shift an unstable matrix by k such that (A − kI) is asymptotically stable. Before presenting the modified Lyapunov equations, the following Lemmas are noted.

Lemma 6.1. Let k be an arbitrary real number. Then $(\underline{A} - k\underline{I}, \underline{B})$ is a controllable pair if and only if $(\underline{A},\underline{B})$ is a controllable pair.

Lemma 6.2. Let k be an arbitrary real number. Then $(\underline{A} - k\underline{I}, \underline{C})$ is an observable pair if and only if $(\underline{A},\underline{C})$ is an observable pair.

Using the above lemmas, and recalling Eq. (6.54)−(6.55), one can move the jw-axis just to the right of the most unstable pole and solve the following modified balancing Lyapunov equations about this shifted axis:

$$\begin{aligned} \underline{W}_o(\underline{A} - k\underline{I}) + (\underline{A} - k\underline{I})^T\underline{W}_o + \underline{C}^T\underline{C} = 0 \\ \underline{W}_c(\underline{A} - k\underline{I})^T + (\underline{A} - k\underline{I})\underline{W}_c + \underline{B}\underline{B}^T = 0 \end{aligned} \tag{6.65}$$

We can relate this shifting technique to frequency-shifting model reduction techniques such as Routh approximation (Jamshidi, 1983). This procedure is discussed next.

Consider the unstable transfer matrix $\underline{H}(s) = \underline{C}(s\underline{I} - \underline{A})^{-1}\underline{B}$ and form a new transfer matrix

$$\tilde{\underline{H}}(s) = \underline{H}(s + k) \qquad k > 0 \text{ and real}$$

The value of k is chosen sufficiently large so that $\tilde{\underline{H}}(s)$ is asymptotically stable. After computing the reduced-order model $\tilde{\underline{H}}_r(s)$, shift the imaginary axis back in the reduced-order model. Thus

$$\underline{H}_r(s) = \tilde{\underline{H}}_r(s - k)$$

where $\underline{H}_r(s)$ denotes the reduced-order model. The following algorithm summarizes the balanced approach for unstable systems.

Algorithm 6.1: Balanced Approach for Unstable Systems.

1. Shift the unstable matrix $\underline{A}$ by k such that $(\underline{A} - k\underline{I})$ is asymptotically stable.
2. Solve the modified Lyapunov equations described in Eq. (6.65).
3. Compute the balancing transformation using the results of step 2 and Laub's method (Laub, 1980).
4. Perform the similarity transformation on the original unstable system matrices $\underline{A}$, $\underline{B}$, and $\underline{C}$ to yield $\hat{\underline{A}}$, $\hat{\underline{B}}$, and $\hat{\underline{C}}$.
5. Partition the system $S(\underline{A},\underline{B},\underline{C})$ appropriately in accordance with Moore's second-order modes.

Example 6.2.5. This example illustrates the use of Algorithm (6.1). Consider the system $S(\underline{A},\underline{B},\underline{C})$ given by

$$\underline{A} = \begin{bmatrix} 0 & 1 & 0 \\ 0 & 0 & 1 \\ -60 & -8 & 9 \end{bmatrix} \qquad \underline{B} = \begin{bmatrix} 0 \\ 0 \\ 1 \end{bmatrix} \qquad \underline{C} = [50 \quad 15 \quad 3]$$

The eigenvalues of $\underline{A}$ are -2, 5, and 6, which indicates that the system is unstable. Using Algorithm (6.1) and letting k = 7 one gets

$$\underline{A} = \begin{bmatrix} 6.5320 & -0.8998 & -0.1211 \\ -0.8998 & 4.5530 & -0.7719 \\ -0.1211 & 0.7719 & -2.0850 \end{bmatrix} \qquad \underline{B} = \begin{bmatrix} -3.003 \\ -2.484 \\ -0.3891 \end{bmatrix}$$

$$\underline{C} = [-3.003 \quad 2.484 \quad -0.3891]$$

$$\underline{D} = \text{diag } (9.641, 1.26, 0.0833)$$

Based on the above realization, the second-order model is given by Moore's direct-elimination technique as

$$\underline{F} = \begin{bmatrix} 6.532 & -0.8998 \\ -0.8998 & 4.553 \end{bmatrix} \qquad \underline{G} = \begin{bmatrix} -3.003 \\ -2.484 \end{bmatrix} \qquad \underline{H} = [-3.003 \quad 2.484]$$

The second-order model has two unstable eigenvalues, 5.976 and 5.109, which compares favorably with the original system. Note that the residual system has an eigenvalue at -2.085.

6.2.2   Perturbation

The basic concept behind perturbation methods is the approximation of a system's structure through neglecting certain interactions within the model that lead to lower order. From a large-scale system modeling point of view, perturbation methods can be considered as approximate aggregation techniques (Jamshidi, 1983). There are two basic classes of perturbations applicable for large-scale systems modeling purpose—"weakly coupled" models and "strongly coupled" models. This classification is not universally accepted, but a great number of authors have adopted it; others use the terms nonsingular and singular perturbations.

   Weakly Coupled Models. In many industrial control systems certain dynamic interactions are neglected to reduce computational burdens for system analysis, design, or both. Examples of such practice are in chemical process control and space guidance (Kokotovic, 1972), where different subsystems are designed for flow, pressure, and temperature control in an otherwise coupled process. However, the computational advantages obtained by neglecting weakly coupled subsystems are offset by a loss in overall system performance. In this section the weakly coupled models for large-scale systems are introduced.

   Consider the following large-scale system split into k linear subsystems:

$$
\begin{bmatrix} \dot{\underline{x}}_1 \\ \dot{\underline{x}}_2 \\ \cdot \\ \cdot \\ \cdot \\ \dot{\underline{x}}_k \end{bmatrix}
=
\begin{bmatrix}
\underline{A}_1 & \varepsilon\underline{A}_{12} & & & & \varepsilon\underline{A}_{1k} \\
\varepsilon\underline{A}_{21} & \underline{A}_2 & \varepsilon\underline{A}_{23} & \cdot & \cdot & \varepsilon\underline{A}_{2k} \\
\cdot & & & & & \cdot \\
\cdot & & & & & \cdot \\
\cdot & \cdot & \cdot & \cdot & \cdot & \varepsilon\underline{A}_{k-1,k} \\
\varepsilon\underline{A}_{k1} & \cdot & \cdot & \varepsilon\underline{A}_{k,k-1} & & \underline{A}_k
\end{bmatrix}
\begin{bmatrix} \underline{x}_1 \\ \underline{x}_2 \\ \cdot \\ \cdot \\ \cdot \\ \underline{x}_k \end{bmatrix}
$$

$$
+
\begin{bmatrix}
\underline{B}_1 & \varepsilon\underline{B}_{12} & \cdot & \cdot \\
\varepsilon\underline{B}_{21} & \underline{B}_2 & & \\
\cdot & & & \cdot \\
\cdot & & & \cdot \\
\cdot & & & \underline{B}_k
\end{bmatrix}
\begin{bmatrix} \underline{u}_1 \\ \underline{u}_2 \\ \cdot \\ \cdot \\ \underline{u}_k \end{bmatrix}
\qquad (6.66)
$$

where ε is a small positive coupling parameter and $\underline{x}_i$ and u_i are ith subsystem state and control vectors, respectively, and all $\underline{A}$ and $\underline{B}$ matrices are assumed to be constant. A special case of Eq. (6.66) is when k = 2; this has been called "ε-coupled" system by Kokotovic (1972), i.e.,

$$\begin{bmatrix} \dot{x}_1 \\ \dot{x}_2 \end{bmatrix} = \begin{bmatrix} A_1 & \varepsilon A_{12} \\ \varepsilon A_{21} & A_2 \end{bmatrix} \begin{bmatrix} x_1 \\ x_2 \end{bmatrix} + \begin{bmatrix} B_1 & \varepsilon B_{12} \\ \varepsilon B_{21} & B_2 \end{bmatrix} \begin{bmatrix} u_1 \\ u_2 \end{bmatrix}$$

$$(6.67)$$

It is clear that when $\varepsilon = 0$, the above system decouples into two sub-systems:

$$\dot{\hat{x}}_1 = A_1 \hat{x}_1 + B_1 \hat{u}_1$$

$$\dot{\hat{x}}_2 = A_2 \hat{x}_2 + B_2 \hat{u}_2 \qquad\qquad (6.68)$$

which correspond to two approximate aggregated models (Jamshidi, 1983), one for each subsystem. In this way the computation associated with simulation, design, etc., will be reduced drastically, especially for large-scale system order n with k larger than 2.

The research regarding weakly coupled systems has taken two main lines. The first line is to set $\varepsilon = 0$ in Eq. (6.67) and try to find a quantitative measure of the resulting approximation when, in fact, $\varepsilon \neq 0$ in an actual condition. Bailey and Ramapriyan (1973) have provided conditions that would give an estimation on the loss in the optimal performance in a linear state regulator formulation of Eq. (6.67). Furthermore, they have presented conditions for criteria of weak coupling, a similar task to that of Milne (1965), whose results will be presented here without proofs.

Consider a coupled A matrix as presented in Eq. (6.67) and assume that A_1, A_{12}, A_{21}, and A_2 are of $n_1 \times n_1$, $n_1 \times n_2$, $n_2 \times n_1$, and $n_2 \times n_2$ dimensions, respectively with $n = n_1 + n_2$ being the order of the original large-scale coupled system. Furthermore, let

$$\lambda_i \{A_1\} = \{\hat{\lambda}_1, \hat{\lambda}_2, \ldots, \hat{\lambda}_{n_1}\} \qquad i = 1, 2, \ldots, n_1$$

$$\lambda_j \{A_2\} = \{\hat{\lambda}_{n_1+1}, \ldots, \hat{\lambda}_n\} \qquad j = n_1 + 1, \ldots, n \qquad (6.69)$$

$$\lambda_k \{A\} = \{\lambda_1, \lambda_2, \ldots, \lambda_n\} \qquad k = 1, 2, \ldots, n$$

be the eigenvalues of diagonal submatrices A_1, A_2, and matrix A. Let us postulate that the moduli of the eigenvalues of A_1 and A_2 are widely separated from each other. Without any loss of generality one can take $|\lambda_j\{A_2\}| \ll |\lambda_i\{A_1\}|$. Let the eigenvalues of A_1 be on or outside a circle with radius $R = \min |\lambda_i\{A_1\}|$, $i = 1, 2, \ldots, n_1$ and the eigenvalues of A_2 be on or inside a circle with radius $r = \max |\lambda_j\{A_2\}|$, $j = n_1 + 1, n_1 + 2, \ldots, n$, as shown in Figure 6.5. If the following conditions are satisfied, then the system is said to be weakly coupled (Milne, 1965; Aoki, 1971):

(i) $(r/R) \ll 1$ $\qquad\qquad (6.70)$

(ii) $\dfrac{(n_1 \varepsilon_{12} \varepsilon_{21})}{R^2} \ll 1$ $\qquad\qquad (6.71)$

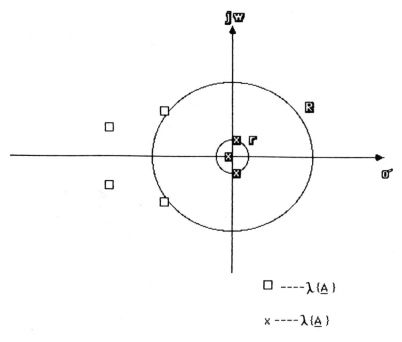

Figure 6.5   A demonstration of the slow and fast eigenvalues and the
"separation ratio."

where $\varepsilon_{12} = \max |(\underline{A}_{12})_{i,j}|$ and $\varepsilon_{21} = \max |(\underline{A}_{21})_{k,\ell}|$ for i,j =
1, 2, . . ., n_1 and k, ℓ = 1, 2, . . ., n_2. The term (r/R) is called
"separation ratio;" ε_{12} and ε_{21} represent the maximum of the moduli of the
elements of $\underline{A}_{12}$ and $\underline{A}_{21}$ submatrices, respectively.

Example 6.2.6.   Consider a sixth-order $\underline{A}$ matrix with the following
structure:

$$
\underline{A} =
\begin{bmatrix}
0 & 1 & 0 & 0 & 0 & 0 \\
0 & 0 & 1 & 0.2 & 0.05 & 0 \\
-400 & -170 & -23 & 0.05 & -.02 & 0 \\
0 & -.01 & -0.02 & 0 & 1 & 0 \\
0 & 0.01 & -0.015 & 0 & 0 & 1 \\
0 & 0 & 0 & -0.04 & -0.53 & -1
\end{bmatrix}
$$

Here we wish to check for weak coupling.
     The eigenvalues of $\underline{A}$ are -9.94, -8.08, -4.97, -0.0816, -0.7924, and
-0.52, which indicate that the first three are much farther away from
the jω-axis than the last three, which can be considered as dominant. For
two 3 × 3 diagonal submatrices, it can be seen that submatrices $\underline{A}_1$ and $\underline{A}_2$
are both in companion forms with the following eigenvalues:

$$\lambda_i\{\underline{A}_1\} = \{-5, -10, -8\} \qquad \lambda_j\{\underline{A}_2\} = \{-0.1, -0.8, -0.5\}$$

implying that r = 0.8, R = 5, and (r/R) = 0.16, which is much smaller than 1; hence condition Eq. (6.70) holds. The values of ε_{12} and ε_{21} are 0.2 and 0.02, respectively, and the quantity

$$(n_1 \varepsilon_{12} \varepsilon_{21})/R^2 = 0.00048 \ll 1$$

Therefore, it is concluded that a system with the given $\underline{A}$ matrix is weakly coupled.

The second line of research regarding the weakly coupled systems has been to exploit it in an algorithmic fashion to find an approximate optimal feedback gain through a MacLaurin's series expansion of the accompanying Riccati matrix in the coupling parameter ε. It has been shown that retaining k terms of the Riccati matrix expansion would give an approximation of order 2k to the optimal cost (Kokotovic, 1972). This approximate solution of the Riccati matrix can be used for near-optimum design of large-scale systems (Jamshidi, 1983). The remainder of this section is devoted to strong coupling of large-scale systems.

Strongly Coupled Models. Strongly coupled systems are those whose variables have widely distinct speeds. The models of such systems are based on the concept of "*singular perturbation*," which differ from regular perturbation (weakly coupled systems) in that perturbation is to the left of the system's state equation, i.e., a small parameter multiplying the time derivative of the state vector. In practice, many systems, most of them large, possess fast-changing variables displaying a singularly perturbed characteristic. A few examples were given earlier; others include power systems where the frequency and voltage transients vary from a few seconds in generator regulators, shaft-stored energy, and speed governor motion to several minutes in prime mover motion, stored thermal energy, and load voltage regulators. Similar time-scale properties prevail in many other practical systems and processes, such as industrial control systems, e.g., cold rolling mills, biochemical processes, nuclear reactors, aircraft and rocket systems, and chemical diffusion reactions.

Consider a singularly perturbed linear dynamic system described by

$$\dot{\underline{x}}(t) = \underline{A}_1 \underline{x}(t) + \underline{A}_{12} \underline{z}(t) + \underline{B}_1 u(t) \qquad \underline{x}(t_0) = \underline{x}_0 \qquad (6.72)$$

$$\varepsilon \dot{\underline{z}}(t) = \underline{A}_{21} \underline{x}(t) + \underline{A}_2 \underline{z}(t) + \underline{B}_2 \underline{u}(t) \qquad \underline{z}(t_0) = \underline{z}_0 \qquad (6.73)$$

and if A_2 is nonsingular, Eq. (6.72) and Eq. (6.73) will become

$$\dot{\hat{\underline{x}}}(t) = (\underline{A}_1 - \underline{A}_{12}\underline{A}_2^{-1}\underline{A}_{21})\hat{\underline{x}} + (\underline{B}_1 - \underline{A}_{12}\underline{A}_2\underline{B}_2)\hat{u} \qquad (6.74)$$

$$\hat{\underline{z}}(t) = -\underline{A}_2^{-1}\underline{A}_{21}\hat{\underline{x}} - \underline{A}_2^{-1}\underline{B}_2\hat{u} \qquad (6.75)$$

Equation (6.74) is an approximate aggregated model for Eq. (6.72)–(6.73), which in effect means that the n eigenvalues of the original system are approximated by the k eigenvalues of $(\underline{A}_1 - \underline{A}_{12}\underline{A}_2^{-1}\underline{A}_{21})$ matrix in Eq.

(6.74). This observation follows the same line of argument when discussing conditions for weakly coupled systems considered by Milne (1965),
Aoki (1971), Arak (1978), Bailey and Ramapriyan (1973).

   Boundary Layer Correction. It is noted that in this model reduction,
i.e., going from Eq. (6.72)–(6.73) to Eq. (6.74) the initial condition z_0
of z(t) is lost and the values of $\hat{z}(t_0)$ and $z(t_0)$ = z_0 are in general different; the difference, as shown in Figure 6.6, is termed a left-side "boundary
layer," which corresponds to the fast transients of Eq. (6.72)–(6.73)
(Jamshidi, 1983). To investigate this phenomenon, which in effect explains
under what conditions $\hat{x}$ and $\hat{z}$ approximate x and z, let control u be zero
in Eq. (6.72)–(6.73) and let the error between z and $\hat{z}$ be defined by

$$\eta(t) = z(t) - \hat{z}(t) = z(t) + A_2^{-1}A_{21}\hat{x}(t) \tag{6.76}$$

and choose a matrix $E_1(\varepsilon)$ so that when

$$\eta(t) = z(t) + A_2^{-1}A_{21}x(t) + \varepsilon E_1(\varepsilon) x(t) \tag{6.77}$$

is substituted in Eq. (6.72) and Eq. (6.73), with u = 0, η separates from
x as

$$\dot{x}(t) = (A_1 - A_{12}A_2^{-1}A_{21} + \varepsilon E_2) x(t) + A_{12}\eta(t) \tag{6.78}$$

$$\varepsilon\eta(t) = (A_2 + \varepsilon E_3)\eta(t) \tag{6.79}$$

It can be shown that there exists an ε^* such that E_i = $E_i(\varepsilon)$, i = 1, 2, 3
bounded over $[0, \varepsilon^*]$ (Kokotovic et al., 1976). As $\varepsilon \to 0$, the eigenvalues
of Eq. (6.79) would tend to infinity very much like $\lambda\{A_2/\varepsilon\}$ would.

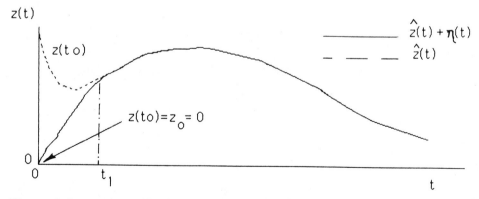

Figure 6.6   A graphic description of the boundary layer of a fast
variable in a singularly perturbed system.

Now a new time variable τ, called "stretched time scale," is defined:

$$\tau = \frac{(t - t_0)}{\varepsilon} \tag{6.80}$$

where $\tau = 0$ at $t = t_0$ and $dt = \varepsilon d\tau$. For a change from t to τ, the system in Eq. (6.79) will become

$$\frac{d\underline{n}(\tau)}{d\tau} = (\underline{A}_2 + \varepsilon \underline{E}_3)\underline{n}(\tau) \tag{6.81}$$

which continuously depends on ε, and at $\varepsilon = 0$ it becomes

$$\frac{d\underline{n}(\tau)}{d\tau} = \underline{A}_2 \underline{n}(\tau) \tag{6.82}$$

with initial condition

$$\underline{n}(0) = \underline{z}(t_0) - \hat{\underline{z}}(t_0) \tag{6.83}$$

Equations (6.82) and (6.83) constitute the so-called boundary layer correction for $\underline{z}(t) = \hat{\underline{z}}(t) + \underline{n}[(t - t_0)/\varepsilon]$. From the above formulation it can be shown that the "slow" and "fast" states x(t) and z(t) are

$$\underline{x}(t) = \hat{\underline{x}}(t) + \underline{O}(\varepsilon) \tag{6.84}$$

$$\underline{z}(t) = \hat{\underline{z}}(t) + \underline{n}(\tau) + O(\varepsilon) \tag{6.85}$$

where $\underline{O}(\varepsilon)$ is a so-called "large-O"-order of ε and is defined as a function whose norm is less than $k\varepsilon$ with k being a constant. It is noted that the boundary layer correction is only significant for the first few seconds away from t_0 and reduces to zero after $t = t_1$ sec as an exponential decay in $\tau = (t - t_0)/\varepsilon$ (Kokotovic et al., 1976). Figure 6.6 shows the boundary layer phenomenon for the fast state $\underline{z}(t)$.

6.3 HIERARCHICAL CONTROL

The notion of a large-scale system, as it was briefly discussed in Section 6.1, may be described as a complex system composed of a number of constituents or smaller subsystems serving particular functions and shared resources and governed by interrelated goals and constraints. Although interaction among subsystems can take many forms, one of the most common is hierarchical, which appears somewhat natural in economic management, organizational, and complex industrial systems such as steel, oil, and paper. Within this hierarchical structure, the subsystems are positioned on levels with different degrees of hierarchy. A subsystem at a given level controls or coordinates the units on the level below it and is, in turn, controlled or coordinated by the unit on the level immediately above it. Figure 6.7 shows a typical hierarchical (multilevel) system. The highest-level coordinator, sometimes called the supremal coordinator, can be thought of as the board of directors of a corporation, while another

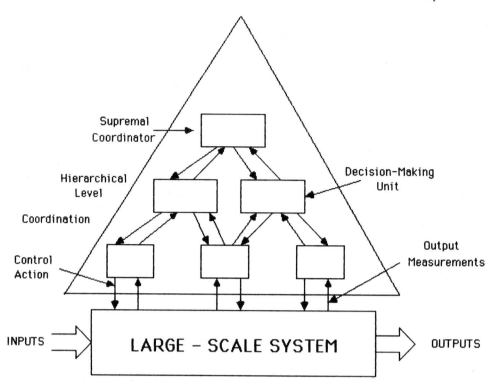

Figure 6.7   A hierarchical (multilevel) control strategy for a large-scale system.

level's coordinators may be the president, vice-presidents, directors, etc. The lower levels can be occupied by plant manager, shop managers, etc., while the large-scale system is the plant itself.   In spite of this seemingly natural representation of a hierarchical structure, its exact behavior has not been well understood mainly because relatively little quantitative work has been done on these large-scale systems.   Mesarovic et al. (1970) presented one of the earliest formal quantitative treatments of hierarchical (multilevel) systems (Jamshidi, 1983).

There is no uniquely or universally accepted set of properties associated with the hierarchical systems.   However, the following are some of the key properties:

1.   A hierarchical system consists of decision-making components structured in a pyramid shape (Figure 6.7).
2.   The system has an overall goal which may (or may not) be in harmony with all its individual components.
3.   The various levels of hierarchy in the system exchange information (usually vertically) among themselves iteratively.
4.   As the level of hierarchy goes up, the time horizon increases; i.e., the lower-level components are faster than the higher-level ones.

Based on the above discussion, one can make a tentative conclusion that a successful operation of hierarchical system is best described by two

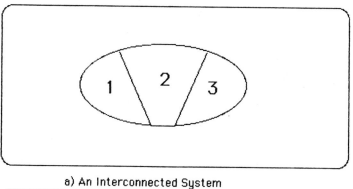

b) An Interconnected System

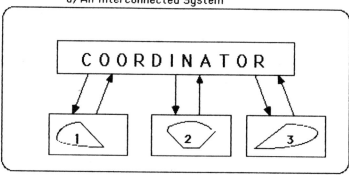

b) Hierarchically Structured System

Figure 6.8   A graphical demonstration of "decomposition" and "coordination" concepts.   (a) An interconnected system.   (b) Hierarchically
structured system.

processes known as decomposition and coordination.   A pictorial representation of these two notions is shown in Figure 6.8.   In summary, then, the
basic notions behind *hierarchical control* are (a) to *decompose* a given
large-scale system into a number of small-scale subsystems, and then
through the (b) *coordination* of these subsystems' solutions, feasibility
and optimality of the overall solution is achieved through a multilevel
iterative algorithm.

6.3.1   Goal Coordination—Interaction Balance

Consider a linear large-scale system described by the following state
equation:

$$\dot{\underline{x}} = \underline{A}\underline{x} + \underline{B}\underline{u} \qquad \underline{x}(t_0) = \underline{x}_0 \tag{6.86}$$

with its cost function, to be minimized,

$$J = \frac{1}{2}\underline{x}^T(t_f)\underline{F}\underline{x}(t_f) + \frac{1}{2}\int_{t_0}^{t_f} (\underline{x}^T\underline{Q}\underline{x} + \underline{u}^T\underline{R}\underline{u})\ dt \tag{6.87}$$

where F, Q $\geqslant$ 0, R > 0, t_0, t_f are initial and final values of time, and x_0 is the initial state. All the remaining terms correspond to the usual linear systems theory (Kwakernaak and Sivan, 1972) optimal control problem. Assume that the order n of system of Eq. (6.86) is too large (say n > 200) to be able to solve this problem as is. It is assumed that Eq. (6.86) can be decomposed into N subsystems described by

$$\dot{\underline{x}}_i = \underline{A}_i \underline{x}_i + \underline{B}_i \underline{u}_i + \underline{z}_i \qquad \underline{x}_i(t_0) = \underline{x}_{i0} \qquad i = 1, \ldots, N \qquad (6.88)$$

where $\underline{z}_i$ defined as

$$\underline{z}_i = \sum_{\substack{j=1 \\ j \neq i}}^{N} \underline{G}_{ij} \underline{x}_j \tag{6.89}$$

describes the ith subsystem's interaction with the remaining N − 1 subsystems. The original system's interaction is reduced to the optimization of N subsystems, which collectively satisfy Eq. (6.88)−(6.89) while minimizing

$$J = \sum_{i=1}^{N} \left\{ \frac{1}{2} \underline{x}_i^T(t)\underline{F}_i\underline{x}_i(t) + \frac{1}{2}\int_0^{t_f} \left[\underline{x}_i^T(t)\underline{Q}_i\underline{x}_i(t) + \underline{u}_i^T(t)\underline{R}_i\underline{u}_i(t) \right. \right.$$

$$\left. \left. + \underline{z}_i^T(t)\underline{S}_i\underline{z}_i(t) \right] dt \right\} \tag{6.90}$$

where $\underline{Q}_i$ and $\underline{F}_i$ are $n_i \times n_i$ positive semidefinite matrices, $\underline{R}_i$ and $\underline{S}_i$ are $m_i \times m_i$ and $n_i \times n_i$ positive definite matrices with

$$n = \sum_{i=1}^{N} n_i \qquad m = \sum_{i=1}^{N} m_i \tag{6.91}$$

The physical interpretation of the last term in the integrand of Eq. (6.90) is difficult at this point. In fact, the introduction of this term is to avoid singular controls.

In this decomposition of a large interconnected linear system the common coupling factors among its N subsystems are the "interaction" variables $\underline{z}_i(t)$, which, along with Eq. (6.88) and Eq. (6.89), is called "global" and is denoted by S_G. The following assumption is considered to hold. The global problem is replaced by a family of N subproblems coupled together through a parameter vector $\underline{\alpha} = (\underline{\alpha}_1, \ldots, \underline{\alpha}_N)^T$ and denoted by $s_i(\underline{\alpha})$, i = 1, . . ., N. In other words, the global system problem is "imbedded" into a family of subsystem problems through an imbedding parameter α in such a way that for a particular value of $\underline{\alpha}^*$, the subsystems $s_i(\underline{\alpha}^*)$, i = 1, 2, . . ., N yield the desired solution to S_G. In terms of hierarchical control notation, this imbedding concept is nothing but the notion of coordination, but in mathematical programming problem terminology, it is

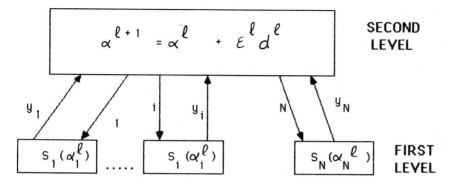

Figure 6.9   The two-level goal coordination structure for dynamic systems.

denoted as the "master" problem.   Figure 6.9 shows a two-level control structure of a large-scale system.   Under this strategy, each local con-troller i receives $\underline{\alpha}_i^\ell$ from the coordinator (second-level hierarchy), solves $S_i(\underline{\alpha}_i^\ell)$, and transmits (reports) some function y_i^ℓ of its solution to the coordinator.   The coordinator, in turn, evaluates the next updated value of α, i.e.,

$$\underline{\alpha}^{\ell+1} = \underline{\alpha}^\ell + \varepsilon^\ell \underline{d}^\ell \tag{6.92}$$

where ε^ℓ is the ℓth iteration step size, and the update term $\underline{d}^\ell$ is de-fined from the following conjugate gradient expression:

$$\underline{d}^{\ell+1}(t) = \underline{e}^{\ell+1}(t) + \gamma^{\ell+1}\underline{d}^\ell(t) \qquad 0 \leqslant t \leqslant t_f \tag{6.93}$$

where

$$\gamma^{\ell+1} = \frac{\displaystyle\int_0^{t_f} [\underline{e}^{\ell+1}(t)]^T \underline{e}^{\ell+1}(t)\,dt}{\displaystyle\int_0^{t_f} (\underline{e}^\ell)^T \underline{e}^\ell \,dt} \tag{6.94}$$

and $\underline{d}^0 = \underline{e}^0$.   The vector $\underline{e}(t)$ is commonly known as the "interaction error," defined by

$$\underline{e}_i[\underline{\alpha}(t),t] = \underline{z}_i[\underline{\alpha}(t),t] - \sum_{\substack{j=1 \\ j\neq i}}^{N} \underline{G}_{ij}\underline{x}_j[\underline{\alpha}(t),t] \tag{6.95}$$

The relations of Eq. (6.94)–(6.95) constitute the so-called "second-level" or "coordinator" problem. The "first-level" or "subsystem" problem, represented by Eq. (6.88)–(6.90), constitutes a linear regulator problem and is handled by a Riccati formulation (Jamshidi, 1983). By virtue of Figure 6.9, each time the coordinator passes down a new (updated) coordination vector $\underline{\alpha}^* = (\underline{\alpha}_1^{*T} \; \ldots \; \underline{\alpha}_N^{*T})$ the coordinator checks to see whether the interaction error among subsystems has sufficiently diminished, at which time it is said that "interaction balance" is achieved; i.e., the normalized interaction error:

$$\left(\sum_{i=1}^{N} \int_0^{t_f} \left\{ \underline{z}_i - \sum_{\substack{j=1 \\ j \neq i}}^{N} \underline{G}_{ij}\underline{x}_j \right\}^T \left\{ \underline{z}_i - \sum_{\substack{j=1 \\ j \neq i}}^{N} \underline{G}_{ij}\underline{x}_j \right\} dt \right) / \Delta t \qquad (6.96)$$

is sufficiently small. Here Δt is the step size of integration.

6.3.2 Interaction Prediction

An alternative approach in optimal control of hierarchical systems which has both open- and closed-loop forms is the interaction prediction method, which avoids second-level gradient-type iterations. Consider a large-scale linear interconnected system which is decomposed into N subsystems, each of which is described by

$$\underline{\dot{x}}_i(t) = \underline{A}_i\underline{x}_i(t) + \underline{B}_i\underline{u}_i(t) + \underline{z}_i(t) \qquad \underline{x}_i(0) = \underline{x}_{io} \qquad i = 1, 2, \ldots, N \qquad (6.97)$$

where the interaction vector $\underline{z}_i$ is

$$\underline{z}_i(t) = \sum_{\substack{j=1 \\ j \neq i}}^{N} \underline{G}_{ij}\underline{x}_j(t) \qquad (6.98)$$

The optimal control problem at the first level is to find a control $u_i(t)$ that satisfies Eq. (6.97)–(6.98) while minimizing a quadratic cost function:

$$J_i = \frac{1}{2}\underline{x}_i^T(t_f)\underline{Q}_i\underline{x}_i(t_f) + \frac{1}{2}\int_o^{t_f} \{\underline{x}_i^T(t)\underline{Q}_i\underline{x}_i(t) + \underline{u}_i^T(t)\underline{R}_i(t)\underline{u}_i(t)\} \, dt \qquad (6.99)$$

This problem can be solved by first introducing a set of Lagrange multipliers $\underline{\alpha}_i(t)$ and costate vectors $p_i(t)$ to augment the "interaction" equality constraint Eq. (6.98) and subsystem dynamic constraint of Eq. (6.97) to the cost function's integrand; i.e., the ith subsystem Hamiltonian is defined by

$$H_i = \frac{1}{2} \underline{x}_i^T(t) \underline{Q}_i \underline{x}_i(t) + \frac{1}{2} \underline{u}_i^T(t) \underline{R}_i \underline{u}_i(t) + \underline{\alpha}_i^T \underline{z}_i$$

$$- \sum_{\substack{j=1 \\ j \neq i}}^{N} \underline{\alpha}_j^T \underline{G}_{ji} \underline{x}_i + \underline{p}_i^T (\underline{A}_i \underline{x}_i + \underline{B}_i \underline{u}_i + \underline{z}_i) \tag{6.100}$$

Then, utilizing the necessary conditions of optimality, using a Riccati formulation,

$$\underline{p}_i(t) = \underline{K}_i(t) \underline{x}_i(t) + \underline{g}_i(t) \tag{6.101}$$

and simplifying the resulting TPBV (two-point boundary value) problem, one obtains

$$\dot{\underline{K}}_i(t) = -\underline{K}_i(t) \underline{A}_i - \underline{A}_i^T \underline{K}_i(t) + \underline{K}_i(t) \underline{S}_i \underline{K}_i(t) - \underline{Q}_i \tag{6.102}$$

$$\dot{\underline{g}}_i(t) = -[\underline{A}_i - \underline{S}_i \underline{K}_i(t)]^T \underline{g}_i(t) - \underline{K}_i(t) \underline{z}_i(t) + \sum_{\substack{j=1 \\ j \neq i}}^{N} \underline{G}_{ji}^T \underline{\alpha}_j^T(t) \tag{6.103}$$

whose final conditions $K_i(t_f)$ and $g_i(t_f)$ follow from

$$\underline{p}_i(t_f) = \partial \left[\frac{1}{2} \underline{x}_i^T(t_f) Q_i \underline{x}_i(t_f) \right] / \partial \underline{x}_i(t_f) = \underline{Q}_i \underline{x}_i(t_f) \tag{6.104}$$

and Eq. (6.101), i.e.,

$$\underline{K}_i(t_f) = Q_i \qquad g_i(t_f) = 0 \tag{6.105}$$

Following this formulation, the first-level optimal control becomes

$$\underline{u}_i(t) = -\underline{R}_i^{-1} \underline{B}_i^T \underline{K}_i(t) \underline{x}_i(t) - \underline{R}_i^{-1} \underline{B}_i^T \underline{g}_i(t) \tag{6.106}$$

which has a partial feedback (closed-loop) term and a feedforward (open-loop) term. Two points are made here. First, the solution of the differential matrix Riccati equation which involves $(n_i + 1)n_i/2$ nonlinear scalar equations is independent of the initial state $x_i(0)$. The second point is that unlike $\underline{K}_i(t)$, $\underline{g}_i(t)$ in Eq. (6.103), by virtue of $\underline{z}_i(t)$, is dependent on $\underline{x}_i(0)$.

The second-level problem is essentially updating the new coordination vector. For this purpose, using partial derivatives, the additively separable Lagrangian would result in the following second-level coordination procedure at the $(\ell + 1)$th iteration:

$$\begin{bmatrix} \underline{\alpha}_i(t) \\ \\ \underline{z}_i(t) \end{bmatrix}^{\ell+1} = \begin{bmatrix} -\underline{p}_i(t) \\ \\ \displaystyle\sum_{\substack{j=1 \\ j\neq i}}^{N} \underline{G}_{ji}\underline{x}_j(t) \end{bmatrix}^{\ell} \tag{6.107}$$

The interaction prediction method is formulated by the following algorithm.

Algorithm 6.3 Interaction Prediction.

Step 1: Solve N independent differential matrix Riccati Eq. (6.102) with final condition Eq. (6.105) and store $\underline{K}_i(t)$, i = 1, 2, . . ., N.

Step 2: For initial $\underline{\alpha}_i^\ell(t)$, $\underline{z}_i^\ell(t)$ solve the "adjoint" Eq. (6.103) with final condition Eq. (6.105). Evaluate and store $\underline{g}_i(t)$, i = 1, 2, . . ., N.

Step 3: Solve the state equation

$$\dot{\underline{x}}_i(t) = [\underline{A}_i - \underline{S}_i\underline{K}_i(t)]\,\underline{x}_i(t) - \underline{S}_i\underline{g}_i(t) + \underline{z}_i(t) \qquad \underline{x}_i(0) = \underline{x}_{i0} \tag{6.108}$$

Step 4: At the second level, use the results of steps 2 and 3 and Eq. (6.107) to update the overall interaction error

$$e(t) = \left(\sum_{i=1}^{N}\int_0^{t_f} \left\{ \underline{z}_i(t) - \sum_{j=1}^{N} \underline{G}_{ij}\underline{x}_j(t)\right\}^T \left\{ \underline{z}_i(t)\right.\right.$$

$$\left.\left. - \sum_{\substack{j=1 \\ j\neq i}}^{N} \underline{G}_{ij}\underline{x}_j(t)\right\} dt \right) / \Delta t \tag{6.109}$$

It must be noted that, depending on the type of digital computer and its operating system, subsystem calculations may be done in parallel and that the N matrix Riccati equations at step 1 are independent of $x_i(0)$, and hence they need to be computed once regardless of the number of second-level iterations in the interaction prediction algorithm [Eq. (6.107)]. It is further noted that unlike the goal coordination methods, no $z_i(t)$ term is needed in the cost function, which is intended to avoid singularities.

Below, two examples are used to illustrate the goal coordination (interaction balance) and interaction prediction algorithms.

Example 6.3.1. Consider a two-reach model of a river pollution control problem (Jamshidi, 1983)

$$
\dot{\underline{x}} = \left[\begin{array}{cc|cc} -1.32 & 0 & 0 & 0 \\ -0.32 & -1.2 & 0 & 0 \\ \hline 0.90 & 0 & -1.32 & 0 \\ 0 & 0.9 & -0.32 & -1.2 \end{array}\right] \underline{x} + \left[\begin{array}{c|c} 0.1 & 0 \\ 0 & 0 \\ \hline 0 & 0.1 \\ 0 & 0 \end{array}\right] \underline{u} \qquad (6.110)
$$

where each reach (subsystem) of the river has two states: x_{11} is the concentration of basic oxygen demand (BOD), x_{12} is the concentration of dissolved oxygen (DO), and its control $\underline{u}_1$ is the BOD of the effluent discharge into the river. For a quadratic cost function

$$
J = \frac{1}{2} \int_0^5 (\underline{x}^T \underline{Q} \underline{x} + \underline{u}^T \underline{R} \underline{u}) \; dt \qquad (6.111)
$$

with $\underline{Q} = \text{diag}\ \{2,4,2,4\}$ and $\underline{R} = \text{diag}\ \{2,2\}$, it is desired to find an optimal control which minimizes Eq. (6.111) subject to Eq. (6.110), $\underline{x}(0) = (1\ 1\ -1\ 1)^T$, using goal coordination.

The first two-level problems are identical, and a second-order matrix Riccati is solved by integrating using a fourth-order Runge-Kutta method for $\Delta t = 0.1$. The interaction error for this example reduced to about 10^{-5} in 15 iterations, as shown in Figure 6.10.

Below, is a CAD example using program "*INTPRD*" (interaction prediction) of *LSSPAK/PC*.

CAD Example 6.3.2. INTRPRD solves a two-subsystem hierarchical control system using the method of interaction prediction.

Optimization via the interaction prediction method.
    Total no. of second level iterations = 4
    Error tolerance for multilevel iterations = .001
    Order of first subsystem n1 = 2
    Order of second subsystem n2 = 2
    Order of first subsystem control vector r1 = 1
    Order of second subsystem control vector r2 = 2

Matrix $\underline{A}1$
    0.200D+01   0.100D+00
    0.200D+00 −.100D+01

Matrix $\underline{B}1$
    0.100D+01
    0.100D+00

Matrix $\underline{R}1$
    0.100+D+01

Matrix $\underline{G}12$
    0.100D-01   0.000D+00
    0.100D+00 −.500D+00

LAGRANGE MULTIPLIER LAMBDA-1
    Lm1 (1): .5
    Lm1 (2): .5

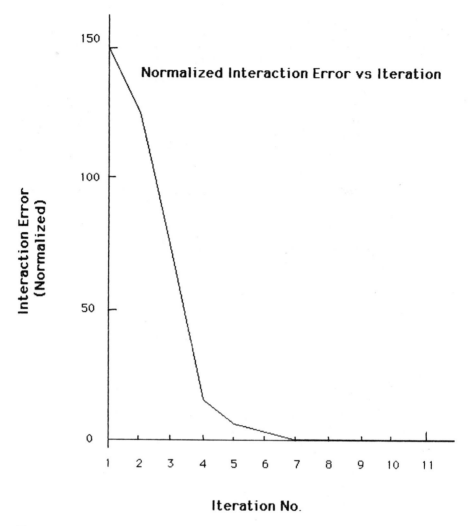

Figure 6.10   Normalized interaction error versus iteration for CAD
Example (6.3.1).

Initial conditions $\underline{x}2$
    $x2(1)$: 1
    $x2(2)$: $-.5$

Initial interaction vector $\underline{z}1$ for subsystem 1
    0.100
    0.3500

The Riccati matrix for each subsystem must be solved using the routine
RICRKUT and the coefficients of the Riccati matrix upper diagonal expressed in a polynomial approximation.

K(1,1) = A(1,1) + A(1,2)*t + A(1,3)*t**2 + · · · + A(1,n+1)*t**n

K(1,2) = A(2,1) + A(2,2)*t + A(2,3)*t**2 + · · · + A(2,n+1)*t**n

.
.
.

K(m,m) = A(m,1) + A(m,2)*t + A(m,3)*t**2 + · · · + A(m,n+1)*t**n

where m = (subsystem order*) (subsystem order + 1)/2 and n = polynomial approximation order.

Matrix $\underline{K}$1

0.444D+01	0.320D+00	−.126D+01
0.900D-01	0.700D-02	−.270D-01
0.500D+00	0.340D-01	−.141D+00

Matrix $\underline{K}$2

0.287D+01	−.526D-01	0.242D+01
−.100D+00	0.160D+00	−.540D-01
0.730D+00	0.118D+00	−.830D+00

Matrix $\underline{A}$2

0.100D+01	0.500D-01
−.250D+00	−.120D+01

Matrix $\underline{B}$2

0.500D+00
0.250D+00

Matrix $\underline{R}$2

0.200D+01

Matrix $\underline{G}$21

0.500D-01	0.150D+00
0.000D+00	−.200D+00

LAGRANGE MULTIPLIER LAMBDA-2
Lm2(1): .75
Lm2(2): .75

Initial conditions $\underline{x}$1
x1(1): −1
x1(2): .1

At second-level iteration no. 1 interaction error = 0.143D+00
At second-level iteration no. 2 interaction error = 0.125D-02
At second-level iteration no. 3 interaction error = 0.966D-03

6.4   DECENTRALIZED CONTROL

As mentioned in Section 6.1, in many large-scale systems the notion of "centrality" does not hold. Under such conditions, the system remains structurally intact while its output information is shared among N controllers, called "decentralized," which collectively contribute to control

the system. Therefore, the basic difference between decentralized and
hierarchical control is the following. In hierarchical control, the system is
decomposed into subsystems whose solutions are *coordinated* by a higher-
level controller, e.g., a supervisory-type control. In decentralized control,
on the other hand, the system's output measurement is shared among a finite
number of controllers which collectively help control the system.

The main motivation behind decentralized control is the failure of conventional methods of centralized control theory. Some fundamental techniques, such as pole placement, state feedback, optimal control, state
estimation, and the like, require complete information from all system sensors for the sake of feedback control. This scheme is clearly inadequate
for feedback control of large-scale systems. Because of the physical configuration and often high dimensionality of such systems, a centralized
control is neither economically feasible nor even necessary. Therefore,
in many applications of feedback control theory to linear large-scale systems some degree of restriction is assumed to prevail on the transfer of
information. In some cases, a total decentralization is assumed; i.e.,
every local control u_i is obtained from the local output y_i and possible
external input v_i. In others, an intermediate restriction on the information is possible.

In this section one major problem related to the decentralized structure
of large-scale control systems is addressed. This problem has to do with
finding a state or an output feedback gain whereby the closed-loop system has all its poles in "*stabilization*" (Seraji, 1978). Alternatively, the
closed-loop poles of controllable system may be preassigned through the
state or output feedback.

6.4.1   The Stabilization Problem

Consider a large-scale linear TIV (time-invariant) system with N local
control stations (channels):

$$\dot{\underline{x}}(t) = \underline{A}\underline{x}(t) + \sum_{i=1}^{N} \underline{B}_i\underline{u}_i(t) \qquad (6.112)$$

$$\underline{y}_i(t) = \underline{C}_i\underline{x}, \; i = 1, \, 2, \, \ldots, \, N \qquad (6.113)$$

where $\underline{x}$ is an $n \times 1$ state vector, and u_i and y_i are $m_i \times 1$ and $r_i \times 1$
control and output vectors associated with the ith control station, re-
spectively. The original system control and output orders m and r are
given by

$$m = \sum_{i=1}^{N} m_i \qquad\qquad r = \sum_{i=1}^{N} r_i \qquad (6.114)$$

The decentralized stabilization problem is defined as follows: Obtain N
local output control laws, each with its independent dynamic compensator

$$\underline{u}_i(t) = \underline{H}_i \underline{z}_i(t) + \underline{K}_i \underline{y}_i(t) + \underline{L}_i \underline{v}_i(t) \qquad (6.115)$$

$$\dot{\underline{z}}_i(t) = \underline{F}_i \underline{z}_i(t) + \underline{S}_i \underline{y}_i(t) + \underline{G}_i \underline{v}_i(t) \qquad (6.116)$$

so that the system Eq. (6.112)–(6.113) in its closed-loop form is stabilized. In Eq. (6.115)–(6.116) $z_i(t)$ is the $n_i \times 1$ output vector of the ith compensator, $\underline{v}_i(t)$ is the $m_i \times 1$ external input vector for the ith controller, and matrices $\underline{H}_i$, $\underline{K}_i$, $\underline{L}_i$, $\underline{F}_i$, $\underline{S}_i$ and $\underline{G}_i$ are $m_i \times n_i$, $m_i \times r_i$, $m_i \times m_i$, $n_i \times n_i$, $n_i \times r_i$, and $n_i \times m_i$, respectively. Alternatively, the problem can be restated as follows: Find matrices $\underline{H}_i$, $\underline{K}_i$, $\underline{L}_i$, $\underline{F}_i$, $\underline{S}_i$, and $\underline{G}_i$, $i = 1, 2, \ldots, N$ so that the resulting closed-loop system described by Eq. (6.112)–(6.113) has its poles in a set £, where £ is a nonempty symmetrical open subset of complex s-plane. It is clear that the membership of a closed-loop system pole λ implies its complex conjugate λ^* in a prescribed manner.

6.4.2 Fixed Modes and Polynomials

The notions of fixed polynomials and fixed modes are generalizations of the "centralized" systems pole placement problem, in which any uncontrollable and unobservable mode of the system must be stable (Jamshidi, 1983) to the decentralized case. The idea of fixed modes for decentralized control was introduced by Wang and Davison (1973) and used extensively the general servomechanism problem by Davison (1976).

Consider the decentralized stabilization problem described by Eq. (6.112)–(6.113). The dynamic compensator Eq. (6.115)–(6.116) may be rewritten in compact form:

$$\underline{u}(t) = \underline{H}\underline{z}(t) + \underline{K}\underline{y}(t) + \underline{L}\underline{v}(t) \qquad (6.117)$$

$$\dot{\underline{z}}(t) = \underline{F}\underline{z}(t) + \underline{S}\underline{y}(t) + \underline{G}\underline{v}(t)$$

where

$$\underline{H} \triangleq \text{block-diag}\{\underline{H}_1, \underline{H}_2, \ldots, \underline{H}_N\} \qquad \underline{K} \triangleq \text{block-diag}\{\underline{K}_1, \underline{K}_2, \ldots, \underline{K}_N\}$$

$$\underline{L} \triangleq \text{block-diag}\{\underline{L}_1, \underline{L}_2, \ldots, \underline{L}_N\} \qquad \underline{F} \triangleq \text{block-diag}\{\underline{F}_1, \underline{F}_2, \ldots, \underline{F}_N\}$$

$$\underline{S} \triangleq \text{block-diag}\{\underline{S}_1, \underline{S}_2, \ldots, \underline{S}_N\} \qquad \underline{G} \triangleq \text{block-diag}\{\underline{G}_1, \underline{G}_2, \ldots, \underline{G}_N\}$$

$$(6.118)$$

and

$$\underline{u}^T(t) \triangleq \{\underline{u}_1^T(t) \vdots \cdots \vdots \underline{u}_N^T(t)\} \qquad \underline{z}^T(t) \triangleq \{\underline{z}_1^T(t) \vdots \cdots \vdots \underline{z}_N^T(t)\}$$

$$\underline{y}^T(t) \triangleq \{\underline{y}_1^T(t) \vdots \cdots \vdots \underline{y}_N^T(t)\} \qquad \underline{v}^T(t) \triangleq \{\underline{v}_1^T(t) \vdots \cdots \vdots \underline{v}_N^T(t)\}$$

$$(6.119)$$

If the control Eq. (6.117)–(6.118) is applied to Eq. (6.112)–(6.113), the following augmented system results:

$$\begin{bmatrix} \dot{\underline{x}}(t) \\ \dot{\underline{z}}(t) \end{bmatrix} = \begin{bmatrix} A + BKC & BH \\ SC & F \end{bmatrix} \begin{bmatrix} x(t) \\ z(t) \end{bmatrix} + \begin{bmatrix} BL \\ G \end{bmatrix} v(t) \qquad (6.120)$$

where

$$\underline{B} = [\underline{B}_1 \vdots \cdots \vdots \underline{B}_N] \underset{m_1 \qquad\qquad m_N}{} \quad \text{and} \quad \underline{C} = \begin{bmatrix} \underline{C}_1 \\ \vdots \\ \underline{C}_N \end{bmatrix} \begin{matrix} \} \, r_1 \\ \\ \} \, r_N \end{matrix} \qquad (6.121)$$

As mentioned earlier, the problem is to find the control laws Eq. (6.117)–(6.118) so that the overall augmented system Eq. (6.120) is asymptotically stable. In other words, by way of local output feedback, the closed-loop poles of the decentralized system are required to lie in the left half of the complex s-plane. The following definitions and theorem provide the ground rules for this problem.

Definition 6.1. Consider the system (C,A,B) described in Eq. (6.112)–(6.113) and integers m, r, i = 1, 2, . . ., N in Eq. (6.114). Let the m × r gain matrix $\underline{K}$ be represented as a member of the following set of block-diagonal matrices:

$$\underline{K} = \begin{bmatrix} K | K = \begin{bmatrix} m_1\{ & \underline{K}_1 & & & \\ & & \underline{K}_2 & & \\ & & & \ddots & \\ & & & & \underline{K}_N \}m_N \\ & & & & \underbrace{\qquad}_{r_N} \end{bmatrix} \end{bmatrix} \qquad (6.122a)$$

where dim $(\underline{K}_i) = m_i \times r_i$, i = 1, 2, . . ., N. Then the "fixed polynomial" of $(\underline{C},\underline{A},\underline{B})$ with respect to $\underline{K}$ is the greatest common divisor (gcd) of the set of polynomial $|\lambda I - A - BKC|$ for all $K \in \underline{K}$ and is denoted by

$$\phi(\lambda; \underline{C},\underline{A},\underline{B},\underline{K}) = \gcd\{|\lambda\underline{I} - \underline{A}-\underline{B}\underline{K}\underline{C}|\} \qquad (6.122b)$$

Definition 6.2. For the system $(\underline{C},\underline{A},\underline{B})$ and the set of output feedback gains $\underline{K}$ given by Eq. (6.122), the set of "fixed modes" of $(\underline{C},\underline{A},\underline{B})$ with respect to $\underline{K}$ is defined as the intersection of all possible sets of the eigenvalues of matrix (A + BKC), i.e.,

$$\Lambda(\underline{C},\underline{A},\underline{B},\underline{K}) = \bigcap_{K \in \underline{K}} \lambda(\underline{A} + \underline{B}\underline{K}\underline{C}) \qquad (6.123)$$

where $\lambda(\cdot)$ denotes the set of eigenvalues of $(\underline{A} + \underline{B}\underline{K}\underline{C})$. Note also that K can take on the null matrix; hence the set of "fixed modes" $\Lambda(\cdot)$ is contained in $\lambda(\underline{A})$. In view of Definition (6.1), the members of $\Lambda(\cdot)$, i.e., the "fixed modes," are the roots of the "fixed polynomial" $\phi(\cdot;\cdot)$ in Eq. (6.122); i.e.,

$$\Lambda \ (\underline{C},\underline{A},\underline{B},\underline{K}) = \{\lambda \,|\, \lambda \epsilon s \quad \text{and} \quad \phi(\lambda, \ \underline{C},\underline{A},\underline{B},\underline{K}) = 0\} \tag{6.124}$$

where s denotes a set of values on the entire complex s-plane.

The fixed modes of centralized system $(\underline{C},\underline{A},\underline{B},\underline{\tilde{K}})$, where $\underline{\tilde{K}}$ is m × r, correspond to the uncontrollable and unobservable modes of the system. The following theorem provides the necessary and sufficient conditions for the stabilizability of a decentralized closed-loop system.

Theorem 6.2 (Wang and Davison, 1973). For the system $(\underline{C},\underline{A},\underline{B})$ in Eq. (6.112)–(6.113) and the class of block-diagonal matrices K in Eq. (6.122a), the local feedback laws Eq. (6.115)-(6.116) would asymptotically stabilize the system, if and only if, the set of fixed modes of $(\underline{C},\underline{A},\underline{B},\underline{K})$ is contained in the open (left-half plane) s-plane; i.e., $\Lambda(\underline{C},\underline{A},\underline{B},\underline{K})\epsilon s$, where s is the open LHP s-plane.

Example 6.4.1. Consider a system

$$\underline{\dot{x}} = \begin{bmatrix} 0.5 & 0 & 1 & 0.75 \\ 0.1 & 1.2 & 0 & 0.1 \\ 0 & 0 & -1 & 0 \\ 0 & 0 & 0.4 & 0.75 \end{bmatrix} \underline{x} + \begin{bmatrix} 0.85 & 0 \\ 0 & 1 \\ 0 & 1.25 \\ 1 & 0 \end{bmatrix} \underline{u}$$

$$\underline{y} = \begin{bmatrix} 1 & 0 & 0 & 0 \\ 0 & 0 & 0.5 & 1 \end{bmatrix} \underline{x}$$

It is desired to find the fixed modes, if any, of this system. A program was written to evaluate $(\underline{A} + \underline{B}\underline{K}\underline{C})$ and find $\lambda(\underline{A} + \underline{B}\underline{K}\underline{C})$ for any K. Here are the results of three iterations. The eigenvalues of A are $\lambda(\underline{A}) = (0.5, \ 1.2, \ -1, \ 0.75)$, which are the diagonal elements of $\underline{A}$. For three arbitrary mxr = 2 × 2 dimensional $\underline{K}$ matrix,

$$\underline{K}_i = \begin{bmatrix} k_1 & 0 \\ 0 & k_2 \end{bmatrix} = \left\{ \begin{bmatrix} 2.248 & 0 \\ 0 & 32.458 \end{bmatrix}, \begin{bmatrix} 1.2458 & 0 \\ 0 & 4.258 \end{bmatrix}, \right.$$

$$\left. \begin{bmatrix} 2.3588 & 0 \\ 0 & -2.146 \end{bmatrix} \right\} \quad i = 1, \ 2, \ 3$$

the respective eigenvalues of (A+BKC) are

$$\lambda(\underline{A} + \underline{B}\underline{K}_1\underline{C}) = (1.0375 \pm j1.7866, \ 2.037, \ 1.20)$$

$$\lambda(\underline{A} + \underline{B}\underline{K}_2\underline{C}) = (1.966 \pm j1.196, \ 3.577, \ 1.20)$$

$$\lambda(\underline{A} + \underline{B}\underline{K}_3\underline{C}) = (1.66 \pm j0.759, \ -2.41, \ 1.20)$$

Clearly, the system has a fixed mode at $\lambda = 1.2$, and hence, according to Theorem (6.2), this system *cannot* be stabilized by decentralized control with dynamic compensators.

6.4.3   Dynamic Compensation

One of the earliest efforts in dynamically compensating centralized systems is due to Brasch and Pearson (1970) using output feedback. The problem can be briefly stated as follows: Consider the system

$$\dot{\underline{x}}(t) = \underline{A}\underline{x}(t) + \underline{B}\underline{u}(t) \tag{6.125}$$

$$\underline{y}(t) = \underline{C}\underline{x}(t) \tag{6.126}$$

It is desired to find a dynamic compensator

$$\dot{\underline{z}}(t) = \underline{F}\underline{z}(t) + \underline{S}\underline{y}(t) \tag{6.127}$$

$$\underline{u}(t) = \underline{H}\underline{z}(t) + \underline{K}\underline{y}(t) \tag{6.128}$$

so that the closed-loop system

$$\dot{\underline{x}}(t) = (\underline{A} + \underline{B}\underline{K}\underline{C})\underline{x}(t) + \underline{B}\underline{H}z(t) \tag{6.129}$$

has a prescribed set of poles. For the case of finding the dynamic compensator, let n_c and n_0 be the smallest integers such that

$$\text{rank}[\underline{B}, \underline{A}\underline{B}, \ldots, \underline{A}^{n_c-1}\underline{B}] = n \tag{6.130}$$

$$\text{rank}[\underline{C}^T, \underline{A}^T, \underline{C}^T, \ldots, (\underline{A}^T)^{n_0-1}\underline{C}^T] = n \tag{6.131}$$

Now, for convenience, let $\eta = \min(n_c, n_0)$, and $\Lambda_\eta = \{\lambda_1, \lambda_2, \ldots, \lambda_{n+\eta}\}$ be a set of arbitrary complex numbers with the only restriction being that for each λ_i with $\text{Im}(\lambda_i) \neq 0$, a pair of complex conjugate pair $\lambda_i = \text{Re}(\lambda_i) \pm j\text{Im}(\lambda_i)$ is contained in Λ_η. Let us define the following augmented triple $(\underline{C}_\eta, \underline{A}_\eta, \underline{B}_\eta)$:

$$\underline{A}_\eta = \left[\begin{array}{c|c} \underline{A} & \underline{0} \\ \hline \underline{0} & \underline{0} \end{array}\right] \begin{array}{l} \}n \\ \}\eta \end{array} \qquad \underline{B}_\eta = \left[\begin{array}{c|c} \underline{B} & \underline{0} \\ \hline \underline{0} & \underline{I} \end{array}\right] \begin{array}{l} \}n \\ \}\eta \end{array} \qquad \underline{C}_\eta = \left[\begin{array}{c|c} \underline{C} & \underline{0} \\ \hline \underline{0} & \underline{I} \end{array}\right] \begin{array}{l} \}r \\ \}\eta \end{array}$$

$$\underbrace{\qquad}_{n} \; \underbrace{\qquad}_{\eta} \qquad\qquad \underbrace{\qquad}_{m} \; \underbrace{\qquad}_{\eta} \qquad\qquad \underbrace{\qquad}_{n} \; \underbrace{\qquad}_{\eta} \tag{6.132}$$

Then the following theorem gives an existence of an output feedback gain for proper pole placement.

Theorem 6.3.   Let (C,A,B) be a controllable system and let the triple $(\underline{C}_\eta, \underline{A}_\eta, \underline{B}_\eta)$ be defined by Eq. (6.132) with $\eta = \min(n_c, n_0)$ and a set of prescribed poles Λ_η. The proof of this theorem with the aid of the properties of cyclic matrices can be found in Brasch and Pearson (1970).

The above theorem, the canonical structure theorem of Kalman (1963), and the above decentralized stabilization problem, which have been considered by Wang and Davison (1973), can be used to find a dynamic stabilizing compensator for a large-scale system under decentralized control.   Consider the set of N dynamic compensator Eq. (6.115)–(6.116) and the triple $(\underline{C}_\eta, \underline{A}_\eta, \underline{B}_\eta)$ in Eq. (6.132); define a real constant $(m + \eta) \times (r + \eta)$ dimensional K_η, i.e.,

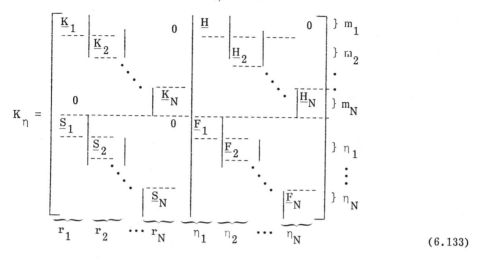

$$(6.133)$$

where $\underline{K}_i$, $\underline{H}_i$, $\underline{S}_i$, $\underline{F}_i$ are submatrices of $(m_i \times r_i)$, $(m_i \times \eta_i)$, $(\eta_i \times r_i)$, and $(\eta_i \times \eta_i)$ dimensions, respectively, defined in Eq. (6.115)–(6.116), m and r are defined in Eq. (6.114), and

$$\eta = \sum_{i=1}^{N} \eta_i$$

The following proposition summarizes the decentralized control pole placement problem.

Proposition 6.1. Consider the triplets $(\underline{C}, \underline{A}, \underline{B})$ and $(\underline{C}_\eta, \underline{A}_\eta, \underline{B}_\eta)$ defined above and the set of block-diagonal K in Eq. (6.122a). Then for any set of integers $\eta_1, \eta_2, \ldots, \eta_N$ with $\eta_i \geqslant 0$, the following two "fixed polynomials" are identical:

$$\phi(\lambda;\ \underline{C}, \underline{A}, \underline{B}, \underline{K}) = \phi(\lambda;\ \underline{C}_\eta, \underline{A}_\eta, \underline{B}_\eta, \underline{K}_\eta) \tag{6.134}$$

where $\underline{C}_\eta$, $\underline{A}_\eta$, $\underline{B}_\eta$ are defined earlier. In other words, the greatest common divisor of det $(\lambda I - \underline{A} - \underline{B}\underline{K}\underline{C})$ and det $(\lambda I - \underline{A}_\eta - \underline{B}_\eta\ \underline{K}_\eta\ \underline{C}_\eta)$ are the same. The proof of this proposition is given by Wang and Davison (1973). The result of this proposition and a matrix identity are used to place poles in decentralized output feedback controllers through dynamic compensation; this is illustrated below.

Example 6.4.1. Let us consider

$$\dot{\underline{x}} = \begin{bmatrix} 0 & 0 \\ 1 & -2 \end{bmatrix} \underline{x} + \begin{bmatrix} 0.2 & 0 \\ 0 & 0.5 \end{bmatrix} \underline{u} \qquad \underline{y} = \begin{bmatrix} 0 & 5 \\ 2 & 0 \end{bmatrix} \underline{x}$$

It is desired to find a decentralized stabilizing output control such that a prescribed set of eigenvalues is achieved.

For this example $\lambda(\underline{A}) = (0, -2)$ and the matrix $\underline{A} + \underline{BKC}$ is

$$\underline{A} + \underline{BKC} = \begin{bmatrix} 0 & k_1 \\ 1 + k_2 & -2 \end{bmatrix} \qquad (6.135)$$

and $\eta = \min(n_c, n_o) = 1$. Hence $\underline{A}_\eta$, $\underline{B}_\eta$, $\underline{K}_\eta$, and $\underline{C}_\eta$ are

$$\underline{A}_\eta = \underline{A}_1 = \begin{bmatrix} 0 & 0 & 0 \\ 1 & -2 & 0 \\ 0 & 0 & 0 \end{bmatrix} \qquad \underline{B}_\eta = \underline{B}_1 = \begin{bmatrix} 0.2 & 0 & 0 \\ 0 & 0.5 & 0 \\ 0 & 0 & 1 \end{bmatrix}$$

$$\underline{C}_\eta = \underline{C}_1 = \begin{bmatrix} 0 & 5 & 0 \\ 2 & 0 & 0 \\ 0 & 0 & 1 \end{bmatrix} \qquad \underline{K}_\eta = \underline{K}_1 = \begin{bmatrix} k_1 & 0 & h_1 \\ 0 & k_2 & 0 \\ s_1 & 0 & f_1 \end{bmatrix} \qquad (6.136)$$

where k_1, k_2, s_1, h_1, and f_1 are unknowns which can be found for a desired pole placement of the two decentralized controllers. The closed-loop matrix $\underline{A} + \underline{BKC}$ is given by Eq. (6.135), and $\underline{A}_1 + \underline{B}_1\underline{K}_1\underline{C}_1$ is obtained:

$$\underline{A}_1 + \underline{B}_1\underline{K}_1\underline{C}_1 = \begin{bmatrix} 0 & k_1 & 0.2h_1 \\ 1 + k_2 & -2 & 0 \\ 0 & 5s_1 & f_1 \end{bmatrix} \qquad (6.137)$$

which for this special case of $\eta = \eta_1 + \eta_2 = 1 + 0 = 1$ [Eq. (6.136)] is expressed by

$$\underline{A}_1 + \underline{B}_1\underline{K}_1\underline{C}_1 = \begin{bmatrix} \underline{A} + \underline{BKC} & \underline{B}_1 h_1 \\ s_1\underline{C}_1 & f_1 \end{bmatrix}$$

which is in agreement with Eq. (6.120) for $v(t) = 0$ and a comparison between Eq. (6.135) and Eq. (6.137). The problem is to find k_1, k_2, h_1, s_1, and f_1 such that the augmented system has a preassigned set of poles for decentralized system and compensator, say:

$$\Lambda_1 = \{-1 \pm j2, -1\} \qquad (6.138)$$

To achieve this, note the following matrix identity:

$$\det \begin{bmatrix} M_1 & M_{12} \\ M_{21} & M_2 \end{bmatrix} = \det(M_1) \cdot \det(M_2 - M_{21} M_1^{-1} M_{12})$$

and apply it to

$$\det(\lambda \underline{I} - \underline{A}_1 - \underline{B}_1 \underline{K}_1 \underline{C}_1) = \det(\lambda \underline{I} - \underline{A} - \underline{B}\underline{K}\underline{C})$$

$$\det \left[(\lambda - f_1) - (0 \quad 5s_1)(\lambda \underline{I} - \underline{A} - \underline{B}\underline{K}\underline{C})^{-1} \begin{pmatrix} 0.2h_1 \\ 0 \end{pmatrix} \right] \qquad (6.139)$$

Choosing k_1 and k_2 such that the first two poles are appropriately placed, $k_1 = 1$ and $k_2 = -6$. The remaining unknowns may be obtained by setting the latter part of Eq. (6.139) equal to zero for $\lambda = \lambda_3 = -1$, i.e.,

$$\det \left[(\lambda - f_1) - (0 \quad 5s_1)(\lambda I - A - BKC)^{-1} \begin{pmatrix} 0.2h_1 \\ 0 \end{pmatrix} \right]$$

$$= \det \left(\lambda - f_1 - \frac{-5h_1 s_1}{\lambda^2 + 2\lambda + 5} \right) \Bigg|_{\lambda = -1} = 0$$

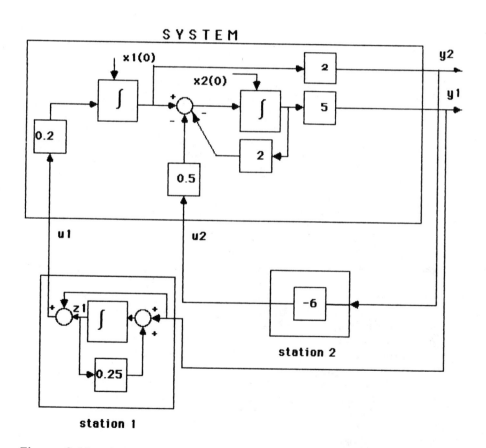

Figure 6.11   A two-controller decentralized structure for Example (6.4.1).

or for arbitrary $s_1 = h_1 = 1$, $f_1 = 0.25$. Thus the decentralized dynamic compensator controllers are

$$\dot{z}_1 = 0.25z_1 + y_1 \qquad\qquad z_1(0) = 0$$

$$u_1 = z_1 + y_1$$

$$\dot{z}_2 = 0 \qquad\qquad\qquad\qquad z_2(0) = 0$$

$$u_2 = -6y_2$$

Figure 6.11 shows a block diagram for the two-controller decentralized structure. Note that while station 1 represents a dynamic compensator, station 2 is a simple static feedback compensator. Note also that the eigenvalues of the combined system are not exactly those specified by Eq. (6.138) for systems with no compensation. In fact, for this decentralized system the closed-loop poles turned out to be $\lambda(.) = \{-0.375 \pm j1.9, -1.0\}$, which are not far from the desired locations.

6.5   STRUCTURAL PROPERTIES OF LARGE-SCALE SYSTEMS

6.5.1   Stability

The high dimensionalities, nonlinearities, and complexities of interconnections in large-scale systems provide computational and analytical difficulties not only in modeling, control, or optimization, but just as severe in the most fundamental issue in any system—stability. As the dimension of the system increases, the problem of assessing its stability becomes much more difficult. When the stability of a large-scale system is of concern, one basic approach, consisting of three steps, has prevailed: (a) decompose a given large-scale system into a number of small-scale subsystems, (b) analyze each subsystem using the classical stability theories and methods, and (c) combine the results leading to certain restrictive conditions on the interconnections and reduce them to the stability of the whole. This approach as been termed "composite system method" by Araki (1978) and many others.

One of the earliest efforts regarding the stability of composite systems is due to Bailey (1966). It assumes a Lyapunov function for each subsystem, and then, using the theory of vector Lyapunov function (Siljak, 1972a), the stability of the composite system is checked (Moylan and Hill, 1978). Alternatively, others (Araki and Kondo, 1972; Michel and Porter, 1972) have constructed a scalar Lyapunov function as a weighted sum of the Lyapunov functions of the individual subsystems. This line of work has given rise to the so-called "Lyapunov methods."

An alternative approach, known as "input-output methods," describes each subsystem by a mathematical relation or an operator on functional space and then functional analysis methods are employed (Moylan and Hill, 1978). Porter and Michel (1974) and Cook (1974) consider the situation where subsystems have finite gain and are conic, which is a generalization of Zames' (1966) single-loop results.

A basic issue involved in interconnected systems stability is the question: How large should the magnitudes and strength of interactions be before the stability of the composite system would be affected? Furthermore, as mentioned by Sandell et al. (1978), in some systems a strong coupling between various subsystems has a major contribution to stability. Such issues lead us to connective stability, which is essentially the extension of stability in the sense of Lyapunov to take into account the structural perturbations (Siljak, 1972a,b, 1978).

The stability of large-scale systems has been one of the most dominant qualitative topics on the subject, and the literature on it is indeed vast. It is beyond the scope of this chapter to go into the subject in any further detail.

6.5.2 Controllability and Observability

Two of the most fundamental issues in control systems are the notions of "controllability" and "observability." As stated in Chapter 2, a system is said to be "completely state controllable" if it is possible to find an unconstrained control vector u(t) that would transfer any initial state $x(t_0)$, for any t_0, to any final state $x(t_f)$, say origin, in a finite time interval $t_0 \leqslant t \leqslant t_f$ (Perkins and Cruz, 1969). Observability, on the other hand, is related to the determination of the state from the measurement of output. A system is said to be "completely observable" at any t_0 if it is possible to determine $x(t_0)$ by measuring y(t) over the interval $t_0 \leqslant t \leqslant t_1$.

The bulk of research in the controllability and observability of large-scale systems falls into these main problems: controllability (and observability) of composite systems, controllability (and observability) of decentralized systems, structural controllability, and controllability of singularly perturbed systems. However, owing to lack of space, only a brief overview of these topics is given here. Interested readers may consult other references, e.g., Jamshidi (1983).

The controllability and observability of composite (series, parallel, and feedback) systems were first considered by Gilbert (1963) who studied the controllability and observability of the system in terms of those of the subsystems. The controllability (and observability) of large-scale systems under decentralized information structure is another area of research. The foundations of this controllability were laid by many, including Corfmat and Morse (1976), who investigated the effect of decentralized output feedback control on the controllable and observable subspaces of each control station and obtained conditions for pole assignability under local feedback. Perhaps one of the best treatments of this subject is that of Kobayashi et al. (1978).

In "structural controllability," introduced by Lin (1974), the controllability of the pair (A,b) through the properties of system structure is determined through the graph of (A,b). The structural controllability has been further considered by many other authors, including Shields and Pearson (1976) and Corfmat and Morse (1975). Graph theoretical concepts have also been used by Davison (1976) for composite systems.

The other area of research is the controllability of singularly perturbed systems, whose model reduction was discussed in Section 6.2. The controllability of such systems is checked in terms of the controllability of "slow" and "fast" subsystems. This problem has been considered by Kokotovic and Haddad (1975) and Chow and Kokotovic (1976) for linear systems and by Sannuti (1977) for linear and nonlinear systems.

6.6   SUMMARY AND CONCLUSIONS

The object of this chapter has been to given an introduction to a special class of systems—"*large-scale systems.*" Several issues pertaining to large-scale systems could have been addressed in this chapter. However, owing to lack of space, we chose to address the most fundamental issues: modeling and control. Each of these issues deserves at least one chapter by itself.

Large-scale systems is a class of systems that have a great deal of promise. However, many key issues are still undetermined. In the area of modeling, new efforts are necessary. Ideas of aggregation and perturbation are both rather old in nature. More work is also needed on structural properties such as controllability and observability. Perhaps the most pressing issue in large-scale systems is application of the present theory to important complex industrial systems. It is hoped that this brief introduction has been able to expose the reader to a very important and still relatively new field.

PROBLEMS

6.1.   Consider a system

$$\dot{x} = \begin{bmatrix} -1 & 0 & -1 \\ 0 & -2 & -1 \\ 1 & 0 & -10 \end{bmatrix} x + \begin{bmatrix} 1 & 0 \\ 0 & 1 \\ 1 & 2 \end{bmatrix} u$$

and find an aggregated model using regular aggregation. Try both schemes. Are these aggregations dynamically exact?

6.2.   Repeat Problem 6.1 for modal aggregation.

6.3.   Show that the aggregation matrix that guarantees exact aggregation is given by

$$C = \begin{bmatrix} \nu_1 \\ \hline \nu_2 \\ \hline \vdots \\ \hline \nu_k \end{bmatrix}$$ where ν_i is the ith left eigenvector of the A matrix.

6.4.   Consider a system

$$\dot{x} = \begin{bmatrix} -1 & -1 & 0 & 0 & 1 \\ 0.2 & -1 & 0.2 & -1 & 0 \\ 0 & -0.5 & -0.2 & -5 & -10 \\ 1 & 1 & -0.1 & -10 & -20 \\ 0 & -2 & 1 & -20 & -25 \end{bmatrix} x + \begin{bmatrix} 1 & 0 \\ 1 & 1 \\ 0 & 1 \\ 1 & 2 \\ 0 & 2 \end{bmatrix} u$$

$$\underline{y} = \begin{bmatrix} 1 & 0 & 0.2 & 0.5 & 1 \\ 0 & 1 & 0.1 & 0 & 0.5 \end{bmatrix} \underline{x}$$

Use chained aggregation to find two aggregated models.

6.5. An unstable system is described by

$$\underline{\dot{x}} = \begin{bmatrix} 0 & 1 & 0 & 0 \\ 0 & 0 & 1 & 0 \\ 0 & 0 & 0 & 1 \\ -0.5 & -1 & -10 & 3 \end{bmatrix} \underline{x} + \begin{bmatrix} 1 & 0 \\ 0 & 0 \\ 1 & 1 \\ 1 & 1 \end{bmatrix} \underline{u}$$

$$\underline{y} = \begin{bmatrix} 1 & 0 & 0 & 1 \\ 0 & 1 & 0 & 0 \end{bmatrix} \underline{x}$$

Find a reduced-order model using the balanced approach.

6.6. Repeat Problem 6.5 for the following stable system:

$$\underline{\dot{x}} = \begin{bmatrix} -1 & 0 & 1 \\ 0 & -2 & 1 \\ 0 & 0 & -1 \end{bmatrix} \underline{x} + \begin{bmatrix} 1 & 1 \\ 0 & 1 \\ 1 & 1 \end{bmatrix} \underline{u}$$

$$\underline{y} = \begin{bmatrix} 1 & 0 & 0 \\ 2 & 2 & 1 \end{bmatrix} \underline{x}$$

6.7. Is the system of Problem 6.4 weakly coupled?

6.8. Letting the following system be dependent on two parameters:

$$\underline{\dot{x}} = \begin{bmatrix} -1 & 0 & a \\ 0 & 0 & -1 \\ b & 0 & -1 \end{bmatrix} \underline{x}$$

For what range on a and b is this system weakly coupled?

6.9. A singularly perturbed system is described by

$$\underline{\dot{x}} = -1/2 \, x + z \qquad x(0) = 1$$
$$\varepsilon \, \underline{\dot{z}} = -x - 2z \qquad z(0) = 1$$

Find a boundary layer for it.

6.10. Consider a two-subsystem problem

$$
\begin{bmatrix} \dot{x}_1 \\ \dot{x}_2 \\ \dot{x}_3 \\ \dot{x}_4 \end{bmatrix} = \begin{bmatrix} -1.2 & 0 & | & 0 & 0.1 \\ -0.5 & -2 & | & 0 & 0.2 \\ \hline 0.5 & 0.5 & | & -2 & 0 \\ 0 & 0 & | & -0.5 & -1 \end{bmatrix} \begin{bmatrix} x_1 \\ x_2 \\ \hline x_3 \\ x_4 \end{bmatrix}
$$

$$
+ \begin{bmatrix} 0.2 & | & 0 \\ 0.1 & | & 0 \\ \hline 0 & | & 0.2 \\ 0 & | & 0.2 \end{bmatrix} \begin{bmatrix} u_1 \\ \hline u_2 \end{bmatrix}
$$

$$
x(0) = (-1 \quad 0 \mid -1 \quad 0)^{-T}
$$

Use the goal coordination algorithm to find a hierarchical controller using a quadratic cost function with $Q_i = I_2$, $R_i = 1$, $i = 1, 2$, $\Delta t = 0.1$, and $t_f = 2$.

6.11. Repeat Problem 6.10 using the interaction prediction algorithm.

6.12 Determine the fixed modes of the following system:

$$
\dot{\underline{x}} = \begin{bmatrix} -1 & 1 \\ 0 & -1 \end{bmatrix} \underline{x} + \begin{bmatrix} 1 & 1 \\ 0 & 1 \end{bmatrix} \underline{u}
$$

$$
\underline{y} = \begin{bmatrix} 1 & 0 \\ 0 & 2 \end{bmatrix} \underline{x}
$$

Is the system stabilizable under decentralized control?

6.13. Find a decentralized stabilizing output controller for the system

$$
\dot{\underline{x}} = \begin{bmatrix} 0 & 1 \\ 0 & -1 \end{bmatrix} \underline{x} + \begin{bmatrix} 1 & 1 \\ 0 & 1 \end{bmatrix} \underline{u}
$$

$$
\underline{y} = \begin{bmatrix} 1 & 0 \\ 0 & 2 \end{bmatrix} \underline{x}
$$

$$
\underline{y} = (1 \ -1) \begin{pmatrix} x_1 \\ x_2 \end{pmatrix} + \underline{u}
$$

REFERENCES

Aoki, M. (1971). Aggregation, in *Optimization Methods for Large-Scale Systems . . . with Applications* (D. A. Wismer, ed.), McGraw-Hill, New York, Chapter 5, p. 191.

Araki, M. (1978). Stability of large-scale nonlinear systems—Quadratic-order theory of composite-system method using M-matrices, *IEEE Trans. Aut. Cont. AC-21*, 254.

Araki, M. and Kondo, B. (1972). Stability and transient behavior of composite nonlinear systems, *IEEE Trans. Auto. Cont. AC-17*, 537.

Bailey, F. N. (1966). The application of Lyapunov's second method to interconnected systems, *SIAM J. Contr. 3*, 443.

Bailey, F. N. (1966). The concept of aggregation in system stability analysis, in *Asilomar Conf. on Circuits, Systems and Computers*, Asilomar, CA.

Bailey, F. N. and Ramapriyan, H. K. (1973). Bounds on suboptimality in the control of linear dynamic systems, *IEEE Trans. Aut. Cont. AC-18*, 532.

Brasch, F. M. and Pearson, J. B. (1970). Pole placement using dynamic compensators, *IEEE Trans. AC. AC-15*, 34.

Chow, J. H. and Kokotovic, P. V. (1976). A decomposition of near-optimum regulators for systems with slow and fast modes, *IEEE Trans. Aut. Cont. AC-21*, 701.

Cook, P. A. (1974). On the stability of interconnected systems, *Int. J. Control 20*, 407.

Corfmat, J. P. and Morse, A. S. (1975). *Structurally Controllable and Structurally Canonical Systems*, Internal Report. Dept. Engr. and Appl. Science. Yale University, New Haven, CT.

Davison, E. J. (1966). A method for simplifying linear dynamic systems, *IEEE Trans. Aut. Cont. AC-12*, 119.

Davison, E. J. (1968). A new method for simplifying linear dynamic systems, *IEEE Trans. Aut. Cont. AC-13*, 214.

Davison, E. J. (1976). Connectability and structural controllability of composite systems, in *Proc. IFAC Symp. Large Scale Systems*, Udine, Italy, p. 241.

Enns, D. F. (1984). Model Reduction for Control System Design, Ph.D. Dissertation, Dept. of Aeronautics and Astronautics, Stanford University, Stanford, CA.

Huseyin, O. (1977). A counter-example on "On the controllability of composite systems," *IEEE Trans. Aut. Cont. AC-22*, 683.

Jamshidi, M. (1983). *Large-Scale Systems-Modeling and Control*, Elsevier, North Holland, New York.

Jamshidi, M. and Banning, R. (1985). *LSSPAK/PC Users' Guide—A CAD Package for Large-Scale Systems*, CAD Lab., Tech. Rep. 85-01, University of Mexico, Albuquerque, NM.

Jamshidi, M. and Malek-Zaverei, M. (1986). *Linear Control Systems—A Computer-Aided Approach*, Pergamon Press, Oxford, England.

Kalman, R. E. (1963). Mathematical description of linear dynamical systems, *S.I.A.M. J. Control 1*, 132.

Kobayashi, H., Hanafusa, H., and Yoshikawa, T. (1978). Controllability under decentralized information structure, *IEEE Trans. Aut. Cont. AC-23*, 182.

Kokotovic, P. V. (1972). Feedback design of large linear systems, in *Feedback Systems* (J. B. Cruz, Jr., ed), McGraw-Hill, New York, p. 99.

Kokotovic, P. V. and Haddad, A. H. (1975). Controllability and time-optimal control of systems with slow and fast modes, *IEEE Trans. Aut. Cont. AC-20*, 111.

Kwakernaak, H. and Sivan (1972). *Linear Optimal Control Systems*, Wiley, New York.

Laub, A. J. (1980). Computation of balancing transformations, *Proc. 1980 JAAC*.

Lin, C. T. (1974). Structural controllability, *IEEE Trans. Aut. Cont. AC-19*, 201.

Mesarovic, M. C., Macko, D., and Takahara, Y. (1970). *Theory of Hierarchical Multilevel Systems*, Academic Press, New York.

Michel, A. N., and Porter, D. W. (1972). Stability analysis of composite systems, *IEEE Trans. Aut. Cont. AC-17*, 111.

Milne, R. D. (1965). The analysis of weakly coupled dynamic systems, *Int. J. Control 2*, 171.

Moore, B. C. (1981). Principle component analysis in linear systems: controllability, observability, and model reduction, *IEEE Trans. Aut. Contr. AC-26*.

Moylan, P. J. and Hill, D. J. (1978). Stability criteria for large-scale systems, *IEEE Trans. Aut. Cont. AC-23*, 143.

Perkins, W. R. and Cruz, J. B. (1969). *Engineering of Dynamic Systems*, Wiley, New York, p. 418.

Porter, D. W. and Michel, A. N. (1974). Input-output stability of time-varying non-linear multiloop feedback systems, *IEEE Trans. Aut. Cont. AC-19*, 422.

Sandell, R., Varaiya, P., Athans, M., and Safonov, M. G. (1978). Survey of decentralized control methods for large scale systems, *IEEE Trans. Aut. Cont. AC-23*, 108.

Sannuti, P. (1977). On the controllability of singularly perturbed systems, *IEEE Trans. Aut. Cont. AC-22*, 622.

Santiago, J. J. and Jamshidi, M. (1986). On the extensions of the balanced approach for model reduction, *Int. J. Cont. Theory Adv. Tech.*

Seraji, H. (1978). A new method for poleplacement using output feedback, *Int. J. Control 28*, 147f-155.

Shields, R. W. and Pearson J. B. (1976). Structural controllability of multi-input linear systems, *IEEE Trans. Aut. Cont. AC-21*, 203.

Siljak, D. D. (1972a). Stability of large-scale systems, *Proc. 5th IFAC World Congress*, Paris, France.

Siljak, D. D. (1972b). Stability of large-scale systems under structural perturbations, *IEEE Trans. Syst. Man and Cybern. SMC-2*, 657.

Tse, E. C. Y., Medaqnic, J. V., and Perkins, W. R. (1978). Generalized Hessenberg transformations for reduced-order modeling of large-scale systems, *Int. J. Control 27*.

Wang, S. H. and Davison, E. J. (1973a). On the controllability and observability of composite systems, *IEEE Trans. Aut. Cont. AC-18*, 74.

Zames, G. (1966). On input-output stability of time-varying non-linear systems, Parts I and II, *IEEE Trans. Aut. Cont. AC-11*, 228-238, 465.

Part II
Software/Hardware Systems

7

Data Structures and Databases

7.1   INTRODUCTION

The purpose of this chapter is to provide an understanding of the important fundamentals of data-handling techniques. Such understanding and techniques provide a link between the physical entities and the simulated ones of the system being studied, as discussed earlier, especially in Chapters 2 and 3. Individuals involved in modeling and simulation are required to understand the problem domain (i.e., the domain of the system to be studied) and then map it to the simulation domain. This mapping requires creative association between physical subsystems (components or objects) and the model. The properties of an object should be properly reflected in the attributes of the corresponding entity. The overview of data structures and their operations enables one to gain insight into correspondence between the two and appreciate the "abstraction" process.

The terms "model" and "simulation" used in this chapter follow closely the definitions given in Section 1.2.1. Readers are encouraged to refer to that section for clarification.

This chapter is divided into five sections: Introduction, Basic Data Structures, Advanced Data Structures, Databases, and Conclusion and Trends. In Section 7.1, the relationship between data handling and the nature of problems is emphasized. Sections 7.2 and 7.3 cover the fundamentals of data structures and their handling techniques. Section 7.4 describes the database concept, its current and potential applications. Section 7.5 presents a few concluding remarks and the future trends of data handling and databases.

The examples used in Sections 7.1 through 7.3 are all written in Pascal. Some of the examples are also described in Fortran to benefit those familiar with Fortran.

As mentioned in Chapter 1, a computer is basically a binary device. It uses binary numbers to represent all information it processes. The elementary memory unit is called a "bit," an abbreviation for "binary digit." It maps a binary digit to a hardware device which behaves like a binary digit in binary operations. To most users this level of knowledge is not important. It is taken for granted that computers can be used to process numbers.

However, it is exactly at this point that we want to start our discussion about data handling in computers. No one will question that a computer can process numerical data because its name implies this capability. We want to emphasize that it can also process nonnumerical data.

Numerical data are represented in binary forms in computers. A decimal number 10 will be represented by 1010 in computer memory with some leading zeros. The basic memory unit of a computer is a "word." A word contains a fixed number of binary digits. The largest integer that can be represented in an eight-digit decimal number is 99999999. Similarly, there is a maximum that can be placed in a computer word.

Nonnumerical data, on the other hand, are encoded letter by letter. A string of letters is stored in a string of computer words. Each letter is represented by an encoded word. Since any alphabet is finite in size, it is not a problem to encode them in a word. The string "letter" can be stored in six computer words as D3C5E3E3C5D9 (in hexadecimal representation). Therefore, any nonnumerical data can also be represented by a string of encoded computer words.

The above discussion refers to the single-valued numerical and non-numerical data. There are times when single-valued data must be grouped together in order to represent some meaningful entity: the location of a point in a three-dimensional space is usually represented by three numbers (x,y,z). These three single-valued numbers must be organized in a specific fashion to represent that point in space. The mailing address is normally arranged in the order of name, number and street name, city and state. The above examples indicate the need to organize single-valued data to represent some physical entity. This is one reason for which data structures are discussed.

The need to organize single-valued data may also appear even when each datum represents a certain physical entity and has its significance. For instance, a set of temperatures is recorded for future processing. It is more convenient to group all the readings and process them together than one at a time (Dale, 1983). Such data need to be organized into some structure before processing. This is a different reason than the one described earlier. However, both are valid and imperative reasons that one should look into data structures.

Two major reasons to organize data into specific structures are: (a) to facilitate the processing of data, and (b) to place relevant information in close proximity for efficient access. Another reason is that the nature of problems may demand that data be structured. This is further explained below.

When matrices are manipulated to reflect some physical phenomena, it is natural to require a data structure for storing matrices. Otherwise, it is very awkward, if not impossible, to manipulate each individual component of a matrix. Just imagine naming all the components one by one for a 10×10 matrix, not even to mention processing them!

Personnel record is frequently mentioned in daily business. Its manipulation also demands special data structure. Usually, it contains at least several different fields, such as name, address, age, department, social security number, pay scale, and marriage status. Since they belong to the same person, all this information should be grouped together. Without proper structure for such information, it will be difficult to retrieve relevant items. A data structure called "record" is designed for this type of problem (Wirth, 1976).

When data structures match the problem under study, one is able to focus on the development of an algorithm to solve the problem. By providing data structures to match problems, programming languages enable us to focus our attention on algorithm development without being distracted by the "nitty gritty" details of data manipulation.

7.2. BASIC DATA STRUCTURES

In a computer, "1" is represented in at least three different ways. It can be a letter, or an integer, or a real number. Each is encoded differently. When it is encoded as a letter, it can only be compared or concatenated. Any other operation is not allowed. If it is encoded in interger form, then all the integer operations can be used on it.

Computers code whole numbers differently than numbers with fractional parts. Whole numbers are represented exactly, other numbers approximately. Thus, the range of fractional numbers representable in a computer is increased. However, this is at the expense of accuracy. Although the numbers 1 and 1.0 are equal mathematically, they are encoded differently in a computer. This is why one gets inequality when they are compared.

Additionally, the numbers 1 and 0 may also represent True and False in a computer. When this is the case, subtracting or adding two values in the set {True, False} does not make sense. Even though they are whole numbers, the operations of whole numbers are not valid. Therefore, we have a new type of values called Boolean.

For each of the four basic types of data—character type, integer type, real type, and Boolean type—there is a set of corresponding operations. There are two different sets of arithmetic operators: one for whole numbers, another for fractional numbers. For Boolean values, there is the set of logical operators: AND, OR, NOT. Similar to arithmetic operators, there are two different sets of relational operators. This is because relational operators are built on arithmetic operators to obtain comparison results; such results can only be Boolean. Most of these operators and data formats are built in to all computers. One does not need to be concerned with them. However, such basic knowledge helps in using computers more effectively.

7.2.1   Elements

All programming languages support some basic "data types" (Wegner, 1980). Usually supported basic data types are integer, real, character, Boolean, and sometimes enumerated type. These basic data types are the fundamental units which serve as building blocks for other structures.

Whole numbers are called integers. They consist of a sequence of digits only, which may be preceded by a + or − sign. If no sign precedes the integer, it is considered to be a positive integer. Integers may not contain any decimal point, dollar sign, or comma.

Real numbers can be represented in decimal and/or exponential notation. In decimal notation, the real number must have a decimal point. The exponential form is similar to the commonly used scientific notation.

Characters are the set of individual symbols used in a language. They include upper/lowercase letters, digits, the space symbol, and other special symbols such as +, =, −, (,), [,], {, }, >, <, *, ;.

Boolean values are the truth values TRUE and FALSE and are used for binary conditions. If a datum is declared Boolean, then it can assume only one of the two values at any particular time. It is analogous to the switch in and "on or off" position.

In most programming languages, variables must be associated with a specific type. When this association is accomplished, it is called "binding."

7.2.2   Arrays

For problems that do not require much data, one can declare each one individually and use them in the computer program. However, when a problem requires manipulation of many similar data at the same time, it is infeasible to name each one individually. Situations such as finding the hottest day in the last month may be solved without declaring 30 variables as Tempday1, Tempday2, . . ., etc. Since each recorded temperature is associated with one day of the month, it is natural to place it in an array. Each basic element in this array is to hold the value of a recorded temperature of a day.

An array is a collection of the same type of elements. The above example can be described in Pascal by

```
var  Temperature  :  real;
     Tempday  :  array[1..31] of Temperature;
```

The equivalent Fortran declaration is

```
Real Tempday (31) or
```

```
Dimension Tempday (31)
```

In this way, one can access the temperature of day 5 of the month by using the notation Tempday [5].

There are two major components in an array: the name and the index. The general form may look like this:

Tempday [I]

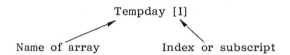

Name of array   Index or subscript

The name identifies a group of elements of the same type. The index allows one to specify one of the elements in the group. Using a variable, such as I in this case, for the index enables us to access all the individual elements by changing the value of the index variable. Arrays are useful when combined with loop constructs in programming languages.

For example, if we are to find the hottest day of the month the following program fragment can be used. Assume the variable Hottestday is declared to be of the type Temperature, and Tempday[1] through Tempday[31] are filled with the temperature readings of 31 days.

```
Hottestday  :=  Tempday[1];
  I  :=  2 ;
while  ( I <= 31 ) do
   begin
   if Hottestday < Tempday[I]
     then Hottestday  :=  Tempday [I];
   I  :=  I + 1
   end
```

The Fortran version looks like

```
      Hotday = Tempday(1)
      Do 123 I = 2,31
      If Hotday >= Tempday(1) then goto 122
      Hotday = Tempday(I)
122   Continue
123   Continue
```

The arrangement of the array Tempday is shown in Figure 7.1.

7.2.3 Tables (Multidimensional Arrays)

Some situations require more complicated data structures than one-dimensional arrays. For instance, the income tax table used to figure our annual income tax can be more easily modeled by a data structure called "multidimensional array" than by several separate arrays. This is illustrated in Figures 7.2 and 7.3.

Figure 7.2 depicts a sample of the actual income tax table for the year 1986. One can view this information as four separate arrays—Single, MarriedFileJoint, MarriedFileSeparate, and HeadofHousehold—as in Figure 7.3. The way to access the contents of the four separate arrays is cumbersome. A better way to access the information is to use a table.

```
type Status =
         (Single,MarriedFileJoint,MarriedFileSeparate,
             HeadofHousehold);
     TaxBracket =  (41000,41050,41100,41150,41200,
                 41250,41300,41350,41400,41450,
                 41500,41550);
```

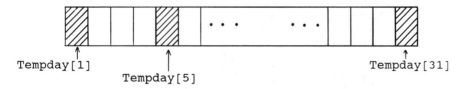

Figure 7.1   Arrangement of the array Tempday.

1986 Tax Table—*Continued*

If line 7(1040EZ), line 19(1040A), or line 37(1040) is-		And you are—				If line 7(1040EZ), line 19(1040A), or line 37(1040) is-		And you are—			
At least	But less than	Single	Married filing jointly *	Married filing sepa- rately	Head of a house- hold	At least	But less than	Single	Married filing jointly *	Married filing sepa- rately	Head of a house- hold
		Your tax is—						Your tax is—			
41,000						**44,000**					
41,000	41,050	9,707	7,773	11,818	9,018	44,000	44,050	10,847	8,763	13,078	10,068
41,050	41,100	9,726	7,790	11,839	9,036	44,050	44,100	10,866	8,780	13,099	10,086
41,100	41,150	9,745	7,806	11,860	9,053	44,100	44,150	10,885	8,796	13,120	10,103
41,150	41,200	9,764	7,823	11,881	9,071	44,150	44,200	10,904	8,813	13,141	10,121
41,200	41,250	9,783	7,839	11,902	9,088	44,200	44,250	10,923	8,829	13,162	10,138
41,250	41,300	9,802	7,856	11,923	9,106	44,250	44,300	10,942	8,846	13,183	10,156
41,300	41,350	9,821	7,872	11,944	9,123	44,300	44,350	10,961	8,862	13,204	10,173
41,350	41,400	9,840	7,889	11,965	9,141	44,350	44,400	10,980	8,879	13,225	10,191
41,400	41,450	9,859	7,905	11,986	9,158	44,400	44,450	10,999	8,895	13,246	10,208
41,450	41,500	9,878	7,922	12,007	9,176	44,450	44,500	11,018	8,912	13,267	10,226
41,500	41,550	9,897	7,938	12,028	9,193	44,500	44,550	11,037	8,928	13,288	10,243
41,550	41,600	9,916	7,955	12,049	9,211	44,550	44,600	11,056	8,945	13,309	10,261

Figure 7.2   Income tax table.

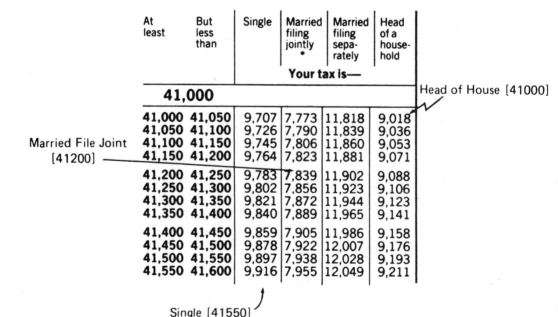

Figure 7.3   Arrays to represent table.

These declarations set the stage that Status is an enumerated type with four entries. And TaxBracket has 12 entries with the specific values. Assume the above is already declared in a program as

```
TaxTable = array[TaxBracket,Status] of integer;
```

This TaxTable can be accessed by assigning values to some variables before retrieving a specific entry. Assume FilingStatus and Bracket are declared to be the proper types; the following program segment can be used to access entries of the TaxTable.

```
FilingStatus := Single;
Bracket := 41250;
    •
    •
    •
Tax := TaxTable[Bracket,FilingStatus];
    •
    •
```

Another way to declare the table is

```
type Entry = array [Status] of integer;
     TaxTable = array [TaxBracket] of Entry;
```

Arrays with dimensions greater than 2 can be declared similarly. They can also be declared in a straightforward manner:

```
MatrixA = array[1..length,0..height,−2..width] of
              real;
```

In this case, the indices must be consecutive integers. Using enumerated type in data structures allows for more readable code.

7.2.4 Records

We have noticed that arrays and tables (or any other multidimensional array) can only accommodate data of the same type. When dealing with simulations that require several items of information grouped together, each of a different type, arrays or tables cannot be used. Most languages support the data structure called "record," which is designed for problems with heterogeneous data types (Horowitz and Sahni, 1982).

Consider a simplified library information system. A minimum amount of information is needed for each book. Let us assume the minimum information includes Title, Author, CallNumber, Price, Borrower, and DueDate. The item Price is the replacement cost in case the book is lost. These six items belong to one book and could be stored together as follows:

```
type Book = record
            Title : string1;
            Author : string2;
            CallNumber : string3;
            Price : real;
            Borrower : string2;
```

```
            DueDate : date;
        end;
    var BestSeller : Book;
```

The first part declares a data type called Book by specifying its individual
fields. The last line indicates that an actual variable named BestSeller is
defined and it takes the form of a Book. Assume that string1, string2,
string3, and date are properly defined before the record is declared.
BestSeller.Title will contain the title of the bestseller, and BestSeller.Price
stores its replacement cost. The record is depicted in Figure 7.4.

To store the books in the entire library, one may declare an array of
books.

```
    var Library : array[1..20000] of Book;
```

Library[1] contains the information of the first book in the library.
Library[10000] contains the information of the 10,000th book. By proper
modification and searching, one can keep the entire library information on
his/her fingertips.

7.2.5   Sets

A set is a mathematical concept and is a collection of objects. Using this
concept can simplify many detailed considerations in an algorithm. Languages
that support this data structure usually put some restrictions on the structure to facilitate language implementations. For example, Pascal supports a
set all of whose elements are of the same type. In addition, there is an
implementation-dependent limit on the size of a set. This size limit is not
found in the mathematical concept (Habermann and Perry, 1983). Let us
illustrate the use of set by an example:

```
    type Alphabet = set of "A".."Z";
    var Vowels : Alphabet;
         .
         .
         .
    Vowels := ["A", "E", "I", "O", "U"];
```

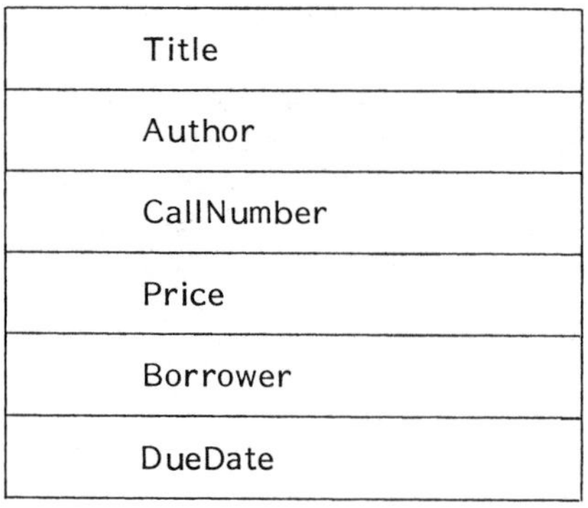

Title
Author
CallNumber
Price
Borrower
DueDate

Figure 7.4   Record of book.

In the declaration, Vowels is declared as a set of alphabet. Later, it is assigned a value of a set containing five vowels. There is a group of operators associated with the data type sets. The relation IN is an important relational operator. Suppose it is desired to find out the number of vowels in a text. The following program segment will be useful:

```
program count(Test,output);
type Alphabet = set of "A".."Z";
var Vowels : Alphabet;
    Ch, Let : char;
    Test : text;
    NumberofVowels : integer;
begin
    Vowels := ["A", "E", "I", "O", "U"];
    NumberofVowels := 0;
    read(Test,Ch);
    while not eof (Test) do
        begin
          if Ch IN Vowels then
            NumberofVowels := NumberofVowels+1;
          read(Test,Ch)
        end;
   writeln("There are" NumberofVowels,"vowels")
end.
```

Compare the following two statements:

"if Ch In Vowels then ..."   (1)

"if (Ch = "A") or (Ch = "E") or (Ch = "I") or (Ch = "O") or
(Ch = "U") then . . ."   (2)

One may begin to appreciate the powerful relational operator in sets.

Set union, set difference, and set intersection must be in a language to support such structured data type. The standard relational operators defined between sets must also be available.

The above-mentioned structured data types, such as arrays, tables, records, and sets, are the basic data structures built in addition to the fundamental data types. These structures are static by nature. They must be allocated and arranged beforehand. One can use them effectively if the information required can be packaged in these structures and their sizes are known in advance.

However, not all problems can be easily solved using these static structures. Consider, for instance, an internal sort algorithm. It may not be wise to choose a static data structure for this algorithm, since one needs to set the upper limit for the static structure beforehand. If the limit is too large, much memory is wasted for most sorting problems. If the limit is too small, the algorithm will not work for problems with large data sets. In addition, static structures are not suitable for sorting because it involves inserting and shifting data. Let us use an example to illustrate this point.

If the list shown in Figure 7.5 is to be sorted, one needs to move the contents of the list back and forth many times as indicated. This is due to the fact that the static order of the list is fixed. The location in the

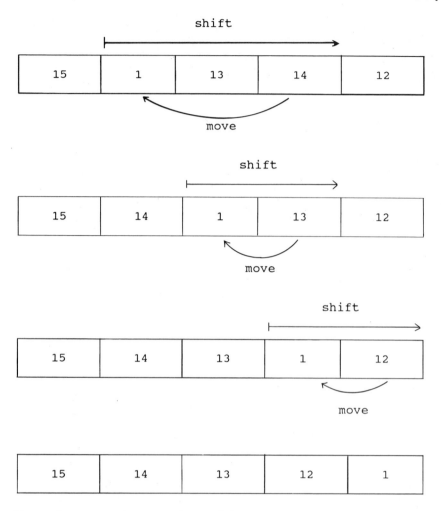

Figure 7.5   Sort in array type of data structure.

list also signifies the order of the resulting list. This is not, however, the case with dynamic structures, which is described next.

7.3   ADVANCED DATA STRUCTURES

The concept that a data item can be linked to another and that each can be created and discarded at will has been developed to overcome the problems facing static data structures. One of the simplest structures is the linked list depicted in Figure 7.6; see Lewis and Smith (1976). This is called dynamic data structure because the location of a data item does not determine its order and the locations can be created and disposed of at will. Dynamic data structures are useful in simulation studies where data sets change frequently. Customer-server types of simulations usually do not have a preset size for the serving queue. Dynamic data structures can be

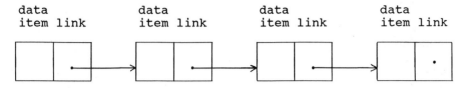

Figure 7.6   Linked list.

properly applied in this situation.   In this context, let us examine the sorting problem (see Figure 7.7).

There is no data movement involved in the sorting algorithm.   The only
changes are in the links in the structure.   Assume the linked list on top is
the original order.   If it is to be sorted in a descending order, there is no
need to move the data items around.   This is because the order of an item,
in a linked list, is not determined by its location, but rather by its links.
When the links are rearranged to become as shown in the bottom list of
Figure 7.7, data are ordered in a descending fashion.   To find the order,
one travels from the header of the linked list.   Each data item visited
through the links appears in the proper order.   In the bottom list of Figure 7.7, the items visited are 15, then 14, then 13, then 12, and finally 1.
The only difference between the two lists is that their links are different.
The data items are still in the same places.

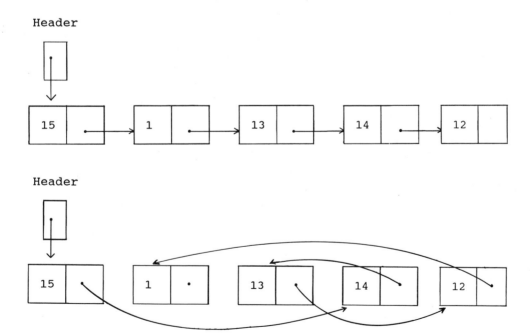

Figure 7.7   Linked list example.

7.3.1   Linked Lists

Data items connected by links are called linked lists. Data items can be of
any declared data structures; they can be simple values or complicated
structures. Let us use the library information system discussed earlier to
illustrate this structure. Refer to Figure 7.8. A new book can be added
to the library by setting up a new record and linking it to the book list.

The information on a specific book can be obtained by checking the name
field and following the links until it matches or reaches the end of the list
(if the book is not in the library). The following program segment (in
Pascal) is for the search of a book by traversing the link. Note that there
are no pointer variables provided in Fortran, and if this program segment is
to be implemented in Fortran, quite a few changes will have to be made.

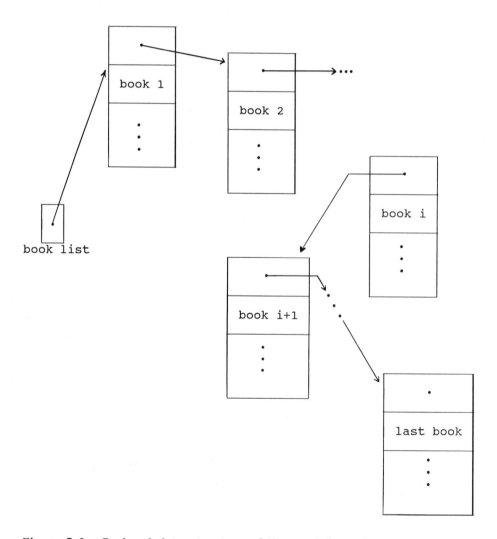

Figure 7.8   Declared data structure of library information system.

```
type
    BookPointer = ^BookRecord;
    BookRecord = record
                    Next : BookPointer;
                    Title : string1;
                    Author : string2;
                    CallNumber : string3;
                 end;

var
    FirstPointer,CurrPointer : BookPointer;
    BookName : string1;
    Found : Boolean;
```

FirstPointer is assumed to point to the first book record, as shown in Figure 7.9. To locate a particular book, we use an auxiliary pointer Curr-Pointer that begins at the first record and moves through the list by traversing the link fields

```
CurrPointer := CurrPointer^.Next
```

This statement enables us to travel to the next record.

```
Found := false;
CurrPointer := FirstPointer;
while (CurrPointer<>nil) and (not Found) do
    if CurrPointer^.Title = BookName then
            Found := true
    else
        CurrPointer := CurrPointer^.Next;
```

The while loop is terminated when the desired book is found, or when the end of the list is reached and the record is not found. In the former case, CurrPointer points to the desired record. In the latter case, CurrPointer contains nil. If the book is found, then the information can be displayed as

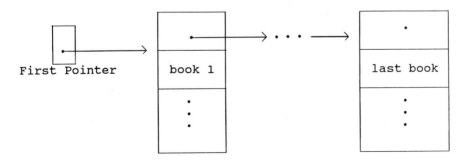

Figure 7.9   Example of FirstPointer.

```
if CurrPointer <> nil then
   writeln("The book is found, . . .")
  else
   writeln("The following book is not found");
printstring(BookName)
```

Here, printstring is assumed to be a procedure that can print an array of characters.

Linked lists can be classified in several ways: by number of links in each item, by usage, and by topology. When they are classified by usage, two prominent structures stand out: stacks and queues. In terms of topology, there are trees and graphs. In terms of number of links, singly linked and doubly linked lists are commonly used.

7.3.2   Singly and Doubly Linked Lists

When it is necessary to travel both directions in a list, it is imperative that two-way links be established. This kind of list is called doubly linked list (Pfaltz, 1977). It is a special kind of multiply linked list, a list in which each node has two or more links. Figure 7.10 is an example of a doubly linked list. When data are in a linked list and one needs to find out whether a certain item is in the list or not, then it would be convenient to work on a doubly linked list. Otherwise, it becomes mandatory that any search will have to start from the beginning of the singly linked list. In applications involving searching or insertion of a list, it is always advisable to use a doubly linked list for data structures.

7.3.3   Stacks and Queues

Stacks and queues (Lewis and Smith, 1976) are frequently used in simulations. They are usually implemented by linked-list structures. A stack is a data structure in which elements can only be accessed from one end. The other end remains untouched unless the stack becomes empty. The addition of a new element, or deletion of an existing element, is always referenced from the same end of the list. This is why it is also called Last-In-First-Out (LIFO) structure. The tray holder in a cafeteria has this property. Only the top tray is accessible. When it is removed, the one below rises to the top. Two operations called PUSH and POP are associated with this data structure. PUSH inserts an element as the first and POP removes the first element from the list. Figure 7.11 shows the effects of PUSH and POP.

The first structure in Figure 7.11, Top, contains 75, 60, 14, and 16. The second structure is the result of pushing an item 3 onto the stack Top. The command used is Push(Top,3). The third structure is the result of

Next

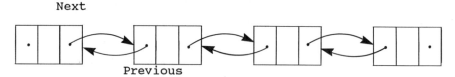

Previous

Figure 7.10   Doubly linked list.

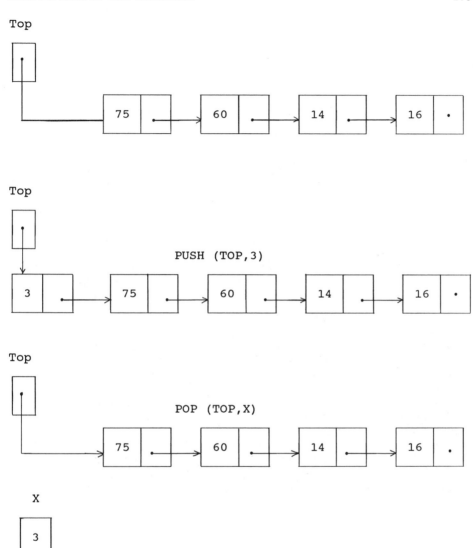

Figure 7.11   Push and Pop of a stack.

popping one item from the stack (which becomes the original stack). The command is Pop(Top). There is no need to specify the item to be popped because it is not ambiguous.

The queue is a structure in which elements are entered at one end and removed from the other end. Service lines are typical examples of this structure. The front person in the line (in a bank, at a movie theater window, etc.) gets to be served first. He/She leaves the line after service (DELETED). New customers are expected to line up at the end of the line (INSERTION). Since both ends can be accessed, two external links are needed: FRONT and REAR.

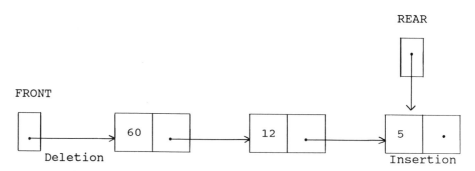

Figure 7.12   Queue.

In Figure 12, the arrows represent links. In this structure there are two ends. Front points to the first customer to be served. Rear points to the end of the line where new customers line up for service.

7.3.4   Trees and Graphs

Linked lists containing more than two links per node can be arranged into a structure called a tree. Figure 7.13 is a binary tree—each node has at most two links pointing to next-level nodes. Each node is pointed by a link and points to at most two other nodes. No single node in a tree can be pointed by two different links. The structure is called a tree because it resembles an upside-down tree (Pfaltz, 1977). The nodes that do not point to any other nodes are called "leaves."

Binary trees are very special because they are the simplest type of trees, yet they exhibit all the necessary qualities of trees. They are studied most thoroughly by mathematicians and computer scientists. Important operations on trees include traversal and restructuring of trees

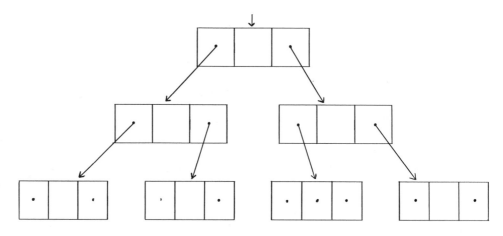

Figure 7.13   Binary tree.

(Wirth, 1976). Traversal on a tree is the operation of following from node to node with the purpose of visiting every node of the tree by following some order (Horowitz and Sahni, 1982). Restructuring (Aho et al., 1974; Wirth, 1976) will change the apparent structure of a tree while maintaining its certain property in order to enhance specific processing.

Many human activities can be best analyzed by using the concept of trees. Decision trees are used to describe all the possibilities to facilitate analysis. The eight-coins problem is a typical example. (Given eight coins, one of which is a counterfeit and has a different weight than the others, and given a balance, what is the minimum number of comparisons required to determine the false coin?) Game trees are used in many different games to help players analyze situations and their next moves (Horowitz and Sahni, 1982).

Graphs are generalized trees dropping the restriction that no node can be pointed to by more than one node. Typically, graphs are used to represent salesmen's traveling routes among cities, routing of flights for airline passengers, and other transportation problems.

7.4 DATABASES

In Sections 7.2 and 7.3, we discussed various data structures, their representations, and some of the algorithms to implement those data structures. For any problem, a programmer must specify a data structure and then decide how to implement it. Since a data structure is required in solving almost any problem, it is feasible to implement it once and let every programmer use it. A collection of interrelated data and a set of programs to access that data are called a database management system (DBMS). The collection of interrelated data is usually referred to as the database. Once a database is created, it is possible for several users to use that data. One user may need only one part of the data; a second user may need another part of the data; a third user may need data used by the first two users. In this case, these three users have different views of the data. The purpose of a DBMS is to provide a convenient and efficient environment for storing information into and retrieving information from the database for all users.

7.4.1 Purpose of Database Systems

Sources of data are the basic necessities of information processing in an organization. "Organization" is simply a convenient generic term for any reasonably large-scale commercial, scientific, technical, or other operation. Any organization must necessarily maintain a lot of data about its operation. In conventional data-processing approach, an organization develops a program or a number of programs for each application. As a typical example, we will consider a savings-and-loan association. A part of the association keeps information about all customers and savings accounts maintained at the association. The savings account and customer records are kept in permanent system files. A number of application programs are used to manipulate these files. One program may be used just to credit or debit an account; another may be designed to add new accounts; a third program may be used to find the balance of an account; and so on. These application programs are created in response to the association's needs.

New application programs may be added as the need arises. As time goes by, the number of files and application programs is expected to grow.

Such files are likely to have different formats, and the programs may be
written in different languages.

The environment described above is a typical file-processing system
supported by a conventional operating system (Chapter 9). Permanent
records are stored in various files, and different programs are written to
retrieve information from and add records to the proper files. The environment, however, has the following disadvantages:

  Data redundancy: It is unavoidable that some data elements are used
    in a number of applications. Since data are required by multiple
    applications, they are often recorded in multiple data files and thus
    stored repeatedly. This situation is called "data redundancy" and
    often leads to problems concerning the integrity of data.
  Limited data sharing: When data files are implemented as separate units,
    it is difficult to relate the data items across the applications.
  Data isolation: Since data scattered in various files may be in different
    formats, it is not an easy task to write programs to retrieve the
    appropriate data.

These difficulties, among others, have prompted the development of
database management systems for large organizations. Database systems
have been successfully exploited in large organizations, and several books
have been published on this subject. See, for example, Atre (1980);
Korth and Silverschatz,(1986), Date (1977), Wiederhold (1977), Cardenas,
(1979), and Everest (1986). These books mainly emphasize the use of a
database management system in an organization. With the advent of microcomputers, database systems have also been developed for small business
systems and have been very successful.

Consider an example of the use of databases in a simulation environment, a VLSI system design. In VLSI design, a designer goes through a
multilevel design process (Mukherjee, 1986). The lowest level in the design
process is the layout-layer, which is an interface between the physical
world and the electrical world. This layer serves as the link between the
circuit and the fabrication process that builds the circuit. The logic or
circuit-level of the design process uses an abstraction of the underlying
electrical circuit in which the currents and voltages are limited to discrete
levels. The layout-layer is derived from the logic-level of the design
process. At the present time, there is a VLSI layout system (Ousterhout
et al., 1985; Scott et al., 1985) which allows users to specify the design
using layout-layer. A cell in the layout-layer consists of an interconnected
network of rectangles of various sizes and colors. A user starts at the
lowest level of the cell design and hierarchically builds the layout of a
circuit. The layout system maintains a hierarchical database of cells defined by linked-list data structures. The user views the design as a large
interconnecting network of rectangles of various sizes and colors. This design is checked online automatically for any design rule violation by another
program which runs in the background. The design is simulated for its electrical properties, timing, and performance. This reported VLSI layout system automatically creates a database of the design and the user interactively
designs the system. Since the design is hierarchically done, any design
can subsequently be used in other designs. If a design is modified, the
database is updated to reflect such modification. This database is used to
create the different views for circuit simulation and design rule checker.

Thus, the proliferation of the use of database systems in various applications including simulations is observed. With these motivations, the fundamental concepts of a database system are presented.

7.4.2 Overview of Database Systems

A database is a collection of related data about a system with multiple users. Multiple users may have different "views of data." As an example, the designer's view of data in the VLSI system design discussed earlier is rectangles of various sizes and colors. These rectangles are interrelated by the interconnecting network of the design. For circuit simulation, the data are viewed as nodes with their capacitances, resistances, and transistor information. A major purpose of a database system is to provide users with their views of data. That is, the system hides certain details of how the data are stored and maintained. However, in order for the system to be usable, data must be retrieved efficiently.

A database system is created by using database schemes, that is, the one concerned with the way in which the data are viewed by individual users. The system consists of a number of functional components, such as file manager, database manager, query processor, language processor, and others. A casual user occasionally will query the system to retrieve some information. A naive user may use the system through the application program requiring access to the database system. The application programmer may access the database system through operating system calls for manipulating the database system. Basic concepts and functional components of a database system are explained in the following.

Data Abstraction. The concern for efficiency leads to the design of complex data structures for the representation of data in the database. Since database systems are often used by non-computer-oriented individuals, the complexity of the data structure should be hidden from the users. This is accomplished by defining three levels of abstraction at which databases may be viewed. These are:

1. Physical level: This is the lowest level of abstraction, at which one describes "how" the data is actually stored. The complex low-level data structure is described in detail at this level.
2. Conceptual level: This is the next level of abstraction at which one describes "what" data are actually stored in the database and the relationship that exists among data. This level describes the entire database in terms of a small number of relatively simpler structures.
3. View level: This is the highest level of abstraction at which one describes only part of the entire database. To simplify the interaction with the system, the view level of abstraction is defined for the user. Many views may be provided for the same database.

The interrelationship among these three levels of abstraction is shown in Figure 7.14.

As an example of data abstraction, the rectangles of the VLSI layout are considered again and illustrated using the concept of data types in programming languages. In a Pascal-like language the rectangle record may be declared as follows

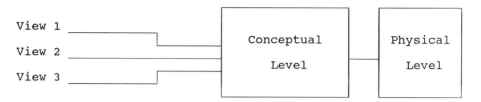

Figure 7.14   Interrelationship among three levels of abstraction.


```
    TYPE    rectangle = RECORD
                          name : string;
                          x1 : integer;
                          y1 : integer;
                          x2 : integer;
                          y2 : integer;
                          color : color;
                          resistance : real;
                          capacitance : real;
                  END;
```

This defines a record called rectangle with eight fields. The first field defines the name of the rectangle. The next four fields define the coordinates
of the rectangle, and the last three fields define the characteristics of the
rectangle. At the physical level, these fields can be described as a block
of storage locations. At the conceptual level, each cell is described
by a record as specified above. At the view level, the coordinates and the
color of the rectangle are enough for the designer during the design
process.

    Data Independence. In a conventional file-processing environment,
data format, location, and method of access are important. Changes in any
of these items may affect the application program. The ability to use the
database without knowing the representation details is called data independence. In database systems, three levels of abstraction have been defined. In this context, there are two levels of data independence, as
follows:

1. Physical data independence is the ability to modify the physical
    scheme without causing application programs to be rewritten.
    Modifications at the physical level are occasionally necessary in order
    to improve the database system performance.
2. Logical data independence is the ability to modify the conceptual
    scheme without causing the application program to be rewritten.
    Modifications at the conceptual level are necessary whenever the
    logical structure of the database is altered.

    Data Models. In order to describe the structure of a database, we need
to define the concept of data models. A data model is a collection of conceptual tools for describing data, data relationship, data semantics, and
data constraints. A number of data models have been proposed. In this
sequel, a few of the widely known ones are described.

1. Relational model: the relational data model is relatively new. A relational database consists of a collection of tables, each of which is assigned a unique name. Although tables are a simple, intuitive notion, there is a direct correspondence between the mathematical concept of relation and the concept of a table. A row in a table represents a relationship among a set of values. A table is then a collection of such relationships. As an example, consider the rectangular table of Figure 7.15. It has six attributes: name, coordinates x1, y1, x2, y2, and color. For each attribute, there is a set of permitted values, called the domain of that attribute. In general, the rectangular table will contain only a subset of the set of all possible rows. Mathematically, a relation is defined to be a subset of a Cartesian product of a list of domains. Since tables represent relations, table and row are referred to as relation and tuple, respectively. The database schema, the logical design of the database, is the relation scheme that corresponds to the type definition in a programming language. There may be several relation schemes in a database. In the above example, the relation scheme in the rectangle table can be denoted as

rectangle-scheme = (name,x1,y1,x2,y2,color)

A database instance is the data in the database at a given instant; thus, a variable in programming languages corresponds to the concept of an instance of a relation.

Frequently it is found that there is one attribute with values that uniquely identify the tuples of the relation. For example, "name" in rectangle relation uniquely identifies the tuple. This key is said to be the primary key to the relation. Not every relation will have a single-attribute primary key. However, every relation will have some combination of attributes to uniquely identify a tuple in the relation. Sometimes, there may be more than one key such that each key uniquely identifies the relation. These keys are candidate keys; any one of them can be chosen as the primary key for the relation.

The user of the database system may query these tables, insert new tuples, delete tuples, and update tuples. There are several languages for expressing these operations. The tuple relational calculus and the domain relational calculus are nonprocedural languages, whereas the relational algebra is a procedural language. Discussion of these languages is beyond the scope of this chapter, and readers are referred to Korth and Silverschatz (1986) and Date (1977). The formal languages provide a concise

Name	x1	y1	x2	y2	color
in	95	271	105	282	blue
metal12	22	−10	28	0	blue
vdd	13	388	23	399	purple
metal13	88	−10	93	0	blue
gnd	16	131	26	142	purple

Figure 7.15   Example of relational data model.

language for representing queries. There are some commercial products that provide "user friendly" query languages. Three influential commercial languages, namely, SQL, Quel, and QBE, are described in Korth and Silverschatz (1986).

2. Network data model: A network database consists of a collection of records which are connected with each other through links. A link is an association between two records. The network model differs from the relational model in that data are represented by collections of records, and relationships among data are represented by links.

A data-structure diagram is a scheme for a network database. The diagram consists of two basic components: boxes, which correspond to record types, and lines, which correspond to links. It specifies the overall logical structure of the database.

In the late 1960s, several commercial database systems based on the network model emerged. These systems were studied extensively by the Database Task Group (DBTG) within the CODASYL group. For further information readers are referred to Atre (1980), Korth and Silverschatz (1986), Date (1977), Cardenas (1979), and Everest (1986).

3. Hierarchical data model: A hierarchical database consists of a collection of records that are connected with each other through links similar to the network model except that the records are organized as a collection of trees rather than arbitrary graphs. A tree-structure diagram is the scheme for a hierarchical database. Such a diagram consists of two basic components: boxes, which correspond to record types, and lines, which correspond to links. It specifies the logical structure of the database. For each diagram, there exists a single instance of a database tree, and the root of such a tree is a dummy node. The children of that node are actual instances of the appropriate record type. Each such instance may, in turn, have several instances of the appropriate record types.

Two influential hierarchical database systems are IBM's Information Management System (IMS) and MRI's system 2000. Readers are referred to the list of references on databases for further information.

Data Definition Language. In order to translate a data model into an operational system, it has to be described in a form that lends itself to implementation. The overall design of the database is called the database scheme. Schemes are changed infrequently, if at all. A database scheme is defined by a special language called a data definition language (DDL). The compilation of DDL statements is a set of tables that are stored in a file called the data dictionary. The data dictionary stores information about data, such as origin, description, relationship to other data, usage, etc. It is a database itself, having data about data. This specifies the implementation details of database schemes which are usually hidden from the users.

Data Manipulation Language. A data manipulation language (DML) is one that enables users to access or manipulate data. There are two types of DML: procedural and nonprocedural. The user of a procedural language is required to specify "what" data are needed and "how" to get the data, whereas in a nonprocedural language, one is only required to specify "what" data are needed. For this reason, a nonprocedural language is easier to learn than a procedural one. It should be noted that the code generated from the nonprocedural language may not be very efficient.

Database Manager. A database manager is a program that provides the interface between the low-level data, in the database, and the application programs and queries submitted to them. Several functions are performed by the database manager.

1. Data manipulation. The database manager is responsible for storing, retrieving, and updating data into the database by interacting with the file manager of the operating system (Chapter 9).
2. Security and integrity enforcement. A user need not have access to the entire database. The database manager administers the security requirements of a database system. In addition, it checks the consistency constraints specified in the database system.
3. Concurrency control. When several users update the database system, the database manager controls the interaction of concurrent users to avoid the inconsistency in the database.
4. Recovery and backup. It is the responsibility of the database manager to detect system failure such as disk crash, software failure, etc., and restore the database to the state that existed prior to failure. This is usually achieved through the initiation of backup and recovery procedures.

Database Administrator. The database administrator (DBA) has central control over the system. The DBA creates the database using the scheme definitions and controls access to the system by performing certain functions, including scheme definition, storage structure and access method definition, scheme and physical organization modification, authorization for data access, and integrity constraint specification.

7.4.3 Database Applications

Database systems are extensively used in commercial- and business-oriented applications mainly in data processing. However, there are many applications outside the realm of data processing, which benefit from the use of a database system. They are essential in systems simulations, especially for large systems. Some of the simulation-related applications are described here.

1. Design simulation databases: In computer-aided design (CAD), a large amount of data must be stored to represent the item being designed and simulated. The data describing a design are interrelated in a complex manner. Several versions of the designs need to be saved. As an example, in VLSI system design (Ousterhout et al., 1985); (Scott et al., 1985), the circuits are designed hierarchically. Such hierarchical structure is then simulated for its electrical properties, timing, and performance(s).

In CAD design environment, an object may be any device or system being designed. In general, the objects may be large and have a relatively complex internal structure. The designers are the users of the design database. They frequently interact with the database, using graphical input/output devices such as a graphics terminal. They move components, insert components, etc. Because of the nature of a design process, text-based languages are not sufficient for the design database user. Many current systems are extensions of existing commercial or research database systems. Certain design applications require operation on an entire object. Examples of such applications include testing, cost estimation, simulation,

and design rule checking. Other applications may involve several objects
at a lower detail level.

2. Environment modeling: Databases are being used as components of
systems designed to automate software development and to simplify user
interaction with a computer system. These require extensions of data
models.

7.5 CONCLUSIONS AND TRENDS

In this chapter we have presented data structures and database concepts.
It is important to note that in a modeling and simulation environment one
needs to identify the specific activities and the corresponding stimuli that
feed the activities. Systems' input(s) and the internal information need to
be arranged in some structure to facilitate their processing. Therefore,
data structures and databases come to the frontline consideration in any
simulation study.

In the area of data structures, we have observed that a technique
called "abstraction" is applied to various data to come up to the few classes
of structures. By concentrating on the important attributes of an item and
ignoring the less essential features, one accomplishes the abstraction process.
Thus, we are able to focus on the common parts. Data structuring essentially uncovers classes of structures that can be used effectively to solve
problems.

The basic data structures and the advanced data structures are presented in Sections 7.2 and 7.3. These are by no means exhaustive, but
they provide some basic understanding required to get into further studies.

Similarly, database concepts are introduced in Section 7.4. A VLSI
example is used to illustrate the importance of database in an industrial
setting. Many concepts are not covered here, such as the relationships
between databases and knowledge bases, how graphics benefit from databases, and how expert systems relate to databases. We believe the fundamentals of databases will enable one to understand those related topics,
but first digesting those concepts is essential.

We conclude by pointing out a recent significant development in both
the data structure and databases area (Stefik and Bobrow, 1986). In the
area of programming research, the programming activity traditionally has
been viewed as either action-oriented or object-oriented. In the action-
oriented view, programs (algorithms) are the fundamental entities and objects
are auxiliary entities on which programs operate. In the object-oriented
view objects are the fundamental entities and actions are auxiliary entities
needed to describe the behavior of objects (Ingalls, 1981). The action-
oriented view is appropriate to scientific computations where the complexity
lies in the algorithm and data structures may disappear once the result has
been computed. The object-oriented view, on the other hand, is appropriate
to data management, embedded computer, and other applications where data
structures represent the state of a system that exists over a period of time
and interacts with its environment. Languages like Algol 60 and Fortran
support the action-oriented view, while Cobol and database languages support
the object-oriented view.

In the object-oriented view, the objects are passive elements. It has
been found that for certain kinds of problems, like simulation, it is better
to view objects as active objects. Now the object may receive input

messages from an input stream and transform them into output messages in an output stream. The programming activity that is based on this point of view has been called "object-oriented programming."

Objects are entities that combine the properties of procedures and data since they perform computations and save local state (Rentsch, 1982). Uniform use of objects contrasts with the use of separate procedures and data in conventional programming.

The concept of "object" is derived from the concept of data abstraction. By combining the abstract structure of an item and its legal operations into a single entity, one is able to view the problem from an elevated plane, thus allowing more concentration on the problem domain considerations.

It is natural for modeling and simulation to adopt the concept of object-oriented programming to solve problems. It is envisioned that object-oriented design will affect the future of modeling and simulation techniques.

Owing to this development, several research organizations have started research projects to develop "object-oriented" database management systems. The evaluations of such systems are not completed at this time. However, we can anticipate some significant changes in terms of data handling and database manipulation in the near future.

REFERENCES

Aho, A. V., Hopcroft, J. E., and Ullman, J. D. (1974). *The Design and Analysis of Computer Algorithms*, Addison-Wesley.

Atre, S. (1980). *Data Base: Structured Techniques for Design, Performance, and Management*, Wiley.

Cardenas, A. F. (1979). *Data Base Management Systems*, Allyn and Bacon.

Dale, N., and Orshalick, D. (1983). *Pascal*, Heath.

Date, C. J. (1977). *An Introduction to Database Systems*, Addison-Wesley.

Elson, M. (1975). *Data Structures*, Science Research Associates.

Everest, G. C. (1986). *Database Management: Objectives, System Functions & Administration*, McGraw-Hill, New York.

Habermann, A. N. and Perry, D. E. (1983). *Ada for Experienced Programmers*, Addison-Wesley.

Horowitz, E. and Sahni, S. (1982). *Fundamentals of Data Structures*, Computer Science Press.

Ingalls, D. (1981). Design Principles Behind Smalltalk, *BYTE*, August 1981.

Korth, H. F. and Silverschatz (1986). *Database System Concepts*, McGraw-Hill, New York.

Lewis, T. G. and Smith, M. Z. (1976). *Applying Data Structures*, Houghton-Mifflin, Boston.

Mukherjee, A. (1986). *Introduction to nMOS & CMOS VLSI Systems Design*, Prentice-Hall, Englewood Cliffs, NJ.

Ousterhout, J. K. et al. (1985). The magic VLSI layout system, *IEEE Design and Test.* 2(1), 19–30.

Pfaltz, J. (1977). *Computer Data Structures*, McGraw-Hill, New York.

Rentsch, T. (1982). Object oriented programming, *SIGPLAN Notices 17*, (9), Sept. 1982.

Scott, W. S., et al. (1985). *VLSI Tools: More Works by the Original Artists*, Report No. UCB/CSD 85/225, Computer Science Division, University of California at Berkeley, Berkeley, CA.

Stefik, M. and Bobrow, D. (1986). Object-oriented programming: Themes and variations, *AI Magazine* 6(4), Winter 1986.

Wegner, P. (1980). Programming languages—Concepts and research directions, in *Research Directions in Software Technology* (P. Wegner, ed.).

Wiederhold, G. (1977). *Database Design*, McGraw-Hill, New York.

Wirth, N. (1976). *Algorithms + Data Structures = Programs*, Prentice-Hall, Englewood Cliffs, NJ.

8
Programming Languages

8.1   INTRODUCTION

Most present-day simulation studies are conducted with digital computers,
using programming languages. The languages may be general purpose or
especially designed for modeling and simulation. The arguments for and
against the use of either general-purpose languages or simulation languages
have persisted from the earliest days to the present time. Both categories
of languages have evolved as perceived needs have evolved, and thus
many helpful features are now available for use in developing computer
software—of which simulation software is a particularly complex subset.

In this chapter, we shall first review the history of programming
language development to the present time, including general purpose and
simulation languages (Section 8.2). The two principal categories of programming language features, namely, data structures and control structures,
are considered in Section 8.3. Whereas Chapter 7 covers data structures
in general, our focus in Section 8.3 is on the support provided by programming languages for representing data and data relationships, especially
for simulation applications. The subject of Section 8.4 is modern features
for expressing flow-of-control within a program, with special emphasis on
time control in simulation languages. In Section 8.5 we shall discuss current programming language support environments and give a brief overview
of processors for programming languages—i.e., compilers, interpreters,
and translators.

Section 8.6 presents a discussion of criteria by which a programming
language may be judged, for general programming use and for simulation
use. Section 8.7 applies the criteria of section 8.6 in providing an assessment of a selected set of programming languages. Section 8.8 seeks to
give a perspective on current trends in the evolution of simulation languages
and support systems.

Earlier chapters (1, 2, and 3) have explored the basic concepts and
examples of discrete-event simulation and continuous simulation, and sub-
section 1.3.2 presented a brief account of the evolution of computers.
Thus, this chapter assumes a fundamental grasp of simulation concepts

and techniques. Appendix B provides a brief summary of many available
simulation languages and may serve as a convenient reference.

8.2   PROGRAMMING LANGUAGE EVOLUTION

A programming language is a notation for use in describing computational
processes (i.e., algorithms and data structures). The earliest programming
languages, by this definition, would have been "natural languages"—
languages used for ordinary communication. The lack of preciseness of
natural languages has led scientists, mathematicians, and engineers through
the ages to devise precise notations for expressing their ideas and for
describing sequences of steps for solving a problem. Horowitz (1984)
describes early algorithms written on clay tablets, found by archaeologists
near the ancient city of Babylon (near the modern city of Baghdad, Iraq).
One example is a procedure for determining the time required to double
an investment of shekels, given a certain annual interest.

Ada Augusta, Countess of Lovelace, the daughter of the poet Lord
Byron, is called the first programmer, because of her work with Charles
Babbage. Babbage designed two machines for computation during 1820–
1850, the "Difference Engine" and the "Analytical Engine." The Difference
Engine was based on the theory of finite differences; the Analytical Engine
used many concepts included in present-day digital computers. The
Analytical Engine didn't use the stored program concept—the Countess
used "operations cards" to describe a desired sequence of operations to the
machine and a "variable card" containing the necessary data. The U.S.
Department of Defense (DoD) has honored Ada Augusta by naming the
language Ada, developed under DoD sponsorship, in her honor. The interested reader may refer to Horowitz (1984) for further information on
Babbage and the Countess Lovelace.

With the advent of digital computers, instructions and data were de -
scribed in the native language of the computer, i.e., machine language.
Thus, strings of zeros and ones were often used to instruct machines operating on binary numbers. Assembly languages occurred next, providing
symbolic representations of the machine-language instructions, a great aid
to writing, reading, and modifying computer programs. Although assembly
languages are still in use, the percentage of programming tasks employing
assembly language has steadily declined since the late 1950s, and now they
are used very infrequently, principally for very specialized, time-critical
applications.

So-called "high-level" programming languages began to appear in the
1950s. Of the many languages developed in the 1950s and 1960s, Sammet
(1976) summarizes 13 languages considered to be especially significant:

> APT (Automatically Programmed Tools); 1956. The first language
> for a specialized application area.
>
> Fortran (FORmula TRANslation); 1956. The first higher-level
> language to be widely used. It opened the door to practical usage
> of computers by large numbers of scientific and engineering
> personnel.
>
> Flowmatic; 1956. The first language suitable for business data
> processing and the first to have heavy emphasis on an "English-
> like" syntax.

IPL-V (Information Processing Language V); 1958. The first—
and also a major—language for doing list processing.

Comit; 1957. The first realistic string-handling and pattern-
matching language; most of its features appear (although with
different syntax) in any other language attempting to do any
string manipulation.

Cobol (COmmon Business-Oriented Language); 1960. One of
the most widely used languages on an absolute basis, and the most
widely used for business applications. Technical attributes include
real attempts at English-like syntax and at machine independence.

Algol 60 (ALGOrithmic Language); 1960. Developed for specifying
algorithms—primarily numerical. Introduced many specific features
in an elegant fashion and, combined with its formal syntatic definition,
inspired most of the theoretical work in programming languages and
much of the work on implementation techniques. More widely used in
Europe than in the United States.

Lisp (LISt Processing); 1960; Introduced concepts of functional
programming combined with facility for doing list processing. Used
by many of the people working in the field of artificial intelligence.

Jovial (Jules Own Version of IAL); 1960. The first language to
include adequate capability for handling scientific computations,
input/output, logical manipulation of information, and data storage
and handling. Most Jovial compilers were written in Jovial.

GPSS (General-Purpose Systems Simulator); 1961. The first
language to make simulation a practical tool for people.

Joss (JOHNNIAC Open-Shop System); 1964. The first interactive language; it spawned a number of dialects, which collectively
helped to make time sharing practical for computational problems.

Formac (FORmula MAnipulation Compiler); 1964. The first
language to be used fairly widely on a practical basis for mathematical problems needing formal algebraic manipulation.

APL/360 (A Programming Language); 1967. Provided many
higher-level operators, which permitted extremely short algorithms
and caused new ways of looking at some problems.

As can be seen from this list, simulation languages have been used
since very early days (e.g., GPSS). Of course, general-purpose languages
were, and still are, used for simulation studies; FORTRAN, for example,
is used extensively for simulation purposes, and FORTRAN code is used
(and generated, in some cases) by numerous simulation processors.
BASIC, PL/I, and SNOBOL, all of which appeared in the 1960s, continue
to have a considerable impact. The SIMSCRIPT II language (Kiviat,
Villanueva, and Markowitz, 1973) was designed for discrete-event simulation use, but had many general-purpose features as well; it appeared in
the 1960s, as did the GASP II simulation language (Pritsker and Kiviat,
1969).

The appearance of ALGOL 60 (Naur, 1963) constitutes a "watershed" in
a sense, in the design of programming languages and the development of
compilers for them. The nested block structure of ALGOL, and its emphasis on declaration of variables, have become common in modern programming languages. As stated earlier, ALGOL 60 was not extensively used in
the United States, but had a marked influence on the development of subsequent languages. Pascal (Jensen and Wirth, 1974) was developed in the

1968-1971 time frame. Ada's planning began in 1975 by the U.S. Department of Defense; production compilers have only recently become available. Both Pascal and Ada employ the concept of strong-typing: each variable must be declared by type before it is used in computation, and only expressions of the variable's type may be used to assign values to the variable. Simula, an Algol-based language for discrete-event simulation, appeared in the 1960s (Dahl and Nygaard, 1966). Like Algol 60, Simula has not been used extensively in the United States, but has had significant influence on the development of languages. We shall pursue this point further in Section 8.3, relative to the "class" feature of Simula.

As can be seen from the foregoing discussion, the 1960s was a period of extremely high activity and creativity in the development of new general-purpose and simulation programming languages. Continuous-simulation languages were advancing rapidly in this period. In 1965, DSL/90 appeared, followed by CSMP/360 in 1966. In 1967, Simulation Councils, Inc. (SCi) published a description of the SCi Continuous-System Language (CSSL) (SCi Software Committee, 1967). This was the first attempt to define the desirable characteristics of a simulation language. CSSL-III was the first attempt to develop a simulation language based on the CSSL report. A PL/I-based language for partial differential equations, PDEL, was also developed in the 1960s. Dynamo, a language for use in analyzing Systems Dynamics models, was developed about 1970. A good overview of the early continuous-simulation languages, and a discussion of the chronology of developments, may be found in Spriet and Vansteenkiste (1982), along with references to definitive documentation; they also provide a summary of the more recent continuous-simulation languages.

ACSL (Mitchell and Gauthier, 1981) and CSSL-IV (Nilsen, 1976) are more recent, currently prominent continuous-simulation languages, based on the SCi CSSL standard. Spriet and Vansteenkiste (1982) provide a comparison of ACSL and CSSL-IV. Among other continuous-simulation languages in current use are DAREP and DARE/INTERACTIVE, developed by Korn and associates, and EASY5, a Boeing product for simulating with nonlinear differential or difference equations. "Catalog of Simulation Software" (1985) contains a summary of an extensive list of simulation software products, categorized by microcomputer software, and minicomputer and mainframe software. All the languages mentioned in this paragraph, and others, are summarized there, including most of the currently available discrete-event simulation languages. For each language listed, a brief description is given, along with the name of an individual to contact for further information, a list of computers and operating systems for which the language is implemented, and an approximate cost of lease or purchase. An updated version of this "catalog" has been published annually in recent years, in the October issue of *Simulation*.

GASP IV was developed by Pritsker (Pritsker, 1974), adding a continuous modeling and simulation capability to the earlier discrete-event capability of GASP. It was the first widely disseminated, well-documented combined discrete/continuous-simulation language. It is FORTRAN-based—the user writes subroutines in FORTRAN; the GASP system subroutines provide such features as time control, random-variate generation, report generation, and plotting capabilities. GASP IV (as well as the earlier GASP versions) also provides an orderly conceptual framework for developing a simulation model. In the late 1970s, Pritsker and associates developed Q-GERT (Pritsker, 1977), a language for modeling systems expressed as networks of nodes and branches. In 1979, the FORTRAN-based language SLAM (Pritsker and Pegden, 1979) was introduced, combining GASP IV and Q-GERT into one system, thus allowing for multiple discrete-event model orientations within a single framework and retaining the combined discrete/continuous capability of GASP IV. SLAM II (Pritsker, 1984) is an improved version of SLAM. TESS supports interactive terminal use of SLAM II (Standridge, 1985).

Pegden, who participated with Pritsker in the development of SLAM, released SIMAN in 1984 (Pegden, 1984). SIMAN is "SLAM-like" in many respects—it supplies multiple discrete-event modeling orientations, supports combined discrete/continuous simulation, and is FORTRAN-based. It contains some features especially designed for modeling manufacturing systems and employs the experimental frame concept of Zeigler (Zeigler, 1976)—separating the model per se from a description of a specific experiment to be conducted by use of the model. Chapter 15 is devoted to manufacturing systems applications using SIMAN.

SIMSCRIPT II.5 (Kiviat, Villanueva, and Markowitz, 1973) was modified in the 1970s to include the process model orientation (Russell, 1974), along with the event orientation that was previously provided. The concept of model orientation will be explored in Section 8.4.

GPSS V is the current IBM-supplied version (Gordon, 1975). GPSS/H (Henriksen, 1983a) is the "state-of-the-art" GPSS with improved features and execution-time performance. Henriksen is now in the process of a complete redesign of GPSS, to result in GPSS/85 (Henriksen, 1985). Schriber (1974) is a popular book for use in learning GPSS.

The general-purpose programming language C (Kernighan and Ritchie, 1978) is "coming on strong" in the middle 1980s. C is a much more "permissive" language than Pascal or Ada—it doesn't, for example, enforce strong typing. Developed at Bell Laboratories, C has been extensively used for systems programming (e.g., the UNIX operating system and its utilities, and numerous compilers). It is also gaining in use for a wide spectrum of applications.

Whereas Pascal and C are of modest size, by intent, as to the number of features, Ada (*Ada Reference Manual*, 1980) is a very large language, relative to concepts and syntax (perhaps five or six times the size of Pascal), and thus compilers have been slow to appear. Ada was designed by DoD to support the development of "embedded systems"—in which computers monitor and control other components of an application system. Pascal, C, and Ada all have their place and will no doubt all be with us for a long time to come. Feuer and Gehani (1984) provide a good collection of papers comparing and assessing these three languages.

It is impossible to mention all the languages currently available and in use, and new languages are being developed constantly, both for general applications and for special-purpose uses, such as simulation. There are some perceptible trends, which are discussed briefly in Section 8.8.

8.3 PROGRAMMING LANGUAGE DATA-RELATED FEATURES

Programming languages describe computational processes, to computers and to humans. The two primary categories of programming language features are data and control—data serving as operands for computation, control determining the order in which operations are to occur. In this section we shall consider programming language features for managing data; control features will be considered in Section 8.4.

8.3.1   Data Structures

A data structure is a data object that contains other data objects as its components; the component data objects may be elementary (e.g., a real number, a Boolean value, or values of other "built-in" types of a language) or another data structure. The general data structures considered here are arrays, records, character strings, files, linked lists, sets, queues, and stacks.

The "array" is the most commonly occurring data structure in programming languages. An array is a collection of a fixed number of components of the same type. Thus, by definition, an array may not be composed of reals and integers intermixed. The EQUIVALENCE statement in FORTRAN allows intermixing of reals and integers in storage, but each array in FORTRAN has a single data type. All modern general-purpose programming languages and simulation programming languages support the use of arrays— they are called matrix savevalues in GPSS. Arrays serve as building blocks for developing other data structures—e.g., linked lists—in languages that do not directly provide the desired data structures. Arrays vary from vectors (single-dimension arrays) to an essentially unlimited number of dimensions in many modern languages.

A "record" is a data structure consisting of a fixed number of components of different types. An array is sometimes called a homogeneous data structure, since its components are all of the same type; a record is thus a heterogeneous data structure. Pascal supplies records as a basic feature; e.g., the code segment

```
var job:   record
              arrival_time:   real;
              service_time:   real;
              resource_reqt:   integer;
              handling_flag:   boolean
           end;
```

is an example of a record declaration in Pascal; the declaration would be similar in Ada. The "fields" of the record can be referenced, for example, as in

```
A := job.service_time
```

for the above declaration. Note that the strong-typing nature of Pascal requires that variable A be of type real. In the languages C and PL/I, this data structure is called a "structure." A record may contain other data structures as components, including other records, arrays, etc. The record may serve as a basic feature in developing simulation models using Pascal and Ada (see, for example, Bryant, 1980; Sheppard, Friel, and Reese, 1984).

"Character strings" (or just strings) are a stock-in-trade for creating computer programs to solve many application problems. They are used, for example, as input to a program to select a certain program option to be performed and in the output from a program to present results in a readable form. FORTRAN 77 provides a declaration for character strings; e.g., the declaration

CHARACTER*10 C1, C2, C3

indicates that the variables C1, C2, and C3 are to accommodate strings of 10 characters each. FORTRAN 77 also provides good features for manipulating character strings, such as concatenating strings (forming a single string from two or more strings) or extracting a substring. Most languages provide some capability for character strings, but the ease of use varies. For example, the new Pascal standard (Cooper, 1983) includes no string data type as a primitive of the language and no operations directly designed for manipulating character strings. Ada does provide such built-in features.

A "file" is a data structure that ordinarily resides in secondary storage and may exist for a greater duration than the program that created it. Files are used for storing program text, programs prepared for execution, data for input to a program, simulation results obtained during execution, and various other data. Most languages do not have a file data type as part of the language per se; Pascal is an exception—the data type "text" in Pascal pertains to a sequential file of records whose components are characters.

"Linked lists" serve an important role in implementing programs for data handling. The concept of linked lists involves a number of "nodes," each of which may contain one or more data fields and one or more "link" fields to provide access (via "pointers"—i.e., addresses) to other nodes. Linked lists are sometimes implemented by use of arrays in languages that have no built-in pointer type. In such an implementation, for example, a two-dimensional array may be used, with rows of values consisting of one or more data fields and one value that is the number of the "next" row in the linked list. GASP IV, implemented in FORTRAN, implements linked lists in this way. Such languages as Pascal, Ada, and C provide built-in features to support the dynamic allocation of storage for nodes (i.e., at execution time) and provide a pointer data type.

The use of linked lists can improve the efficiency of storage use. A node may be inserted between existing nodes without moving the nodes, as would be necessary with a linear sequence stored in an array. Deletion of a node involves changing pointers in one or two remaining nodes and putting the deleted node space onto an "available space" list. Again, no movement of information is necessary, as would be the case using a linear sequence in an array. The event list (or event set, or event calendar) of discrete-event simulation languages is generally implemented as a linked list with the elements (event notices) ordered by increasing time of execution. User chains in GPSS are just linked lists with elements ordered in specified ways.

The "set" data structure of Pascal is based on mathematical concepts—supporting membership tests, permitting insertion and deletion of single values, and supporting such fundamental set operations as union, intersection, and set difference. SETL (Dewar, 1979) is a language whose very nature is based on the use and manipulation of sets. There are only a limited number of general-purpose languages that directly support the set feature—Ada, for example, does not.

The set concept is widely used in simulation models. SIMSCRIPT II.5 supplies a built-in set concept, considered to be an attractive feature by its users. Generally speaking, the data structures that are called sets do not satisfy the mathematical definition of set—in particular, that no ordering be imposed on the elements. In SIMSCRIPT, for example,

orderings are imposed on the elements of a set; thus, technically, SIMSCRIPT
sets are really sequences of elements. This is true of event lists as well—
which are frequently called event sets.

"Queues" occur in many real-world systems (e.g., cafeteria lines, cars
in an assembly line waiting for the next activity, jobs in a computer's
"ready queue"). Queueing models are discussed in Section 3.5. Queue
implementations are often called FIFO lists (for "first-in-first-out").
Queues may readily be represented in programming languages by linked
lists (using "front" and "rear" pointers to nodes); in discrete-event simulation languages, built-in features are generally provided for representing
and manipulating queues. For example: SLAM II has a QUEUE mode;
GPSS has a QUEUE block; the set feature of SIMSCRIPT II.5 may be used
to effect queue representations.

The "stack," another prevalent data structure, is often called a LIFO
("last-in-first-out") list. Stacks are easily implemented by use of linked
lists or by use of arrays.

As mentioned earlier in this section, languages supply various data
types. The most common are integer, real, Boolean (or "logical"), and
character. FORTRAN 77 also supports a complex data type. These types
are adequate for most purposes, especially if address variables (i.e.,
pointers) are supported.

8.3.2   Data Referencing Environment

The approach taken to "nonlocal referencing" of variables varies by language. Most languages provide linkage to actual parameters when a procedure is called, either "by reference" (the address is passed) or "by
value" (the value of the actual parameter is passed; thus the called subprogram cannot alter the value of the parameter in the caller's storage).

An approach to sharing of nonlocal storage is supported by the
"explicit specification" approach, exemplified by COMMON in FORTRAN—
the programmer specifies that certain variables are to be known and
shared by two or more software components (main program, subroutines,
functions). In "block-structured" languages, such as ALGOL, Pascal,
Ada, C, Simula, and **SIMSCRIPT II.5,** "global" variables may be known and
shared by all components of a program, without explicitly so specifying in
each component; e.g., in SIMSCRIPT, a variable appearing in a DEFINE
statement in the preamble is global, as is any variable declared in the main
program in Simula or Pascal. In SIMSCRIPT and C, there is only one level
of nonlocal referencing—i.e., global variables. But in ALGOL, Pascal,
and Ada, procedures (and begin...end blocks in ALGOL) may be "nested"
to any depth; thus, variable references in a nested procedure (e.g., in
PROCA) may pertain to variables that are declared in PROCA or in any
procedure containing PROCA. This approach to nonlocal referencing in
block-structured languages is called "static scoping." With static scoping,
in all executions of a procedure, its references to a certain variable will
always be identified with a certain declaration of that variable—specifically
to the innermost declaration of a variable of that name, in the nesting of
procedures (and main program) containing the reference. For example,
suppose we have the following program structure, where PROCA is nested
within PROCB.

```
   . . .
PROCEDURE PROCB ...
   INTEGER I, J;
      . . .
   PROCEDURE PROCA ...
     INTEGER J, K;
        . . .
     K : = I + J;
        . . .
   END PROCA;
      . . .
END PROCB;
   . . .
```

Then in the statement K : = I + J within PROCA, J and K are variables declared locally in PROCA; but I is one of the variables declared in PROCB, since PROCA has no declaration for a variable with the name I. Note that, although PROCB has a declaration for a variable with name J, it cannot be referenced within PROCA since PROCA declares a variable of the same name.

In contrast to static scoping, a number of languages use the "dynamic scoping" approach; examples are LISP, APL, and SNOBOL. In these languages, a reference to a variable that is not declared locally in a subprogram is satisfied by a declaration in the subprogram (or main program) which was most recently activated (i.e., called), is still active, and which contains a declaration for the variable. For example, if MAIN calls PROCA, PROCA calls PROCB, and PROCB calls PROCC, then if in PROCC the statement A : = B + C occurs, and B is not declared in PROCC, then B is the variable B declared in PROCB if there is such a declaration, otherwise the variable B declared in PROCA if there is such a declaration, otherwise the variable B declared in MAIN; failing all of these, of course, use of B is invalid in PROCC. Note that, in the static scoping case, the order of calls would have no effect whatsoever on the meaning of a given reference, while in the dynamic scoping case the meaning depends entirely on the order of calls—presuming that the variable is not declared locally. The very nature of a language is greatly affected by this design decision. In the static scoping case a reference to a variable is always satisfied by the variable in a specific declaration; thus, the reference is always of a fixed, predetermined data type. In the dynamic scoping case each execution of a procedure could result in a variable reference having a different data type, depending on the order of calls and which declaration of a variable is used to satisfy the reference. As could be inferred, more execution time is involved in nonlocal referencing in dynamic scoping languages than in static scoping languages.

An idea that evolved during the 1970s, and which is having increasing impact in programming language design and software design, is the concept of "abstract data type." The fundamental idea is that a user-defined data structure and its allowable operations are implemented within a module, with the calling modules being allowed to "know" (i.e., access) only the interface to the structure and its operations, but not the specific aspects of the structure's implementation. A desirable result is that such an implementation

can be altered to improve efficiency with no impact on a calling program,
so long as the interface is not altered. This concept for user-defined data
types is analogous to the concept of user-defined operations, which have
been available for many years through use of subprograms. In both cases,
the caller has a precisely defined interface for use, but is not permitted
(and does not need) access "inside" the implementation. The "class" concept of Simula was the forerunner of this idea; Ada's "package" supports
the concept. Thus, we have an example of a feature of a discrete-event
simulation language that has had significant impact on modern general-
purpose programming languages.

The reader interested in the status of data representation and management capabilities in modern programming languages is referred to Horowitz
(1984) and Pratt (1984).

8.4   PROGRAMMING LANGUAGE CONTROL STRUCTURES

8.4.1 Control Structures for General Use

All programming languages provide the means to specify flow-of-control,
corresponding to decisions, iterations, etc., in the problem solution
method embodied in the program. The basic approaches are: sequence,
selection, iteration, and recursion.

The concept of "sequence" simply means that, unless explicit instructions to the contrary are given, statements are to be executed in sequential
order as they appear. "Selection" refers to mechanisms for decision making,
i.e., conditional statements. Examples in common use are the IF-THEN-ELSE
and IF-THEN-ELSEIF-ELSE constructs (e.g., in FORTRAN 77, Ada, and C)
and the CASE construct, which is increasingly more prevalent in modern
languages (e.g., in Pascal and Ada; called "switch" in C). Earlier languages relied principally on the GOTO for control transfer following a test,
mirroring the machine language jump instruction. Provision of control structures, such as those mentioned above, has resulted in diminished reliance
on the GOTO and thus has resulted in code that is more readable and easier
to debug and maintain. Emphasis on use of such control structures is one
facet of "structured programming." There is no single accepted definition
of structured programming; however, it pertains to designing and coding
software with emphasis on communicating with a human reader (Myers,
1976). Prior to the advent of structured programming, the principal emphasis was on developing code to reduce execution time. Although this is
still a concern, the high cost of personnel time compared to machine time,
coupled with high-quality code optimization capabilities in available compilers,
has resulted in the shift in emphasis.

"Iteration" refers to looping. Popular forms of iterative statement structures are for the FOR structure, the WHILE-DO structure, the REPEAT-
UNTIL structure, and the LOOP-REPEAT structure. Many variations and
combinations of these constructs may be found in available languages.

"Recursion" is provided in many modern languages—providing the
capability for a subprogram to call itself, either directly or indirectly.
SIMSCRIPT II.5 supports recursion, as do Ada, Pascal, C, and many
others. FORTRAN 77 does not support recursion.

Subprogram invocations (i.e., calls) also constitute a form of flow-of-
control management—control passes to the subprogram body and then returns to the statement following the call statement.

"Exception handling" is a feature for altering control flow less often found in programming languages; it is somewhat analogous to a hardware interrupt—the occurrence of any of certain prespecified conditions causes control to be interrupted and passed to a block of code associated with the interrupt condition. Ada provides support for exception handling, as does PL/I. Exceptions need not be error conditions; the idea is that "exceptional conditions" can better be handled by such a mechanism, without "cluttering" the code with a great many tests on conditions that do not pertain directly to the main logical flow of the program.

"Concurrency management" is becoming increasingly important in the design and use of programming languages. The concept is that two or more tasks associated with a problem solution may execute concurrently, requiring synchronization of the tasks in certain predetermined situations (e.g., one task must finish writing data to a file before another task can read it). For many years such processing has been done, but in earlier years it was achieved by calls to the operating system rather than through the use of programming language syntax. PL/I was one of the earliest general-purpose languages to support concurrency directly; Ada and Modula-2 (Wirth, 1982) are recent general-purpose languages supporting concurrency directly.

Discrete-event simulation languages have for many years represented concurrency in systems—i.e., activities that occur in parallel. For example, in simulations using a GPSS model, several transactions may move through a process during the same time frame (simulation time); in simulations based on GASP IV or SIMSCRIPT II.5 models, event routines may "simultaneously" process a ship arriving at a dock, another ship being loaded, and a third ship leaving the dock. The "executive program" of a discrete-event simulation language system achieves the effect of this "simulated parallelism."

"Distributed processing" (the use of two or more central processing units, connected only by communication lines, in the solution of a single problem) has brought about much greater complexity in the management of concurrency. Much progress has been made in determining workable methods for distributed problem solutions, and much research is under way. Distributed processing is attracting interest from simulationists—the Society for Computer Simulation sponsored a conference in 1985 on this subject (see Reynolds, 1985).

The concept of "coroutine" provides another approach to control flow management between procedures; rather than the usual caller-callee approach, coroutines may share control equally; given a group of coroutines, say A, B, and C, then if A is processing, it may suspend its execution and invoke B, with A "remembering" the location where it should resume; B may process for a time, and then invoke C; C may then cause A to resume, etc. Simula was a "trailblazer" in this regard as well. Its class mechanism provides the coroutine capability. BLISS is another language providing the coroutine capability (Wulf, Russell, and Habermann, 1971).

A "functional" programming language (also called an "applicative" language) is a language in which expressions made up of function calls are the basic "building blocks" for programs—as opposed to statements, which constitute the basic unit for programs in most languages. In contrast to these functional programming languages, the more common languages are called "imperative" languages. In an imperative language, a program describes specifically how a solution is to be achieved (principally by changing the state of variables through assignment). On the other hand, functional languages achieve their effect by applying functions, either sequentially or recursively. LISP is the best-known functional language. Other, more recent functional languages are Backus' FP (for functional programming) (Backus, 1978) and Dennis' dataflow language VAL (see Horowitz, 1984). The decreasing cost of hardware, coupled with increased interest in parallelism in computing, suggests that interest in functional languages will increase in the coming years.

"Object-oriented" languages and object-oriented programming are attracting increasing interest. An object-oriented language is one in which data elements are active and have some of the characteristics usually associated with programs. In more conventional languages, data elements are passive. Smalltalk is the best-known (and original) object-oriented language (see Horowitz, 1984). Software design approaches are developing, based on the object orientation (see, for example, Wiener and Sincovec, 1984).

8.4.2 Time Control in Simulation Languages

A subcategory of flow-of-control that is of special interest to simulationists is time control and management. Much of the decision making as to control transfer and module activation during simulation executions is based on "simulation time." It is interesting to note that the representation of parallelism in a simulation, relative to the simulation clock, is very similar in concept to the management of parallelism in computer systems, relative to "real time." Thus, simulation languages were the earliest languages to deal with the issue of parallelism.

In continuous simulations, the simulation clock is updated successively by the amount of the integration interval, as integration occurs at each time step. Depending on the integration method used, the interval size may be adjusted for accuracy as the simulation proceeds. In addition, a time interval (the "communication interval") is specified for recording values of variables describing simulation status; whenever this interval has been achieved by successive updates of the clock, control is transferred to the portion of the program that accomplishes the recording function. Further, discrete points in time may be prespecified, at which certain actions are to be taken. This may be accomplished by causing the immediately preceding "continuous" time interval to coincide with this discrete time and transferring control to the appropriate code location.

Discrete-event simulation languages do not, in general, employ the concept of fixed time intervals, but rather successively update the clock to the time of the next (in time) "significant event" that is to occur. Thus, any period of time containing no events pertinent to the simulation is skipped over in a single clock update. There are three fundamental approaches to time control in discrete event languages: (a) event scheduling, (b) activity scanning, and (c) process interaction (these are described later in this section). Many writers, however, omit activity scanning from their discussions, but we shall summarize it briefly. The terms "approach," "strategy," "orientation," and "method" are all used synonymously, by different writers. We shall use the term "strategy" in our discussion. It is an interesting aspect of discrete-event simulation languages that the strategy chosen determines how the modeler must view a system that is to be modeled; the term "world view" is used to describe the perspective of a system one assumes in using a given language. Hooper (1986) gives a detailed comparison of the three strategies; the following summary is extracted from that article.

Event-scheduling strategy	Activity-scanning strategy	Process Interaction strategy
GASP (II, IV)	AS	GPSS (/360, V, /H)
SIMSCRIPT (I.5, II, II.5)	CSL ECSL	Q-GERT SIMSCRIPT II.5
SLAM (and SLAM II)	ESP	SLAM (and SLAM II)
SIMAN	SIMON	SIMAN SIMULA

Figure 8.1   Some simulation languages by strategy.

Figure 8.1 shows a number of languages categorized by strategy. Note that SIMSCRIPT II.5, SLAM, and SIMAN appear in both the event-scheduling and process interaction columns—meaning that these languages provide the user both of these strategies for use. CSL and SIMON (an entirely different language than SIMAN) are among the better-known activity scanning languages; languages in this category have been used primarily in Europe.

With any one of the strategies, when the next event is selected for processing, control is transferred to the corresponding "modeling routine" (block of code), which is executed to model appropriate changes in model state. Basic to the strategies are the concepts of conditional and unconditional events. An "unconditional event" is eligible to be executed when its scheduled clock time occurs—it depends entirely on time. A "conditional event" may depend on additional conditions other than time.

The "event-scheduling" (ES) strategy involves a succession of unconditional events over time; i.e., the world view of a language or model incorporating the ES strategy is that the system's operation constitutes a succession of unconditional events. As to how this strategy is implemented: the ES time control procedure selects from the event list the event notice having the earliest occurrence time (with ties being resolved by assigned or default priority), updates the simulation clock to that time, and invokes the corresponding event routine. Any condition testing, other than on clock time, must occur within event routines. Events are chosen and processed successively until termination time. As an example of how events and their corresponding event routines are related, let us consider GASP IV (Pritsker, 1974). In a discrete-event model using GASP IV, the modeler supplies the event routines and a subroutine called EVNTS, having a form similar to the following:

```
        SUBROUTINE EVNTS(IX)
        INTEGER IX
        GOTO (101,102,103...),IX
101     CALL EVTRT1...
        GOTO 999
102     CALL EVTRT2
        GOTO 999
```

```
103   CALL EVTRT3...
      GOTO 999
      . . .
999   RETURN
      END
```

The approach is that each event notice has stored with it an integer event number, designating its corresponding event routine: the value of 1 corresponds to event routine EVTRT1, 2 corresponds to EVTRT2, etc. Thus if an event notice currently being processed has an event number of 3, then EVNTS will be called (by a GASP-supplied executive routine) with IX having the value 3; thus EVTRT3 will be called from EVNTS. The names EVTRT1, etc., are merely representative. They should, in practice, be names meaningful to the modeler—such as ARRIVE, DEPART, etc.

The "activity-scanning" (AS) strategy chooses the next event based on both scheduled time and condition testing. The basic concept is activity, which is (conceptually) a system state transition requiring a period of time; an activity is usually represented as two distinct events which mark the beginning and end of the activity. The world view of a language or model based on the AS strategy is that a system consists of components that engage in activities, subject to specified conditions. In an implementation, each "active-type" component (or "mobile" component) of a system may have an associated activity routine which, when executed, models active phases of the active-type component. Also, for each active-type component, the time at which the component may next be considered for activation is maintained. A condition routine corresponds to each active-type component, to determine whether conditions have been met for activation. The AS time control procedure scans activities in priority order for time eligibility and other activation conditions, executing the activity routine of the first component whose activation conditions are met. When an activation occurs, the scan starts over again in priority order; this process continues until termination time. In recent years, greater emphasis has been given to explicit scheduling of events in languages employing the AS approach, in order to reduce execution time involved in condition testing.

The "process interaction" (PI) strategy has characteristics related to both the event-scheduling and activity-scanning strategies. The world view of a language or model based on the PI strategy is that the components of a system progress through a sequence of steps (referred to as a process). Each step may consist of a condition segment and an action segment; execution of the condition segment determines whether execution of the action segment may occur. For example, in GPSS one could have the sequence:

```
      . . .
SEIZE CPUA
      . . .
RELEASE CPUA
      . . .
```

The idea of the SEIZE statement is to obtain exclusive use of the "facility" called CPUA. However, if CPUA is currently in use (i.e., by a transaction elsewhere in the program), the requesting transaction must wait until CPUA is available. Thus, there is an implicit condition in the SEIZE statement.

The RELEASE statement is essentially unconditional—the only check is whether the current transaction is the current user of the facility.

There are various implementation approaches that might be used to achieve the process interaction strategy. The following description is based on the approach generally employed in GPSS implementations and is very similar to Zeigler's approach in his "Process Interaction Prototype" (Zeigler, 1976); he uses a "Current Activations List" and a "Future Activations List," corresponding respectively to the CEL and FEL of the following descriptions.

The PI time control procedure uses two event lists: a future events list (FEL), containing event notices for events scheduled for execution at a later clock time, and a current events list (CEL), containing event notices for events that are already eligible from the standpoint of time to be executed, but whose other conditions may not yet have been met. Each event notice contains an indication of its component's current step location in a process. When time is advanced, all events scheduled for current time are moved from the FEL to the CEL. Then a CEL scan occurs; this consists of evaluating each entry's condition routine to determine whether the corresponding component may move to the next step; if so, the step's action segment is executed. A component moves through as many successive steps as possible (i.e., as long as time need not advance, and condition segments evaluate true). When a component "stops" (owing to time or other conditions), the scan resumes with the next CEL entry; when no CEL component can move, time is advanced.

The significance of strategy is considerable: it determines the fundamen-the modeler is still constrained by the strategy chosen for representing a given portion of a model. And, if a general-purpose language is used for de-concepts consistent with the world view of the specific language. If a language includes more than one world view—which is increasingly more common—the modeler is still constrained by the strategy chosen for representing a given portion of a model. And, if a general-purpose language is used for developing a model, one or more of the strategies as described will probably be used.

SLAM II and SIMAN are notable examples of combined discrete/continuous languages—simulation languages that provide the user the capability to use both discrete and continuous modes, in combination if desired. Extra complexity is introduced in achieving time control for such languages. To synchronize the two separate components of a model, capabilities must be available to (a) permit a discrete event to effect a change in the continuous component (i.e., alter "state variable" values), and (b) allow the achievement of specified conditions in the continuous-model component to cause the occurrence of discrete events. These features are implemented in SLAM and SIMAN by (a) permitting a state variable to be directly altered from the discrete component, and (b) providing the means to specify conditions for occurrence of a "state event" (i.e., by specifying a threshold value for a variable) and the means to alert the discrete component that the state event has occurred; control will then be transferred to the discrete component of the model.

From the foregoing discussion of time control procedures in simulation languages, it is evident that flow-of-control is much more complex in simulation languages than in special-purpose languages. There is a significant advantage to a modeler, however, in using a simulation language, in that time control is a built-in-feature; if a general-purpose language is used, the modeler must develop the time control code. There are many other attractive built-in features of simulation languages; we shall explore this further in Sections 8.6 and 8.7.

8.5   SUPPORT ENVIRONMENTS FOR
PROGRAMMING LANGUAGES

During the 1980s increasing emphasis is being placed on "programming
environments"—i.e., on the support tools available to aid the software development process.   Some support tools are language-related—e.g., the
Ada program support environment (*Ada Reference Manual*, 1980; Sommerville,
1985), which consists of such tools as a run-time support system, database
primitives, peripheral device interfaces, a compiler for Ada, an editor, a
loader, a debugger, and a configuration manager.   The Unix Programmers
Workbench is a programming environment for use with the Unix operating
system (Ivie, 1977).   In most settings, however, it is likely that there is
a collection of support tools from various sources that may be used for
several languages and with various operating systems.

Sommerville (1985) points out that current thinking suggests building
programming environments around a database system, so that all tools
output information to and extract information from the database.   Howden
(1982) categorizes tools in a software development environment by software
life-cycle activities, in particular, by (a) requirements tools and techniques, (b) design tools and techniques, (c) coding tools and techniques,
(d) verification tools and techniques, and (e) management tools and techniques.   In category a, he includes such aids as SADT, PSL/PSA, HIPO
charts, and SREM.   In category b, he includes automated data dictionaries,
Jackson's method, PDLs, Nassi-Schneiderman diagrams, simulation languages,
and others.   In category c, he mentions source code control tools and text
editors, among others.   In category d, file comparators, test harnesses, test
coverage analyzers, and machine-readable test plans are among the items
listed.   Category e support tools include PERT and CPM charts, among
others.   Howden (1982) also mentions various other tools, such as data
flow analyzers, control flow analyzers, interface checkers, performance
monitors, source code debugging tools, cross-reference tools, and test data
generators.   The reader is referred to Howden (1982) for references to
the tools mentioned in the above summary.

Significant progress has been achieved in designing and developing
support environments for simulation use.   Standridge (1985) gives an
overview of TESS, which may be used with SLAM II.   The support capabilities of TESS include (a) graphically building SLAM networks, (b) forms
entry of control and data, (c) database management of user-defined model
parameters, inputs, and simulation-generated data, (d) preparation of reports and graphs, (e) analysis of simulation results, and (f) animation of
simulation runs.   Planning is underway to make TESS available to users of
GPSS/H.

Pegden and his associates (Systems Modeling Corporation) have developed
around SIMAN a set of integrated tools that provide for graphical building
of models, data collection during simulation runs, graphical preparation of
simulation results, and simulation of concurrent animation (Pegden, 1983—
also see Chapter 15 of this book).   They also have developed "Cinema,"
which includes SIMAN, animation software, a graphics board, a color
graphics monitor, and a mouse.   They characterize the product as a combination of SIMAN, color graphics, and an "intelligent-user interface."

Before ending this discussion on support environments, it is useful
to briefly discuss programming language processors—especially compilers
and interpreters.   The fundamental distinction between compilers and

interpreters is as follows: a "compiler" translates a program written in a high-level language (e.g., FORTRAN, SIMSCRIPT) to an assembly language or machine language. An "interpreter" (generally speaking) translates a program in a high-level language to an intermediate form (e.g., reverse polish, PCODE) and then executes directly from the intermediate form. There is, thus, no output program from an interpreter; the interpreter provides execution directly, whereas the compiler outputs an equivalent program which is then linked, loaded, and executed. In general, run-time execution is faster with a compiler, whereas the translation aspect is less complex (and thus takes less time) with an interpreter. As an example, previous generations of GPSS have been interpreted; GPSS/H has been speeded up greatly by use of compilation (Henriksen, 1983a). BASIC and SNOBOL are languages that are usually interpreted; Pascal and FORTRAN are usually compiled.

The term "translator" usually refers to a software component used to translate one high-level language to another high-level language—although technically a compiler is a special kind of translator. A language "pre-processor" is a translator that recognizes special constructs over and above a given language and translates them to features of the standard language. An example would be a preprocessor for a superset of FORTRAN 77 allowing a WHILE-DO construct, which would translate each WHILE-DO to standard FORTRAN 77 constructs (i.e., using GOTO's). Another example would be a translator for Simula, whose output would be a standard ALGOL 60.

Other commonly available language processors are assemblers, linkers, and loaders. The reader may obtain a good overview of language processors from Aho, Sethi, and Ullman (1986).

Expert systems are beginning to play a role in support environments, especially for simulation studies. We shall consider this subject briefly in Section 8.8.

8.6 CRITERIA FOR PROGRAMMING LANGUAGE SELECTION

A bewildering array of languages is available for any project involving computers. Occasionally the choice of language is dictated for a project— e.g., because the firm's contract specifies use of a certain language, or because only one language is available currently, and time doesn't permit obtaining another. But, in general, a real choice is available, and in this case the choice should be carefully considered. In a great many application areas, including simulation, a choice is available between general-purpose languages and application-specific languages. The choice between general-purpose languages and simulation languages often comes down to weighing the benefits to be obtained from the built-in simulation-specific features against the often-greater generality and computational power of a general-purpose language. This dichotomy has been addressed by some language designers, in providing simulation languages with broad general-purpose computational features (e.g., Simula and SIMSCRIPT), and by developing sets of simulation-specific procedures for use with a general-purpose language (e.g., GASP IV, for use with FORTRAN).

There are a number of functions that must be performed with any simulation model and that should be supported by a simulation language. The necessary functions may be stated, in summary form as:

1. Modeling the dynamics of a system
2. Modeling a system's state
3. Performance data recording and analysis

The exact meaning of these aspects varies depending on whether the simulation is continuous, discrete-event, or combined. In the continuous case, for example, a system's dynamics may be achieved by use of integration of differential equations and executing blocks of code as specified time values are reached. In the discrete-event case, dynamics are effected by moving the clock ahead to the next event in the event set and executing code to model the associated state transition (frequently including generation of random variates). The exact approach depends on a language's strategy (or strategies).

A great deal is involved in modeling a system's state, but in the final analysis, the system state at any given time is represented by entries in data structures. Thus, the ease with which system states may be expressed depends on the ease with which entities may be created and terminated, their attributes and relationships expressed, and entities grouped according to their attributes and relationships.

With regard to recording and analyzing data, such capabilities as automatic collection of certain types of data, and the means to easily generate graphs and charts, are very important.

In a larger sense, it is important that a simulation language aid the modeler by providing "conceptual guidance" (Shannon, 1975)—i.e., provide a conceptual framework for development of the model. This provides a significant advantage over using a general-purpose language and having no such guidance for organizing and developing the model.

In somewhat more detail, we could thus expand the three functions listed above into the set of criteria for measuring simulation-specific features of a language, as shown in Figure 8.2.

Modeling a System's Dynamics
   Framework for modeling and simulation (providing
         support for conceptualizing/organizing a model)
   Time management features supplied
     Discrete event methods
       Event scheduling
       Activity scanning
       Process Interaction
     Continuous
     Combined discrete/continuous

Modeling a System's State
   Range of data structures and operations supplied (especially linked
         lists, ordered sets, entities with attributes and relationships)
   Random variable generation features

Performance Data Collection and Analysis Features
   Features for specifying and collecting performance data
   Features for obtaining standard reports
   Features for designing special reports

Figure 8.2   Simulation-specific criteria for evaluating programming language features.

To provide effective programming support, a simulation language should have considerable flexibility and computational power. For example, relative to modeling a system's dynamics, it is important to have a full range of features (control flow, arithmetic computation, data handling) as provided by a good general-purpose language.

There are numerous features that could profitably characterize all programming languages—whether general-purpose or simulation languages. These features may be divided into the categories of (a) design aspects and (b) environmental aspects. Figure 8.3 summarizes some important criteria against which all programming languages may be measured.

Now, one who is considering various languages for use in a simulation study may measure their features against the criteria listed in Figures 8.2 and 8.3. An important general criterion for programming languages is naturalness for a given application. Since our emphasis here is primarily on simulation studies, it follows that a language that measures up poorly against the simulation-specific criteria of Figure 8.2 will not be especially "natural" for this application; on the other hand, a language that exhibits most of the simulation-specific criteria may rate poorly overall if some of the general criteria (of Figure 8.3) are not met.

Software development has become an extremely complex and error-fraught endeavor, as software systems have become larger and more complex. The field of software engineering has been developed since 1968 to attempt to solve the prevailing problems of late, over-budget software, characterized by numerous errors and difficult to understand and modify. Large simulation systems are examples of very complex software. Thus to as great an extent as possible, the proven approaches of software engineering should be employed in the development of simulation software, including emphasis on activities throughout the "software life-cycle." This implies having languages that support stepwise refinement and structured programming, are readable, and have conceptually simple, yet powerful, features. The criterion of "support for the software development process" of Figure 8.3 is intended to summarize such aspects of languages. Wiener and Sincovec (1984) state the opinion that languages that offer support in the following areas provide the basis for constructing reliable and maintainable software: readability, modules for modular software construction, separate compilation with strong cross-reference checking, the control of side effects, data hiding, data abstraction, structured control of flow, dynamic memory

---

Language Design Aspects
    Understandable language concepts
    Understandable language syntax
    Computational/representational power and flexibility
    Support for the software development process

Environmental Aspects
    Availability
    Programming support environment
    Computer run-time
    Portability
    Standardization

Figure 8.3   General criteria for evaluating programming language features.

management, type consistency checking between various subprograms, and run-time checking.

The quality of support environments is becoming more and more important and relates directly to the goal of developing error-free software more quickly—i.e., of making programmers more productive. This aspect of languages is discussed in Section 8.5. This will no doubt be an increasingly important criterion by which simulation languages are measured.

There is no best language, in any absolute sense. Each project has its own priorities and resource limitations (e.g., human skills, time, money), and thus each language selection activity gives rise to a different set of "weighting factors" for each of the criteria of Figures 8.2 and 8.3. One should be neither too ready to abandon an "old friend"—i.e., a language that has proven useful over time—nor too reluctant to seriously consider the merits of new tools and approaches that could ease a difficult task.

The reader who wishes to pursue further the ideas discussed here may find information on general-purpose language features in Horowitz (1984) and Pratt (1984). Shannon (1975), Banks and Carson (1984), Shub (1980), and Law and Kelton (1982) discuss simulation-specific language features. The subject of language support to the software development process is treated at length by Ghezzi and Jazayeri (1982). The implications of discrete-event language strategy for model development and execution are considered in Hooper (1986).

8.7 ASSESSMENT OF SOME LANGUAGES FOR USE IN DISCRETE-EVENT SIMULATION

Using the criteria of Section 8.6, we shall now give an assessment, in summary form, of comparative strengths of various general-purpose and simulation languages. The descriptive words used for assessment are: none, poor, fair, good, very good, and excellent. In the case of time management features, availability is indicated by use of yes or no. The set of languages chosen for this assessment is by no means encompassing, but does collectively exhibit a broad range of qualities. The eight languages chosen are Pascal, Ada, FORTRAN 77 (all general-purpose languages), SIMSCRIPT II.5, SLAM, SIMAN, GPSS V, and GPSS/H. No attempt is made here to assess continuous-simulation languages, primarily because they adhere in the main to the SCi standard (SCi Software Committee, 1967) and thus are more "closely clustered" in features than is true of languages used for discrete-event simulation. The reader interested in a comparison of continuous-simulation languages is referred to Spriet and Vansteenkiste (1982).

Figure 8.4 provides assessments of the eight languages. These assessments should not be taken too literally; they are comparative, and the adjectives used are intended to reflect comparative strengths—not necessarily absolute strengths. And, of course, such assessments are subjective to a great extent, based on one's background and experience. With such reservations considered, it is hoped that the assessments may provide some useful insights.

8.8 TRENDS

As can be inferred from earlier sections of this chapter, the number of languages from which a simulationist may choose is large and increasing. But the quality and power of languages are noticeably increasing as well. Older languages are being updated, and considerable cross-fertilization is occurring. FORTRAN, for example, is gradually being revised to incorporate features considered important in other languages; SIMSCRIPT II.5 was updated to include the process orientation; GPSS/H constitutes a significant updating of GPSS, with the planned GPSS/85 going much further (Henriksen, 1985).

The emphasis on software engineering techniques for addressing the many pitfalls in developing large, complex software systems has impacted language design. Greater emphasis is being placed on "readability" and "writeability." Strong typing characterizes a number of languages (e.g., Pascal, Ada), and simulationists are looking seriously at strong typing languages for simulation. In February 1984 the Conference on Simulation in Strongly Typed Languages was held (see Bryant and Unger, 1984). As may be seen from the papers in the proceedings and references in those papers, considerable activity is under way in this area.

Distributed simulation is growing in importance as techniques and methodologies for distributed processing mature. As mentioned earlier (in Section 8.4.1), a conference on this subject was recently held (see Reynolds, 1985).

Prototyping in an area of increasing importance in the development of large software systems. Prototypes of software systems are (generally speaking) "throwaway" representations of software systems, developed for use in early experimentation and analysis, with the intent to improve the software specifications. "Rapid prototyping" refers to quick development of prototypes—implying existence of supporting tools and methodologies. Discrete-event simulation is beginning to play a role in conjunction with prototyping tools and methods. For example, the Jade Project at the University of Calgary is developing an environment for distributed prototyping and discrete-event simulation (Unger et al., 1984).

The ready availability of powerful personal computers (PCs) has brought about the development of a sizable number of simulation languages and tools for PCs. This can be a very valuable approach, especially with a PC being used as a "user work station" with the capability to "upload" large simulations to a "mainframe" computer for execution.

Henriksen (1983b) suggests what the characteristics of simulation tools of the future may be. Working from a common "knowledge base" would be a number of tools—including a model editor, an input preparation aid, a statistics collection facility, an experimental design facility, and an output definition facility, along with program editor, compiler, and runtime support; i.e., these tools would be integrated relative to a knowledge base.

Expert systems offer promise in easing the development of simulation models and interpretation of simulation results. Shannon, Mayer, and Adelsberger (1985) provide a summary of the likely potential of expert systems with regard to simulation. O'Keefe (1986) makes some observations

GENERAL LANGUAGE CRITERIA	Ada	FORTRAN 77	Pascal
LANGUAGE DESIGN ASPECTS			
Understandable Language Concepts	excellent	excellent	excellent
Understandable Language Syntax	excellent	very good	excellent
Computational/Representational Power and Flexibility	excellent	excellent	very good
Support for the Software Development Process	excellent	good	excellent
ENVIRONMENTAL ASPECTS			
Availability	fair	excellent	excellent
Programming Support Environment	excellent	fair	fair
Computer Runtime	very good	excellent	good
Portability	excellent	excellent	excellent
Standardization	excellent	excellent	excellent
SIMULATION-SPECIFIC CRITERIA			
MODELING A SYSTEM'S DYNAMICS			
Framework for Modeling and Simulation (providing support for conceptualizing/organizing a model)	poor	poor	poor
Time Management Features Supplied Discrete Event Methods			
Event Scheduling	no	no	no
Activity Scanning	no	no	no
Process Interaction	no	no	no
Continuous	no	no	no
Combined Discrete/Continuous	no	no	no
MODELING A SYSTEM'S STATE			
Range of Data Structures and Operations Supplied (especially linked lists, ordered sets, entities with attributes and relationships)	good	poor	good
Random Variable Generation Features	poor	poor	poor
PERFORMANCE DATA COLLECTION AND ANALYSIS			
Features for Specifying and Collecting Performance Data	poor	poor	poor
Features for Obtaining Standard Reports	poor	poor	poor
Features for Designing Special Reports	fair	fair	poor

Figure 8.4   Assessment of some languages for use in simulation.

GPSS/H	GPSS/V	SIMAN	SIMSCRIPT II.5	SLAM
fair	fair	very good	very good	very good
fair	poor	good	very good	good
good	fair	excellent	excellent	excellent
fair	poor	good	good	good
fair	fair	fair	fair	fair
very good	poor	excellent	very good	excellent
good	poor	good	good	good
excellent	excellent	excellent	excellent	excellent
good	good	good	good	good
excellent	excellent	excellent	excellent	excellent
no	no	yes	yes	yes
no	no	no	no	no
yes	yes	yes	yes	yes
no	no	yes	yes	yes
no	no	yes	yes	yes
good	fair	excellent	excellent	excellent
good	good	excellent	excellent	excellent
good	good	excellent	excellent	excellent
excellent	excellent	excellent	good	excellent
good	poor	good	good	good

concerning the viability of expert systems for effective use in simulation projects. Reddy, Fox, Husain, and McRoberts (1986) report on a knowledge-based simulation system for use as part of a factory management system. The expert system described (KBS) uses heuristics to evaluate the output from the simulation program. KBS offers facilities for automatic generation of model scenarios, goal-oriented instrumentation, cause-and-effect analysis, and interactive model building.

The interactive interfaces being used in the available support systems, including some of the expert systems, represent an important trend. This approach can go a long way toward helping achieve model credibility through increased understanding, can greatly aid interpretation of results, and thus can expedite achievement of effective simulation studies.

In concert, the current trends reflect a quantum increase in capability for simulation. It is a time of high activity and innovation in design and development of simulation tools and methods, and all indications are that the momentum is still increasing. Thus, the future appears very bright for simulation practitioners and researchers.

PROBLEMS

8.1. Compare the features of FORTRAN 77 with an early version of FORTRAN (to see how the language has evolved).

8.2. (a) List two advantages that FORTRAN 77 has relative to Pascal.

(b) List two advantages that Pascal has relative to FORTRAN 77.

8.3. The ALGOL-like program fragment

while a>b do
    ...
endwhile

can be rewritten using conditional and unconditional jumps. Compare the two approaches in terms of readability and writeability.

8.4. Separate compilation of modules is the usual approach with FORTRAN programs. Why is separate compilation more difficult with Pascal?

8.5. Select another general-purpose language and another simulation language than those assessed in Figure 8.4 and assess them according to the criteria used there.

8.6. Take a small simulation problem and sketch it in process-oriented form and in event-oriented form. Use one of SLAM, SIMAN, and SIMSCRIPT II.5 as your guide.

REFERENCES

Ada Reference Manual (1980). United States Department of Defense, Washington, D.C.

Aho, A. V., Sethi, R., and Ullman, J. D. (1986). *Compilers: Principles, Techniques, and Tools*, Addison-Wesley, Reading, MA.

Backus, J. (1978). Can programming be liberated from the von Neumann style? *Comm. ACM 21*(8), 613–641.

Banks, J. and Carson, J. S., II (1984). *Discrete-Event System Simulation*, Prentice-Hall, Englewood Cliffs, NJ.

Bryant, R. M. (1980). SIMPAS: A simulation language based on Pascal, *Proceedings of the 1980 Winter Simulation Conference*, pp. 25–40.

Bryant, R. and Unger, B. W. (1984). Simulation in strongly typed languages: Ada, Pascal, Simula. . ., Society for Computer Simulation, San Diego, CA.

Catalog of Simulation Software (1985), *Simulation 45*(4), 196–209.

Cooper, D. (1983). *Standard Pascal User Reference Manual*, Norton, New York.

Dahl, O. J. and Nygaard, K. (1966). SIMULA—An ALGOL-based simulation language, *Comm. ACM 9*(9), 671–678.

Dewar, R. B. (1979). *The SETL Programming Language*, Courant Institute of Mathematical Sciences, New York University, New York.

Feuer, A. and Gehani, N. (1984). *Comparing and Assessing Programming Languages: Ada, C, Pascal*, Prentice-Hall, Englewood Cliffs, NJ.

Ghezzi, C. and Jazayeri, M. (1982). *Programming Language Concepts*, Wiley, New York.

Gordon, G. (1975). *The Application of GPSS V to Discrete System Simulation*, Prentice-Hall, Englewood Cliffs, NJ.

Henriksen, J. O. (1983a). "State-of-the-Art-GPSS," in *Proceedings of the 1983 Summer Computer Simulation Conference*, pp. 918–933.

Henriksen, J. O. (1983b). The integrated simulation environment (simulation software of the 1990's), *Operations Res. 31*(6), 1053–1073.

Henriksen, J. O. (1985). The development of GPSS/85, in *Proceedings of the Annual Simulation Symposium*, Tampa, FL, pp. 61–77.

Hooper, J. W. (1986). Strategy-related characteristics of discrete event languages and models, *Simulation 46*(4), 153–159.

Horowitz, E. (1984). *Fundamentals of Programming Languages*, 2nd ed., Computer Science Press, Rockville, MD.

Howden, W. E. (1982). Contemporary software development environments. *Comm. ACM 25*(5), 318–329.

Ivie, E. L. (1977). The Programmers' Workbench—A machine for software development, *Comm. ACM 20*(10), 746–753.

Jensen, K. and Wirth, N. (1974). *Pascal User Manual and Report*, 2nd ed., Springer-Verlag, New York.

Kernighan, B. W. and Ritchie, D. M. (1978). *The C Programming Language*, Prentice-Hall, Englewood Cliffs, NJ.

Kiviat, P. J., Villanueva, R., and Markowitz, H. M. (1973). *Simscript II.5 Programming Language*, 2nd ed., CACI, Los Angles, CA. Los Angeles, CA.

Law, A. M. and Kelton, W. D. (1982). *Simulation Modeling and Analysis*, McGraw-Hill, New York.

Mitchell, E. E. and Gauthier, J. S. (1981). *ACSL: Advanced Continuous Simulation Language—User Guide/Reference Manual*, Mitchell and Gauthier Assoc., Concord, MA.

Myers, G. J. (1976). *Software Reliability: Principles and Practices*, Wiley, New York.

Naur, P. (1963). Report on the Algorithmic Language ALGOL 60, *Comm. ACM 6*(1), 1–17.

Nilsen, R. N. (1976). *CSSL-IV: The Successor to CSSL-III*. Young Lee and Associates, Torrance, CA.

O'Keefe, R. M. (1986). Simulation and expert systems—A taxonomy and some examples, *Simulation 46*(1), 10–16.

Pegden, C. D. (1983). Introduction to SIMAN, in *Proceedings of the 1983 Winter Simulation Conference,* pp. 231–241.

Pegden, C. D. (1984). *Introduction to SIMAN,* Systems Modeling Corporation, State College, PA.

Pratt, T. W. (1984). *Programming Languages,* 2nd ed., Prentice-Hall, Englewood Cliffs, NJ.

Pritsker, A. A. B. (1974). *The GASP IV Simulation Language,* Wiley, New York.

Pritsker, A. A. B. (1977). *Modeling and Analysis Using Q-Gert Networks,* Halsted Press, New York.

Pritsker, A. A. B. (1984). *Introduction to Simulation and SLAM II,* 2nd ed., Halsted Press, New York.

Pritsker, A. A. B. and Kiviat, P. J. (1969). *Simulation with GASP II,* Prentice-Hall, Englewood Cliffs, NJ.

Pritsker, A. A. B. and Pegden, C. D. (1979). *Introduction to Simulation and Slam,* Halsted Press, New York.

Reddy, Y. V. R., Fox, M. S., Husain, N., and McRoberts, M. (1986). The knowledge-based simulation system, *IEEE Software 3*(2), 26–37.

Reynolds, P. (1985). *Distributed Simulation 1985,* in *Proceedings of the Conference on Distributed Simulation 1985* (P. Reynolds, ed.), Society for Computer Simulation, San Diego, CA.

Russell, E. C. (1974). *Simulating with Processes and Resources in SIMSCRIPT II.5,* CACI, Arlington, VA.

Sammet, J. E. (1976). Programming languages, in *Encyclopedia of Computer Science* (A. Ralston and C. L. Meek, eds.), Petrocelli/Charter, New York, pp. 1169–1174.

Schriber, T. J. (1974). *Simulation Using GPSS,* Wiley, New York.

SCi Software Committee (1967). The SCi continuous-system simulation language, *Simulation 9,* 281–303.

Shannon, R. E. (1975). *Systems Simulation: The Art and Science,* Prentice-Hall, Englewood Cliffs, NJ.

Shannon, R. E., Mayer, R., and Adelsberger, H. H. (1985). Expert systems and simulation, *Simulation 44*(6), 275–284.

Sheppard, S., Friel, P., and Reese, D. (1984). Simulation in Ada: An implementation of two world views, in *Simulation in Strongly Typed Languages: Ada, Pascal, Simula, . . . ,* Society for Computer Simulation, San Diego, CA.

Shub, C. M. (1980). Discrete event simulation languages, in *Proceedings of the 1980 Winter Simulation Conference,* Vol. 2, pp. 107–124.

Sommerville, I. (1985). *Software Engineering,* 2nd ed., Addison-Wesley, Reading, MA.

Spriet, J. A. and Vansteenkiste, G. C. (1982). *Computer-Aided Modelling and Simulation,* Academic Press, New York.

Standridge, C. R. (1985). Performing simulation projects with the extended simulation system (TESS), *Simulation 45*(6), 283–291.

Unger, B., Birtwistle, G., Cleary, J., Hill, D., Lomow, G., Neal, R., Peterson, M., Witten, I., and Wyvill, B. (1984). Jade: A simulation and software prototyping environment, in *Simulation in Strongly Typed Languages: Ada, Pascal, Simula . . . ,* Society for Computer Simulation, San Diego, CA, pp. 77–83.

Wiener, R., and Sincovec, R. (1984). *Software Engineering with Modula-2 and Ada*, Wiley, New York.

Wirth, N. (1982). *Programming in Modula-2*, Springer-Verlag, New York.

Wulf, W., Russell, D. B., and Habermann, A. N. (1971). BLISS: A language for systems programming, *Comm. ACM 14*(12), 780–790.

Zeigler, B. P. (1976). *Theory of Modelling and Simulation*, Wiley, New York.

9
Operating Systems

A knowledge of operating system principles is important to the developer or user of model simulations. Many of the problems encountered in the design of operating systems are also encountered in the design of simulation programs, and the techniques used there are directly applicable to simulation development. Moreover, an understanding of the basic operating system concepts, such as segmentation, paging, input/output (I/O), and scheduling, is necessary in order to develop simulations that will have reasonable performance when running under the control of an operating system.

As simulation size and complexity have increased, there has been more and more interest in developing "distributed simulations," or simulations divided into parts which are running concurrently on more than one central processing unit (CPU). The problems of controlling and synchronizing the execution of concurrently running processes were first encountered in the development of multiuser operating systems. It is believed that an understanding of the implementation of primitive synchronization operations in the operating system is helpful in designing simulations that use these operations.

9.1   NATURE OF AN OPERATING SYSTEM

In order to understand the relationship of simulations to operating systems, one should first consider the question: "What is an operating system?" The most fundamental characteristic is that an operating system is simply a computer program, or a collection of computer programs whose purpose is to manage the resources (such as I/O devices, memory, processor time) of a computer system and to control the execution of "user" programs on that system.

In order to understand the need for the various operating system functions, it is helpful to consider the motivation for early operating systems and their growth as computer systems have become more powerful. For a more thorough treatment of this subject, see Peterson and Silberschatz (1985).

The first computer systems did not have an operating system of any kind. The user of the machine acted as both programmer and operator, and frequently the programs were written at an assembly language (or even at a machine language) level. The computer system was scheduled so that a user was given the exclusive use of the machine for a period of time. Although these machines were very expensive, the performance of the largest machines was not as great as that of many modern microcomputers.

Any computer system devoted to a single user has the characteristic that the machine itself may be idle for considerable periods of time while the user is thinking about what to try next or how to fix the last programming error. The primary motivation for the development of operating systems came not from any consideration for the programmer, but simply from trying to increase the efficiency and utilization of the computer hardware by decreasing the idle time of the system.

The first steps taken to reduce the idle time of the machine were to replace the single user as the operator of the machine and to allow a stack of jobs to be presented to the system by a system operator so that as soon as one job was completed the next would be read in from a queue of waiting jobs.

The developed "batch" system decreased the idle time of the machine, but introduced new complications for both the programmer and the system designer. Even in these simple systems, some form of protection had to be added to the system to prevent one user's programs from interfering with the execution of the next program in the queue. If all the programs were presented to the system as a large stack of card decks, something had to keep one user from accidentally reading a portion of the next program as input data. This requires the addition of system I/O routines to read the input data instead of allowing the user direct access to I/O. The system input routine could be programmed to look for special end-of-job cards placed between each user's program decks. After one of the separator cards was read, further reading of cards would not be allowed for that job.

If a program contained an infinite loop, something had to be done to detect this so that the program could be terminated, allowing the next job to be run. This required the addition of system clocks to time the execution of programs and special system software to monitor the clock and remove jobs whose time has expired.

Job scheduling could be easily performed by the operator who orders the card decks belonging to each job in the stack of decks presented to the card reader according to their priority and order of arrival.

As memory capacities grew, it became possible to consider allowing more than one job in memory at a time, so that while one was busy with an I/O to a peripheral device (such as a printer), the next could be using the CPU to perform computations. This overlapping of I/O with processing further decreased the idle time of the CPU and resulted in a more efficient system. More complications resulted, however. The two or more jobs in the memory of the machine at the same time had to be protected from each other, so that one could not accidentally read or write into a memory location belonging to the other program. This resulted in the need for some form of partitioning or other division of memory by the system hardware and software to allow protection of one program's memory from the other programs. In the event that one program attempted to read or write a memory cell belonging to another program, the hardware would detect this and would interrupt the execution of the offending instruction before the damage was done; system software could then terminate the offending program.

The concept of allowing more than one program into the memory of a single machine and overlapping the execution of one program with the I/O of another is known as "multiprogramming." (This term should not be confused with "multiprocessing," which is the use of more than one CPU in a computer system.) Ideally, two jobs would run so that when one was doing I/O the other was computing, and the I/O system and the CPU would both be busy all the time. Unfortunately, this is seldom the case. There will be times when both want to output information or both have extensive computing to do. By increasing the number of jobs in memory, the probability is raised that at least one job is ready to use the CPU at all times. If more than one job is ready to use the single CPU or any other non-sharable system resource simultaneously, some form of scheduling has to be performed to determine which of the competing programs will gain the use of the resource. The number of jobs in the system at one time is known as the "degree of multiprogramming." As the degree of multiprogramming is increased, the probability that all jobs will be waiting for I/O is decreased, and the utilization of the expensive CPU is increased.

Another motivation for multiprogramming came from the need for time sharing. Microprocessors were not generally available until the late 1970s. Before that time any need for interactive computing had to be met by allowing multiple user's jobs into the memory of a shared computer system at the same time. The CPU would be allocated to one person's job for a brief period known as a time slice, and then it would be switched to the next job for a brief time and eventually would get back to the first job again. Each user would get small bursts of computer time at hopefully frequent intervals. It would appear to each interactive user as if his program were running independently on a machine that was considerably slower than the actual system. Again, there would be a need for special system software, the scheduler, to control the allocation of time slices to the various users' programs.

The size of memory limits the degree of multiprogramming by limiting the number of jobs that may be simultaneously in memory. Virtual memory was introduced as a mechanism for increasing the degree of multiprogramming beyond the number of jobs that may be contained in available memory. Only the portion of the job's data and code that is considered likely to be needed remains in memory, and the remainder is placed in disk storage. If a reference is made to a non-memory-resident location, the program is interrupted, the memory location is retrieved from disk, and then the program execution is resumed. The portion of the operating system that decides which portions of a job's code and data will remain resident in memory and accomplishes the retrieval of non-memory-resident locations is the memory manager.

The brief summary given above points out several of the basic operating system functions. The operating system has to provide protection of a program from other programs running in the machine. It has to provide some form of scheduling for the jobs competing for processor time and for other nonshared system resources such as the printer and to provide memory management. It also has to provide basic I/O services for the user programs and handle interrupts from I/O devices and the system clock.

9.1.1   View as a Resource Manager

The preceding description of the functions of an operating system leads to the view of an operating system as a collection of managers which manage the various system resources: the device manager, the processor manager, the memory manager, and the file system manager.

Device Manager. The device manager is responsible for controlling the system's peripheral devices. This manager contains the device drivers which translate requests for I/O into hardware-specific commands for a particular I/O device, such as a disk or printer. The device manager also contains the much higher-level function of queueing I/O requests for devices, such as the disk, that are shared between jobs running on the system. The waiting requests may have to be scheduled, or ordered, in a way to provide efficient use of the I/O device. Another function of the device manager is to establish ownership of peripheral devices, such as the printer or tape drives, which may not be shared between jobs.

Sometimes a higher level of manager, the I/O manager, is included which controls the allocation of user I/O requests to the specific device manager that actually carries out the I/O operation.

Processor Manager. Central-processing-unit time has to be split between jobs on the system in a manner consistent with the system goals of adequate response time for interactive jobs and reasonable performance on large number-crunching applications, such as batch simulations. Frequently these goals are nonconsistent, and compromises must be made.

Scheduling of jobs takes place at several levels. The highest level of scheduling determines the order in which jobs will even be allowed into the system, while the lowest level of scheduling determines the allocation of CPU time between jobs currently resident in memory. The actual allocation of the CPU to a given job is called dispatching, and the portion of the operating system is the dispatcher.

The CPU is actually switched between all the programs running in the system. Each of the programs running independently in the system is called a process. The processor manager that controls the allocation of the actual processor to the processes is frequently called the "process manager."

Memory Manager. In multiprogramming, several jobs are allowed in the memory of the system at the same time. The memory manager keeps track of the memory not in use, decides where to place a given job in memory, and allocates space in memory for each job. As the memory manager allocates memory to new jobs, and old jobs complete execution and leave the system, unused "holes" will form in memory between the remaining jobs. Some of these holes may be too small to hold any new job, and in effect these memory locations are wasted. Memory is considered an extremely valuable resource in most computer systems, and considerable effort is expended to prevent this "fragmentation" of unused memory.

If the computer system uses virtual memory, the memory manager decides which portion of the job will be made resident initially and controls the retrieval of nonresident code or data locations from the disk.

File Manager. Computer systems require some form of inexpensive bulk storage, such as disks, where user programs and data files may be kept for later recall. Off-line devices such as magnetic tapes are inadequate for storage of programs for interactive users as it takes too long to retrieve and mount a magnetic tape on the system. Magnetic tape drives are also too expensive to tie up with tapes from interactive users while they figure out what to do next.

The file manager allocates space on the disk for user files and maintains directories or catalogs showing the location and size of each of these files. The directories can be as simple as a linear list of all of the files on a given floppy disk, or as complex as tree-structured directories for a multi-user system.

These are the four major managers of an operating system. Each will be described in greater detail in the following sections. The concept of concurrency is of particular importance to distributed simulations. The methods and concepts used in the device managers are of importance to simulations requiring interfaces to special hardware devices such as displays.

9.1.2   View as an Interface Between the User and the Machine

Another view of an operating system is that it is an interface between the hardware of the system and the user. If one were given a "bare" computer system to use, there would have to be a large amount of software written just to perform the simplest of tasks.

Even as the sole user of a computer system there are many functions that have to be performed which would be difficult to include in the user's program. The system clock has to be kept up to date. Some program has to be resident in the system to load the user's program into the machine. On most systems, just reading data from a floppy disk or other I/O device is a complex task.

The basic computer hardware does not provide a friendly environment in which to program. There needs to be a central core of software available in the machine at all times just to provide basic services for even the simplest user program. This "core" of programs comprises the operating system and its associated system programs. It successfully hides the characteristics of the hardware from the user program and makes it appear to the user that all that is required to read from the floppy disk is to call the read routine. If one needs the time of day, it is available by simply requesting the time from the appropriate system routine.

Thus, the system as seen by the user is a considerably simpler system than the basic hardware of the system. We can say that the basic hardware of the system combined with the operating system provides a new "virtual machine" to the user: a simpler, easier-to-use machine, one with a number of relatively high-level features available.

9.1.3   View as a Hierarchy

The concept of a virtual machine, introduced in the preceding section, provides another view of operating system software: the view as a hierarchy of programs.

The first multiprogramming operating systems were much larger software efforts than had previously been attempted. They took much longer to develop than anticipated, and most were highly unreliable. There were several reasons for this. One reason was that the problems of debugging such large programs in a multiprocessing environment had not be considered.

Some early attempts constructed the operating system as one large monolithic program, so that it was difficult, or impossible, to test a single component of the operating system without the rest of the operating system present. The problems in testing such systems led to an understanding of the importance of writing large programs as a collection of independently testable modules. Although the importance of module independence was

easily appreciated, it was not understood how to program an operating
system (or any other large program) so that the modules could be independently tested. It appeared that most of the operating system had to be
present in order for any portion to run at all.

Another problem in the testing of programs running on multiprocessing
systems stems from the inherent nonrepeatability of certain types of errors.
When a single program is running in a dedicated system that does not use
interrupts, the exact execution sequence of the program is repeatable, so
that the program may be run as many times as necessary to find an error,
and the error will be observed each time the program is run. When I/O devices interrupt the execution of the CPU at the completion of an I/O operation, the exact timing of the interrupt is not repeatable. Hence, if the
interrupt-handling portion of the operating system has programming errors
that are observed on a particular computer run, the error may not be repeatable on subsequent runs and testing will be extremely difficult. We now have
introduced a particularly insidious type of error to computer programming:
errors that are dependent on execution timing.

The same problem of nonrepeatability may occur owing to the nature of
multiprogramming itself. If the protection between programs running in the
system is not complete, there may be circumstances when one program
changes a memory location belonging to another program. The effects of
such an error will be dependent on the relative locations in memory of each
program, and on the relative progress toward completion of each program
when the error occurs. It may not be possible to duplicate the exact mix
of programs and their exact state of execution in the system at the time of
an observed error, and as a result it will not be possible during testing
to reconstruct the error.

Dijkstra (1968) demonstrated a hierarchical technique for operating system development which allowed the development and testing of the system in
small increments. This concept has been found to be extremely important
to the development of any large computer program and is especially applicable
to the development of large simulations.

Starting with the bare system hardware, there is a minimum core of programs that must be present just to allow even the simplest program to run.
System interrupts must be handled, and some manner of scheduling must be
present to allocate processor time between the various processes in the system. Other functions of the process manager must also be present in the
core, such as process creation, termination, and some form of interprocess
synchronization and communication.

Once this basic core of the operating system has been written and tested,
the next level of capability can be added to the system. The new parts of
the operating system would have not only the capabilities of the basic hardware available, but also the capabilities added by the core of programs already written. The system hardware plus the basic core of programs provides a new virtual machine with more of the capabilities needed to develop
the next portion of the operating system. This core is small enough to be
adequately tested, so that its correctness can be assured before proceeding
with the next level. Each new level could be considered to be a new
layer surrounding the previous layers, like the layers of an onion, as shown
in Figure 9.1. A new layer is not independent of the lower layers, but the
lower layers have already been tested and their correctness is not affected
by the addition of the new layer. An excellent description of this layering
may be found in Lister (1979).

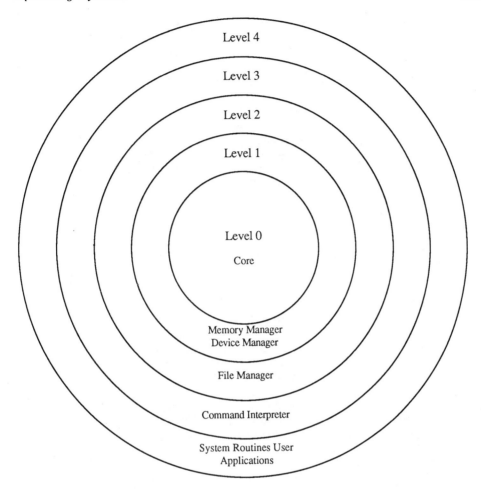

Figure 9.1   Example of a possible hierarchy for operating system functions.

The next level of capabilities to be added could include the basic I/O system, the device managers, and the memory manager. The addition of this second level now gives the system the capability of memory allocation and basic I/O operations. The next layer could add higher-level functions using the basic functions provided by lower levels. The file manager would be implemented at this level. The next layer could contain the command interpreter, the program that actually reads in user commands and interprets them into calls of the functions available at lower levels. User programs and various system programs such as compilers would be considered the last layer in the hierarchy and could use all the capabilities of any lower layer(s).

The layers of software described above form a hierarchy in the sense that any layer can use features at an equal or lower level. No layer is allowed to use features at a level higher than its own level. The actual placement of the given operating system components into the various levels differs from operating system to operating system.

It is important to preserve the hierarchical structure of the layers when implementing a large simulation. All the layers below the application layer could be regrouped into a single central core from the viewpoint of the simulation. In other words, the simulation may request services from lower layers, but no lower layer may call a routine in the simulation.

Is it possible, as a user, to violate this basic hierarchical structure? It is not too uncommon for real-time simulations to "trap" (or intercept) and redirect basic operating system functions or I/O interrupts to special code in the simulation. This is logically equivalent to having the operating system request services from the user program, a clear violation of the hierarchical structure. The danger is that any error in this code affects the correctness not only of the simulation, but also of the operating system itself and any other programs running on the system. Hence, all the testing of the operating system is now suspect, and errors of a nonrepeatable type may have been introduced into the system.

9.1.4   Operating System Software and System Utility Software

Compilers, editors, assemblers, and other similar programs are needed by all system users. They are not a true part of the operating system as they are not a part of any of the operating system managers. From the standpoint of the operating system, these programs are no different from any other application program and are run at the same level as a user's application program. Since this utility software is needed by all users, it is generally considered to be part of the computer system, rather than belonging to any particular user, and hence these programs are known as system programs.

Commonly used subroutines for mathematics functions, graphics, and similar applications are placed in libraries available to all users and are also frequently considered to be system routines.

9.2   EFFECT OF THE OPERATING SYSTEM ON SIMULATION

The operating system acts as the manager of all system resources: memory, processor time, I/O devices, and the file system. It provides the basic environment in which the simulation software runs and also controls the environment in which the simulation software is developed.

As a manager of processor time, it decides when a job will execute and thus controls the basic speed at which the simulation is run. As a manager of the system's I/O devices, the design of the I/O manager affects the basic throughput of the simulation's I/O requests. If the memory manager does not allocate a sufficient amount of memory to a simulation, performance can be significantly affected.

Thus, even though the developer of a simulation programmed in a high-order simulation language may not interface in a direct way with the operating system, its design may have a significant effect on the operation and performance of the simulation.

The development environment seen by a simulationist is to a large extent determined by the operating system and its system utilities. The scheduling policies affect the basic system responsiveness to interactive requests, and the file system determines the basic organization of data and programs stored in the system. The ease with which information and files may be shared between users is controlled by the memory manager and the file system. The limitations on physical memory allocated to running programs may limit the responsiveness of large programs such as compilers and simulation programs by requiring an excessive number of mass storage accesses to retrieve nonresident data and code.

9.3 CONCURRENCY, PROCESS SYNCHRONIZATION, AND SCHEDULING

A single CPU that is switched rapidly between two or more processes may give an appearance of concurrency, known as simulated concurrency. Each process appears to be running on a separate virtual machine which is slower than that of the actual CPU. At any instant of time the CPU is actually executing instructions from only one process, and there is never any true concurrent execution of processes.

When two processes are running on a multiprocessor system so that one is running on one processor at the same time that the second process is running on another processor, we have true concurrency in the execution of two processes. (Multiprocessor systems are discussed in Chapter 16.) Processes that are running with either simulated or true concurrency are known as concurrent processes.

If two concurrent processes have no way to interact through the sharing of any resource such as memory, the processes are independent, and it may be shown that the calculations made by the two processes are independent of the relative timing of their execution.

On the other hand, if the concurrent processes under consideration both belong to a single simulation with multiple processes, interaction between the processes is necessary in order to produce useful results. One must coordinate or synchronize the progress of the simulation in each process and also communicate results from one portion of the simulation to another. For example, current simulation time must usually be known to all the distributed processes. Techniques for communication of data values between processes, and for process synchronization using a shared memory, will be discussed in this section.

9.3.1 Effects of Scheduling on Simulation Software

There are three types of scheduling: short-term scheduling, medium-term scheduling, and long-term scheduling. Short-term scheduling concerns the switching of the CPU between processes. Medium-term scheduling controls the moving (swapping) of a program's data and code between the secondary memory on disk and the main memory in order to make room for another program. Long-term scheduling controls the original allocation of memory and peripherals for the program.

We have been using the terms "job" and "program" rather loosely. A job will be considered to be the collection of everything that a user does between logging on and logging off. Thus, a job might consist of several editing, compiling, and running sequences for an interactive session. A program is considered to be a collection of code that is loaded into the machine and executed as a unit. A process, in this context, is a CPU executing a stream of instructions in a program. Since it appears to a process that it is running on the CPU independently of other processes, it is said to be running on its own "virtual" CPU. It is possible that there are several processes running the instructions of a single program.

The long-term scheduler (or job scheduler) has little effect on the simulation, as it controls only its initial loading. Simulations of large applications may, however, experience difficulty attaining the resources needed without being assigned a priority over the many smaller programs running in the computer system.

Some operating systems have a medium-term scheduler. The medium-term scheduler controls the moving, or swapping, of an entire program between main memory and secondary memory (usually disk or drum storage) in order to make room for another program. If an insufficient amount of main storage is available, considerable systems overhead may be incurred owing to excessive movement of data and code between primary and secondary memory. This problem may be solved on some systems by locking the program in memory.

Short-term scheduling controls the allocation of the CPU between the processes that are ready to execute. A process may be considered to have three primary states: ready, blocked, and running. The process may be "running" if the processor has been allocated to it and it is currently executing, or it may be "ready" to run but waiting for processor allocation. If the process has just done an I/O operation, the process may be "blocked," waiting for completion of the I/O. It is also possible for a process to be blocked waiting for some type of event from another process.

The states may be illustrated as circles, as shown in Figure 9.2, with transitions between the states drawn as arcs connecting the states. Two additional states are also shown: a suspended state when execution has been suspended by the medium-term scheduler, and a hold state for a job waiting to be allowed into the system initially by the job scheduler.

A process moves from the running state to the blocked state when it is waiting for the completion of an I/O operation or is waiting for another process. When the cause of blocking is removed, the process is placed in the ready state, rather than in the running state, as another process is probably now running. At the expiration of the time slice for the currently running process, the running process is moved to the ready state. Then, a portion of the short-term scheduler, known as the dispatcher, selects one of the ready processes for allocation of the CPU, placing it in the running state.

In a single-processor system there can be at most one process in the running state at any time. In a multiprocessor system there may be as many running processes as there are processors in the system. The ready process may be grouped by processor so that the CPU is cycled through the processes in its group, or the CPU may be allocated to any ready process.

If a running process completes its processing, it will terminate and leave the system, perhaps allowing the long-term scheduler to admit another process into the system from the hold state.

The short-term scheduler may have a great impact on a multiprocess simulation, as it determines when a process will be run. The act of switching the CPU from one process to another requires considerable overhead in terms of CPU time. Registers must be saved and restored, accounting information must be recorded for the process that has been executing, and other operations, such as placing this process in the ready state and selecting the next ready process for execution, must be completed. This

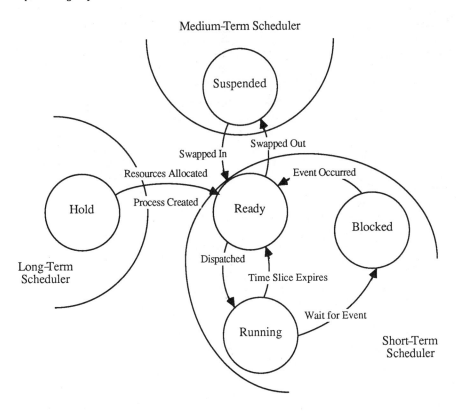

Figure 9.2   Process states.

switching time may be of the order of 1 msec, even on medium-scale computers.

Unfortunately, high overhead or not, this frequent switching is necessary to provide adequate response to the very short requests from an interactive user. Otherwise, if frequent switching were not used, even a short request would have to wait while the CPU completed other longer requests. The switching overhead results in an effective loss in the useful computational power of the system, which may be quite detrimental to a computationally intensive simulation trying to run in real time. The optimal scheduler for the simulation would minimize switching overhead by allowing each process to run until it became blocked waiting for another process or for an I/O operation. This may be accomplished on some systems by giving the process a higher priority than any other competing one. On single processor systems, locking out timer interrupts prevents the operating system from ever gaining control until the user process relinquishes it. This approach has the disadvantage that vital system functions may be prevented from executing properly and time on the system clock may not be properly maintained.

Long periods of execution by compute-bound processes will appear to the interactive user as large gaps with no responses to requests. Thus, the scheduling requirements for real-time simulation are not compatible with the rapid process-switching requirements for interactive jobs. This makes

it difficult to mix the two kinds of processing on the same system. Batch
and non-real-time simulations may be more easily mixed with interactive
users, as the speed of execution may not be of great importance.

9.3.2   Desirability of Concurrency in Simulations

Many simulations tax the capabilities of today's small- to medium-scale computer systems. Frequently the costs may not allow the dedication of a large-
scale computer for simulation, and so there has been a trend toward the use
of multiple, small- to medium-scale or specialized CPUs in simulations. The
resulting use of concurrency on these systems has emphasized the importance of understanding and developing techniques for process synchronization and communication.

Programming considerations may make the use of multiple processes
desirable even within a single CPU. The architecture of the system may
limit the memory used by any one process to an amount less than that required for the entire simulation. For example, the addressing limitations of
many micro- and minicomputer systems limited the memory visible to a single
process to 64K, even if the computer system had much more memory. Also,
if the computer is dedicated to a given simulation, it may be desirable to
break the simulation into multiple processes in order to allow computation to
continue while I/O operations are being performed. Finally, it is desirable
from the standpoint of software engineering and testing to allow logically
separate functions to be performed by independent program modules having
completely controlled interactions. These modules may be tested independently and combined together as building blocks with a flexibility that is highly
desirable.

Simulation languages such as GPSS present the user with simulated concurrency which provides some of the features mentioned above. The
actual code is run, however, as a single process. Recent operating systems and programming languages such as Modula 2 and Ada make it
especially convenient to express a problem to be solved as a collection of
independent processes.

9.3.3   Problems with Concurrent Processes Sharing Data

If two or more processes are sharing data items, some form of protection
must be provided to prevent access to data items by one process while they
are being updated by another process. Even very simple operations, such
as updating a counter or inserting an item into a queue, need this kind of
protection, as illustrated by the following example.

Consider a simulation of a grocery store in which a simple counter,
"ncust," is used to track the number of customers in the store. Arrivals
of customers are handled by an arrival process which increments the counter
whenever a customer arrives. Departures from the store are simulated by
a "checkout" process which decrements the counter when a customer leaves
the store. Figure 9.3 shows the portions of the code in each of the two
processes that deal with updating the counter "ncust."

These code sections appear innocuous enough; however, there is a
potential problem with simultaneous access to the shared variable "ncust"
while it is in the process of being updated. Figure 9.4 shows the actual
machine language operations that might be performed by the increment and
decrement operations of Figure 9.3. Assume that both processes are running
simultaneously in two separate CPUs, and that "ncust" is equal to 5. If both

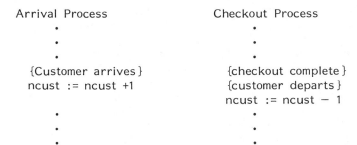

Figure 9.3   Portions of code from processes using counter ncust.

processes execute line 1 at the same time, both will load the current value of "ncust," or 5, into their respective registers. The arrival process will increment its register to 6, and the checkout process will decrement its register to 4. The two processes will then proceed to store their answers back into the shared memory location containing "ncust." One of the two processes will attempt to store 6 while the other attempts to store 4 in the same location!

Shared memories are built to prevent truly simultaneous loads or stores into a single location. Two simultaneous requests to read or write a single memory location will be arbitrated by the hardware so that one follows the other. Unfortunately, this doesn't help the above problem. Whichever process stores its answer first will have the answer changed by the other process. The final answer stored will be either 6 or 4, depending on which process stores its answer last. Either answer, of course, is incorrect, as one arrival and one departure should cancel out, leaving "ncust" at a value of 5.

The problem is caused by two processes trying to accomplish an update on a shared variable during the time periods that overlap. If either process had completed its update before the other started, there would have been no problem. The problem can occur for any update operation that requires multiple memory accesses for completion, allowing another process to interfere before the update is complete. Notice that the problem can occur on a single processor system, if the time slice for a process expires in the middle of an

{ncust := ncust + 1}
    1A   Load register R1 with value of location ncust
    2A   Increment register R1
    3A   Store result back in location ncust

{ncust := ncust − 1}
    1B   Load register R1 with value of location ncust
    2B   Decrement register R1
    3B   Store result back in location ncust

Figure 9.4 Machine code for incrementing and decrementing ncust.

update sequence, allowing another process to change the shared variable while an update is under way.

One way of describing the problem is to say that the updating of the shared data item is not an indivisible operation. There are two accesses to memory required to complete the operation (one load and one store), and another process could look at or change the shared data item between these two accesses. If other processes need to be prevented from accessing the shared data while a process is in a section of code, such as the update code, that section is called a "critical region." If no other process may access the shared data while the critical region is being executed, we say that mutually exclusive access to the data has been provided, or that we have "mutual exclusion" on the use of the data.

The apparent method for providing mutual exclusion is to use a flag variable that indicates whether or not an update is in progress. Any process would check the flag to see whether another update was in progress before setting the flag and then beginning its own update, as shown in Figure 9.5. This is in principle the correct approach; however, there are two complications. First, the operations of testing the flag and then beginning the update after setting the flag have an identical problem, needing the same kind of protection we were attempting to provide for the update. Two processes could simultaneously test the flag, both decide the flag was clear, both set the flag, and both begin their update operations. We now need a flag to protect the flag test operation!

The second problem occurs after the flag has been tested and it has been correctly determined that an update is in progress. If the test is simply repeated until the flag is clear, as in the program of Figure 9.5, the CPU is not being allowed to do any useful work during this waiting time. This form of waiting is called a "busy wait." Not all updating operations are as simple as incrementing and decrementing a counter, and some operations, for example updating an entry in a shared database, may take considerable time. Thus, it is not desirable for one CPU to remain essentially idle for this time period. Instead, the process should be placed in a blocked state and a new process dispatched. As soon as the offending update is complete, the blocked process will be moved to the ready state.

Software solutions to this mutual exclusion problem for multiple processes have been found, but will not be presented in this book. Most require having a separate flag for each process to indicate whether the process is in its critical region. The trick to providing mutual exclusion is to have the process set its flag before testing the flags of other processes. This technique alone will assure mutual exclusion, but special attention has to be paid to ensure a property called "fairness." A fair implementation of critical-region protection makes sure that any process attempting to enter

```
REPEAT
  test flag
UNTIL flag is clear
flag := 1
    .
    .
    .
```

Figure 9.5   First attempt at using a flag to protect a critical region.

its critical region cannot be prevented from entering in a reasonable amount of time by other processes looping through their critical regions. Any method, whether fair or not, which assures mutual exclusion is said to be a "safe" method. A good description of these problems and some solutions may be found in Ben-Ari (1982).

Though software solutions to the mutual-exclusion problem have been found, considerable overhead is incurred in testing all the flags. Most modern computer systems provide a hardware instruction that provides the capability to both set and test the flag as one indivisible operation. This instruction may be used to establish a critical region for testing the flag variable and subsequently basing a decision on its value.

9.3.4 Methods for Controlling the Access to Shared Data

It was shown in the preceding section that some implementation is needed that will provide for mutually exclusive access to shared data. Even very simple operations in a program that has multiple processes need this capability. Any operation on shared data that involves making a change in the data based on a previous sample of that data must be placed in a critical region. While the process is executing the instructions in its critical region, all other processes must be prevented from gaining access to the shared data. This will allow the process to take an action based on its initial sample of the data with the assumption that the data have not been changed.

Implementations for critical-region protection typically provide the user with routines similar to SIGNAL and WAIT, whose use is shown in Figure 9.6. These routines provide critical-region protection through the use of special flag variables called "semaphores." A separate semaphore is used for each set of shared data that is to be protected against simultaneous access. WAIT is called before entering the critical region or, in other words, before first accessing the shared data. SIGNAL is called to indicate that the routine is leaving the critical region and is no longer referencing the shared data. Pseudocode for the implementation of WAIT and SIGNAL is shown in Figure 9.7.

Waiting on a semaphore puts a process to sleep if the flag indicates the critical region is in use by changing the process state to blocked and allocating the CPU to another process that is in the ready state. A queue of all the processes waiting on each semaphore is maintained by the operating system. If the critical region is not in use, the flag is set to indicate that

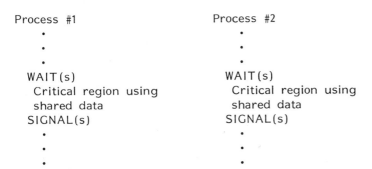

Figure 9.6 Mutual exclusion.

```
WAIT(s)
  IF s.flag = 1 THEN s := 0
  ELSE put process in blocked state waiting for semaphore s; dispatch
  another ready process

SIGNAL(s)
  IF any process is blocked on semaphore s THEN
    select one for removal from the queue of blocked process and place in
    the ready to run state
  ELSE s := 1
```

Figure 9.7   Binary semaphore.

the critical region is now busy, and the process continues into its critical region.

Signaling a semaphore will wake up one of the processes that is blocked on the semaphore by removing it from the queue of blocked processes and changing its state to the ready state. If no process is blocked on the semaphore, SIGNAL simply ensures that the flag indicates the critical region is not in use and returns.

WAIT and SIGNAL are sometimes called P and V, respectively. These routines place the calling process in the blocked state to prevent useless busy waiting. Since process queueing is not normally available at the user application level, SIGNAL and WAIT must be implemented as part of the operating system. As these routines are needed by other operating system routines, they are frequently included as part of the operating system's core, or kernel.

The WAIT and SIGNAL routines discussed so far each use a flag variable, the semaphore, which keeps track of whether the critical regions are in use. As this flag has only two states, one and zero, or not-busy and busy, the semaphore is known as a "binary semaphore." Pseudocode for a possible implementation of WAIT and SIGNAL which uses the test and set instruction is given in Figure 9.8. The test-and-set instruction is assumed to be a function that returns the value TRUE if the flag indicates the critical region is busy and FALSE if the flag indicates the critical region is free. In either case, after test-and-set is invoked, the flag will be set to indicate that the critical region is busy. Test-and-set is in actuality a hardware-implemented instruction that performs the testing and setting of the flag as an indivisible operation.

Note that there is a busy wait as the first operation in both SIGNAL and WAIT as part of the code that prevents more than one process from accessing the flag variable while it is being tested, thus establishing the WAIT and SIGNAL routines themselves as critical regions. Any process will be in only one of these critical regions for the time it takes to execute the small number of instructions necessary to implement the routines. Thus, this busy wait can be guaranteed to last only for the time it takes to execute a few instructions and is considered to be an acceptably efficient implementation. Busy waits for the implementation of user critical regions are not considered acceptable, however, as the length of the user critical region can't be guaranteed. In a single-processor system, a process entering a busy wait, waiting for another process to clear the critical region, would cycle through the busy wait until at least the end of its time slice, as no other process is

```
Procedure Wait (s:  binary-semaphore)
  Begin {Loop until test-and-set is false}
    While test-and-set(s.crflag) Do {Nothing};
    {Now in critical region protected by s.cr-flag}
    {Code for Wait from Figure 9.7}
      IF s.flag = 1 THEN s := 0
      ELSE put process in blocked state waiting for semaphore s; dispatch
      another ready process
    s.crflag := 1 {clear critical region flag}
  End
```

```
Procedure Signal (s:  binary-semaphore)
  Begin {Loop until test-and-set is false}
    While test-and-set Do {Nothing}
    {Now in critical region protected by s.cr-flag}
    {Code for Signal from Figure 9.7}
      IF any process is blocked on semaphore s THEN select one for removal
      from queue of blocked processes and place in the ready to run state
      ELSE s := 1
    s.crflag := 1
  End
```

Figure 9.8   Signal and Wait.

executing to free the critical region. The implementation of WAIT and
SIGNAL in the operating system can prevent this problem by not allowing
time slices to expire while any process is executing SIGNAL or WAIT. This
can be accomplished by simply disabling interrupts while in these short
critical regions.

User programs are capable of causing processes to be blocked through
the use of WAIT and also of waking up blocked processes through the use
of SIGNAL. As a result, WAIT and SIGNAL are used to synchronize the
execution of processes, in addition to their normal use for the protection of
critical regions and provision of mutual exclusion. This use is illustrated
in Figure 9.9. Process A needs to wait until process B has completed a
certain task T. Process A waits on semaphore s, which has been initialized
to 0. When process B finishes task T it wakes up process A by performing
the SIGNAL on semaphore s. If process A is not yet waiting on s, then

```
VAR s: binary semaphore, initial value 0

Process A                        Process B
  •                                •
  {Wait for Process B to           •
  complete task T}                 •
  WAIT(s)                          •
  •                                •
  •                                {Code for Task T}
  •                                SIGNAL(s)
  •                                •
```

Figure 9.9   Use of WAIT and SIGNAL for process synchronization.

when it finally reaches the WAIT, s will be 1 and process A will continue as it should.

9.3.5   Methods for Communicating Between Processes

The basic method for communication between processes is through the use of shared data. It was shown in the last section that SIGNAL and WAIT may be used to protect the shared data against simultaneous access.

Process synchronization is another form of process communication, where one process needs to wait for data to become available from another process. Again, it has been shown that SIGNAL and WAIT may also be used for this purpose.

The two forms may be combined together. One or more processes, usually referred to as the consumer processes, are waiting for data from another set of processes, the producer processes. A producer process places its results into a shared data buffer area and signals the consumer processes that data are available. This wakes one of the waiting consumer processes, which removes the data items from the buffer and processes them. The addition and removal of data items in the shared buffer must be inside a critical region, which may be protected by a semaphore. Any real program probably has a limited buffer space in which to store data, and the producer processes must not continue to store data items when the buffer is full.

The problem just described is commonly encountered in concurrent programming and is called the "bounded buffer" problem. A more complete discussion may be found in Ben-Ari (1982). This problem may be solved using only binary semaphores; however, it is quite easy to make implementation errors if there are multiple consumer and producer processes. The trickiness of the implementation comes from the fact that if signals get ahead of waits on a binary semaphore, the signals are lost, as binary semaphores have no counters to remember the number of signals, only a binary flag to remember whether there have been any signals or not.

Another type of semaphore, the counting semaphore, is very useful for this problem. A pseudocode implementation of counting semaphores is shown in Figure 9.10. A semaphore counter is used in place of the binary flag of the binary semaphore. This counter is incremented on SIGNALs and decremented on WAITs. A process is blocked only if WAIT is called when the counter is zero. Thus the counter can be considered to be the count of the excess of SIGNALs over WAITs if the counter was initially zero. Conversely, the counter may be considered to keep track of the number of

```
WAIT(s)
  IF s.count ≤ 0 THEN
    place the process in blocked state on semaphore s and
    dispatch another process
  ELSE s.count := s.count − 1

SIGNAL(s)
  If any processes are blocked or semaphore s THEN
    select one and place in the ready state
  ELSE s.count := s.count + 1
```

Figure 9.10   Pseudocode implementation of counting semaphores.

WAITs that may be done before blocking will occur if the counter is initially set to a positive integer value.

Considering our bounded-buffer problem, we may use one counting semaphore to keep track of how much buffer space is left. The producer processes must WAIT on this semaphore before attempting to store an item in the shared data buffer. Consumer processes SIGNAL this semaphore to indicate that more room is available in the buffer after removing an item. A second counting semaphore may be used to indicate the availability of data items. Producers will SIGNAL this semaphore when an item is placed into the buffer, and consumers will WAIT on this semaphore when they are ready to process a new item. Of course, the addition and removal of data items from the shared buffer must be performed in a critical region provided by a binary semaphore. Figure 9.11 shows a possible implementation of this problem. There are three semaphores. A binary semaphore "s" to protect accesses to the shared buffer, a counting semaphore "room" to indicate the amount of space available in the shared buffer, and another counting semaphore "items" to represent the number of items that have been produced by the producer process.

It is important to observe that semaphores are data structures belonging to the operating system routines WAIT and SIGNAL. They are not stored in the user's data area, and the user has no way to inquire about their value. The user of a semaphore may do nothing with the semaphore but initialize it or use it as an argument for WAIT and SIGNAL. The actual argument of the WAIT and SIGNAL calls is not the semaphore itself, but is probably an integer pointer into an array of semaphores local to the WAIT and SIGNAL routines. This is very important, as it is a common mistake in using semaphores to attempt to test the value of the semaphore in an IF

Producer Processes
      .
      .
                                          {Produce an item in local storage}
WAIT(room)   {Wait for room in shared data buffer}
WAIT(s)   {Protect shared data area}
          {Move item into shared buffer}
SIGNAL(s)   {Free data area}
SIGNAL(items)   {Signal that new item is available}
      .
      .

Consumer Processes
      .
      .
WAIT(items)   {Wait for item to be produced}
WAIT(s)   {protect shared buffer}
          {Remove item from shared buffer and place in local store}
SIGNAL(s)   {Free shared data area}
SIGNAL(room)   {More room in buffer now that item is
      .   removed}
      .

Figure 9.11   Use of counting semaphores for the bounded-buffer problem.

statement. For instance, it is tempting to try to eliminate one of the two counting semaphores from the bounded-buffer implementation, noting that room and items always sum to the available buffer space. However, it is not possible to retrieve the value of either semaphore; hence this approach may not be taken. One may, of course, maintain one's own counters to keep track of the number of available items, or the amount of room, but the implementation turns out to be considerably more difficult than that of Figure 9.10.

The bounded-buffer problem is probably the most common problem encountered in communication between processes. There are, unfortunately, some rather unsatisfactory characteristics to the solution using WAIT and SIGNAL given in Figure 9.11. First, if many users are writing programs to produce or consume data, it is only a matter of time before someone forgets to include one of the all-important WAIT or SIGNAL operations. This can cause errors to show up, not only in the process with the error, but strange effects may appear in other processes which were previously checked out. If the counters are inaccurate, data may be retrieved before they are available, or one may wait forever for data to be produced that have really already been produced.

There is an approach to reducing the risk of inappropriate use of WAIT and SIGNAL taken by many recent programming languages. Ada, Concurrent Pascal, and Modula2 all allow the packaging of all access routines for shared data with the shared data itself into a "module" or "package" that is separate from the user program. Any program that wishes to add or remove an item from the shared data buffer does so by simply calling an access routine that is part of the shared data package, such as ADD or REMOVE. All of the code to provide access protection for the data is coded only once and debugged only once, as a part of the access routine. This concept was first introduced as "monitors" and is discussed in Hoare (1974) and Brinch Hansen (1973).

This concept was later expanded in scope with Ada "rendezvous" construct. In this construct, the data and its access routines are packaged together as in the monitor. However, the access routines are only run by a separate process which may wait for a rendezvous with a calling process in much the same way that a person may wait at a street corner to pick up a package from a friend. When both processes have met at the rendezvous point, the access routine's process proceeds to give the data to the calling process, and the calling process returns. The use of the rendezvous is discussed in Ben-Ari (1982).

The second problem with the implementation of the bounded-buffer problem in Figure 9.11 is the difficulty of implementing priorities for the various producer and consumer processes. A property of WAIT and SIGNAL is that no guarantee is given as to the discipline of the process queues. Though on many computers the queues are FIFO, or first in first out, there is nothing in the standard definition of these routines that guarantees any special queue ordering. In order to allow each process to have a user-controlled queue priority, it is necessary to write an interface routine to the system's WAIT and SIGNAL routines. This interface routine will not call WAIT or SIGNAL on a single semaphore, but will have to have an array of semaphores, one for each process that may WAIT. Each process will be made to wait on its independent semaphore. When the interface SIGNAL routine is called, it can choose which waiting process to wake up by calling the system SIGNAL on the appropriate semaphore. This use of arrays of semaphores to accomplish user priorities is discussed in Habermann (1976).

Though modern languages are implementing advanced process control constructs, WAIT and SIGNAL are all that is available on many machines, and an understanding of these routines is important as the provide the basis through which the more advanced constructs may be implemented.

9.4   FILE SYSTEMS

Programs and data are stored in a computer system in areas called "files." A storage area for these files is needed that allows the retrieval of programs and data in a reasonable amount of time. Main memory is too expensive to have the volume of memory required, and the high access speeds and word-by-word access ability of main memory are not necessary. Some form of mass storage device is needed (such as disk storage) which is inexpensive enough to allow for the storage of files and at the same time allow retrieval of the information fast enough to satisfy the requirements of interactive and batch programs.

It is characteristic of today's storage devices that the faster the access, the more expensive is a unit of storage. The storage devices of a computer system may be considered to form a hierarchy. At one end of the hierarchy is main memory, which has fast access for any individual word, but is expensive and therefore limited in size. Next in the hierarchy might be disk storage, which has slower access rates than memory and can only read and write data in chunks called "sectors," but is inexpensive enough to hold a reasonable number of frequently used files. At the far end of the hierarchy is magnetic tape. Tape is a very inexpensive form of storage, but access time is very long. To retrieve a file from tape, first the tape must be found in the tape library and mounted on the machine after a tape drive becomes available, and then the tape must be searched until the file is found. The access for files on tape is not fast enough to be reasonable for interactive use, but it is quite reasonable for semipermanent storage of files not currently being used.

Storage devices may also be categorized by the types of access that are possible. Main memory has the characteristic that one word may be retrieved with the same speed as any other word (ignoring the effects of the memory cache) and is said to have true random access capabilities.

We have been considering access time, or the time to reach a particular data item on a storage device. Another important parameter of a storage device is its transfer rate, or the rate at which data may be moved into main memory once it has been reached. Surprisingly, the transfer rates of fast magnetic tape units are comparable to those of disks.

The data on disk are read by magnetically sensitive heads mounted on moving arms. To read a particular data item from disk, first the arm must be moved to the correct position or track and then the disk has to turn until the data of interest pass under the head. Hence, disks have a form of random access, however much slower than main memory. Also, most disks cannot read or write an arbitrary number of words, but must read or write a number of words equal to a multiple of the sector size.

A data item on magnetic tape may not be read without scanning through the tape until the item is reached. Devices, such as tapes, for which the data must be accessed in the same order that they were written are said to

provide "sequential access" to the data. Tapes also have the characteristic
that data are not written or read as individual words, but in groups of
words called a "physical record."

The file manager is responsible for storage and retrieval of files on the
system's storage devices. It must allocate space on the devices for the
creation of new files, and must be able to find existing files.

The system's file manager has a great impact on the programming environment, as it controls the organization and access to the user's files.
Once a simulation is running in the system, the file manager has little impact on performance unless the simulation has extensive file use. Frequently, the only use of the file system is to log the results of the simulation. However, a brief discussion of file organization and access methods
is given next to allow a basic understanding of the handling of user file requests and the characteristics of file systems that should be considered
when selecting a system for simulation.

9.4.1   Organization of the System's Files

A directory is kept containing the location of each file on an I/O device.
A large disk may hold thousands of files, so the directory must be organized to allow the efficient retrieval of information about a file. At the
same time, a large simulation project may itself involve hundreds of subprogram and data files, and these files need to be organized to allow the
developer to easily keep track of the files and to group them in meaningful
ways.

Most file systems group the files by user and possibly into subgroups
by project for each user. Many newer systems have adopted a true
hierarchical structure for the file system so that the user is allowed to develop any desired grouping of files. This is accomplished by allowing any
directory to contain entries that are directories in addition to the entries
for individual files.

The basic structure of the data on magnetic tape is different from that
of data on disk. Many older systems used basically different formats for
files on different types of devices. Different system commands and interface
routines were used to create files on the different devices, causing problems
for the user when files were moved from one device to another. There has
been a trend in recent systems to have "device independence" so that the
physical structure of the storage device is invisible to the user. This concept may be extended even to devices such as printers. Even though a
file may not be easily retrieved from a printer, a file may be written to the
printer using the same interface procedures that are used to write a file to
a disk.

9.4.2   Access Methods and I/O Throughput

Data logging from simulation programs needs only to use sequential writes
to a file, and many simulation users will not need to use anything besides
sequential access to simulation files. However, interactive retrieval of data
from the log file for later analysis may well involve the need for random
access to the data. Some simulations may require such large quantities of
input data that the data must be kept in files even during simulation runs.
If, for example, the data represent time-ordered inputs to the simulation,
then sequential access is adequate. On the other hand, if the data are
tabular, then random access may be required to reach a given entry.

(Further details about basic data structures may be found in Section 7.2, and database organization is covered in Section 7.5.)

The types of access required by a simulation may be of great importance, as I/O performance can vary considerably with different designs of the file manager and of the file structure. The speed of access depends not only on the speed of a physical I/O device, but also on how hard it is to determine the location of a desired data item on an I/O device.

Many of the design considerations for interactive programming are not consistent with those of fast production runs containing many random accesses. Interactive programming requires convenience of organization and use of files. Production program runs usually care less about the convenience, but speed is of overwhelming importance. Simulation programs may be at either end of the spectrum. Many simulation activities involve frequent modifications to the models, and programming convenience is much more important than speed of execution. Simulations may be pushed to the limit on speed, but may or may not actually do any real-time I/O.

In order to read information from a file, the entry for the file itself must be found in the directory, and the file must be "opened" for input. If a simulation requires rapid switching between input files, the speed of directory searches may be important, but usually this is not the case.

The unit of information read by the user in a read or write request is a "logical record." The unit of information written to a physical I/O device is a "physical record." Frequently several logical records are grouped together, or buffered, by the file manager before they are actually sent to the physical I/O device. Therefore, each I/O request for a write, for example, made by the user does not necessarily involve a real I/O request to a device, but may only involve moving the data to a buffer area in memory. When enough data have been written to fill the buffer, an actual I/O operation is performed.

Several factors determine the time required to complete the I/O operation. First, the location on the physical disk must be determined, and possibly additional space has to be allocated on the disk for the file. After waiting for one's turn to access the disk, the disk heads must move to the correct location, the disk must rotate to the correct position, and finally the data may be transferred from the buffer to the disk.

"I/O throughput" is a term that describes the overall rate at which I/O is performed and includes many factors, as described above. One factor of special importance is that the data location must be determined before a physical I/O operation can begin. The difficulty in determining this location depends, of course, on the organization of the data in the file itself, but it also depends on how space was allocated on the disk to the file.

The simplest method of space allocation is contiguous allocation. When a new file is opened for writing, all the space needed for the file is reserved as one big contiguous chunk. Unfortunately, as files are created and deleted, empty spaces will form that are too small to be effectively used. This is called "external fragmentation." Periodically all the files on the disk will have to be moved up to eliminate the wasted spaces, in an operation called "packing" or "crunching" the disk, a very time-consuming operation for a large disk.

Other methods of file space allocation involve gathering up pieces of available disk space, wherever they may be found, and keeping some form of a map that shows where the different parts of the file are to be found. The amount of information in these maps may become so large that the map itself is stored on the disk, so now it may be necessary to read the map just to find where to read the data, increasing the number of I/O operations required. There are several schemes for containing the information in these maps. A detailed description may be found in Peterson and Silberschatz (1985).

If the physical records within a file are of fixed length, it is possible to calculate the relative location in the file for any record. If the file is stored in a contiguous area, it is then possible to calculate the physical location of any record in the file simply by knowning the record length and the location of the first record in the file. This makes random access very simple and fast.

If, on the other hand, the file is not contiguous, the map will have to be consulted to determine how to find the nth record. The time required to find the record will depend on the complexity and structure of the map itself, and also on the number of I/O operations that are required to retrieve the necessary information from the map.

Other capabilities may be required of the file system. It may be necessary to insert or remove records from an existing file, or to extend the file without rewriting the entire file. These capabilities are difficult to achieve in the contiguous file. File space allocation methods that are adequate for some of these features are not as adequate for others. Contiguous allocation allows very fast random or sequential access but does not have the flexibility to easily implement other features. The correct choice of an allocation method depends on the application.

9.5   INPUT/OUTPUT

The previous section described some of the characteristics of the file manager. The file manager handles requests from the user application for I/O to files. The file manager then issues requests to the device manager for I/O to physical devices, such as the printer or the disk. It is the device manager that actually performs I/O operations on the specified device. The combination of the file manager and the device manager is frequently referred to as the "I/O system."

9.5.1   Use of the I/O System by Simulation Software

Normally, simulations use the I/O system by executing ordinary read, write, open, and close requests written in a high-level language for files on the regular system devices.

Sometimes, especially in real-time applications, simulations are interfaced to special hardware, such as special display devices or actual hardware from the system being simulated. A special hardware device is interfaced to an operating system by adding device-specific software, called a "device driver" to the operating system's device manager. The device driver acts as an interface between the special-purpose device and the I/O system. It transforms special control requirements for the device into standard I/O requests. Data and control functions are sent to the device from the simulation as standard I/O requests to the file manager.

Direct I/O by simulation programs to the device manager, bypassing the file manager, is used by some simulations to eliminate the overhead of the file manager, and also to allow special types of I/O requests needed by an unusual I/O device.

9.5.2 Handling of a Simulation's I/O Requests by the Operating System

In order to understand the interrelationship of the file manager and the device manager, it is useful to follow the sequence of events when a simulation program issues an open, close, read, or write request in a high-level language. The request is first sent to a library routine for that language, which transforms the request into the form required by the file manager. The file manager processes the request and either handles it through the existing buffer area in memory or decides to do a physical I/O operation such as a read or write to the disk. The file manager issues an I/O request to the device manager. Chances are that the disk is busy with a request from another user, and so the device manager places the I/O request in a queue of I/O requests waiting for the disk. This process is placed in the blocked state, waiting for completion of the I/O request. Another process is selected from the ready list by the dispatcher, and the CPU continues running the new process.

Eventually the current I/O operation is completed, and the waiting I/O request is selected from the queue of waiting requests. The request for an I/O operation is given to the disk driver program, which actually sends the proper commands to the disk hardware. When this operation is completed the disk driver notifies the device manager, which causes the requesting process to be removed from the blocked state and to be put back on the ready list. When the process is finally dispatched, the file manager returns control to the user program, which resumes execution. An excellent description of this sequence of events and the data structures used by each of the managers is presented in Lister (1979).

9.5.3 Software Problems in Interfacing Special-Purpose Hardware to Real-Time Simulation Systems

The only portion of the whole I/O system that knows about the exact nature of a specific hardware device is the device driver for that device. In order to add a new special-purpose I/O device to an operating system, it is usually only necessary to add a new device driver to the system and a new entry into the list of system device drivers. The addition of a new device driver to some systems requires relinking the operating system, or at least the device manager, in a process called generating a new system. Some operating systems will actually let a new device driver be added during a regular program run without requiring generation of a new system. This feature saves a lot of time, as the generation of some operating systems takes many hours.

Many operating system vendors provide a "skeleton" device driver which may be modified to meet specific needs. Another approach is to take an existing device driver for a similar device and modify it. The exact structure of device drivers differs considerably from system to system, but most micro- and minicomputers have the general structure described below. But before discussing the structure of the device driver itself, let's consider the way that an I/O operation is performed on a typical simple I/O device such as a printer.

Normally, one communicates an I/O request to a device by observing and setting bits in special registers belonging to the device. In some systems the device registers are mapped into specific memory locations and the programmer accesses these registers by simply loading or storing to the appropriate memory location. In other words, there are no special I/O instructions; normal instructions use the device registers as if they were regular memory locations. This type of I/O is known as "memory-mapped I/O" and is used on many small- and medium-scale computer systems. On other systems there are special I/O instructions for accessing the device registers. These device registers are, in reality, inside the hardware of the device interface, and the device can also observe and set bits in these registers.

A simple character-at-a-time printer may be used as an example of a simple I/O device. It could have as few as two registers: a status register and a data register. The status register contains one or more bits indicating the current status of the printer, such as out of paper, busy printing, and ready for more data. This register may also contain bits that the programmer may set or clear to disable interrupts for the device. To print a character on such a printer one sets the interrupt disable bit to the desired value, waits for the bit that says the printer is ready for a new character, and then stores the character to be printed in the printer's data register, which will cause it to be printed. When the printer is finished with that character, it will interrupt the CPU so that it may be fed another character to print if any are available.

This printer would be interfaced to an operating system by writing a device driver. The device driver may be considered to have four basic parts: (a) the initialization routine, (b) the I/O start routine, (c) the device interrupt handler, and (d) the device completion routine.

The initialization routine is called only once when the operating system is first loaded, so that any device initialization, such as enabling or disabling interrupts or initializing any local variables in the device driver, may be done.

The I/O start routine is called by the device manager when an I/O operation is requested, such as to print a line. The device driver sends the first character to the printer and then returns to the device manager waiting for a printer interrupt to signal that the printer is ready for a new character.

The device interrupt handler is called when the printer signals its readiness to print another character by causing an interrupt. The device interrupt handler first checks to see whether the last character on the line has been printed; then if it hasn't, the device interrupt handler sends the next character to the printer and returns, waiting for the next interrupt. If the last character has been printed, the device interrupt handler jumps to the device completion routine.

The device completion routine resets any local variables and possibly disables printer interrupts, since there are no more characters to be printed. It then signals the device manager that the I/O operation is complete, which in turn notifies the file manager, which in turn allows the user to resume execution.

Devices such as disks are considerably more complex to handle than the printer described above, but many typical special-purpose I/O devices needed by simulations have simple interfaces, similar in complexity to the printer described above.

9.6 MEMORY MANAGEMENT

Limitations on the amount of main memory available have always been a major factor in the design of computer systems. As described earlier, it is desirable to increase the amount of available main memory in order to increase the degree of multiprogramming and also to make it possible to run large programs without the need for user-defined overlays. Considerable effort has been expended in developing system designs with virtual memory that efficiently run programs requiring more memory than is actually available.

The effort to efficiently utilize memory has considerable effect on the performance of large simulation programs. An understanding of the virtual memory concepts used in today's computer systems is necessary in order to properly control the execution of a large program in the virtual memory environment.

9.6.1   Concepts of Virtual Addressing and Virtual Memory

The terms virtual addressing and virtual memory are often confused. When a program is written, compiled and linked, it is linked as if it were to load contiguously starting at a fixed location in memory. For example, assume that the program is linked to load in location 1000 and is 3000 locations long. In reality, other programs are probably already occupying those memory locations, so the program must be able to execute when loaded at some other unused location in memory, such as location 3500. The hardware of the CPU will adjust the addresses of all instructions referencing memory by adding the constant 3500 - 1000 = 2500 to all addresses generated by the program, as the program executes. Thus, when the program tries to access a variable that it thinks is in location 2000, it will really access location 2000 + 2500 = 4500. This is the location where the variable is really located. The moving of a program from its intended place in memory is called "relocation," and when it happens as the program runs, it is known as "dynamic relocation." When the program is addressing different locations in memory than the ones actually coded in the program, we have introduced the concept of "virtual addressing." The addresses generated by the program seem to be real, the program does work, but they are really fake, or "virtual," addresses, as they are different from the actual memory locations accessed by the program. We call the set of all virtual addresses occupied or referenced by the program the "virtual address space" of the program. Each virtual address in the virtual address space of the program is "mapped" into a unique address in "physical," or actual memory. The set of all physical addresses that correspond to an address in the virtual address space is called the "physical address space" of the program. Thus, the physical address space is the set of all real-memory locations used by the program.

When the program is loaded into consecutive physical memory locations, relocation may be handled as described by a single constant stored in a single register known as the base register. Many programs are logically composed of several pieces, and it would be easier to load such a program if the system could place each piece, called a "segment," independently in a separate unused spot in memory. This way, smaller unused portions of memory could be utilized. When an instruction in the program now references a memory location, it will have to specify which segment, as well as the location within the segment, or the segment offset. A separate relocation constant will be needed for each segment to indicate the location in real memory where the segment has been loaded. Systems using this scheme are said to be using "segmentation."

The structure of a virtual address in a system with segmentation is shown in Figure 9.12. Recall that the virtual addresses is the address the program

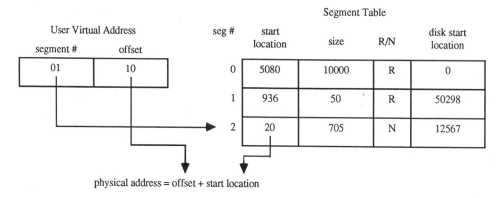

Figure 9.12   Calculation of physical address from user's virtual address in a virtual memory system with segmentation.

specifies and is now composed of two parts. The first part specifies which segment is to be addressed, and the second part is the location within the segment. Machines having large word lengths actually break up a word specifying an address into two parts, as shown in Figure 9.12. Machines having smaller word lengths, such as microprocessors, frequently use two address words, one for the segment and one for the offset. Some systems will assume which segment is desired if it is not specified by the instruction, and only one word is used for the address. In any case, a separate relocation constant is needed for each segment, and these are frequently arranged into a "segment table," as shown in Figure 9.12. For the moment, ignore the right two columns of the table (R/N and disk start location), which are concerned with virtual memory on disk.

Each segment used by the program is referred to by an integer. This integer serves as an index into the table to select the proper row that contains the entry for the given segment. Hence, segment number 5 will select the row containing the information for segment 5, such as the starting location of that segment. The table also contains the length of each segment, which may be used to prevent accidental accesses beyond the end of the segment.

Most programs contain far more code and data than are actually needed at any given time. For example, a simulation frequently contains code related to error printouts which is not executed under normal conditions. If the code and data are arranged into logical groups, with each group in a separate segment, it is likely that some of the segments could be absent from memory and would seldom be missed. Perhaps some segments are only needed to initialize the simulation and are not needed after the simulation has run a while. If these segments were available in a reasonable time from the disk, they could be removed from memory. For years a similar technique was used, under user control, to overlay sections of code that did not have to be in memory simultaneously. Most programmers found it very difficult to design and control the overlays. Overlays were retrieved from disk in their original form, with all variables having initial values. This made it even more difficult to program, as all local storage would be reinitialized whenever the overlay was retrieved from disk.

It was necessary to make the removal and reinstatement of segments invisible to the user. The swapping of segments to and from disk should be handled by the hardware automatically when any location in the segment is accessed, with all local variable sintact. A system designed in this way is said to have "virtual memory," as it appears to have more memory than it really does. Figure 9.13 demonstrates a system running a program with three segments named SEGA, SEGB, and SEGC. SEGA and SEGC are resident in physical memory, and all three segments are stored on disk. In theory, if the segments such as SEGB in the figure are loaded into memory automatically upon access, the user doesn't even need to know that the segment wasn't in memory.

Automatic loading of segments may be accomplished using the segment table, which has to be accessed by the hardware on every instruction with a virtual address, to contain a flag, shown as the R/N entry in the table in Figure 9.12, which indicates whether the segment being referenced is resident or nonresident. If the segment isn't resident, an interrupt, called

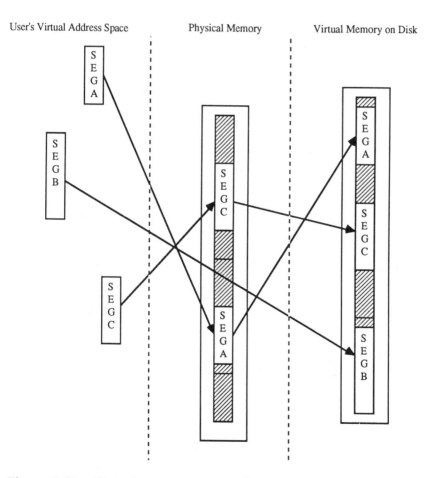

Figure 9.13   Virtual memory system with segmentation.

a "segment fault interrupt," is caused. This interrupt activates a routine in the memory manager which reads the segment into an available spot in memory. It is possible that there is no room in memory for the segment, in which case another segment will have to be removed to make room. The questions is how to determine which segments are reasonable to remove from memory. The first objective is to make sufficient room for the new segment. Beyond that, any segment that has not been accessed in a reasonable period of time would be a likely candidate. An additional entry could be added to the segment table of Figure 9.12 which indicates whether each segment has been accessed. This entry would be set by the hardware automatically if the segment was accessed during the execution of any instruction. It would be desirable to know the rate of accesses to this segment, but there would be too much overhead in accumulating this information, so the only information recorded is whether or not the segment has been accessed. If the entries were periodically cleared, the entries that were still clear at the time of a segment fault would indicate that at least no accesses had been made since the last periodic clearing, and the corresponding segment could be considered removable.

If a segment has to be removed from memory, the contents of its local variables need to be saved, so the segment could be copied to disk and its location recorded in the segment table along with an indication of non-residence. After the segment to be removed has been placed on disk the new segment may be read into its space. After both segment transfers are complete, the program's execution may be resumed, blissfully unaware that anything has happened, except for a mysterious jump in time.

In order to save I/O operations, the following technique is used. Before executing the program, a copy of all the program's segments are placed on disk, and the location of each segment is recorded in the segment table. When a segment fault occurs, any segment to be removed from memory doesn't really need to be copied back to disk unless locations have been modified in the segment since it was loaded. Thus, another additional entry is made in the segment table shown in Figure 9.12 which records whether or not any segment has been modified since it was loaded into memory. This entry will be updated automatically by the hardware when any location within the segment has been modified. Some computer systems associate a "modified" bit with each block of the main memory itself. These bits could be used to determine which blocks need to be saved on the disk.

It is quite possible, if SEGC or SEGA of Figure 9.13 has been modified since they were loaded, that the copies of these segments on disk are not up to date. If so, then these segments would have to be copied to disk before they could be removed from memory.

We have discussed policies for removing segments from memory, known as "replacement" policies, but which segments will be placed in memory? If segments are only placed in memory when they have been accessed and a page fault occurs, we have "demand segmentation."

If the segment table were stored in memory, any memory access made by an instruction would actually take two accesses: One for the segment table access and one for the data itself. If a computer system allowed only a few segments, it would be feasible to contain the entire table in registers. Many systems allow too large a number of segments to allow the storage of all of the segment table in special high-speed registers; however, system performance would be unacceptable if the number of memory accesses were doubled by the use of a segment table. To solve this problem, only the most recently used entries in the table are kept in a set of registers known as a translation-look-aside buffer, or sometimes known simply as a cache. If the entry for a given segment is in the look-aside buffer, very little overhead is required to obtain the segment table information. If, however, the information is not in the look-aside buffer and has to be retrieved from memory, there is the considerable overhead of an extra memory access.

Segmentation shares a major problem with file space allocation discussed previously: external fragmentation. It is possible to have a lot of little pieces of unused memory between the resident segments so that there is quite an unused space available, but none of the pieces are big enough to hold a segment. Since memory is considered a scarce resource, this may be a serious problem. A fair amount of work is required to keep track of all the unused portions of memory and their sizes, and searching this list to find room for a segment also requires considerable overhead. Most systems do not actually attempt to keep track of memory space to the nearest word, as shown in Figure 9.12, but allocate memory only in pieces of a minimum size, which is a power of two. This allows removal of a few bits from each of the entries in the segment table, as segment locations and sizes are all powers of two. For example, if the sizes were all multiples of 256, the trailing eight bits would always be zero and hence would not need to be stored.

Paging was introduced to largely eliminate the problems of external fragmentation and to simplify memory allocation to programs. Instead of dividing a program in logical pieces, just hack the program into fixed-size pieces called "pages." Memory is also divided into pieces of the same size called "frames" to distinguish them from the pieces of the program itself. Now, any page of the program can fit into any available frame of memory. Each page of the program has an entry in a "page table" which shows where that page is located in physical memory by specifying the frame number in memory. The size entry does not have to be in a page table, as all pages are of the same size, although some computers still have a size entry which tabulates the last-used location in each page.

If virtual memory is to implemented, each page has an image on the disk, and entries must be added to the page table to indicate whether a resident page has been modified and whether the page has been accessed since the last clearing of access flags. Figure 9.14 shows an example of a system running a program requiring five pages. The page table for this program is shown in Figure 9.15, which also demonstrates the method for formulating a physical address given a user virtual address. Since pages are contiguous in the user virtual address space, it is unnecessary to specify a page number for any address. If the page size is an even power of two, it takes a specific number of bits to contain the offset into the page. For example, if the page size is 512 bytes, it takes nine bits to hold the offset within the page, which could vary between 0 and 511. The rest of the bits in the word represent the page number. If the computer has a 16-bit address, the top seven bits represent the page number. This page number is used as an index into the page table which gives the frame number in memory. The offset into the page is the same as the offset into the frame, so the physical address is formed by concatenating the frame number on the front of the offset.

Just because the user address is limited to 16 bits in this example places no limit on the number of bits in the frame number, so by using a page table one can reference addresses requiring more than 16 bits. This

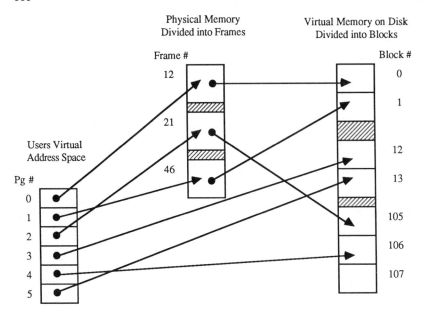

Figure 9.14   Virtual memory system with paging.

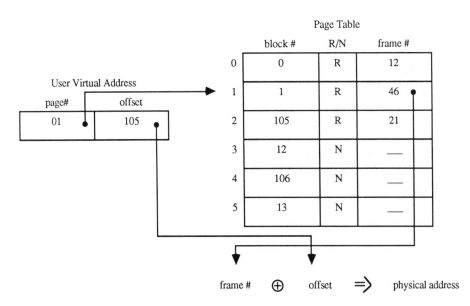

Figure 9.15   Calculation of physical address from user's virtual address in virtual memory system with paging.

explains why memories on 16-bit computers may be quite large, even though a user may specify only a 16-bit address in the range from 0 to 64K. Note that the total size of the user address space is still limited to 64K in this case, but many users may be running, each mapped by the page table into a memory much larger than 64K.

A more detailed coverage of segmentation and paging may be found in Madenick and Donovan (1974) and Deitel (1984). The MULTICS operating systems was developed emphasizing file sharing, and extended the concept of segmentation to the file system. A description of MULTICS may be found in Corbato (1962) and (1965). The memory management techniques used in Digital Equipment Corporations VAX computers are described in Kenah and Bate (1984).

9.6.2 Effects of Segmentation and Paging on Simulation Software

Virtual memory systems are proclaimed to do away with any limits to program size. If the total number of pages of resident memory is insufficient to allow the main computational parts of a simulation to remain resident, there will be constant I/O to bring pages to and from the disk. Many systems have a parameter that determines the maximum number of resident pages allowed. Frequently this number must be considerably increased to allow reasonable performance for a large simulation. The number of pages of the simulation that must be resident to prevent excessive page swapping is called the simulation's "working set size." If the working set size is greater than the maximum allowed number of resident pages, the swapping can become so intensive as to hurt the performance of the entire system, not just the simulation. Real-time requirements may require locking the pages of the simulation into memory for the duration of the run.

Paging and segmentation are both schemes to increase the number of programs that may run simultaneously in memory by efficiently utilizing the available memory. This is done at the expense of a loss of performance to any program experiencing page faults or segment faults. Simulations tend to be large, computationally intensive programs. Memory prices are decreasing to the point that it may be feasible to simply buy enough memory to contain the entire simulation so that it may be locked into memory, effectively defeating the paging or segmentation capabilities and their inherent loss of performance.

Systems that have segmentation may have a limit on maximum segment size. This is especially true in systems with a small word length. In 16-bit computers, the maximum segment size is usually 64K. This limit may not be harmful at all, if program and data units may be divided into logical segments of this size or smaller. If data arrays exceed 64K, there may be overhead introduced by switching segments when moving from one array element to the next.

9.7 SUMMARY

As the manager of all system resources, the operating system has a significant impact on the speed at which simulation software runs. Services are requested from the operating system's managers for file and device I/O, memory management functions, and process synchronization. Special devices are interfaced to the operating system by adding device drivers to the operating system's device manager. However, simpler and more powerful methods for interfacing between applications, such as simulations, and the operating system need to be developed.

There is a recent trend toward developing distributed simulations, or simulations with processes running concurrently on more than one CPU. The development of efficient methods for synchronizing the execution of distributed processes is a topic of much current research effort. An example of recent research in this area may be found in Zhixin et al. (1986), where an unusual method of process synchronization called "time-warp" is described.

Stankovic (1984) presents six areas of current research in distributed computer systems and points out the need for coordinated research in the areas of distributed computer systems, distributed databases, and distributed system communication. Distributed simulations involve all these areas, and there is a need for research into new and efficient techniques for ensuring the integrity of data throughout the system, and for dividing and coordinating the work to be done among the simulation's processes. As new techniques for distributed simulations are discovered, there will be a corresponding need for changes in the operating system design to support these techniques.

REFERENCES

Ben-Ari, M. (1982). *Principles of Concurrent Programming*, Prentice-Hall Englewood Cliffs, NJ.

Brinch Hansen, P. (1973). *Operating System Principles*, Prentice-Hall Englewood Cliffs, NJ.

Corbato, F. J. and Vyssotsky, V. A. (1965). Introduction and overview of the MULTICS system, *Proc. AFIPS Fall Joint Computer Conf. 27*, 185.

Corbato, F. J., Merwin-Daggett, M., and Daley, R. C. (1962). An experimental time sharing system, *Proc. AFIPS Fall Joint Computer Conf. 21*, 335.

Deitel, H. M. (1984). *An Introduction to Operating Systems*, Addison-Wesley, Reading, MA.

Dijkstra, E. W. (1968). The structure of the T. H. E. multiprogramming system, *Comm. ACM. 11*, 341.

Habermann, A. N. (1976). *Introduction to Operating System Design*, Science Research Associates, Chicago.

Hoare, C. A. R. (1974). Monitors: An operating system structuring concept, *Comm. ACM 17*, 549.

Kenah, L. J. and Bate, S. F. (1984). *VAX/VMS Internals and Data Structures*, Digital Press, Bedford, MA.

Levy, H. M., and Eckhouse, R. H. (1980). *Computer Programming and Architecture: The VAX*. Digital Press, Bedford, MA.

Lister, A. M. (1979). *Fundamentals of Operating Systems*, Springer-Verlag, New York.

Madnick, S. E. and Donovan, J. J. (1974). *Operating Systems*, McGraw-Hill, New York.

Peterson, J. L. and Silberschatz, A. (1985). *Operating System Concepts*, Addison-Wesley, Reading, MA.

Stankovic, J. A. (1984). A perspective on distributed computer systems, *IEEE Trans. on Computers C-33*, 1102.

Zhixin, D., Heying, Z., Lomow, G., and Ungar, B. (1986). Distributed systems in simulation database concurrency control using time warp, in *Proceedings of the 2nd European Simulation Congress*, Antwerp, Belgium.

10

Hardware and Implementation

10.1   INTRODUCTION

The heart of any digital simulation system is the digital computer. The
simulation is carried out either on a general-purpose computer or on a
special-purpose computer dedicated to the particular simulation. Types of
computers used in simulation include *digital* and *analog* computers, as discussed in Section 1.3.1. A digital computer processes the data in digital
or discrete form and produces results in the digital form. An analog computer, on the other hand, processes continuous signals.

The primary function of a digital computer is *processing* of the *data*
input to it, to produce the results. The processing follows an *algorithm*,
which is a sequence of steps to achieve the *result* from the input data. The
algorithm when coded using a computer language forms the *program* executed
by the computer. In the case of a simulation, the model parameters form
the data input, and the simulator (program) is executed by the digital computer to process the input to produce the simulation results.

Analog computers are capable of processing analog signals directly. It
should be noted that the real world is usually analog in nature, and analog
computers tend to be faster and more versatile than digital computers in
simulating dynamic real-world systems. The components of analog computers
are amplifiers and integrators, and they are programmed by interconnecting
these components by patch chords. This programing mode is rather inconvenient and less flexible compared to that of digital computers. We shall
now provide a brief comparison of the two types of computers with respect
to simulation.

10.1.1   Analog Versus Digital Computers

Analog computers provide a "natural way" for simulation (Kobayashi, 1981)
and have been used extensively either purely as analog systems or as hybrid systems in conjunction with digital computers. The reasons for
naturalness of analog computers in simulation are: real-time execution,
retention of functionality, parallel execution, continuity of data and
processing, and natural man-machine interaction.

As mentioned in Section 1.3.1, analog computers can simulate at real-time speeds. The speed of a simulation can be altered through scaling, by selecting faster or slower integrators without actually changing the programs (i.e., patched wires). The processor speed and the complexity of programs dictate the speed of simulation using digital computers. Careful tailoring and control of programs through a real-time clock would be needed to bring the real-time speed to digital simulations.

A system is often represented in terms of a block diagram, as discussed in Section 2.2.4. Analog computers allow implementing a model, without altering its functionality as exhibited by the block diagrams. In digital simulation, the functionality may have to be altered to suit the computational characteristics of the machine, especially in a uniprocessor environment where the computation is completely serial. The parallel simulation paths in functional block diagrams can be easily retained in analog computers. A serialization of computation may be needed to suit the serial processing mode of digital uniprocessors.

When the simulation involves continuous input(s) and requires continuous output(s) to interface with the real world, analog computers provide a natural processing mode and are faster since data transformation (from analog to digital to analog form) is not needed, as when using digital computers. The data transformation steps could also introduce accuracy problems.

The recent advances in digital hardware technology have increased the speed and cost effectiveness of digital computers. Various architectures have been devised to accommodate parallel processing, thereby increasing the speed of simulation and making real-time simulation practical. As such, the orientation of this chapter is toward digital computers in simulation. We introduce single-processor systems in this chapter. Chapter 16 describes the multi-processor architectures. Section 10.2 provides a model for the digital computer itself. Section 10.3 details the organization and programming of a simple hypothetical computer. Section 10.4 introduces the various classes of computer architectures. This is followed by a discussion of the use of the digital computer as a component of the simulation system. The hardware components that perform analog-to-digital and digital-to-analog conversion are described in Section 10.5. Signal sampling and quantization effects are presented in Section 10.6. Section 10.7 details the implementation aspects, and an example simulation system description is the object of Section 10.8.

10.2   DIGITAL COMPUTER MODEL (Baer, 1980; Shiva, 1985)

The basic components of any digital computer system are: the physical devices that constitute the *hardware* and the set of programs that are executed on the hardware that constitute the *software*.

10.2.1   Hardware

Figure 10.1 shows the four major hardware blocks of a general-purpose computer system: the memory unit, the arithmetic-logic unit, the input/output unit, and the control unit. Programs and data reside in the memory unit. The arithmetic-logic unit processes the data taken from the memory unit and places the results into the memory unit. The control unit retrieves the instructions from the memory resident programs one at a time,

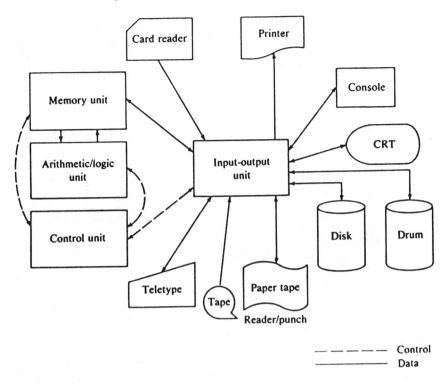

Figure 10.1   A typical computer system.   (From Shiva, 1985.   Courtesy of Little, Brown Company.)

decodes them, and directs the arithmetic-logic unit to perform the processing called for by the instruction. The control unit oversees the operation of all the other units of the machine. The input unit is used to input data (and programs) into the memory, and the output unit is used to transfer the processed data to the external devices. This is a general organization. In practical systems the input/output units transfer data into and out of the arithmetic-logic unit and/or the memory.

Several input/output devices are shown in Figure 10.1. Devices such as card reader and paper tape reader are strictly input devices, and others such as the printer and paper tape punch are strictly output devices. The other devices shown can be used as an input or an output device. The magnetic tape, disk, and drums are also used as the secondary memory devices to expand the capacity of the system's primary memory.

The digital computer hardware can be modeled at various levels of detail (Shiva, 1979). The following levels have been used:

1. Processor-memory-switch (PMS), where the components are the blocks depicting the processors, memories, and switches that interconnect them
2. Instruction set level, which is a programmer's model wherein each processor instruction is described
3. Register transfer level, where the elements of description are the registers and the flow of data and instruction is depicted in detail

4. Gate level, in which the register level details are expanded to logic components such as AND, OR, and NOT gates and flipflops
5. Circuit level, in which the logic components of level 4 are replaced by circuit components such as resistors, capacitors, and transistors
6. Mask level, in which the circuit components are in terms of mask geometries that fabricate the component on the silicon real estate of an integrated circuit.

The model of the general-purpose digital computer provided in this section is essentially at the PMS level. We shall provide a programmer's model of a computer system in the next section.

10.2.2   Software   (Shiva, 1985; Calingaert, 1979)

The hardware components of a computer system are electronic devices in which the basic unit of information representation is either a 0 or a 1, corresponding to the two states of the electronic signal. For instance, in one popular integrated-circuit technology (TTL-Transitor Transistor Logic), 0 is represented by 0 volt and 1 is represented by +5 volts. Programs and data must therefore be represented using this binary alphabet of 0 and 1 within the hardware. Programs written using only these binary digits (or bits) are *machine language programs* (see Section 1.3.3). At this level of programming, operations such as ADD, SUBTRACT, etc., are each represented by a unique string of 0s and 1s and the hardware is designed to recognize such strings and interpret them when decoded in the control unit. Programming at this level is tedious since the programmer is required to have an intimate knowledge of the hardware structure of the machine.

The next higher level of programming is the *assembly language level*. In this level, symbols such as ADD, SUB, etc., are used to denote the operations; the operands (data) are also identified by symbolic names such as X, Y, DAT1, etc. The assembly language program is first translated into the machine language level, before it is executed by the machine. The translator is a program known as the *assembler*. Programming at this level also requires a detailed knowledge of the machine structure. The operations permitted tend to be primitive and depend on the machine structure. Thus each processing step needs to be broken into the primitive set of operations allowed by the instruction set of the machine. Each computer system will have its own instruction set, and hence the programs at this level are not portable to other systems. Chapter 8 focuses in more detail on programming languages.

Use of *high-level* programming languages such as FORTRAN, COBOL, and Pascal further reduces the requirement of the detailed knowledge of the computer structure. Each processing step is now represented by a high-level language statement which is independent of the machine structure. A *compiler* program is used to translate the high-level program into the machine level. A separate compiler is needed for each high-level language on the machine. High-level language programs can be easily ported to other machines.

Figure 10.2 shows the sequence of operations that occurs once a program is developed. The program, written in either the assembly or a high-level language, is called the *source* program. The source program is translated into the machine language level by either an assembler or a compiler. This translated program is the *object code*. The object code ordinarily resides in an intermediate memory device such as a magnetic disk or tape. A

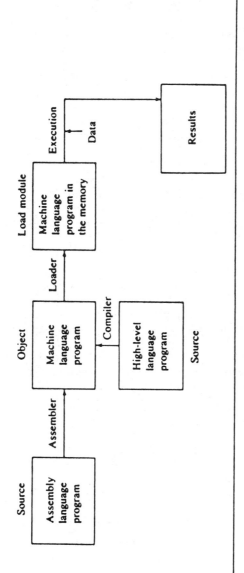

Figure 10.2   Program translation and execution.   (From Shiva, 1985.   Courtesy of Little Brown Company.)

loader program loads the object code into the appropriate parts of the machine's memory unit. This program in the memory is *executed* to produce the results. The data required during execution are either available in the memory by a previous operation or supplied by the external input devices during execution.

Operations such as selecting the appropriate compiler, loading the object code into the memory, starting and stopping the execution, accounting of the computer system usage, and resource (memory, peripheral, etc.) usage management, etc., are automatically done by the computer system. A set of supervisory programs that perform these operations is usually provided by the machine manufacturer. This set of programs is called the *operating system* of the machine. (Chapter 9 is devoted to operating systems.) The operating system receives the commands from the user through a command language and manages the overall operation of the computer system. Figure 10.3 is a simple rendering of the complete hardware and software environment of a general-purpose digital computer system.

10.2.3   Firmware

The primary memory of the machine is composed of *random access memory (RAM)* devices, in which data can be *written into* or *read from* any memory location randomly and consumes the same time no matter which memory location is used. Ordinarily, all the application programs and data reside in RAM. Some of the most often used program segments may also reside in another type of memory called *read only memory (ROM)*. As the name implies, the content of ROM can only be read and cannot be written into during the program execution. Special devices known as ROM programmers are used to write data and programs into ROMs in an offline mode. The set

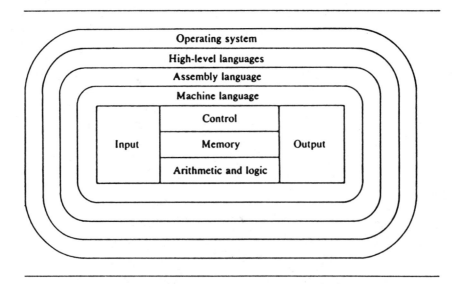

Figure 10.3   Hardware and software components of digital computer system. (From Shiva, 1985. Courtesy of Little, Brown Company.)

of programs and data residing in the ROM is termed the *firmware* of the system. For instance, the control sequences that are used in the control unit to decode and execute instructions are usually stored in a control ROM. Depending on the complexity of the system, either all or part of the operating system and application programs may be ROM-resident.

10.3 A HYPOTHETICAL COMPUTER
(Adapted from Shiva, 1985)

The purpose of this section is to introduce the general structure of the digital computer and illustrate assembly level language programming using a simple hypothetical computer (AHC). Although AHC appears to be very primitive, its organization reflects the basic features of most complex modern machines. The instruction set is primitive but complete enough to write powerful programs. AHC is introduced from a programmer's point of view. As such, the hardware level details are minimized. A prior knowledge of binary number system is assumed.

10.3.1 Organization

The hardware components of AHC are shown in Figure 10.4. We shall assume that AHC is a 16-bit machine and hence the unit of data manipulated by and transferred between various registers of the machine is 16-bits long. A *register* is a temporary data storage device. AHC is a binary, stored-program computer. That is, the data and programs are represented within the machine in binary form and programs are stored in machine memory

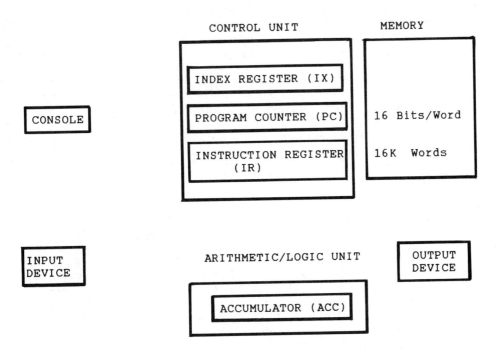

Figure 10.4 AHC hardware components.

before they are executed. The memory is a RAM organized into several
words each 16 bits long. Each memory word has an *address*. It is
assumed that the memory contains 64K words ($1K = 2^{10} = 1024$). To read
the contents of a memory word, the address of the word is provided to the
memory along with the READ request; to write a 16-bit data into the memory
word, the data and the address of the memory word into which the data are
to be written are provided to the memory along with the WRITE request.
READing does not alter the memory word content, whereas WRITing changes
the content of the memory word to the new data.

The programs are stored in the memory along with the required data.
During the program *execution*, each instruction from the program is first
fetched into the control unit register, called instruction register (IR), and
the circuitry around IR decodes the instruction. The control unit generates
the appropriate control signals to perform the operation called for by the
instruction. During the *execute* phase, the operations are performed. The
fetch-execute phases put together are known as the *instruction cycle* of the
machine. A program counter (PC) register in the control unit contains the
address of the instruction in memory to be executed next. At the end of
each fetch phase, the content of PC is incremented so that the control unit
can fetch the next sequential instruction after the execution of the current
instruction. The machine thus executes instructions in the sequential order
unless the content of PC is altered during the execution of an instruction.
The housekeeping instructions such as Jump and Halt alter the content of
PC.

The arithmetic-logic unit of the machine contains a 16-bit accumulator
(ACC). This register accumulates the results of arithmetic and logic operations.

A 16-bit index register (IX) is used in the manipulation of addresses.
We shall describe its function further later in this section.

The *console* permits the operator to interact with the machine. Through
the console, any memory location can be examined and altered, programs
can be loaded into the machine memory, and PC can be initialized. The
console also contains the power ON/OFF and START/STOP switches. To
execute a program, the operator first loads the program (and data) into the
memory, sets the PC to the address of the first instruction in the program,
and STARTs the machine. During the execution, additional data is input
through the input device and the results are forwarded to the output device.

AHC has one input device that can transfer a 16-bit quantity into the
ACC and has one output device that can transfer the content of ACC to the
output media.

10.3.2   Data Format

AHC memory is an array of 64K, 16-bit words. Each word will contain
either a 16-bit data item or an instruction. The exact interpretation of
what the 16-bit quantity is depends on the phase the machine is in when the
particular word is examined. If a word is retrieved during the fetch
phase, its content is interpreted as an instruction, whereas it would be
interpreted as data (or address) if retrieved during the execute phase.
At the machine and assembly programming levels the programmer should be
aware of the data and program segments in the memory and should make
certain that a data item is not encountered during the phase in which the
machine is accessing an instruction, and vice versa.

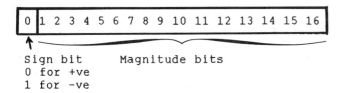

Figure 10.5   AHC data representation format.

Figure 10.5 shows the data format for AHC. The most significant bit is the sign bit (0 for positive; 1 for negative). Since the binary point is assumed to be to the right of the least significant bit, only integer numbers can be represented and the numbers range from $+(2^{15} - 1)$ to $-(2^{15} - 1)$. If the binary point is assumed to be to the right of the sign bit, a fixed-point fraction representation would have resulted, with the range of fractions from $+(1 - 2^{-15})$ to $-(1 - 2^{-15})$. Scientific simulation-oriented machines normally use a floating-point data representation.

10.3.3   Instruction Set

Table 10.1 lists the complete instruction set of AHC. There are 16 instructions. The first column of the table shows the symbolic operation code, or the *mnemonic*, for each of the instructions allowed. The second column lists the corresponding binary patterns for these operations (Opcodes). The last column describes the function of each instruction. An assembly level language programmer uses the mnemonics in coding and the assembler translates the mnemonics into the binary opcodes.

LDA and STA enable data transfer between ACC and the memory. ADD and SUB are arithmetic instructions. BRU, BIP, and BIN are for program flow control. LDX and STX transfer data between memory and the IX. TIX and TDX enable program loop control and IX manipulation. RWD and WWD are input and output instructions, respectively. SHL multiplies the ACC by 2 and SHR divides it by 2. HLT stops the machine.

Note that the first 11 instructions are single address instructions since they refer to a memory word either as an operand (LDA, STA, ADD, SUB) or as an address (BRU, BIP, BIN, LDX, STX, TIX, TDX). The last five instructions do not require a memory reference and hence are zero-address instructions. The first four of these instructions use the ACCumulator as the operand.

Typically, it takes longer to address a memory word and retrieve or write data from/to it compared to a data transfer between two registers. Hence, the memory references should be minimized to achieve faster program execution. Instruction sets with more than a single address in each instruction are used on modern machines.

10.3.4   Instruction Format

Each instruction in AHC occupies 16 bits. An instruction word has three fields, shown below:

0 1 2 3	4	5 6 7 8 9 10 11 12 13 14 15
OPCODE	↑ INDEX FLAG	ADDRESS

Table 10.1   AHC Instruction Set

Mnemonic	Opcode	Description	
LDA	0000	ACC ← (MEM)	Load ACC from MEM
STA	0001	MEM ← (ACC)	Store ACC in MEM
ADD	0010	ACC ← (ACC) + (MEM)	Add to ACC
SUB	0011	ACC ← (ACC) − (MEM)	Substract from ACC
BRU	0100	Branch Unconditional	
BIP	0101	Branch if (ACC) > 0	
BIN	0110	Branch if (ACC) < 0	
LDX	0111	IX ← (MEM)	Load Index register
STX	1000	MEM ← (IX)	Store Index register
TIX	1001	IX ← (IX) + 1 Branch if (IX) = 0	Test Index Increment
TDX	1010	IX − (IX) − 1 Branch if (IX) ≠ 0	Test Index Increment
RWD	1011	Read a word into ACC	
WWD	1100	Write a word from the ACC	
SHL	1101	Shift ACC left once	
SHR	1110	Shift ACC right once	
HLT	1111	Halt	

MEM is a symbolic address of a memory word.   Parentheses indicate
"content of."   ← indicates "transferred to."

The first field contains the four-bit opcode.   Bit 4 is the index flag and is
set to 1 if an index register reference is made; otherwise it is set to 0.
The remaining 11 bits indicate a memory address.

With 11 bits for the address one can directly address only 2^{11} or 2K
words.   The IX can be used to extend the addressing range from 2K to
64K, ad discussed later in this section.

The index flag and address fields are not required for representing
zero-address instructions.   Nevertheless, it is assumed that a zero-address
instruction also uses a 16-bit word and bits 4 through 15 are ignored.   In
practice, a variable-length format is used for instruction representation, in
order to optimize the memory size needed for program representation.

The assembler translates each instruction in the program to the above
16-bit format, and the control unit interprets the instruction when fetched,
based on the first field (opcode).

10.3.5   AHC Assembler

An assembly language program consists of a sequence of statements (instructions) coded in mnemonics and symbolic addresses.   Each statement consists
of four fields:

LABEL	MNEMONIC	OPERAND	COMMENTS

The *label* is a symbolic address denoting where the instruction itself is located in the memory. It is not necessary to provide a label for each statement in the program. Only those statements that are referenced elsewhere in the program need a label. When provided, the label is a set of alphanumeric characters usually beginning with an alphabetic character. The assembler associates this symbolic address to an actual (*absolute*) address in the memory. The mnemonic field contains 1 of the 16 mnemonics from the instruction set (Table 10.1). In addition to these, there will be some *pseudooperations* coded in this field. These pseudooperations are the instructions to the assembler itself and hence do not translate into an executable instruction. Four typical pseudoinstructions are:

ORG Indicates the beginning address of the program and is always the first instruction in the program.

END Indicates the physical end of the program and is always the last instruction.

BSS Reserves a block of storage. The operand field of this instruction indicates the number of words to be reserved.

BSC Reserves memory words and inserts constants provided as operands into those words

The operand field consists of symbolic addresses, absolute addresses, and IX designation. Typical operands are:

Operand	Description
35	Absolute address 35
35,X	Absolute address 35, indexed
M	Symbolic address M
M,X	M indexed
Z+1	Address of the word next to Z
Z-1	Address of the word before Z
Z+4,X	Z+4 indexed
Z—P	Z and P are symbolic addresses; Z-P yields an absolute address

The *comment field* is an optional field and usually starts with a special character such as "." or "*". It does not affect the instruction in any way and is ignored by the assembler. A complete statement can usually be designated as a comment by using a special character as the first character of the statement.

An assembly language program thus consists of two types of instructions: executable instructions, which translate into the machine code, and pseudoinstructions (assembler directives), which aid the assembler and are not executable. Care must be taken by the programmer to make sure that an assembler directive is not in the execution sequence.

We shall now illustrate various features of assembly language programming with the following examples.

Example 10.3.1.

Instruction	Meaning
LDA M	Load the ACC with the content of memory word M. Content of M is not altered but that of ACC is.
LDA M,X	ACC-[M+(IX)] Content of IX is first added to the memory address M to generate a new address. The data at this new address are loaded into ACC.

The second instruction above shows the indexing feature. By changing the content of IX, the same instruction can be used to access different memory words. Usually the memory words just before or just after the address M are accessed by decrementing or incrementing IX, respectively, at the end of each load before LDA is executed again. The following example illustrates this feature.

Example 10.3.2. The assembly language program shown below accumulates four integers located beginning at M and leaves the result in SUM and is equivalent to the following FORTRAN program:

```
        DATA M/45,23,-8,7/
        SUM = 0
        DO 30 I = 1,4
30      SUM = SUM + M(I)
```

```
                ORG     0
0       BEGIN   LDX     THREE
1               LDA     M
2       LOOP    ADD     M,X
3               TDX     LOOP
4               STA     SUM
5               HLT
6       THREE   BSC     3
7       M       BSC     45,23,-8,7
11      SUM     BSS     1
                END     BEGIN
```

The first column of numbers above are for reference only and not part of the program. The index register is first loaded with a 3. The number at M is now loaded into ACC. The ADD instruction adds the number at M + 3 to ACC. TDX decrements IX by 1. Since it is not zero, the machine jumps to LOOP; adds the number at M + 2; decrements IX, loops; adds the number at M + 1; decrements IX, which is now 0, and hence the result in the ACC is stored at SUM.

The indexing can be used to extend the addressing range. With 11 bits in the address field, only 2K of memory words can be directly addressed. When indexing is used, the content of IX is added to the direct address. Since IX is 16 bits long, a 16-bit number can be added to the 11-bit direct address, resulting in a 16-bit address and thereby addressing 64K of memory.

In the above program, the numbers in the first column are the memory addresses assigned by the assembler to the instructions in the program.

The first instruction is at 0 because of ORG 0, and subsequent instructions each occupy one memory word. The BSC directive defining the block of numbers M requires four words since there are four operands for BSC. SUM consumes only one word, and the content of the word is not defined, whereas the content of the word THREE is 3. BEGIN identifies the first executable instruction and is the operand of the END command; thereby the assembler passes this beginning address to the loader and PC is initialized to 0 just before program execution.

When the program assembly is complete, all the instructions will be in binary form. The machine language version of the above program is:

```
                ORG   0
 0   BEGIN   LDX   THREE      0111  0   00000000110
 1           LDA   M          0000  0   00000000111
 2   LOOP    ADD   M,X        0010  1   00000000111
 3           TDX   LOOP       1010  0   00000000010
 4           STA   SUM        0001  0   00000001011
 5           HLT              1111  -   -----------
 6   THREE   BSC   3          0000  0000  0000  0011
 7   M       BSC   45,23,-8,7 0000  0000  0010  1101
 8                            0000  0000  0001  0111
 9                            1000  0000  0000  1000
10                            0000  0000  0000  0111
11   SUM     BSS   1          ----  ----  ----  ----
             END   BEGIN
```

Locations 0 through 5 contain instructions. Locations 6 through 11 contain data; "-" indicates undefined data bits.

We have provided a brief introduction to digital computer structure. The books listed in the reference section should be consulted for further details. Although high-level programming languages are used in writing simulation programs, knowledge of assembly level language is useful in tuning the timing characteristics of programs. It is possible to control the execution time of the program and also optimize the program for fast execution when it is coded in assembly language. High-level language programs do not give this flexibility.

10.4   TYPES OF DIGITAL COMPUTER SYSTEMS

A digital computer system can be classified as a microcomputer, a mini-computer, a large-scale computer, or a supercomputer. It is difficult to list a set of characteristics that help place a given system into one of these classes, since the progress in hardware and software technologies has blurred the distinctive characteristics. For instance, a microcomputer of the eighties can provide the same processing capability of a large-scale system of the sixties or the minicomputer system capabilities of the seventies. Table 10.2 lists some characteristics of contemporary computers. As can be seen, there is considerable overlap in the characteristics of these classes. Architectural features found in large-scale machines eventually appear in mini- and microsystems, and hence the concepts described in this chapter apply equally well to all classes of computers.

Physical size probably is the most distinguishing feature. The process-ing unit of supercomputers and large-scale machines occupies one or more

Table 10.2 Characteristics of Contemporary Computers

Characteristic	Supercomputer	Large-scale computer	Minicomputer	Microcomputer
CPU size	Not easily portable	Not easily portable	About 2' × 3' rack	8" × 12" board
Cost of minimum system	>$1M	>$100K	$15–90K	$40–500
Word size (bits)	64	32–64	16–32	4–32
Typical processor cycle time (μsec)	0.01	0.05–0.75	0.2–1	1
Typical application	High volume scientific	General purpose	Most dedicated some general purpose	Dedicated (controller)

equipment racks and that of a minicomputer system is relatively small. The processing unit of a microcomputer is an integrated circuit chip, and the whole system fits on a circuit board.

The minimum system configuration of a microcomputer system is the processing unit (CPU) chip itself; that of a mini is usually the CPU and around 64–256 kilowords of memory and a system console; that of a large-scale system is the CPU, several megawords of memory, and a console; and the supercomputer requires another mini or large-scale machine as its front end.

10.4.1 Architecture Classification

Figure 10.6 shows the four logical classes (Flynn, 1972) of computer architectures. The classification depends on the number of *instruction streams* and *data streams* found in the system. A system can exhibit any or all of these classes of operation. As such, this classification does not clearly classify modern day computers into nonoverlapping classes. We shall use these classifications to introduce the architectural concepts found in contemporary computer systems.

The microcomputer system described in the previous section is a single instruction, single data stream (SISD) system. This class of architectures covers all the scalar machines that do not provide for parallel processing of data and typically are uniprocessor machines.

The single instruction, multiple data stream (SIMD) architectures contain a single control processor that controls several data processors. Each data processor will be executing the same instruction at any given time, although on its own data stream. This is the class of *vector processors* useful in parallel computation of vector-oriented data. If the computations can be adopted to vector processing, a processing step that would take N^2 time units in a scalar processor can be accomplished in N time units in an SIMD with N data processors. (See Figure 10.6b.)

The multiple instruction, single data stream (MISD) architectures are known as *pipeline* machines. Almost all modern computer systems adopt pipelining, to gain speeds either during instruction fetch and decode cycle or in their arithmetic logic units. (See Figure 10.6c.)

The multiple instruction, multiple data stream (MIMD) classification encompasses systems ranging from simple systems with a CPU and a data channel to networks of computers. In a system with a CPU and a data channel, the CPU initiates the data channel and continues with its processing activities while the channel is busy gathering data; thereby two instruction streams are at work at the same time on two data streams. Systems with more than one processor operating on independent data units are *multiprocessor* (MIMD) systems.

As can be seen, the last three classifications above enhance the speed of the computer system by adopting the parallel processing concept. The *coprocessor* concept introduced by some microcomputer systems (Intel 8086/ 8087, for instance) is MIMD in nature. In these systems, the coprocessor takes over the floating-point-oriented computations while the main processor continues its processing of fixed-point data. In an *array processor* system the host processor bundles the computations and parcels them to one or more specialized processor arrays. The processors are special-purpose processors that can execute a special set of operations fast (Floating Point Systems FP120 series, for example). When more and more processing

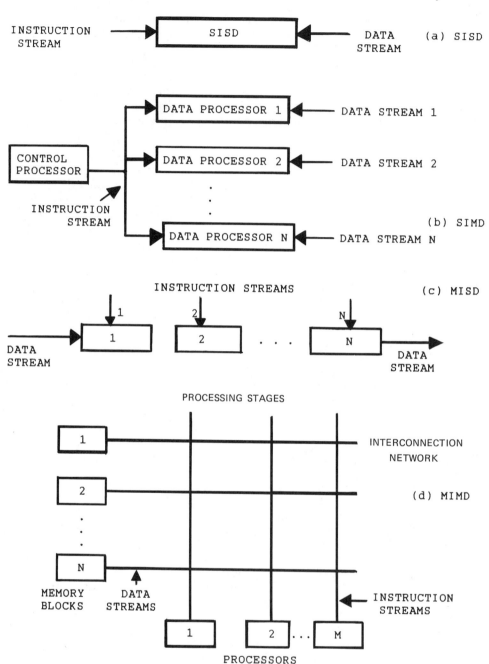

Figure 10.6   Architecture classes.   (a) SISD.   (b) SIMD.   (c) MISD.
(d) MIMD.

activities are transferred to the array processors, the resulting system is a *peripheral processor* system. In such a system, the peripheral processors perform a designated set of operations while the main or the host processor performs the computation-intensive functions. When the peripheral processors are made general enough to handle all computations, the system becomes a true MIMD system. A *distributed processing* system is an MIMD system in which not only the data and processing, but also the control, are distributed among the processors. Supercomputers, such as CDC STAR-100 and Texas Instruments' Advanced Scientific Computer (ASC), are pipeline machines. The CRAY series of supercomputers exhibit all four modes of operation described above (Stone, 1980; Hwang and Briggs, 1984). Chapter 16 further describes the multiprocessor systems.

10.5 A/D AND D/A CONVERTERS

The control of a physical system with a digital computer requires that data taken from the system which is normally in analog form be converted into digital form. Therefore, a device that continuously converts analog information into equivalent digital words is needed. For this purpose samples are taken from the system outputs at specific times and converted to digital word using analog-to-digital (A/D) converters. On the other hand, the computed digital signals in the digital processor need to be converted into analog form using digital-to-analog converters (D/A), before being applied to the actual physical system. Chapter 11 utilizes such converters in implementing digital control systems. A large number of such devices have been developed and are in use. We shall consider next examples of A/D and D/A converters.

10.5.1 Analog-to-Digital Converters

An A/D converter usually consists of two sections: a sample-and-hold section followed by an A/D converter section. The purpose of a sample-and-hold section is to take a sample from the input analog signal and hold it for a short time to allow the A/D converter to complete its function before the next sample is taken. Figure 10.7 shows a typical sample-and-hold device (zero-order hold) using an operational amplifier. During the sampling interval the sampler switch closes momentarily, and the capacitor C charges up and tracks the input analog voltage x(t). When the input signal is disconnected by the sampler, the capacitor continues to hold, for a short time,

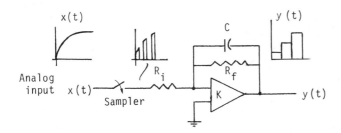

Figure 10.7 A sample-and-hold device. R_i = input resistor. R_f = feedback resistor. K = OP amplifier.

the input signal value immediately before disconnection. In practice, the
output of the hold device is not constant in response to an input signal,
but rather decays exponentially with a large time constant. The effectiveness of the zero-order hold as a device, to stretch the input pulses into a
rectangular wave of a certain width, depends on the sampling frequency
(Kuo, 1963). The zero-order hold function improves by increasing the
sample rate.

The A/D converter function is to convert the sampled analog signal,
which is a piecewise constant, into its equivalent digital signal. For this
purpose, the sampled input signal is divided into a number of finite levels
which can be represented by an n-bit digital word. There are many different forms of A/D converters. One typical A/D converter using a counter
is shown in Figure 10.8. A continuous sequence of equally spaced clock
pulses is applied to an AND gate. The AND gate is normally closed, but
it stays open as long as the output of the comparator is high. The number
of clock pulses passing through the gate is recorded by a binary counter.
The number of recorded pulses increases linearly in time, and the binary
number representing the counter output is converted to an analog signal
(v_d) through a D/A converter. Signal v_d is continuously compared with
the input signal y (y is the output signal from the sample-and-hold device).
As long as the sampled analog input signal y is greater than v_d, the comparator output is high and the counter receives the clock pulses continuously.
When v_d exceeds y, the comparator output drops and the counting process
stops, as shown in Figure 10.8b. The counter output at this time is
approximately equal to the input signal $y = v_d$ and can be read out as a
digital word representing the analog signal at the time of the sampling.

10.5.2   Digital-to-Analog Converters

A D/A converter is a device that accepts a digital word or code as an input
signal and converts it into an analog voltage or current signal. The input
digital word is usually an n-bit binary or binary-coded decimal (BCD) word
in the form $(a_{n-1}a_{n-2} \cdots a_0)$, where a_{n-1} is the most significant bit (MSB)
and a_0 is the least significant bit (LSB). The D/A converter output v_{out}
is proportional to the input binary code. In general v_{out} is given by a
weighted linear combination of the digital bits in the form

$$v_{out} = (a_{n-1}2^{-1} + a_{n-2}2^{-2} + \cdots + a_0 2^{-n})v_{cc} \qquad (10.1)$$

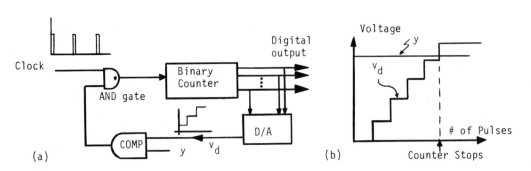

(a)

(b)

Figure 10.8   An A/D converter with a counter.

where v_{cc} is the reference voltage. Consider, for example, a four-bit input binary code $(1010)_2$. The D/A converter output voltage for the given binary code is

$$v_{out} = (1 \times 2^{-1} + 0 \times 2^{-2} + 1 \times 2^{-3} + 0 \times 2^{-4})v_{cc} = 0.625 \, v_{cc} \text{ volts}$$

There are also different forms of D/A converters. One simple parallel D/A converter circuit is shown in Figure 10.9. The blocks M_0, M_1, ..., M_{n-1} are 2×1 multiplexers (MUX). Each MUX has two inputs, one output, and one control terminal. The MUXs act as two-way electronic switches, transferring signal from a selected input terminal directly to the output terminal. The input is selected by a proper signal applied to the control terminal. The input binary word to the D/A converter controls the operation of the MUXs. When an input bit is 1, the corresponding MUX resistor is connected to the reference voltage v_{cc}. A zero-bit input connects the MUX resistor to ground. The output analog voltage is proportional to the input digital word. The D/A converter output is positive for a negative reference voltage v_{cc}. For converting both positive and negative binary numbers represented in sign-magnitude form, the MSB that represents the sign of the number can be used to select the sign of the reference input voltage v_{cc}.

Example 10.5.1. Consider a four-bit D/A converter similar to the one shown in Figure 10.9. A five-bit binary word in the form $N = (a_4 a_3 a_2 a_1 a_0)$ is applied to the D/A converter. The MSB represents the sign of the input number $(0 = "+", 1 = "-")$, and the reference voltage is $v_{cc} = \pm 12$ volts. Assume that $a_1 = a_0 = 1$ and all other bits including, the sign bit, are zero. The output signal is given by

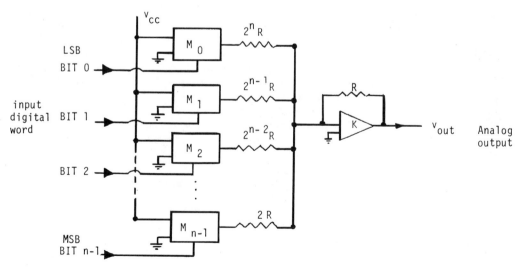

Figure 10.9   A D/A converter.

$$v_{out} = +(8\,a_3 + 4\,a_2 + 2\,a_1 + a_0)\,v_{cc}/16 = +36/16 \text{ volts}$$

The MSB selects the sign of the output signal.

10.6   SAMPLING/QUANTIZATION EFFECTS

The quality of representation of a continuous signal by the sampled and digitized signal is determined by several factors such as sampling rate, sampling pulse width, and quantization accuracy. These factors can cause unacceptable distortions in the sampled signal if the sampling system is not properly designed. In the following sections we explore the sampling and quantization effects on the sampled data system performance.

10.6.1   Sampling Effects

The basic problem in discrete-time processing of analog signals is how to sample a continuous signal to be able to reconstruct the signal from its samples after processing. The number of samples taken from an analog signal (sample rate) for a digital system directly affects the system performance and cost of the design. The sampling theorem developed by Shannon and Nyquist states that in order to recover a signal from its samples, the sample rate must be at least twice the signal bandwidth. This minimum sample rate prevents distortion due to aliasing (overlapping of frequency spectrum) (see Katz, 1981). In practice, however, the theoretical lower bound of sampling rate is not sufficient for a specific time response. This is mainly due to the plant dynamics and system open-loop behavior between the samples. Increasing the sample rate helps the discrete signal converge to its corresponding analog signal more accurately. In practical applications, a sampling rate of 3 to 4 or even as high as 20 times the bandwidth of the analog signal is not uncommon (Franklin, 1980). The performance factors that provide a lower limit to the sample rate are as follows:

1. Time response requirement: For practical purposes the theoretical lower limit of sampling at twice the highest frequency contained in the input signal is not sufficient for a desired time response in a system. The lower bound is limited by the specification to track a certain input command signal. The higher sampling rate reduces the delay in the system response for a sudden change in the input signal and results in a much smoother system output.

2. Sensitivity: In almost all real implementations the system model that is used for simulation and control differs more or less from the actual physical system. The system parameters are subject to variations any time during the processing. Digital systems tend to be more sensitive to inherent modeling errors and parameter variations compared to their continuous counterparts. This increase in sensitivity is due to the open-loop behavior of the system between the samples. The system sensitivity to modeling errors and parameter variations increases as the sample rate is reduced.

3. External disturbances: A digital system must be able to accommodate the external disturbances acting on the system at least to some extent. The disturbance accommodation aspect of an analog system will be degraded when it is implemented digitally. This is due to the fact that the sampled values do not coincide with the continuous signal all the time except at the sampling instant. If we choose a very high sample rate compared to the frequencies

contained in the disturbance input, the performance loss is expected to be insignificant for the digital system. On the other hand, if the sampling interval is very long compared to the frequencies contained in the disturbance, the system response to disturbance can be very poor. Therefore, the effect of sample rate in the disturbance accommodation property of a digital system is an important factor in selecting the proper sample rate for a system.

In general, the main factor that influences the designer to select a lower sample rate for a digital control system is the cost of the design. Fewer samples mean more time for computation between samples and hence slower and much cheaper computers for controller implementation or more computing capability for a given control computer. The lower sample rate also effects the cost of A/D and D/A converters. A lower sample rate results in a slower signal conversion rate and therefore less costly A/D and D/A converters.

In summary, the system performance and cost of the design are two main factors that influence a designer in selecting a sample rate for a digital system. Reducing the sample rate degrades the system performance and increasing the sample rate results in a more costly design. In selecting a proper sample rate for an application, the designer is often forced to make a compromise. The best sample rate for a digital control system is the slowest possible rate that meets all performance requirements.

10.6.2 Quantization Effects

In a digital control system the system outputs and variables are sampled at certain specific times, and a digital word is formed that represents the amplitude of the signals at that time. Real numbers and coefficients are also represented by digital values with finite accuracy. As a result, not only are the variables and system states used in the control processor not exact and accurate, but also the control dynamics are not represented exactly as designed. This means the processor is handling a slightly different system compared to the actual system. This introduces an additional performance loss in overall system operation.

In the previous section we described the effects of sampling rate on system performance and showed that selection of a proper sample rate minimizes the possible performance degradation. In this section we shall explore the effects of quantization on system performance. The conversion from a continuous sampled signal to a digital signal requires that the signal voltage be divided into a finite number of levels which can be represented by an n-bit digital word. With an n-bit digital word 2^n different voltage levels can be defined. The number of bits needed in an A/D conversion depends on the range of signal levels a system must operate with and the conversion accuracy for a certain application. Theoretically, for an exact conversion an infinite word length is required. Therefore, a finite word length introduces an additional error (quantization noise) in the system operation. The quantization noise N_q is defined by the difference between a sampled signal and its corresponding quantized signal. Figure 10.10 illustrates the quantization error for an A/D converter. The ratio of peak signal power P to mean quantization noise N_q is a function of the digital word length (assuming uniform quantization and uniform distribution of the signal voltage). Thus,

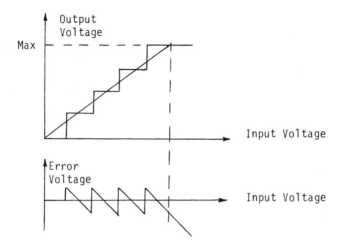

Figure 10.10   Quantization error.

$$\frac{P}{Nq} = 2^n - 1 \tag{10.2}$$

or in db

$$\frac{P}{Nq} = 6n \text{ for } n > 2 \tag{10.3}$$

(see Ahmad and Natarajan, 1983). An increase in the word length reduces quantization noise, and for a proper choice of word length, performance loss due to quantization can be insignificant. Chapter 11 discusses the effects of finite word length in system performance in detail.

10.7   IMPLEMENTATION ASPECTS

One of the major difficulties facing designers in system simulation is the actual implementation of the analog or digital system algorithm. The credibility of a simulation and the accuracy of its results are greatly affected by the selection of the digital or analog hardware, word length, memory size, sampling rate, speed of simulation, etc. Some of these problems were discussed earlier in this chapter and will be considered in more detail in Chapter 11. Here we present a general approach for analog or digital system implementation and computer programming.

10.7.1   Analog Simulation

Analog computers have been used for continuous-time system simulation and in hybrid simulation (where the analog computer interacts with the digital hardware). Chapter 4 discusses simulation techniques in more detail. In this section we mainly explore different techniques for converting a given continuous time algorithm to an analog computer program and other practical aspects of analog simulation, such as amplitude and time scaling.

10.7.2 Analog Computer Programming

The continuous-time systems are usually represented either in time domain by a set of first-order differential equations (state equations) or in the frequency domain by a transfer function. If the system is represented by its state equations, it can be directly implemented on an analog computer. The system state variable diagram is the same as the analog computer simulation diagram. The state variables are often chosen to be the integrator outputs. For a continuous system defined by its overall transfer function, the computer simulation diagram can be derived four different ways: (1) direct programming, (2) parallel programming, (3) cascade programming, and (4) M programming. The basic elements used in all of these methods are integrators, potentiometers, amplifiers, and summers. Differentiators are avoided in analog and digital simulations since they are very sensitive to the noise that might be present in the signals (D'Azzo and Houpis, 1981).

To illustrate the different programming techniques, consider an nth-order linear continuous-time single input/output system with an overall transfer function

$$\frac{Y(s)}{U(s)} = \frac{b_m s^m + b_{m-1} s^{m-1} + \cdots + b_1 s + b_0}{s^n + a_{n-1} s^{n-1} + \cdots + a_1 s + a_0} \tag{10.4}$$

where $Y(s)$ is the system output, $U(s)$ is the input, and $n \geqslant m$. a_i and b_i are constant coefficients.

Direct Programming. For direct programming Eq. (10.4) is cross-multiplied and all the terms containing s are moved to the left-hand side of the equality in the form

$$(s^n + a_{n-1} s^{n-1} + \cdots + a_1 s)Y(s) - (b_m s^m + \cdots + b_1 s)U(s)$$

$$= b_0 U(s) - a_0 Y(s) \tag{10.5}$$

Now, $Y(s)$ can be computed from Eq. (10.5) by performing the following two steps successively: dividing both sides by s, followed by moving all the terms not containing s to the right-hand side of the equation. After n such operations, $Y(s)$ can be obtained in the form

$$Y(s) = 1/s\{\cdots + 1/s\{b_1 U(s) - a_1 Y(s) + 1/s[b_0 U(S) - a_0 Y(s)]\}\cdots\} \tag{10.6}$$

The analog computer diagram derived form Eq. (10.6) is shown in Figure 10.11. The integrator outputs x_1, x_2, . . ., x_n may be assumed as system state variables. The initial values for integrators can be computed from the system given initial conditions. This method is also known as rectangular programming method.

Parallel Programming. In parallel programming, the transfer function Eq. (10.4) is expanded as a sum of some simpler factors in the form

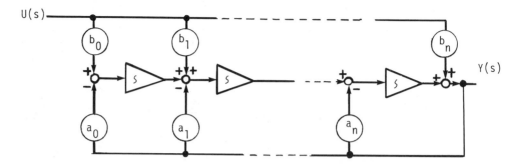

Figure 10.11   Direct programming.

$$\frac{Y(s)}{U(s)} = M_0(s) + M_1(s) + \cdots + M_p(s) \qquad (10.7)$$

where, for the case of a system with distinct denominator roots (characteristic roots), $M_i(s)$ is a first- or second-order factor in the form

$$M_i(s) = \frac{Y_i(s)}{U(s)} = \frac{b_i}{s + a_i}$$

or

$$M_i(s) = \frac{Y_i(s)}{U(s)} = \frac{c_i s + d_i}{s^2 + a_i s + b_i} \qquad (10.8)$$

The computer diagram for the transfer function Eq. (10.4) can be derived by programming each M(s) factor in Eq. (10.7) separately and combining them in parallel, as shown in Figure 10.12. The computer diagrams for typical $M_i(s)$ blocks are given in Figure 10.12a and b.

Cascade Programming. For programming the transfer function Eq. (10.4) using the cascade programming method, it is factored out into product of some simpler terms

$$\frac{Y(s)}{U(s)} = P_0(s) \cdot P_1(s) \cdots P_q(s) \qquad (10.9)$$

where $P_i(s)$ is in general a first- or second-order factor similar to those given by Eq. (10.8). The cascade computer program derived from Eq. (10.9) is shown in Figure 10.13, where $P_i(s)$ blocks are similar to the first- or second-order blocks given in Figure 10.12.

M Programming. In computer programming using the M method, the numerator and denominator of Eq. (10.4) is divided by s^n (the highest-degree term appears in the denominator). Eq. (10.4) now has the form

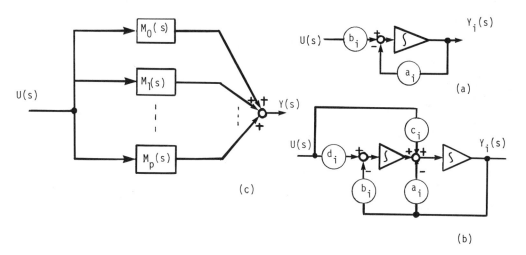

Figure 10.12   Parallel programming.

$$\frac{Y(s)}{U(s)} = \frac{b_m s^{m-n} + b_{m-1} s^{m-1-n} + \cdots b_1 s^{1-n} + b_0 s^{-n}}{1 + a_{n-1} s^{-1} + \cdots + a_1 s^{1-n} + a_0 s^{-n}} \qquad (10.10)$$

or

$$Y(s) = (b_m s^{m-n} + b_{m-1} s^{m-1-n} + \cdots + b_1 s^{1-n} + b_0 s^{-n}) M(s) \qquad (10.11)$$

where

$$M(s) = U(s)/(1 + a_{n-1} s^{-1} + \cdots + a_0 s^{-n}) \qquad (10.12)$$

It follows from Eq. (10.12) that

$$M(s) = U(s) - a_{n-1} s^{-1} M(s) - \cdots - a_0 s^{-n} M(s) \qquad (10.13)$$

The computer diagram obtained from Eq. (10.11) and (10.13) is shown in Figure 10.14. The simulation diagram requires only two summers regardless of the order of the system.

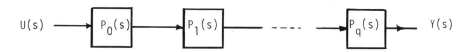

Figure 10.13   Cascade programming.

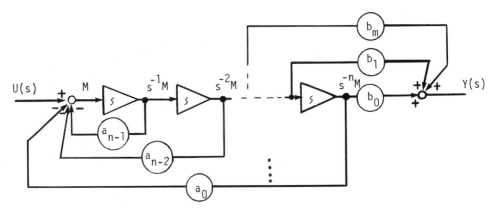

Figure 10.14   M programming.

Example 10.7.1. Consider a third-order single input/output linear continuous-time system with an overall transfer function

$$\frac{Y(s)}{U(s)} = \frac{s + 1}{s^3 + 9s^2 + 26s + 24}$$

Direct programming method. Following the procedure given in this section, the system output is expressed by an equation similar to Eq. (10.6):

$$Y(s) = 1/s\{-9\ Y(s) + 1/s\{U(s) - 26Y(s) + 1/s[U(s) - 24Y(s)]\}\}$$

The analog computer program for the example system is shown in Figure 10.15.

Parallel programming method. For parallel programming, the example system transfer function can be written as a sum of three first-order transfer functions:

$$\frac{Y(s)}{U(s)} = \frac{-0.5}{s + 2} + \frac{2}{s + 3} + \frac{-3/2}{s + 4}$$

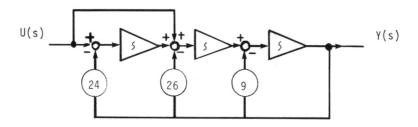

Figure 10.15   Simulation using direct programming.

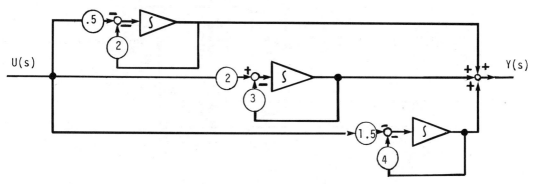

Figure 10.16   Simulation using parallel programming.

Now, each of the first-order terms can be simulated separately using the
direct programming method. The overall system analog diagram can be obtained by parallel combination, as shown in Figure 10.16.

Cascade program method. The example system transfer function can
be written as a product of three first-order factors as follows:

$$\frac{Y(s)}{U(s)} = \left(\frac{s + 1}{s + 2}\right) \cdot \left(\frac{1}{s + 3}\right) \cdot \left(\frac{1}{s + 4}\right)$$

Thus, the overall system can be assumed to consist of three first-order
systems connected in cascade. The overall system diagram is given in
Figure 10.17.

M Programming method. The example transfer function can be written
in the following form:

$$\frac{Y(s)}{U(s)} = \frac{s^{-3} + s^{-2}}{1 + 9s^{-1} + 26s^{-2} + 24s^{-3}}$$

Assuming

$$M(s) = \frac{U(s)}{1 + 9s^{-1} + 26s^{-2} + 24s^{-3}}$$

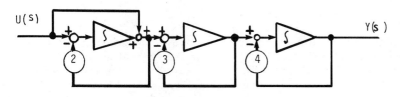

Figure 10.17   Simulation using cascade programming.

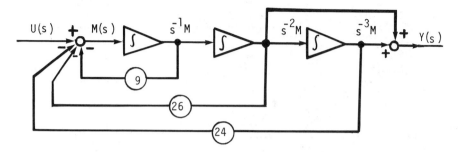

Figure 10.18   Simulation using M programming.

then

$$Y(s) = (s^{-3} + s^{-2})M(s)$$

The computer program using the M method is shown in Figure 10.18.

10.7.3   Amplitude and Time Scaling

Two types of scaling are often required for analog system simulation, namely, amplitude and time scaling. The system variables in the analog computer are required to stay within the computer reference voltages to avoid overloading amplifiers. In general, amplitude and time scaling are very time-consuming and often require estimation of the range of values for system variables and several trial-and-error tests before obtaining an acceptable scaling. Scaling is not a significant problem for digital simulation and in most cases is not required. Amplitude scaling is performed on the system-dependent variables to transform them into analogous machine variables in such a way that all the maximum values of the variables stay within the machine positive and negative reference voltages. Amplitude scaling is also necessary to make the system coefficients small enough to be in the range of values that can be set on the analog computer potentiometers. Time scaling, on the other hand, transforms the real-time simulation into machine-time simulation or, as referred to by Spriet and Ghislain (1982), human-time simulation. Time scaling can be used to speed up the simulation or slow it down for recording or analysis of the simulation results. For almost all analog simulations, amplitude or time scaling or both are required and should be carried out before executing the program. There are different scaling techniques that a designer might use to scale his simulation program. To illustrate the amplitude and time scaling, consider the system defined by differential equation

$$\frac{dy}{dt} + 0.05y = u(t)$$

where u(t) is a unit step input. The corresponding simulation program is given in Figure 10.19a. The coefficients of the differential equation can be set on the pots, but the maximum value of y exceeds machine limits. For a zero slope, y is equal to 20.0. Assuming y_M, $\dot{y}_M$, . . ., etc., are the

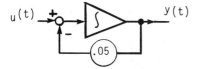

(a)

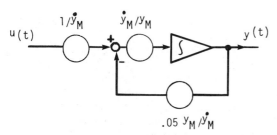

(b)

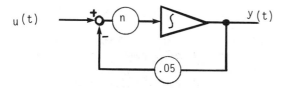

(c)

Figure 10.19   Amplitude and time scaling.

expected peak values for y, $\dot{y}$, . . ., etc., we define a set of normalized variables as follows:

$$x = \frac{y}{y_M}, \quad \dot{x} = \dot{y}/\dot{y}_M, \quad \cdots$$

The normalized variables x, $\dot{x}$, . . . are nondimensional variables lying within the machine voltage limits where

$$\frac{dx}{dt} = \left(\frac{d}{dt}\right)\left(\frac{y}{y_M}\right) = \frac{1}{y_M}\left(\frac{dy}{dt}\right)$$

Assuming a gain of unity on the integrator, we have

$$x = -k \int \dot{x} \, dt$$

where k is an unknown factor that needs to be evaluated. We may write the above equation as

$$\frac{y}{y_M} = -k \int \frac{\dot{y}}{\dot{y}_M} \, dt$$

or

$$y = -k \left(\frac{y_M}{\dot{y}_M} \right) \int \dot{y} \, dt$$

This equality exists if $k = \dot{y}_M/y_M$. The amplitude-scaled diagram is shown in Figure 10.19b. For speeding the simulation up or down we replace the real time with $T = nt$ in the system equation where n is the time scale factor. n can be less than or larger than unity, depending on what type of time scaling one needs to perform. The effect of time scaling in the example system simulation diagram is shown in Figure 10.19c. Note that time scaling affects only the integrator input pots. More discussion of magnitude and time scaling may be found in Section 4.8.

10.7.4   Digital Simulation

Digital systems are usually represented in two different forms, state space form and transfer function form. In state space representation, system dynamics are defined by a set of difference equations, which can be directly programmed on a digital computer or realized by digital hardware, such as delay units, adders, inverters, etc. The state variables for the digital system are often assumed to be the outputs of the delay units. There are various implementation forms for digital systems when they are defined by a transfer function in z^{-1}. This section briefly explores different techniques that are often used to derive simulation diagrams for the digital systems. Chapter 11 discusses in more detail various implementation approaches and their effects on system performance and simulation error; simulation of discrete-time systems is the subject of Section 4.5. The simulation diagram for a digital system defined by its transfer function may be derived in three different ways: (a) direct programming, (b) parallel programming, and (c) cascade programming.

   Direct Programming. The transfer function of a digital system is the ratio of two polynomials in z^{-1} given by

$$\frac{Y(z)}{U(z)} = G(z) = \frac{\displaystyle\sum_{i=0}^{n} p_i z^{-i}}{1 + \displaystyle\sum_{j=1}^{n} q_j z^{-j}} \tag{10.14}$$

Equation (10.14) may be written as

$$Y(z) \left(1 + \sum_{j=1}^{n} q_j z^{-j} \right) = U(z) \sum_{i=0}^{n} p_i z^{-i} \tag{10.15}$$

or

$$Y(z) = U(z) \sum_{i=0}^{n} p_i z^{-i} - Y(z) \sum_{j=1}^{n} q_j z^{-j} \qquad (10.16)$$

Equation (10.16) can now be written as

$$Y(z) = p_0 U(z) + z^{-1}\{p_1 U(z) - q_1 Y(z) + z^{-1}\{p_2 U(z) - q_2 Y(z)$$
$$+ \cdots + z^{-1}\{p_n U(z) - q_n Y(z)\}\} \cdots \} \qquad (10.17)$$

The digital simulation diagram follows from Eq. (10.17) and is shown in Figure 10.20. The state variables $x_1(k)$, $x_2(k)$, ..., $x_n(k)$ are the outputs of the delay units. The difference equation for software realization is given by

$$y(k) = p_0 u(k) + p_1 u(k - 1) + \cdots + p_n u(0) - q_1 y(k - 1)$$
$$- \cdots - q_n y(0) \qquad (10.18)$$

Parallel Programming. For parallel programming the transfer function is first decomposed into partial fractions. Thus Eq. (10.14) is written in the form

$$\frac{Y(z)}{U(z)} = M_1(z) + M_2(z) + \cdots + M_r(z) \qquad (10.19)$$

$M_i(z) = Y_i(z)/U(z)$ is a first-order or second-order term in the form

$$M_i(z) = a_i/(1 + b_i z^{-1})$$

$$M_i(z) = (c_i + d_i z^{-1})/(1 + b_i z^{-1} + a_i z^{-2}) \qquad (10.20)$$

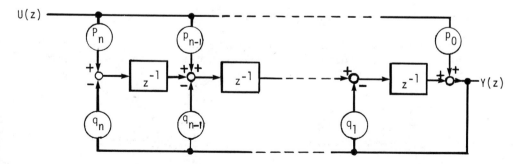

Figure 10.20   Direct programming.

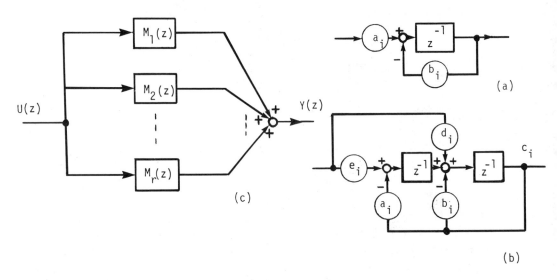

Figure 10.21   Parallel programming.

The simulation diagrams for the above terms are shown in Figure 10.21a and b. Figure 10.21c shows the overall simulation diagram for the given transfer function. The state variables are the outputs of the delay units.

Cascade Programming. In cascade programming the system transfer function Eq. (10.14) is expressed as a product of simple first- or second-order factors:

$$\frac{Y(z)}{U(z)} = G(z) = K \cdot M_1(z) \cdot M_2(z) \ldots \cdot M_r(z) \qquad (10.21)$$

where

$$M_i(z) = (1 + a_i z^{-1})/(1 + b_i z^{-1})$$

or

$$M_i(z) = (1 + a_i z^{-1} + b_i z^{-2})/(1 + c_i z^{-1} + d_i z^{-2}) \qquad (10.22)$$

Figure 10.22 shows the hardware realization for Eq. (10.21).

Example 10.7.2. To illustrate the various ways of deriving the digital simulation programs, consider a closed-loop system with this overall transfer function:

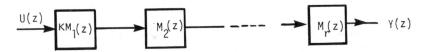

Figure 10.22   Cascade programming.

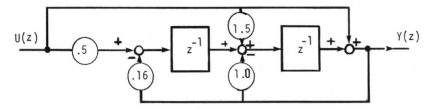

Figure 10.23   Simulation using direct programming.

$$\frac{Y(z)}{U(z)} = G(z) = \frac{(1 + 0.5z^{-1})(1 + z^{-1})}{(1 + 0.2z^{-1})(1 + 0.8z^{-1})}$$

Direct programming.   Following the procedure given in this section, the example system output can be derived from the given transfer function in a form similar to Eq. (10.17).

$$Y(z) = U(z) + z^{-1}\{1.5U(z) - Y(z) + z^{-1}[0.5U(z) - 0.16Y(z)]\}$$

The block diagram in Figure 10.23 shows the details of digital simulation of the example system using the direct programming method.

Parallel programming.   The example system discrete transfer function can be expanded as follows:

$$\frac{Y(z)}{U(z)} = 1 + [0.4/(z + 0.2)] + [0.1/(z + 0.8)]$$

or

$$\frac{Y(z)}{U(z)} = 1 + [0.4z^{-1}/(1 + 0.2z^{-1})] + [0.1z^{-1}/(1 + 0.8z^{-1})]$$

Now, the example system is assumed to consist of three lower-order parallel systems, each defined by the different terms in the expansion. The overall system output is the sum of the outputs of its individual subsystems, as shown in Figure 10.24.

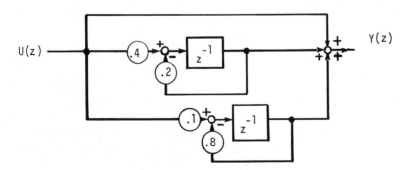

Figure 10.24   Simulation using parallel programming.

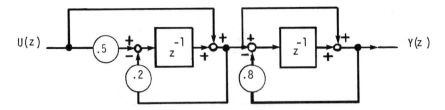

Figure 10.25   Simulation using cascade programming.

Cascade programming.   For cascade programming, the example system transfer function can be written as a product of two first-order factors in the form

$$\frac{Y(z)}{U(z)} = [(1 + 0.5z^{-1})/(1 + 0.2z^{-1})][1 + z^{-1})/(1 + 0.8z^{-1})]$$

Therefore, the original system can be simulated by two first-order systems in cascade.   The overall simulation diagram using the direct programming method for each subsystem is shown in Figure 10.25.

10.8   EXAMPLE OF DIGITAL CONTROL SYSTEM USING MICROCOMPUTER

In this section we shall consider the digital control of a single-link robot arm.   The design objective is to control the arm in such a way to follow exactly the system input torque commands.   For controller implementation, the Intel 2920 signal processor is used.   The 2920 processor is a single-chip microcomputer which has an on-board program memory, scratchpad memory, and I/O circuitry, including A/D and D/A converters.   The 2920 processor is designed to process real-time analog signals.   The programming of 2920 is simple and can be directly derived from the designed system block diagram, using 2920 short assembly language routines.   For further information about the 2920 signal processor unit readers are referred to the 2920 design handbook (1980).

10.8.1   Robot Arm Model

A single-section robot arm is shown in Figure 10.26a.   The robot arm can only rotate about a shaft along the Z-axis, which is perpendicular to the horizontal X,Y-plane.   A simple linear model for the arm is given by Asada and Slotine (1986) in the form

$$\tau(t) = \ell^2 \frac{m}{3} \ddot{\theta}(t) + B\dot{\theta} \qquad (10.23)$$

where $\theta(t)$ is the position of the arm with respect to the X-axis at time t, m is the mass of the arm, which is assumed to be uniformly distributed, ℓ is the length of the arm, B is the joint friction, and $\tau(t)$ is the input torque about the Z-axis.   The open-loop system analog diagram is given in Figure 10.26b.   For servo "command control design" purposes, the input positioning command $\theta_c(t)$ is modeled according to Johnson (1976) by

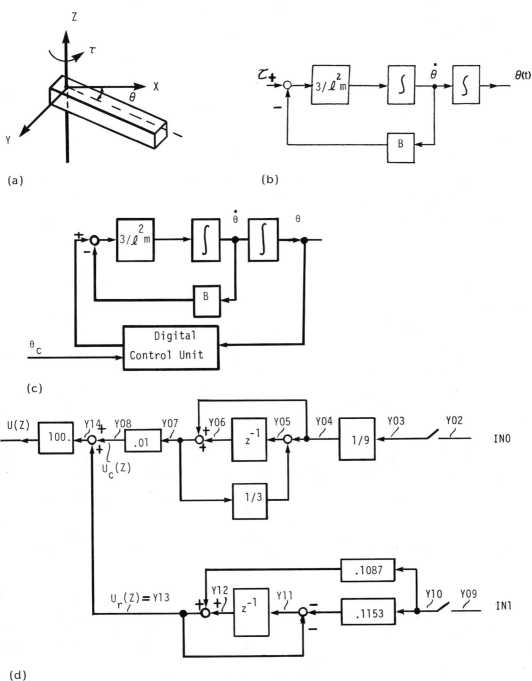

Figure 10.26 A single-section robot arm. (a) Robot arm. (b) Analog diagram of the arm. (c) Arm system with digital controller installed. (d) Digital control unit.

$$\theta_c(t) = q \tag{10.24}$$

where q is an unknown constant coefficient that could randomly jump from one value to another, representing the changes in the input command. Thus, it can be assumed that θ_c is governed by a differential equation in the form

$$\dot{\theta}_c = \delta(t) \tag{10.25}$$

where $\delta(t)$ is a randomly arriving random intensity impulse representing the jumps in q. The servo controller for the robot arm has two different functions. These functions are regulating the arm operation and tracking the unknown input commands. Therefore, the total control action for the robot arm can be written as a sum of these two actions as follows:

$$u(t) = u_c(t) + u_r(t) \tag{10.26}$$

where $u_c(t)$ has the task of tracking the input commands and $u_r(t)$ is to regulate the arm about the desired set point. To implement the servo controller it is required to estimate the arm states $\theta(t)$, $\dot{\theta}(t)$ and the assumed command process states. Hence, to simplify the computations we use the numerical values m = 1, b = 1, and ℓ = 1. Also, we assume that the system states are accessible and only input command states need to be estimated.

The command estimator has the form

$$\dot{\hat{\theta}}_c(t) = -10\hat{\theta}_c(t) + 10\theta_c \tag{10.27}$$

where $\hat{\theta}_c$ is the estimate of the command input θ_c. The servo controller for the arm can be designed using standard regulator design techniques (see, for example, Franklin, 1980) and in general has the following form:

$$u(t) = u_c(t) + u_r(t) = k_1\theta_c + k_2\theta + k_3\dot{\theta} \tag{10.28}$$

where k_1, k_2, and k_3 are constant controller gains. For the numerical values used in this example, the controller gains are $k_1 = 1/3$, $k_2 = -1/3$, and $k_3 = 0.56$.

10.8.2   Analog/Digital Simulation

The robot arm block diagram is shown in Figure 10.26b. The designed servo-controller algorithm Eq. (10.28) will be implemented on Intel 2920 signal processor. The 2920 processor will interact with the analog system (robot) to control the movement of the arm. Figure 10.26c shows the robot arm system along with the digital controller.

For digital computer implementation of the servo controller, first we replace θ_c in Eq. (10.28) by its estimate $\hat{\theta}_c$ from Eq. (10.27). Then, the continuous-time control function is transformed to its equivalent discrete-time equation using bilinear transformation in s-domain in the form

$$s = \left(\frac{2}{T}\right)\left(\frac{z-1}{z+1}\right) \tag{10.29}$$

The discrete-time controller for a sample period of T = 0.1 sec is given by

$$U(z) = U_c(z) + U_r(z) \qquad (10.30)$$

where

$$U_c(z) = [(1/9)(1 + z^{-1})/(1 - z^{-1}/3)]\theta_c(z)$$

$$U_r(z) = [(10.87 - 11.53z^{-1})/(1 + z^{-1})]\theta(z) \qquad (10.31)$$

The discrete-time control system diagram is shown in Figure 10.26d.

10.8.3   Programming 2920 Signal Processor

The 2920 signal processor executes its programs at typically 13,000 times a second with 10-MHz clock and full program memory. For a sample rate of 0.1 sec the program execution needs to be slowed down. This can be done by adding a clock routine which controls the timing of the sampling process. The 2920 can process up to four input signals and up to eight output signals. A limited number of operations are available in 2920, such as ADD, SUB, LDA, etc., and in most cases it is required to scale system coefficients and parameters. The 2920 program simulating the designed robot arm control unit of Figure 10.26d is given below. For more information about programming of the 2920 processor, readers are referred to the 2920 design handbook. The robot arm closed-loop system response for a step and ramp positioning commands are shown in Figures 10.27a and b.

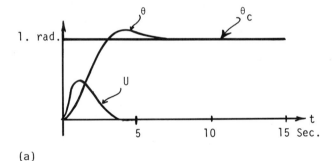

(a)

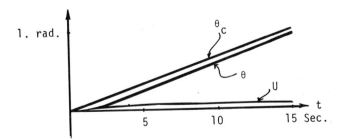

(b)

Figure 10.27   Robot closed-loop response for an input positioning command. (a) Step command input. (b) Ramp command input.

Computer Program

```
              *** CLOCK ROUTINE ***
              *************************

 1. SUB     Y00, KP1, R11, NOP
 2. LDA     DAR, Y00, R00, NOP
 3. LDA     Y01, Y00, R00, NOP
 4. LDA     Y00, KP7, R00, CNDS

         *** POSITIONING COMMAND INPUT ***
         ****************************************

 5. SUB     DAR, DAR, R00, INO     ;CLEAR DAR FOR INO
 6. INO
 7. INO
 8. INO
 9. INO
10. NOP
11. NOP

         *** A/D CONVERSION ROUTINE ***
              AND COMPUTING
            COMMAND CONTROL SIGNAL
         ************************************

12. CVTS                          ;CONVERT SIGN BIT
13. ADD     DAR, KM2, R00, CND6
14. NOP
15. NOP
16. CVT7                          ;CONVERT BIT 7
17. NOP
18. NOP
19. CVT6
20. NOP
21. NOP
22. CVT5
23. NOP
24. SUB     Y08, Y08, R00, NOP    ; START COMM. ROUTINE
25. LDA     Y08, Y07, R07, CVT4
26. ADD     Y08, Y07, R09, NOP
27. ADD     Y08, Y07, R12, NOP
28. LDA     Y07, Y06, R00, CVT3
29. ADD     Y07, Y04, R00, NOP
30. LDA     Y05, Y04, R00, NOP
31. ADD     Y05, Y07, R02, CVT2
32. ADD     Y05, Y07, R04, NOP
33. ADD     Y05, Y07, R06, NOP
34. LDA     Y04, Y03, R04, CVT1
35. ADD     Y04, Y03, R05, NOP
36. ADD     Y04, Y03, R07, NOP
37. CVTO
38. NOP
39. NOP
```

```
              *** SAMPLE-AND-HOLD ROUTINE ***
              *************************************

40.   LDA     Y02, DAR, R00, NOP
41.   LDA     DAR, Y01, R00, NOP
42.   LDA     Y03, Y02, R00, CNDS
43.   LDA     Y06, Y05, R00, CNDS

              *** FEEDBACK SIGNAL INPUT ***
              *********************************

44.   SUB     DAR, DAR, R00, IN1     ;CLEAR DATA FOR IN1
45.   IN1
46.   IN1
47.   IN1
48.   IN1
49.   NOP
50.   NOP

                  *** A/D CONVERSION ***
                    AND COMPUTING
                  FEEDBACK CONTRL SIGNAL
              *****************************

51.   CVTS                          ;CONVERT SIGN BIT
52.   ADD     DAR, MM2, R00, CND6
53.   NOP
54.   NOP
55.   CVT7                          ;CONVERT BIT 7
56.   NOP
57.   NOP
58.   CVT6
59.   NOP
60.   SUB     Y13, Y13, R00, NOP    ;START REG. ROUTINE
61.   LDA     Y13, Y12, R00, CVT5
62.   ADD     Y13, Y10, R04, NOP
63.   ADD     Y13, Y10, R05, NOP
64.   ADD     Y13, Y10, R06, CVT4
65.   SUB     Y13, Y10, R11, NOP
66.   SUB     Y13, Y10, R13, NOP
67.   SUB     Y11, Y11, R00, CVT3
68.   SUB     Y11, Y10, R04, NOP
69.   SUB     Y11, Y10, R05, NOP
70.   SUB     Y11, Y10, R06, CVT2
71.   SUB     Y11, Y10, R08, NOP
72.   SUB     Y11, Y10, R09, NOP
73.   SUB     Y11, Y13, R00, CVT1
74.   SUB     Y14, Y14, R00, NOP
75.   LDA     Y14, Y13, R00, NOP
76.   ADD     Y14, Y8,  R00, CVT0
77.   NOP
78.   NOP
```

*** SAMPLE–AND–HOLD ROUTINE ***
****

```
 79.  LDA    Y09, DAR, R00, NOP
 80.  LDA    DAR, Y01, R00, NOP
 81.  LDA    Y10, Y09, R00, CNDS
 82.  LDA    Y12, Y11, R00, CNDS
```

*** OUTPUT ROUTINE ***
****

```
 83.  SUB    DAR, DAR, R00, NOP    ; CLEAR DAR
 84.  LDA    DAR, Y14, R00, NOP    ; OUTPUT TO DAR
 85.  NOP
 86.  NOP
 87.  NOP
 88.  NOP
 89.  NOP
 90.  OUT0                                          ;OUTPUT TO OUT0
 91.  OUT0
 92.  OUT0
 93.  OUT0
 94.  OUT0
 95.  OUT0
 96.  OUT0
 97.  NOP
 98.  NOP
 99.  NOP
100.  NOP
101.  NOP
102.  NOP
103.  NOP
104.  NOP
105.  NOP
106.  NOP
107.  NOP
108.  NOP
109.  NOP
110.  NOP
111.  NOP
112.  NOP
113.  NOP
114.  NOP
115.  NOP
116.  EOP                                           ; RETURN TO LOCATION 1.
117.  NOP
```

PROBLEMS

10.1. Study the architectural characteristics of a large, a mini-, and a microcomputer you have access to.

10.2. Write AHC assembly language programs to perform the following:

(a) P = X − Y
(b) P = X * Y
(c) P = X/Y

where P, X, and Y are memory locations. Note that, multiplication is repeated addition of multiplicand to itself and division is a repeated subtraction of divisor from dividend.

10.3. Input 10 integers; store the integers in memory starting at location NUM; store the sum of positive integers (only), at location PSUM.

10.4. Find the maximum integer in the block of integers at NUM in Problem 10.3.

10.5. Sort the numbers at NUM from Problem 10.3, in increasing order of magnitude. (HINT: Program for Problem 10.4 can be called 10 times to arrange the numbers in the required order.)

10.6. Certain applications require that the integers be larger than the values that can be represented in 16 bits. Then, two consecutive words are used to represent each integer. Write a program to add two such integers. What changes are needed to the AHC instruction set, to accomplish this addition?

10.7. Obtain the simulation diagram for each transfer function using direct, parallel, cascade, and M realization methods.

(a) $G(s) = 10(s + 2)/s(s + 1)(s + 3)$
(b) $G(s) = k/s(s + 1)(s + 5)$
(c) $G(s) = K(s^2 + 2s + 1)/s(s^2 + s + 1)$
(d) $G(s) = K(s - 1)/s^3(s + 5)$

10.8. Consider the transfer function

$$G(z) = (3 + 5z^{-1} + 5z^{-2})/(1 + z^{-1} + 4z^{-2} - 16z^{-3})$$

Determine the digital simulation diagram (a) using the cascade method, (b) using the direct method, (c) using the parallel method.

10.9. Determine the simulation diagram for the discrete filter

$$G(z) = 1/(z + 5)^2(z + 0.1)$$

Use cascade programming and draw the block diagram.

10.10. Repeat Problem 10.10 using direct programming.

10.11. Consider a continuous-time system defined by the state equations

$$\dot{x}_1 = x_2$$

$$\dot{x}_2 = x_3$$

$$\dot{x}_3 = -2x_1 - 3x_2 + u$$

where x_1, x_2, and x_3 are the system state variables and u is the input signal. Determine the simulation diagram.

10.12. Draw the simulation diagram representing a discrete-time system defined by

$$y(t + 2T) - 5y(t + T) + 6y(t) = u(t)$$

where T is the sampling period. (Hint, use z^{-1} to simulate one step time delay.)

10.13. Repeat Problem 10.12 using

$$y(t + 3T) - 4y(t + 2T) + 2y(t + T) = u(t + T) + 2u(t)$$

REFERENCES

Ahmad, N. and Natarajan, T., (1983). *Discrete-Time Signals and Systems*, Reston, Reston, VA.

Asada, H. and Slotine, J. E. (1986). *Robot Analysis and Control*, Wiley, New York.

Baer, J. L. (1980). *Computer Systems Architecture*, Computer Science Press, Rockville, MD.

Calingaert, P. (1979). *Assemblers, Compilers and Program Translation*, Computer Science Press, Rockville, MD.

D'Azzo, J. J. and Houpis, C. H. (1981). *Linear Control System, Analysis, and Design*, McGraw-Hill, New York.

Flynn, M. (1972). Some computer organizations and their effectiveness, *IEEE Trans. Computers C-21*, 948–960.

Franklin, G. F. (1980). *Digital Control of Dynamic Systems*, Addison-Wesley, Reading, MA.

Hwang, K. and Briggs, F. A. (1984). *Computer Architecture and Parallel Processing*, McGraw-Hill, New York.

Intel 2920 Analog Signal Processor Design Handbook (1980).

Johnson, C. D. (1976). Theory of disturbance—Accommodating controllers, control and dynamic systems, in *Advances in Theory and Applications*, Vol. 12, Academic Press, New York.

Katz, P. (1981). *Digital Control Using Micro-processors*, Prentice Hall, Englewood Cliffs, NJ.

Kobayashi, Y. (1981). Design of a real time, interactive, parallel simulation computer, Ph.D. Dissertation, Ohio State University.

Kuo, B. C. (1963). *Analysis and Synthesis of Sampled-Data Control Systems*, Prentice Hall, Englewood Cliffs, NJ.

Shiva, S. G. (1979). Hardware description lanaguages—A tutorial, *Proc. IEEE 79*, 1605–1615.

Shiva, S. G. (1985). *Computer Design and Architecture*, Little, Brown, Boston.

Sloan, M. E. (1983). *Computer Hardware and Organization*, Science Research Associates, Chicago.

Spriet, J. A. and Vansteenkiste, G. C. (1982). *Computer Aided Modeling and Simulation*, Academic Press, New York.

Stone, H. S., ed. (1980). *Introduction to Computer Architecture*, Science Research Associates, Chicago.

Part III
Applications

11

Digital Control Systems

11.1 INTRODUCTION

The contemporary and economical digital computer is being used in many
and varied environments for the purpose of improving performance across
a wide spectrum of applications. Specific environments reflecting this
operational performance include those found in aircraft systems, automobile
manufacturing, medicine, and process control. Such applications are diverse
and cover a wide spectrum of requirements. The attributes of the general
digital computer, including accuracy, sensitivity, versatility, decision logic ,
size, and flexibility, permit it to function very well in these various environments. In particular, the use of a minicomputer or microprocessor to
control the dynamic response of a physical system model, as presented in
Chapter 2, is an interesting application. In this chapter the discussion
focuses on the theoretical design, implementation, and simulation of a digital
controller through the use of a computer.

Models for physical systems usually consist of a linear time-invariant
set of equations. This type of model, as introduced in Chapters 1 and 2,
provides for easier derivation of control system theory as well as development of physical controllers. This chapter focuses on the design and implementation of discrete (or digital) controllers as compared to continuous
controllers. Specific modeling and analysis techniques include methods involving the time domain and the frequency domain with s-plane (Laplace),
z-plane, and w-plane representations.

Although this chapter presents a large collection of design and implementation techniques for a digital controller, the obvious limitation of space
suggests that additional references be consulted for details and subtler
aspects of digital control system design and implementation. Some current
texts of interest include Houpis and Lamont (1985), Phillips and Nagle
(1984), Iserman (1981), Jacquot (1981), Katz (1981), Franklin and Powell
(1980), Kuo (1980), Biberro (1977), and Cadzow and Martens (1974). For
understanding of continuous-domain controller design D'Azzo and Houpis
is suggested.

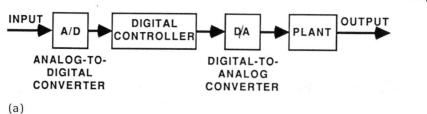

(a)

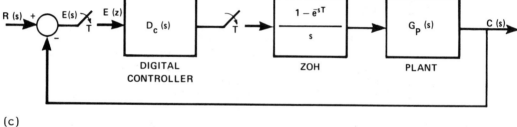

(b)

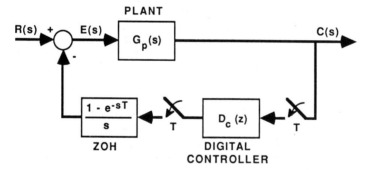

(c)

(d)

Figure 11.1   Digital control system structures.   (a) General digital control system.   (b) Open-loop cascade compensation.   (c) Closed-loop cascade compensation.   (d) Closed-loop feedback compensation.

11.2 BASIC DIGITAL CONTROL STRUCTURES

The general diagram of a digital control system is presented in Figure 11.1a. It consists of four blocks; the analog-to-digital converter (A/D), the digital controller, the digital-to-analog converter (D/A), and the continuous plant model that is to be controlled. Expanded discussion of each of these elements is presented in this and the following sections. The structure of a controller is defined as "open-loop" (Figure 11.1b) or "closed-loop" (Figure 11.1c or d) depending on the inclusion of a "feedback" path from output to input. This figure also indicates the conceptual structure of a digital controller for the various organizations. An open-loop "cascade" controller is defined by Figure 11.1b, a closed-loop cascade controller is shown in Figure 11.1c, and a closed-loop "feedback" controller is depicted in Figure 11.1d. The design techniques associated with these structures vary because of the resulting transfer functions, as discussed in the following sections.

The continuous signal, R(s) or E(s) as shown in Figure 11.1, is sampled with time period T and, thus, converted to a digitized signal through the use of an A/D converter. The digital controller then manipulates the signal in the form of discrete digits. The resulting digitized value is transformed into a continuous signal through the use of the converter. This signal then directly controls the plant.

The A/D converter is usually modelled as an ideal impulse sampler, as presented in Figure 11.2a, which involves a series of unit impulses defined as $\delta(t - kT) = \delta_T(t)$. The general sampled signal x*(t) is depicted in Figure 11.2c resulting from sampling the continuous signal x(t) (Figure 11.2b).

The D/A converter is represented as a sample-and-hold device (see Figure 11.1b); i.e., the resulting controller digitized signal is converted to an analog value which is held for one sampling period T. This nonlinear

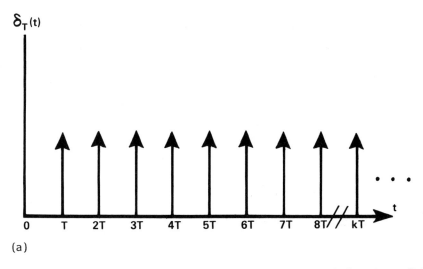

(a)

Figure 11.2 Sampling process. (a) Ideal sampler unit impulse. (b) Continuous signal x(t). (c) Sampled signal x*(t). (d) Effect of ZOH.

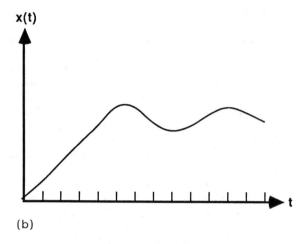

(b)

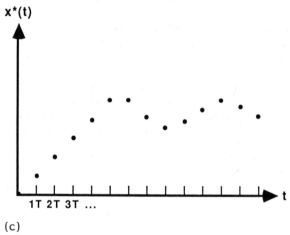

(c)

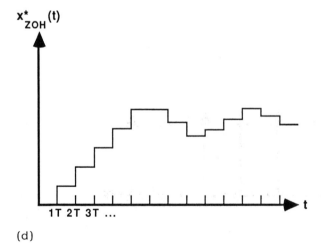

(d)

Figure 11.2   (Continued)

operation is called a zero-order-hold (ZOH), as shown in Figure 11.2d. The ZOH can be modeled by the functional representation $(1 - e^{-sT})/s$, as used in Figure 11.1. Higher-order "holds" use derivative information and are thus more difficult to implement.

The intent of this chapter is to use the general structures presented in Figure 11.1 to present the theory and techniques for the generation of digital controllers. This discussion thus focuses on digital algorithm design, associated modeling characteristics of the converters (A/D, D/A), transformation from the algebraic equation of the digital controller into a computer software program, integration of the program into a digital control system (a sampled-data system), and associated system simulation.

11.3 DESIGN APPROACHES

Various methods are available for modeling a plant and formulating a digital controller. This section presents a capsule discussion of current methods with emphasis on generation of the desired control ratio between input and output. It is assumed that the reader is familiar with the standard s-plane (Laplace) formulation and transformation approaches. The Z-transform discussion of Section 2.3 is extended, since it is a fundamental aspect of digital controller design.

11.3.1 Modeling

The foundation for the development of models of digital controllers is based on linear systems theory, as presented in Chapter 2. One mathematical structure of linear time-invariant difference equations is of the form

$$c(kT) = \sum_{i=1}^{n} b_i c(kT - iT) + \sum_{i=0}^{n} a_i r(kT - iT) \tag{11.1}$$

where r is the input, a_i's and b_i's are constants, and T in seconds is the sampling period ($\omega_s = 2\pi f_s = 2\pi/T$ is the sampling frequency). Note that c/r is defined as the control ratio, which directly relates to the generation of the desired system performance.

Note that Shannon's sampling theorem states that the sampling frequency or rate should be greater than twice the highest frequency in the input signal r(t) if the input signal is to be reconstructed from the samples (Oppenheim and Schafer, 1975). If this condition is met, there are no higher-frequency signals that "aliase" into lower-frequency signals due to sampling. It is assumed throughout this chapter that the digital control developer analyzes the input signal for proper concurrence with this constraint.

Z-Transform. In the development of continuous controllers, linear differential equations representing the plant are Laplace-transformed for ease of manipulation and analysis. In the development of digital controllers the Z-transformation (see Section 2.3) is used to perform the same function. The Z-transformation can be defined as a mapping from the s-plane into the z-plane using the definition $z = e^{Ts}$. Consider the sampling of a continuous

signal x(t) as a modulation process (see Figure 11.2) using the impulse
train

$$x*(t) = \sum_{k=0}^{n} x(kT)\,\delta(t - kT)$$ (11.2)

where x* is defined as the sampled signal x(t). Taking the one-sided
(t > 0) Laplace transform of x* yields

$$X*(s) = \sum_{k=0}^{n} x(kT)e^{-kTs}$$ (11.3)

If $X*(s)\Big|_{z=e^{Ts}} = X(z)$ by definition, then

$$X(z) = \sum_{k=0}^{n} x(kT)z^{-k}$$ (11.4)

This power series is defined as the Z-transform of the discrete signal
x(kT). As discussed in Chapter 2, note that s can be considered as a
differential operator in the continuous domain and z^{-1} can be defined
similarly as a delay operator in the discrete domain. Equation (11.4)
permits a direct mapping from the discrete-time domain to the z-domain.
Using inverse mappings usually in the form of partial fraction expansions
of X(z) and table searches, a discrete-time domain equation (difference
equation) in kT can be derived. Formal inverse mappings exist and use,
for example, contour integration in the z-plane or power series expansion
by long division (Section 2.3). Note also that the independent time variable kT can be replaced, in all time domain equations, by k since T is
a constant. Thus, the abstract time dimension k is shifted depending on
the specific value of T.

Transforming Eq. (11.1) to the z-plane formulation results in

$$\frac{C(z)}{R(z)} = \sum_{i=0}^{n} a_i z^{-i} \Bigg/ \left[1 - \sum_{i=1}^{n} b_i z^{-i} \right]$$ (11.5)

In reality, the objective is to design a digital controller to meet the desired performance as constrained by plant dynamics. The resulting z-plane
controller is transformed into a difference equation, which is then implemented on a digital computer.

As shown in Figure 11.3, the impact of sampling on a continuous signal is reflected in the defining of the primary strip whose frequency
boundaries are, by definition, $\pm\omega_s/2$ or $\pm\pi/T$. In addition, complementary
strips have frequency boundaries at multiples of the primary strip
boundaries. This phenomenon is due to the periodic nature of z as a function of s = $\sigma + j\omega$:

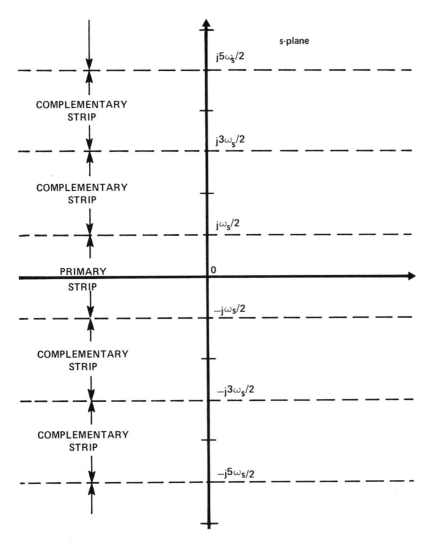

Figure 11.3   Location of primary and complementary s-plane strips for mapping onto z-plane.

$$z = e^{Ts} = e^{T(\sigma + j\omega)} = e^{T(\sigma + j\omega \pm j2\pi n/T)} \tag{11.6}$$

along the imaginary axis in the s-plane for n = 0, 1, ...   Thus, poles and zeros of the s-plane transfer function outside the primary strip are "folded" into the primary strip, possibly causing undesired performance. As in the aliasing situation (Shannon), careful analysis of the plant dynamics must be considered in order to determine the effect of folded poles and zeros.

   Mapping of the s-plane primary strip left of the imaginary axis (see Figure 11.4a) into the z-plane results in a "unit circle," as shown in Figure 11.4b.   The primary strip to the right of the imaginary axis is mapped into the region outside the unit circle, with the s-plane primary

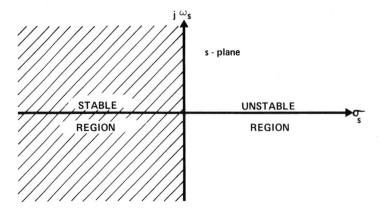

(a)

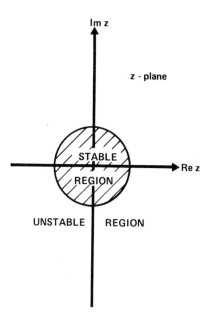

(b)

Figure 11.4   s-plane to z-plane mapping.   (a) s-plane primary strip.
(b) z-plane unit circle.

strip imaginary axis becoming the unit circle boundary. Note that the "complementary strips," as shown in Figure 11.3, are each mapped onto the unit circle. Thus, poles and zeros (roots) of the s-plane transfer function outside the primary strip map into the unit circle due to folding. These folded poles and zeros act as if they were originally in the primary strip, thus changing system performance. This phenomenon can also easily be shown by analyzing the above equations in the frequency domain ($s = j\omega$; Fourier transform) and relates directly to Shannon's sampling theorem for input signals.

The z-plane unit circle is the domain in which performance characteristics are studied. This process is similar to the continuous-domain analysis that is done in the left-half s-plane for stable analog controller designs. For example, constant frequency loci ω_1 and ω_2 correspond to radial lines, as can be derived from Eq. (11.6) and shown in Figure 11.5a. For constant σ_1 and σ_2, the z-plane equivalents are concentric circles, as depicted in Figure 11.5b. In particular, each constant s-plane damping coefficient (a line through $s = 0$) becomes a helix (logarithmic spiral) starting at $z = +1$ and ending at $z = 0$ (Figure 11.5c).

As in the s-domain, similar transformation theorems and properties hold in the z-domain. Examples include arithmetic (addition, subtraction, multiplication), translation (delay by mT), convolution, and initial-value and final-value theorems, as discussed in Section 2.2. The use of these theorems allows the required manipulation of z-domain transfer functions.

Approximations to the exact Z-transformation permit ease of manipulation, as compared to generating the explicit closed-form solution to a power series in z. For example, the Tustin transformation (Tustin, 1947) is defined as

$$s^n = \left[\frac{2}{T} \frac{z-1}{z+1} \right]^n \tag{11.7}$$

which is an approximation to $s^n = [(1/T) \ln z]^n$. Thus, an s-plane transfer function can easily be transformed into a z-plane transfer function by substitution. Of course, the Tustin transformation is an approximation, and thus one should determine its region of "goodness." Analysis of this approximation, using an approximate 5% tolerable error in associated s-plane root location, suggests the following rectangular s-plane region for using the Tustin transformation: left-half s-plane roots should have imaginary values less than $\pm j0.6/T$ and have real values greater than $-0.1/T$. All dominant transfer function poles and zeros should not be outside this region by definition for a "good" Z-transform Tustin approximation. Note that the specific boundary values are functions of T such that the selection of the sampling time can guarantee that all s-plane roots are within the acceptable region. Of course, small values of T may not allow practical implementations. Depending on the allowable tolerance of s-plane root movement (warping), the region suggested above could be enlarged or decreased in size.

Another view of the Tustin transform gives further insight to its performance. Since (1/s) is the integral operator, inverting Eq. (11.7) results in an integral approximation, namely a trapezoid integration rule in the discrete z-domain. Other approximations to the log function represent possible transformations, examples being first-order backward or forward difference equations.

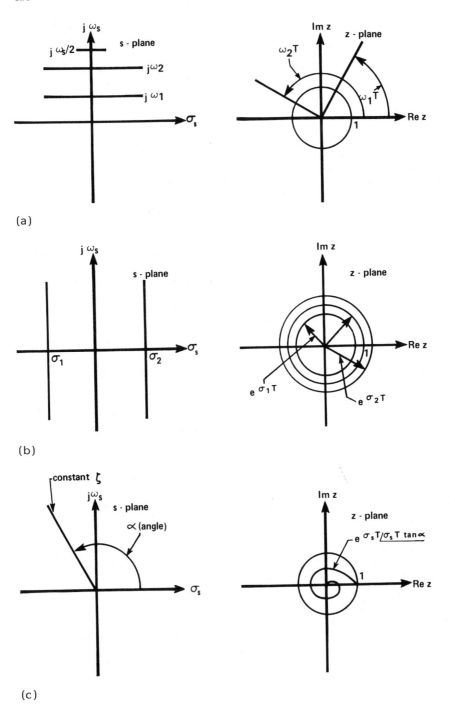

Figure 11.5   s-plane loci mapping to z-plane.   (a) Constant frequency.
(b) Constant real values of s.   (c) Constant damping coefficient.

The w-plane approach to digital control design uses the inverse of the Tustin technique. The W-transformation is defined as

$$w = \frac{2}{T} \frac{z - 1}{z + 1} \qquad \text{or} \qquad z = \frac{Tw + 2}{-Tw + 2} \qquad (11.8)$$

as shown in Figure 11.6. If 2/T is not present, this is the standard "bilinear" transformation. Using the W-transformation, a z-plane transfer function can be mapped into the w-plane using direct substitution. The w-plane is an approximation to the s-plane because of the relationships discussed earlier with the bilinear transformation. In the w-plane, all the standard continuous-domain design techniques are utilized, without having to develop new design techniques for the z-plane assuming the "good" Tustin region contains all dominant roots.

State-Space. State-space methods for modeling digital control systems or sampled-data systems use vector formulation for the system states and matrix transformations to generate next states (Houpis and Lamont, 1985). Section 2.2 of this book develops continuous state-space models, and the discrete-time models are presented in Section 2.3. Simulation techniques for state-space models are discussed in Chapter 4.

In general, the vector-matrix state and output equations for the continuous case are

$$\underline{\dot{x}}(t) = \underline{A}\underline{x}(t) + \underline{B}\underline{u}(t) \qquad \text{state} \qquad (11.9)$$

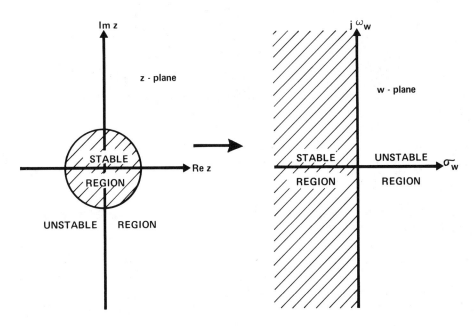

Figure 11.6   z-plane to w-plane mapping.

$$\underline{y}(t) = \underline{C}\underline{x}(t) + \underline{D}\underline{u}(t) \qquad \text{output} \tag{11.10}$$

where the components of the n-dimensional vector x, i.e., x_i, are defined as the state variables. The elements of the r-dimensional input vector u are u_j, and $\underline{y}$ is the system output vector with components y_q with dimension p. Thus,

$\underline{A}$ is defined as the n-by-n dimensional plant matrix
$\underline{B}$ is defined as the n-by-r dimensional control matrix
$\underline{C}$ is defined as the p-by-n dimensional output matrix
$\underline{D}$ is defined as the p-by-r dimensional feedforward matrix

Equations (11.9) and (11.10) represent a linear-time-invariant (LTI), multiple-input-multiple-output (MIMO) continuous-system model. Chapter 2 shows how different vector-matrix formulations can be developed from scalar differential equation models, depending on the selected relationships between state-variables and output variables. Example formulations include phase-variable, canonical-variable, physical-variable forms (D'Azzo and Houpis, 1981) and reduced-order models.

The formulation of the discrete state-space equations is also presented in Chapter 2, with the general time-invariant structure being

$$\underline{x}(kT + T) = \underline{F}\,\underline{x}(kT) + \underline{H}\,\underline{u}(kT) \qquad \text{state} \tag{11.11}$$

$$\underline{y}(kT + T) = \underline{C}\,\underline{x}(kT) + \underline{D}\,\underline{u}(kT) \qquad \text{output} \tag{11.12}$$

where $\underline{F}$ has dimension n by n and $\underline{H}$ has dimension n by r. Again, the state-vector $\underline{x}$ has dimension n by 1, and the control $\underline{u}$ dimension is r by 1. The equation for the output $\underline{y}$ is essentially the same, since it is an algebraic equation in both the continuous and discrete models. Thus, the dimensions for Eq. (11.10) are the same as for Eq. (11.12).

The continuous state-vector model can be transformed to the Laplace domain (s-domain) using the techniques shown in Chapter 2. Similarly, the discrete state-vector model can be transformed into the z-domain, yielding

$$\underline{X}(z) = [z\underline{I} - \underline{F}]^{-1} z\, x(0) + [z\underline{I} - \underline{F}]^{-1}\underline{H}\underline{U}(z) \tag{11.13}$$

$$\underline{Y}(z) = (\underline{C}[z\underline{I} - \underline{F}]^{-1}\underline{H} + \underline{D})\underline{U}(z) \qquad \text{with } \underline{x}(0) = 0 \tag{11.14}$$

The rationale for using this approach is to solve the equation for $\underline{x}$ resulting in the explicit equations

$$\underline{x}(k) = \underline{\Phi}(k,0)\,\underline{x}(0) + \sum_{j=0}^{k-1} \underline{\Phi}(k, j + 1)\underline{H}\underline{u}(j) \tag{11.15}$$

$$\underline{y}(k) = \underline{C}\underline{\Phi}(k,0)\,\underline{x}(0) + \underline{C}\sum_{j=0}^{k-1} \underline{\Phi}(k, j + 1)\underline{H}\underline{u}(j) + \underline{D}\underline{u}(k) \tag{11.16}$$

where $\underline{\Phi}(k,0)$ is called the "state-transition matrix" and satisfies the following relationships:

$$\underline{\Phi}(k,0) = \underline{F}^k$$

$$\underline{\Phi}(0,0) = \underline{I}$$

$$\underline{\Phi}(k,m)\ \underline{\Phi}(m,0) = \underline{\Phi}(k,0)$$

$$\underline{\Phi}(k,m) = \underline{\Phi}^{-1}(m,k) \tag{11.17}$$

Again, kT has been replaced with k. Similar equations exist for the continuous case, as shown in Chapter 2.

The solution to the discrete-system model, as shown in the previous equations, can be used to simulate the system and evaluate system performance if the control vector $\underline{u}$ is known. Open-loop performance can be found by setting the control to zero. To determine an appropriate control vector, a matrix performance criterion, or index, is chosen as a function of $\underline{x}$ and $\underline{u}$. The minimization or maximization of such a criterion as constrained by the system equations is developed in the next section. Questions of state-vector controllability and observability also arise and are important in developing acceptable controllers (Chapter 2).

11.3.2 Design Methods

Design methods for the development of digital controllers are essentially of two major types. The first consists of designing a continuous controller based on the continuous model of the plant using conventional procedures. This analog controller is then discretized for computer implementation. Certain aspects of the discrete domain must be reflected in the continuous plant model before standard design procedures are employed. The other method involves discretizing of the plant and designing the digital controller directly in the discrete domain. Relationships between this discrete-system model and associated theory must be integrated with real-world implementation constraints and physical reality. In this section, a variety of approaches for each general method are introduced.

Although the use of just gain in the cascade or feedback digital controllers of Figure 11.1 may provide acceptable performance, most applications require the design of a dynamic controller or compensator to generate specified system performance. Acceptable performance can be defined in the time domain, frequency domain, or combination of characteristics. A dynamic digital controller is usually developed using trial-and-error procedures for either design methods. Efficiently using a trial-and-error procedure requires the designer to be familiar with computer-aided design tools for control analysis and synthesis, as summarized in Appendix A.

Indirect Method (Digitization). The initial design of the continuous controller requires that the poles and zeros of the plant lie within the "good" Tustin region. Using this design technique requires that a linear model for the digital-to-analog (D/A) block, including the sampler in Figure 11.1, be formulated in the s-plane. Note that the ZOH function $(1 - e^{-Ts})/s$ must be approximated. Modeling these functions is accomplished through the use of the Padé approximation (Houpis and Lamont, 1985). The most predominant model is described by

$$G_{D/A} = \frac{2}{Ts + 2} \qquad (11.18)$$

and $G_{D/A}$ is then used as part of the plant model.

The exact transfer function of the controller D_c is found by using standard continuous-design procedures such as root locus or Guillemin-Truxal (D'Azzo and Houpis, 1981). The root-locus approach requires that zeros and poles be added in lead/lag configurations. In the Guillemin-Truxal method an overall system transfer function (control ratio) is initially generated that incorporates the desired performance constraints. The controller is generated based on this desired system transfer function through algebraic manipulations that also involve the given plant transfer function.

Once an analog controller is generated, the associated time and frequency responses permit comparison with the desired performance characteristics (rise-time, peak-overshoot, settling-time, . . .), as presented in Chapter 2. If performance values are not met, the compensator design step is iterated with a perturbed set of transfer function coefficients, and the design process continues until an acceptable continuous controller is found. At this point, the s-plane controller can be transformed into the z-plane, resulting in a digital controller. This transformation can be accomplished using the exact Z-transformation or the Tustin approximation. The selection depends on the locus of the s-plane roots and the availability of tables, as discussed in Section 11.3.1. Both might be attempted for comparison of different controllers. Next, simulate or test the overall system with the current digital controller. If performance objectives are not met, return to continuous controller design step for better selection of poles and zeros. Finally, transfer $D_c(z)$ into a difference equation for computer implementation. Note that there are additional implementation considerations, such as finite word length and compensator structure, which are discussed in Section 11.4.

Direct Method. The direct development of a digital controller requires that the plant including the ZOH be transformed into the z-domain first before designing a compensator. Design occurs in the z-plane using root-locus techniques or other methods. A particular approach is to select a parameterized version of a discrete-lag, lead, or lag-lead controller network, i.e.,

$$D_c(z) = K(z - a)/(z - b/beta) \qquad (11.19)$$

where beta > 1 results in a lead digital compensator and beta < 1 yields a lag digital compensator. Using this compensator model, the values of "a" (zero) and "b" (pole) that meet performance objectives may be found through trial-and-error.

Note that the existence of an extensive body of knowledge regarding design approaches in the z-plane is limited. Also, the accuracy requirements are much more extensive than in the s-plane owing to the exponential function, an example being the number of digits required to express the s-plane roots and z-plane roots; i.e., $s = -0.234, -0.235$ transforms to $z = +0.791362, +0.790571$ for $T = 1$. For $T < 1$, considerably more digits are required in general.

Another approach is to transform the z-domain model of the plant including the ZOH into the w-plane. Assuming that the poles and zeros of the plant lie within the good Tustin region, s-plane root-locus design techniques can be used to design a compensator since the w-plane has the same structure as the s-plane. Implementation constraints still must be included, however. Many examples of all these approaches to designing a digital controller are presented in the references cited in Section 11.1.

Another aspect of digital compensator development is the analysis of stability. Section 2.4 presents an introduction to this topic. Beyond the obvious root-locus stability regions in the s-, z-, and w-planes, through factorization, one can easily determine whether roots exist in unstable areas of these domains. Examples include Jury's test and Routh's stability criterion (Section 2.4). The use of a control CAD tool is also a fast way of determining whether the overall system is stable and to what extent (stability margins). For state-space models, Section 6.5 presents a discussion of stability properties based on system structure.

11.3.3 Cascade Compensation

Figure 11.1c represents a block diagram of a closed-loop cascade digital controller. As suggested in the previous section, various methods can be employed in designing a digital compensator to meet performance specifications. In this section examples of these approaches are presented. First, the Guillemin-Truxal approach is used based on a desired system transfer function. Next, a lead compensator is developed to meet modified performance specifications.

Consider the following desired system transfer function (control ratio) that is to be used in the digitization (indirect) technique:

$$\frac{C(s)}{R(s)} = \frac{2.0}{(s + 1)(s^2 + 1.7s + 2.0)} \tag{11.20}$$

This model was developed from the required classical performance specifications of rise-time, peak-time, peak-value, and settling time for a step input. Note that the generation of this model for control ratios of order greater than 2 from a given set of performance specifications is not an easy task. Considerable experience and trial-and-error are required. Use of CAD tools and simulation is suggested.

The emphasis here is on minimizing peak-value (overshoot) and controlling damping. Using the Tustin transformation with T = 0.1 sec (a priori selection), the resulting z-domain control ratio from Eq. (11.20) is

$$\frac{C(z)}{R(z)} = \frac{0.000218(z + 1)^3}{(z - 0.9048)(z - 0.9128 \pm j0.1037)} = N(z)/D(z) \tag{11.21}$$

Note that all the s-plane roots are within the good Tustin region (-1, $\pm j6$). The given plant model $G_p(s)$ for this example problem is

$$G_p(s) = \frac{2.0}{s^2 + s + 1} \tag{11.22}$$

with figures of merit shown in Table 11.1. The complete plant transfer function $G_h(s)$, including the ZOH, becomes, from Figure 11.1c,

Table 11.1   Resulting Figures of Merit for Design Problem with Unit-Step
Input

Controller type	Rise-time	Peak-time	Peak-value	Settling-time
Original plant	0.99	2.37	1.31	7.74
Guillemin-Truxal	2.18	4.50	1.01+	4.80
Lead compensator	0.29	0.96	1.26	3.05

$$G_h(s) = G_{ZOH}(s)G_p(s) \tag{11.23}$$

or with the Tustin transformation

$$G_h(z) = \frac{0.001188(z + 1)^3}{z(z - 0.9477 \pm j0.1254)} \tag{11.24}$$

Using the Guillemin-Truxal approach, the control equation is

$$D_c(z) = \frac{N(z)}{[D(z) - N(z)]G_h(z)} \tag{11.25}$$

and the resulting digital controller is

$$D_c(z) = \frac{1.8z(z - 0.9477 \pm j0.1254)}{(z - 1)(z - 0.8659 \pm j0.1197)} \tag{11.26}$$

which can now be transferred into a difference equation for computer implementation using the delay operator.

Note that using this straightforward algorithmic approach, a physically
unrealizable controller may result, since the order of the numerator could
be higher than the denominator. To rectify the situation, one approach is
to replace all the numerator terms (z + 1) with their dc gain constant of
2. Another refinement includes elimination of controller high-frequency
ringing caused by z-plane poles near z = −1. In this case, the poles
(z + b) are replaced by the DC gain value (1 + b) resulting in no significant performance degradation since most plants have low-frequency band-
widths.

The design objective is now modified to reduce the rise-time through a
continuous cascade lead compensator of the form

$$D_c(s) = \frac{s + 0.6}{s + 60.0} \tag{11.27}$$

Using Eq. (11.22) and (11.27) to develop the system transfer function
yields

$$\frac{C(s)}{R(s)} = \frac{5000(s + 0.6)}{(s + 0.9604)(s + 3.28)(s + 15.56)(s + 61.20)} \tag{11.28}$$

The values of the standard lead network of Eq. (11.19) were tuned, based on root-locus analysis, using control CAD tools (see Appendix A); i.e., move the loci of s-plane roots by adding a cascade compensator in order to achieve a reduced rise-time value.

If a smaller peak-value was the main objective, as in the previous Guillemin-Truxal design, the root-locus approach indicates that a higher-order compensator is required. This is shown in the previous example.

Even though the pole of the lead network is outside the Tustin region, its effect on system performance is minimal because of the dominant plant poles.

The digital controller using Eq. (11.27) is

$$D_c(z) = 0.2575(1 - 0.9417z^{-1})/(1 + 0.5z^{-1}) \qquad (11.29)$$

which is simulated using control CAD tools producing the results in Table 11.1.

If the direct method is used with a lead compensator the value of beta is $0.9417/0.5 = 1.88$ [see Eq. (11.19)], which is within the suggested range. To fine-tune the digital controller, simulation with various values of beta and the gain may improve the performance using this trial-and-error procedure. It is suggested that numerous design approaches be tried to determine the "best" controller to implement.

In designing a specific digital control system, the form of the controller relates directly to the requirements objective, as shown above. This phenomenon occurs also in the w-domain design method, where the standard root-locus approach is used. Because of the good Tustin approximation used in this example, the w-plane design would proceed in a very similar direction with similar results expected.

To precisely determine the parameters of the digital controller transfer function, the trial-and-error procedure for pole-zero placement and gain selection can be tedious. An efficient method for fine-tuning a designed controller can be accomplished by defining an error function E in the frequency domain, relating current and desired values of magnitude and phase at selected frequency values; i.e.,

$$E = \sum_{i=1}^{N} [C/R_{mag}(j\omega_i) - C/R'_{mag}(j\omega_i)]^2 W_{mag}(j\omega_i)$$

$$+ \sum_{k=1}^{N} [C/R_{angle}(j\omega_k) - C/R'_{angle}(j\omega_k)]^2 W_{angle}(j\omega_k)$$

$$(11.30)$$

where C/R' is the desired frequency domain system response. This approach is a constraint minimization problem with the form of the controller transfer function being the constraint. If r second-order compensators are used in cascade, then there are $4r + 1$ unknowns. A gradient vector approach is used to solve this problem (Rattan, 1980) usually with a numerical-differentiating computer program owing to the transcendental functions of "e." The selection of the weighting coefficients, W_{mag} and

W_{angle}, is a critical aspect of successful design using this technique.
Usually specific frequency bands are given more weight than others, depending on the application, i.e., control bandwidth versus noise (disturbance) filtering.

11.3.4   Feedback Compensation

Developing a digital controller in the feedback path of Figure 11.1d is generally not as easy as designing a cascade controller and requires more trial-and-error iterations. The use of feedback controllers can considerably enhance the system performance in high-accuracy situations. Applications include vehicle control (aircraft, automobiles), process control (chemical, power systems), disturbance rejection systems, and other real-time environments. The selection of cascade (feedforward) or feedback controller design depends on the application, the performance requirements, the designer's experience, and the availability of appropriate design and simulation tools.

As indicated previously, a continuous-feedback controller can be designed and then discretized, or the plant discretized and a controller designed in the z-plane or w-plane. Essentially, the approach is to select a priori a feedback controller transfer function form and then determine coefficients' values using a trial-and-error procedure. The Guillemin-Truxal technique can also be employed. In either case, the feedback compensator is usually of lower order than a cascade controller with similar performance. This is another reason for selecting a feedback controller design.

Many systems involve rapid tracking or disturbance rejection of plant output signals. In these cases a feedback implementation probably has better performance characteristics, especially with regard to stability and control of response time (rise-time).

In particular, the dominant poles of the plant of Section 11.3.3 [see Eq. (11.22)] are generally more controllable if feedback compensation is employed instead of first-order cascade lead compensators.

The digitization technique in this case again requires an approximation to the ZOH function to design a digital controller from continuous-domain design. With the use of the Padé approximation, the controller $D_c(s)$ can be associated with the ZOH. Specifying a particular form for $D_c(s)$, such as first-order or second-order results in selection of pole-zero values based on root-locus analysis. Simulation again can fine-tune the specific values, which can have substantial effects, even in the least significant bit locations, when poles or zeros are close to the +1 point in the z-plane. The Guillemin-Truxal digitization technique can also be employed using a similar approach to that in Section 11.3.3.

In the Guillemin-Truxal direct design approach, the assumed control ratio is again defined as

$$\frac{C(z)}{R(z)} = \frac{N(z)}{D(z)} \tag{11.31}$$

From Figure 11.1d, the output of the overall system is

$$C(z) = \frac{GR(z)}{1 + G_h(z)D_c(z)} \tag{11.32}$$

where

$$G_h(z) = G_{ZOH}(z) \, G_p(z) \qquad (11.33)$$

If the combined function GpR(z) can be decomposed in a given problem and $G_p(z)R(z)$, assuming the input R(z) is known, then the control ratio becomes

$$\frac{C(z)}{R(z)} = \frac{G_p(z)}{1 + G_h(z)D_c(z)} \qquad (11.34)$$

Thus,

$$D_c(z) = \frac{G_p(z)D(z) - N(z)}{G_h(z)N(z)} \qquad (11.35)$$

Using the same evaluation criteria presented in Section 11.3.3, the various zeros and poles in D_c can be generated and analyzed.

The root-locus approach for the feedback controller direct design can use a lead, lag, or lead-lag structure similar to the cascade case. A second-order transfer function could also be formulated (real or imaginary roots). In either case, the open-loop transfer function $G_h(z)D_c(z)$ is generated and root-locus placement of the poles and zeros of D_c is carried out using z-plane or w-plane techniques (see Houpis and Lamont, 1985). The specific gain and pole-zero values define a digital controller that can be simulated with fine tuning and eventual computer implementation.

In general, the form of a feedback digital controller is simpler than that of a cascade controller in a closed-loop system with similar performance. The design effort, on the other hand, is probably more intensive and inter-active with a CAD package.

11.3.5 PID Controllers

Special forms of lead and lag compensators have been used in the process-control industry for many years and are called proportional-integral-deriva-tive (PID) controllers. Other controllers were presented briefly in Section 2.4.1. PID controllers are used to maintain process "set-points," i.e., to perform the transitions from point-to-point with proper damping or control system responses. The design of a PID controller is usually approached from either the frequency-domain or time-domain performance point of view with appropriate performances matrices. The form of the continuous PID controller is given by

$$D_c(s) = K_p + K_i/s + K_d s$$

$$= \frac{K_d s^2 + K_p s + K_i}{s} \qquad (11.36)$$

where k_p is the proportional gain, K_i is the integral gain, and K_d is the derivative gain. The proportional gain defines the response to be obtained

due to a set-point step input. The integral term in conjunction with negative feedback allows the output steady-state error to become zero for step
and ramp inputs. Damping is controlled by the derivative term.

Various combinations of the three operators are used, such as PI
(proportional-integral) and PD(proportional-derivative) for various purposes.
The PI controller is a lag filter with a pole at s = 0, as can be seen from
Eq. (11.36) with K_d zero. In the frequency domain, the PI compensator
has relatively high low-frequency gain (low-pass filter) and therefore reduces steady-state errors and improves stability margins, which is characteristic of a lag compensator.

The PD controller, on the other hand, is a lead controller with a pole
at infinity which improves system stability and increases frequency bandwidth and therefore the response time. High relative gain is accomplished
in this case at high frequencies. The gain instability at very high frequencies must be modified in the pure PD model with additional poles.

The PID controller with proper selection of gain coefficients can combine the attributes of the PI and PD compensators to represent a lag-lead
filter. Of course, careful selection of K_p is required since it influences
the cutoff frequencies for both the lag and lead portions of the PID controller. Note, also, that for very high frequencies the gain with the
present model increases without bound similar to the PD controller. Thus,
to design a stable system, additional poles must be added to generate a
high-frequency rolloff. Because the high gain is caused by the derivative
term, these rolloff poles are usually associated with it. With this correction, the derivative operation is designed for a smaller-frequency bandwidth.
Thus, the modified continuous PID model becomes

$$D_c(s) = K_p + \frac{K_i}{s} + \frac{K_d s^2}{s^n + b_{n-1}s^{n-1} + \cdots} \qquad (11.37)$$

The order of the denominator of the derivative term depends on the desired rolloff characteristics at high frequency. The added poles are usually
selected to be outside the frequency bandwidth of interest and contribute
little phase-lag at the frequencies of interest.

To design a PID controller, the required performance characteristics
are determined from the desired control application. Rise-time, damping,
peak-overshoot, and final-value are used to determine the PID gain values
with a root-locus approach. Using frequency and root-locus design techniques, an efficient process controller can be designed in a relatively short
time.

By simulating a PID controller and tuning the gain values, a useful
controller can also be developed. This technique is useful when the plant
parameters are not known explicitly and some search is required to determine appropriate gain settings.

For a digital version of a PID controller, Eq. (11.36) can be transferred to the z-domain after the gains are determined using continuous-
domain design methods (digitization), or the design could be developed in
the z-plane (direct). Both approaches, as presented in Section 11.3.2,
can generate effective PID controllers.

The discrete version of the PID controller can be derived from Eq.
(11.36) by approximating both the derivative and integral functions, yielding

$$D_c = \frac{K(z^2 + a_1 z + a_2)}{z(z - 1)} \tag{11.38}$$

where $K = K_p + TK_i/2 + K_d/T$, $a_1 = (TK_p - TK_i/2 + 2K_d)/TK$, and $a_2 = K_d/TK$. The approximations used in this derivation are trapezoidal numerical integration (Tustin) and a first-difference approximation to the derivative. Letting $K' = KK_p$, where K_p is the gain of the plant, yields

$$K_d = a_2 TK'$$

$$K_i = \frac{-2a_1 K'}{T} + \frac{4K_d}{T}$$

$$K_p = K' - \frac{K_d}{T} - \frac{TK_i}{2} \tag{11.39}$$

Considering the previous plant of Eq. 11.22 with $K_p = 0.001188$, the open-loop z-domain transfer function becomes

$$\frac{K'(z^2 - a_1 z + a_2)(z + 1)^3}{z^2(z - 1)(z^2 - 1.8959z + 0.9052)} \tag{11.40}$$

If a_1 is set equal to 1.8959 and a_2 is 0.9052, the plant complex poles are effectively canceled, resulting in

$$K' (z + 1)^3 / z^2(z - 1) \tag{11.41}$$

The root-locus approach is used to determine the appropriate gain K' in order to achieve the desired system performance (see Section 11.3.3). The following parameters for this problem were generated using the control CAD simulation tool ICECAP: risetime < 0.3, peak-value = 1.017, peak-time = 0.7, settling-time = 1.5, and final-value = 1.000. These values are better than those in Table 11.1 since a second-order compensator allowed canceling plant complex poles. As before, fine tuning through gain adjustment or small pole-zero perturbations is possible with interactive simulation CAD tools. The closed-loop control ratio (transfer function) in this case is

$$\frac{C(z)}{R(z)} = \frac{0.04762(z + 1)^3}{(z + 0.1599)(z + 0.4847 \pm j0.2508)} \tag{11.42}$$

In this example the steady-state error is zero for a unit step input. If a ramp steady-state error is desired, then the order of the numerator in Eq. (11.41) must be at least one less than the denominator order. In this case different values of K', a_1, and a_2 are chosen.

For PI and PD controllers the previous approach may be used with the appropriate gain being set to zero before pole-zero placement. The use of CAD simulation tools permits easy design, analysis, and evaluation

of controllers, especially PID structures, with approximate plant parameter
values.

11.3.6   State-Space Controllers

The vector-matrix equations of Section 11.3.1 as presented represent the
general state-space model. Using this model and a performance objective,
a controller structure can be derived. The method of controller derivation
depends on the specific performance objective selected and the mathematical
techniques employed. In particular, the objective may involve development
of a feedback controller to make the digital control system output insensitive
to plant parameter variations (Houpis and Lamont, 1985). Also, "optimal"
controllers can be derived from a performance index or cost function and
the state-space equations. Examples of cost functions are minimum energy,
minimum movement (displacement), and minimum fuel. The specific meaning
of a cost function depends on the physical attributes associated with the
state-space variables, as discussed in Section 11.3.1.

In this discussion, the linear quadratic optimal control approach is
chosen which uses the quadratic cost function

$$J = 1/2 \sum_{k=0}^{N-1} [\underline{x}(k)\underline{Q}\underline{x}^T(k) + \underline{u}(k)\underline{R}\underline{u}^T(k)] + 1/2\,\underline{x}(N)\underline{S}\underline{x}^T(N) \qquad (11.43)$$

which is to be minimized. N is the finite number of time samples. $\underline{Q}$, $\underline{R}$,
and $\underline{S}$ are symmetrical matrices, and $\underline{x}^T$ is the transpose of $\underline{x}$. This
quadratic form permits ease of mathematical manipulation, as is shown in
the following controller development. Note that the design approach results
in a controller (control law) of the form

$$\underline{u}(k) = -\underline{K}(k)\underline{x}(k) \qquad (11.44)$$

In order to obtain the structure of $\underline{K}$, the general calculus-of-variations
technique is used. The cost function J is augmented with the state equation through the use of arbitrary (unknown) Lagrangian multipliers, $\underline{\lambda}(k)$,
yielding

$$J' = J + \underline{\lambda}^T(k+1)][\underline{F}\underline{x}(k) + \underline{H}\underline{u}(k) - \underline{x}(k+1)] \qquad (11.45)$$

The objective is to find $\underline{x}$ and $\underline{u}$ such that the new J formulation is
minimized. Note that the addition of the state equation to J did not change
the minimization problem since zero was added.

To find the minimum value, J' is differentiated with respect to $\underline{x}$ and $\underline{u}$,
resulting in

$$dJ' = \sum_{k=0}^{N-1} \{\underline{x}^T(k)\underline{Q}d\underline{x}(k) + \underline{u}^T(k)\underline{R}d\underline{u}(k) + \lambda^T(k+1)[\underline{F}d\underline{x}(k)$$

$$+ \underline{H}d\underline{u}(k) - d\underline{x}(k+1)]\} + \underline{x}^T(N)\underline{S}d\underline{x}(N) \qquad (11.46)$$

Since dJ' must vanish for arbitrary differentials, each of the differential term coefficients must be zero, resulting in the following equations:

$$\lambda(k) = \underline{Q}\underline{x}(k) + \underline{F}^T\underline{\lambda}(k + 1) \tag{11.47}$$

$$u(k) = -\underline{R}^{-1}\underline{H}^T\underline{\lambda}(k + 1) \tag{11.48}$$

$$\underline{\lambda}(N) = \underline{S}\underline{x}(N) \tag{11.49}$$

Analyzing the backward difference, Eq. (11.47) yields the general form of the solution for λ as

$$\underline{\lambda}(k) = \underline{P}(k)\underline{x}(k) \tag{11.50}$$

Substituting this form into the previous equations yields the discrete Riccati equation for P:

$$\underline{P}(k) = \underline{F}^T[\underline{P}(k + 1) - \underline{P}(k + 1)\underline{H}(\underline{H}^T\underline{P}(k + 1)\underline{H} + \underline{R})^{-1}\underline{H}^T\underline{P}(k + 1)]\underline{F} + \underline{Q} \tag{11.51}$$

with $P(N) = S$ from Eq. (11.49) and (11.50). The controller equation using Eq. (11.48) is

$$\underline{u}(k) = \underline{R}^{-1}\underline{H}^T(\underline{F}^T)^{-1}[\underline{P}(k) - \underline{Q}]\underline{x}(k) \tag{11.52}$$

The inverse of $\underline{R}$ must exist in Eq. (11.52), and thus $\underline{R}$ must be positive definite. Also, $\underline{Q}$ must be positive semidefinite for a solution to exist. Note also that Q and $\underline{R}$ matrices were assumed to be constant over time. However, the same formulation can be used to develop similar equations with these matrices as functions of k. The modified cost function represents changes in relative weighting between the state and controller as a function of discrete time [see Eq. (11.43)]. The important aspects of the optimal control approach is the ration between the elements of Q and $\underline{R}$ and the elements of $\underline{S}$.

Selection of the values for the matrix elements is a difficult task since these values indirectly relate to physical performance, i.e., rise-time, peak-time, settling-time. It is suggested that the initial approach assume that the off-diagonal elements are zero. Various simulations of the overall system with the optimal controller structure can be used to determine acceptable matrix values.

Other approaches to developing the optimal solution use the dynamic programming method (principle of optimality—Bellman) and the maximum principle (Phillips and Nagle, 1984).

The numerical solution to the equations presented in this section is easily found through the use of CAD tools, as discussed previously.

11.3.7 Quantitative Feedback Technique

The qualitative feedback synthesis technique (QFT) is a graphical method that uses feedback to achieve closed-loop system response within perform-ance tolerances, despite plant uncertainty and input disturbances (Horowitz and Sidi, 1972). The range of plant uncertainty and the output performance specifications are quantitative parameters in the frequency-domain design process. The fundamentals of the design method are presented in relation to a single-input, single-output (SISO) design. The multiple-input, multiple-output (MIMO) design procedure can also be generated using the same fundamentals (Horowitz, 1979, 1982). The computational aspects of the development process require a digital computer for design and analysis since the complexity of the design approach necessitates a multitude of symbolic and numerical calculations.

Continuous Problem. The general (SISO) problem involves a plant transfer function, with uncertain parameters (gain, poles, and zeros) known only to be members of bounded sets. Design specifications dictate the desired response of the plant to inputs and expected disturbances. The problem is to design a controller that forces the plant output to satisfy performance tolerances over the range of plant uncertainty and disturbance inputs.

The basic SISO control loop structure is shown in Figure 11.7, where r(t) is the command input to the system and d'(t) and d(t) are disturbance inputs that are to be attenuated. G_p is the plant transfer function, the characteristics of which are not precisely known. The compensator D_c and the prefilter F are to be designed to force the system output c(t) to be an element of an acceptable response set, despite the uncertainty in the plant and the disturbance inputs. The problem is significant in many practical applications.

Access to r(t) and c(t) allows the use of the two degree-of-freedom structures of Figure 11.7 and provides the designer with two independent compensator elements F(s) and $D_c(s)$.

To approach a solution, define the open-loop transfer function or the loop transmission as $L = D_c G_p$, where D_c is assumed to include the ZOH. A continuous design approach is presented first. There are three transfer functions of interest in Figure 11.7, which relate to the system output to the command and disturbance inputs as follows:

$$T_R = \frac{C(s)}{R(s)} = \frac{FD_c G_p}{1 + D_c G_p} = \frac{FL}{1 + L} \tag{11.53}$$

$$T_D = \frac{C(s)}{D(s)} = \frac{1}{1 + D_c G_p} = \frac{1}{1 + L} \tag{11.54}$$

$$T_D' = \frac{C(s)}{D(s)} = \frac{G_p}{1 + D_c G_p} = \frac{G_p}{1 + L} \tag{11.55}$$

The design specifications, or closed-loop system response tolerances, describe the upper and lower limits for an acceptable output response to a desired input or disturbance. Any output response between the two bounds is assumed acceptable. The response specifications must be determined prior to applying a design method. Typically, response specifications are given in the time domain, such as the classical figures of merit or as a bounded region in the time domain. Additional similar bounds are needed if other inputs are to be considered.

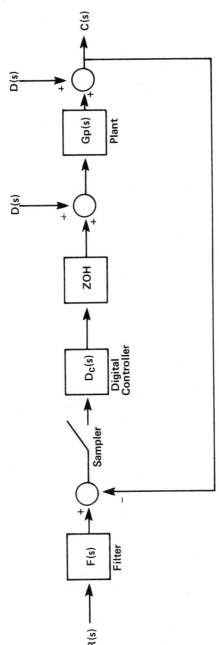

Figure 11.7 QFT SISO control structure.

The design techniques chosen is a frequency-domain approach; therefore, the time-domain specifications must be translated to bounds in the frequency domain. Desired control ratios are modeled to satisfy these frequency-domain performance specifications using the root-locus pole-zero placement method.

The principal steps of the procedure as as follows:

1. Translation of the time-domain specifications on the output C into tracking and disturbance control ratios and subsequent determination of the tracking and disturbance bounds
2. Derivation of the bounds on the loop transfer function $L(j\omega)$ from the tracking and disturbance bounds
3. Synthesis of the loop transfer function $L(j\omega)$ from the results of step 2
4. Derivation of the prefilter $F(j\omega)$
5. Modification of $L(j\omega)$ and $F(j\omega)$, if necessary, to achieve performance objectives.

For response to a step input, a control ratio example of $T_R = L/(1+L)$ for a range-of-plant uncertainty can be read directly from the "curved scales." In a first-order model (Figure 11.8), the range-of-plant gain is from A to

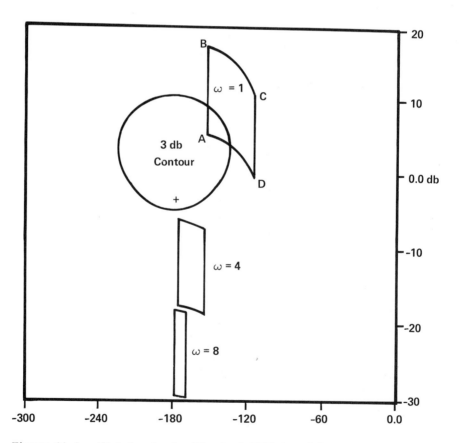

Figure 11.8   Nichols chart with plant QFT templates.

B or equivalently from C to D, and the pole uncertainty is BC or AD for $\omega = 1$. An illustrated example is presented later. The 3-db contour is a bound for high frequencies to avoid selecting closed-loop poles of T_R near the imaginary axis. These poles would cause oscillatory disturbance responses. A more extensive rationalization is presented later. Any point corresponding to the magnitude and angle of T_R on the curved scales provides a point corresponding to the magnitude and angle of L on the horizontal and vertical scales. This correspondence between L and T_R on the Nichols chart is very important.

The Nichols chart can also be used for the disturbance response. Recall that $T_D = 1/(1 + L)$. Using the transformation $L = 1/m$, the system control ratio due to the disturbance becomes $T_D = m/(1 + m)$, which is of the same form as $T_R = L/(1 + L)$ (Horowitz and Sidi, 1978). One could design the inverse of the loop transmission, $m = 1/L$, directly on the Nichols chart. However, by turning the Nichols chart upside down, reflecting the vertical angle of L lines about the $-180°$ line (i.e., -190 becomes -170, -210 becomes -150, etc.), and reversing the signs on all magnitude lines, the chart can be used directly to design L. Note that the horizontal and vertical lines still correspond to the magnitude and angle of L, and the curved magnitude lines correspond to the magnitude of $(1 + L)$. For design purposes, only the magnitude of $(1 + L)$ is required. Therefore, the curved angle lines on the chart can be ignored. In practice, the transformation is merely implied by turning the Nichols chart upside down and modifying the scales as described.

A plant template is a plot on the Nichols chart of the range of uncertainty in the plant at a given frequency. Consider the example plant $G_p(s) = K/s(s + p)$, where the gain K is in the range $2 < K < 8$, and the location range of the second pole is given by $0.5 < p < 2.0$. An infinite number of possible plant models exist owing to the variation of the parameters K and p. Each parameter is a member of a set with finite boundaries. Likewise, the magnitude and phase angle of all possible plants lie within finite boundaries when plotted at a given frequency.

The plant template is obtained by plotting on the Nichols chart $\text{Lm}[G_p(-j\omega)]$ versus $\text{Ang}[G_p(j\omega)]$ for all possible $G_p(j\omega)$'s, over the range of uncertainty at a given frequency, Lm being logmagnitude and Ang meaning angle. Note that only the outer edges of the template need be calculated. The plant transfer functions at the boundaries are found by holding one parameter constant at a boundary value, i.e., set $K = 2$, and varying "p" in increments from 0.5 to 2.0. The frequency response at $\omega = 1$ for the G_p's obtained provides a set of points from A ($K = 2$, $p = 0.5$) to D ($K = 2$, $p = 2$) on the Nichols chart, as shown in Figure 11.8. The process is continued until the complete template is formed. For example, for $p = 0.5$, varying K from 2 to 8 to obtain the line from A ($p = 0.5$, $K = 2$) to B ($p = 0.5$, $K = 8$). Templates are needed for a number of frequencies taken at regular intervals, such as every octave. A set of templates is shown to demonstrate the change in size and location of the range of uncertainty in G_p for $\omega = 1$, 4, and 8.

To facilitate the shaping of the loop transmission, a reference or "nominal" plant transfer function is required. This nominal plant G_0 is nothing more than a reference plant to be used in the definition and shaping of the nominal loop transmission $L_0 = D_c G_0$. There are no rules or constraints on the selection. It doesn't even have to be from the set of possible plants, but it is usually convenient to choose it such that it lies at a recognizable point on the templates. It is convenient, as in the previous example, to select G_O such that it lies at the lower left-hand corner of the templates. This choice keeps the bounds on L_O as near the center of the Nichols chart as possible. Once selected, the G_O point should be marked on each template (see, for example, point A ($\omega = 1$)) in Figure 11.8. For our example, the plant described by $G_O = 2/(s + 0.5)$ is chosen as the nominal plant.

Owing to plant uncertainty, the frequency response of the output $C(j\omega)$ for a step-input can vary from the upper bound (T_U) to the lower bound (T_L) at specified frequencies. In general, the allowable relative change in $C(j\omega)$ at a given frequency ω_i is expressed as

$$\Delta Lm[C(j\omega_i)] = Lm[T_U(j\omega_i)] - Lm[T_L(j\omega_i)] \tag{11.56}$$

The relative change in the output, on the other hand, is related to the control ratio as follows:

$$\Delta Lm[C(j\omega_i)] = \Delta Lm[T(j\omega_i)] = \Delta Lm \frac{L(j\omega_i)}{[1 + L(j\omega_i)]} \tag{11.57}$$

Likewise, the relative change in $L(j\omega)$ is equal to the relative change in the plant:

$$\Delta Lm[L(j\omega_i)] = \Delta Lm[G_P(j\omega_i)] \tag{11.58}$$

The variation in G_p arises, as mentioned above, because of parameter uncertainty. Thus, the problem is to find an L such that the relative change requirements on the closed-loop response are satisfied for the entire uncertainty range of plant models. The design specifications state the requirements on the closed-loop response $C(j\omega)$ and thus on $T(j\omega)$ as given by Eq. (11.57). Constraints on the loop transmission $L(j\omega)$ are now developed.

The relative uncertainty in L is equal to the range of uncertainty in G_p by Eq. (11.58). As described earlier, the plant template is a plot on the Nichols chart of the range of uncertainty in G_p at a given frequency. Because $Lm(L) = Lm(G_p) + Lm(D_c)$ and also $Ang(L) = Ang(D_c) + Ang(G_p)$, a template may be translated (but not rotated) horizontally or vertically on the Nichols chart, where horizontal and vertical translations correspond to the angle and magnitude requirements on $D_c(j\omega)$, respectively, at a given frequency.

With the template corresponding to $\omega = 1$ of Figure 11.8, translate it to position 1 shown in Figure 11.9. Since the template represents the range of uncertainty in G_p, the area now covered by the template corresponds to the variation in L and in T due to model uncertainty. Recall the correspondence between L and T on the Nichols chart. Using the curved magnitude contours, i.e., contours of constant $Lm[T(j\omega)]$, read the maximum and minimum values of T covered by the template. If the difference between the maximum and minimum values is greater than the allowable variation in T at the frequency $\omega = 1$, shift the template vertically, as shown in Figure 11.9, until the difference is equal to $Lm[T(j1)]$ (to position 2). Conversely, if the difference is less than that allowed, move the template vertically, downward, until equality is obtained. Note that

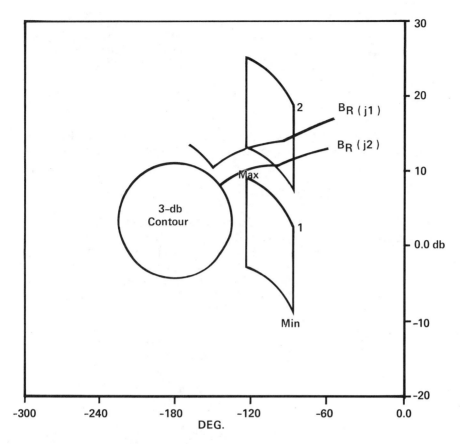

Figure 11.9 $L_0(j\omega_i)$ bounds on Nichols chart.

max and min occur in different places along the template perimeter. When the position of the template achieves equality (position 2 of the example), mark the nominal point G_0 (lower left-hand corner) of the template on the Nichols chart. The marked point corresponds to a bound on the magnitude and phase angle values of $L_0(j1)$ read from the horizontal and vertical scales of the Nichols chart. The nominal loop transmission $L_0(j\omega_i)$ is given by

$$L_0(j\omega_i) = D_c(j\omega_i)G_0(j\omega_i) \qquad (11.59)$$

Repeat the process horizontally across the chart for different values of Ang(L_0). The points marked on the chart form a curve $B_R(j\omega_i)$ representing the boundary of $L_0(j\omega_i)$ at the given frequency of the template. As long as $L_0(j\omega_i)$ lies outside or above the boundary $B_R(j\omega_i)$, the variation in T due to the uncertainty in G_p is less than or equal to the relative change in T allowed by the design specifications. Repeat this derivation for various frequencies, using the corresponding plant templates to obtain a series of bounds on $L_0(j\omega_i)$, as shown in Figure 11.9.

Likewise, the step-disturbance response specification is converted to bounds on $L_0(j\omega)$. Only output disturbances defined by Eq. (11.54) are considered in this development. In order to effectively reject the disturbance, the following inequality must be satisfied:

$$\left| \frac{1}{1 + L(j\omega)} \right| \leqslant |B(j\omega)| \qquad\qquad (11.60)$$

where $B(j\omega)$ is the magnitude of the boundary T_D. Converting the magnitudes to decibels and rearranging terms, the inequality can be expressed as

$$Lm[1 + L(j\omega)] \geqslant -Lm[B(j\omega)] \qquad\qquad (11.61)$$

A template is placed on the inverted Nichols chart such that its lowest point rests directly on the contour of constant $Lm[1 + L(j\omega)]$ equal to $-Lm[B(j\omega)]$ at the frequency ω_i for which the template is drawn. The point G_0 is marked and the template slid along the same contour forming a bound $B_D(j\omega_i)$ for L_0. Bounds are formed for each frequency using each template. Using the rectangular (Lm L) grid, transcribe the bounds $B_D(j\omega_i)$ on L_0 onto the upright Nichols chart, which already contains the

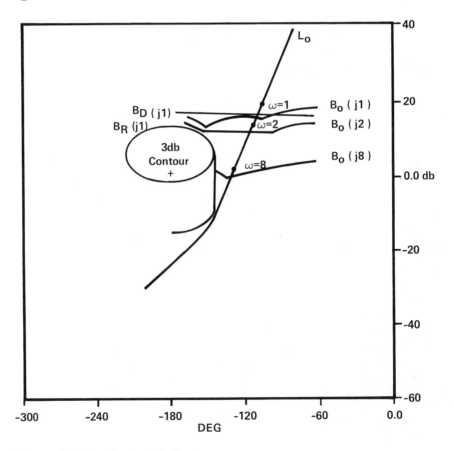

Figure 11.10   Nominal $L_0(j\omega_i)$.

command response bounds $B_R(j\omega)$ on L_0, as shown in Figure 11.10. For each frequency of interest, erase the lower of the two bounds, where the remaining bound is labeled $B_0(j\omega)$. The point here is that the worst case (most restrictive) bound must be used in shaping L_0.

The universal frequency (UF) bound ensures that the loop transmission L has positive phase and gain margins, whose values depend on the "oval" of constant magnitude chosen (see Figure 11.10). As the frequency increases, the plant templates become narrower and can be considered vertical lines as ω approaches infinity. The allowable variation in T increases with frequency also. The result is the bounds of $L_0(j\omega_i)$ tend to become a very narrow region around the 0-db, $-180°$ point (origin) at high frequency. To avoid placing closed-loop poles near the $j\omega$ axis, resulting in oscillatory disturbance response, a UF bound is needed. With increasing frequency the bounds on L_0 approximately follow the ovals encircling the origin. In Figure 11.10, the contour of constant magnitude equal to 3 db is used.

From the templates corresponding to high frequency, the template is found with the greatest vertical displacement v in db. The displacement v may be accurately determined by finding the maximum change in $Lm[G_p(j\omega)]$ in the limit as ω goes to infinity. Translate the lower half of the 3-db oval down the length of the template, thus obtaining the UF bound.

The shaping of a nominal loop transmission conforming to the boundaries of L_0 is a crucial step in the design process. A minimum bandwidth design has the value of L_0 on its corresponding bound at each frequency. In practical designs, the goal is to have the value of L_0 occurring above the corresponding bounds, but as close as possible to keep the bandwidth to a minimum. Figure 11.10 shows a practical design for L_0. Note, any right-half-plane poles or zeros of G_0 must be included in L_0 to avoid any attempt to cancel them with zeros of G_0. This suggested starting point in the design of L_0 avoids any implicit cancellation of roots in determining D_c.

The compensator D_c is obtained from the relation $D_c = L_0/G_0$. If the L_0 found above does not contain the roots of G_0, then the compensator D_c must cancel them. Note that cancellation occurs only for purposes of design using the nominal plant transfer function. In actual implementation, exact cancellation does not occur (nor is it necessary) since G_p can vary over the entire uncertainty range.

Design of a proper L_0 only guarantees that the variation in $T(j\omega)$ is less than or equal to that allowed. The purpose of the prefilter is to position $Lm[T(j\omega)]$ within the frequency-domain specifications. For the example of magnitude the frequency response must lie within the specified bounds which are drawn in Figure 11.11, i.e., T_U and T_L.

One method for determining the bounds on the prefilter F is as follows. Place the nominal plant of the $\omega = 1$ template on the Nichols chart where $L_0(j1)$ point occurs. Record the maximum and minimum values of $Lm(T)$ (1.2 and 1.0 in the example), obtained from the curved magnitude contours. Compare the values found above to the maximum and minimum values allowed by the frequency-domain specifications at $\omega = 1$ (0.7 db and -0.8 db). Determine the range, in db, that $Lm(T)$ must be raised or lowered to fit within the bounds of the specifications.

For example, at $\omega = 1$, the actual $Lm(T)$ must be within $[Lm(T_0) = 0.7$ db$] > Lm[T(j1)] > [Lm(T_L) = -0.8$ db$]$. But, from the plot of L_0, the actual range of $Lm(T)$ is 1.2 db $> Lm[T(j1)] > 1.0$ db. To lower $Lm[T(j1)]$ from the actual range to the desired range, the prefilter $Lm(F)$ is required such that $(0.7 - 1.2$ db$) > Lm[F(j1)] > (-0.8 - 1.0$ db$)$, or -0.5 db $> Lm[F(j1)] > -1.8$ db (see Figure 11.11). The process is repeated for each frequency corresponding to the templates used in the design of L_0. Bounds

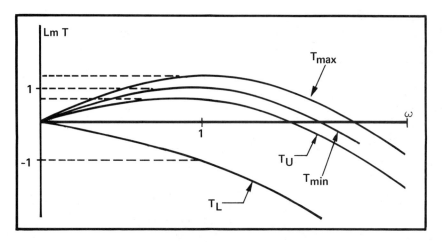

Figure 11.11   Frequency bounds for prefilter F.

of F, $[Lm(T_o) - Lm(T_{max})] > Lm(F) > [Lm(T_L) - Lm(T_{min})]$, can be
plotted as a function of frequency. By use of the straight-line approximation, F is determined such that its magnitude lies within these bounds.

The single-loop design is complete with the design of F. The system
response is guaranteed to remain within the bounds of the design specifications, provided the uncertainty in G_p stays within the range assumed at
the beginning of the design process.

The QFT technique is entirely based in the frequency domain and makes
considerable use of the Nichols and Bode plots. Much of the designing can
be done by graphical methods and use of CAD tools (see Appendix A).

QFT Digitization Approach. The feasibility of using the digitization
approach is discussed for extending the continuous-domain control system
designs to the discrete domain. The example design considered is a transport aircraft with continuous fixed compensators for three different flight
conditions. For the transport aircraft, the compensators are discretized
using two transformation techniques after they are modified to reflect the
effects of the sampler and ZOH. Simulation results for the transport aircraft show that the Tustin transformation provides acceptable results using
a sampling rate of 40 Hz (Coucoules et al., 1986).

The longitudinal mode aircraft equations are represented in input-output
matrix form, where the input and output vectors are, respectively,

$$\underline{\delta} = \begin{bmatrix} \delta_e \\ \delta_{sb} \\ \delta_t \end{bmatrix} \qquad\qquad \underline{C} = \begin{bmatrix} h \\ \theta \\ u \end{bmatrix} \qquad\qquad (11.62)$$

The input variables are defined, respectively, as elevator deflection,
speed brake deflection, and percent engine thrust. The output vector
elements are altitude, pitch angle, and velocity (see Figure 11.12).

The desired response models are used to develop bounds in the con-
tinuous-frequency domain which are constraints on the loop transmission
function L for each channel, as discussed in the previous section. Figure
11.13 shows an example of a loop transmission shaped to meet bounds at
various frequencies. In fact, this is the continuous-domain nominal loop
transmission designed for loop two (tracking loop) of the transport aircraft
longitudinal mode.

As can be seen from Figure 11.13, the loop transmission almost coincides
with the universal high-frequency boundary (with only a few degrees of
separation over a frequency range of about 75 rad/sec.) Noting that
$L = D_c G_p$, it is apparent that any change in the compensation that results
in additional lag could take the loop transmission beyond the boundary region. If this occurs, and depending on the severity of the additional lag,
some plants in the parameter space may exhibit less than desired performance or can become unstable. Since the intent in the digital system is to
maintain the frequency response characteristics of the original continuous-
domain loop transmission, it is clear that the low-pass filter (lag) characteristic of the ZOH by itself can have an undesirable effect.

The original s-domain QFT compensators $D_c(s)$ can be modified by the
Padé approximation to obtain

$$D_c(s) = [G_{A/D}(s)]^{-1} D_c(s) \qquad (11.63)$$

The net affect is that the gain is modified by a factor of T and phase
lead is added to the compensation. Thus, based on the digitalization
approach, the loop transmission is

$$L = [G_{A/D}(s)]^{-1} D_c(s) G_p(s) \qquad (11.64)$$

A multitude of techniques are used to select the sample rate of a
sampled-data system, many of which often depend on the selected design
technique. For both the lateral and longitudinal modes of the transport
AC a sample rate of 40 Hz is selected and is about five times the maximum
closed-loop bandwidth. For the transport, this sample rate proves to be
quite satisfactory.

Two methods for obtaining $D_c(z)$ from $D_c(s)$ are the standard Z-transform and the Tustin transformation. The standard Z-transform generated
the proper z-domain poles but had mapped the zeros to the same location
as the poles. The discrete frequency response of $D_{c1}(z)$ (see Figure 11.12),
for the lateral mode of transport AC is essentially flat in the frequency
range defined by the primary strip $\omega_{s/2}$ using the Z-transform. Similar
results are obtained with all the other compensators.

Each s-domain QFT compensator has a high-frequency complex pole
pair for the purpose of noise rejection. All the poles, except $D_{c1}(s)$ and
$D_{c3}(s)$, of the transport longitudinal mode are warped significantly since
they are outside the acceptable Tustin region. This warping has no real
effect since these poles were generally at or well above the sampling frequency. $D_{c1}(s)$ for the transport lateral mode had several real poles that
were warped onto the left-half z-plane real axis within the unit circle,
thereby severely limiting their intended lag effect on the loop transmission.

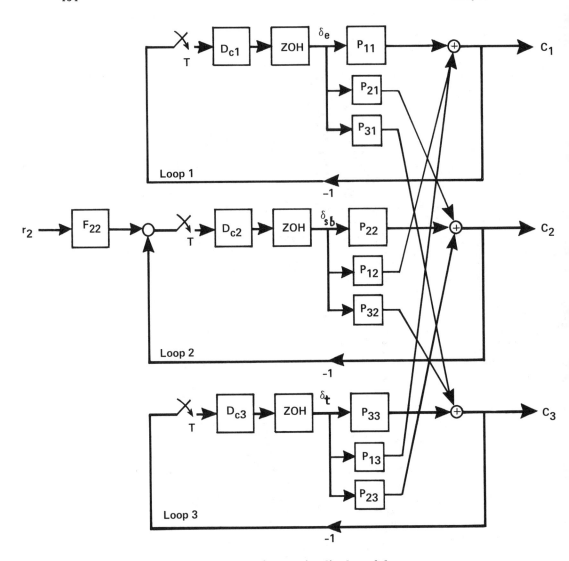

Figure 11.12   Transport Aircraft Longitudinal model.

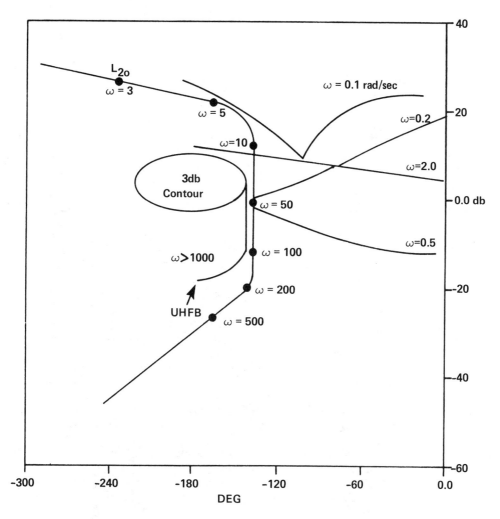

Figure 11.13   Bounds and nominal loop transmission for transport longitudinal mode.

This warping had no effect on the robustness or performance because all the bounds are still met. In other words, the loop transmission moved to the right owing to the loss of the lag effect but still is within all the bounds.

A single maneuver is simulated for the rigid- and elastic-body aircraft. For the rigid body, flight condition (FC) 3 input was a 4° pitch angle command, which rises to 4° in 1 sec and remains at 4° thereafter, whereas for FCs 1 and 2, a similar input was used but the command only rises to 1°. For the elastic-body simulations, the input for all three FCs is a 1° pitch command, which rises to 1° in 1 sec.

Simulation results are obtained using a processor emulation subroutine which permits specification of word length (Coucoules, 1985). The transport's longitudinal mode responses are obtained using a processor configuration consisting of a 32-bit wordlength with 17 bits to the right of the binary point and rounding. Several simulations using various configurations of binary point placement and truncation/rounding generate the qualitative results of Table 11.2. Rounding provides a larger range of acceptable performance. A similar analysis for the lateral mode reveals that a minimum of 26 bits produces the same results.

It is possible, but less likely, that a 16-bit microprocessor can achieve acceptable results for the longitudinal mode. Implementing a majority of the compensator gain outside the digital controller might allow a 16-bit microprocessor, especially considering that the magnitude of the gain for one of the longitudinal compensators is higher than the maximum lateral compensator gain. The pitch angle response is shown in Figure 11.14. Figure 11.15 shows the perturbation altitude response, and Figure 11.16 shows the perturbation velocity response. Note that both the rigid and elastic body responses for all three flight conditions are shown on the same figure. The simulation indicates that all performance specifications are met. Examination of the tracking response curves shows that the desired response shape has been achieved. Also, comparison of the finite wordlength responses with the continuous system responses shows only a small amount of variation.

Use of the Tustin transformation and digitization model to extend existing QFT analog designs to the discrete domain provide acceptable system

Table 11.2   Transport Longitudinal Mode Processor Configuration Study

Wordlength	Bits right of binary point	System performance	
		Rounding	Truncation
32	25	Unacceptable	Unacceptable
	20	Unacceptable	Unacceptable
	18	Acceptable	Unacceptable
	17	Acceptable	Acceptable
	14	Acceptable	Acceptable
	12	Acceptable	Unacceptable
	10	Acceptable	Unacceptable

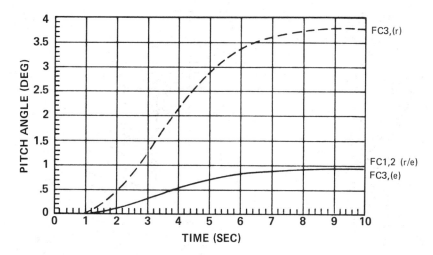

Figure 11.14   Transport pitch angle responses for various FC's.   r = rigid body model, e = elastic body model.

performance when all real compensator zeros are located within $s > -2/T$. This criterion may apply to real compensator poles as well, depending on the location of the Nichols chart bounds.   Thus, the empirical results of this investigation show that if the criterion is met, the Tustin approximation effectively accounts for the problems associated with sampling, finite wordlength, and ZOH delay inherent in a digital control system.   If the above criterion is not met, then the Tustin approximations limit the degree of achievement of the performance specifications.   Improving the performance

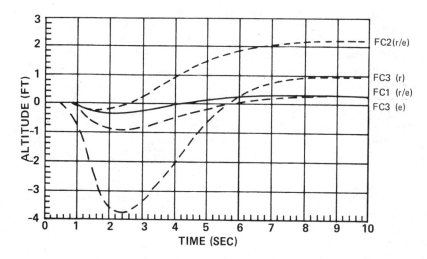

Figure 11.15   Transport altitude responses.

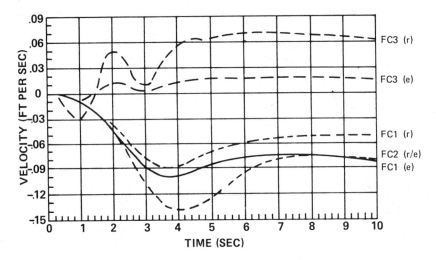

Figure 11.16   Transport velocity responses.

requires the application of "cut-and-try" methods. In this situation, the
system performance guarantees inherent in the QFT design technique are
lost. Using the extension method, realizable sample times (40–60 HZ) and
processor wordlengths (16–32 bits) provide acceptable performance for the
resulting QFT digital control systems (Coucoules et al., 1986). Note that
the standard Z-transformation is not a useful tool in extending QFT analog
control system designs to the discrete domain because of pole/zero transformation characteristics.

For those cases where the warping of the poles and zeros is significant
it is best to implement the QFT design technique directly in the discrete
domain.

Direct QFT Approach (Horowitz and Liao, 1985). As a general approach,
the w-plane model is generated from the z-plane model in order to use the
analysis techniques employed in the continuous QFT Method. The assumption
is that a sampling time T or sampling frequency ω_s is selected a priori.
However, the selection of T is another degree of freedom in the direct QFT
approach and is one of the basic design parameters to be determined.

Since a discrete model is required, the continuous plant and ZOH are
transformed to the z-plane with the same equations for plant uncertainty
as Eq. (11.54) and (11.55). The loop transmission becomes L(z) =
$G_{A/D}(z)G_p(z)$, which is transformed into the w-plane. From the previous
section, the specifications on $Lm[C(j\omega_i)]$ suffice to determine the time-
domain response in a stable minimum-phase system where the right-half-
plane zeros, if they exist, are known (Horowitz, 1979). This constraint
applies to the direct QFT digital design as well. As before, the time-
domain specifications are first translated into frequency-domain requirements before the controller design proceeds. The desired closed-loop
control ratios (transfer functions) are determined as before as a function
of frequency. These derived specifications are used to develop the loop
transmission bounds, with a resulting controller structure generated using
the techniques discussed previously.

The direct QFT approach for digital controllers is therefore similar to the continuous technique, the difference being the use of the w-domain instead of the Laplace-frequency domain although the design methods used in the w-domain are essentially the same as those in the s-domain. Also, the method of ZOH modeling and sampling frequency selection is different. With the disturbance rejection constraint included, the larger of the two bounds (uncertainty, disturbance) is used. The references cited in this section should be consulted for further insight.

11.4   IMPLEMENTATION

Implementation of a digital controller requires detailed environmental evaluation, study of timing considerations, development of relationships between theoretical and physical controllers, and analysis of accuracy requirements as impacted by finite word length. This section deals with the association of these phenomena, with a focus on overall digital controller performance. However, some of the analysis techniques require a solid foundation in statistical mathematics, complex variables, and numerical analysis. Various references, such as those cited in Section 11.1, provide the mathematical details.

11.4.1   Effect of Finite Word Length

In the theoretical generation of digital controller coefficients the assumption of infinite accuracy (infinite word length) was a subtle element of the theory. The various arithmetic operations (addition, subtraction, multiplication) involved in executing the difference equation controller also are impacted by finite wordlength (see Section 10.4). In realizing a controller, one of the effects of the computer's finite wordlength is to change the frequency and time response characteristics of the theoretical controller, which may be inconsistent with performance objectives. This phenomenon is due to the movement in poles and zeros of the realized controller and the arithmetic manipulation errors (rounding or truncation). The changes in performance can be developed by using various approaches, such as simulation (digital, hybrid), analysis of pole/zero movement, coefficient perturbations, and controller structural analysis.

Specific errors resulting from input-signal quantization (conversion) are deterministically modeled as $1/2$ of the value of the least significant bit. Assuming a number representation with the binary point to the left of the most significant bit, a 10-bit A/D, for example, has a maximum quantization error of $\pm(2^{-10+1})/2 = \pm0.0009765$. In general, an error of $\pm2^{-M}$ is expected, where M is the total number of bits. A rounding A/D quantizer has been assumed in this error model. If the A/D uses a truncation operation, then the range of the quantization error becomes $(-2^{-M+1},\ 0)$.

Another consideration is selection of the binary number representation within the computer. Most commercial computers use sign-plus-two's complement (2's complement) representation, although sign-plus-magnitude and sign-plus-one's complement can be implemented (Houpis and Lamont, 1985). The specific representation causes interval errors similar to those above. For example, using 2's complement, the truncation error range is $(-2^{M+1},\ 0)$ and rounding error range is $(-2^{-M},\ 2^{-M})$.

The error ranges due to rounding and truncation quantization can also be modeled with probability density functions, specifically uniform PDFs (see Chapter 3). Using this error model the expected value m due to rounding is zero, and for truncation it is -2^{-M}. The variance σ for both cases is $2^{2(-M+1)}/12$. Using the expected value, variance, and the controller transfer function $D_c(z)$, a relationship can be found between input quantization and controller output error (expected value and variance);

$$m_{output} = m_{input} D_z(z)\Big|_{z=1}$$

$$\sigma^2_{output} = \sigma^2_{input}(1/2\pi j) \oint D_c(z)D_c(z^{-1})z^{-1} \, dz \qquad (11.65)$$

The solution of the integral equation requires complex variable knowledge, although approximations exist (Houpis and Lamont, 1985).

Within a computer, arithmetic errors are generated because of finite wordlength. For a specific number, its representation and the selected quantization technique (rounding, truncation) define the range of maximum error. Integer-fixed-point arithmetic is assumed in the error models discussed earlier, although a floating-point model can easily be developed since the number of mantissa bits and exponent bits is usually fixed. The range of fixed-point errors is identical to that for A/D conversion and also can be formulated as a statistical model using uniform PDfs. The impact of an arithmetic error, at a particular point in the solving of the difference equation, relates to the controller output by a transformation relationship. This relationship is part of D_c, the overall controller transfer function. To find the specific relationship, the precise implementation structure of the compensator must be known.

Various implementation forms for digital controllers (compensators, filters) have been developed which have different finite wordlength error performance. The general realization categories are direct, cascade, and parallel (see Chapter 10). The direct method usually includes four structure subclasses, defined as 1D, 2D, 3D, and 4D. The 1D structure for second-order filters has the following form, which possesses only n time-delay elements (n = 2):

$$s(k) = r(k) - b_1 s(k - 1) - b_2 s(k - 2)$$

$$c(k) = a_0 + a_1 s(k - 1) + a_2 s(k - 2) \qquad (11.66)$$

The 2D structure requires (n + 1) equations and has the form

$$c(k) = a_0 + q_1(k - 1)$$

$$q_1(k) = a_1 r(k) - b_1 c(k) + q_2(k - 1)$$

$$q_2(k) = a_2 r(k) - b_2 c(k) \qquad (11.67)$$

For 3D, the equation requires 2n time delays and is of the form

$$c(k) = a_0 r(k) + a_1 r(k - 1) + a_2 r(k - 2) - b_1 c(k - 1) - b_2 c(k - 2)$$

$$(11.68)$$

The 4D form has 2n equations:

$$p_0(k) = r(k) + p_1(k - 1)$$

$$c(k) = a_0 p_0(k) + q(k - 1)$$

$$p_1(k) = -b_1 p_0(k) - b_2 p_0(k - 1)$$

$$q(k) = a_1 p_0(k) + a_2 p_0(k - 1) \tag{11.69}$$

The various equations for all structures should be calculated in the order presented. This approach limits the effect of the computational delay. Figure 11.17 represents a 1D structure. The other structures are similar.

The various models of direct structures differ only in positioning of the multiplication operators (using difference equation coefficients that are real) before or after a delay element z^{-1}. The output error due to a specific multiplication or addition operator can be found by deriving the transfer function from that point to the controller's output. Figure 11.17 also presents the transfer functions for various error sources in a 1D structure.

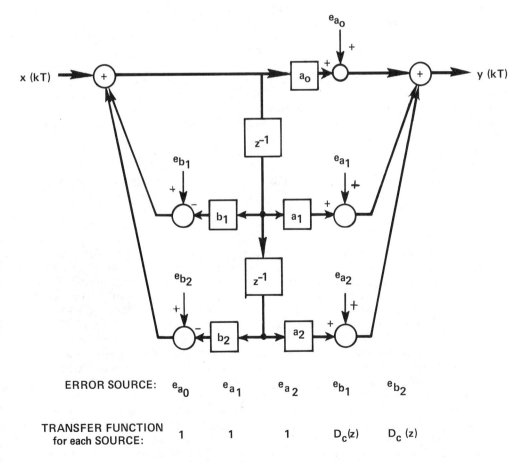

ERROR SOURCE:	e_{a_0}	e_{a_1}	e_{a_2}	e_{b_1}	e_{b_2}
TRANSFER FUNCTION for each SOURCE:	1	1	1	$D_c(z)$	$D_c(z)$

Figure 11.17   1D structure with multiplication error sources.

In addition to the direct structures above, the transfer function numerator and denominator polynominals can be factored, with the resulting complex roots represented directly in a compensator structure. The associated forms are defined as 1X and 2X. The 1X structure is derived from the following second-order factored form:

$$D_c(z) = a_o + \frac{1/2(g_3 + ig_4)}{z - g_1 - ig_2} + \frac{1/2(g_3 - ig_4)}{z - g_1 + ig_2} \tag{11.70}$$

The compensator difference equations for the 1X canonical form are

$$c(t) = a_0 + s_2(k - 1)$$

$$s_1(k) = g_1 s_1(k - 1) - g_2 s_2(k - 1) + g_4 r(t)$$

$$s_2(k) = g_1 s_2(k - 1) + g_2 s_1(k - 1) + g_3 r(t) \tag{11.71}$$

The transpose of the 1X structure generates the 2X form (Phillips and Nagle, 1984).

Cascade and parallel structures are usually implemented with second-order forms for ease of implementation. Chaining the output(s) of one filter, or compensator, to another in a serial fashion defines a cascade implementation. Parallel implementation, on the other hand, requires each second-order structure to accept the same input, with every output being added together to form the controller output.

The selection of poles and zeros to be assigned to a specific second-order filter is known as pole-zero pairing (Phillips and Nagle, 1984). Because of the sensitivity issue discussed previously, a pole (zero) should be paired with the pole (zero) that is farthest away. Poles and zeros that are close together should then be combined to generate a second-order filter component with reduced gain and sensitivity.

Proper scaling is an important aspect of achieving desired performance and minimizing error propagation in both cascade and parallel implementation. In particular, the objective is to utilize the complete dynamic range of each second-order form by shifting data and limiting error propagation. The selection of second-order filter gains can be achieved by analysis of worst-case amplitude values and simulation.

Finite wordlength also changes the position of the poles and zeros because of the change from the theoretical coefficients to the implemented coefficients. Specific root changes can be found by generating an equation relating the known coefficient quantization error to a specific pole or zero-position error with a first-order difference approximation (Oppenheim and Schafer, 1975). The form of this equation is a function of the difference between each pair of poles. If two poles are close together, then even for a relatively small coefficient realization error, large variations can occur in realized pole locations. This phenomenon makes sense when one realizes that roots of a polynomial are functions of the coefficients.

Limit-cycle phenomena is another situation that can occur owing to a finite-wordlength limit-cycle phenomena. In limit cycling the output of the controller oscillates because of the nonlinear quantization operation. Various bounds (see references cited in Section 11.1) have been developed to determine whether limit-cycles would impact specific performance characteristics such as settling-time or plant fatigue due to a small oscillating input. With any model a considerable analysis effort is required in order to determine the exact impact of quantization error on system performance, including limit-cycles and error ranges.

Various simulation methods can provide evaluation of system performance (see Chapter 4). The simulated controller should be represented by implementing the D_c wordlength and the plant with a longer wordlength for more accuracy (better reflection of the real continuous physical world). In addition, the plant output should be calculated (at least) at a rate 10 times faster than the controller for good representative results for the same reason. Hybrid simulations permit the continuous plant to be simulated on an analog computer and the digital controller realized with different microprocessor systems. Also, CAD packages, as mentioned previously, can generate characteristic values, such as overshoot, damping, settling-time, and frequency bandwith.

11.4.2 Computational Delays

The theoretical designed digital (z-plane) controller does not include the conversion and computational delays in the system model. This approach is reasonable if the sampling period T is much greater than the sum of the A/D, computer, and D/A delays. The implication of the theoretical controller is that the controller's output, at time kT, is available instantaneously at the same time the input r(kT) is applied [see Eq. (11.1)]. Thus, if the additive delay from the three sources above is much less than T, the compensator resulting from the theoretical design process is expected to be close to the form of the overall implemented controller. If this constraint is not met, the delay operator e^{-Ts} or z^{-1} must be an explicit multiplier in the controller transfer function model. The design method then must include this term in the analysis and development of an acceptable controller.

Of course, the digital controller can be designed without regard to conversion and computation delay time. In this case, simulation and tuning can play an important role in meeting system performance objectives. By tuning or tweaking the coefficients of the programmed digital controller, an acceptable time and/or frequency response may be achieved. Simulation, as mentioned previously, is an integral aspect of this engineering approach.

11.4.3 Software

The development of a software program for a digital controller is in reality a rather simple task. The solution to the difference equation can be reflected in an algorithm programmed in a higher-order language like FORTRAN, Pascal, or Ada or in assembly language (see Appendix B). The input/output routines for signal conversion again are rather straightforward, and most manufacturers provide example programs for setting T, initializing the conversion process, and starting the sampling process. Using the unique functionality of the overall process (initialization, input, difference equation computation, and output), specific software modules can be defined using the concepts of structured programming (see Chapter 8).

The more difficult aspects of the software development process relate to the generation of a complete set of documentation records. Examples include specification listings, data flow diagrams, software design structures, commented code, testing results, and operational experience. The extensive list of documentation, whether in English, graphical forms, data dictionaries,

or simple listings, is required in order to easily fix errors (bugs), extend
the current configuration for new applications, and understand the system
operation from a user's point of view.   Also, detailed decisions related to
specific data structure selection (see Chapter 7) for coefficients, partial
products, and past and current values generate different control structures
(algorithms).

The process of software development alluded to is usually called software engineering (Jensen and Tonies, 1979), i.e., the efficient and effective generation of functioning software through consistent and complete
documentation.   Many design techniques exist (Houpis and Lamont, 1985)
and are acceptable if used properly and interfaced to other methods to
meet the documentation objectives.   The effort requires a disciplined approach
to software development, i.e., requirements first, structural design, algorithm design, code, test, integration, and operation and maintenance.   This
process is called the software life-cycle.   However, in the design of digital
controllers, one important dimension usually not considered in simulation is
the real-time aspect.   As mentioned in Section 11.3.2, the additive computation and signal conversion time should be relatively small as compared
to the overall sampling time T.   Yet T must be small enough to permit
sampling of input signals within the given bandwidth (Shannon's theory).
Thus, T is bounded from above and below and should be selected using
good engineering analysis.   Walter (1984), Andrews (1982), and Auslander
and Sagues (1981) present various examples of software designs for digital
controllers.

In many digital control systems, not one but many modes and degrees
of freedom are involved with multiple controllers.   To determine the proper
sequence of conversion and compensator computation requires elaborate
analysis.   A real-time operating system is usually employed to sequence the
various operations (see Chapter 9).   Different types of such operating
systems are available with different data structures and control structures.
Commercial real-time operating systems have a variety of data structures
and also include interrupt mechanisms for initializing and starting each sampling window.   Interrupts can be generated from a real-time clock, watch-
dog timer clock, exception handlers, or other event occurrences that are
part of the overall digital control system.   Multiple sampling techniques can
also be implemented in such a system, but require considerable more analysis than the straightforward single sampling frequency discussed.

11.5   SUMMARY

The general approaches for designing a digital controller include the
digitization technique and the direct discrete-domain method.   Both approaches require the proper modeling of the ZOH and sampler along with
the given plant transfer function.   Various approximations, involving the
s-plane, z-plane, and w-plane design techniques, require that the designer
know explicitly the impact of pole and zero locations on system performance
through the use of simulation.

The a priori selection of a cascade or feedback controller permits
efficient design using the Guillemin-Truxal approach or pole-zero placement
using the root-locus method in any domain.   Feedback compensation requires more trial-and-error analysis.   Thus, cascade compensation is
suggested as the first attempt in meeting performance objectives.   Lead-lag

and PID compensator structures provide initial models for developing a controller design, and, as appropriate, all should be attempted before finalizing a design, depending on time and resources available.

Fine tuning of digital controller coefficients can be achieved by minimizing a mean-square error function of the magnitude and angle differences between the desired and current design using gradient search. Another approach is to use simulation with slight variations in controller parameters to improve system performance with selected inputs.

The quantitative feedback technique (QFT) provides additional design capability by including in the design problem uncertainties in the plant model as well as input disturbances. The approach requires transformation of time-domain system-performance objectives into the frequency-domain. The associated trial-and-error procedure requires simulation and testing.

Two important aspects of implementing a digital controller are the effect of finite wordlength and the structure of the associated compensator. In practice, second-order structures are used to develop generic digital control systems. Finite wordlength impacts the implemented difference equation, the D_C coefficient values, and the selection of converters (A/D, D/A). Analysis and emulation of various configurations are suggested.

Selection of a programming language can also influence system performance in terms of storage requirements, execution time, and accuracy. When considering the real-time burden of digital controllers, analysis of timing relationships of various programmed controllers must be considered and compared with actual operation. Simulation provides one effective tool for such an approach.

Simulation and the use of CAD tools are essential to the development of contemporary digital controllers. Computer capabilities are continually improving, and designers should always be on the search for better and more extensive hardware and software to solve their analysis and synthesis problems.

PROBLEMS

11.1. From Figures 11.1b, c, and d, derive the closed-loop transfer functions (control-ratio) equations C/R as functions of G and H.

11.2. Derive the s-plane to z-plane mapping loci using the exact Z-transformation, the Tustin approximation, and the first-difference model. Derive the z-plane to w-plane bilinear mapping loci.

11.3. Develop a "good" Tustin region for a 1% error tolerance. Generate a "warping" equation relating s-plane root locus to z-plane root locus through the use of the Tustin approximation. Plot results. Is the Tustin approach a good transformation for the derivative operation?

11.4. Derive the Padé approximation for the ZOH as shown in Eq. (11.8). Discuss other possible approximations to the transcendental delay formulation.

11.5. Generate a digital cascade controller for a plant with transfer function $4/s(s + 1)$ using the direct and digitization approaches.

For a unit-step input, the peak-value should be less than 5% of the input, peak-time less than 4 sec, and settling-time less than 6 sec. Compare simulation results and discuss the Tustin region as related to selected sampling period.

11.6. Using the PID structure of Eq. (11.38), generate a controller based on the model and performance requirements of Problem 11.5.

11.7. For the plant of Problem 11.5, generate a feedback controller using the same specifications. Compare to cascade controller form, simulation performance, and sampling time.

11.8. Develop the equivalent state-space equations from Eq. (11.15) and (11.16) for the time-varying linear model. In this model the $\underline{F}$, $\underline{H}$, $\underline{C}$, and $\underline{D}$ matrices of Eqs. (11.11) and (11.12) are functions of discrete time kT.

11.9. For the linear time-varying model, generate the equivalent equations to Eq. (11.51) and (11.52).

11.10. Using the state-space controller of Eq. (11.52), determine the state-space response through Eq (11.11) and (11.12). Simulate response using identity matrices and relate performance to classical metrics (rise-time, peak-time, . . .) for each state-vector element.

11.11. Consider the selection of an appropriate wordlength for Problems 11.4 and 11.5 using sign-plus-2's complement representation. Use a simulation if possible to validate selection.

11.12. Derive the Guillemin-Truxal controller equations [Eq. (11.25) and (11.35)] for the digital closed-loop cascade and feedback methods.

11.13. Generate a continuous QFT design for the plant of Eq. (11.22) assuming the gain varies 100% and the poles can vary over a range ±200%. Repeat the design for a digital controller using the direct approach in the w-plane. Simulate both systems and discuss performance in terms of classical metrics. Determine wordlength ranges for this problem based on analysis and simulation.

11.14. Expanding on Problem 11.13, add an output disturbance D of a unit step-input. Develop a digital controller using the QFT technique. Discuss performance through analysis and simulation results.

11.15. Derive the 1D, 2D, 3D, and 4D versions [Eq. (11.66–11.69)] of the second-order compensator direct implementation and develop graphical representations similar to Figure 11.17 (1D structure). Generate error equations for the multiplication operations within each structure.

11.16. Derive the 1X form [Eq. (11.71)] ano 2X form for a second-order compensator, develop graphical representations, and generate multiplication error functions.

11.17. Develop computer programs for the six second-order compensator representation forms and evaluate program performance (accuracy, run-time, storage requirements). Use assembly language, if possible, and a high-level language.

11.18. Discuss the integration of the programs of Problem 11.17 into a selected commercial "real-time" operating system. If possible, execute and evaluate performance.

REFERENCES

Andrews, M. (1982). *Programming Microprocessor Interfaces for Control and Instrumentation*, Prentice-Hall, Englewood Cliffs, NJ.

Auslander, D. M. and Sagues, P. (1981). *Microprocessors for Measurement and Control*, McGraw-Hill, New York.

Biberro, J. (1977). *Microprocessors in Instruments and Control*, Wiley, New York.

Cadzow, J. A. and Martens, H. (1974). *Discrete-Time and Computer Control Systems*, Prentice-Hall, Englewood Cliffs, NJ.

Coucoules, J. S. (1985). Study of the effects of discretizing quantitative feedback theory using analog feedback designs, unpublished thesis, Air Force Institute of Technology, Wright-Patterson AFB, Dayton, Ohio, December.

Coucoules, J. S., Houpis, C. H., Lamont, G. B. (1986). Study of the effects of discretizing QFT analog control system design, 25th Control and Decision Conference, Athens, Greece, December.

D'Azzo, J. J. and Houpis, C. H. (1981). *Linear Control System Analysis and Design: Conventional and Modern*, 2nd ed., McGraw-Hill, New York.

Franklin, G. F. and Powell, J. D. (1980). *Digital Control of Dynamic Systems*, Addison-Wesley, Reading, MA.

Horowitz, I. (1979). Quantitative synthesis of uncertain multiple-input multiple-output feedback systems, *Int. J. Control 30(1)*, 81–106.

Horowitz, I. (1982). Improved design technique for uncertain multiple-input multiple-output feedback systems, *Int. J. Control 36(6)*, 977–988.

Horowitz, I. and Liao, Y. K. (1985). Quantitative feedback design for sampled-data systems, Report of Weizmann Institute of Science, Israel.

Horowitz, I. and Sidi, M. (1972). Synthesis of feedback systems with large plant ignorance for prescribed time-domain tolerances, *Int. J. Control. 16*, 287.

Horowitz, I. and Sidi, M. (1978). Optimal synthesis of nonminimum-phase feedback systems with parameter uncertainty, *Int. J. Control 27*, 361–368.

Houpis, C. and Lamont, G. B. (1985). *Digital Control Systems, Theory, Hardware, Software*, McGraw-Hill, New York.

Iserman, R. (1981). *Digital Control Systems*, Springer-Verlag, New York.

Jacquot, R. (1981). *Modern Digital Control Systems*, Dekker, New York.

Jensen, R. W. and Tonies, C. C. (1979). *Software Engineering*, Prentice-Hall, Englewood Cliffs, NJ.

Katz, P. (1981). *Digital Control Using Microprocessors*, Prentice-Hall, Englewood Cliffs, NJ.

Kuo, B. C. (1980). Digital Control Systems, Holt, Rinehart and Winston, New York.

Oppenheim, A. V. and Schafer, R. W. (1975). *Digital Signal Processing*, Prentice-Hall, Englewood Cliffs, NJ.

Phillips, C. L. and Nagle, H. T. (1984). *Digital Control System Analysis and Design*, Prentice-Hall, Englewood Cliffs, NJ.

Ragazzini, J. R. and Franklin, G. F. (1958). *Sampled-Data Control Systems*, McGraw-Hill, New York.

Rattan, K. S. (1980). Digitization of existing continuous-data control systems, Air Force Flight Dynamics Laboratory, Air Force Wright Aeronautical Laboratories, Wright-Patterson, AFB, Ohio, *Report AFWAL-TM-80-105-FIGC*.

Truxal, J. G. (1955). *Automatic Feedback Control System Synthesis*, McGraw-Hill, New York.

Tustin, A. (1947). A method of analyzing the behavior of linear systems in terms of time series, *JIEE (London)*, *94* (part IIA).

Tzafestas, S. G. ed. (1985). *Applied Digital Control*, North Holland, New York.

Walter, C. (1984). Control software specification and design: An overview, *IEEE Computer 17(2)*.

12

Robotics

12.1   INTRODUCTION

Robot manipulators were introduced into industry about 20 years ago with
the idea of replacing teleoperators (Paul, 1983), which allow a human operator to work from a distance.   The robot manipulator has its origin in both
teleoperators and numerically controlled machine tools.   The first industrial
generation of robots consisted of "pick-and-place" robots, initially used in
the unloading of diecasting machines.   A sequence of positions was previously recorded, and the task executed was in replaying these positions
by acting on the joint axes separately.   Thus, no interaction existed between the robot motions and the task.   The application field of the robot
type was extended to other industrial processes.   The robot then became
a component of transfer lines used in automation.

The second generation of robots (Ernst, 1961) consisted of robot
manipulators equipped with sensors allowing the robot control based on
task data.   Initially, these sensors consisted of switches set with respect
to the path of the task.   They stopped the joint motions when the foreseen
positions were reached.   Then, in 1967, the first vision system associating
a television camera with a computer allowed the robot to identify objects
in real time.   The vision processing system identified the position and the
orientation of these objects in terms of homogeneous transformation; here,
the control was based on high-level languages, but the coordination between
the joint motions was not considered.   Later, this problem was successfully
solved at MIT and at Stanford University (Feldmann, 1971).

The third generation of robots consists of robots controlled by hand-eye
systems, able to modify their own trajectory according to changes during
the actual task.

At the present time, robots of all these types are still utilized in industry in order to perform handling, painting, arc and spot welding, cutting,
and assembly work (Groover et al., 1986).

In Section 12.2, the mechanical structure of robots is presented and the
sensors, actuators, and computers used with the robots are described.   The
usual methods of programming used in robotics are introduced.

In Section 12.3, we develop the homogeneous transformations allowing the building of mathematical models. Then they are successively applied to kinematic, Jacobian, and dynamic models.

In Section 12.4, several control techniques are presented, in particular the resolved acceleration control and the adaptive control.

In Sections 12.5 and 12.6, we describe the main uses of a robot (welding, painting, handling) and the current directions of research in robotics.

12.2 BACKGROUND

12.2.1 Description and Functional Structure of a Robot

The functional structure of a robot can be divided into four parts: the mechanical structure, the sensors, the actuators and mechanical transceivers, and the computer; see Snyder (1985) and Fu et al. (1987).

The Mechanical Structure of a Robot. A robot consists of a chain of n rigid links (Gorla and Renaud, 1984). Each link i is attached to a previous link (i − 1) and a following link (i + 1) by a joint. In robotics, only prismatic and revolving joints are used because the actuators do not allow other motions. The prismatic joints are such that adjacent links translate linearly to each other along the joint axis, while the revolving joints allow adjacent links to rotate with respect to each other about the joint axis. Then the link i motion with reference to the link (i − 1) depends on only one variable, rotation θ_i or translation d_i. In general (except for mobile robots), the link 0 is fixed to the robot base. The terminal link n carries a gripper or a pincer or a tool and is also called the end effector (or hand) of the robot. The location of an object in space is determined by six independent variables or six degrees-of-freedom (d.o.f.), three of which indicate the position and the other three the orientation. If a task is performed in space without constraints, 6 d.o.f. are necessary. But if the task is performed in a plane, only 3 d.o.f. are required. The d.o.f. of a robot manipulator are the independent parameters required to fix the place of the end effector. This number is not always equal to the joint variable number. For example, two prismatic joints operating in the same direction are equivalent to only one prismatic joint.

A redundant robot mainpulator is a robot that has a larger d.o.f. number than that of the d.o.f. task number. In this case, it will be difficult to eliminate problems. In general, n, the d.o.f. number of a robot is equal to 6, but for less efficient robots n = 5 or n = 4.

If the joint axes of the final three links are concurrent, both position and orientation of the end effector can be decoupled. Only the position of the common point of three final axes is defined by the first three joint variables of the robot, and the orientation of the end effector, in relation to this point, is determined by three final joint variables.

Different structures exist for the first three links: only three prismatic joints directly give the position of end effector in Cartesian coordinates (14% of the robots) and the structure is called PPP; one revolving joint followed by two prismatic joints having orthogonal axes (RPP and PRP structure) give the position of end effector in the cylindrical coordinate system (47%); two revolving joints followed by one prismatic joint (RRP structure) give the position in the spherical coordinate system (13%). A robot that contains only revolving joints gives the position in the anthro-poid coordinate system (25%).

The SCARA structure (RRRP), which has recently appeared in Japan, merits attention, because this simple and stiff structure allows actuators and reducers to be coupled directly to a common axis, which is also the joint axis.

The joint variables of the robot are limited between both initial and final values which depend on the robot structure. As a result, the number of points that can be reached by the end effector is also limited. The set of points that can be reached is called the working space of the robot. Generally, it is a volume with a complex form calculated from the limit values of the joint variables. For a task to be workable, all the paths constitut-ing that task must be contained within the working space.

Sensors. All the sensors used in robotics belong to one of the follow-ing three classifications: sensors for measurements of joint variables, sen-sors located on the end effector, sensors for perception of the immediate environment.

In robot control the joint variables interfere either directly or through their velocity with control laws. Thus, four types of physical variables will have to be measured: angular and linear positions and angular and linear velocities.

The positional measurements are performed by digital or analog methods. In the digital method, the classical optical encorder is frequently used for incremental and absolute measurements. These devices are accurate (up to 40,000 steps/round) and reliable, for they give a digital signal possibly encoded with a security code. Their cost is smaller than that of analog devices, but the information is lost when the power supply is switched off. The analog measurement system groups the angle sensor and the digital-to-analog-converter in a single box. Ohmic or inductive potentiometers, re-solvers (angle) or inductosyn (linear) are used to operate the angle sensor.

The velocity sensors are conceived in the same manner. The digital velocity sensors measure the rotation or displacement velocities either based on pulses counted during a fixed time or based on the measurement of the time required to count the pulses. These techniques require a de-lay which does not exist for the analog velocity sensors. Indeed, the analog signal is produced from a D.C. tachometer generator. But digital analog conversion time delays the emission of the output signal. These analog sensors are very accurate and very fast, but their cost is high.

The sensors located on the end effector are proximity sensors and force sensors. The proximity sensors (ultrasonic transducer, infrared telemeter, or pneumatic system) indicate to the robot computer the exact position of the object between the fingers of the gripper. These operate only in the final phase of the trajectory and allow the grasp to be effected correctly. The force sensors (piezocrystal, constraint gauges) are used to measure and limit the force exerted by the gripper on the object or the compliance force operating on the wrist.

The sensors for perception of working environment consist of television cameras and charged coupled device (CCD) cameras. The scene analysis was first emphasized in two dimensions (2D systems); then only one camera was required. The 2D-vision system are currently used in industrial

processing, and several manufacturers propose systems that are very effective in real time. But 3D-vision processing is more difficult on both
theoretical and practical levels. First, the 3D-vision needs at least two
cameras; second, the complexity of algorithm is high, and accordingly
real-time processing becomes a problem; Groover et al. (1986).

For mobile robots, ultrasonic transducers attached to the robot base allow
the measurement of the robot's position with reference to fixed obstacles

Actuators. The actuators must move the links about their joints. Only
two motions are possible: rotation and translation. Two types of actuators
realize these motions, the rotative motors (electric or hydraulic) and the
hydraulic jack. But the transformation of actuator motions is made possible
by a mechanical transmitter. Thus, a rotation motion can be transformed
into displacement motion by a screw-nut system, while a translation motion
is turned into rotation motion by a connecting rod-crank system.

As a general rule, the electrical motors used in robotics are too fast
and have too small a torque to directly operate on the joint's axis. In
this case the joint is driven by an actuator through a gear reduction of
ratio r, which divides the actuator velocity by r, multiplies the actuator
inertia by r^2, and also multiplies the actuator torque by r. The main reducer used in robotics is the "harmonic drive," which is compact and
light.

Robots equipped with electrical motors are less powerful than hydraulic
robots, but they are also more dynamically accurate. Several motor types
are adequate provided that their inertia is small and their velocity is high,
or their torque is high and their velocity is very small. These conditions
are required by the gear reducer. Almost always they are D.C. motors with
a very large or very small ratio (rotor length)/(rotor diameter) to minimize
the inertia. A.C. motors of the type reluctance variable synchronous
motors (high torque, small velocity) or autocontrolled synchronous machines
(with permanent magnets) are seldom used. The stepping motor offers the
possibility of open-loop position control, because it turns a constant angle
(the "step") for each pulse received. One advantage of this motor is to sup-
press the position and velocity sensors, but a major disadvantage exists if
one or several "steps" are lost during the motion.

Hydraulic robots are operated by means of a rotative or linear hydraulic
jack. The control system of a hydraulic jack is the servo-valve, which
controls the oil flux in the room by an electrical signal. The main advantage
of a hydraulic jack is the very high ratio power/mass which allows the
actuators to generate. large forces or torques while occuppying a small
amount of space. Besides, the incompressibility of oil gives a very high
stiffness to the hydraulic robot at stop position. The main disadvantage
of hydraulic actuators is their bad dynamics and the presence of considerable
Coulomb friction.

Computers. The robot operation requires a highly efficient computer
to perform several tasks: data logging, analog-to-digital conversion, real-
time control, coordinate transformation, interpolation, trajectory generation,
etc. Since not all robots use all these functions, a large variety of computers are encountered. The simplest robots utilize a programmable controller specialized in sequential processing. Next, the classical industrial
robots which use "teaching by doing" and point-to-point techniques require

monoprocessor computers with significant capabilities of computation and storage. Finally, multiprocessor computers (see Chapter 16) using parallel computation are required for modern robots. At the present time, only parallel computation allows the autoadaptive control in real time. The vision processing is always carried out by a special processor, connected and synchronized to the main robot computer.

Example 12.2.1 (Fages, 1984)

PUMA 560		d.o.f:	6 (R.R.R.R.R.R)
Base rotation	Axis 1	2 × 320°	Loading capacity: 2.6 kg
Shoulder rotation	Axis 2	250°	Actuators: electrical motors
Elbow rotation	Axis 3	270°	Velocity of hand: 1m/s
Wrist rotation	Axis 4	300°	in point-to-point technique
Wrist bend	Axis 5	210°	
Hand rotation	Axis 6	532°	Power supply: 1.5 KVA
			Manufacture: Unimation, Inc.

Computer PDP 11/45: programming language: VAL—use: assembly

12.2.2 Programming of Industrial Robots

Various methods are used to carry out the programming of industrial robots. The more complex methods have recourse to high-level languages and take into account the forces, torques, synchronization, collision-free path planning, camera vision, etc. On the other hand, a useful technique in industrial applications is the "teaching-by-doing" technique, which does not require complex programming languages.

Teaching-by-Doing Technique. The teaching-by-doing method is based on the record of the joint variables in terms of time, when the end effector of the robot is manually led by a human operator through the points of a trajectory. Thus, all the points of the trajectory are expressed in terms of robot variables. Also, the angular and linear velocities can be recorded. If the robot has 6 d.o.f., each point of the trajectory corresponds to 12 values of the joint variables and their derivatives. Two usages of these data are possible: either to consider these values as inputs to the servomechanisms of the robot (in order to play back the same trajectory) or to use them in dynamic control algorithms of the robot. The first usage is the most common.

Occasionally, a human operator carries the robot hand along the desired path. Here a master-slave technique is said to be used. The master consists of a geometrical copy of the actual robot. The dimensions of the copy are exactly those of the robot (or with fixed and known ratio). The position and velocity sensors are strictly identical to those of the robot. The output signals of the copy's sensors feed servo-inputs of the robot. Thus, the motions of master along the trajectory are copied by the actual robot. The master structure is very light, so it is easy for the operator to lead the hand through all the points of the desired path. For playback utilization only the actuator control voltages of the slave are recorded.

Sometimes the points of the trajectory are reached by moving each joint independently by means of keys. The number of points thus recorded is very limited because the operation takes a long time.

Use of High-Level Programming Languages. Many high-level programming languages have been developed in the United States, Japan, and Europe over the last 15 years, and there is practically a command language for each kind of robot. While the robots developed from numerically controlled machine tools, the first robot programming languages were derived from machine tool languages. These languages operate especially on the actuators (position and velocity) by means of perforated tapes using interpolation subroutines; Groover et al. (1986).

Next the object of languages was to control the motions of the end effector. This assumes the capacity of the control program to transform the trajectory coordinates into joint variables in real time. Then the only specifications to give to the program are the positions and orientations of the hand along the trajectory. These data can be modified during the operation by a vision system or an interactive system. The environment model, i.e., the position and orientation of the objects surrounding the robot, is also furnished to the robot computer.

The WAVE programming language is the prototype of this family. R. Paul originally developed this language at Stanford University in 1972 (Paul 1977). Other languages operating on the end effectors have been created; examples are AL (1974), POINTY (1977), VAL (1979).

Other languages have also been developed to undertake specific tasks, such as assembly work, handling, etc. These are called goal-oriented languages, and their specifications are defined as operations such as to screw, to insert, etc. Thus, Auto-pass was developed in the IBM Research Center by Liebermann and Wesley (1977), AML by Taylor et al. (1982), another by Gini and Gini (1975) in Italy, RAPT by Popplestone et al. (1978) at Edinburgh University, PAL at Purdue University, and L. M. by Latombe (1979) in France. Note that PAL and L. M. are based on Pascal language.

Coordination of Motion. If each motion is left free, each actuator stops after the joint variable change has taken place. But joint variable changes have different durations, and each actuator stop causes a little shock. Thus, the displacement of the end effector from one point to another is discontinuous and is accompanied by vibrations. In order to avoid this undesirable effect, it is convenient to stop all the actuators at the same time. In other words, all the joint variable changes must have the same duration. Assume a robot with six revolving joints and θ_i and Ω_i are the angle and angular velocity, respectively, of joint i. Going from point 1 to point 2 along the trajectory, the six corresponding angular changes of the six joints are

$$\Delta \theta_i = \theta_i^2 - \theta_i^1 = \Omega_i T_i \qquad i = 1, 2, \ldots, 6$$

where the angular velocity Ω_i of joint i is assumed maximum and constant (except during acceleration and deceleration phases) and T_i is the time necessary for joint i to go from point 1 to 2. Then,

$$T_{max} = \max_i \frac{(\Delta\theta_i)}{(\Omega_i)} \qquad i = 1, \ldots, 6$$

defines the longest time to go from point 1 to 2; the velocities Ω_i must be adapted in order that the motions of each joint will have the same duration T_{max}.

Simulation and CAD to Carry the Robot into Effect. The simulation of the robot motions and its environment on a television screen is of great help to engineers in the industrial setting. Thus, collision-free paths can be studied without dangerous and expensive experiments. It also allows optimization of the motions and feed systems (Catier, 1984).

The CAD graphic program represents the robot image in 3D on the screen from either the geometrical or the kinematic model of the robot. Each variation of some joint variable modifies the image so that the robot drawing appears to move. The environment model is also required because the position and orientation of the pieces to be handled must be taken into account. The control language of the robot is used during the simulation; thus, when the task is correctly performed on the screen, the actual pro-gram is ready. Some CAD programs compute the torques and forces of the robot in order to design mechanical characteristics of the links, actuators, or reducers. IBM developed the EMULA system from AML control robot language for the IBM 3575 robot. At Stanford University, a robot has been achieved and associated with the acronym system in order to simulate the PUMA 500 robot. In France the CATIA system has been developed by Dassault Systemes.

12.3   MODELING OF ROBOTS

12.3.1   Mathematical Background

Introduction. Robot control uses mathematical models based on analytical transformations which are necessary for kinematical description of robot manipulators. Homogeneous transformation, in particular, allows vectors and frames to be changed by rotation and translation with only one matrix operator (Paul, 1982). This form can be used effectively in the formulation of kinematic and dynamic problems in the robotics field.

Rotation Matrices. Consider a reference coordinate system R_0 (O_0, O_0x_0, O_0y_0, O_0z_0) and a coordinate system R_1 (O_0, O_0x_1, O_0y_1, O_0z_1) initially the same as R_0. The R_1 system is rotated at a θ angle about the O_0x axis (Figure 12.1). If M is a fixed point represented by $a_0b_0c_0$ and $a_1b_1c_1$ with respect to R_0 and R_1 coordinate systems, respectively, then

$$(a_0 b_0 c_0)^T = \begin{bmatrix} 1 & 0 & 0 \\ 0 & \cos\theta & -\sin\theta \\ 0 & \sin\theta & \cos\theta \end{bmatrix} (a_1 b_1 c_1)^T = \underline{Rot}\,(\theta,x)\,(a_1 b_1 c_1)^T$$

The $\underline{Rot}$ (θ,x) matrix is called the rotation matrix. In the same manner, if the R_1 coordinate system is rotated about the O_0y or O_0z axis, then

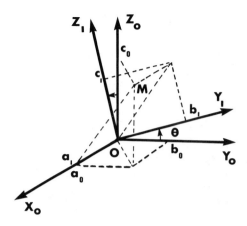

Figure 12.1   Rotation about x-axis.

$$\underline{Rot}(\theta,y) = \begin{bmatrix} \cos\theta & 0 & \sin\theta \\ 0 & 1 & 0 \\ -\sin\theta & 0 & \cos\theta \end{bmatrix} \qquad \underline{Rot}(\theta,z) = \begin{bmatrix} \cos\theta & -\sin\theta & 0 \\ \sin\theta & \cos\theta & 0 \\ 0 & 0 & 1 \end{bmatrix}$$

Now, if R_1 is chosen as the reference system and R_0 as the rotated co-ordinates system, the same relative positions for R_1 and R_0 are obtained by rotating R_0 about O_0x_1 axis at a $-\theta$ angle. In this case the $a_0 b_0 c_0$ and $a_1 b_1 c_1$ values do not change, and

$$(a_1 b_1 c_1)^T = \underline{Rot}(-\theta,x)\ (a_0 b_0 c_0)^T$$

Note that $\underline{Rot}(-\theta,x) = \underline{Rot}^T(\theta,x)$ because the sign of the $\cos\theta$ and "1" diagonal terms are invariable. The result is that

$$\underline{Rot}(-\theta,x) = \underline{Rot}^T(\theta,x) = \underline{Rot}^{-1}(\theta,x)$$

As a general rule, the columns of rotation matrix are the projections of the three unit vectors i', j', and k' of R_1 onto the three unit vectors i, j, and k of R_0 successively: the first column consists of three components of the O_0x_1 axis direction according to the R_0 reference system, the second column consists of O_0y_1 axis direction components, and the third column consists of O_0z_1 axis direction components (Lee, 1982).

The inverse rotation matrix is obtained by the projection of i, j, k (unit vectors of R_0) onto i', j', and k' (unit vectors of R_1). This results in the following transposed rotation matrix.

$$\underline{Rot}(\theta) = \begin{bmatrix} i'\cdot i & i'\cdot j & i'\cdot k \\ j'\cdot i & j'\cdot j & j'\cdot k \\ k'\cdot i & k'\cdot j & k'\cdot k \end{bmatrix}$$

$$\underline{Rot}(-\theta) = \underline{Rot}^{-1}(\theta) = \begin{bmatrix} i \cdot i' & | & i \cdot j' & | & i \cdot k' \\ j \cdot i' & | & j \cdot j' & | & j \cdot k' \\ k \cdot i' & | & k \cdot j' & | & k \cdot k' \end{bmatrix} = \underline{Rot}^T(\theta)$$

Where (.) represents the dot product.

<u>Product of Rotation Matrices.</u> Let the product of several rotation matrices be

$$H = \underline{Rot}(\theta_o x_o)\underline{Rot}(\theta_1 y_1)\ \underline{Rot}(\theta_2 Z_2) \cdots \underline{Rot}(\theta_i x_i) \cdots \underline{Rot}(\theta_q z_q)$$

Each partial product: $\underline{Rot}(\theta_o x_o)$; $\underline{Rot}(\theta_o x_o) \cdot \underline{Rot}(\theta_1 y_1)$; $\underline{Rot}(\theta_o x_o)$, $\ldots$, $\underline{Rot}(\theta_i x_i)$, etc. describes the orientation of intermediate coordinate frames $R_1,\ R_2,\ \ldots,\ R_{i+1},\ \ldots,\ R_{q+1}$, with respect to the R_o reference system:

$$R_o \xrightarrow{\underline{Rot}(\theta_o x_o)} R_1 \xrightarrow{\underline{Rot}(\theta_1 y_1)} R_2 \xrightarrow{\underline{Rot}(\theta_2 z_2)} \cdots \xrightarrow{\underline{Rot}(\theta_i x_i)} R_{i+1}$$

$$\xrightarrow{\quad} \cdots \xrightarrow{\underline{Rot}(\theta_q z_q)} R_{q+1}$$

The $(i + 1)$ frame is obtained by rotating the ith frame about the specified axis at a θ_i angle.

The elements of columns of $\underline{Rot}(\theta_i x_i)$ are the directions of x_{i+1}, y_{i+1}, and z_{i+1} axes of R_{i+1} frame, defined in R_i coordinate system.

The elements of columns of $\underline{H}$ matrix are the directions of x_{q+1}, y_{q+1}, and z_{q+1} axes of R_{q+1} frame expressed in the reference coordinate system. The x_{q+1}, y_{q+1}, and z_{q+1} axes will be called n (normal), O (orientation), and a (approach), respectively, as is usually the case in robotics.

$$\underline{H} = \begin{bmatrix} n_x & O_x & a_x \\ n_y & O_y & a_y \\ n_z & O_z & a_z \end{bmatrix}$$

where n_x, n_y, n_z; O_x, O_y, O_z; and a_x, a_y, a_z are the projections of unit vectors n, O, and a, respectively, onto the $O_o x_o$, $O_o y_o$, and $O_o z_o$ axes of R_o.

The inverse matrix $\underline{H}^{-1}$ is the transposed matrix of $\underline{H}$:

$$\underline{H}^{-1} = \{\underline{Rot}(\theta_o x_o)\underline{Rot}(\theta_i y_1) \cdots \underline{Rot}(\theta_q z_q)\}^{-1}$$

$$= \underline{Rot}^{-1}(\theta_q z_q) \cdots \underline{Rot}^{-1}(\theta_1 y_1)\underline{Rot}^{-1}(\theta_o x_o)$$

$$= \underline{Rot}^T(\theta_q z_q) \cdots \underline{Rot}^T(\theta_i y_1)\underline{Rot}^T(\theta_o y_o) = \underline{H}^T$$

All the products of rotation matrices can be expressed in terms of three rotations about the x, y, or z axis. Several solutions are possible: Euler angles, roll pitch, yaw angles (see Problem 12.1).

<u>Example 12.3.1: Euler Angles.</u> The three Euler angles ϕ, θ, Ψ are defined in Figure 12.2.

$$\underline{H} = \underline{Rot}(\phi z_o) \cdot \underline{Rot}(\theta y_1) \cdot \underline{Rot}(\Psi z_2)$$

$$\underline{H} = \begin{bmatrix} C\phi & -S\phi & 0 \\ S\phi & C\phi & 0 \\ 0 & 0 & 1 \end{bmatrix} \cdot \begin{bmatrix} C\theta & 0 & S\theta \\ 0 & 1 & 0 \\ -S\theta & 0 & C\theta \end{bmatrix} \cdot \begin{bmatrix} C\Psi & -S\Psi & 0 \\ S\Psi & C\Psi & 0 \\ 0 & 0 & 1 \end{bmatrix} = \begin{bmatrix} n_x & O_x & a_x \\ n_y & O_y & a_y \\ n_z & O_z & a_z \end{bmatrix}$$

$$\begin{bmatrix} (C\phi C\theta C\Psi - S\phi S\Psi) & -(C\phi\, C\theta S\Psi + S\phi C\Psi) & C\phi S\theta \\ (S\phi C\theta C\Psi + C\phi S\Psi) & -(S\phi\, C\theta S\Psi - C\phi C\Psi) & S\phi S\theta \\ -S\theta C\Psi & S\theta S\Psi & C\theta \end{bmatrix} = \underline{H} \qquad (12.1)$$

where $C = \cos$, $S = \sin$. If the elements of the H matrix are numerically defined, it is possible to calculate ϕ, Ψ, and θ. For example: $\tan\phi = a_y/a_x$; $\tan\Psi = O_z/-n_z$; $\tan\theta = \sqrt{a_x^2 + a_y^2}\,/\,a_z$.

<u>Rotation About a Given Axis.</u> If the rotation is done around a given axis which is not a coordinate axis of R_O, the complexity of the rotation matrix expression increases considerably.

Let us take a $\underline{O\,M}$ vector rotating at a θ angle about a $O_o\lambda$ axis (Figure 12.3) and with $\underline{k}$ a unit vector along this axis (with k_x, k_y, and k_z

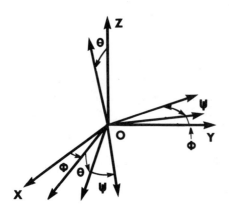

Figure 12.2   Euler angles.

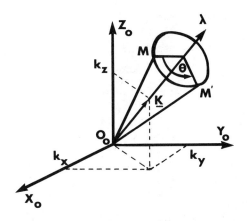

Figure 12.3   Rotation about a given axis.

components in R_o coordinate system).   After rotation, $\underline{O}_o\underline{M}$ becomes $\underline{O}_o\underline{M}'$ and

$$\underline{O}_o\underline{M}' = \underline{Rot}(\theta, O_o\lambda)\underline{O}_o\underline{M}$$

The $O_o\lambda$-axis can be assumed to be the z-axis of a coordinate frame obtained by several rotations from R_o.   We call $\underline{H}$ the product of rotation matrices representing these successive rotations and R_1, which is the coordinate frame obtained from R_o by $\underline{H}$ transformation.   Let x_o, y_o, z_o and x_1, y_1, z_1 be the coordinate of M point in R_o and R_1, respectively, then

$$\underline{O}_o\underline{M} = \underline{X}_o = (x_o\ y_o\ z_o)^T \qquad \underline{X}_1 = (x_1\ y_1\ z_1)^T \qquad \underline{X}_o = \underline{H}\underline{X}_1$$

Thus

$$\underline{X}_1 = \underline{H}^{-1}\underline{X}_o$$

The rotation around the $O_o\lambda$ axis for the $O_o M$ vector is equivalent to the rotation around the z-axis of R_1.

$$\underline{X}_o' = \underline{O}_o\underline{M}' = \underline{Rot}(\theta, O_o\lambda)\underline{X}_o = \underline{H}\ \underline{Rot}(\theta, z_1)x_1 = \underline{H}\ \underline{Rot}(\theta, z_1)\ \underline{H}^{-1}\ \underline{X}_o$$

where $\underline{Rot}(\theta, z)X_1 = X_1'$.
Thus, $\underline{Rot}(\theta, O_o\lambda) = \underline{H}\ \underline{Rot}(\theta, z_1)\ \underline{H}^{-1} = \underline{H}\ \underline{Rot}(\theta, z_1)\underline{H}^T$ where

$$\underline{H} = \begin{bmatrix} n_x & O_x & a_x \\ n_y & O_y & a_y \\ n_z & O_z & a_z \end{bmatrix} \quad \text{and} \quad \underline{Rot}(\theta, z_1) = \begin{bmatrix} \cos\theta & -\sin\theta & 0 \\ \sin\theta & \cos\theta & 0 \\ 0 & 0 & 1 \end{bmatrix}$$

We obtain the final expression

$$\underline{Rot}(\theta, O_o\lambda) = \begin{bmatrix} k_x^2(1-\cos\theta) + \cos\theta & \big| k_yk_x(1-\cos\theta) - k_z\sin\theta \big| \\ k_xk_y(1-\cos\theta) + k_z\sin\theta & \big| k_y^2(1-\cos\theta) + \cos\theta & \big| \\ k_xk_z(1-\cos\theta) - k_y\sin\theta & \big| k_yk_z(1-\cos\theta) + k_x\sin\theta & \big| \end{bmatrix}$$

$$\left. \begin{array}{c} k_zk_x(1-\cos\theta) + k_y\sin\theta \\ k_zk_y(1-\cos\theta) - k_x\sin\theta \\ k_z^2(1-\cos\theta) + \cos\theta \end{array} \right] \qquad (12.2)$$

where $\underline{k} = [k_x\ k_y\ k_z]^T = a$.

This result is independent of n and O components. Indeed, an infinity of R_1 frames exist that satisfy the previous assumptions. It is only defined by $O_o\lambda$ direction.

The $\underline{Rot}(\theta, O_o\lambda)$ matrix has all the properties of rotation matrices, especially $\underline{Rot}^{-1}(\theta, O_o\lambda) = \underline{Rot}(-\theta, O_o\lambda) = \underline{Rot}^T(\theta, O_o\lambda)$.

The $\underline{k}$ vector of $O_o\lambda$ axis and the θ rotation angle can be established from elements of the $\underline{Rot}(\theta, O_o\lambda)$ matrix, if these are numerically defined (Problem 12.2).

<u>Homogeneous Transformations</u>. The homogeneous coordinates of a point M consist of four scalar obtained by adding the 1-scalar to three cartesian orthogonal coordinates x, y, z of point M. Then, the homogeneous coordinates of point M are: (x y z 1). In like manner, a position vector $p = (p_x\ p_y\ p_z)^T$ is represented in the homogeneous coordinate system by a four-dimensional vector $\underline{p}^* = (p_xp_yp_z1)^T$.

The homogeneous components of a vector representing a direction are obtained by adding "0" to three classical components of this direction. Then the direction $\underline{V}$ is given by $\underline{V} = (V_x\ V_y\ V_z\ 0)^T$, where $V_x\ V_y\ V_z$ are the projections of the $\underline{V}$-direction vector onto the $O_x\ O_y\ O_z$ axes.

In these conditions the rotation matrices expressed in a homogeneous coordinate system are 4×4 matrices.

$$\underline{H} = \begin{bmatrix} n_x & O_x & a_x & 0 \\ n_y & O_y & a_y & 0 \\ n_z & O_z & a_z & 0 \\ 0 & 0 & 0 & 1 \end{bmatrix}$$

Thus,

$$\underline{Rot}(\theta, O_x) = \begin{bmatrix} 1 & 0 & 0 & 0 \\ 0 & C & -S & 0 \\ 0 & S & C & 0 \\ 0 & 0 & 0 & 1 \end{bmatrix} \quad \text{and} \quad \underline{Rot}(\theta, O_y) = \begin{bmatrix} C & 0 & 0 & 0 \\ 0 & 1 & S & 0 \\ -S & 0 & C & 0 \\ 0 & 0 & 0 & 1 \end{bmatrix}$$

where $C = \cos\theta$ and $S = \sin\theta$.

The homogeneous transformations have been set up in order to express the coordinate transformations by means of matrix-vector product. Let H be such a homogeneous transformation; $\underline{X}_0$ and $\underline{X}_1$ are vectors expressed with respect to R_o and R_1 coordinate systems, respectively. Thus:

$$\underline{X}_o = \underline{H}\,\underline{X}_1 \qquad \text{or} \qquad \{x_o\ y_o\ z_o\ 1\} = \underline{H}\ \{x_1\ y_1\ z_1\ 1\}^T$$

where $\underline{H}$ can represent translations or rotations assembled in the same single operator.

To this end, the translation is expressed in matrix form; for example, the translation at a, $-$ b along o_x and o_y axis, respectively, is therefore given by

$$\underline{\text{Trans}}\ (a,\ -b,\ 0) =
\begin{bmatrix}
1 & 0 & 0 & a \\
0 & 1 & 0 & -b \\
0 & 0 & 1 & 0 \\
0 & 0 & 0 & 1
\end{bmatrix}
=
\left[
\begin{array}{c|c}
\underline{I}_3 & \begin{matrix} a \\ -b \\ 0 \end{matrix} \\
\hline
\underline{0}^T & 1
\end{array}
\right]$$

where $\underline{I}_3$ is the 3×3 unit matrix, and $\underline{0}^T$ is a null element vector.

$$\underline{X}_o = \text{Trans}\ (a,\ -b,\ 0)\underline{X}_1 \qquad \{x_o\ y_o\ z_o\ 1\}^T = \{x_1 + a),\ (y_1 - b),\ z_1, 1\}^T$$

As a general rule, the $\underline{H}$ matrix is partitioned into two parts: a 3×3 rotation matrix and a translation or position vector. The fourth row only consists of three "0" and one "1".

$$H =
\left|
\begin{array}{c|c}
\text{Rot}\,(\theta,\text{axis}) & \begin{matrix}\text{Position}\\ \text{vector}\end{matrix} \\
\hline
\underline{0}^T & 1
\end{array}
\right|
=
\left|
\begin{array}{c|c|c|c}
\underline{n} & \underline{o} & \underline{a} & \underline{p} \\
\hline
& \underline{0}^T & & 1
\end{array}
\right|\ \underline{p}$$

The homogeneous transformation matrix can be viewed as resulting from a product of translation and rotation matrices:

$$\underline{H} = \underline{\text{Trans}}\ (\underline{p})\ \underline{\text{Rot}}\ (\theta,\ \text{axis})$$

It is noted that this product is not commutative.

The inverse homogeneous transformation matrix is

$$\underline{H}^{-1} = \{\underline{\text{Trans}}(\underline{p})\ \underline{\text{Rot}}(\theta,\text{axis})\}^{-1} = \underline{\text{Rot}}^T(\theta,\text{axis})\,\underline{\text{Trans}}(-\underline{p})$$

where

$$\underline{\text{Rot}}(\theta,\text{axis}) =
\begin{vmatrix}
n_x & o_x & a_x & 0 \\
n_y & o_y & a_y & 0 \\
n_z & o_z & a_z & 0 \\
0 & 0 & 0 & 1
\end{vmatrix}
\qquad
\underline{\text{Trans}}(\underline{p}) =
\begin{vmatrix}
1 & 0 & 0 & p_x \\
0 & 1 & 0 & p_y \\
0 & 0 & 1 & p_z \\
0 & 0 & 0 & 1
\end{vmatrix}$$

and

$$
\underline{H}^{-1} = \begin{vmatrix} n_x & n_y & n_z & -p.n \\ O_x & O_y & O_z & -p.O \\ a_x & a_y & a_z & -p.a \\ 0 & 0 & 0 & 1 \end{vmatrix}
$$

where $-p.n$, $-p.O$, $-p.a$ are the dot products of vector p with unit vectors n, O, and a.

12.3.2   Kinematic Modeling of Robots

Introduction. As stated earlier, a robot consists of several rigid elements called links, connected by prismatic or revolving joints. Each link is numbered with a number i ($i = 0, \ldots, n$). One extremity of the kinematic chain is fixed to a robot base (link 0) and the other end (link n) supports an effector, or a tool, and is free. Six degrees-of-freedom are required to specify the location of a coordinate frame attached to the effector: the position of origin is determined by a vector p and the orientation can be expressed by three angles (Euler angles, for example).

In order to describe the motion of the effector, it is necessary to define a reference coordinate system fixed to the robot base and a moving intermediate coordinate frame for each link. The relative motion between two links is specified by an angular or linear joint variable. Each joint axis i is placed at the connection of link i and link (i - 1). A relationship expressed in terms of only one joint variable q_i exists between the locations of two adjacent coordinate frames bonded to link (i − 1) and link i, respectively. A homogeneous transformation matrix, specified by the structure and the dimensions of each link, can be used to describe the geometrical relationship between the coordinate frame locations. Then, it is necessary to define a systematic method to establish the coordinate systems for each link. The method proposed by Denavit and Hartenberg (1955) is frequently used.

Denavit-Hartenberg Convention. An iterative method allows the engineer to attach an orthonormal cartesian coordinate system R_i to each link i of the robot:

$$
R_i = (O_i, \underline{x}_i \ \underline{y}_i \ \underline{z}_i)
$$

where O_i is the origin, and $\underline{x}_i$, $\underline{y}_i$, $\underline{z}_i$ are the unit vectors along the principal axes of R_i. See Figure 12.4.

The R_i frame is fixed on the link i, at the joint (i + 1) connecting the link i to link (i + 1).

The z_i axis is the axis of motion of joint (i + 1); if joint (i + 1) is rotational, z_i is the axis of rotation between link i and link (i + 1), and if joint (i + 1) is translational, z_i lies along the center line of link (i + 1).

The joint variables are defined as

$$
q_i = \sigma_i \theta_i + (1 - \sigma_i) d_i \tag{12.3}
$$

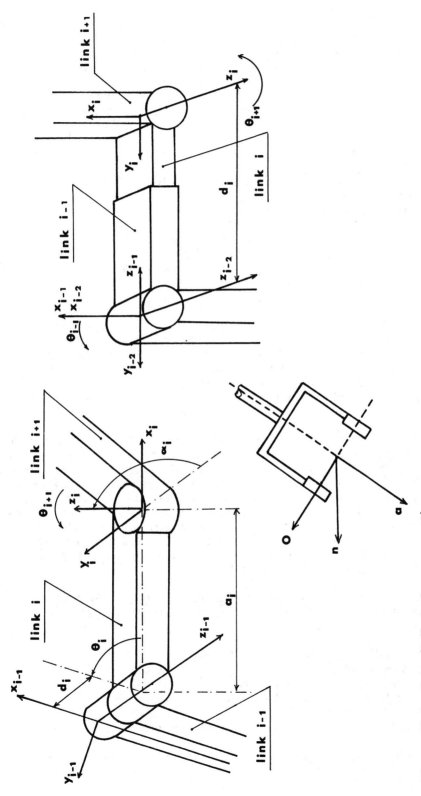

Figure 12.4   Denavit-Hartenberg convention.

where θ_i is a rotation angle about z_{i-1}, the axis of joint i connecting link (i −1) to link i, if joint i is rotational ($\sigma_i = 1$). d_i is the linear displacement of link i along the z_{i-1} axis of link i, if joint i is prismatic ($\sigma_i = 0$).

The O_i origin of R_i is generally placed at the intersection of the z_{i-1} and z_i axes, or at the intersection of the z_i axis and the common normal between the z_{i-1} and z_i axes. If z_{i-1} and z_i axes are parallel, an arbitrary common normal is chosen using symmetry properties of link i.

The x_i axis lies along the common normal between z_{i-1} and z_i axes. When they are concurrent or colinear, x_i axis is specified by the cross-product $(\underline{z}_{i-1}) \times (\underline{z}_i) = \underline{x}_i$, or it has an arbitrary direction. Likewise, the y_i axis is defined as $\underline{y}_i = (\underline{z}_i) \times (\underline{x}_i)$, which completes the right-hand coordinate system.

The R_0 frame is the reference coordinate system and is attached to the supporting base of the robot. The z_0 axis is the first motion axis of the link 1 for the joint variable q_1.

The R_n system fixed to the final link specifies the orientation of the hand. For R_n, the x_n y_n z_n axes are replaced, respectively, by n, O, and a unit vectors (normal, orientation, approach).

The relative position of R_i from R_{i-1} is fully described by four so-called Denavit-Hartenberg (DH) parameters: θ_i, d_i, a_i, and α_i defined as follows:

θ_i is the rotational joint angle around the z_{i-1} axis in the counter-clockwise direction. It is the angle between x_{i-1} and x_i axes.
d_i is the distance between the x_{i-1} and x_i axes along the z_{i-1} axis.
a_i (length of link i) is the projection of $O_{i-1}O_i$ onto the x_i axis, or the shortest distance between the z_{i-1} and z_i axes.
α_i (twist of link i) is the angle between the z_{i-1} and z_i axes about the x_i axis in a counterclockwise direction.

Only two DH parameters are always constant: a_i and α_i. The angle θ_i is variable when the joint i is a revolving joint, and in this case d_i is constant. The distance d_i is variable when the joint i is prismatic, and in this case θ_i remains constant. The DH parameters allow us to define the moving R_i frame using translations and rotations with respect to the z_{i-1} and x_i axis only. The y_i axis is never utilized. This restriction means that the robot has to perform a special initial configuration in order to assign the coordinate frames according to DH convention. This particular configuration is the "zero" of the robot because in this case all the joint variables will be zero. See Figure 12.4.

<u>Homogeneous Transformation Matrix for a Robot.</u> Using DH parameters, the assignment of the R_i frame from the R_{i-1} coordinate system can be split up into four transformations:

1. Rotation of angle θ_i about the z_{i-1} axis
2. Translation of distance d_i along the z_{i-1} axis
3. Translation of a distance a_i along the x_i axis
4. Rotation of an angle α_i about the x_i axis

Let $\underline{A}_i$ be the product of the previous transformation matrices, that is:

$$\underline{A}_i = \underline{Rot}(\theta_i, z_{i-1}) \cdot \underline{Trans}(0,0,d_i) \cdot \underline{Trans}(a_i,0,0) \cdot \underline{Rot}(\alpha_i, x_i)$$

$$\underline{A}_i = \begin{bmatrix} \cos\theta_i & -\sin\theta_i\cos\alpha_i & \sin\theta_i\sin\alpha_i & a_i\cos\theta_i \\ \sin\theta_i & \cos\theta_i\cos\alpha_i & -\cos\theta_i\sin\alpha_i & a_i\sin\theta_i \\ 0 & \sin\alpha_i & \cos\alpha_i & d_i \\ 0 & 0 & 0 & 1 \end{bmatrix} \qquad (12.4)$$

For a revolving joint, θ_i changes and a_i, d_i, and α_i remain constant. For a prismatic joint, the R_{i-1} and R_i coordinate systems have the same origin; then $a_i = 0$ and d_i is variable.

The first three columns of $\underline{A}_i$ matrices consist of the projections of $\underline{x}_i$, $\underline{y}_i$, and $\underline{z}_i$ unit vectors of R_i onto x_{i-1}, y_{i-1}, and z_{i-1} axes of R_{i-1}; the fourth column is the position of O_i with respect to the R_{i-1} coordinate system.

Using $\underline{A}_i$ matrices, it is possible to specify the position and orientation of the R_n moving coordinate frame with respect to the R_o fixed-reference coordinate system.

$$R_o \xrightarrow{\underline{A}_1} R_1 \xrightarrow{\underline{A}_2} R_2 \cdots \xrightarrow{\underline{A}_i} R_1 \xrightarrow{\underline{A}_{i+1}} \cdots \xrightarrow{\underline{A}_n} R_n$$

(with T_n spanning from R_o to R_n)

Noting that $\underline{T}_n$ is the homogeneous transformation matrix which expresses the orientation and position of the end effector with respect to the reference coordinate system, we obtain

$$T_n = \underline{A}_1\underline{A}_2\underline{A}_3 \cdots \underline{A}_i\underline{A}_{i+1} \cdots \underline{A}_n = \left[\begin{array}{ccc|c} n & O & a & p \\ \hline & \underline{O}^T & & 1 \end{array}\right]$$

$$(12.5)$$

We denote $i\underline{T}_j$, the homogeneous transformation matrix, to expresses the location of the R_j coordinate frame with respect to the R_i coordinate frame.

$$R_o \xrightarrow{\underline{A}_1} \cdots \xrightarrow{\underline{A}_i} R_i \xrightarrow{\underline{A}_{i+1}} \cdots \xrightarrow{\underline{A}_j} R_j \xrightarrow{\underline{A}_{j+1}} \cdots \xrightarrow{\underline{A}_n} R_n$$

(with $i\underline{T}_j$ spanning from R_i to R_j)

$$(12.6)$$

$$i\underline{T}_j = \underline{A}_{i+1}\underline{A}_{i+2} \cdots \underline{A}_{j-1}\underline{A}_j$$

If all the DH parameters are known, it is possible to calculate the n, O, a, and p components in the matrix T_n for the imposed values of q_i. In this case, the position and the orientation of the robot's hand are specified as a function of q_i variables. On the other hand, if the position and orientation of the end effector are imposed by a matrix X, it is possible to calculate q_i variables, solving the matrix equation

$$\underline{T}_n = \underline{X} \qquad (12.7)$$

This problem is more complex because of the $\underline{T}_n$ determination from specified joint variables. The set of matrices $\underline{A}_i$ and $\underline{T}_n$ forms the kinematic or geometrical modeling of the robot manipulator.

Example 12.3.2. We wish to determine the direct kinematic model of a SCARA type of robot: 4C06 robot, a product of SCEMI Co. The end effector motion depends on 4° freedom: three rotations and one translation. The four joint variables θ_1, θ_2, θ_3, and d_4 and the assignment of coordinate frames R_0, R_1, R_2, R_3, and R_4 are shown in Figure 12.5. The reference coordinate system is chosen so that the matrix A_1 is as simple as possible.

The joint motions are always defined about or along the z-axis. Table 12.1 gives the DH parameters. The ith row of Table 12.1 indicates the transformation parameters from the R_{i-1} frame to the R_i frame. The homogeneous transformation matrices are obtained from Table 12.1 and Eq. (12.4). In this case we have

$$\underline{A}_1 = \begin{bmatrix} c_1 & -s_1 & 0 & a_1 c_1 \\ s_1 & c_1 & 0 & a_1 s_1 \\ 0 & 0 & 1 & 0 \\ 0 & 0 & 0 & 1 \end{bmatrix} \qquad \underline{A}_2 = \begin{bmatrix} c_2 & -s_2 & 0 & a_2 c_2 \\ s_2 & c_2 & 0 & a_2 s_2 \\ 0 & 0 & 1 & d_2 \\ 0 & 0 & 0 & 1 \end{bmatrix}$$

$$\underline{A}_3 = \begin{bmatrix} c_3 & -s_3 & 0 & 0 \\ s_3 & c_3 & 0 & 0 \\ 0 & 0 & 1 & 0 \\ 0 & 0 & 0 & 1 \end{bmatrix} \qquad \underline{A}_4 = \begin{bmatrix} 1 & 0 & 0 & 0 \\ 0 & 1 & 0 & 0 \\ 0 & 0 & 1 & -d_4 \\ 0 & 0 & 0 & 1 \end{bmatrix}$$

Table 12.1   Transformation Parameters from the R_{i-1} Frame to the R_i frame

	θ_i	d_i	a_i	α_i	σ_i
R_1	θ_1	0	a_1	0	1
R_2	θ_2	d_2	a_2	0	1
R_3	θ_3	0	0	0	1
R_4	0	$-d_4$	0	0	0

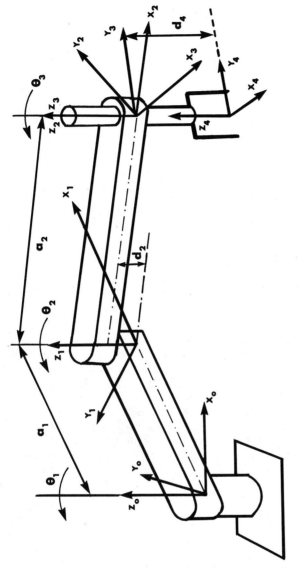

Figure 12.5 Robot 4C06.

$$
{}^{2}\underline{T}_4 = \begin{bmatrix} c_3 & -s_3 & 0 & 0 \\ s_3 & c_3 & 0 & 0 \\ 0 & 0 & 1 & (-d_4) \\ 0 & 0 & 0 & 1 \end{bmatrix}
\qquad
{}^{1}\underline{T}_4 = \begin{bmatrix} c_{23} & -s_{23} & 0 & a_2 c_2 \\ s_{23} & c_{23} & 0 & a_2 s_2 \\ 0 & 0 & 1 & (d_2 - d_4) \\ 0 & 0 & 0 & 1 \end{bmatrix}
$$

$$
\underline{T}_4 = \underline{A}_1 \underline{A}_2 \underline{A}_3 \underline{A}_4
\qquad
{}^{1}\underline{T}_4 = \underline{A}_2 \underline{A}_3 \underline{A}_4
\qquad
{}^{2}\underline{T}_4 = \underline{A}_3 \underline{A}_4
\qquad
{}^{3}\underline{T}_4 = \underline{A}_4
$$

$$
\underline{T}_4 = \begin{bmatrix} c_{123} & -s_{123} & 0 & a_2 c_{12} + a_1 c_1 \\ s_{123} & c_{123} & 0 & a_2 s_{12} + a_1 s_1 \\ 0 & 0 & 1 & d_2 - d_4 \\ 0 & 0 & 0 & 1 \end{bmatrix}
\qquad
\begin{aligned}
a_1 &= 400 \\
a_2 &= 275 \\
d_2 &= 100
\end{aligned}
$$

$$(12.8a)$$

where s_{ijk} and c_{ijk} represent $\sin(\theta_i + \theta_j + \theta_k)$ and $\cos(\theta_i + \theta_j + \theta_k)$, respectively.

12.3.3 Kinematic Equations Solution

The position and orientation of the end effector can be specified by means of of six parameters corresponding to 6 d.o.f. of a rigid body in 3D-space: the three coordinates of the origin and the three parameters define the orientation of the coordinate frame attached to the hand with respect to the reference coordinate system. We have already presented the use of the Euler angles to define the orientation of a coordinate system. Let $p = \{p_x\ p_y\ p_z\ 1\}^T$ be the specified position of origin of R_n frame attached to the hand, and $\phi\ \theta\ \Psi$ the three Euler angles of R_n with respect to the R_0 reference frame. Then from Eq. (12.1) and (12.7), we obtain

$$
\underline{T}_n = \begin{bmatrix} c\phi\, c\theta\, c\Psi - s\phi\, s\Psi & -(c\phi\, c\theta\, s\Psi + s\phi\, c\Psi) & c\phi\, s\theta & p_x \\ s\phi\, c\theta\, s\Psi + c\phi\, s\Psi & -(s\phi\, c\theta\, s\Psi - c\phi\, c\,\Psi) & s\phi\, s\theta & p_y \\ -s\theta\, c\Psi & s\theta\, s\Psi & c\theta & p_z \\ 0 & 0 & 0 & 1 \end{bmatrix}
\qquad (12.8b)
$$

Solving the kinematic equation consists of the discovery of the q_i variable for a specified location of the hand. Three methods are suggested here to calculate this: a direct method based on identification of matrix elements two by two, a variational method using the Newton-Raphson optimization technique, and an iterative numerical method using the inverse Jacobian matrix of the robot. Only the first method is convenient for real-time computation because the others require several iterations and consequently are very slow.

Direct Analytical Solution. For a specified location of the end effector of an $\underline{n}$ - d.o.f robot, the numerical values of the $\underline{T}_n$ matrix elements are imposed, and each A_i matrix [Eq. (12.5)] contains only the variable q_i. It is possible to isolate q_1, q_2, . . ., q_n successively in the following manner:

$$\underline{T}_n = \underline{A}_1 \quad \underline{A}_2 \cdot \cdot \cdot \cdot \cdot \cdot \cdot \cdot \cdot \cdot \cdot \cdot \underline{A}_n$$

$$\underline{A}_1^{-1} \ \underline{T}_n = \underline{A}_2 \quad \underline{A}_3 \cdot \cdot \cdot \cdot \cdot \cdot \cdot \cdot \cdot \cdot \cdot \cdot \underline{A}_n = {}^1\underline{T}_n$$

$$\underline{A}_{n-1}^{-1} \cdot \cdot \cdot \cdot \cdot \cdot \cdot \cdot \underline{A}_1^{-1} \ \underline{T}_n = {}^{n-1}\underline{T}_n \qquad (12.9)$$

The matrix elements of the right-hand side of the first equation are either zeros, constants, or functions of q_1 . . . q_n variables. All of the elements of the $\underline{T}_n$ matrix are zeros or constants. By identifying each element of the two matrices, it is possible to find simple relations giving one or several angular variables.

The matrix elements of the left-hand side of the second equation are only dependent on q_1 since $\underline{T}_n$ is numerically specified. The ${}^1\underline{T}_n$ matrix elements consist of zeros, constants, or trigonometrical functions of q_2 . . . q_n.

Solving the first matrix equation generally gives q_1 and sometimes other variables. The calculation of the q_2 variable is achieved in the same way, using the third matrix equation, where q_1 is determined earlier. The other variables are calculated by the same method.

In some cases, this procedure is very difficult to carry out, according to Lee (1982) (for the PUMA robot, for example). In this case the solution is obtained in two steps. Indeed, the end effector is often moving about three concurrent axes, in which case R_4, R_5, and R_6 have the same origin (DH convention) situated at the end point of link 3. The position of this point is easy to determine from the product of the matrices $\underline{A}_1$, $\underline{A}_2$, $\underline{A}_3$ and from the position of the end effector. Note that variables q_4, q_5, and q_6 do not interfere with the calculation of q_1, q_2, and q_3. This calculation makes up the first step. During the second step, the q_4, q_5, and q_6 joint variables are solved for using the orientation of the hand.

Solvability and Computational Aspect. Solving of the inverse kinematic problem generally gives several solutions. Given a position and orientation of the end effector specified by r parameters (r = 1, 2, . . ., 6), let n be the degrees of freedom of the robot. Thus,

If n < r, then no solution exists.
If n = r, then several sets of solutions exist.
If n > r, then infinite solutions are available.

The solutions for the inverse kinematic problem can be easy to obtain with a computer working in real time. They generally require no more than 30–40 multiplications, 20–25 additions, and 15–20 transcendental function calls.

For all these calculations, the numerical values of angles are more accurate if the determination is based on the function $\tan^{-1}$ (indeed, $-\infty < \tan\theta_i < +\infty$, while $-1 \leqslant \cos$ or $\sin \leqslant +1$). If the tangent function is calculated from an a/b quotient, it is easy to specify θ by the FORTRAN function ATAN 2, and we have

$0 \leqslant \theta \leqslant 90°$ for a > 0 and b > 0 $90° \leqslant \theta \leqslant 180°$ for a > 0 and b < 0

$-180° \leqslant \theta \leqslant -90°$ for a < 0 and b < 0 $-90° \leqslant \theta \leqslant 0°$ for a < 0 and b > 0

Example 12.3.3. The inverse kinematic solution for the 4C06 robot is required. The end effector location is specified by

$$
\underline{X} =
\begin{vmatrix}
-0.87 & 0.5 & 0 & 210 \\
-0.5 & -0.87 & 0 & 364 \\
0 & 0 & -1 & -80 \\
0 & 0 & 0 & 1
\end{vmatrix}
=
\begin{vmatrix}
n_x & O_x & a_x & p_x \\
n_y & O_y & a_y & p_y \\
n_z & O_z & a_z & p_z \\
0 & 0 & 0 & 1
\end{vmatrix}
= \underline{T}_4
\qquad (12.10)
$$

Solving the previous equation, the values of θ_1, θ_2, θ_3, and d_4 joint variable, required for the hand to reach the given position and orientation, can be found.

We directly obtain from Eq. (12.10) and (12.8a)

$$d_2 - d_4 = p_z = -80 \qquad d_4 = 80 + d_2$$

$$\theta_1 + \theta_2 + \theta_3 = \tan^{-1}\left[\frac{n_y}{n_x}\right] \quad \text{or} \quad \tan^{-1}\left[\frac{-O_x}{O_y}\right]$$

$$\theta_1 + \theta_2 + \theta_3 = 210° \text{ or } 30°$$

$$p_x = a_2 c_{12} + a_1 c_1 \qquad c_{12} = \frac{p_x - a_1 c_1}{a_2}$$

$$p_y = a_2 s_{12} + a_1 s_1 \qquad s_{12} = \frac{p_y - a_1 s_1}{a_2}$$

$$p_x c_1 + p_y s_1 = \frac{p_x^2 + p_y^2 + a_1^2 - a_2^2}{2a_1} = k$$

The solution of this trigonometrical relation is

$$\theta_1 = 2 \tan^{-1}\left(\frac{p_y}{p_x + k} \pm \frac{\sqrt{p_y^2 + p_y^2 - k^2}}{p_x + k}\right) = \begin{cases} 99° \\ 21° \end{cases}$$

From the above relations

$$T_{12} = \frac{s_{12}}{c_{12}} = \frac{p_y - a_1 s_1}{p_x - a_1 c_1}$$

$$\theta_1 + \theta_2 = \tan^{-1} \frac{p_y - a_1 s_1}{p_x - a_1 c_1} = \begin{cases} -6°4 \text{ for } \theta_1 = 99°; \ \theta_2 = -105°4 \\ 126°4 \text{ for } \theta_1 = 21°; \ \theta_2 = +105°4 \end{cases}$$

The θ_3 angle is easily deduced from $\theta_1 + \theta_2 + \theta_3 = 210°$ or $30°$ and for $\theta_1 = 99°$, $\theta_2 = 105°4$, then $\theta_3 = 216°4$ or $36°4$; and for $\theta_1 = 21°$, $\theta_2 = 105°4$, then $\theta_3 = 83°6$ or $-96°4$. The set of solutions requires 24 multiplications, 24 additions/subtractions, eight divisions, and 16 transcendental function calls.

Computer-Aided Setting of the Kinematic Model. Standardization of homogeneous transformation matrices by means of DH parameters gives the possibility of writing the kinematic model of a robot symbolically using a computer. The elements of $\underline{T}_n$ or $^i\underline{T}_n$ matrices are obtained using symbolic calculation based on the laws of matrix product and applied to symbolic expressions. All the languages convenient for string processing can be used. It must be noted that the automatic reduction of complex trigonometric expressions is a very difficult task. Only artificial-intelligence-oriented programs can achieve this aim. If low-level languages are used, the desired reduced expressions are not obtained.

Despite these disagreements (which can be minimized) the computer-aided method for the setting of the kinematic model is very useful because it avoids tedious work and eliminates miscalculations.

12.3.4 Jacobian Model (Gorla and Renaud, 1984; Paul, 1982)

Mechanical background

Linear and angular velocities. The linear velocity V of the extremity of a vector resulting from a rotating motion about an axis Oz is determined as

$$V = O'M.\Omega = OM \sin\alpha\Omega$$

where

$$\Omega = \frac{d\theta}{dt}$$

or in vector form

$$\underline{V} = \underline{\Omega} \times \underline{OM} \tag{12.11}$$

If $|\underline{OM}| = \rho$ and varies with time, the linear velocity along the trajectory of point M can be represented by two orthogonal components $\underline{V}_R$

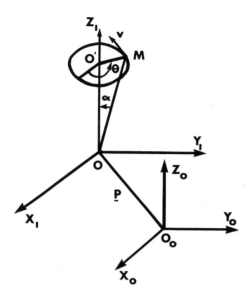

Figure 12.6   Linear velocity of a rotating vector.

and $\underline{V}_T \cdot \underline{V}_R$ lies along $\underline{OM}$, and $\underline{V}_T$ is tangent, at M, to a circle situated in a plane orthogonal to $\overline{O_z}$ and containing M (see Figure 12.6). Now,

$$\underline{V} = \underline{V}_R + \underline{V}_T = \frac{d\rho}{dt} \cdot \frac{\overline{OM}}{\rho} + \underline{\Omega} \times \underline{OM}$$

Now the oxyz coordinate system is translated by a vector $\underline{p}$ from the $O_0 x_0 y_0 z_0$ fixed-reference coordinate frame and the translation velocity is $\underline{V}_e = d\underline{p}/dt$. The linear velocity of point M with reference to the base coordinate system is then

$$\underline{V} = \underline{V}_e + \frac{d\rho}{dt} \cdot \frac{\overline{OM}}{\rho} + \underline{\Omega} \times \underline{OM} \tag{12.12}$$

The angular velocity of vector O_1M and vector O_0O_1 with reference to the base coordinate $(O_0 x_0 y_0 z_0)$ are called absolute angular velocity $\underline{\Omega}_a$ and carrying velocity $\underline{\Omega}_e$, respectively. The angular velocity of the O_1M vector with reference to the moving coordinate system is termed relative velocity $\underline{\Omega}_r$ (see Figure 12.7).

Then we have

$$\underline{\Omega}_a = \underline{\Omega}_r + \underline{\Omega}_e \tag{12.13}$$

The angular velocity is unchanged in the course of a translation motion.

Linear and angular acceleration. The linear acceleration Γ is expressed by means of the derivative of $\underline{V}$ with respect to time; thus, by using the cross-product, we obtain

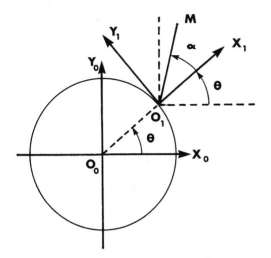

Figure 12.7   Angular velocity of a rotating vector.

$$\Gamma = \frac{d\underline{V}}{dt} = \frac{d}{dt}(\underline{\Omega} \times \underline{OM}) = \dot{\underline{\Omega}} \times \underline{OM} + \underline{\Omega} \times (\underline{\Omega} \times \underline{OM}) \tag{12.14}$$

where $|\underline{OM}|$ is a constant.

If $|\underline{OM}|$ varies with time, the extremity of vector $\underline{V}$ goes through four displacements. Two of these displacements are as above, caused by variation in $\underline{V}_T$; the other two are caused by variation in $\underline{V}_R$.

$$\Gamma = \frac{d\underline{V}_R}{dt} + \frac{d\underline{V}_T}{dt}$$

$$\Gamma = \left(\frac{d^2|\underline{OM}|}{dt^2}\right)\frac{\underline{OM}}{|\underline{OM}|} + \underline{\Omega} \times \left(\frac{d|\underline{OM}|}{dt}\right) \cdot \frac{\underline{OM}}{|\underline{OM}|} + \dot{\underline{\Omega}} \times \underline{OM}$$

$$+ \Omega \times \left(\frac{d|\underline{OM}|}{dt}\right) \cdot \frac{\underline{OM}}{|\underline{OM}|} + \underline{\Omega} \times (\underline{\Omega} \times \underline{OM})$$

$$\Gamma = \dot{\underline{\Omega}} \times \underline{OM} + 2\underline{\Omega} \times \left(\frac{d|\underline{OM}|}{dt}\right) \cdot \frac{\underline{OM}}{|\underline{OM}|} + \underline{\Omega} \times (\underline{\Omega} \times \underline{OM}) + \frac{d^2|\underline{OM}|}{dt^2}$$

$$\cdot \frac{\underline{OM}}{|\underline{OM}|} \tag{12.15}$$

If the origin O moves with a velocity $\underline{V}_e$, the linear acceleration $\dot{\underline{V}}_e$ will be added to Eq. (12.15).

The second and third terms in Eq. (12.15) are Coriolis acceleration and centrifugal acceleration, respectively.

The same method is used for the angular acceleration $\dot{\underline{\Omega}}_a$:

$$\dot{\underline{\Omega}}_a = \dot{\underline{\Omega}}_e + \dot{\underline{\Omega}}_r$$

But the displacement of the extremity of vector $\underline{\Omega}_r$ is caused here by the increase in the length of the $\underline{\Omega}_e$. Then,

$$\dot{\underline{\Omega}}_r = \frac{d|\underline{\Omega}_r|}{dt} \cdot \frac{\underline{\Omega}_r}{|\underline{\Omega}_r|} + \underline{\Omega}_e \times \underline{\Omega}_r$$

and

$$\dot{\underline{\Omega}}_a = \dot{\underline{\Omega}}_e + \frac{d|\underline{\Omega}_r|}{dt} \cdot \frac{\underline{\Omega}_r}{|\underline{\Omega}_r|} + \underline{\Omega}_e \times \underline{\Omega}_r \qquad (12.16)$$

<u>Definition of the Jacobian Model (Whitney, 1972).</u> The position and orientation of the end effector is specified by

$$\underline{T}_n(q) = \underline{X}$$

where $\underline{q} = (q_1 \, q_2 \, \cdots \, q_n)^T$ is a vector representing the n generalized joint variables.

If q increases by $\underline{dq}$, the hand of the robot is displaced very slightly by the amount dp along the trajectory, and there is a very small rotation dα about the k axis. The components $\underline{dp}$ and $\underline{kd\alpha}$ are calculated in terms of $\underline{dq}$ by

$$(dp_x, dp_y, dp_z, k_x d\alpha, k_y d\alpha, k_z d\alpha)^T = \begin{vmatrix} \dfrac{\partial p_x}{\partial q_1} & \cdots\cdots & \dfrac{\partial p_x}{\partial q_n} \\[2mm] \dfrac{\partial (k_z \alpha)}{\partial q_1} & \cdots\cdots & \dfrac{\partial (k_z \alpha)}{\partial q_n} \end{vmatrix} \cdot \begin{vmatrix} dq_1 \\ \cdot \\ \cdot \\ \cdot \\ dq_n \end{vmatrix}$$

$$(12.17)$$

or

$$(\underline{dp}, \, \underline{kd\alpha})^T = \underline{J}(q) \, \underline{dq}$$

The $\underline{J}(q)$ matrix is called the Jacobian matrix of the robot. This matrix is dependent on q, i.e., the position and orientation of each joint. The q_i values set is called robot configuration.

The direct calculation of $\underline{J}(q)$ can be undertaken by noting that the $\underline{p}$ components constitute the fourth column of the $\underline{T}_n$ matrix and that $\underline{kd\alpha}$ is obtained from $\underline{T}_n$ by means of Eq. (12.2). The direct calculation poses many difficulties, and other methods are more convenient.

The Jacobian model has several uses:

Considering the different elements $\underline{dp}$, $\underline{kd\alpha}$, and $\underline{dq}$ as small finite variations $\underline{\delta p}$, $\underline{k\delta\alpha}$, and $\underline{\delta q}$, it is possible to define an incremental motion for the end effector.

By introducing the time derivative, the Jacobian model becomes

$$[\dot{\underline{p}}, (\underline{k\dot{\alpha}})]^T = \underline{J}(q)\dot{\underline{q}} \qquad (12.18)$$

Thus, we obtain the instantaneous kinematic model. This model is useful to calculate the velocities of the end effector from joint velocities, and to achieve resolved rate control and resolved acceleration control.

Jacobian Matrix Calculation. Beginning with the Jacobian matrix with respect to the reference frame, and assuming all the q_j variables to be fixed, except q_j ($j = i$), the angular and linear velocities of R_n frame, attached to the end effector, are easily obtained from Eq. (12.18) and (12.13) (see Figure 12.8).

If joint i is a revolute, then:

$$\underline{V}_{ni} = (\underline{z}_{i-1} \, \dot{\theta}_i) \times \underline{O_{i-1}O_n}$$

$$\underline{\Omega}_{ni} = \underline{z}_{i-1} \, \dot{\theta}_i \qquad\qquad (12.19)$$

If joint i is a prismatic joint then

$$\underline{V}_{ni} = \underline{z}_{i-1} \, \dot{d}_i \qquad \text{and } \underline{\Omega}_{ni} = 0$$

where $\underline{O_{i-1}O_n} = \underline{p}_n - \underline{p}_{i-1}$; $\underline{p}_n$ and $\underline{p}_{i-1}$ are the fourth columns of the $\underline{T}_n$ and $\underline{T}_{i-1}$ matrices, respectively, and $\underline{z}_{i-1} = \underline{a}_{i-1}$ is the third column of the $\underline{T}_{i-1}$ matrix.

The hand velocities result from the sum of the contributions of each joint. In this case we obtain

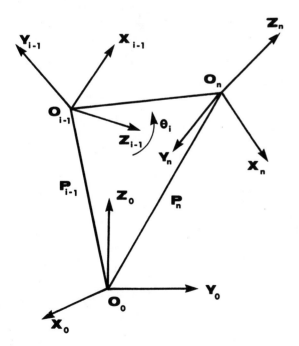

Figure 12.8   Determination of the velocity of R_n frame.

$$\underline{V}_n = \sum_{i=0}^{i=n} \underline{V}_{ni} \qquad \qquad \underline{\Omega}_n = \sum_{i=0}^{i=n} \underline{\Omega}_{ni}$$

$$[\underline{V}_n, \underline{\Omega}_n]^T R_o = \begin{vmatrix} \sigma_i \{\underline{z}_{i-1} \times (\underline{p}_n - \underline{p}_{n-1})\} + (1 - \sigma_i)\underline{z}_{i-1} \\ \\ \\ \\ \sigma_i \underline{z}_{i-1} \\ \\ \end{vmatrix} \begin{vmatrix} \dot{q}_1 \\ \dot{q}_2 \\ \vdots \\ \dot{q}_n \end{vmatrix}$$

$$\underbrace{}_{\text{ith column}} \qquad (12.20)$$

where σ_i and q_i are defined in Eq. (12.2), and $\underline{z}_{i-1}$, $\underline{p}_n$, and $\underline{p}_{i-1}$ are expressed with respect to the fixed-reference coordinate frame.

Jacobian Matrix with Respect to Moving Coordinate System. The above calculation is now undertaken with respect to the R_{i-1} frame; that is

$$\underline{z}_{i-1} = \{0 \ 0 \ 1\}^T \quad O_{i-1}O_n = ({}^{i-1}p_{nx}, \ {}^{i-1}p_{nx}, \ {}^{i-1}p_{ny}, \ {}^{i-1}p_{nz})^T$$

where ${}^{i-1}\underline{p}_n$ is the fourth column of the ${}^{i-1}\underline{T}_n$ matrix which gives the position and orientation of the R_n frame with respect to the R_{i-1} frame.

The angular $\underline{\Omega}_n$ and linear $\underline{V}_n$ velocities obtained in this way can be projected onto the axes of any R_j moving coordinate system. Thus, the hand rates will be described in the R_j frame, and in particular for the R_n frame attached to the end-effector, the linear velocity components are the projections of $(\underline{z}_{i-1} \ \dot{\theta}_i) \times ({}^{i-1}\underline{p}_n)$ onto the n, 0, and a directions, i.e., the invariable dot products; for a revolute joint

$$\underline{n} \cdot [\underline{z}_{i-1} \ \dot{\theta}_n \times {}^{i-1}\underline{p}_n] \qquad \qquad \underline{O} \cdot [\underline{z}_{i-1} \ \dot{\theta}_i \times {}^{i-1}\underline{p}_n]$$

$$\underline{a} \cdot [\underline{z}_{i-1} \ \dot{\theta}_i \times {}^{i-1}\underline{p}_n]$$

and for a prismatic joint

$$\underline{n} \cdot \underline{z}_{i-1}\dot{d}_i \qquad \underline{O} \cdot \underline{z}_{i-1}\dot{d}_i \qquad \underline{a} \cdot \underline{z}_{i-1}\dot{d}_i$$

where $\underline{n}$, $\underline{O}$, and $\underline{a}$ are expressed with respect to the R_{i-1} frame. Then

$$(\underline{V}_n, \underline{\Omega}_n)\ {}^T\!R_n \;=\; \begin{bmatrix} \sigma_i \begin{bmatrix} -\,n_x p_y + n_y p_x \\[4pt] -\,O_x p_y + O_y p_x \\[4pt] -\,a_x p_y + a_y p_x \\[4pt] n_z \\[4pt] O_z \\[4pt] a_z \end{bmatrix} + (1 - \sigma_i) \begin{bmatrix} n_z \\ O_z \\ a_z \\ 0 \\ 0 \\ 0 \end{bmatrix} \end{bmatrix} \cdot \begin{bmatrix} \dot{q}_1 \\ \vdots \\[6pt] \vdots \\ \dot{q}_n \end{bmatrix}$$

$$\underbrace{\hphantom{\begin{bmatrix} -\,n_x p_y + n_y p_x \\ n_z \end{bmatrix}}}_{\text{ith column}}$$

$$\hfill (12.21)$$

where $\underline{n}$, $\underline{O}$, $\underline{a}$, and p are the columns of the ${}^{i-1}T_n$ matrix.

This formulation is useful when the robot is controlled by a camera.

On the other hand, $(\underline{V}_6,\ \underline{\Omega}_6)^T\!{}_{Ro}$ can be easily obtained from $(\underline{V}_6,\ \underline{\Omega}_6)^T\!{}_{R_6}$ using

$$(\underline{V}_6\ \underline{\Omega}_6)^T\!{}_{Ro} \;=\; \left|\begin{array}{c|c} \rho_6 & 0 \\ \hline 0 & \rho_6 \end{array}\right| \cdot \left|\begin{array}{c} \underline{V}_6 \\ \hline \underline{\Omega}_6 \end{array}\right|_{R_6}$$

where ρ_6 is the rotation submatrix in the $\underline{T}_6$ homogeneous transformation matrix for a 6-d.o.f. robot.

Note that the Jacobian matrix has the following particular form, if the last three joints have concurrent axes.

$$\underline{J} \;=\; \left|\begin{array}{c|c} A & 0 \\ \hline B & C \end{array}\right| \hfill (12.22)$$

<u>The Inverse Jacobian Matrix</u>. In order to determine the δq joint variable variations (or velocities) from a specified δp displacement (or linear velocity) and a specified $\delta \alpha$ rotation (or angular velocity), it is necessary to invert the Jacobian matrix. Thus

$$\underline{\delta q} \;=\; \underline{J}^{-1}(q)(\underline{\delta p},\ \underline{\delta \alpha})^T$$

Next, we shall discuss the direct symbolic inversion of the $\underline{J}(q)$ matrix and the differentiation of the inverse kinematic model, which are two convenient methods.

<u>Direct Symbolic Inversion of $\underline{J}(q)$</u>. As a general rule, this problem is very difficult to solve owing to the complexity of $\underline{J}(q)$ matrix elements. But

this difficulty is decreased if $\underline{J}(q)$ has the form Eq. (12.22); in this case inverse Jacobian is as follows:

$$\underline{J}^{-1}(q) \;=\; \left| \begin{array}{c|c} \underline{A}^{-1} & 0 \\ \hline -\underline{C}^{-1}\underline{B}\underline{A}^{-1} & \underline{C}^{-1} \end{array} \right| \tag{12.23}$$

where $\underline{A}$, $\underline{B}$, and $\underline{C}$ are 3×3 sub-matrices for a 6-d.o.f. robot. Then the inversion of a 6×6 matrix requires inversion of two 3×3 matrices. But this method assumes that the $\underline{J}(q)$ matrix is not singular, i.e.,

$$\det \underline{J}(q) \neq 0 \qquad \text{or} \qquad \det \underline{A}.\det \underline{C} \neq 0$$

The q_i values, which result in a zero det J, form the so-called singular configuration of the robot. These are obtained from

$$\det \underline{A} = 0 \qquad \text{and} \qquad \det \underline{C} = 0$$

In general, the calculation of det $\underline{J}(q)$ is very difficult.

Differentiation of the Inverse Kinematic Model. The kinematic model provides the q vector using the equation $T_n(q) = \underline{X}$. Each q_i component is a function of $\underline{p}$, $\underline{n}$, $\underline{O}$, and $\underline{a}$ vectors. The derivative of the expressions with respect to time gives $\underline{dq}$ in terms of $\underline{dp}$, $\underline{dn}$, $\underline{dO}$, and $\underline{da}$.

As a general rule, the θ_i joint variables are expressed as

$$\theta_i = \tan^{-1}\frac{N}{D}$$

or

$$\tan\theta_i = \frac{N}{D}$$

The derivative of the tan function gives $d\theta_i$ in terms of N and D:

$$d\theta_i = \frac{DdN - NdD}{D^2 + N^2} \tag{12.24}$$

12.3.5   Dynamic Models of Robots

The kinematic modeling of robots allows the engineer to compute the position and orientation of the end effector without having to refer to the forces and torques operating in the robot. The kinematic equations are used to describe the position and orientation of a hand that is moving very slowly along a trajectory.

The equation of motion takes into account the dynamic parameters of the robot, such as mass, inertia, and viscous damping. This allows calculating the torques and forces that are produced by the actuators in order to impose accelerations and velocities on the hand of the robot along a desired trajectory. Robot control requires the use of dynamic models.

Several approaches are available to model the robot dynamics. The Lagrange-Euler formulation is the oldest method (Paul, 1982). Despite many improvements (Hollerbach, 1980), this formulation remains computationally inefficient, but it is very useful to the control law design because it directly leads to state-space equations. More recently, the Newton-Euler formulation was expanded with a view to obtaining recursive dynamic equations adapted to real-time computation. Efficient algorithms which were derived from these equations by Luh et al. (1980a) allow compact and fast computations. But the control law design is difficult with the Newton-Euler formulation because the equation structure is not adaptable to the state-space model. These two formulations are presented next.

Lagrangian Formulation of Robot Dynamics. The derivation of the equations of motion for the joint variables of a robot utilizes the Lagrangian function defined as $L = K_e - P_e$, where K_e is kinetic energy and P_e is potential energy. The generalized forces B_i, which are applied to joint i, are obtained from this Lagrange-Euler relation:

$$B_i = \frac{d}{dt}\left[\frac{\partial L}{\partial \dot{q}_i}\right] - \frac{\partial L}{\partial q_i} \tag{12.25}$$

Lagrangian Function Calculation

Kinematic energy. Let M be a point on the link i and $^i r$ be the position vector of M with respect to the R_i moving coordinate system. r, the position vector of point M, with respect to the R_0 fixed-reference coordinate system, is calculated from

$$\underline{r} = \underline{T}_i \, ^i\underline{r}$$

where $\underline{T}_i = \underline{A}_1 \underline{A}_2 \ldots \underline{A}_i$. The velocity of point M, with respect to the R_0 frame, is calculated when only joint i is moving by obtaining

$$\left[\frac{d\underline{r}}{dt}\right]_i = \frac{\partial T_i}{\partial q_i} \, ^i\underline{r} \, \dot{q}_i = \underline{v}$$

$[dr/dt]_i$ is the contribution of joint i to the velocity of point M. Each joint situated between the robot base and link i brings contribution $(\underline{dr}/dt)_j$, j = 1, 2, . . ., i, and the resulting velocity is then

$$\frac{d\underline{r}}{dt} = \sum_{j=1}^{j=i} \frac{\partial T_i}{\partial q_j} \, ^i\underline{r} \, \dot{q}_j \tag{12.26}$$

Let dm_i be the differential mass located in the M point. The differential kinetic energy is then

$$dK_i = \frac{1}{2} \underline{v}^2 dm_i$$

Since $v^2 = T_r(vv^T)$ where T_r is the trace operator, then

$$dK_i = \frac{1}{2} T_r \left[\sum_{j=1}^{i} \sum_{k=1}^{i} \left(\frac{\partial T_i}{\partial q_j} \right) (^i\underline{r}\, dm_i \,^i\underline{r}^T) \left(\frac{\partial T_i}{\partial q_k} \right)^T \dot{q}_j \dot{q}_k \right] \qquad (12.27)$$

Integrating dK_i in the complete volume of link i yields

$$K_i = \int_i dK_i = \frac{1}{2} T_r \left[\sum_{j=1}^{i} \sum_{k=1}^{i} \left(\frac{\partial \underline{T}_i}{\partial q_j} \right) \underline{J}_i \left(\frac{\partial \underline{T}_i}{\partial q_k} \right)^T \dot{q}_j \dot{q}_k \right] \qquad (12.28)$$

where $\underline{J}_i$ is the pseudo inertia matrix of link i.

For a robot with n d.o.f., the total kinematic energy is as follows:

$$K_e = \sum_{i=1}^{n} (K_i + K_{a_i}) = \frac{1}{2} \sum_{i=1}^{n} \left\{ T_r \left[\sum_{j=1}^{i} \sum_{k=1}^{i} \left(\frac{\partial \underline{T}_i}{\partial q_j} \right) \underline{J}_i \left(\frac{\partial \underline{T}_i}{\partial q_k} \right)^T \dot{q}_j \dot{q}_k \right] \right.$$

$$\left. + I_{a_i} \dot{q}_i^2 \right\} \qquad (12.29)$$

where K_{a_i} and I_{a_i} are the kinematic energy and the moment of inertia (or moving mass) of the actuator a_i, respectively.

Potential energy. Let $^i r_i$ be the position of the center of mass G_i of link i with respect to the R_i coordinate frame. Its position vector $\underline{r}_i$, with respect to the R_0 reference system, is given by

$$\underline{r}_i = \underline{T}_i \,^i\underline{r}_i$$

The potential energy P_i of a link i having a mass m_i in the gravitational field is expressed by

$$P_i = - m_i \, \underline{g}^T \, \underline{T}_i \,^i\underline{r}_i \qquad (12.30)$$

where g is the acceleration vector of gravity. Then

$$P_E = -\sum_{i=1}^{n} m_i \underline{g}^T \, \underline{T}_i \,^i\underline{r}_i \qquad (12.31)$$

The Lagrangian function is therefore

$$L = \sum_{i=1}^{n} \left\{ \frac{1}{2} \left[\sum_{j=1}^{i} \sum_{k=1}^{i} T_r \left(\frac{\partial T_{-i}}{\partial q_j} J_j \frac{\partial T_{-i}}{\partial q_k} \right)^T \dot{q}_j \dot{q}_k + I_{a_i} \dot{q}_i^2 \right] + m_i g^T T_{-i} r_{-i} \right\}$$

(12.32)

Generalized Forces Calculations. Following the successive calculation of $\partial L / \partial \dot{q}_i$, $\frac{d}{dt}(\partial L / \partial \dot{q}_i)$ and $\partial L / \partial q_i$, the generalized force for the joint i is

$$B_i = \sum_{j=1}^{n} \left[\sum_{k=1}^{j} T_r \left(\frac{\partial T_{-j}}{\partial q_k} J_{-j} \frac{\partial T_{-j}}{\partial q_i} \right)^T \ddot{q}_k + \sum_{k=1}^{j} \sum_{m=1}^{j} T_r \left(\frac{\partial^2 T_{-j}}{\partial q_k \partial q_m} \right. \right.$$

$$\left. \left. \times J_{-j} \frac{\partial T_{-j}^T}{\partial q_i} \right) \dot{q}_k \dot{q}_m - m_j g^T \frac{\partial T_{-j}}{\partial q_i} {}^j r_{-j} \right] + I_{a_i} \cdot \ddot{q}_i$$

(12.33)

This equation can be expressed in matrix form as

$$B_i = \sum_{j=1}^{n} D_{ij} \ddot{q}_j + I_{ai} \ddot{q}_i + \sum_{J=1}^{n} \sum_{k=1}^{n} D_{ijk} \dot{q}_j \dot{q}_k + D_i$$

(12.34)

where

$$D_{ij} = D_{ji} = \sum_{\substack{p=max \\ (i,j)}}^{n} T_r \left[\frac{\partial T_{-p}}{\partial q_j} J_{-p} \frac{\partial T_{-p}}{\partial q_i}^T \right]$$

$$D_{ijk} = \sum_{\substack{p=max \\ (ijk)}}^{n} T_r \left[\frac{\partial^2 T_{-p}}{\partial q_j \partial q_k} J_{-p} \frac{\partial T_{-p}^T}{\partial q_i} \right]$$

$$D_i = - \sum_{p=i}^{n} m_p g^T \frac{\partial T_{-p}}{\partial q_i} {}^p r_{-p}$$

Note that the joints j located behind the joint i (j < i) do not contribute to the expression of B_i.

The n generalized forces can be viewed as the n components of a vector $\underline{B}$. In this case, a nonlinear matrix differential equation represents the dynamics of the robot as

$$\underline{D}(q)\,\ddot{\underline{q}} + \underline{f}[(\dot{q}_i\,\dot{q}_j),q] + g(q) = \underline{B} \tag{12.35}$$

where $\underline{D}(q)$ is an $n \times n$ matrix of inertia depending on the instantaneous configuration of the robot, $\underline{f}$ is an $n \times 1$ vector corresponding to Coriolis and centrifugal forces, and $\underline{g}(q)$ is an $n \times 1$ vector representing the gravity loading terms. The term $V\dot{q}$ expressing the viscous damping may be added to Eq. (12.35).

The structure of Eq. (12.31) is well adapted to the state variables formulation, which allows us to design control laws and makes the simulation of the dynamics of the robot easier. But the complexity of computing the dynamics in the Lagrangian-Euler formulation causes difficulties for real-time computation.

Newton-Euler Formulation of the Dynamics. The second formulation is Newton-Euler, which is based on direct calculations of forces and torques exerted on each link of the robot. A recursive formulation is used for calculation of the velocities and accelerations.

Kinematics of the links. The R_i and R_{i+1} coordinate frames are assigned to link i and i + 1 according to the Denavit-Hartenberg convention. The origins O_i and O_{i+1} of the R_i and R_{i+1} frames are located by the position vectors $\underline{p}_i$ and $\underline{p}_{i+1}$ relative to the origin of R_o frame (see Figure 12.9).

Let $\underline{v}_i$, $\underline{\omega}_i$ be linear and angular velocities of R_i coordinate frame with reference to R_o system, $\underline{v}_{i+1}$, $\underline{\omega}_{i+1}$ be linear and angular velocities of R_{i+1} with reference to R_o, and $\underline{\omega}_s$ be angular velocity of R_{i+1} with reference to R_i coordinate frame. Then Eq. (12.13)–(12.16) give

$$\underline{v}_{i+1} = (d\underline{s}/dt)_{Ri} + \underline{\omega}_i \times \underline{s} + \underline{v}_i$$

$$\dot{\underline{v}}_{i+1} = (d^2\underline{s}/dt^2)_{Ri} + \dot{\underline{\omega}}_i \times \underline{s} + 2\underline{\omega}_i \times (d\underline{s}/dt)_{R_i} + \underline{\omega}_i \times (\underline{\omega}_i \times \underline{s}) + \dot{\underline{v}}_i$$

$$\underline{\omega}_{i+1} = \underline{\omega}_i + \underline{\omega}_s \qquad \dot{\underline{\omega}}_{i+1} = (d\underline{\omega}_s/dt)_{R_i} + \underline{\omega}_i \times \underline{\omega}_s$$

where $\times$ denotes the cross-product, $(d/dt)_{R_i}$ denotes the time derivative with respect to the R_i coordinate system, and s is the position vector locating O_{i+1} in relation to O_i with respect to the R_o frame. Since the DH convention imposes the motions to be realized about or along the z-axis, we obtain

$$\underline{\omega}_s = \sigma_{i+1}\underline{z}_i\,\dot{q}_{i+1} \qquad (d\underline{\omega}_s/dt)_{R_i} = \sigma_{i+1}\underline{z}_i\,\ddot{q}_{i+1} \tag{12.36}$$

$$\underline{\omega}_{i+1} = \underline{\omega}_i + \sigma_{i+1}\underline{z}_i\dot{q}_{i+1} \qquad \dot{\underline{\omega}}_{i+1} = \dot{\underline{\omega}}_i + \sigma_{i+1}[\underline{z}_i\ddot{q}_i + \underline{\omega}_i \times (\underline{z}_i\dot{q}_{i+1})]$$

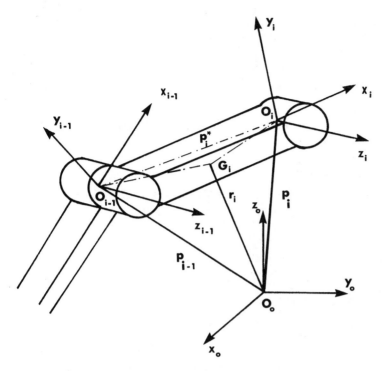

Figure 12.9   Determination of the kinematics of R_i frame.

The vector $\underline{s} = \underline{p}_{i+1}^*$ expresses the position of the O_{i+1} point relative to the O_i origin and results from the addition of $\underline{d}_{i+1}$ and $\underline{a}_{i+1}$ translation vectors, as specified by the DH convention; thus,

$$(ds/dt)_{R_i} = \sigma_{i+1}\underline{\omega}_s \times \underline{p}_{i+1}^* + (1 - \sigma_{i+1})\underline{z}_i\dot{q}_{i+1}$$

$$(d^2\underline{s}/dt^2)_{R_i} = \sigma_{i+1}(d\underline{\omega}_s/dt)_{R_i} \times \underline{p}_{i+1}^* + \underline{\omega}_s \times (\underline{\omega}_s \times \underline{p}_{i+1}^*)$$
$$+ (1 - \sigma_{i+1})\underline{z}_i\ddot{q}_{i+1}$$

Substituting these expressions in $\underline{v}_{i+1}$ and $\dot{\underline{v}}_{i+1}$ equations, we obtain

$$\underline{v}_{i+1} = \underline{\omega}_{i+1} \times \underline{p}_{i+1}^* + (1 - \sigma_{i+1})(\underline{z}_i\dot{q}_{i+1} + \underline{\omega}_{i+1} \times \underline{p}_{i+1}^*) + v_i \qquad (12.37)$$

$$\dot{\underline{v}}_{i+1} = \dot{\underline{\omega}}_{i+1} \times \underline{p}_{i+1}^* + \underline{\omega}_{i+1} \times (\underline{\omega}_{i+1} \times \underline{p}_{i+1}^*) + \dot{\underline{v}}_i + (1 - \sigma_{i+1})[\underline{z}_i\ddot{q}_{i+1}$$
$$+ 2\underline{\omega}_{i+1} \times (\underline{z}_i\dot{q}_{i+1})] \qquad (12.38)$$

Equations (12.36)–(12.38) express the velocities and accelerations of the R_{i+1} coordinate system, with respect to the base coordinates, in terms of velocities and accelerations of the R_i frame; they consist of a recursive formulation of the kinematics of the robot arm.

Dyna ic Equations. Obtaining dynamic equations is based on application of the fundamental dynamic principle; that is,

$$\underline{F}_i = \frac{d(m_i \hat{\underline{v}}_i)}{dt} = m_i \hat{\dot{\underline{v}}}_i \quad \text{and} \quad \underline{N}_i = \frac{d(\underline{J}_i \omega_i)}{dt} = \underline{J}_i \dot{\omega}_i + \underline{\omega}_i \times \underline{J}_i \underline{\omega}_i$$

where

m_i = total mass of link i

$\hat{\underline{r}}_i$ = position vector of center of mass G_i of link i

$\hat{\underline{v}}_i = \dfrac{d\hat{\underline{r}}_i}{dt}$

$\underline{F}_i$ = total external vector force exerted on link i

$\underline{N}_i$ = total external vector moment exerted on link i

$\underline{J}_i$ = inertia matrix of link i about its center of mass

$\underline{\omega}_i$ and $\dot{\underline{\omega}}_i$ = angular velocity and acceleration of center of mass calculated from Eq. (12.36)

All the vectors and matrix $\underline{J}_i$ are expressed with respect to the R_o reference frame.

The movement of G_i is derived from Eq. (12.36)−(12.38) in terms of the movement of the R_i frame by considering a coordinate frame located in G_i and deduced by a translation from R_i.

Let

$\underline{p}_{i-1}$ = vector position of origin O_{i-1} of R_{i-1} frame

$\underline{p}_i^*$ = vector distance from O_{i-1} to O_i

$\hat{\underline{s}}_i$ = vector position of the center of mass from O_i

All the vectors are expressed with respect to the R_o reference frame. Since the point G_i is fixed in the R_i frame, $(d\hat{\underline{s}}_i/dt)_{R_i} = 0$ and $(d^2\hat{\underline{s}}_i/dt^2)_{R_i} = 0$, even if the joint $(i - 1)$ is a prismatic joint. Thus

$$\hat{\underline{v}}_i = \underline{\omega}_i \times \hat{\underline{s}}_i + \underline{v}_i \qquad \hat{\dot{\underline{v}}}_i = \dot{\underline{\omega}}_i \times \hat{\underline{s}}_i + \underline{\omega}_i \times (\underline{\omega}_i \times \hat{\underline{s}}_i) + \dot{\underline{v}}_i$$

The force $\underline{f}_i$ exerted on link i by link $(i - 1)$ balances the dynamic force $\underline{F}_i$ and the force $\underline{f}_{i+1}$ exerted on link i by link $i + 1$; i.e.,

$$\underline{f}_i = \underline{F}_i + \underline{f}_{i+1} = m_i \hat{\dot{\underline{v}}}_i + \underline{f}_{i+1} \tag{12.40}$$

The same balance exists for the moments. Thus,

$$\underline{n}_i = \underline{N}_i + \underline{n}_{i+1} + (\text{moments exerted by } \underline{f}_i \text{ and } \underline{f}_{i+1} \text{ forces})$$

where $\underline{n}_i$ and $\underline{n}_{i+1}$ are the moments exerted on link i by link (i −1) and (i + 1), respectively:

$$\underline{m}_i = \underline{n}_{i+1} + \underline{N}_i - (\underline{p}_{i-1} - \hat{\underline{r}}_i) \times \underline{f}_i + (\underline{p}_i - \hat{\underline{r}}_i) \times \underline{f}_{i+1}$$

Since

$$\underline{r}_i - \underline{p}_{i-1} = \underline{p}_i^* + \hat{\underline{s}}_i \qquad \text{and} \qquad \underline{p}_i - \underline{p}_{i-1} = \underline{p}_i^*$$

Then

$$\underline{n}_i = \underline{n}_{i+1} + \underline{p}_i^* \times \underline{f}_{i+1} + (\underline{p}_i^* + \hat{\underline{s}}_i) \times \underline{F}_i + \underline{N}_i \qquad (12.41)$$

The generalized force B_i exerted by the actuator at joint i is the sum of the projection of f_i or n_i onto the z_{i-1} axis and the viscous damping force or moment, respectively. The Coulomb friction forces can also be added:

$$B_i = \sigma_i \, (\underline{n}_i^T \cdot \underline{z}_{i-1}) + (1 - \sigma_i)\underline{f}_i^T \cdot \underline{z}_{i-1} + b_i\dot{q}_i + c_i\text{sign}(\dot{q}_i) \qquad (12.42)$$

where b_i and c_i are the coefficients of the viscous damping and Coulomb frication, respectively.

Equations (12.36)–(12.38) are often called forward equations and Eq. (12.40)–(12.42) backward equations.

The Computational Point of View. The forward and backward equations are referenced to the base coordinate system. Under these conditions, each inertia matrix J_i depends on the robot configuration and changes in the orientation of link i; this increases the complexity of computation. An important simplification for computing the generalized force B_i is to use link i's own coordinate system. The 4 × 4 matrices $\underline{T}_i$ developed in the preceding sections transform the expression of components of any vector, with respect to the R_i coordinate system, into components with respect to the R_0 base coordinate system.

Let $\underline{\rho}_i$ and $^i\underline{\rho}_j$ be the 3 × 3 submatrices expressing a pure rotation in the $\underline{T}_i$ and $^i\underline{T}_j$ matrices. Then,

$$(\underline{\omega}_{i+1})_{R_0} = \rho_i(\underline{\omega}_{i+1})_{R_i}$$

and

$$(\underline{\omega}_{i+1})_{R_i'} = \underline{\rho}_i^T(\underline{\omega}_{i+1})_{R_0}$$

where R_i' is a frame deduced from R_i by translation into the point O_0.

Then, the set of equations can be transformed in this way. Thus, the inertia matrices J_i are fixed in each link frame. These transformations result in an efficient reduction in the numbers of addition and multiplication and real-time computation becomes impossible.

12.4 CONTROL OF ROBOTS

The purpose of robot control is to achieve a prescribed motion for the robot hand along a desired trajectory with a specified orientation and velocity. Several stages in control design are required in order to solve this problem:

1. Transformation from variables defining the trajectory to the joint variables
2. Determination of torque and force inputs for the actuators achieving the prescribed joint variables
3. Setting of control laws and controllers

Kinematic, Jacobian, and dynamic models which have been described above must be taken into account. But the differential equation models are non-linear and highly coupled, because the dynamic parameters vary with the position of the joint variables.

The most primitive solution neglects the coupling terms and the variations of the inertia matrix in the motion equations. Many industrial robots are controlled by conventional servo mechanisms based on former assumptions. But they move at slow speeds, vibrate, and are imprecise. Another approximate method is to generate the motion from a joint variables set stored in the computer's memory and to apply this to the inputs of servo mechanisms. This is known as control through a look-up table. Other methods have been suggested in order to improve the efficiency of robot control. The best known of these are "resolved motion rate control" (Whitney, 1969), "near-minimum time control" (Kahn and Roth, 1971), the computed torque technique, predictive control, decoupled control (Horowitz and Tomizuku, 1981), and adaptive control (Dubowsky and Desforges, 1979; Lee, 1982; Lee and Chung, 1984; Fu, 1987; Snyder, 1985).

12.4.1 Joint Variable Determination

The determination of the joint variables and their derivatives is accomplished by either experimental techniques or analytical calculations. Experimental techniques are based on "teaching-by-doing" methods. The robot itself transforms the trajectory coordinates into joint variables when the end effector is carried along the specified trajectory. The obtained joint variables $q(t_k)$ and $\dot{q}(t_k)$ will be used in the motion equations and represent discrete values of $\underline{q}$ and $\dot{q}$ at sampling times t_k. On the other hand, the analytical calculation of $\underline{q}$ and $\dot{q}$ needs to use the inverse kinematic model developed in Section 12.3.

The q vector can also be calculated in the recursive sense. Convergence of this method depends on initialization of q_k and the singularities of the $\underline{J}(q_k)$ matrix.

12.4.2 Positional Control (Luh, 1983)

The aim of positional control is to ensure that the robot goes from one specified point to another traveling along a straight line. The trajectory consists of several segments of a straight line separated by specified corner points. The robot control has to make sure that the end effector passes exactly through the specified corner points. The robot is positionally controlled by joint variables q_i, which are obtained by one of the techniques explained above. A classical positional servo mechanism for each joint ensures this task. Besides, no coordination exists between joint motions, because each actuator works at its maximum velocity. Control is generally undertaken by PID controllers, and the torques exerted by the actuators do not take into account any variations in inertia. In order to reduce the storage of excessively large numbers of computed points, a functional representation of the trajectory can be used.

12.4.3 Resolved Motion Rate Control (Whitney, 1969)

In positional control, if the number of specified points in a desired path increases, the length of the segments of straight line decreases. But when these segments become very short, the distance between adjacent points is an increment $d\underline{X} = (dp_x, dp_y, dp_z, d\phi, d\theta, d\psi)^T$. The corresponding increments of joint variables can then be calculated using the inverse Jacobian model:

$$dq = \underline{J}^{-1}(\underline{q}) \; d\underline{X}$$

But if the positional control is preserved, the hand will stop at every corner point! This effect may be eliminated by controlling the joint variable velocities. In this case, it is necessary to add a velocity control loop to each servo mechanism.

Whitney used the RMRC with a redundant manipulator. Then vector q has a higher dimension than $\underline{X}$, and $\underline{J}^{-1}(\underline{q})$ does not exist. In order to utilize the superfluous d.o.f., an optimality criterion can be minimized during the motion with possible constraints.

The torques and forces exerted on the robot joints are modeled as simple inertial torques:

$$\tau_i = \underline{I}_i \ddot{q}_i$$

where $\underline{I}_i$ is the inertia matrix of the robot. This modeling assumes that the velocities are very slow, that the Coriolis and centrifugal effects are negligible, and that $\underline{I}_i$ is a constant matrix.

If $\underline{q}_d$ and $\dot{\underline{q}}_d$ are the desired position and velocity vectors, respectively, then

$$\tau = \underline{I}\ddot{q} = \underline{K}_o(\underline{q}_d - \underline{q}) + \underline{K}_v(\dot{\underline{q}}_d - \dot{\underline{q}})$$

$\underline{K}_p$ and $\underline{K}_v$ are 6 × 6 constant gain matrices and $\underline{q}$ and $\dot{\underline{q}}$ are measurable.

12.4.4 Resolved Acceleration Control (Luh et al., 1980b)

This method is an extension of RMRC, but now the desired accelerations are taken into account, and the forces and torques are determined according to the Newton-Euler formulation.

Position and orientation errors are defined as 3 × 1 vectors:

$$\underline{e}_p = \underline{p}_d(t) - \underline{p}(t) \qquad \underline{e}_o = \frac{1}{2}(\underline{n} \times \underline{n}_d + \underline{O} \times \underline{O}_d + \underline{a} \times \underline{a}_d)$$

$$(12.43)$$

where $\underline{p}$ is the current position vector of the end effector and $\underline{p}_d(t)$ is the desired position. Then $\underline{n}$, $\underline{O}$, and $\underline{a}$ give the current orientation of the axes of the coordinate frame attached to the hand, and $\underline{n}_d$, $\underline{O}_d$, and $\underline{a}_d$ give the desired orientations.

Let $\underline{X}$ and $\underline{\dot{X}}$ be the 6×1 vectors defined as

$$\underline{X} = (p_x p_y p_z, \, \theta_x \theta_y \theta_z)^T \qquad \underline{\dot{X}} = (V_x V_y V_z, \, \omega_x \omega_y \omega_z)^T \qquad (12.44)$$

where V_x, V_y, V_z are the components of the linear velocity vector of moving coordinate frame attached to the hand, and ω_x, ω_y, ω_z are those of the angular velocity vector. Then,

$$\underline{\dot{X}} = \underline{J}(\underline{q}) \, \underline{\dot{q}}$$

By differentiating this relation, we obtain

$$\underline{\ddot{X}} = \underline{J}(\underline{q}) \, \underline{\ddot{q}} + \underline{\dot{J}}(\underline{q}) \, \underline{\dot{q}} \qquad (12.45)$$

The $\underline{\dot{J}}(\underline{q})$ matrix may be calculated either from Eq. (12.36)–(12.38) with $\dot{\omega}_0 = 0$ and $\underline{\dot{V}}_0 = (00g)^T$ or directly by differentiation of the $\underline{J}(\underline{q})$ matrix. The actuator exerts torques and forces on the joint which generate angular and linear displacements, velocities, and accelerations. In order to reduce the errors to zero, torques and forces are determined in terms of calculated accelerations, measured velocities, and displacements. These calculated accelerations consist of the desired accelerations and a corrective term which is proportional to errors and which satisfies:

$$\underline{\dot{V}}(t) = \underline{\dot{V}}_d(t) + k_V \, (V_d(t) - V(t)) + k_p(\underline{p}_d(t) - \underline{p}(t))$$

$$\underline{\dot{\omega}}(t) = \underline{\dot{\omega}}_d(t) + k_V \, \{\underline{\omega}_d(t) - \omega(t)\} + k_p \underline{e}_o(t)$$

or

$$\underline{\ddot{X}}(t) = \underline{\ddot{X}}_d(t) + k_V \, \{\underline{\dot{X}}_d(t) - \underline{\dot{X}}(t)\} + k_p \underline{e} \qquad (12.46)$$

where $\underline{e} = \{\underline{e}_p, \, \underline{e}_o\}^T$. Then,

$$\underline{\ddot{q}} = \underline{J}^{-1}(\underline{q}) \, \{\underline{\ddot{X}}_d(t) - k_V(\underline{\dot{X}}_d - \underline{\dot{X}}) + k_p \underline{e} - \underline{\dot{J}}(\underline{q})\underline{\dot{q}}\}$$

$$\qquad = k_V \underline{\dot{q}} + \underline{J}^{-1}(\underline{q}) \, \{\underline{\ddot{X}}_d - k_V \underline{\dot{X}}_d + k_p \underline{e} - \underline{\dot{J}}(\underline{q})\underline{\dot{q}}\} \qquad (12.47)$$

where k_V and k_p are scalar constants.

The position error vanishes if the values of k_V and k_p are such that the characteristic roots of the equation obtained from Eq. (12.47):

$$\frac{d^2(p_d - p)}{dt^2} + \frac{k_V d(p_d - p)}{dt} + k_p(p_d - p) = 0$$

have negative real parts.

The $q(t)$ and $\dot{q}(t)$ vectors are measured directly and the kinematic model is determined from $q(t)$. The error vector $\underline{e}$ is calculated from Eq. (12.43), and the computation of acceleration vector $\ddot{q}$ can be achieved by using Eq. (12.47). The torques and forces are determined by using the forward and backward equations. The Newton-Euler formulation associated with the RAC technique permits real-time computation.

12.4.5 The Computed Torque Technique

The computed torque technique uses the Lagrangian-Euler formulation for the dynamic model of the robot. The basic idea of this method is to express, in Eq. (12.35), the acceleration vector $\ddot{q}$ as a sum of the desired acceleration $\ddot{q}_d$ and a corrective term which is proportional to errors. Then Eq. (12.35) becomes

$$\underline{B} = \underline{D}_c(\underline{q})[\ddot{\underline{q}}_d + \underline{k}_v(\dot{\underline{q}}_d - \dot{\underline{q}}) + \underline{k}_p(\underline{q}_d - \underline{q})] + V_c\dot{\underline{q}} + f_c(\dot{\underline{q}},\underline{q}) + g_c(\underline{q})$$
(12.48)

where $\underline{D}_c(\underline{q})$, $\underline{V}_c$, $f_c(\dot{\underline{q}},\underline{q})$ and $g_c(\underline{q})$ are the calculated counterparts of $\underline{D}(q)$, $\underline{V}$, $\underline{f}(q,q)$ and $\underline{g}(q)$; $\underline{k}_v$ is a 6×6 velocity feedback gain matrix; and k_p is a 6×6 position feedback gain matrix.

Then the $\underline{q}$ vector converges to $\underline{q}_d$ if $J_c = J$, $V_c = V$, $f_c = f$, $g_c = g$. By comparing Eq. (12.35) and (12.48) we obtain

$$D(q) \{\ddot{q}_d - \ddot{q} + k_v(\dot{q}_d - \dot{q}) + k_p(q_d - q) = 0$$

if $\underline{e} = \underline{q}_d - \underline{q}$ is the joint position error and $D(q)$ is nonsingular. In this case we have

$$\underline{\ddot{e}} + k_v\underline{\dot{e}} + k_p\underline{e} = 0$$
(12.49)

If the $\underline{k}_v$ and k_p matrices are chosen so that the solution of Eq. (12.49) is asymptotically stable, then $\underline{e}$ approaches zero asymptotically.

But the torques and forces can be calculated from the Newton-Euler formulation instead of the Lagrangian formulation. The desired acceleration $\ddot{q}_d$ is introduced by substituting $\ddot{q}_i$ in the forward and backward equations:

$$\ddot{\underline{q}}_i = \ddot{\underline{q}}_{di} + \sum_{j=1}^{n} k_{v_{i_j}} (\dot{\underline{q}}_{dj} - \dot{\underline{q}}_j) + \sum_{j=1}^{n} k_{pij}(\underline{q}_{dj} - \underline{q}_j)$$

$$\ddot{\underline{q}}_i = \ddot{\underline{q}}_{dj} + \sum_{1}^{n} k_{vij}\underline{e}_j + \sum_{1}^{n} k_{pij}\underline{e}_j$$

where k_{vij} and k_{pij} are the derivative and positional feedback gains for joint i, respectively. These error acceleration terms generate a correlation torque which compensates for the uncertain parameters and errors due to the mechanism (Luh et al., 1983).

12.4.6   Adaptive Control   (Lee and Chung, 1985)

Efficient control methods take into consideration any changes in the inertia
matrix and in the interaction forces during the motion of the hand along
the trajectory.   Adaptive control allows a significant improvement in robot
performance.   Several approaches have been suggested, such as those of
Lee (1982), Dubowsky and Desforges (1979), Koivo and Guo (1983), and
Vukobratovic and Potkonjak (1982), which take into consideration the
changes in the dynamic parameters of the payload.

We present here the adaptive control method of Lee and Chung based on
the linear perturbation equations in the vicinity of a nominal trajectory.   The
nonlinear equations obtained from the Lagrange-Euler formulation are linearized
about the desired path and give a linearized perturbation system.   A 2n-dimen-
sional state vector is defined for a robot with n d.o.f. from Eq. (12.35) as

$$\underline{D}(q)\ddot{q} + \underline{f}(\dot{q},q) + g(q) = B$$

$$\underline{x}^T(t) = \{\underline{q}^T(t), \dot{\underline{q}}^T(t)\} = (q_1 \ldots q_i \ldots q_n, \dot{q}_1 \ldots \dot{q}_i, \ldots, \dot{q}_n)$$

$$= (x_1 \ldots x_n, x_{n+1} \ldots x_{2n})$$

and an n-dimensional input vector U(t) as

$$\underline{U}^T(t) = (B_1 B_2 \ldots B_n) = (U_1 \ldots U_n)$$

Equation (12.35) can be rewritten in the state-space representation as

$$\dot{\underline{X}}(t) = \underline{h}\{\underline{X}(t), \underline{U}(t)\} \tag{12.50}$$

$$\dot{x}_i(t) = x_{n+i}(t)$$

or

$$\dot{x}_{n+i}(t) = h_{n+i}(X) + b_{n+i}(X)\underline{U}(t) \qquad \text{for } i = 1, \ldots, n \tag{12.51}$$

where $h_{n+i}(X)$ is the ith component of the vector $-\underline{D}^{-1}(q) \{\underline{f}(\dot{q}.q) + g(q)\}$
and $b_{n+i}(X)$ the ith row of matrix $D^{-1}(q)$.   The nominal states $X_n(t)$ are
obtained from the nominal values of q and $\dot{q}$ on the trajectory.   The Newton-
Euler equations of motion give the nominal inputs $\underline{U}_n(t)$ required to make
the robot hand travel along the nomal path.   Then

$$\dot{\underline{X}}_n = \underline{h}[\underline{X}_n(t), \underline{U}_n(t)] \tag{12.52}$$

The linearization of Eq. (12.52) is achieved by using the Taylor expansion
series about the nominal path:

$$\delta\dot{\underline{X}}(t) = \{\underline{\nabla}_x \underline{h}\}_n \delta\underline{X}(t) + \{\underline{\nabla}_u \underline{h}\}_n \delta\underline{U}(t)$$

where $\{\underline{\nabla}_x \underline{h}\}_n$ and $\{\underline{\nabla}_u \underline{h}\}_n$ are the Jacobian matrices of h evaluated at the nominal state $X_n(t)$ and the nominal input $U_n(t)$, respectively, therefore

$$\delta \underline{X}(t) = \underline{X}(t) - \underline{X}_n(t) \qquad\qquad \delta \underline{U}(t) = \underline{U}(t) - \underline{U}_n(t)$$

Let $\{\underline{\nabla}_x \underline{h}\}_n = \underline{A}(t)$ and $\{\underline{\nabla}_u \underline{h}\}_n = \underline{C}(t)$. Then,

$$\delta \underline{\dot{X}}(t) = \underline{A}(t)\ \delta \underline{X}(t) + \underline{C}(t)\ \delta \underline{U}(t) + \underline{e}(t) \tag{12.53}$$

where $\underline{e}(t)$ is an error term including the bias and modeling errors of the robot; $\underline{A}(t)$ and $\underline{C}(t)$ are complex and change during the course of the nominal trajectory. Only the parameter identification techniques allow the engineer to identify $\underline{A}(t)$ and $\underline{C}(t)$.

The $\delta \underline{U}(t)$ vector is called the perturbation torque vector and can be calculated on the basis of the optimal control theory in order to reduce $\delta \underline{X}(t)$ to zero at any time.

Identification of the parameters of $\underline{A}(t)$ and $\underline{C}(t)$ with a digital computer needs the discretization of Eq. (12.53) as

$$\underline{X}(\{k + 1\})T = F(kT)X(kT) + \underline{G}(kT)\underline{U}(kT) + \underline{W}(kT) \qquad k = 0, 1, \ldots \tag{12.54}$$

where T is the sampling period and $\underline{U}(kT)$ a piecewise constant input vector; $F(kT)$ and $\underline{G}(kT)$ are $2n \times 2n$ and $2n \times n$ matrices, respectively. Thus $6n^2$ parameters must be identified. An identification technique that is well adapted to this problem is the recursive real-time least-square parameter identification scheme.

If $F(kT)$ and $\underline{G}(kT)$ are known through the parameter identification scheme, a control law for $\delta \underline{U}(t)$ will allow the engineer to reduce $\delta \underline{X}$ to zero. A one-step performance index that satisfies any constraints is chosen:

$$\underline{H}(k) = \frac{1}{2} [\underline{X}^T(k + 1)\ \underline{Q}\ \underline{X}(k + 1) + \underline{U}^T(kT)\ \underline{R}\underline{U}(kT)]$$

Equation (12.54) makes up the constraint; $\underline{Q}$ is $2n \times 2n$ semipositive definite weighting matrix, and $\underline{R}$ is $n \times n$ positive definite weighting matrix. The optimal control solution $\underline{U}^*(t)$ is obtained by the classical minimization of H(k).

$$\underline{U}^*(t) = -\{\underline{R} + \underline{\hat{G}}^T(k)\underline{Q}\underline{\hat{G}}(k)\}^{-1}\underline{\hat{G}}^T(k)\ \underline{Q}\underline{\hat{F}}(k)\underline{X}(k)$$

where $\hat{G}(k)$ and $\hat{F}(k)$ are the matrices obtained by the real-time least-square identification algorithm. The Newton-Euler formulation of dynamic equations and the recursive real-time least-square identification algorithm require relatively short computation times. The control algorithm computation is slower, but the nominal torque computation can be performed in parallel with the perturbation torque computation.

12.5   APPLICATIONS

12.5.1   Robot Vision

A vision system is a noncontact sensor type that permits a closed feedback loop in order to control the hand position. Its main functions are to identify the objects and to determine their position and orientation. Industrial robots use a two-dimensional representation of three-dimensional objects. Indeed, this technique is very efficient and requires a short processing time. On the other hand, the perspective views of three-dimensional objects require a relatively long processing time.

The identification and determination of the position and orientation of objects are based on several techniques. At first, detection of the edges, based on image contrast, was studied and used. Next, an analysis of connecting images was developed from a run-length binary coding scheme and was efficient for one-line operations. Statistical methods were also used. At present, 2D vision systems are produced by several manufacturers. These machine vision systems operate on the 2D projection of a 3D object, with 16 levels of gray scale.

12.5.2   Welding by Robot

Use of the welding robot has five main advantages: improved workers' conditions, increased quality and consistency, an increase in productivity, and computerization of the industry. Two welding techniques are possible with a robot: spot welding or resistance welding and arc welding.

Resistance welding is a process in which the heat for coalescing the two metals is produced by the resistance of the joint to the passage of an electric current. Automatic control of current, timing, and electrode force does not cause difficulties. The robot control is a positional point to point control with unsophisticated algorithms. Arc welding is a very complex joining process. It requires metal, flux, and the addition of gas. The arc is positioned between a consumable electrode and the workpiece, which becomes part of the electric circuit. An inert gas is applied separately. In robotized welding, special techniques (pulsed or high-frequency current) have been studied to bait the arc, which differ from the manual methods.

The welding process requires the electrode to travel along the joint at a constant speed. The robot will then be controlled by velocity feedback loops allowing a fixed orientation of the electrode with reference to the joint and a constant velocity along the trajectory.

Any adaptive robot is able to track a seam with a seam-tracking sensor. The following categories of sensors exist: probe systems, electro-optic vision systems, and intensity sensors which measure change in the welding current.

12.5.3   Product Assembly Automation

The positional control of industrial robots is not accurate, and assembly tasks require precise position control. In order to compensate for this disadvantage, sensors of force and touch must be used. Positionally controlled motion is possible only if large tolerances can be accepted. But if industrial robots are force-controlled, many assembly tasks become feasible. Another technique uses a compliant wrist provided with strain gauges indicating electrical signals from which the forces and moments are computed. But the computation is complex and slow. The best method is to equip the joint with torque or force sensors and to control the joint torque directly.

12.5.4   Other Applications

Robotized Water Cutting. The tranditional methods of trimming or bandsawing used in the aircraft industry in order to cut aluminum sheet metal parts are inefficient for advanced composites. Indeed, bandsawing requires a second manual operation of fettling and cleaning of the cutting edge. The water-jet cutting system eliminates these dust problems. This technique is easily robotized, but the velocity of the water jet along the cut trajectory must be kept strictly constant. Water-jet cutting also allows industry to cut leather for the manufacture of boots.

Laser cutting. The energy produced by laser beams melts the material, and the molten metal is removed from the cutting edge by a gas jet. The laser systems used are generally gas lasers or CO_2 lasers. The laser beam serves strictly as a controlled heat source for the processing of materials. The robot does not carry the laser system, which is too large, but it handles the workpiece and makes it follow a programmed path. The main advantage of this technique is that the cutting speed is much faster than those of the other systems.

Painting and Handling. Robots have frequently been used in this field for several years. The trajectories do not need high precision, and point-to-point position control or teaching-by-doing techniques are largely adequate.

12.6   CURRENT DIRECTIONS OF RESEARCH IN ROBOTICS

12.6.1   Mobile Robots

At the present time, mobile robots have two fields of application: discovery of space and parallel traveling synchronized with the movement of an assembly line. In the future, other possible applications for mobile robots are: robots for industrial cleaning of grounds (airport, stations, etc.), robots for underwater metal discovery, and robots for internal inspection of fissures in nuclear reactors. In order to do this, the robot must be able to localize its position with respect to fixed and recognized marks in order to specify its motion with reference to the task. This localization is obtained by camera vision or ultrasonic systems. But for the cleaner robot, the environment is fixed and recognized, in which case a teaching-by-doing method is possible in combination with a safety system avoiding the possibility of collision with unforeseen hindrances. For both other applications, only telecontrol by human operators using camera vision is practical at present.

12.6.2   Vision and Scene Analysis in Conjunction with the Robot

In the vision research field, one of the main problems is scene analysis in three dimensions on-line. At present, the algorithm complexity is very high and the real-time computations are very difficult or impossible. Solution of this problem would allow classifying of pieces in bulk by a robot. Another vision problem is that of the robot keeping track of moving objects in the case of fast displacements. Vision is a very large field of research, of which robot vision is only a part.

12.6.3   Artificial Intelligence

Artificial intelligence is above all used in robotics when the robot environment is changing and uncertain. Then the arm motions must be determined by the robot computer itself. Studies in this area are based on the simulation of the occurrence of events or scenes analyzed by the vision system.

For instance, if there are several alternative paths around a hindrance, the simulation results indicate to the computer which of these is best suited to fulfill the task.

12.6.4   Programming

High-level languages must allow users to be able to control a robot without having to worry about its dynamic and kinematic parameters. Thus, more complex tasks could be carried out by using several robots simultaneously. Rapid progress has been made in this domain, and every year it accelerates still further; the reader is encouraged to consult with the recent text by Groover et al. (1986).

PROBLEMS

12.1. Obtain the matrix $\underline{H}$ for the roll (ϕ), pitch (θ), and yaw (ψ) angles defined in Figure 12.10. If the entries of the matrix $\underline{H}$ are numerically specified, calculate the angles ϕ, θ, and ψ. [See Section 12.3, Equation (12.1).]

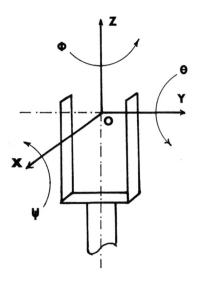

Figure 12.10   Roll, pitch, and yaw angles.

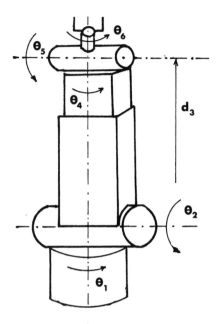

Figure 12.11   Robot polar 6000.

12.2. For the rotation matrix $\underline{Rot}(\theta,k)$ [Eq. (12.2)], express the angle θ and the unit vectors k_x, k_y, and k_z in terms of components of n, O and a vectors.

12.3. For the Polar 6000 Robot, define the fixed and moving coordinate systems attached to each link, according to the Denavit-Hartenberg convention. The robot position that defines the origin of the variables is shown in Figure 12.11. Calculate the matrix $\underline{T}_6$ of this robot.

12.4. For the robot of Problem 3, solve the kinematic model and write the joint variables as functions of the position p and the orientation of the hand, specified by $\underline{n}$, $\underline{O}$, and $\underline{a}$ vectors [Ex. (12.3.3)].

12.5. For the robot of Problem 3, obtain the Jacobian Matrix $J_6(q)$ with respect to the reference coordinate frame [Eq. (12.20)].

12.6. For the 4C06 Robot (Section 12.3.2, Figure 12.5), calculate the Jacobian matrix $J_4(q)$ and the inverse Jacobian matrix $\underline{J}_4^{-1}(q)$.

12.7. The Lagrange-Euler formulation, used to calculate the torques applied to the joints, leads to expressions containing partial derivatives of matrices $\underline{T}_i$, such as $\partial T_j/dq_i$. Show that $\partial T_j/\partial q_i$ can be expressed by means of a matrix product as

$$\partial T_j/\partial q_i = T_{i-1} Q_i^{i-1} T_j$$

where Q_i is defined by $\partial \underline{A}_i / \partial q_i = Q_i \underline{A}_i$. For a rotational joint and a prismatic joint, obtain the Q_i expression.

12.8.   The resolved acceleration control method uses Eq. (12.45), which contains the derivative $\dot{\underline{J}}(\underline{q})$ term. From Eq. (12.36) and (12.38) write the relation

$$(\dot{\underline{V}}_n, \dot{\underline{\Omega}}_n)^T = \underline{J}(\underline{Q})\ddot{\underline{q}} + \dot{\underline{J}}(\underline{q})\dot{\underline{q}}$$

and directly obtain $\dot{\underline{J}}(\underline{q})$ without derivation of $\underline{J}(\underline{q})$.

REFERENCES

Catier, E. (1984).   La CAO des robots, *Electron. Industr.*, No. 72, 103–106.

Denavit, J. and Hartenberg, R. S. (1955).   A kinematic notation for lower-pair mechanisms based on matrices, *J. Appl. Mechanics, June*, 215–221.

Dubowsky, S. and Desforges, D. T. (1979).   The application of model reference adaptive control to robotic manipulators, *Trans. ASME J. Dyn. Syst. Measurement Control. 101*, 193–200.

Ernst, H. A. (1961).   A computer-operated mechanical hand, Sc.D. Thesis, MIT, Cambridge, MA.

Fages, G. (1984). *Robots, le guide de l'utilisateur*, Editions Techniques d'Alsace, Strasbourg, France.

Feldmann, J. (1971).   The use of vision and manipulation to solve the puzzle, *Proceedings of 2nd Int. Joint Conf. on Artificial Intelligence*, London, pp. 359–364.

Fu, K. S., Gonzalez, R. C., and Lee, C. S. G. (1987). *Robotics Control, Sensing, Vision, and Intelligence*, McGraw-Hill, New York.

Gini, G. and Gini, M. (1975).   Control of intelligent robots and goal oriented languages, *Industr. Robots 2*, 67–74.

Gorla, B. and Renaud, M. (1984). *Modèles des robots manipulateurs. Application à leur commande*, Cepadues Editions, Toulouse, France.

Groover, M. P., Weiss, M., Nagel, R. N., and Odrey, N. G. (1986). *Industrial Robotics: Technology, Programming, and Applications*, Mc-Graw-Hill, New Your.

Hollerbach, J. M. (1980).   A recursive Lagrangian formulation of a manipulator dynamics. *IEEE Trans. SMC 10*, 730–736.

Horowitz, R., and Tomizuku, M. (1981).   An adaptive control scheme for mechanical manipulators, *Proceedings 20th IEEE Conf. on Decision and Control.* pp. 271–276.

Kahn, M. E. and Roth, B. (1971).   The near-minimum time control of open-loop articulated kinematic chains, *J. Dyn. Sys. Measurement Cont. 101*, 193–200.

Koivo, A. J. and Guo, T. H. (1983).   Adaptive linear controller for robotic manipulators, *IEEE Trans. A.C., 28(2)*, 162–171.

Latombe, J. C. (1979).   Une analyse structurée d'outils de programmation pour la robotique industrielle, *Séminaire International IRIA*, Paris, France.

Lee, C. S. G. (1982).   Robot arm kinematics dynamics and control, *IEEE Computer*, pp. 62–77.

Lee, C. S. G. and Chung, M. J. (1984). An adaptive control strategy for mechanical manipulators, *IEEE Trans. A. C. 29(9)*, 837–840.

Lee, C. S. G. and Chung, M. J. (1985). Adaptive perturbation control with feed forward compensation for robot manipulators, *Simulation, March*, 127–136.

Liebermann, L. I. and Wesley, M. A. (1977). Auto-Pass: An automatic programming system for computer controlled mechanical assembly, *IBM J. Res. Dev. 21*, 321–333.

Luh, J. Y. S. (1983). An anatomy of industrial robots and their controls, *IEEE Trans. A.C. 28(2)*, 133–152.

Luh, J. Y. S., Walker, M. W., and Paul, R. P. (1980a). On line computational scheme for mechanical manipulators, *Trans. J. Dyn. Syst. Measurement Control, 102*, 69–76.

Luh, J. Y. S., Walker, M. W., and Paul, R. P. (1980b). Resolved-acceleration-control of mechanical manipulators, *IEEE Trans. A.C. 25(3)*, 468–474.

Luh, J. Y. S., Fischer, W. D., and Paul, R. P. (1983). Joint torque control by a direct feedback for industrial robot, *IEEE Trans. A.C. 28(2)*, 153–161.

Paul, R. P. (1977). Wave: A model based language for manipulator control, *Indust. Robot 4*, 10–17.

Paul, R. P. (1982). *Robot Manipulators: Mathematics Programming Control*, MIT Press, Cambridge, MA.

Paul, R. P. (1983). The early stages of robotics, *Proceedings of Symposium IFAC-IFIP Real Time Digital Control*, Guadalajara, Jalisco, Mexico, Pergamon Press, London.

Popplestone, R. J., Ambler, A. P., and Bellos, I. (1978). RAPT: A language for describing assemblies, *Industr. Robot 5*, 131–137.

Synder, W. E. (1985). *Industrial Robots: Computer Interfacing and Control*, Prentice-Hall, Inc., Englewood Cliffs, NJ.

Taylor, R. H., Summers, P. D., and Meyer, J. M. (1982). AML: A manufacturing language, *Int. J. Robotics Res. 1(3)*, 19–41.

Vukobratovic, M., and Potkonjak, V. (1982). *Dynamics of Manipulation Robots*, Communications and Control Engineering Series, Springer-Verlag, Berlin.

Whitney, D. E. (1969). Resolved motion rate control of manipulators and human prosthesis, *IEEE Trans. M.M.S. 10(2)*, 47–53.

Whitney, D. E. (1972). The mathematics of coordinated control of prosthetic arms and manipulators, *J. Dyn. Syst. Measurement Control, Dec.*, 303–309.

13
World Modeling
Concepts and Applications

13.1 INTRODUCTION

In the present-day world it is unwise to study and analyze the problems of a single country or region in isolation from those of other countries or regions. The existence of links of international character in both socio-political and economic areas makes the changes in policies of a single country, especially an industrially developed one, felt throughout the world. Even a small nation, or group of far less developed countries, with monopolistic advantage in any particular economic sector can generate large enough shocks to humble economic giants, as happened in the 1970s through the "oil strategy" of the OPEC countries. The world's resource base and ecosystems, population and environment are dependent on each other. Over the years, with advancement in communication and transportation facilities, the interactions among the nations have become stronger and closer, thus making them more interdependent. Therefore, a natural result of this phenomenon has been to study and analyze this complex and multidirection interaction through models to shed light on the intricacies of the system.

13.2 BIRTH OF WORLD MODELS

The recognition of an economic reality like "global interdependence" motivated the Club of Rome, which consists of a group of individuals representing different specialties, to identify a "problématique" in 1968 which was defined as a complex of interrelated problems: the recent growth and apparent persistence of a major gap in living standards between the richest and the poorest people; the shrinking of the nonrenewable resource base of the globe; the increasing interrelatedness of issue areas once perceived to be self-contained; mankind's ever-growing ability to affect its larger environment in frequently unanticipated and undesirable ways; and the inability of current social and political institutions to deal with these problems (Peccei, 1977). One purpose of the Club of Rome was to sponsor activities that would develop the tools to analyze and understand the problématique.

Another important development on this line was the adoption of
"Declaration and Program of Action for the Establishment of a New International Economic Order" in April 1974, by the Sixth Special Session of the
U.N. General Assembly. This focused attention on deepening relations of
interdependence between the industrially advanced and the developing
countries, generally known as the north-south relationship.

All these activities at the world level drew attention to the fact that
worldwide economic security was to be attained and nurtured to avoid any
country or region running into extraordinary crises. This called for a
better understanding of the total economy of the world and spurred activities that led to the development of many global economic models. The
models have been developed over the years by different groups of people
located in various parts of the world, including the United States, Japan,
the United Kingdom, Argentina, etc. Some models have been developed by
such world bodies as the World Bank, the United Nations, the International
Monetary Fund, etc. Quite often the economic models included food, agriculture, energy, population, and environment sectors, but there were also
attempts to develop separate models to analyze the problems individually.
Many of these models developed over the years cannot be included here
because of lack of proper documentation.

The world models, as expected, are complex and are designed to
address more long-term problems over long time horizons. Being policy-
oriented computer models, they suffer from the same drawbacks as any
other computer models. These models differ from each other mainly in
their structure. The models have been designed on the principle of system dynamics, multilevel, hierarchical systems, optimization, economics and
econometrics, input-output, or a combination of some of these. This is
not surprising mainly because of the bias and expertise of those developing
such models and the important roles they have played in their structures.
One should refer to Chapter 6 to learn more about the theory of hierarchical control and Chapter 14 for a discussion of various types of econometric
models.

The issues that have been focused on and can be analyzed by these
models are many, encompassing the fields of economy, demography, natural
resources and environment, ecosystems, defense and nuclear arms, poverty
and vulnerability of nations, political and sociological-cultural philosophy,
to mention a few of the important, critical, and not well-fathomed problems
of the globe.

There are many world models that have been developed in different
parts of the world. The most popular ones are discussed next. The first
global model to be conceived and developed during the years 1968-1973 was
project Link (Ball, 1973). Pioneering work was done by Forrester (1971)
with a model called World 2 to initiate a new family of global models
sponsored by the Club of Rome. A subsequent version of this model was
named World 3, the findings of which were published in a popular book,
The Limits to Growth (Meadows et al., 1972, 1974). This was followed by
other models that were presented through publications such as *Mankind at
the Turning Point* (Mesarovic and Pestel, 1974a), describing the results
of the Mesarovic-Pestel world model [World Integrated Model (WIM)], and
Catastrophe or New Society? (Herrera et al., 1976), describing a Latin
American model known as the Bariloche model. A model called Future of
Global Interdependence (FUGI) was developed in Japan by Kaya and his
group (Kaya et al., 1977; Kaya, 1978). The World Bank responded to

these modeling efforts by developing SIM-LINK (IBRD, 1975). A model named SARUM (SARU, 1977, 1978) was built by Systems Analysis Research Unit of the Department of Environment and Transport in Great Britain. The United Nations' world model was designed and built by a group headed by Leontieff, and the results have been published in *The Future of the World Economy* (Leontieff et al., 1977). Modeling efforts have continued through the years, and newer versions of the world models have been built and presented. There have been publications describing newer versions, more detailed than the older ones. Technical documents of World 3 are contained in a report called *Dynamics of Growth in a Finite World* (Meadows, 1977). More information on WIM can be found in a paper by Mesarovic (1979). Complete technical reports of the Bariloche model are contained in a Bariloche Foundation report (1978). The most recent information on World Bank modeling activities can be found in Carter (1978) and Gupta et al. (1979).

There are also models that have focused their attention mainly on only one sector of economy. *MOIRA—Model for International Relations in Agriculture* (Linneman et al., 1979) has been developed in The Netherlands; it deals with food and hunger. Another world model on agriculture is the U.S. Department og Agriculture's grain-oilseeds-livestock model (GOL) (Rojko et al., 1978). Global energy models have been developed by the U.S. Department of Energy (Shaw et al., 1978) and the Stanford Research Institute (SRI, 1977). The United Nations Industrial Development Organization's world industry cooperation model (UNIDO, 1980) deals with industrial development around the world. An environment model examining CO_2 contents in the atmosphere has been presented by IEA/OARU (1984).

One can also find a good number of reviews on world models. *Models of Doom* gives a good technical critique of World 3 (Cole et al., 1973). In *Growth and Its Implications for the Future* (1974) one can find discussions of "The Limits to Growth." Reviews of more well-known world models can be found in Richardson (1977), Sundaram and Krishnayya (1979), Barry Hughes (1980) and Meadows et al. (1982). Models on agriculture have been reviewed by Meadows (1977) and U.S. General Accounting Office (1977). *The Global 2000 Report to the President* (Barney, 1980) gives a comparative study of the usage and findings of World 3, WIM, Bariloche model, and U.N. model. *Groping in the Dark* (Meadows et al., 1982) is an interesting publication that describes the various models that were presented at the Sixth IIASA (International Institute for Applied Systems Analysis) Symposium in Global Modeling.

13.3 APPROACHES TO WORLD MODELING

World models are computerized mathematical simulations designed to study the world's economic, physical, and political systems on a relatively long-term basis. Objectives of the models and the expertise of their builders have significantly influenced the designs and structures of these models and modeling paradigms. The structures range from treating the whole world as one geographical region to 38 regions, and from dividing the economy into a few sectors to as many as 48 sectors. Depending on the objectives, the models are biased toward economy, agriculture, or environment, etc. Time horizons for these models have also widely differed, the shortest being 3 years and the longest 200 years.

13.3.1   Regions and Sectors

Ideally in a world model, every sovereign nation should be treated as an
individual entity. So should be the case with each type of good or service
so far as production and consumption are concerned. Unfortunately, the
corresponding data and the number of variables and parameters will be
huge, and storing and manipulating them in a model will be an impossible
task even with present-day large, fast computers. In addition to this,
the computer time required in such simulations will be cost-prohibitive.
Also, there are many small nations for which it will be difficult to get any
data to obtain meaningful results. So, to increase the efficiency of the
model, the world is divided into many regions based on geographical
proximity; cultural, sociological, and economic similarities; and political systems, e.g., United States, Western Europe, Southern Europe, Eastern
Europe, Southeast Asia, Latin America, etc. The goods and services are
also grouped into sectors or categories based on their nature, e.g., agriculture, energy, materials, manufacturing, etc.

13.3.2   Variables in Model

The objectives of a specific model play a role in the selection of variables
and factors to be included in the model for analysis. The constraints and
initial conditions that are required to facilitate proper analysis of the system, and to help attain the objectives, influence the choice of the model's
variables and the parameters. For example, if the model is used to analyze
the pollution of atmosphere by the year 2000, the types of energy source,
demand and supply of energy, existing reserves of the energy types,
their discovery and extent of processing, price, elasticity, etc., will have
to be included in the system. Consequently, certain causal relationships
will feature in the logic of the model. Sometimes, though, lack of knowledge and ability to estimate accurately will force the modeler to exogenize
some variables to help in creating different problems and situations in world
or regional economy.

13.3.3   Demand, Supply, and Equilibrium

There will be demand for consumption of each type of goods or service.
It has to be produced and supplied to the consumers. The consumers and
producers will have to look after their own interests, which means they
will try to maximize their own benefit. Transactions between the producers
and consumers will take place at the "right price." In other words, total
quantity produced will be equal to that consumed at the right price, and,
thus, there will be equilibrium.

Depending on the economic philosophy of the modeler, the model can be
demand-driven where demand forces the supplier to supply, and the system may assume that there is no constraint on volume of supply. The
model can be just the opposite, i.e., supply-driven. It is a well-known
fact that demand or supply of any item depends on price. Using this,
some models do try to attain economic equilibrium in every time period,
when the demand for any commodity is equal to the supply, for a price
called the "equilibrium price." This is attained in one of two ways. In
the first case, the model forces demand and supply to be equal without
using any mechanism based on economic principles. The larger value,
between supply and demand, is reduced to become equal to the smaller
value, and this demand (equal to supply) is reallocated among different

consumers (or producers). Price plays no role. In the second procedure , price and elasticities of demand and supply are included in the process of obtaining equilibrium, which is attained usually through an iterative procedure, and the resulting equilibrium price is obtained along with the demand that is equal to supply. The model can use any of these procedures at the regional or world level. At the world level, for any good, the total demand (import demand) in the demanding nations should be equal to total supply (export supply) from the supplying nations.

In real-life situations, it is difficult to get an equilibrium every time period. Some models use the notion of dynamic equilibrium or partial equilibrium. Here, the model always chases the point of equilibrium over time. In other words, the iterative process always tends to move the system toward equilibrium but may not attain it in all attempts. The modeler is satisfied as long as the system is trying to converge on the equilibrium point without being too far from it; i.e., there is no large discrepancy between supply and demand. This is a type of satisfying approach as proposed by Herbert Simon.

13.3.4 Dynamic and Static Models

World modelers cannot just obtain results for 1 year and be content. Analysis of the world economy requires the model to shed light on the nature and extent of change over time. Models can be dynamic in the sense that changes over a time horizon are analyzed using the resulting state of one period as the initial state of the following period. Often the outcome of an economic decision will be realized after a time lag, especially in the case of capital investments. Such dynamics, of course, make the system more complex to deal with. To avoid this, some models are built to be static models where all changes over time are given exogenously. The reader should refer to Chapter 1 for discussion of static and dynamic models. Also, to have a better understanding of various types of systems, such as continuous-time, discrete-time, etc., of system dynamics concepts, one should refer to Chapter 2.

13.4 WORLD MODELS

Most world models deal with the problems of economies. Economies, of course, cannot stand alone and be analyzed without referring to the aspects of regional interdependence. As a result, all economic models do have submodels looking into population, agriculture, energy, environment, etc. There are models where other aspects of the world society play more important roles than economies. In this section, the models are grouped into functional areas, and the details of structure and objectives of some of those models are presented. The selected functional areas are economics, agriculture, food, and environment.

13.4.1 World Economic Models

The models, such as World 3 (Meadows et al., 1972, 1974), LINK (Ball, 1973), Mesarovic-Pestel (Mesarovic and Pestel, 1974b), SARUM (SARU, 1977, 1978), World Bank (IBRD, 1975; Carter, 1978, Gupta et al., 1979), and UN (Leontieff et al., 1977), that emphasize economics belong to this category from the point of view of structure. They vary, sometimes radically, but variables and parameters representing gross national product, value added, income, expenditure, supply, demand, price, input-output co- efficients, investment, import, export, economic equilibrium, etc., are found in abundance in these models. The Bariloche model (Herrera et al., 1976) is also included here, though it is actually a type by itself because of its main concern about the life-span of human beings.

World 3 Model. This model was designed and developed at MIT by a group led by Forrester and Meadows in the early 1970s. Based on the principle of systems dynamics, this model uses the Dynamo simulation language. Since the publication of its findings in 1972, there has been no concerted effort to modify or to develop a new generation of this model. At present, the model is being used more in the academic world.

The purpose of the model was to gain insight into the workings of the world system and the limits and constraints within which the society has to operate. The study attempted to identify and understand the various elements that dominate other factors in the evaluation and progress of the world system and to assess the effects and limiting behavior of those elements under different conditions. The model treated the world as a single geographical unit and ignored the differences in social, political, and economic structures among the nations and regions of the world.

The model analyzes the important interrelations among the various aspects of the world system through five major areas represented by sub- models (Figure 13.1).

1. Population: This submodel looks into the problems related to population, which is divided into four cohorts (age groups). The effects of industrialization, family income, health services, avail- ability of food, and pollution on the birth and death rate are analyzed. The population growth is determined by the birth and death rate. This submodel interacts with other submodels through factors such as industrialization, family income, etc., as mentioned above. It is assumed that an increase in food intake, material standard of living, and health services decreases the death rate, while an increase in pollution and crowding has the opposite effect. The birth rate is assumed to go down with an increase in food in- take, pollution, crowding, etc.

2. Economy: This has been divided into three sectors—industrial, agricultural, and service. The output in the industrial section is used for direct consumption and investment. Growth of capital through investment and its decline because of depreciation are analyzed. Labor force and resource usage are some of the inputs to this submodel and are also the linkages between this and other submodels.

3. Agriculture: Here the effects of availability of land, population, degree of pollution, availability of fertilizer, and capital inputs on agricultural production are studied. Food is not divided into dif- ferent types. Capital is assumed to affect the land development and its yield.

4. Nonrenewable resources: This submodel deals with all energy and nonenergy inputs required in the economy submodel. It is assumed that these resources are continually depleted because of their usage by the economy sector, thus requiring more investment

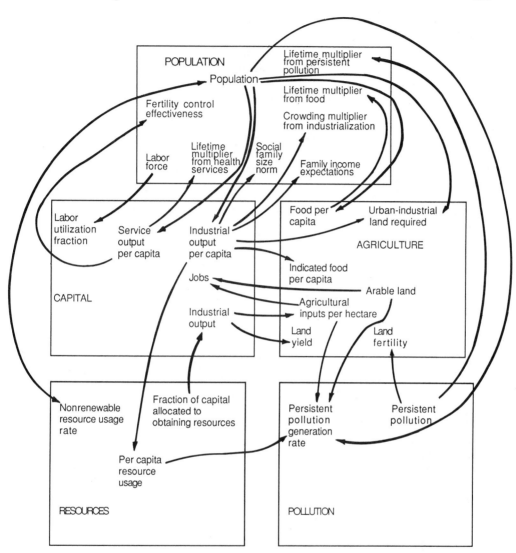

Figure 13.1   Interactions among the five basic sectors of World 3.

in obtaining resources. The resource usage, which is also determined to some extent by population, is considered one of the
determinants of pollution.

5. Pollution: Production in economic and agricultural submodels and
    usage of resources increase pollution, which, in turn, affects the
    fertility of land and human life expectancy. Pollution is slowly
    absorbed by the environment, but large quantities inhibit the
    process, resulting in harmful effects on human beings and land.

In the model, population and capital are assumed to grow exponentially,
but this growth is somehow decelerated by agriculture, nonrenewable resources, and pollution sectors, whose growth rates have limits.

The model, which has a time horizon of 200 years (1900–2100), is supply-driven where supply (production) determines the demand (consumption) in every sector. Of course, over time the level of economic development that affects the desired global consumption determines the allocation of investment among sectors, thus making the model dynamics more demand-driven over a long term. Because of absence of any conflict between demand and supply, there is no procedure to solve the system. The model has been run for 14 different scenarios under various assumptions about technical progress, value changes, social policies, and other exogenous parameters.

Project Link (Ball, 1973). A variety of national and regional economic models have been linked in this model in order to generate forecasts of world trade and payments. The models for the countries or regions have been developed by universities and research institutes of government agencies. These are all econometric models. They represent 25 countries/regions including 13 Organization for Economic Cooperation and Development (OECD) countries and developing countries in Africa, Latin America, the Mideast, and South Asia.

The LINK system has a time horizon of 3 years and, hence, can be used to facilitate simulation over a short term. The economic structure varies from model to model—some are less complex and some are more.

Mesarovic-Pestel Model (World Integrated Model). Criticisms were leveled against the World 3 model mainly because of its treatment of the world as a single geographical unit that ignored the basic social, political, and economic differences between nations and regions. This prompted two groups—one led by M. D. Mesarovic at Case Western Reserve University, Cleveland, and the other by E. Pestel at the Technical University, Hannover (West Germany)—to work together to design and develop the World Integrated Model (WIM) under the auspices of the Club of Rome to give a more disaggregated view of the world and its problems. The model uses the multilevel hierarchical system approach developed by Mesarovic (Mesarovic et al., 1970).

Scope and purpose. The purpose of the model is twofold:

1. The model intends to address the problématique and to analyze the interdependence of national and regional problems. Problems related to supply of adequate amounts of food, energy and its influence on economic development, and, of course, the "economic gap" between the "have and the have-nots" of the world have been addressed by various versions of the model over the years.
2. The second purpose of the model is to make it usable as a policy analysis tool by decision makers to create and assess different scenarios through simulations with alternate sets of inputs.

As will be evident from the following discussion, the structure of the model makes it a very powerful, versatile instrument to achieve the above purposes.

Structure of the model. (a) Regionalization: The WIM divides the world into many regions. The original model had 10 regions, and the most recent version has as many as 38 regions. Even in the original version there were provisions to subdivide any region into subregions to facilitate in-depth analysis of the policy(s) of a particular country from the global perspective. This built-in flexibility gives WIM an added advantage.

(b) Multilevel Hierarchical Structure: The theory of multilevel system has enabled model builders to organize complex systems into relatively simpler interconnected subsystems. Figure 13.2 shows the organization of each regional submodel in WIM. The stratum structure shown on the left side gives superior and inferior levels. A higher level has the controlling power over the ones below it. For example, the technology stratum is affected by the activities in the demographic and economic stratum above it and, in turn, determines the quality of the environment. It should be noted that the presence of arrows with directions from a lower level to a higher level indicates some influence of the inferior level over the superior. For example, the physical laws and resources in the environment do pose some constraints for the technologies. This type of structure clearly emphasizes the fact that analysis for economic, demographic, or technological systems cannot be carried out in isolation from each other because of the existing interactions. Similarly, the group stratum that takes into account the sociological and political processes has to interact with choice of an individual and also with demographic and economic decisions. Thus, the individual stratum enjoys the maximum power in the hierarchy. Representing the higher strata, namely, individual and group stratum involving human elements, has always posed difficulties so far as models are concerned. The WIM has adopted an approach of building analytical models for the lower three levels, which are easy to describe through equations or statements governed by physical and economic laws and the availability of resources. The activities of the top two levels that deal with human

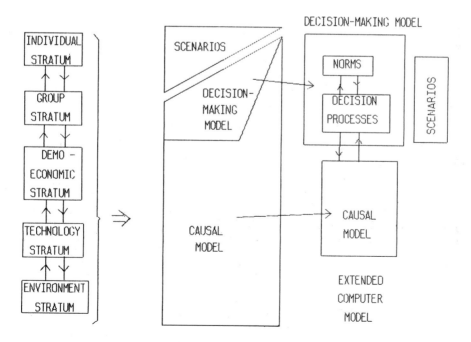

Figure 13.2   Multilevel Hierarchical Structure of WIM.

elements have been modeled through "open loops." The decision maker, as the analyst, introduces various political or sociological factors through exogenous variables or constraints and creates a number of scenarios for quick and efficient implementation of such exogenous factors. The WIM relies heavily on an interactive computer model and a very powerful and versatile software package called ARISTOTLE (SAI, 1986) that can house any code. Figure 13.2 shows this approach with "causal models" and the open-loop part named "scenarios" in column 2. Part of the causal model, as expected, should deal with the decision-making process that is guided by the open-loop part and the activities of the lower strata. Column 3 shows the extended model. The open loop of the upper strata has been partially closed through endogenization in different versions of the model. Many versions have the provision to analyze a region or group of regions through only open loops at the upper level or through partially open loops, and this has enabled analysts to experiment with a broad range of policies.

Figure 13.3 describes the multilevel feature in more detail. For the purpose of illustration, only three submodels, dealing with population, economics, and agricultural production, have been chosen to show the vertical interconnection of strata through some key variables and components. It should be noted that each submodel and its components are much more

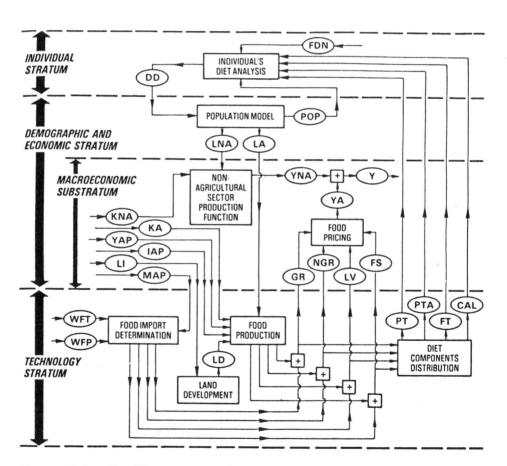

Figure 13.3   Simplified example of vertical interconnection of strata.

complex than that shown here. Each submodel will be discussed in more detail later in this section. At the individual stratum, population POP, the food diet needs FDN, the availability of food in terms of protein PT, animal protein PTA, calories CAL, and fats FT determine the deficiency in diet DD for individuals. The population model, which is part of the demographic and economic stratum, determines the population and labor in agriculture LA and nonagriculture LNA.

In the economic part of this stratum, an overly simplified version of the macroeconomic model is given. Here, the nonagricultural production YNA is determined by using a suitable production function and non-agricultural labor LNA and capital KNA as inputs. In the technology stratum, the land development, food production, and total food availability are shown. The investment in land LI determines the arable land available LD. The land available LD and labor LA, along with investment in agricultural production IAP, available capital KA, and technical inputs like fertilizer, seeds, etc., YAP, serve as input to the food production submodel. The outputs of this submodel are grain GR, nongrain NGR, livestock LV, and fish FS. All the inputs to this submodel, with the exception of LA, are the outputs of the economic submodel. The food produced combined with that imported gives the food available, which decides the economic value of regional agricultural output YA through a pricing mechanism. The imports of the food are determined by food available for world trade WFT, world food price WFP, and the regional economic output allocated for food import MAP. As shown, the total food available determines the diet components PT, PTA, FT, and CAL. The sum of outputs for all production sectors in a region gives a gros national product Y.

Figure 13.4 shows a schematic diagram of the economic model. The MAC of this figure represents the macroeconomic substratum of Figure 13.3.

(c) Major Elements of a Regional Model in the WIM: The WIM consists of various elements, as shown in Figure 13.5. One should take note of the hierarchical system here. The segments dealing with labor and education and aid and loans appear at the top of the hierarchy representing superiority of individual and group decisions, whereas the submodels analyzing materials, energy, food, and machinery form the lowest level to represent technological and environmental strata. Population and economic submodels appear in between to indicate the place of demographic and economic strata in the hierarchy. For every region one can find such a structure. The trade and payments submodel represents the suitable linkage among the regional models and facilitates analysis of the interdependence among the regions or nations. A brief description of each submodel in Figure 13.5 is given below.

(i) Labor and Education: Here the regional demand and supply of labor in four educational categories are computed and then the required expenditures in those education categories are determined after comparing demand and supply of labor. The total demand of labor LT in a region is computed as a function of per capita gross regional product y (gross regional product Y divided by population P). Then the demand of labor in the ith educational category LD_i is obtained by dividing LT proportionately using a factor e_i, which also is a function of Y. Thus, one has

$$LD_i = LT \cdot e_i \cdot Y \tag{13.1}$$

The supply in labor in any educational category is computed from the current population in that category and allowable immigration and emigration

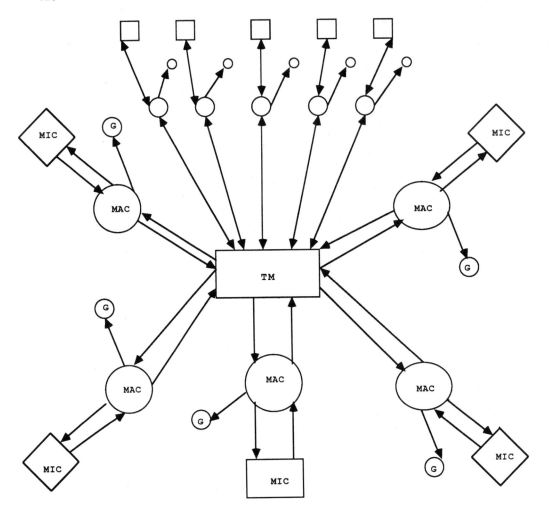

MIC = Regional   Microeconomic   Substratum

MAC = Regional   Macroeconomic   Substratrum

G   = Gross   Regional   Product

TM = Trade   Matrix

Figure 13.4   Economic model.

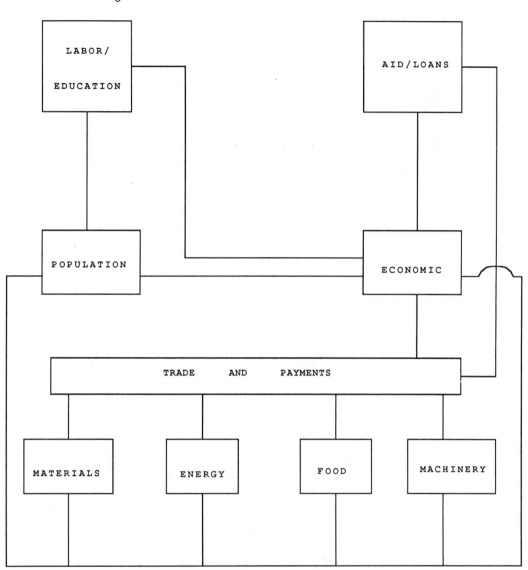

Figure 13.5   Major elements of the world integrated model.

depending on the exogenous political decisions. On the balance between demand and supply, the labor shortage or surplus is computed, and this, when fed back into the economic model, affects the economic output of the region. Depending on the required rate of growth in economy, the requirements in labor are determined and the corresponding desired expenditure in those educational categories is computed. Subsequently, the total desired expenditure on education is computed and that is modified depending on the total educational expenditure allocated in the national economy. One should take note of the dynamic nature of the model because the expenditure in education at any time will affect the labor supply at a future period owing to a time lag.

(ii) Population: The model uses 87 categories of age distribution for population. Fertility distribution and age distribution together determine new births. Similarly, deaths in each category are computed using a mortality distribution and age distribution. The objective of this submodel is to estimate the population in different age categories over time. Births increase the population of the following period at the lowest age category whereas deaths reduce the population in each category. If $P_{t,j}$ is the population in the tth period for jth age distribution, B_t the births, and $D_{t,j}$ the deaths in jth age distribution; then

$$B_t = \sum_j fr_j \cdot P_{t,j} \tag{13.2}$$

$$D_{t,j} = mr_j \cdot P_{t,j} \tag{13.3}$$

$$P_{t,1} = B_t - D_{t,1} \tag{13.4}$$

$$P_{t+1,j+1} = P_{t,j} - D_{t,j} \tag{13.5}$$

where fr_j and mr_j are the fertility rates and mortality rates respectively in jth age distribution. Fertility rate, which is age-specific, is a function of per capita income that is computed in the economic model and certain exogenous factors (e.g., effect of family-planning activities sponsored by the government). The mortality rate depends on medical advances, disease control activities, per capita income, etc. In addition, starvation deaths, which depend on calorie and protein deficiencies, are computed.

(iii) Economy: As expected, this submodel plays the most important role, with variables and factors, both endogenous and exogenous, affecting all other submodels such as energy, agriculture, metals, machinery, and trade. Economy is divided into eight sectors: agriculture, manufacturing, nonmanufacturing, industrial, construction, wholesale/retail, transportation/communication, and services. The model, to start with, computes the available capital and the production capacity in the eight sectors. If $K_{t,s}$ denotes the capital in the sth sector in the tth period, d_s the depreciation rate, and IS_s the investment, then

$$K_{t,s} = K_{t-1,s} (1 - d_s) + IS_s \tag{13.6}$$

The gross regional product is computed either by using a production function using capital and labor as inputs, or by determining the output in each sector by a capital/output ratio and then by summing up the outputs of all the sectors. The initial gross regional product YI can be computed by using the Cobb-Douglas function as follows:

$$YI = m(KT)^{\alpha} (LT)^{\beta} \tag{13.7}$$

where m is the coefficient and α and β the parameters. KT is the total capital and is equal to the sum of sectoral capital $K_{t,s}$, and LT is the total labor.

Using the capital-to-output ratio, Q_s for the sth sector, one can compute YI with t corresponding to the initial period as

$$YI = \sum_s \frac{K_{t,s}}{Q_s} \tag{13.8}$$

Q_s is defined as the capital required to produce unit output in the sth sector. This initial gross regional product is used to determine the initial expenditures, such as private consumption CI, governmental consumption GI, investment II, exports XI, and imports MI, through coefficients for expenditures shares.

Thus,

$$CI = YI \cdot kc$$
$$GI = YI \cdot kg$$
$$II = YI \cdot ki$$
$$XI = YI \cdot kx$$
$$MI = YI \cdot km$$

where kc, kg, ki, kx, and km are the coefficients for consumption, government, investment, export, and import, respectively. Investment depends on savings. The estimates of the government and private consumptions and investment are suitably modified by the final results of the aid-and-loans submodel to give the final values of these expenditures. The initial estimates of exports and imports are modified by what happens in the other submodels, which are physical models, to give the final exports and imports of energy, materials, machinery, food, residual goods, and services. The sum of the final expenditure gives the gross regional product.

In some versions, the targeted growth rate trg in gross domestic product plays an important role in determining the desired output YD. With t, as before, denoting the time period,

$$YD_t = Y_{t-1}(1 + trg) \tag{13.9}$$

A balance is obtained between the supply and demand in the domestic market, and the final demands are computed using Leontief's input/output matrix. The exports and imports are determined by the world trade submodel.

Using the final values, the indicators like rate of economic growth, per capita income, total exports and imports, trade balances, etc., are

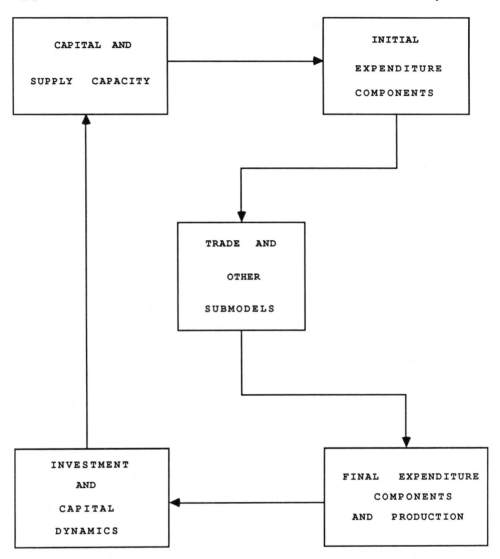

Figure 13.6   Economic submodel.

computed. Figure 13.6 shows the basic structure of the economic sub-model. The block for investment and capital dynamics represents the effect of time on capital availability.

(iv) Energy: This is a commodity or physical submodel using a partial economic equilibrium approach. Energy has been divided into five categories: oil, gas, coal, hydroelectric power, and nuclear energy. The model computes the demand, supply, and price in each category. Statistical estimates are used in determining the energy demand per unit of gross regional product ed; this value, the gross regional product (prices) P, and aggregate demand elasticity E_d, together determine the total energy demand ED for the region, as given in Eq. (13.10). (Elasticity is defined as the percentage change in quantity, e.g., demand or supply, for the corresponding change in price.)

$$ED = (ed) \ Y \ (1 + E_d \cdot P) \tag{13.10}$$

The total energy demand, after being suitably modified by any exogenous factor that is rather political, social, or some other policy oriented than economic, is divided into energy demand by types of energy using elasticities and exogenous variables. The supply of energy is determined through investments and reserves. Current investment depends on past investments and reserves. Current investment depends on past investments with time lags, price, economic growth, production capacity, reserves, and suitable elasticities. The reserves are determined by using new discoveries, which increase reserves, and past production. The current investment coupled with the past capital used gives the current capital in the energy sector. Then, with the help of a capital-output ratio, the production capacity of energy is computed. As mentioned earlier, this model uses a partial equilibrium approach to compute the price of energy while trying to equate total supply and demand as well as modifying the total supply and demand by imports of energy from the world trade submodel. Figure 13.7 is a flow diagram of the energy model.

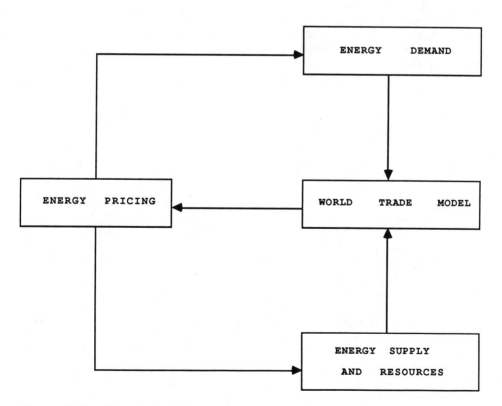

Figure 13.7 Energy submodel.

(v) Agriculture: Agriculture is divided into five categories: grain, nongrain, industrial crops, fish, and livestock. This submodel is a commodity model and uses the partial economic equilibrium approach for balancing supply and demand in each category and computing the corresponding prices. The demand comprises demand of food, including grain, fish, and livestock, demand of feed required by the livestock, and that of industrial input, including nonfood, nonfeed agricultural products. Food demand is computed using per capita income and income elasticity, which depends on change in prices and population. Size and growth in livestock determine the feed demand. The industrial demand depends on that fraction of the manufacturing sector that needs agricultural input. Suitable elasticities to reflect effect of price changes are used here. On the supply side, the production of crop depends on the availability of land and yield from it. Three types of land are considered, namely: dry, irrigated, and grazing. The cultivable land is divided among various crops, depending on the price at that time and elasticities. Yield depends on factors such as type of land, fertilizer used, etc. Using parameters a, b, and c, the following yield function is used:

$$yl = a + b_r[1 - \exp(-k_a/c_r)] \tag{13.11}$$

where k_a is capital investment in agriculture, r is the crop type. Fish production depends on the level of mariculture, price, and elasticities. The production of livestock is a function of size of herd, which depends on the rate of reproduction, and slaughter rate, which is a function of price, elasticities, and sometimes the lack of proper feed. Another aspect of the agricultural production segment deals with the question of investment, which is divided for use of capital and land development. The land developed LN is computed by the following relationship:

$$LN = \frac{I_\ell}{c_\ell} \tag{13.12}$$

where I_ℓ is investment in land development and c_ℓ is the cost per unit land development. Investment is a function of its productivities in different land types and prices of the agricultural outputs. After computing the demand and supply, the submodel goes through the equilibrium process to balance the demand and supply and compute the prices after taking into account the imports and exports of different agricultural products through the world trade submodel. One can construct a flow diagram of the type given in Figure 13.7.

(vi) Materials: This submodel deals with the demand, supply, and prices of materials. This, like the previous two submodels, is also a physical model that uses a partial equilibrium approach to balance demand and supply and to compute the prices. The materials that are considered here are: copper, aluminum, and iron. The demand of these is computed using the output of the economic sectors that use such materials and then is adjusted using prices and elasticities. There are three categories so far as the supply is concerned: primary, secondary, and government stockpiles. The production of primary materials depends on the capacity of the previous period, depreciation, change in production from year to year, and the level of materials reserve. The secondary or recycled supply depends on the recycling fraction, the materials in circulation, and elasticities. Change in government stockpiles is specified exogenously. After computing the total demand and supply of materials, the model considers imports and exports and goes through the equilibrium process.

(vii) Machinery: The total demand of machinery depends on investment. The level of activities in the manufacturing sectors determines the production of machineries. This model deals with the value of machineries rather than the physical quantities. Given the domestic demand and supply of machineries, it goes through the partial equilibrium process after including the actual imports and exports from the world trade submodel.

(viii) Aid and Loans: Interregional financial transactions involving foreign aid and loan are considered in this submodel. These transactions take place in two ways. Donors of financial assistance or loans give the amount to a pool which supplies a certain amount to a receiving region that is trying to achieve a percentage growth rate in its economy. The other method of transaction is a direct method where direct bilateral and, more or less, fixed sums are involved. In the pool total inflow should be equal to the total outlfow and, hence, a balancing mechanism is needed.

(ix) Trade: This submodel, as the name indicates, deals with the international trade linkages among the nations or regions of the world. The import demands and export capacities of the regions are taken into consideration in many trading categories. Different versions of WIM have used different numbers of trading categories, varying from a total of 7 to 15, for energy (oil, gas, coal, hydraulic power, and nuclear power), agriculture (grain, nongrain, industrial crops, livestock, fish), investment goods (machineries), and other goods (manufactured goods, three types of materials). The physical submodels described earlier compute the desired imports and exports of different commodities. The WIM assumes that anything that is not included in those submodels is included in a category called residual. The residual for each region is computed as a fraction of gross regional product using an exogenously supplied coefficient based on past history. The desired imports and exports are then modified by any exogeneously given limits. Imports also are affected by the availability of foreign exchange in the form of aid, loan, or any other international transfers. The world price of a commodity either is set equal to regional prices of one of the exporting regions or is computed as a weighted sum of prices of all exporting regions. The total world import and export of a commodity are determined by adding those for all regions. At the world level, total import must be equal to total export for every commodity, and, hence, the submodel goes through the trade-balancing procedure.

The balance is attained in one of two ways: In one procedure, actual world trade flow is set equal to the minimum of world import and export, and then the regional values of imports or exports, as the case may be, are proportionately reduced so that total world value is equal to the above-mentioned trade flow. In the second procedure, world prices are used in an iterative process to modify the regional import and export volumes so that world import and export are equal for every commodity. In a recent version, a goal-programming approach using regional prices, world prices, and elasticities has been adopted to achieve this equilibrium (Mohanty, 1982).

Figure 13.8 gives the flow of computation in the WIM. The model computes population, labor, economic capacity, and aid and loans first. Then the demand and supply of energy, food, machinery, and materials are calculated along with import demand and export capabilities of a region.

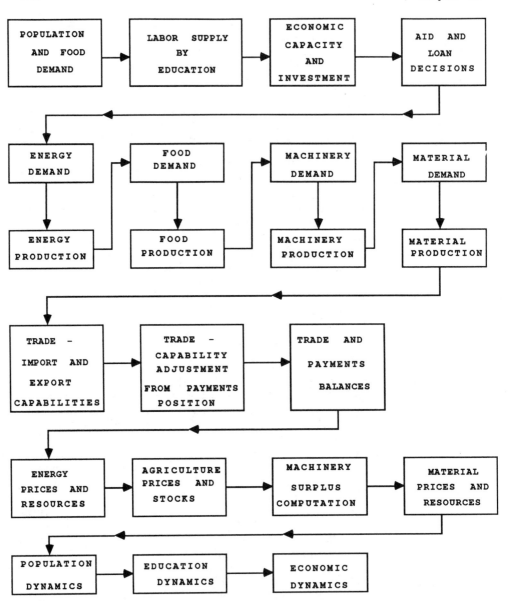

Figure 13.8   Sequential flow of the world integrated model.

The actual volume traded is computed next. Depending on the output of the above steps, the model goes back and adjusts the quantities in energy, agriculture, machinery, and materials and computes their prices. In the final stage, the dynamic character of the model deals with the change or adjustment in population because of adjustments in other submodels, change in labor supply, and final results in the economic models in terms of investments in the coming time periods.

The time horizon for WIM is 1975–2025. The model cycles annually.

Bariloche Model (Herrera et al., 1976). This is a normative model that is concerned with the development and analysis of ways and means by which a "new (ideal) world," void of the current disparities and miseries can be attained. It does not use the current situation of the world as the basis to predict the future. The model is designed to achieve two specific goals:

1. To demonstrate that resources and environment do not impose any serious limits on proper development of an ideal society. The authors carried out an analysis of the current situation in non-renewable resources, energy, and pollution to prove their point.
2. To build a mathematical model to show that the world can move from the present state toward the goals of the ideal society in a reasonable time. The modelers used mathematical optimization as an instrument in this phase to ensure proper allocation of available resources to maximize life expectancy at birth. Figure 13.9 gives the basic structure of the Bariloche model. Nutrition, housing, education, other services and consumer goods, and capital goods are the five economic sectors of the model. The various submodels deal with population, food, urbanization, education, and trade. The world is divided into four regions: developed countries, Latin America, Africa, and Asia. The time horizon for the model is 1960–2060.

The United Nations World Model (Leontief et al., 1977). The World 3 model and its findings, as well as the specific development objectives regarding world economy that were included in the UN's second development decade strategy beginning in 1970, prompted the world organization to sponsor a project to study the world economy in the early seventies through its Center for Development Planning, Projections, and Policies (CDPPP). A group of economists at Brandeis and Harvard Universities, headed by Wassily Leontief, designed and developed the model.

The purpose of this study was to develop a disaggregated multiregional model that can be used as a tool to analyze the impact of various economic policies on the International Development Strategy of the UN. More specifically, one of the basic objectives was to look into the widely shared concerns regarding the material feasibility and social desirability of long-term economic growth.

Structure of the Model. This global model uses input-output analysis as its basic frame. The world has been divided into 15 regions. Figure 13.10 shows a simplified structure of a regional model (Petri, 1977). Each of the

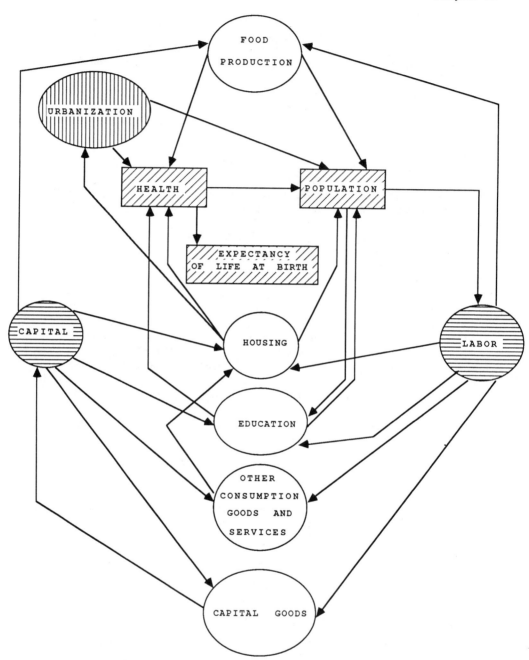

Figure 13.9   Basic structure of Bariloche model.

regional models deals with 175 equations and 229 variables. There are 45 sectors in the economic system. Figure 13.10 also shows the linkages between the regions through the world trade pool. World pool is a unique feature of this model where it is assumed that each country, or region, exports the commodities to a pool and imports various commodities from this pool. Thus, the complexities involved in the bilaterial transactions between two nations are avoided.

The model, on the whole, consists of a set of equilibrium conditions, and the equations are solved simultaneously. To give more freedom to the analyst, the model has more unknowns than equations so that one can choose a subset of variables to have values exogenously assigned. This also helps to avoid determining relationships between variables that are difficult to estimate. Of course, some variables, such as population, resource reserves, environmental standards, etc., are always treated as exogenous variables. The gross domestic product GDP can be either an exogenous variable, which is used to indicate a prespecified target value for any region, or an endogenous one where the model determines a value for GDP with a set of given conditions. As mentioned before, linear methods are used and, hence, many coefficients are estimated through regression in satellite models. Also, nonlinear relationships are approximated by linear ones.

The model divides the population into 5-year cohorts. Mortality depends on cohorts, but fertility is a function of female children born to any woman during her childbearing years, cultural factors in a region or country, and family planning programs. In the economic sector, the U.N. model develops a demand-driven model. Prices do not play any major role in consumption. Production is determined by the demand and input-output matrices. Capital requirements are computed from production and consumption through suitable functional relationships in simultaneous equations. There is no time lag in the features dealing with the usage of capital and investment. In this model, prices affect only the cost of inputs. In the agricultural sector, this model has five products, namely: animal products, high-protein crops, grains, roots, and other. Demand of these is determined through consumption and per capita gross national product, and production is computed using the input/output matrix. Energy and materials are treated in a similar way through input/output structure. The model has a special feature where some products are measured in physical units (e.g., agricultural products, mineral extraction, and pollutants), and others in monetary values.

A major emphasis of the model is on pollution. The 45 sectors mentioned earlier include eight pollutants and five pollution abatement programs. The system determines the level of pollution of the environment in terms of various levels of economic activities in different scenarios after taking into account the effects of pollution abatement programs in a region.

The time horizon for the model is 1970–2000, but it gives only "snapshots" of the results at 10-year intervals. The model has four categories of food, nine categories of mineral and energy, 22 manufacturing categories, and eight pollutants.

13.4.2 Food and Agriculture Models

Most of the world economic models include submodels that deal with agriculture and food. Some of these submodels are quite extensive, but some

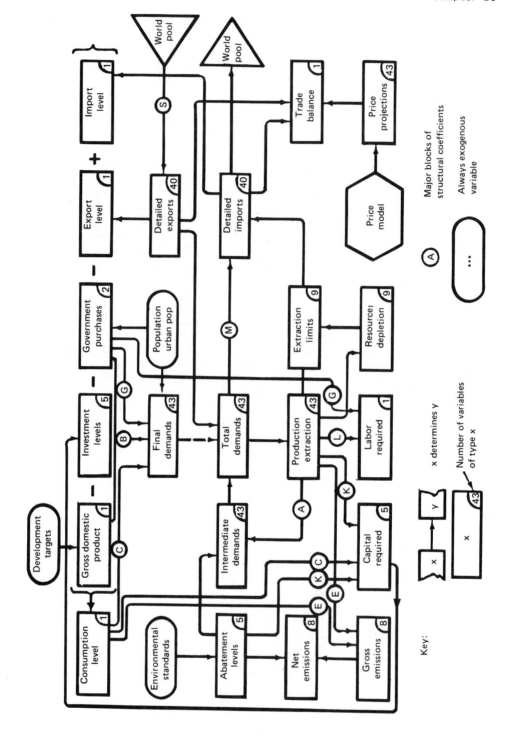

world models have been designed and built only to analyze the problems related to the world situation in food and agriculture. MOIRA and GOL are two such models.

MOIRA (Linneman et al., 1979). The Model of International Relations in Agriculture (MOIRA) was cosponsored by the Club of Rome, the Ministry of Agriculture and Fisheries, the Ministry of Foreign Affairs in the Hague, and the Haak Bastean Kunemann Foundation. The project was located at the Free University of Amsterdam, with another group working at the Agricultural University of Wageningen.

The objectives of the project were to analyze the world food situation, to shed light on current constraints on production and changes in policies that would be needed to relax those constraints, and, of course, to determine the maximum possible production quantity. Special emphasis was to be placed on problems of food production in developing countries. MOIRA surpasses world models in its extent of disaggregation. It is basically an econometric model where elegant statistical estimation procedures have been used extensively. The model analyzes 103 free nations and three centrally planned nations. For each nation there are two consumption sectors and one production sector. All agricultural products are converted into one product, called "consumable protein." The nonagricultural production sector values are supplied exogenously. At the national level there is no sector dealing with energy, but domestic price and balance-of-payment problems are taken into account by the policy makers. Functions giving production and consumption at national and world levels are included in the model, and a mathematical algorithm is set to work to optimize the production decision at the national level while aiming at balancing the total demand and supply at the world market. This equilibrium approach determines the world price of this product. Figure 13.11 shows the structure of the model.

The population in each nation is divided into 12 categories with six income classes. Each nation is also divided into two regions, such as urban and rural. The model carries out sensitivity analysis of many policies provided exogenously and assumes that people starve, not because there is no food, but because they cannot afford to buy food.

13.4.3 Environmental Models

The U.N. world model emphasizes environment and pollution in its structure. Recently, researchers have shown more concern about the effect

---

Figure 13.10   The internal structure of a region in the U.N. global model. Circled capital letters denote major structural coefficients as follows: A, interindustry inputs; B, composition of investment; C, composition of consumption; E, Generation of gross pollution; G, composition of government purchases; K, capital inputs; L, labor in industry; M, import dependence ratios; S, shares of world export markets (Petri, 1977).

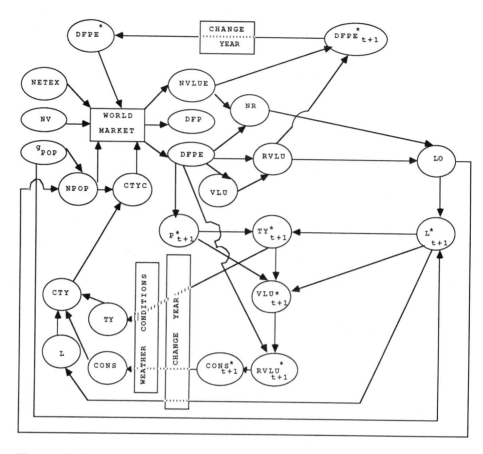

Figure 13.11   Structure of the MOIRA model.

CONS = food consumption in the agricultural sector, per capita
CTY = total supply of food by agriculture to non-agriculture
CTYC = supply of food by agriculture to non-agriculture, per capita of non-agricultural population
DFP = world market price level of unprocessed food, measured as deviation from the base-year value
DFPE = domestic price level of unprocessed food, measured as deviation from the base-year value
DFPE* = desired DFPE
g_{POP} = population growth rate
L = agricultural population
LO = labour outflow from agriculture
NETEX = net food exports of centrally-planned economics
NPOP = non-agricultural population
NR = disposable non-agricultural income per capita, in real terms
NV = (nominal) non-agricultural income per capita
NVLUE = disposable (nominal) non-agricultural income per capita
P = price received by the agricultural producer
RVLU = agricultural income per capita, in real terms
TY = total yield of agricultural production
VLU = (nominal) agricultural income per capita

Asterisks* denote expected values. The time subscript t is suppressed.

of many economic decisions on the atmosphere. A discussion on such a model dealing with usage of energy and resulting CO_2 emission follows.

Long-Term Global Energy—CO_2 Model (IEA/ORAU, 1984). Carbon dioxide (CO_2) is produced as a by-product during fossil fuel combustion, certain specific industrial activities, and, of course, animal respiration. CO_2, while remaining as a nontoxic gas in the atmosphere, exerts a greenhouse effect where the sunlight is allowed to pass on to the earth, but the outgoing heat is obstructed from escaping, thus tending to raise the global temperature. This will result in a change in climate pattern. It is known that combustion of fossil fuel is a major determinant of the CO_2 content in the atmosphere, and, thus, there is a very close association between energy, economics, and environment.

This model deals with the long-term supply and demand of energy and analyzes the energy-economics relationship in the global context and projects the energy usage, which is translated into the use of fossil fuels and the resulting CO_2 emission into the atmosphere.

In the model, the energy is divided into six primary fuel types, namely: oil (conventional oil, shale oil, etc.), gas, solids (coal, biomass), resource-constrained renewables (hydro- and egothermal), nuclear, and solar (solar electric, wind power, tidal power, etc.). The world is divided into nine global regions: United States, Canada, Western Europe, centrally planned Asia, Middle East, Africa, Latin America, and South and East Asia. The model can be used to develop projections for any desired year up to 2100.

Figure 13.12 shows the structure of this model. It has four parts: supply, demand, balance between supply and demand, and CO_2 emissions. For each region the supply of energy in each type is computed. The model has two classes of energy supply technologies: one includes conventional oil, gas, and hydropower and is resource-constrained because of high demand and low availability; the other deals with unconventional oil, gas, coal, nuclear, and solar energy and is considered resource-unconstrained because the amount available is large relative to small potential demand. The supply of six primary energy types is the result of aggregation of eight different supply modes. The energy demand for each region in each category is computed using exogenous inputs such as population, economic activity, technological change, energy prices, and energy taxes and tariffs. The regional gross national product (GNP) is used as a measure of economic activity. The unique regional energy prices are derived from world prices and regional taxes and tariffs. In the energy balance component of this model, the world prices are computed through an iterative process such that the total supply equals the total demand in each energy type at the global level. After the final amounts of energy usage have been determined, the CO_2 emission coefficients are used to assess the CO_2 release for consumption of oil, gas, and coal.

This model is a powerful and flexible tool for analysis of long-term global energy problems. It has the required provisions to formulate different scenarios through many exogenous inputs such as GNP, population, taxes, tariffs in international trade, regional energy policies to reflect certain sociopolitical problems, and the state of technology on a long-term basis. This model can be effectively used in analyzing energy-environment relationships.

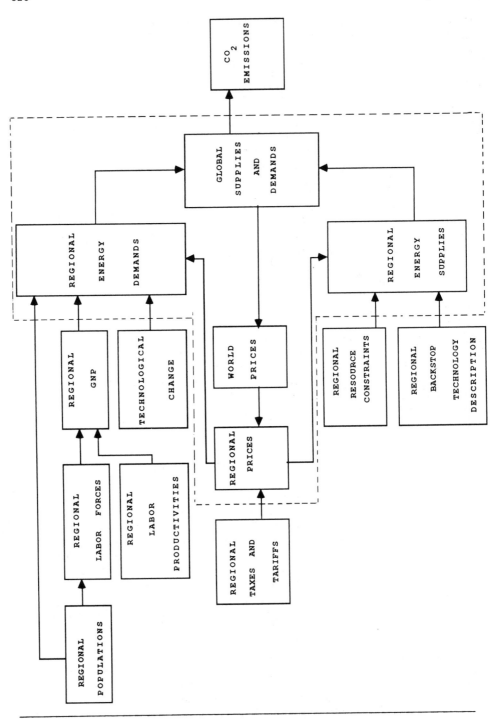

Figure 13.12   IEA/ORAU CO$_2$ emissions model.

13.5 CASE STUDY USING THE WORLD INTEGRATED MODEL*

In this section, the effect of the world economy on a representative multi-national corporation is discussed. This case study has been chosen to analyze the effects on the world market of changes made at national levels because of both economic and political reasons. Simulation studies have been carried out to generate different scenarios. The study aims at understanding and analyzing the impact of such changes at the world level on a single unit such as a multinational corporation. ARISTOTLE (a software package developed by Systems Applications, Inc.) will be the strategist's primary tool and the focal point of the planning process. ARISTOTLE will be used to consolidate information obtained from all sources, to formulate the issues relevant to the company and its divisions, to simulate these conditions and their likely outcomes on a set of integrated computer models, to conduct impact analysis of alternative scenarios, and to generate a consolidated set of projections for the plan. ARISTOTLE will also be used to achieve a proper blend of quantitative analysis and human experience.

13.5.1 AGB'S Sustained-Growth Objective: The Problem

AGB is a multinational corporation with headquarters in the United States and substantial operations in West Germany and Brazil. The company has manufacturing facilities in these countries, with three product lines in both the United States and West Germany and two in Brazil. The U.S. division's products are oil-based; the West German division manufactures transportation vehicle products. Fifty-three percent of total revenue comes from the U.S. operations, with 29% and 16% from the West German and Brazilian operations, respectively.

Currently AGB has a good cash position due to record profits from their Brazilian operations in the late seventies and to high sales volumes in the United States for the last several years. However, management is concerned about future prospects and would like to know how to best use current liquid assests to ensure future growth.

ABG's staff has analyzed the company's past performance, examined data on past economic and business environment, and implemented a set of FORESIGHT models to simulate the company. On the basis of a reference "status quo" or base scenario, the staff has forecast performance for the next four years. These results are illustrated in Figures 13.13 and 13.14. Figure 13.13 illustrates the net income of the three divisions forecast over 16 quarters (4 years) in millions of dollars. Figure 13.14 displays the expected net income and retained earnings (Retained-earn) for the corporation over the next 16 quarters in millions of dollars.

---

*The case study presented in this section and the corporate model were developed by Gerard D'Souza, senior systems analyst at Systems Applications, Inc., Cleveland, Ohio.

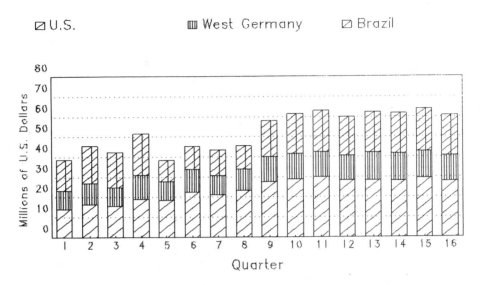

Figure 13.13   Net income of AGB by division.

Unfortunately, it is evident that the company will have difficulty meeting future growth objectives. The German division, which provided much
of the impetus for growth in the late sixties and early seventies, is now
stagnating, although it continues to generate a profit. The Brazilian
division is showing strains from the current Latin American slump. The
forecast high growth of the U.S. division is quite uncertain because of
potential sluggishness and competition in domestic markets. AGB faces
serious problems.

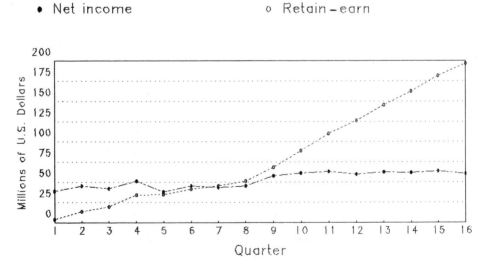

Figure 13.14   Revenue of AGB.

13.5.2 Management Options

AGB's top management has considered several options and selected two as providing the best potential investment for its current liquid assets.

Option A: Foreign Expansion. This option is based on management's expectation for rapid growth of the international market for product Z. Management believes it might be worthwhile to enter this market by establishing a manufacturing plant for Z in Brazil, where favorable treatment and local management expertise are already available. To implement option A, the company could use some of its available cash and borrow the rest on the international financial market.

Option B: Increase U.S. Market Share. This option is based on management's belief that the markets for two of its three U.S. product lines are rapidly expanding, offering an opportunity for the company to increase its market shares for both products, despite increasing competition. This strategy would require an aggressive marketing effort with a special focus on advertising.

Management needs to determine which of these two options, or what combination of the two, would be most suitable for meeting the company's sustained-growth objectives.

Management decides that at least 50% of the available cash is to be kept in reserve, the remainder being allocated in the best possible fashion. However, of this remainder, at least 10% will be allocated to improve market share in the United States, and no more than 40% will be spent on foreign expansion. If worthwhile, the company can borrow additional cash to finance the Brazilian expansion.

13.5.3 Strategy Formulation

The job of the corporate strategic planners is to help formulate a strategy based on the two options under consideration—foreign investment in Brazil and increased market share in the United States—that yields the target growth rate within the restrictions on the use of available cash. The strategists must produce a set of projections showing the impact of the chosen strategy at both the divisional and corporate levels, and they must present an assessment of the risk factors associated with implementation of this strategy.

The process can be summarized in terms of three specific tasks that the strategists must perform during the course of their work.

Task 1: Determine the amount of cash that must be borrowed in order to purchase a plant in Brazil that can achieve production levels of product Z consistent with overall revenue and net income targets.

Task 2: Determine the amount of cash that must be spent on advertising in U.S. markets to achieve the targeted growth of the company's shares in those markets by the end of 4 years.

Task 3: International market (risk) assessment—assess the impact of a range of possible uncontrollable global and regional trends and events on AGB's financial position and plans.

13.5.4   Strategy Mix

For initial analysis purposes, it is decided to treat the two options separately but to combine the results later. Two analysis teams are formed, and
the following specific tasks are completed by each team:

Option A: Foreign Expansion. Possible facilities for the production
of Z are identified, along with detailed cost estimates for construction and
acquisition. Time schedules for phased production are generated with the
help of ARISTOTLE. A marketing strategy, based on product availability
and estimates of product development and market penetration, is developed.
Alternative arrangements for loans are explored. Finally, several
scenarios, based on the availability of various amounts of capital, are developed, documented, and reported to headquarters.

Option B: U.S. Market Expansion. Using the goal-seeking facility of
ARISTOTLE, the staff determines the level of advertising required in the
first quarter in order to increase the market share to an initial target level.
The U.S. division is then asked to comment on both the assumptions and
outcome regarding this initial time period and to suggest any market share
target modifications for subsequent quarters, in view of the state of the
company and the market as indicated by ARISTOTLE. These suggestions
are implemented, and ARISTOTLE provides the outlook for the following
quarter. This iterative process is repeated until the end of the targeted
time period is reached, at which time the scenario is fully developed.
   The overall results for the U.S. division are evaluated and the entire
process repeated to generate a range of alternative scenarios. The results
are again reported to corporate headquarters.

Combining Options. Using the corporate ARISTOTLE model, the two
submodels (the world and corporate models) are integrated to yield consolidated results. Such results are determined for the entire range of
scenarios implied by the U.S. and Brazilian analyses. When a number of
reasonable scenarios are available, top management is consulted for final
modifications.
   After a number of alternative mix strategies are established, ARISTOTLE's
scenario-processing capability is used to select the alternatives that satisfy
the corporate objectives and to rank them with respect to the company's
multiple goals. Top management is again consulted for the final selection
from the set of satisfactory alternatives, using ARISTOTLE's multiple-
criteria selection process.
   Computer simulation of the world and regional economies by ARISTOTLE
results in the predictions presented in Figures 13.15 and 13.16. Figure
13.15 is a forecast of relative oil prices for the years 1986–1989. Figure
13.16, similarly, is a summary of the GNP growth rates for the United
States, West Germany, and Brazil in percentages. These forecasts are
based on a wide variety of assumptions made in WIM. Retained earnings
are projected to increase by 40% and to be sustained even in the face of
reduced U.S. sales due to predicted weakness of the market for the U.S.
division's products. The addition of a Brazilian line will further
strengthen AGB's financial outlook. The relative contribution of each
division to the corporate net income for each scenario is shown in Figures
13.17 and 13.18.

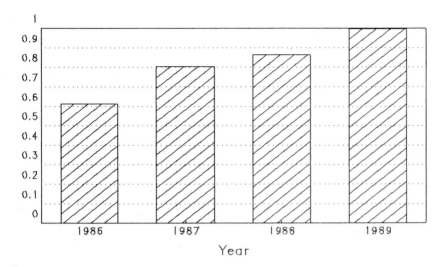

Figure 13.15 Relative world oil price (average).

13.5.5 International Market Assessment

The strategic planning process has succeeded in formulating a strategy to meet AGB's targets based on the options under consideration by management. However, it is also necessary to present management with an assessment of the key risk factors. These are events which, in the judgment of the

Figure 13.16 GNP Rate of host countries.

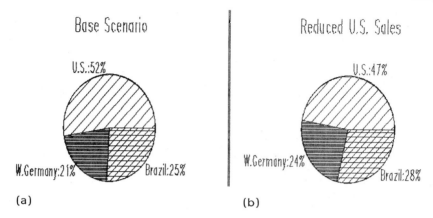

(a)   (b)

Figure 13.17   Net income at quarter 16.   (a) Base scenario.   (b) Reduced
U.S. sale.

strategic planners, management, and outside consultants, may compromise
successful implementation of the strategy and attainment of its objectives.
An assessment of a risk factor must consider the likelihood of its occurrence
and the magnitude of its potential impact.

Three risk factors are of particular concern to AGB in connection with
the recommended mixed strategy:

1. The effect of more stringent oil export conditions by the Organiza-
    tion of Petroleum Exporting Countries (OPEC).

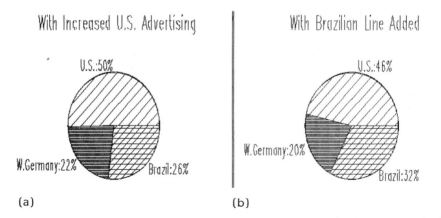

(a)   (b)

Figure 13.18   Net income at quarter 16.   (a) With increased U.S. advertising.   (b) With Brazilian line added.

2. The effect of expected changes in the world automotive market.
3. The impact of tighter international loan restrictions on Brazil.

ARISTOTLE will be used to examine these cases by developing the appropriate scenario for each and evaluating their impact on the recommended strategy.

Oil Production Restraint by OPEC Members.   Up to this point, the scenarios developed on ARISTOTLE have relied on the world model (FORESIGHT/WIM) and the oil model (FORESIGHT/OIL) to establish world oil prices through a free-market supply-and-demand mechanism. The planners and experts at AGB suspect that the future may bring a renewal of concerted restraints on the export of oil by the OPEC nations. They have hypothesized that the restraint will initially take the form of an oil production decrease during 1988. The policy makers at AGB must know how this action will affect oil prices and the performance of the AGB corporation, and how it will fuel future actions on the part of OPEC.

The first step in this investigation entails building an alternative scenario that reflects the assumed OPEC oil restraint policy. Given that OPEC nations reduce production in 1988 by 10% (this figure being the best estimate of experts), FORESIGHT/OIL is run in conjunction with the world model to calculate the resultant market price caused by this artificial oil supply reduction (alternately, the user can override the equilibrium oil price generated by FORESIGHT/OIL with a user-supplied value).

The second step is to rerun the recommended scenario for this first year, 1988, in order to determine the effects of increased oil prices on AGB's performance. The linkage between oil prices and AGB is through its U.S. division, which sells oil-based products. An increase in oil prices actually increases corporate revenue, a result that seems counterintuitive. (Here it is assumed that the profit markup is a percentage of the total cost of a product. Under inelastic consumer demand, increased product price will translate into increased revenues for the oil-based products of the U.S. division.)

After an informed discussion, it is decided that the most likely OPEC action in the second year, 1989, would be to raise prices 15% over 1988 prices while maintaining depressed production levels. Again, the world (Figure 13.19) and the corporate (Figure 13.20) models are run and the outcomes and their policy implications noted.

With ARISTOTLE, the process of scenario generation can be continued for as many time periods as required. Note that this process is interactive and stepwise, allowing for adjustments and a greater comprehension of the underlying trends and processes. Typically, sensitivity analysis is performed on all the studied parameters (here oil supply, oil price, corporate revenues, etc.) in order to determine the strength of the relationship between these parameters.

Increase in the World Automotive Demand.   The strategic planner at AGB will have determined that the world demand for automotive-related products will undergo a rapid growth in the next few years. Specifically, their forecasts indicate that the demand for German automotive products is expected to rise by 10% in 1988, by another 20% in 1989, and thereafter to hold constant. Officials at AGB expect to increase production at their German division in order to satisfy this forecast demand, but they want

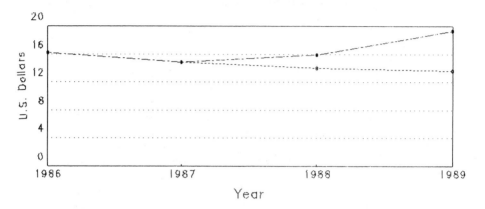

Figure 13.19   World oil prices for two scenarios.

to know exactly how these conditions will impact on AGB's financial
standing.

    Because these objectives are known in advance and involve controllable
factors (e.g., production), the world, corporate, and passenger transport
(FORESIGHT/PTV) models can be run with no scenario generation or

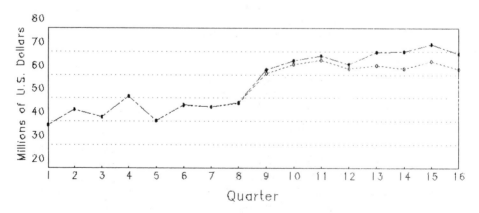

Figure 13.20   Net income of AGB for two scenarios.

interaction necessary. FORESIGHT/PTV and the world model are run
through the entire time span, 1986–1989, with the assumptions on vehicle
demand as outlined above. This model generates the expected imports to
each country and the equilibrium sales price. The corporate model uses
these figures to generate expected corporate revenues (Figure 13.21).

Although this straightforward analysis indicates that AGB should implement an expansion policy, a number of factors have been ignored. For
example, AGB's national and international competitors would probably respond to the expanding market by increasing their own production levels.
Foreign countries may respond to the sudden wave of imports by introducing protective measures, such as tariffs or trade quotas. Scenarios simulating these conditions could readily be developed and explored with the
ARISTOTLE software system. But in these cases an interactive scenario development would probably be pursued because of the more dynamic and re-
active nature of the circumstances.

Imposed Loan Restrictions on Brazil. The final risk factor to be
studied in connection with the recommended strategy is the impact of
slower growth of the Brazilian division owing to tighter international
loan restrictions on Brazil. The Brazilian government's program to repay
its foreign debt requires continuing loans from international banks. There
is some question as to whether these loans will be forthcoming. If not,
economic growth in Brazil is likely to slow, with possible negative effects
on the AGB's investment, production, and sales.

Again, AGB's analysts use ARISTOTLE's scenario formulation features
to conduct this risk assessment. In their best judgment, concerted action
on loan restrictions would probably occur in 1988. Figure 13.22 illustrates
the magnitude of the total loans received by Brazil during 1988 for two
scenarios (in billions of dollars): the "unlimited" case where Brazil

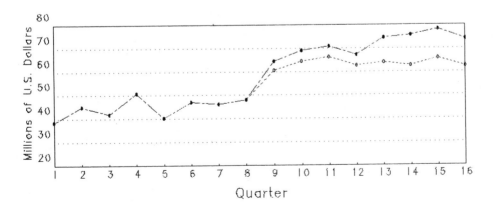

Figure 13.21   Net income of AGB for two scenarios.

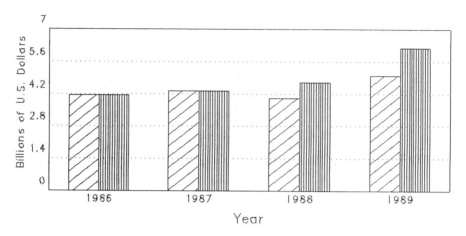

Figure 13.22   Loans to Brazil for two scenarios.

receives the amount it would normally request, and the "restricted" case
where a maximum loan amount is determined by the International Monetary
Fund and the loaning banks.   Loan restrictions for Brazil are considered
separately each year, taking into account the state of the Brazilian economy
and international economic and financial systems as indicated by FORESIGHT
models.   The level of restrictions represent the best judgments of consultants and company experts.

ARISTOTLE, in conjunction with the world model, is then used to determine the impact of loan restrictions on the growth rate of the Brazilian

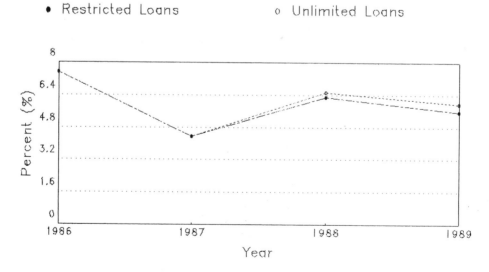

Figure 13.23   GNP rate of Brazil for two scenarios.

• Unlimited Loans to Brazil   o Restricted Loans to Brazil

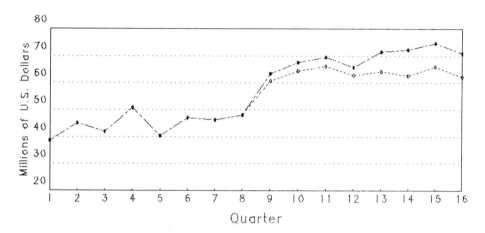

Figure 13.24   Net income of Brazil division for two scenarios.

economy.   Noting the outcome of these actions in 1988 and accordingly adjusting the inputs, the process can be continued for the year 1989 (refer to Figure 13.23 with growth rates measured in percentages).

With the Brazilian economic scenario now fully developed, ARISTOTLE can now run the corporate model in order to assess the impact of the lowered Brazilian GNP on division 3 (Figure 13.24).   From this figure it is clear that the resilience of the recommended strategy is quite good relative to the performance of the Brazilian economy.

• Base
o Base+Reduced US Sales & Increased Advertising+Brazil Line
■ Decrease in Oil Production & Increase in Price
♦ Increased World Automotive Demand
▲ Restricted Loans to Brazil

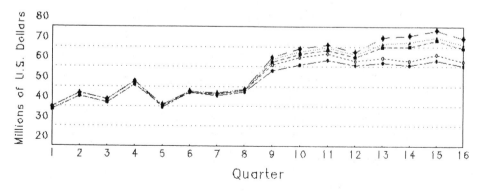

Figure 13.25   Net income of AGB for five scenarios.

Further simulations may be taken in order to develop a range of alternative scenarios more completely encompassing AGB's prospects and risks. A review of the scenarios (judged by net income) developed in this study is presented in Figure 13.25.

13.5.6   Conclusion

Having examined all the reasonable alternatives, having incorporated the best judgments available, and having assessed the most probable risks, the AGB corporation is now prepared to implement the recommended strategy (increased advertising for the U.S. division products and a Brazilian line addition). The economy, market disposition, and corporate performance will be closely monitored and compared to expectations. Deviations will be analyzed, and appropriate strategic responses to unplanned events will be developed using ARISTOTLE. Owing to the integrated, dynamic approach to decision analysis provided by ARISTOTLE, the AGB corporation is believed to be well prepared for the future.

13.5.7   System Features Used in the Case Study

No amount of data can be a substitute for solid judgment.

W. Ruckelshaus, former Director, EPA

Data is 200,000 numbers.   Information is the answer to the right question.

Walter Wriston, former Chairman, Citicorp

The case study demonstrates some of the powerful and versatile features of the ARISTOTLE decision support system and the FORESIGHT model series as they would typically be used in a simulation study. This section briefly describes these features.

ARISTOTLE is the first of a new generation of decision support tools and represents a significant advance over earlier decision support systems. ARISTOTLE has been specifically designed to increase the user's effectiveness and efficiency using modeling techniques to reach appropriate decisions.

ARISTOTLE makes it possible to design, construct, manipulate, execute, and evaluate complex models. It represents a prerequisite to practical (cost-effective) use of world models.

Some of the unique features of ARISTOTLE include:

Integration of subjective and objective factors in decision assessment
Interactive process of scenario formulation
Procedure for the construction of simplified representations (models) at the user-specified levels of complexity in order to relate to the user's intuition
Scenario (options) preprocessing and ranking by goal seeking
Inference of intrinsic preference from a sample of the user's expressed choices
Reasoning capability regarding structure and procedures.

REFERENCES

Ball, R. J., ed. (1973). *International Linkage of National Economic Models*, North Holland, Amsterdam.

Carter, N. (1978). Assumptions and sensitivity analysis for the global framework, Background Paper no. 8. World Bank Economic Analysis and Projections Department, Washington, D.C.

Forrester, J. W. (1968). *Principles of Systems*, Wright-Allen Press, Cambridge, MA.

Forrester, J. W. (1971). *World Dynamics*, Wright-Allen Press, Cambridge, MA.

Foundacion Bariloche (1978). *Complete Technical Report of the Bariloche Model*, Foundation Bariloche, Bariloche.

Gupta, S., Schwartz, A., and Padula, R. (1979). The World Bank model for global interdependence, *J. Policy Modeling*, 1(2).

Herrera, A. (1974). Introduction and basic assumptions of the model, in *Latin American World Model*, Laxenburg, Austria, October 7–10, IIASA, pp. 3–8.

Herrera, A. O., Scolnik, H. D., Chichilnisky, G., Gallopin, G. C., Harday, J. E., Mosovich, D., Oteiza, E., de Romero Brest, G. L., Suarez, C. E., and Talavera, L. (1976). *Catastrophe or New Society? A Latin American World Model*, International Development Research Centre, Ottawa.

IEA/OARU Long Term Global Energy—CO_2 Model (1984). (J. A. Edmonds and J. M. Reilly, contribs.), Oak Ridge National Laboratory, Oak Ridge, TN.

International Bank for Reconstruction and Development (1975). The Simlink model of trade growth for the developing world, World Bank Staff Working Paper no. 220, IBRD, Washington, D.C.

Kaya, Y. (1978). Present status of project FUGI, Faculty of Engineering University of Tokyo, Tokyo.

Kaya, Y., Onishi, A., Ishitani, H., Ishakawa, M., Suzuki, Y., and Shoji, K. (1977). Project FUGI, University of Tokyo, Tokyo.

Leontief, W., Carter, A., and Petri, P. (1977). *The Future of the World Economy*, Oxford University Press, New York.

Linneman, H., De Hough, J., Keyzer, M., and Van Heemst, H. (1979). *MOIRA—Model for International Relations in Agriculture*, North Holland, Amsterdam.

Liu, T. C. (1973). *Chinese and Other Asian Economies: A Quantitative Evaluation*, AER.

Meadows, D. H. (1977). A conceptual overview of agricultural models, in *MOIRA: Food and Agricultural Model*, Laxenburg, Austria, February, IIASA, pp. 151–171.

Meadows, D. H., Meadows, D. L., Randers, J., and Behrens, W. K., III (1972). *The Limits to Growth*, Universe Books, New York.

Meadows, D. L., Behrens, W. W., III, Meadows, D. H., Naill, R. F., Randers, J., and Zahn, E. K. O. (1974). *Dynamics of Growth in a Finite World*, Wright-Allen Press, Cambridge, MA.

Meadows, D. L., et al., (1976). *Toward Global Equilibrium*, Wright-Allen Press, Cambridge, MA.

Mesarovic, M. (1979). World modeling and its potential impact, in *Growth in a Finite World*, (J. Grunfield, ed.), Franklin Institute Press, Philadelphia.

Mesarovic, M., and Pestel, E. (1972). A goal-seeking and regionalized model for the analysis of critical world relationships: The conceptual foundation, *Kybernetes J., 1.*

Mesarovic, M., and Pestel, E. (1974a). *Mankind at the Turning Point,* Dutton, New York.

Mesarovic, M., and Pestel, E., eds. (1974b). Multilevel computer model of world development system, in Proceedings of the Symposium held at IIASA, April 29–May 3.

Mesarovic, M., Macko, D., and Takahara, Y. (1970). *Theory of Hierarchical, Multilevel Systems,* Academic Press, New York.

Mohanty, B. B. (1982). A regionalized goal oriented model for dynamic analysis of world trade, Ph.D. Dissertation, Dept. of Operations Research, Case Western Reserve University, Cleveland, Ohio.

Peccei, A. (1977). *The Human Quality,* Pergamon Press, Oxford.

Petri, P. A. (1977). An introduction to the structure and application of the United Nations world model, *Appl. Mathematical Modelling, 1*(June), 261–267.

Rojko, A., Regier, D., O'Brien, P., Coffing, A., and Bailey, L. (1978). *Alternative Futures for World Food in 1985,* 3 Vol., U.S. Department of Agriculture Economics, Statistics, and Cooperative Service, Washington, D.C.

SAI (Systems Applications, Inc.) (1986). *ARISTOLE—Decision Support, Decision Analysis and Simulation Software System,* SAI, Beachwood, Ohio.

Shaw, M. L., Allen, B. J., Goodhue, L. H., and Hutzler, M. J. (1978). *The International Energy Evaluation System,* Vol. 2: *Technical Documentation,* Logistics Management Institute, Washington, D.C.

Stanford Research Institute (1977). *Generalized Equilibrium Modeling: The Methodology of the SRI-Gulf Energy Model, Final Report,* SRI International, Menlo Park, CA.

Sundaram, A. K., and Krishnayya, J. G. (1979). *Global Models—An Evaluation,* Systems Research Institute, Pune, India.

Systems Analysis Research Unit (SARU) (1977). *SARUM 76 Global Modeling Project,* Departments of the Environment and Transport, London.

Systems Analysis Research Unit (SARU) (1978). *SARUM Handbook,* Departments of the Environment and Transport, London.

United Nations Industrial Development Organization (1980). *The UNIDO World Industry Cooperation Model,* IFIP Working Conference on Global Modelling, Dubrovnik, September 1–5, 1980.

U.S. General Accounting Office (1977). *Food and Agriculture Models for Policy Analysis,* U.S. General Accounting Office, Community and Economic Development Division, Washington, D.C.

Waelbroeck, J., ed. (1976). *The Models of Project LINK,* North Holland, Amsterdam.

14

Economic Systems

14.1 INTRODUCTION

National economies are dynamic in that economic conditions change with
time and are affected by unexpected shocks and governmental controls.
Although the economic principles that govern their dynamics are not as
precise as for physical systems, it is nevertheless possible to model and
simulate national economies with reasonable accuracy. As with the physical
systems discussed in Chapters 1 and 2, a set of ordinary differential equations is used to depict the complex interactions of the economic variables.

Although it is possible to simulate almost any aspect of the national
economy, some difficulties must be overcome. First, the bulk of economics
literature concentrates on illustrative notions that address the effects of
changes in individual variables. Seldom are there quantitative relationships and rarer still are there dynamic relationships suitable for simulation.
To make matters worse, economic views are often inconsistent and contro-
versial. Second, the economic database is vast and diverse, but limited
in significant ways. The data are periodic, usually monthly or quarterly,
noisy, and often biased. Because of the tendency of economic conditions
to fluctuate periodically, as in the 4-year business cycle, the frequency
content of economic variables is concentrated at 1.57 radians/year. This
causes a strong correlation among economic variables, which presents an
obstacle in differentiating between cause and effect. The complexity of a
complete economic system is truly awesome. It is the task of the econo-
metrician to model the particular aspect of the economy of interest with
minimum complexity.

The objective of this chapter is to develop macroeconometric models of
the United States and employ them for forecasting and for studying policies
for regulating inflation and unemployment. In order to achieve this objective, (a) it will be necessary to model the pertinent economic variables,
including those affected directly by government control; (b) it will also be
necessary to estimate the model parameters from historical data; (c) finally,
it will be possible to simulate monetary policy in controlling inflation and
to simulate future economic conditions.

The problem of controlling inflation and unemployment has long been studied and continues to be a concern. Our present understanding of macroeconomics, and in particular the business cycle, is the result of many individual contributions. The modeler of economic systems should take full advantage of this rich lore of economic principles. Much of the differences in points of view of economists is not with regard to structural issues, but rather the relative influence of competing causes. The modeler, then, may consider all candidate causes, letting statistical analysis indicate their relative influence. By repetition of this process for all pertinent variables, the necessary elements for a complete model are obtained. The model structure then is guided by economic principles, with historical data used to estimate and to test the model parameter values.

It was the view of Adam Smith (1776) that prices were a result of supply and demand and that fluctuations in prices had the same cause. Hansen (1964) credits Wicksell in the 1890s as describing the fluctuations of the business cycle as a lightly damped system buffeted by random economic disturbances. Tinbergen (1937) used ordinary differential equations to describe the dynamics of the business cycle for the Netherlands economy. Today's complex macroeconometric models employ a large number of the same type of equations. Keynes (1936) argued for increased government spending to alleviate the chronic unemployment of the Great Depression. His policy has become commonplace. Phillips (1958) studied the relationship between inflation and unemployment using data spanning almost 100 years. Although it has been necessary to modify the Phillips curve, the close relationship between unemployment and inflation remains. Friedman (1968) models production growth as related inversely to the growth of money supply. Friedman argues that irregular and inappropriate regulation of the nation's money supply is the primary cause of changing economic conditions. Anderson and Carlson (1970), of the Federal Reserve Board of St. Louis, have continued to advocate similar views. Okun (1970) models the relationship between changes in production and changes in unemployment during the business cycle, which suggests that less than half the number of workers are laid off that might be justified by reduced production. We see then that by 1970 several studies exist that might contribute to a more comprehensive model of the nation's economic system.

Klein and Goldberg (1955), Dusenberry et al. (1965), and Fromm and Klein (1976) developed rather extensive econometric models for the United States. Similar techniques were used to develop world models, as discussed in Chapter 13. Modern control theory was applied to economic models to analyze and forecast econometric behavior by Vishwakarma (1970), Mehra (1974), Athans (1974), Chow (1974), and Aoki (1976). Although these investigations and the analyses of numerous researchers made significant contributions, there remains a need for refinement, improved accuracy, and greater understanding of the dynamics of the economy by using relatively simple models.

As years passed after the Phillips curve was published, it became increasingly evident that in the long run the Phillips curve shifts its position. Fellner et al. (1963) argue that in the long run inflation and unemployment are independent because of changing expectations with regard to inflation. Taylor and Smith (1984) developed a dynamic relationship linking unemployment and inflation which is consistent with the Phillips curve in the short term and also consistent with the view of Fellner for the long term. Wingrove (1983) related a simple second-order systems model to specific observations concerning economic behavior, such as Friedman's model of production growth and Okun's model of unemployment. Wingrove also studied governmental control in terms of feeding back unemployment rate and production growth.

In this chapter, an econometric model will be developed that is based on these earlier works. The resulting econometric model of the United States will be used to study the control of inflation rate and unemployment rate. A study is made of the effects of using feedback variables to represent governmental control of disturbances representing the random shocks to the economy. Comparisons of different regulatory policies will then be made.

A second area of application of economic simulation to be considered is economic forecasting. A special econometric model is developed based on spectral analysis and is used to study forecasting techniques, ranging from simplistic to complex. The accuracy of the methods is compared to show the degree to which increased sophistication improves forecast accuracy. Finally, it is shown how low-order-systems models produce forecasts that are often superior to the most comprehensive macroeconometric models of the United States.

14.2 ECONOMETRIC MODEL DEVELOPMENT

First, the second-order model of Wingrove (1983) is examined, reformated, and modified to include both the unemployment-inflation relationship and the money supply of Taylor and Smith (1984). The model parameters are obtained from referenced study results and by using least-squares in the form of auto- and cross-correlation functions. The resulting macroeconometric model of the United States is given in a state-space form with stochastic disturbances. The model is then used to study monetary policy, particularly with regard to inflation and unemployment, by employing feedback of various economic variables.

14.2.1 Macroeconometric Model Structure

In Wingrove (1983) a second-order model is used to describe the relationship between money supply growth, interest, unemployment, inflation, and growth in the real gross national product of the United States of America. Wingrove's treatment of linking the elements of the model to the concepts of Friedman, Okun, and Phillips will serve as a starting point for the simulation model.

It is useful to reformat the Wingrove model in a way that doesn't change its dynamics but shows more clearly the casual relationship of the economic variables. It can be seen, for example, in Figure 14.1, that it is the real money supply growth rate RM1% that affects first real interest rates, next production growth rate RGNP%, then unemployment rate, and finally inflation rate. In the Wingrove model, unemployment rate, and inflation are modeled as being directly related through a constant of proportionality, similar to the Phillips curve. Taylor and Smith (1984) show, in Figure 14.2, that such a simple relationship has not been applicable since 1958. The link between unemployment and inflation requires a more complex, dynamic relationship. Taylor and Smith (1984) offer such a model in the form of the transfer function:

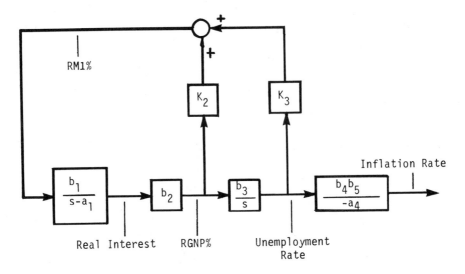

Figure 14.1   The modified Wingrove (1983) model.

$$\frac{\text{Inflation rate}(s)}{\text{Unemployment rate}(s)} = \frac{-4.79(s + 1.55)}{s(s + 3.85)} \tag{14.1}$$

The growth of money supply is assumed to be proportional to real interest rate, real gross national product growth rate, unemployment rate, and inflation rate. A block diagram of the complete model is shown in Figure 14.3. The view is taken that the growth in money supply is due to both inherent economic factors and the actions of the Federal Reserve

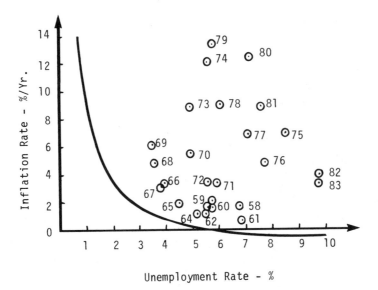

Figure 14.2   Recent economic data compared with the Phillips curve.

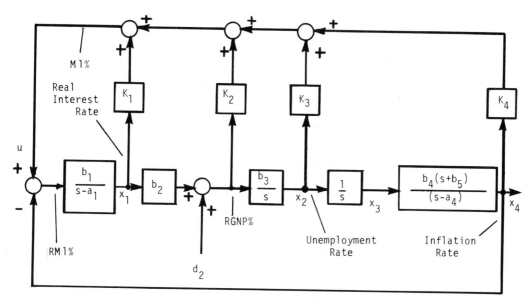

Figure 14.3   Block diagram of macroeconomic model.

Board.   Separation of the two causes will not be attempted.   Because the
Federal Reserve Board can affect money supply growth and is charged with
the responsibility of its control, we represent money supply growth as the
control variable u.   As discussed in Chapter 2, the state-space model
equations are

$$\frac{dx_1}{dt} = a_1 x_1 + b_1 u - b_1 x_4$$

$$\frac{dx_2}{dt} = b_2 b_3 x_1 + b_3 d_2$$

$$\frac{dx_3}{dt} = x_2$$

$$\frac{dx_4}{dt} = a_4 x_4 + b_4 x_2 + b_4 b_5 x_3 \tag{14.2}$$

$$u = K_1 y_1 + K_2 y_2 + K_3 y_3 + K_4 y_4 \tag{14.3}$$

In state-space format we have

$$\frac{d\underline{x}}{dt} = \underline{A}\,\underline{x} + \underline{B}\,\underline{u} + \underline{D}\,\underline{d} \tag{14.4}$$

$$\underline{u} = \underline{K}\,\underline{x} \tag{14.5}$$

$$\underline{y} = \underline{H}\,\underline{x} + \underline{S}\,\underline{d} \tag{14.6}$$

where

$$\underline{x} = \begin{bmatrix} x_1 \\ x_2 \\ x_3 \\ x_4 \end{bmatrix} = \text{State variables}$$

$$\underline{y} = \begin{bmatrix} y_1 \\ y_2 \\ y_3 \\ y_4 \\ y_5 \end{bmatrix} = \begin{matrix} \text{Real Interest rate} \\ \text{Production growth rate} \\ \text{Unemployment rate} \\ \text{Inflation rate} \\ \text{Personal income} \end{matrix}$$

$$\underline{u} = [\text{Money supply growth rate}]$$

$$\underline{A} = \begin{bmatrix} a_1 & 0 & 0 & -b_1 \\ b_2 b_3 & 0 & 0 & 0 \\ 0 & 1 & 0 & 0 \\ 0 & b_4 & b_4 b_5 & a_4 \end{bmatrix} = \text{Stability matrix}$$

$$\underline{B} = \begin{bmatrix} b_1 \\ 0 \\ 0 \\ 0 \end{bmatrix} = \text{Control matrix}$$

$$\underline{H} = \begin{bmatrix} 1 & 0 & 0 & 0 \\ b_2 & 0 & 0 & 0 \\ 0 & 1 & 0 & 0 \\ 0 & 0 & 0 & 1 \\ b_2 & 0 & 0 & 1 \end{bmatrix} = \text{Observation matrix}$$

$$\underline{D} = \begin{bmatrix} 0 & 0 & 0 & 0 \\ 0 & b_3 & 0 & 0 \\ 0 & 0 & 0 & 0 \\ 0 & 0 & 0 & 0 \\ 0 & 0 & 0 & 0 \end{bmatrix} \qquad \underline{S} = \begin{bmatrix} 0 & 0 & 0 & 0 \\ 0 & 1 & 0 & 0 \\ 0 & 0 & 0 & 0 \\ 0 & 0 & 0 & 0 \\ 0 & 1 & 0 & 0 \end{bmatrix}$$

$$\underline{d} = \begin{bmatrix} 0 \\ d_2 \\ 0 \\ 0 \end{bmatrix} = \text{Random disturbance}$$

$$\underline{K} = [K_1 \ K_2 \ K_3 \ K_4] = \text{Gain matrix}$$

Next we shall turn our attention to the subject of estimating the values of these model parameters.

14.2.2 Model Parameter Estimation

The model of interest rate and production growth is based on Friedman's work, as reported by Wingrove (1983). The parameters of the model for real production growth driven by money supply growth were determined by a least-square technique. The simulation results compare well with the historical data in Figure 14.4. The value of a_1 and the product b_1b_2 required some algebraic manipulation because of the reformating of Wingrove's model discussed earlier. The value of b_1 per se was obtained directly from Wingrove. The value of b_3 was based on Okun's law and estimated by Wingrove (1983). The simulated response in Figure 14.5 is seen to compare well with the actual, measured unemployment growth. The parameter values of the inflation rate model are based on the model of Taylor and Smith

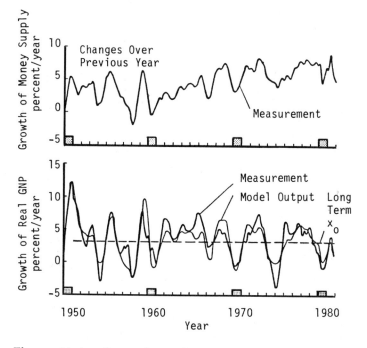

Figure 14.4   Comparison of actual and modeled production growth.

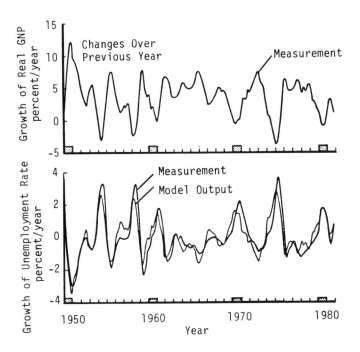

Figure 14.5   Comparison of actual and modeled unemployment rate.

(1984). In Figure 14.6 it can be seen how well this model represents the unemployment/inflation relationship. The money supply growth model used is an expansion of that studied by Wingrove (1983). Feedbacks using real interest rate and inflation rate have been added to production growth and unemployment. The values of the feedback gain parameters were determined using least-squares by first calculating the autocorrelation and cross-correlation functions (Gelb, 1974) of the pertinent economic variables.

Later in the chapter the macroeconometric model dynamics are examined and compared with power spectral densities of economic variables. It proves to be necessary to adjust the parameters b_1, b_3, and b_4 in order for the model's results to match historical data. Both the unadjusted and the adjusted values of the model parameters are summarized below.

Parameter (reference)	Unadjusted values	Adjusted values
a_1 (Wingrove, 1983)	−2.4	−2.4
b_1 (Wingrove, 1983)	−0.1875	−0.28125
b_2 (Wingrove, 1983)	−10.0	−10.0
b_3 (Wingrove, 1983)	−0.4	−0.6
a_4 (Taylor and Smith, 1984)	−3.85	−3.85
b_4 (Taylor and Smith, 1984)	−4.79	−7.185
b_5 (Taylor and Smith, 1984)	1.55	1.55
$K_1{}^a$ (Taylor and Smith, 1984)	0.3461	0.3461
$K_2{}^a$ (Taylor and Smith, 1984)	0.1574	0.1574
$K_3{}^a$ (Taylor and Smith, 1984)	0.4045	0.4045
$K_4{}^a$ (Taylor and Smith, 1984)	0.5014	0.5014

[a] Historical values (1954–1983).

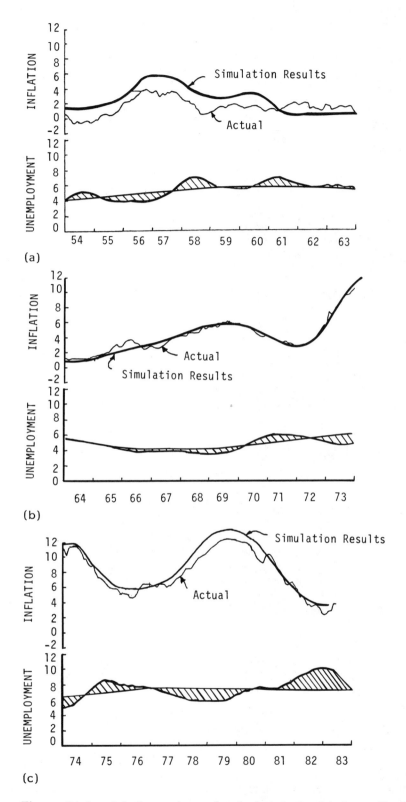

Figure 14.6   (a) Comparison of actual and simulated results for 1954 to 1963.   (b) Comparison of actual and simulated results for 1964 to 1973. (c) Comparison of actual and simulated results for 1974 to 1983.

14.2.3   Dynamic Characteristics

Let us examine the dynamic characteristics of the model by determining the
roots of its closed-loop characteristic equation. Substitution of the control
law into the state equation yields

$$\frac{dx}{dt} = \underline{A}\underline{x} + \underline{B}\underline{u} = \underline{A}\underline{x} + \underline{B}\underline{K}\underline{x} = (\underline{A} + \underline{B}\underline{K})\underline{x}$$

The characteristic equation is

$$|s\underline{I} - \underline{A} - \underline{B}\underline{K}| = 0$$

$$= s^4 + (-a_1 - a_4 - b_1 K_1 - b_1 b_2 K_2)s^3$$

$$+ (a_1 a_4 + b_1 a_4 K_1 + b_1 b_2 a_4 K_2 + b_1 b_2 b_3 K_3 a_4)s^2$$

$$+ (b_1 b_2 b_3 b_4 - b_1 b_2 b_3 b_4 K_4)s$$

$$+ b_1 b_2 b_3 b_4 b_5 - b_1 b_2 b_3 b_4 b_5 K_4 \qquad (14.7)$$

For the values of K_1, K_2, K_3, and K_4 set equal to the values obtained
from historical data we get the following roots:

ROOT(1) = $-0.05330 + 0.5995j$

ROOT(2) = $-0.05330 - 0.5995j$

ROOT(3) = -2.132

ROOT(4) = -4.158

It is clear from examining the undamped natural frequency of the first
two complex roots that its value is too low to correspond to that of the
business cycle. The loop gain of the unadjusted model needs to be increased to be consistent with these results. This result illustrates the
importance of examining the complete system when modeling. Modeling individual elements proved to be inadequate. The loop gain was increased
to make the roots consistent with the business cycle frequency by multiplying b_1, b_3, and b_4 by 1.5. As a result of these adjustments, the characteristic roots of the macroeconometric model becomes

ROOT(1) = $-0.04116 + 1.180j$

ROOT(2) = $-0.04116 + 1.180j$

ROOT(3) = -1.586

ROOT(4) = -4.236

It is now possible to simulate the effects of random shocks to the
economy. If the power spectral density of the disturbance is constant
with frequency, the power density of a response variable equals the corresponding transfer function times its complex conjugate (Gelb, 1974).
Using Eq. (14.4)–(14.6), the transfer functions which relate the economic
variables to the random disturbance d can be expressed as

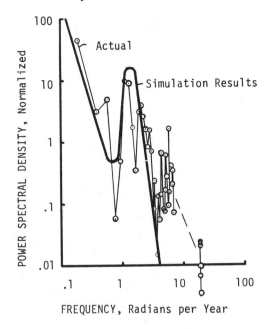

Figure 14.7   Power spectral density of composite inflation rate.

$$\frac{y(s)}{d(s)} = \underline{H}(s\underline{I} - \underline{A} - \underline{BKH})^{-1}(\underline{D} + \underline{BKS}) + \underline{S} \qquad (14.8)$$

In Figure 14.7 the power spectral densities of inflation for the actual economy are compared with those of the model. The similarities of the modeled power spectral density of inflation rate to that of the actual economy support the validity of the model developed. The model is now ready for simulation applications.

14.3   ECONOMIC POLICY APPLICATION

It is the objective of this chapter to model, simulate, and explain the effects of monetary policy, to evaluate different approaches, and to recommend a policy that can better stabilize the economy when buffeted by random disturbances. Of course, much has been written on the subject of economic stability, for example, by Samuelson (1973), Smith (1985), and Dauten and Valentine (1978). In this section we shall apply the results of systems analysis to the macroeconometric model just developed. In this way one hopes to develop an improved monetary policy by studying the effects of different feedback laws on the stability of the economy.

It is the contention of Friedman (1968) that this is best done by keeping the growth of the money supply constant. We can evaluate this policy by making all of the feedback gains K_1, K_2, K_3, and K_4 equal to zero and then examine the characteristic roots of the macroeconometric model.

The roots for this case of constant real money supply growth are

ROOT(1) = 0.1192 + 1.473j

ROOT(2) = 0.1192 − 1.473j

ROOT(3) = −1.858

ROOT(4) = −4.631

The positive real parts of the complex roots correspond to a slightly unstable or undamped business cycle. Although the dynamic instability is slight, the amplitude of the envelope of the oscillations would double every 5.8 years. Slight changes in the macroeconometric model might affect the results, but we must conclude at this time that fixing the growth of real money supply is not the best policy. In an ideal economy that is undisturbed and in equilibrium, money supply growth is constant. But this does not mean that keeping money supply growth constant will cause the economy to be steady, as we have seen.

To illustrate the complex situation facing those charged with controlling inflation, let us examine the dynamics of our model as direct feedback is used to control inflation. Figure 14.8 shows a root-locus plot as the direct feedback gain K_4 is increased (actually larger negative values). One can see how the frequency of the business cycle increases as the feedback gain is increased. Unfortunately, as the static stability of the business cycle is increased, damping is decreased. The roots drift into the right-half plane causing the business cycle to become more and more undamped. It is clear that regulation of money supply based on inflation solely is inadequate. Additional feedback variables must be used in concert to enhance economic stability.

It is useful to examine the effects of each of the feedbacks on the dynamic characteristics of the business cycle. Figure 14.9 is a guide to

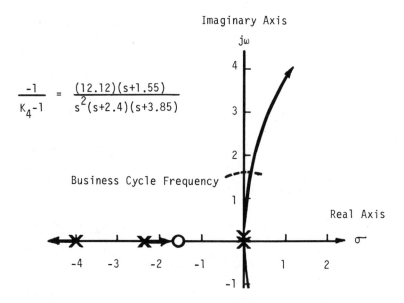

$$\frac{-1}{K_4 - 1} = \frac{(12.12)(s+1.55)}{s^2(s+2.4)(s+3.85)}$$

Figure 14.8   Root-locus plot for K_4, feedback of inflation only.

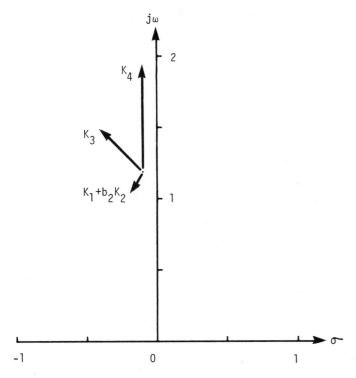

Figure 14.9   Effect of feedback gains on frequency and damping of the business cycle.

the effects of changing each of the feedback gains relative to the historical values discussed earlier. The length of each arrow shows how much the roots for the business cycle change when each is changed by an increment of 1 or -1. Increasing K_4 in the negative direction increases the stability and the frequency of the business cycle. The gains K_1, K_2, and K_3 can effectively add to the damping when changed in combination.

Having examined the effects of increasing the various feedback gains in Figure 14.9, we can now consider increasing both the damping and the frequency of the business cycle. The importance of a higher frequency is that the economy responds to unexpected shocks to a diminished degree. An increase in the frequency of the business cycle of 41% can be expected to cut in half the amount a given shock affects changes in the economic variables. Armed with increased understanding as to the effects of each of the feedbacks, a set of gains has been generated which promises a high degree of static and dynamic stability and therefore a more stable economy than ever before experienced in the United States. It represents an actively controlled economy.

The gain values are

$$K_1 = 1.5$$
$$K_2 = -0.15$$
$$K_3 = 4.0$$
$$K_4 = -1.0$$

The resulting characteristic roots are

ROOT(1) = −0.5648 + 2.674j

ROOT(2) = −0.5648 − 2.674j

ROOT(3) = −1.017

ROOT(4) = −4.947

Figure 14.10 shows power spectral density functions for inflation rate for both the actively controlled economy and the standard case. The same random, uncorrelated shocks are added to both models, as in Gelb (1974). It is striking to see the degree of stabilization that is possible. The root mean square of the excursions in inflation rate is reduced by a factor of five by actively controlling the economy. The period of the actively controlled business cycle would be only 2.35 years instead of 4.0 years for the uncontrolled case. Unemployment rate, production growth, and interest rates would be stabilized to a similar degree at the same time.

It is possibly more convenient to use interest rate per se instead of real interest as a feedback. The above equation for monetary policy would then become

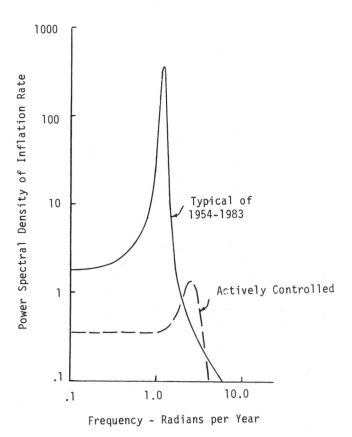

Figure 14.10   Comparison of power spectral density function.

Money supply growth rate = 1.5 * interest rate

$\qquad$ − 0.15 * production growth rate

$\qquad$ + 4.0 * unemployment

$\qquad$ − 2.5 * inflation rate

$\qquad$ + target inflation rate $\qquad$ (14.9)

The results of this study are encouraging, causing one to contemplate the potential value of actively controlling free-market and mixed economies, thereby reducing the degree of hardship that millions now suffer during recessionary periods of the business cycle. It is clear how important simulation of economic systems can be in studying government monetary policy. Let us now address a different application of economic simulation, that of forecasting economic conditions.

14.4 ECONOMIC FORECASTING

The task of forecasting future economic conditions at first appears to be straightforward. Given the past values of economic variables, the goal is to accurately estimate future values based on the past ones. What is not straightforward, however, is what basis or set of assumptions is best in terms of forecasting accuracy. Each set of assumptions corresponds to a particular forecasting technique. The state-space equations discussed in Chapter 2 and in Gelb (1974) are particularly well suited to the study of these forecasting techniques, especially those that fall into the class of linear operators. Discrete-time models are used because of the sampled nature of economic data. The simpler forecasting techniques require only the autocorrelation functions of the historical economic data for formulation and error estimation of the time series being forecasted. The more complicated techniques, however, require a more complete mathematical description of the economy. The approach used here is to assume the economic system is modeled as a dynamic system that is buffeted by random shocks. Noisy observations of this system consist of linear transformations of the state variables. The model structure is that described in Chapter 2 and in Gelb (1974). A variety of forecasting techniques will be studied in the context of this class of dynamic systems.

We shall start by considering the dynamic system given by a state equation and by having a noisy observation related linearly to the state, as

$$\underline{x}(n + 1) = \underline{A}\underline{x}(n) + \underline{F}\underline{w}(n) \qquad (14.10)$$

$$\underline{y}(n) = \underline{H}\underline{x}(n) \qquad (14.11)$$

$$\underline{z}(n) = \underline{y}(n) + \underline{C}\underline{v}(n) \qquad (14.12)$$

where

$\underline{x}$ = the state vector

$\underline{A}$ = the stability matrix

$\underline{F}\underline{w}$ = the random, uncorrelated process noise

$\underline{y}$ = the response vector

$\underline{H}$ = the observation matrix

$\underline{z}$ = the observation vector

$C\underline{v}$ = the observation noise

Correlation functions will be used to compute the expected forecast error. For our purposes of assessing forecasting techniques, the information contained in the historical database will be represented by its auto-correlation functions. That is, the autocorrelation function will be determined from the observed time series minus its average and seasonal variations.

Because both modeling and forecasting depend on information contained in the correlation functions of observation data, it is necessary to develop certain statistical relationships.

The expected forecast error covariance developed by Taylor and Sliwa (1983) is

$$\underline{R}ee(O,K) = \sum_{m=1}^{M} \underline{G}(N - m)\underline{R}zz(O)\underline{G}(N - m)^T + \underline{R}zz(O)$$

$$+ \sum_{j=1}^{M-1} \sum_{m=1}^{M-K} [\underline{G}(N - m + j)\underline{R}zz(j)\underline{G}(N - m)^T$$

$$+ \underline{G}(N - m)\underline{R}zz(j)\underline{G}(N - m + j)^T] - \sum_{m=0}^{M} \underline{G}(N - m)\underline{R}zz(K + m)^T$$

$$- \sum_{m=0}^{M} \underline{R}zz(K + m)\underline{G}(N - m)^T \tag{14.13}$$

where

$\underline{e}$	=	forecast error
$E\{\}$	=	expected value
$\underline{G}()$	=	weighting function of forecasting technique
j	=	general index
J	=	expected mean square forecast error
K	=	time horizon of forecast, months
m	=	general index
M	=	number of past values required of the forecasting technique
N	=	index for the current time or most recent data
$\underline{R}ee$	=	expected forecast error covariance
$\underline{R}zz(O)$	=	autocorrelation function of historical data
$\underline{R}zz(j)$	=	cross-correlation function of historical data

The forecasting performance measure J will consist of a scalar related to the error covariance matrix.

$$J = E\{\underline{e}^T M \underline{e}\} = Tr\{\underline{R}ee(O)M\}$$

where

$$Tr(\underline{A}) = \text{trace of } \underline{A} = \sum_{i=1}^{n} a_{ii} \qquad \text{for } \underline{A}(n \times n) \text{ matrix}$$

Although this formulation is useful in combining different forecast errors into a single measure, in this study the forecast errors will be examined independently. We shall compare forecasting techniques within the framework just developed, but first let us examine the two time series of data that will serve as examples.

14.4.1 Forecasting Applications

Historical values of inflation rate will be used to illustrate and assess several forecasting techniques. Specifically, the monthly Consumer Price Index for the United States for the years 1945 to mid-1983 is the source information that is used. The raw datum considered for this study is monthly annualized percentage change of the index. Figure 14.11 shows the autocorrelation functions for the series. The rapid decrease in the value of the

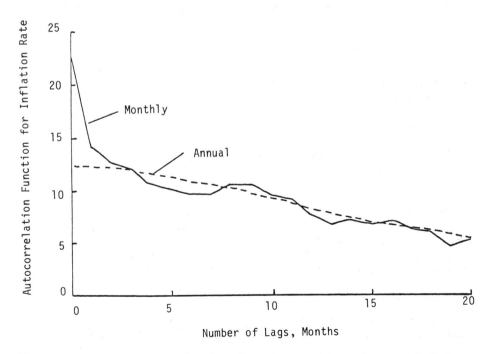

Figure 14.11   Autocorrelation functions for monthly and annual inflation rates.

autocorrelation function in going from "no lag" to a "lag of 1 month" is evidence of the large amount of observation noise in the raw, monthly rates of change.

A more popular and less erratic version of inflation rate is the change in the Consumer Price Index from one year to the next. This is closely approximated by averaging 12 consecutive monthly changes. The filtering effect of the averaging is evident in Figure 14.11, where the raw (monthly) and averaged (annual) autocorrelation functions are compared. It will be seen that another result of the averaging is to improve the accuracy of the forecasts of inflation rate. The same holds true for predicting other economic variables.

14.4.2   "No-Change" Forecast

Perhaps the most simple forecast is to assume "no change," that is, to expect current conditions to extend into the future, especially the immediate future. Expressing this mathematically we have

$$\underline{G}(N) = \underline{I} \tag{14.14}$$

The forecast performance measure is

$$J = Tr\{[\,2\underline{R}zz(O) - \underline{R}zz(K) - \underline{R}zz(-K)]M\} \tag{14.15}$$

Figure 14.12 shows the normalized error in forecasting inflation rate as a function of the horizon time of the forecast. Normalized error is the root-mean-square error divided by the standard deviation of the variable forecast. The no-change forecast is most accurate for very short time horizons. Its accuracy quickly degenerates as the time horizon increases, however.

14.4.3   "Average" Forecast

Another simple forecast is to expect the "average" value of the subject variable to occur in the future. Because we have subtracted the average from $\underline{z}$ at the start, we have

$$\underline{G}(N) = 0 \tag{14.16}$$

The forecasting performance measure is

$$J = Tr[\underline{R}zz(O)\underline{M}] \tag{14.17}$$

In Figure 14.12 the average forecasts for inflation rate are compared with the no-change forecasts. It is better to use the no-change technique for forecasting inflation rate for time horizons of less than 18 months, and 6 months in the case of production growth. The "average" forecasts are better for time horizons longer than these values.

14.4.4   "Exponential" Forecast

A slightly more complicated forecast technique is to assume that the variable of interest will subside to the average (zero, in our case) exponentially; i.e.,

$$\underline{G}(N) = \underline{B}exp(K) \tag{14.18}$$

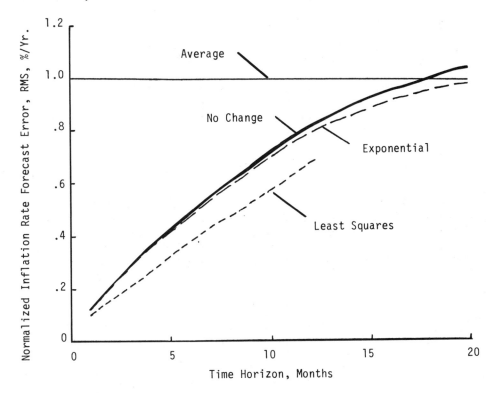

Figure 14.12　Comparison of errors for forecasts of inflation rate.

For cases in which the number of observation variables equals the number of states, the value of $\underline{B}$ is often determined by

$$\underline{B} = \underline{R}zz(1)[\underline{R}zz(O)]^{-1} \tag{14.19}$$

The forecasting performance measure is

$$J = \mathrm{Tr}\{[\underline{B}\exp(2K)\underline{R}zz(O)(\underline{B}\exp(2K))^T + \underline{R}zz(O) - \underline{B}\exp(K)\underline{R}zz(-K)^T$$
$$- \underline{R}zz(K)\exp(K)^T\underline{B}^T]\underline{M}\} \tag{14.20}$$

In Figure 14.12 the forecast error for "exponential" forecasts is compared with those using the no-change and "average" forecasts. Improvement is evident in the forecast accuracy.

14.4.5　"Least Squares" Forecast

An improved forecasting technique results if one takes advantage of past observations of the time series being forecast. The "best" weighting of these observations can be formulated to minimize the sum of the squares of the expected forecast error. Hence the name "least squares" forecast.

The resulting weighting vector is

$$
\begin{bmatrix} \underline{G}(N) \\ \underline{G}(N-1) \end{bmatrix} = \begin{bmatrix} \underline{R}zz(O) & \underline{R}zz(1) \\ \underline{R}zz(1) & \underline{R}zz(O) \end{bmatrix}^{-1} \begin{bmatrix} \underline{R}zz(K) \\ \underline{R}zz(K+1) \end{bmatrix} \tag{14.21}
$$

The forecast error becomes

$$
\underline{R}ee(O) = \underline{R}zz(O) - [\underline{R}zz(K)\ \underline{R}zz(K+1)] \begin{bmatrix} \underline{G}(N) \\ \underline{G}(N+1) \end{bmatrix} \tag{14.22}
$$

Forecast accuracy is plotted as a function of horizon time and compared with results discussed earlier in Figure 14.12. It can be seen that the least-squares forecasts are superior to the no-change, average, and exponential forecasts. In order to assess the value of using past data, Figure 14.13 shows how forecast accuracy improves slightly as additional data are used.

14.4.6 "Low-Order System" Forecast

The next level of complexity and sophistication that will be considered is forecasting based on a linear, time-invariant, low-order econometric model.

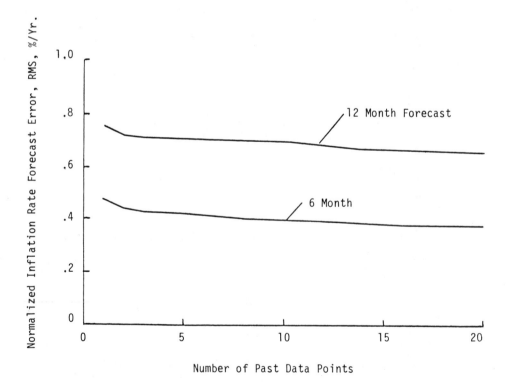

Figure 14.13   Inflation rate forecast error as a function of past data points.

Three parts of the forecasting process will be discussed: (a) modeling, (b) state estimation, and (c) prediction.

The modeling process consists of determining (a) the appropriate order, (b) the elements of the state vector $\underline{x}$, (c) the state transition matrix $\underline{A}$, (d) the state or process noise covariance matrix $\underline{FF}^T$, (e) the observation matrix $\underline{H}$, and (f) the measurement of noise covariance, $\underline{GG}^T$. This list is quite formidable when one considers that the only source of information is the past history of the response variables $\underline{z}(n)$, or, more accurately, the correlation functions $\underline{R}zz(K)$, or spectral density $\underline{PHI}zz(w)$. We shall use the low-order systems model of Taylor (1981). From the same reference we have the following expressions for perhaps the most sophisticated forecasting technique in which a Kalman filter is used to estimate the current state $\underline{x}(N)$.

$$E\{\underline{x}(N)\} = \sum_{n=1}^{w} (\underline{A} - \underline{KHA})^{n-1}\underline{Kz}(N - n + 1) \tag{14.23}$$

where $\underline{S}$ is large enough to assure convergence and

$$\underline{K}(n) = \underline{L}(n)\underline{M}(n)^{-1} \tag{14.24}$$

$$\underline{M}(n) = \underline{HL}(n) + \underline{GG}^T \tag{14.25}$$

$$\underline{L}(n) = [\underline{AP}(n)\underline{A}^T + \underline{FF}^T]\underline{H}^T \tag{14.26}$$

$$\underline{P}(n + 1) = \underline{AP}(n)\underline{A}^T + \underline{FF}^T - \underline{L}(n)\underline{M}(n)^{-1}\underline{L}(n)^T \tag{14.27}$$

The expected forecast error covariance is obtained using the following relationships. First the state error correlation matrix P(N) is obtained solving the above relationship for the steady-state value of P.

Changes in the state error correlation function are obtained from

$$\underline{P}(N + 1) = \underline{AP}(N)\underline{A}^T + \underline{FF}^T$$

$$\underline{P}(N + 2) = \underline{AAP}(N)\underline{A}^T\underline{A}^T + \underline{AFF}^T\underline{A}^T + \underline{FF}^T$$

For the general case:

$$\underline{P}(N + K) = \underline{A}exp\underline{KP}(N)\underline{A}expK^T + \sum_{i=0}^{N-1} \underline{A}^i\underline{FF}^T\underline{A}^{Ti} \tag{14.28}$$

The expected future state values are

$$\underline{E}\{\underline{x}(N + K)\} = \underline{A}^K\underline{E}\{\underline{x}(N)\} \tag{14.29}$$

and the expected observation values are given by

$$\underline{E}\{\underline{z}(N + K)\} = \underline{HA}^K\underline{E}\{\underline{x}(N)\} \tag{14.30}$$

The forecast error covariance matrix is

$$\underline{R}ee(O)(K) = \underline{HP}(N + K)\underline{H}^T \tag{14.31}$$

Figure 14.14 shows the expected error in forecasting inflation rate as a function of the forecast time horizon. One can see that for inflation rate forecasts of only 2 years the expected error reaches 75% of the steady-state error. These results suggest that forecasts of inflation rate more than 2 years into the future are of little value. The simplistic "average" forecast would be almost as accurate.

14.4.7   Large Econometric Model Forecasts

Since the first econometric models of Tinbergen (1937) the econometric models of national economics have grown in sophistication. Perhaps the principal use of these models is to forecast economic conditions such as inflation rate and production growth. In Figure 14.15 errors in forecasting inflation are compared for a number of sources collected by Taylor (1986). One can see that the RMS inflation rate forecast errors made by Chase, Wharton, Data Resources (DRI), University of California at Los Angeles (UCLA), and Investment Analysis Company (IAC) exceed the theoretical errors for the Kalman Filter and Fourier Series forecast techniques. The

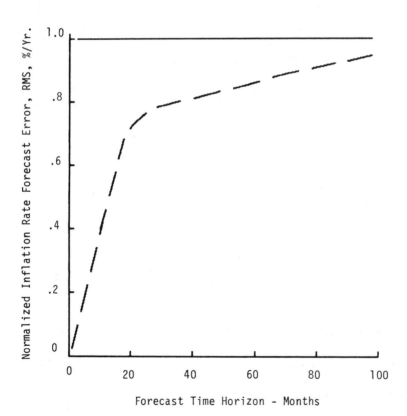

Figure 14.14   Kalman filter forecast error for inflation rate.

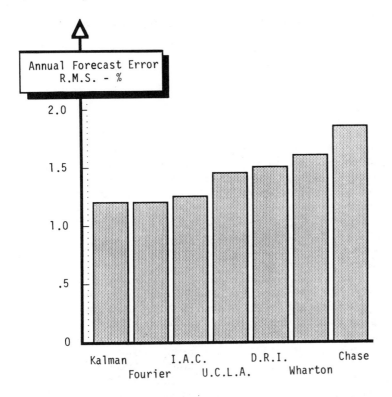

Figure 14.15   Comparison of average errors in forecasting inflation rate.

Fourier Series technique of Taylor and Taylor (1980) uses a low-order
systems model and Fourier analysis for state estimation.

Because the expected forecast errors are a function of the precise
econometric model, there remains the possibility that further work can improve both modeling and forecast accuracy.   At the same time the process
of estimating forecast errors points out the fact that there exists an absolute
limit to forecast accuracy that cannot be exceeded.   This limit to our ability
to forecast is due to the unpredictable nature of shocks to the economy and
errors in generating economic measures.

14.5   CONCLUSIONS

The second-order macroeconometric model of Wingrove and the inflation-
unemployment model of Taylor and Smith are combined to form an improved
model of the economy of the United States for the years 1954 through 1983.
Actions of the Federal Reserve Board in regulating the money supply are
modeled as a controller that feeds back interest, production growth, unemployment rate, and inflation rate.

The stochastic systems model is examined using characteristic roots,
transfer functions, root-locus diagrams, and spectral analysis to ensure it
is a valid representation of the U.S. economy and to illustrate the effects
of different monetary policies.

The difficulties in controlling inflation and unemployment are studied as a feedback synthesis problem. A feedback law for significantly reducing the mean square inflation rate by actively controlling the economy through discretionary monetary policy is generated and studied with respect to the resulting dynamics of the economy.

Feedback of only inflation rate is seen to increase the frequency of the business cycle, but if used exclusively can have the disastrous result of making the economy dynamically unstable. The monetary policy of holding the growth rate of real money supply constant results in a slightly undamped economy which is worsened by random shocks to the economy. Feeding back unemployment rate is of critical importance in damping the business cycle. Feeding back real interest rate is equivalent to feeding back production growth rate in that both contribute to the damping of the business cycle, provided unemployment is also used as a feedback. Feeding back interest, production growth rate, unemployment, and inflation rate resulted in the best system performance and, hence, the most stable economy. The root mean square deviations were reduced by a factor of 5.0 for the actively controlled economy.

A second low-order economic systems model is developed for forecasting applications. Harmonic or spectral analysis and multiple regression are used to determine the parameters of the state space model. Historical economic data are used to determine cross-correlation functions, which are then used to evaluate the accuracy of several forecasting techniques.

The simplistic no-change forecast is seen to be useful, but only for time horizons of a few months. The equally simplistic "average" forecast is seen to be useful for long time horizons. The "exponential" forecast gives reasonably accurate forecasts for a wide range of time horizons. The least-squares forecast gives improved forecasting performance with only slightly increased complexity.

Forecasts based on low-order systems models using (a) Fourier series and (b) Kalman filter techniques are seen to provide the most accurate forecasts. Comparisons of forecasts based on large, complex macroeconometric models show that the accuracy of forecasts using simple, low-order systems models are more accurate. There is, therefore, strong support for economic systems simulation that employs simple, low-order models of the economy.

PROBLEMS

14.1. Derive the characteristic equation for the reformated Wingrove model of Figure 14.1. Compare the resulting function with that derived for the complete model of Figure 14.3 for K_1 and K_4 equal to zero.

14.2. Calculate the amplitude ratio and phase angle for the inflation rate/unemployment rate transfer function for the business cycle having a period of 4 years.

14.3. By using the results shown in Figure 14.9, sketch how the damping of the business cycle can be increased by 200% without changing its frequency, using only K_1 and K_3. Repeat using only K_3 and K_4. Repeat using only K_1 and K_4.

14.4. Based on the forecast errors shown in Figure 14.12, how much does the forecast error increase for the no-change forecast if the time horizon is increased from 5 to 7 months because of delays in obtaining economic data?

14.5. Based on the results shown in Figure 14.13, how much does the forecast accuracy increase by using 1 year's worth of past data, instead of just the most recent value, for a 6-month forecast?

REFERENCES

Anderson, L. C., and Carlson, K. M. (1970). *A Monetarist Model for Economic Stabilization*, Federal Reserve Board, St. Louis.

Aoki, M. (1976). *Optimal Control and System Theory in Dynamic Economic Analysis*, North-Holland, Amsterdam.

Athans, M. (1974). The importance of Kalman filtering methods for economic systems, *Ann. Econ. Social Measurement*.

Chow, G. C. (1974). *Analysis and Control of Dynamic Economic Systems*, Wiley, New York.

Dauten, C. A. and Valentine, L. M. (1978). *Business Cycles and Forecasting*, South-Western.

Dusenberry, J. S., Fromm, G., Klein, L. R., and Kuh, E. (1965). *The Brookings Quarterly Model of the United States*, Rand McNally and North-Holland.

Fellner, W., et al. (1963). *Fiscal and Debt Management Policies; A Series of Research Studies for the Commission on Money and Credit*, Prentice-Hall, Englewood Cliffs, NJ.

Friedman, M. (1968). The role of monetary policy, *Am. Econ. Rev. 58.*

Fromm, G. and Klein, L. R. (1976). The NBER/NSF model comparison seminar: An analysis of results, *Ann. Econ. Social Measurement 5(1)*.

Gelb, A. (1974). *Applied Optimal Estimation*, Analytical Sciences Corp.

Hansen, A. A. (1964). *Business Cycles and National Income*, Norton, New York.

Keynes, J. M. (1936). *The General Theory of Unemployment, Interest, and Money*.

Klein, L. R. and Goldberg, A. (1955). *An Econometric Model of the United States, 1929–1952*, North-Holland, Amsterdam.

Mehra, R. (1974). Identification in control and econometrics; similarities and differences, *Ann. Econ. Social Measurement*.

Okun, A. M. (1970). *The Political Economy of Prosperity*, The Brookings Institution, pp. 132–145.

Phillips, A. W. (1958). The relation between unemployment and the rate of money wage rates in the United Kingdom, 1851–1957, *Econometrica 25*, pp. 283–299.

Samuelson, P. A. (1973). *Economics*, McGraw-Hill, New York.

Smith, A. (1776). *The Wealth of Nations*, Modern Library Edition, 1937, Random House, New York.

Smith, G. (1985). *Macroeconometrics*, Freeman, New York.

Taylor, L. W., Jr. (1981). Forecasting Applications of a Low-Order Systems Model of the United States Economy Based on Structural Analysis, the 10th IFIP Conference on Systems Modeling and Optimization, New York.

Taylor, L. W., Jr. (1986). *The Calculating Investor*, Investment Analysis
    Co., Williamsburg, VA.

Taylor, L. W., Jr., and Sliwa, S. M. (1983). A Statistical Comparison of
    Several Techniques for Forecasting Inflation, 3rd International Symposium
    on Forecasting, Philadelphia.

Taylor, L. W., Jr. and Smith, R. (1984). A Stochastic Systems Approach
    to Controlling Inflation and Unemployment in the United States of
    America, International Federation for Information Processing, Working
    Conference on Stochastic Differential Systems, Baku, USSA.

Taylor, L. W., Jr. and Taylor, D. K. (1980). Economic Forecasting Based
    on a Low-Order Systems Model of the United States Economy, Third
    IFAC/IFORS Conference on Dynamic Modeling and Control of National
    Economies, Warsaw, Poland.

Tinbergen, J. (1937). *An Econometric Approach to Business Cycle Problems.*
    Actualités Scientifiques et Industrielles, Hermann and Co., Editeurs,
    Paris, France.

Vishwakarma, K. P. (1970). Prediction of economic time-series by means of
    the Kalman Filter, *Int. J. Systems Sci.* *1*(1), 25–32.

Wingrove, R. C. (1983). Classical Linear-Control Analysis Applied to
    Business-Cycle Dynamics and Stability, NASA Technical Memorandum
    84366.

15

Manufacturing Systems Simulation Using SIMAN

15.1 INTRODUCTION

This chapter discusses the simulation of manufacturing systems. In particular, the modeling of systems designed to transform raw material into discrete items (e.g., cars, appliances, computers), and the components that compose these items, are presented.

Manufacturing systems represent an important area of application for simulation/modeling. It is estimated that manufacturing constitutes over half of the real gross national product of the United States, and that over one-quarter of the workforce is employed in manufacturing activities (Fairless Memorial Lecture, 1972). Our standard of living, in terms of material goods, depends directly on our ability to efficiently manufacture products; simulation provides an important tool for improving the productivity of these systems.

Manufacturing systems are generally classified, according to production volume, into three types: jobshop production, batch production, and mass production (Groover, 1980; Groover and Zimmers, 1980). Jobshop production involves a wide variety of low-volume items. As a result, the equipment and workers must be flexible and general-purpose. Batch production systems manufacture medium-size lots. They are typically used when the production rate exceeds the demand, and hence the shop manufactures to build up an inventory of the item. As in the case of the jobshop environment, the equipment and workers must be general-purpose. Mass production is characterized by very high volume and involves the continuous manufacture of identical items using dedicated special-purpose equipment.

Mass production lines typically lend themselves to a high degree of automation and, as a result, tend to be highly productive systems. Jobshop and batch production systems, on the other hand, are much more difficult to automate. It is estimated that as much as 75% of all manufacturing is in lot sizes of 50 items or less. Hence, jobshop and batch production systems represent a significant portion of the total manufacturing activity.

Since the introduction of the microprocessor, dramatic changes have been underway in both the equipment and methods used in manufacturing.

Many of these revolutionary changes have centered around the introduction
of flexible manufacturing systems (FMS). These systems provide the advantages of automation without being restricted to a specific product by
employing hardware devices such as robots and automatic guided vehicles
(AGVs). Although these new systems involve massive capital expenditures
by industry, when properly designed and implemented they have tremendous
potential for improving productivity by bringing the advantages enjoyed by
high-volume automated production lines to the much more common small-
batch-size jobshop.

Because of the high investment cost involved with these new manufacturing systems, design and analysis of these systems is a critical issue. It is
important to be able to predict the performance of the system before it is
built. However, dynamic interactions within these systems are sufficiently
complex that intuition and analytical models are typically inadequate for this
purpose. As a result, simulation has emerged as an essential tool in the
design and analysis of these new manufacturing systems.

The importance of the role of simulation in manufacturing systems is
highlighted by the recent introduction of many new simulation languages
specifically designed for this application area. SIMAN is an example of one
such language.* During its first 3 years of use, SIMAN was installed in
more than 1000 institutions throughout industry, government, and universities.

This chapter discusses the ease and applicability of the SIMAN simulation language for modeling manufacturing systems. The objective of this
chapter is to discuss some of the unique aspects involved in the modeling
of manufacturing systems and to illustrate how a manufacturing-oriented
language, such as SIMAN, can greatly simplify the modeling of these systems. The modeling aspects of SIMAN are also introduced. In addition,
the important role that real-time graphic animation can play in a manufacturing simulation project is discussed, and an overview of the Cinema system
for animating SIMAN simulation models is presented. The ideas presented
in the chapter are then illustrated by a case study involving a SIMAN
model and Cinema animation of a modern Japanese flexible manufacturing
system.

15.2   GENERAL-PURPOSE MODELING FEATURES
OF SIMAN

SIMAN is a general-purpose SIMulation ANalysis program for modeling combined discrete-continuous systems (see References, Chapters 2 and 3). The
modeling framework of SIMAN allows component models, based on three distinct modeling orientations, to be combined in a single system model. For
discrete-change systems, either a process or an event orientation can be
used to describe the model. Continuous-change systems are modeled with
algebraic, difference, or differential equations. A combination of these
orientations can be used to model discrete-continuous systems.

---

*Other languages include MAP/1 (Rolston and Miner, 1985), Modelmaster
(General Electric Company, Charlottesville, VA), XCELL (Conway et al.,
1986), and SimFactory (CACI, La Jolla, CA).

Although SIMAN is a powerful general-purpose simulation language which can be used to model a wide range of systems, its real strength lies in its ability to easily model complex manufacturing systems. The reason for its strength in this area is that SIMAN embeds within its general-purpose framework a set of special-purpose constructs specifically designed to simplify and enhance the modeling of manufacturing systems. However, before turning our attention to these special features, we shall begin by overviewing the general-purpose features of SIMAN.

The SIMAN modeling framework is based on the system theoretic concepts developed by Zeigler (1976). Within this framework, a fundamental distinction is stressed between the *system model* and the *experimental frame*. The system model defines the static and dynamic characteristics of the system. The experimental frame defines the experimental conditions under which the model is run to generate specific output data. For a given model, there can be many experimental frames resulting in many sets of output data. By separating the model structure and the experimental frame into two distinct elements, different simulation experiments can be performed by changing only the experimental frame. The system model remains the same.

Given the system model and the experimental frame, the SIMAN simulation program generates output files that record the model state transitions as they occur in simulated time. The data in the output files can then be subjected to various data analyses, such as data truncation and compression, and the formatting and display of histograms, plots, tables, etc. These same data files can also be used to drive a graphical animation of the system. Within the SIMAN framework, the data analysis and display functions follow the development and running of the simulation program and are completely distinct from it. The output file can be subjected to many different data treatments without reexecuting the simulation program. Data treatments can also be applied to sets when performing an analysis based on multiple runs of a model or when comparing the response of two or more systems.

Although SIMAN permits component models to be developed using a process, event, or continuous orientation, we shall focus our attention on the process orientation, in which models are constructed as block diagrams. This orientation is the one best suited for modeling most manufacturing systems. These diagrams are linear, top-down flowgraphs that depict the flow of entities through the system. The block diagram is constructed as a sequence of blocks whose shapes indicate their function. The sequencing of blocks is depicted by arrows that control the flow of entities from block to block through the entire diagram.

These entities are used to represent "things," such as workpieces, information, people, etc., which flow through the real system. Each entity may be individualized by assigning attributes (or variables) to describe or characterize it. For example, an entity representing a workpiece might have attributes corresponding to due date and processing time for the workpiece. As the entities flow from block to block, they may be delayed, disposed, combined with other entities, etc., as determined by the function of each block.

There are 10 different basic block types in SIMAN. The symbol and functions for each of the 10 block types are summarized in Table 15.1.

The OPERATION, HOLD, and TRANSFER blocks are further subdivided into several different block functions, depending on their *operation*, *hold*, or *transfer* type. These types are specified as the first operand of the

Table 15.1   SIMAN Basic Block Types

Name	Symbol	Function
OPERATION		The OPERATION block is used to model a wide range of processes such as time delays, attribute assignments, etc.
TRANSFER		The TRANSFER block is used to model transfers between stations via material handling systems.
HOLD		The HOLD block is used to model situations in which the movement of an entity is delayed based on system status. The HOLD block must be preceded by a queueing facility to provide a waiting space for delayed entities.
QUEUE		The QUEUE block provides a waiting space for entities which are delayed at following HOLD or MATCH blocks.
STATION		The STATION block defines the interface points between model segments and the material handling systems.
BRANCH		The BRANCH block models the conditional probabilistic and deterministic branching of entities.

Table 15.1 (continued)

Name	Symbol	Function
PICKQ		The PICKQ block is used to select from a set of following QUEUE blocks.
SELECT		The SELECT block is used to select between resources associated with a set of following OPERATION blocks.
QPICK		The QPICK block is used to select from a set of preceding QUEUE blocks.
MATCH		The MATCH block delays entities in a set of preceding QUEUE blocks until enties with the same value of a specified attribute resides in each QUEUE.

block and consist of a verb that describes the specific function the block
is to perform. For example, the operation type ASSIGN specifies that the
block is to assign a value to an attribute or variable; the operation type
DELAY specifies that the block is to delay entities.

Each block function of SIMAN is referenced by a *block function name*.
In the use of the QUEUE, STATION, BRANCH, PICKQ, QPICK, SELECT,
and MATCH blocks, each block type performs only one function and the
block function name is the basic block name. However, in the case of the
OPERATION, HOLD, and TRANSFER blocks, each basic block type performs
several different functions. The block function name for each of these
blocks is the operation, hold, or transfer type specified as the first operand
of the block. Blocks are frequently referred to by their function names.
For example, an OPERATION block with the DELAY operation type will be
called a DELAY block.

All the basic block types, including the OPERATION, HOLD, and
TRANSFER types, have operands that control the function of the block.
For example, the CREATE block has operands that prescribe the time between batch arrivals, the number of entities per batch arrival, the number
of entities per batch, and the maximum number of batches to create.

Blocks may optionally be assigned a block *label,* one or more block
modifiers, and a *comment* line. A block label is appended to the lower side
of a block and can consist of up to eight alphanumeric characters. A block
label is used for branching or referencing from other blocks. Block modifiers are special symbols appended to the right or bottom of a block and
either modify or extend the standard function to be performed by the block.
The comment line, if specified, is entered to the right of the block and
serves to document the model.

A block diagram model can be defined in either of two equivalent forms
referred to as the *diagram model* and the *statement model.* The diagram
model is a graphic representation of the system using the 10 basic block
symbols in Table 15.1. The statement model is a transcription of the diagram
model into statement form for input to the model processor program. There
is a one-to-one correspondence between blocks in the diagram model and
input statements in the statement model.

The modeler normally proceeds by first constructing the block diagram,
which is then transcribed into the equivalent statement form for input to the
SIMAN language. The latter step is straightforward and, in fact, has been
automated by Professor Randall Sadowski at Purdue University, as part of
an interactive graphics preprocessor named BLOCKS which he developed for
SIMAN. BLOCKS allows the modeler to graphically build the diagram model
directly on the graphics display of a personal computer. The BLOCKS program then automatically generates the corresponding statement version of
the model for input to the SIMAN language.

To illustrate the general-purpose modeling approach of SIMAN, consider
the simple manufacturing system in which workpieces arrive, are processed
in order on a single machine, and then depart the machine. A schematic
of this example is shown in Figure 15.1. Assume that the parts arrive in

Figure 15.1   Schematic of a simple system.

batches of size 10, with the time between arrivals exponentially distributed with a mean of 20. The processing time on the machine is assumed to be uniformly distributed between one and two.

The block diagram model for this example is shown in Figure 15.2. The workpieces enter the system at the CREATE block. The operands for this block specify that the workpieces enter in batches of 10 and that the inter-arrival time between batches is exponentially distributed. The first number in the EX(1,1) operand denotes that parameter set number 1 in the experimental frame PARAMETERS element holds the mean of the exponential; the second 1 indicates that samples should be taken from random number stream 1. The workpieces continue to the QUEUE block, where they wait in turn to seize a unit of the resource MACHINE. Once a workpiece seizes the MACHINE, it enters the DELAY block, where it is delayed by the processing time UN(2,1). This delay time is taken as a sample from a uniform distribution whose minimum and maximum are given in parameter set 2 of the PARAMETERS element, and which is drawn using random stream 1. Following this delay, the workpiece releases the resource MACHINE, which allows it to be reallocated to workpieces waiting in the QUEUE block. The symbol attached to the bottom of the RELEASE block is called the DISPOSE modifier and models the departure of the workpiece from the system.

An experimental frame for this model is shown in Figure 15.3.

The experimental frame specifies the experimental conditions associated with the model and includes a specification of the parameter values referenced by the distributions in the model, the definition and capacity of the resources employed, a specification of the number and length of each simulation replication, etc. Note that in the RESOURCES element, the capacity of

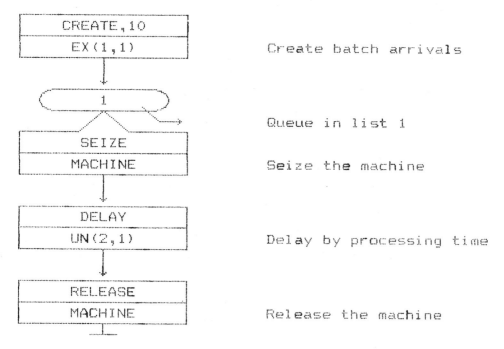

Figure 15.2   Block diagram for simple system.

```
BEGIN;
PROJECT, Single Machine Ex., Pegden, 5/20/86;
DISCRETE, 20, 0, 1;
RESOURCES: 1, MACHINE, 1;
PARAMETERS:  1, 20:              !Interarrival time mean
             2, 1, 2;           Processing min and max
REPLICATE, 1, 0, 480;
END;
```

Figure 15.3   Experiment for simple system.

resource number 1 named MACHINE is set to 1. A second machine is added
to the system simply by increasing this resource capacity to 2.

15.3   CHARACTERISTICS OF MANUFACTURING SYSTEMS

Manufacturing systems have a number of unique characteristics that make
them awkward to model within the framework of a strictly general-purpose
simulation language. These characteristics are present in both the classical
jobshop system and the modern automated version of the jobshop referred
to as an FMS. The most significant of these characteristics include the
following:

   1. Large manufacturing systems are typically comprised of a number
of different work centers or cells. A natural way to model such systems
is to decompose a large system into its workcenters, modeling each workcenter separately, and then combine the workcenter models into an overall
system model. General-purpose simulation languages typically do not provide a logical format for modeling separate workcenters.

   2. Often several workcenters within a manufacturing system are functionally equivalent. As a result, it is possible to develop a single functional
description that can be used to model all similar workcenters within a
system. Again, general-purpose simulation languages generally do not provide any features to exploit this property.

   3. The workpieces that move through a manufacturing system typically
have unique process plans. This means that each entity in the model must
have its own routing sequence through the workcenters as well as its own
setup, processing time, tool requirements, etc. If a general-purpose simulation language is used, a considerable amount of effort is usually consumed
in incorporating logic within the model to maintain process plans on each
workpiece and to control the flow from one workcenter to another.

   4. In most manufacturing systems, there is an uneven distribution of
workload between workcenters. One way that this varying workload is
accommodated is through the use of different operating schedules for the
different workcenters based on their workloads. Hence, it is desirable to
be able to assign each workcenter an operating schedule to follow during
the simulation. Again, this is often awkward to do with general-purpose
simulation languages.

   5. An essential element of most manufacturing systems is the materials-
handling component. This includes devices such as robots, AGVs, conveyors,
power and free monorail systems, etc. These types of devices can be extremely difficult to model. It should be noted that although in theory a
general-purpose simulation language can, given enough effort, accurately

model these characteristics of manufacturing systems, the modeling effort involved can be enormous. This is particularly true when the system of interest contains a major materials-handling component.

15.4 MANUFACTURING MODELING FEATURES OF SIMAN

In this section, we shall describe the special features included in the SIMAN simulation language for modeling the special characteristics of manufacturing systems.

15.4.1 Modeling Workstations

In manufacturing systems, it is frequently desirable to model distinct work-centers within the system. This can be done with SIMAN by employing the STATION block, which defines the beginning of a station submodel. An entity is entered into the station submodel by sending the entity to the STATION block using a TRANSFER block, which is used to represent entity movements between station submodels.

Each station submodel is referenced by a positive integer, called the station number. This number corresponds to a physical location within the system. The station number is an operand of both the STATION block and the TRANSFER block.

When an entity (workpiece) enters a STATION block, the entity's station attribute M is set by SIMAN to the station number of the STATION block. The entity carries this special attribute with it as it proceeds through the sequence of blocks that comprise the station submodel. The entity remains within the station submodel until it is disposed or is sent to a new station submodel via a TRANSFER block.

The block sequence within a station submodel defines the processes through which the entities flow. Such processes normally involve the queueing of entities due to limited resources such as operators, tools, etc.

To illustrate the concept of a workcenter submodel, consider the block diagram submodel shown in Figure 15.4; it contains the frequently occurring sequence QUEUE-SEIZE-DELAY-RELEASE, which can be used to model both single-server and multiserver queueing systems, depending on the capacity for the resource, which is specified in the experimental frame. The work-pieces arriving to this submodel enter the STATION block, proceed through the QUEUE-SEIZE-DELAY-RELEASE blocks, and then enter the ROUTE block. The ROUTE block is a TRANSFER block that routes the workpieces to their next destination. The block sequence in this example is particularly simple and employs only a small subset of the features of SIMAN. Once the modeler/ analyst becomes familiar with the wide range of block functions included in SIMAN, complex workcenters can be modeled with similar ease.

15.4.2 Macro Submodels

One particularly useful feature for modeling workcenters in SIMAN is the macro submodel. This feature permits the development of a similar single macro submodel to represent a set of two or more similar, yet distinct, work-centers. For example, a typical jobshop may consist of several different workcenters (lathes, planers, etc.) that are functionally equivalent and differ only in their number and type, buffer sizes, etc. One can model such a jobshop by constructing a single macro submodel that represents the process encountered by a job at a general jobshop workcenter. This

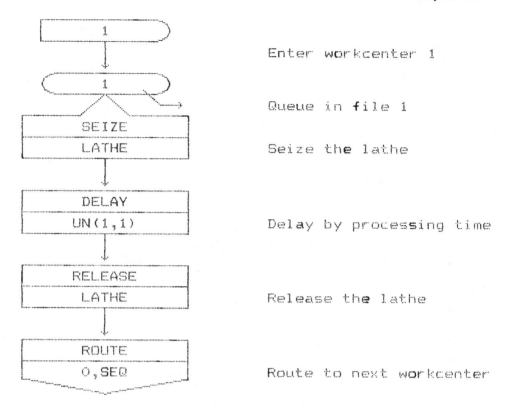

Figure 15.4   Block diagram of a workcenter submodel.

single macro submodel can then be used to model a jobshop of any arbitrary size.

The beginning of a macro submodel is defined by a STATION block. The range of station numbers, represented by the macro submodel, is specified as the operand of the block. An entity is entered into the macro submodel by sending it to the STATION block using a TRANSFER block. All entities sent to a station number in the specified range of the STATION block are processed as arrivals to the block. Upon entering the STATION block, the macro station attribute M of the entity is set by SIMAN to the station number to which the entity was sent.

When a macro submodel is employed, the station attribute is typically incorporated in the operands of one or more of the blocks that follow the STATION block. In this way, the operation of the macro submodel can depend on the station number of the entity. For example, the queue number at a QUEUE block can be specified as a function of M.

The station attribute can also be used to specify a resource to be seized or released through the use of *indexed resources*. An indexed resource allows a single name to be assigned to a set of two or more different resource types, with each resource in the set having its own capacity. The resource types within the set are distinguished by an index appended to the resource name. For example, MACHINE(1) and MACHINE(2) represent two distinct resources with individual capacities. These resources are completely independent, other than sharing the common name MACHINE.

To illustrate the use of indexed resources within a macro submodel, consider the block diagram shown in Figure 15.5. When an entity arrives at the STATION block, the station attribute M of the entity is set to its current station number. Since this macro models stations 1 through 6, M in this case will be between 1 and 6. The entities proceed to the QUEUE-SEIZE combination, where they attempt to seize a unit of MACHINE(M). If all units of the resource are busy, the entity is placed into list number M, where it waits for an idle unit of MACHINE(M). Once a unit is seized, the entity continues to the DELAY block, where it is delayed by its processing time, given by attribute 1. The entity next arrives to the RELEASE block, where one unit of MACHINE(M) is released. The entity is then sent, as before, via the TRANSFER block to its next station.

15.4.3 Visitation Sequences

As illustrated in the previous example, a workpiece is sent to its next workcenter using a TRANSFER block. However, the TRANSFER block must have some way to determine which workcenter is next in sequence for a particular workpiece. In addition, it may be necessary to update one or more attributes of the workpiece to correspond to the processing parameters at that workcenter. For example, the general purpose attribute A(1) was used in our macro submodel, shown in Figure 15.5, to specify

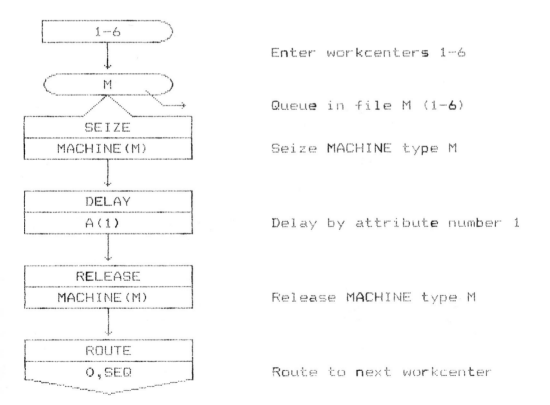

Figure 15.5   Block diagram of a macro submodel.

the processing time on the machine. Thus, an additional function of the TRANSFER block is updating attribute A(1) to correspond to the processing time at the next workcenter.

The workcenter visitation sequence and corresponding attribute update are specified in SIMAN using the SEQUENCES element, which is included as part of the experimental frame. The following SEQUENCES element defines two visitation sequences.

```
SEQUENCES:  1,  5,EX(1,1)  /  3,UN(2,1) :
            2,  4,UN(3,1)  /  2,10,3  /  6,EX(4,1) ;
```

Here, sequence number 1 consists of two workcenter visits. The first visit is to station number 5 and assigns to A(1) a sample from an exponential distribution. The second visit is to station number 3 and assigns to A(1) a sample from a uniform distribution. Sequence number 2 consists of visits to stations 4, then 2, and then 6. The visits to station 4 and 6 make assignments only to attribute 1; the visit to station 2 assigns to A(1) a value of 10 and to A(2) a value of 3.

Each workpiece in the system has two special attributes that are used in conjunction with the SEQUENCES element to determine its next station and attribute update values at the TRANSFER block. The first special attribute is NS, which specifies the number of the visitation sequence that the workpiece is to follow. This value is typically assigned to the entity when it first enters the model. The second special attribute is IS, which keeps track of the current index within the sequence. The index attribute is automatically updated by SIMAN whenever the entity arrives at a TRANSFER block.

Although the example discussed here is relatively simple and involves only a single attribute update, the constructs provided by the SEQUENCES element, in combination with the special attributes NS and IS, are flexible. Additional attributes could be employed to specify setup times, special tool requirements, etc., which might be part of the process plan. In addition, by resetting the value of IS for a given entity within a workcenter submodel, a portion of a given sequence could be repeated or skipped. Likewise, by resetting the value of NS, the sequence which that workpiece follows could be changed.

15.4.4   Resource Schedules

As noted earlier, the workcenters within a manufacturing system often operate according to different work schedules as a result of their differing loads. Within SIMAN, this characteristic can be modeled through the use of the SCHEDULES element which is included in the experimental frame. The SCHEDULES element is used to define a work schedule by specifying a resource capacity over time. A resource within the model can then be directed to follow a given work schedule. For example, the resources in workcenter 1 might be directed to follow schedule number 1 and the resources in workcenter number 2 might be directed to follow schedule number 2.

The following SCHEDULES element defines two different work schedules:

```
SCHEDULES:  1,  1*8,  0*16 :
            2,  1*EX(1,1),  0*UN(2,1);
```

In schedule number 1, the capacity is 1 for 8 time units, then 0 for 16
time units, and then this cycle repeats. In schedule number 2, the capacity is 1 for a duration that is sampled from an exponential distribution,
and th:n 0 for a duration that is sampled from a uniform distribution, and
then '..:: cycle repeats. Note that schedule number 2 could be used to
represent an unscheduled breakdown and repair activity for a resource
such as a machine.

15.4.5   Modeling Materials-Handling Systems

Within a manufacturing system, the movement of entities between workcenters is accomplished by the materials-handling system. This is an extremely critical function in most manufacturing systems. Apple (1977)
notes that the materials-handling function can account for 50–70% of the
production activity.

In simple terms, the function of materials handling is the movement of
materials from one point to another. A large variety of materials-handling
devices have been developed to support this function. Within the materials-
handling literature, these devices are frequently categorized into the
following equipment classes (Apple, 1972, 1977; Blanding, 1978; Boltz,
1958; Muther and Haganas, 1969).

1. Industrial trucks—hand or powered vehicles used for intermittently
    moving items by maneuvering across a fixed surface. Common ex-
    amples are forklift trucks, hand carts, and platform trucks.
2. Cranes, hoists, and manipulators—mechanical devices used for
    intermittently moving items through space. Common examples are
    overhead cranes, jib cranes, industrial robots, hoists, and mono-
    rail devices.
3. Conveyors—gravity or powered devices used in moving items con-
    tinuously and simultaneously along a fixed path. Common examples
    are belt conveyors, bucket conveyors, hook conveyors, and trolley
    conveyors.

From a systems-modeling perspective, the devices in the above three
equipment categories all perform one of two basic movement functions.
The first function, the *transport function*, corresponds to the intermittent
movement of items, one load at a time, along a fixed or varied path. The
term load unit, as applied here, could denote a box, a roll of material, or
a pallet containing a number of items grouped together. The transport
function is performed by devices in the first two equipment categories.
From above, it is noted that devices in the first equipment category perform
the transport function along the ground, whereas devices in the second
category perform the transport function above the ground. Devices in
either of these two equipment categories are called *transporters*.

The second basic movement function, the *convey function*, corresponds
to the continuous and simultaneous movement of items along a fixed path.
The convey function is performed by the *conveyors* listed in the third
equipment category. Special blocks and experimental elements are included
in SIMAN that allow both these basic movement functions to be modeled
in a straightforward manner.

The block functions used to model materials-handling systems employ
the concept of a station submodel, as discussed earlier. All movement
functions are made relative to station numbers assigned to each station

submodel. A materials-handling system is represented in SIMAN as a component of the system model that models the utilization of materials-handling devices to move from one station submodel to another. The travel time for entities between stations is based on the speed of the materials-handling device and the spatial relationship of the origin and destination stations relative to the device. Both are specified by the modeler in the experimental frame.

The generic term *transporter* is used in SIMAN to denote a general class of movable devices that may be allocated to entities. Each transporter device has a specific station location in the system and required time to travel from one station to another. Examples of devices that might be modeled as transporters are carts, cranes, and mechanical manipulators.

The characteristics of each transporter type are specified in the experimental frame and include the transporter name, capacity, distance set number, and initial station position and operational status of each of the transporter units for that type. The transporter name is an arbitrary alphanumeric name assigned by the modeler to each transporter type. The transporter capacity is the number of independent movable units of that transporter type. The *distance set number* is a cross-reference to a matrix containing the travel distances between all pairs of stations that each transporter unit of that type may visit. This distance matrix is specified in the experimental frame.

Transporter units are allocated to entities at HOLD blocks using the REQUEST block type. Once an entity has been allocated a transporter unit at a REQUEST block, the entity can be transported from one station to the next using the TRANSFER block with the TRANSPORT transfer type. The duration of the transport is automatically computed by SIMAN based on the distance to the destination and the speed of the transporter unit. At the end of the transport duration, the entity enters the STATION block of the destination station submodel.

The generic term *conveyor* is used in SIMAN to denote a general class of devices that consist of positioned cells which are linked together and move in unison. Each cell represents a location on the device and can be either empty or occupied. Entities that access the conveyor must wait at the entering station until the specified number of consecutive empty cells are located at that station. The entity then enters the conveyor and the status of the cells is changed from empty to occupied. The entity remains in the cells until the conveyor is exited at the entity destination station.

The representation of conveyor devices as linked cells, which are either empty or occupied, imposes the restriction that items can only enter the device at discrete points along the conveyor. This is representative of discrete spaced conveyor devices, such as bucket, cable, and magnetic conveyors, for which items can enter the device only at fixed positions along the conveyor. However, other devices, such as belts, permit continuous spacing of items along the device. These devices can be reasonably approximated by defining a small spacing (called the cell width) between cells. The number of cells that must be accessed is specified as the length, in cell widths, of the item to be loaded onto the conveyor.

Each conveyor device moves along a fixed path defined by one or more *segments*. A segment is a section of a conveyor path that connects two station submodels. Segments can be connected to form either *open-* or *closed-loop* conveyor paths. A closed-loop path is one in which an item on the conveyor can return to a station by continuing on the device.

An open-loop path is one that is not closed. The segments defining a
conveyor path are specified in the experimental frame.

Conveyor cells are allocated to entities to HOLD blocks with the
ACCESS hold type. Once a conveyor has been accessed, the entity can
be conveyed to its destination workcenter using the TRANSFER block with
the CONVEY transfer type. A conveyor may be stopped and started using
the OPERATION block with the STOP and START operation block types,
respectively.

15.5   CINEMA: AN ANIMATION SYSTEM

It is well known that computer simulation is useful in the design and
analysis of manufacturing systems. It allows an analyst to evaluate the
consequences of design and operating decisions before those decisions
have to be made.

One of simulation's shortcomings has been that outputs from a simulation study typically take the form of summary statistics or simple graphs.
Although these outputs are necessary to measure and draw conclusions on
the performance of the modeled system, they provide little insight into the
dynamic interactions between the components of the system.

Computer animation represents the ideal solution to the problem of
understanding the dynamic behavior of a simulation model. There are
several examples of the use of specialized computer animation for this
purpose (e.g., Healy, 1983). In applications in which animations have
been employed, the benefits of "seeing" the system operation have been
substantial. The two main benefits that have been cited are the following:

1. An animation is an ideal method for verifying the correct operation
    of the model. Subtle errors that might not be apparent from
    standard simulation output become obvious when the system opera-
    tion is displayed graphically.
2. Moreover, the results from simulation studies are sometimes difficult
    to present to convince the management (decision makers). Anima-
    tion is a powerful aid to assure the management that the model
    does, indeed, represent the system being modeled.

A typical animation, however, may require months of specialized
graphics programming as well as special graphics hardware. For most
simulation projects, even the well-recognized benefits of animation do not
justify the added expense.

These considerations provided the motivation for the development of
Cinema, a general-purpose, microcomputer-based animation system designed to work with the SIMAN simulation language. A Cinema animation
is a dynamic display of graphical objects that change location, color, or
shape on a static background to display the dynamic behavior of a SIMAN
simulation model as it is executing. The development of such animation requires no programming skills. Highly detailed animations, which might
otherwise require months of programming, can be developed in a few hours
with Cinema.

The key to the Cinema system is the design of the user interface. The
design is based on developments made at the Xerox Palo Alto Research
Center (PARC) (see Lampson, 1978). The user interacts with the Cinema

system by using a mouse-controlled graphics screen cursor to make selections from a hierarchy of "pull-down" menus. When the mouse-controlled cursor is positioned on a selection and the left button on the mouse is pressed, the selected header will "pull-down," revealing a menu of secondary choices. A selection within the menu is made by moving the mouse-controlled cursor to the desired item and again pressing the left button. Figure 15.6, an actual screen image, illustrates the pull-down menu concept.

On-line help is available on any menu item by moving the cursor to the item and pressing the right button on the mouse. Cinema then displays detailed information on the selected item.

Animations are generated by first constructing a SIMAN simulation model of the system. With minor exceptions, the SIMAN model is constructed without special consideration for whether it will be run with an animation. The Cinema program is then used to construct a corresponding animation layout, which, as stated earlier, is a graphical depiction of the physical components of the system being modeled. The user then executes the SIMAN simulation model in conjunction with the Cinema layout to generate a graphical animation of the system dynamics.

The Cinema system is designed to run on an IBM PC-AT (or its compatible) with 640K bytes of memory, an 80287 coprocessor, and the DOS operating system. Graphics specific hardware includes a high-resolution (1024 × 768 noninterlaced) color graphics board capable of displaying 16 simultaneous colors from a palette of 4096 and a high-resolution, fast-scanning, 19-in. color monitor.

15.5.1   The Animation Layout

An animation layout is a combination of one or more graphical objects that form a representation of the system being modeled. The objects that comprise a layout are one of two types referred to as static and dynamic. The static objects within a layout form a background and represent the portion of the layout that does not change during the animation. In a simulation of a manufacturing system, this might correspond to a sketch of the walls, aisles, posts, etc., of the facility being modeled. The dynamic objects within a layout are superimposed on the static background and represent the objects within the system that change location, color, or shape during execution of the simulation. Workpieces, workers, machines, robots, etc., would be represented as dynamic objects within a layout. Figure 15.7 is an example of an animation layout.

The Static Component. The static component of the layout is constructed using a set of elementary computer-aided drawing functions which allow the user easily to add both graphics and text to the background. The basic graphic drawing functions include line, box, circle, and arc. Pull-down menus are used to set attributes of the current drawing function including color, style, and line width. These basic graphics functions are drawn in "rubberband" mode, allowing the user to view the size and orientation of the object before it is actually added to the layout.

A "sketch function" allows the user to enter any free-form curve by simply moving the graphics cursor along the desired path while holding down the left button on the mouse. The curve is drawn on the screen using the current line color, style, and width. A "fill function" allows the user to color any enclosed region with either solids or patterns.

RESET

PALETTE

FILES

LIBRARIES

LAYOUT

Background
Queue
Variable
Level
Storage
Transfer
Symbols
Save
Print

QUIT...

Figure 15.6 Pull-down menus.

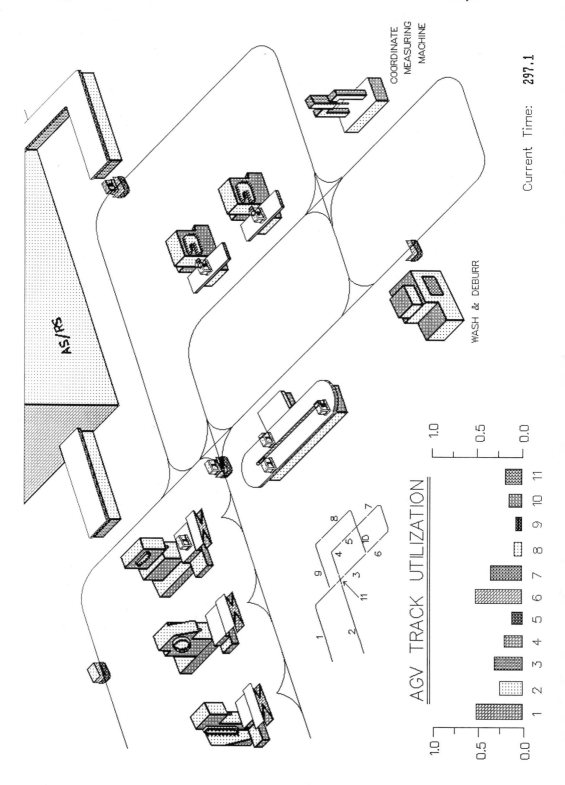

Current Time: **297.1**

Graphics text can be added to the static background in a choice of font, size, color, and orientation that are selected using the pull-down menus. An "erase function," which operates similarly to the sketch function, erases all graphics and text in the path of a mouse-controlled eraser cursor as long as the left button on the mouse is held down. Examples of the various drawing functions are illustrated in Figure 15.8.

The Dynamic Component. A Cinema animation is the dynamic display of objects that change location, color, or shape on a static background in correspondence with changes in a SIMAN simulation model. The changing objects constitute the dynamic component of the animation layout.

The dynamic objects in a Cinema layout are directly tied to specific modeling constructs within the SIMAN simulation language. When SIMAN executes a simulation of a system, it is continually updating an internal representation of the state of the system. The state of the system is defined by the current value of status variables, the number and location of entities within the system, the values assigned to attributes of the entities, the status of the resources within the system, etc.

Each dynamic object in an animation layout is associated with a specific element of the system state as represented by the SIMAN model. For example, one object might be associated with a status variable such as the simulated clock time. Another object might be associated with the status of a SIMAN resource. The dynamic objects in a Cinema layout are automatically updated as the state of the system changes during the simulation. Following is a discussion of the dynamic objects that can be incorporated into a Cinema layout and the association between these objects and specific modeling constructs within the SIMAN simulation language.

Entities. In a SIMAN simulation model, entities represent items that move through the system. An entity is represented in a Cinema layout as a moving and/or changing symbol. The symbol could be a sketch of a workpiece or a partially assembled car. As the entity moves from workstation to workstation within the SIMAN simulation model, its corresponding symbol is automatically moved across the static background. Entity symbols are created by coloring boxes, on a screen-sized grid, using the mouse-controlled cursor. Each box in the grid corresponds to one picture element (pixel) of the actual symbol image. As the symbol is created on the grid, it is displayed in actual size in the upper left corner of the screen (see Figure 15.9).

The symbols are created and stored in an entity symbol library. Symbol libraries are maintained separately from the animation layout and may be saved (stored on disk) and recalled at will.

To establish the association between a graphical entity symbol and a specific entity within the SIMAN model, the user must reserve one of the general-purpose entity attributes in the SIMAN model. The modeler is responsible for using that attribute to assign an entity number to entities in the SIMAN model. In Cinema, a specific symbol from an entity library

---

Figure 15.7   Example animation layout.

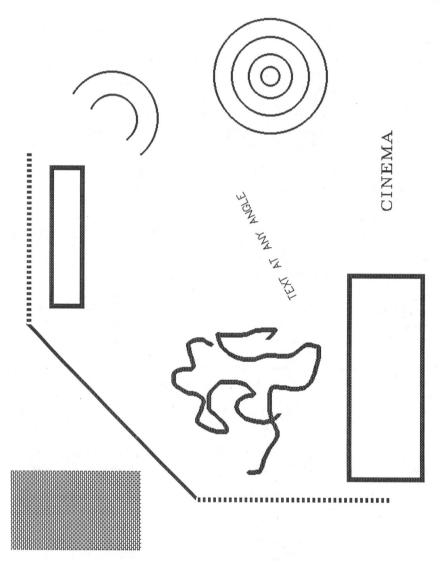

Figure 15.8   Example of static background drawing functions.

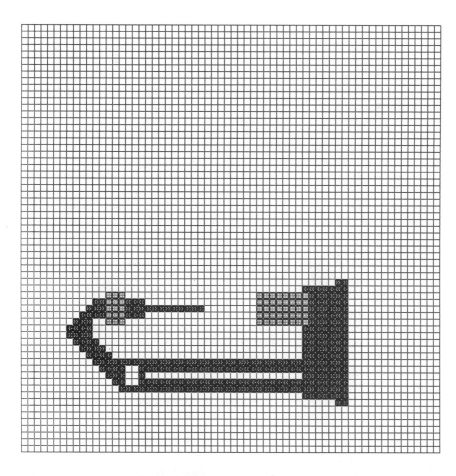

Figure 15.9 Symbol drawing grid.

is then associated with a particular entity number. If the designated
entity attribute changes values in the SIMAN simulation model, the corresponding entity symbol changes in the animation. Consider, for example,
a simulation of an automotive assembly plant. The designated attribute of
an entity arriving at an assembly station might be set to a value that
corresponds to a symbol of a car body without doors. After leaving the
workstation, the associated symbol could be changed to a car body with
doors simply by assigning a new entity number to the designated attribute
in the SIMAN model.

Queues. A Cinema queue is a dynamic representation of an ordered
list of entities residing in a file that is associated with a SIMAN QUEUE
block. The entities residing in the file of an associated QUEUE block in
the SIMAN model might represent workpieces awaiting the availability of a
machine, setup operator, space on a conveyor, etc.

A queue can be added to the layout at any location, and in any length
and orientation, using the mouse-controlled cursor. This graphical representation is then associated with a specific file number in the corresponding
SIMAN simulation model.

When an entity enters a queue in the simulation model, the entity's
symbol is displayed along the corresponding queue symbol at the proper
location relative to the other members of the queue. When an entity exits
the queue, its associated symbol is removed and all following symbols are
moved forward one position. When the queue in the SIMAN model contains
more entities than can be displayed along the fixed length of the graphical
queue, subsequent arrivals to the queue are not displayed. The entities
that are not displayed will eventually become visible as they move forward
into the display portion of the queue when preceding entities exit.

Resources. Resources are used in SIMAN to model limited items in a
system, such as machines and workers. Entities compete for the limited
number of units of a resource and incur queueing delays when enough
units are not available. In a SIMAN model, each resource assumes one of
four possible states: idle, busy, inactive, or preempted.

Resource status changes within a SIMAN model are displayed in a
Cinema animation using resource symbols. The resource is represented ·by
four distinct symbols called the idle, busy, inactive, and preempted symbols. Like entity symbols, resource symbols are created by coloring boxes
on a screen-sized grid and are stored in a resource symbol library that is
maintained separately from the animation layout.

A resource is added to the layout by selecting an idle symbol from the
library and positioning it on the static background, using the mouse. At
that time the user must also associate a SIMAN resource number with that
symbol. When the status of a resource changes in the SIMAN simulation
model, the associated Cinema resource symbol is displayed at the proper
location on the screen to reflect the status change.

Transfers. Distinct workstations within a system can be modeled in
SIMAN as station submodels. When this feature of SIMAN is used, the
movement of entities between workstations is modeled using the TRANSFER
block. This movement of entities between workstations can be animated in
Cinema by defining the station locations on the layout and then defining
the travel paths for the entities between the stations.

Station symbols are placed on the animation layout using the mouse.
Each station symbol in the animation layout is assigned a station number,

which is the same as the corresponding station number in the SIMAN model of the system. Once the station symbols are placed on the layout, the travel paths between stations are defined. This is done by digitizing connecting paths of one or more line segments using the mouse. Three different types of connecting paths are possible in Cinema, corresponding to the three types of TRANSFER blocks (ROUTE, TRANSPORT, CONVEY) in SIMAN.

1. A *route path* is used to define the travel path for simple entity routing between stations. When an entity is routed to a station in the SIMAN simulation model, its associated symbol is continuously moved along the route path defined in the layout, at a speed that is a function of the routing time in the model.

2. A *transporter path* is used to define the travel path for a transporter device that moves entities between stations. A device (cart, AGV, forklift truck, etc.) that moves along this path is represented by a set of symbols which are defined using the mouse, similar to resource symbols. Three symbols may be associated with each transporter device, corresponding to its idle, busy, and inactive states within the simulation model. In addition, locations called *transfer points* may be defined at each station symbol in the layout, where transporters wait when delayed at a station.

3. A *conveyor path* is used to define the travel path for entities that are being moved between stations on a conveyor within the simulation model. In this case, the speed of movement of the entity along the path is controlled by the movement of the conveyor within the simulation model.

Status Variables. While a simulation is executing, SIMAN automatically updates the value of many status variables that define the system state. Examples of these status variables are: the value of simulated time, the number of entities in a queue, and the number of workpieces processed. A representation of any SIMAN status variable can be incorporated into an animation layout using one of four dynamic features in Cinema.

A digital display of the numerical value of a status variable can be added to a layout using a feature called dynamic variables. The user selects the variable; defines the format, size, and color using pull-down menus; and then positions the variable anywhere on the screen, using the mouse.

A second way to display a status variable is with a feature called levels, which is an analog representation of the variable's value. Three different level shapes are included in Cinema: a box, circle, and dial. During an animation, a box or circle fills and empties in response to changes in the value of the associated status variable. The dial is a circular level with a sweep hand that rotates either clockwise or counterclockwise. The shape, fill and empty colors, fill direction, size, and location of a level are all specified using pull-down menus and the mouse cursor.

A feature called global symbols is a third way to display the value of a status variable. Like entity and resource symbols, global symbols are drawn on a screen-sized grid and maintained in a separate global symbol library. To incorporate global symbols into a layout, the user associates selected symbols from a library with designated values of a specified system status variable. For example, one symbol could be displayed when the number in a queue is less than or equal to, say, 10 and a second symbol displayed when the queue length exceeds 10.

Dynamic colors represent a fourth way to display the value of a status variable. The user first associates 1 of the 16 Cinema colors indices with a specific status variable. The user then defines one or more different colors for that index (in the form of different combinations of red, green, blue intensities) and associates each new color with a designated value of the status variable. For example, in the simulation of a steel-making facility, a status variable that represents the temperature of a furnace could be tied to the color index used to draw the furnace in an animation layout. Specific values of the variable (furnace temperatures) could be associated with different combinations of red, green, blue intensities that produce colors ranging from gray to red. As the temperature of the furnace increases in the simulation model, we would see the furnace gradually change from gray to red.

15.5.2   Runtime Features

The Cinema system requires a special version of the SIMAN simulation language which incorporates the additional code required to dynamically update the graphics screen when a simulation is executed with animation. In addition, a series of mouse-controlled pull-down menus are provided for managing the execution of a simulation/animation. Other features in this special version, as well as new features in the general release of SIMAN Version 3.0, allow users to interact with a simulation model while it is executing.

The Cinema-specific version of SIMAN includes the capability to switch between two or more animation layouts that might represent different parts of a simulation model that is executing. In addition, snapshots of the system status and graphics screen can be saved (stored on disk) and recalled later to restore the state of the system to a previous point in simulated time. Users can also control the run speed of an animation by specifying values for parameters that are used to scale simulation time to real screen update time. While the simulation/animation is executing, users can zoom in on the layout by pressing the "+" key and pan to different areas of the zoomed-in layout by pressing one of the arrow keys (up, down, left, right). The "-" key can be pressed to zoom back out.

SIMAN Version 3.0 includes an interactive debugging facility that allows users to interrupt the execution of a model to examine and modify the values of system status variables. For example, machines can be broken down by decreasing the capacity of the corresponding SIMAN resource. A Cinema animation provides immediate visual feedback on the consequences of such changes.

15.6   AN EXAMPLE APPLICATION

15.6.1   Overview of the Problem

The example application is based on a manufacturing system installed at a large Japanese machine tool manufacturer. This new flexible manufacturing system was designed and installed to produce up to 1500 different parts for machine tools and other large machinery. The system modeled has five machine tools (three vertical and two horizontal machining centers), an automatic wash and deburr center, and a coordinate measuring machine.

There are as many as 100 machine pallets and fixtures, with clamps and mounting holes for a variety of parts.

An automated storage and retrieval system (AR/RS) is used to store and deliver raw materials and finished goods. Machine pallets are also recycled within the AS/RS. All in-line buffer is handled by a centrally located carousel. Materials transfer within the FMS is handled by an automated guided vehicle system (AGVS). AGVs pick up delivered raw material on the fixtured machine pallets at the warehouse delivery station, move pallets from station to station within the system, and deliver finished product back to the AS/RS. Idle AGVs and those undergoing battery recharge are sent to a staging area.

The simulation analysis was undertaken to determine whether the proposed AGVS track layout would be feasible to handle the expected production load and how many AGVs would be required. Simplifying assumptions were made concerning the scheduling of parts to the system, part routings (the 1500 parts were divided into four large part families), and tool availability at the machines, since details about the system had not yet been finalized.

15.6.2 The SIMAN Model

In modeling manufacturing systems that use AGVS for materials handling, the SIMAN transporter constructs may be used to track the location, status, and velocity of each individual AGV. In this example model, the AGV track is mapped out by placing stations at control points and intersections on the track, and entering the station numbers and distances between stations into the DISTANCES element of the experimental frame. This information must be augmented by control logic in the model to handle the interaction among the AGVs as they move through the system. The stations divide the track into control zones, each of which is modeled as a resource with a capacity of one, allowing only one AGV in any control zone at a time. The resources are indexed in SIMAN and given the name TRACK. Tables in the experiment define the sequence of stations an AGV must visit to move from any station on the track to any other station and also specify in which track zone each of the stations belongs.

To send an AGV to pick up a part at a station, the model must determine where the AGV is and the sequence of stations it must move through to arrive at the part's station. To start, it must find what the first station between the AGV location and the part station is, seize the track resource for the zone that must be clear before the AGV may move, and then move the AGV through that zone. As the AGV leaves a zone, it releases the TRACK resource for that area, so that another AGV may move into the zone. Unidirectional and bidirectional track segments are modeled using the same method.

When an AGV drops off a pallet at a station, if there are no other requests for it, the model begins to move it to a staging area. During travel to the staging, the AGV may be redirected to pick up a pallet if instructed by the controlling computer to do so at one of the control stations. When an AGV arrives at the staging area, it remains there until it is assigned to pick up a pallet.

Part processing sequences and processing times at the machining centers, wash, and measuring stations are specified in parameter sets in the experimental frame. Each job is also assigned a due date, which is also taken from a parameter set.

Statistics collection is defined in the experimental frame, through the DSTAT element. For this system, the important information centers around AGV traffic in the track zones, which is retrieved from the utilization of the TRACK resources. Also, the AGV utilization is important for determining the required number of AGVs.

15.6.3   The Cinema Layout

The layout created to animate the SIMAN model of this FMS is shown in Figure 15.7. Most of the Cinema layout was created using the background drawing function (line, fill, etc.) in order to create an animation that closely resembles the physical system. To begin, the AGVS track was drawn using lines and arcs in the background, in order to establish the general layout of the system on the animation screen. Machines were drawn next, using lines and fill, as well as the explode function, which allows pixel-by-pixel drawing of sections of the background in a grid similar to that used for symbol drawing. The AS/RS was placed on the layout for reference, even though the control of pallets within the AS/RS is not included explicitly within the SIMAN model.

The AGVs are represented as transporter symbols; the idle symbol is shown in Figure 15.10. The busy symbol is similar, with the top of the AGV cleared to allow space for the entity symbol to be placed. Entities are drawn simply as numbered pallets, with the numbers denoting the part family to which the entity belongs. Family number 4 parts are represented by the symbol in Figure 15.11.

In order to define travel paths for the AGVs, all stations that belong to the AGVS in the SIMAN model are added to the Cinema layout, at the appropriate location on the drawn track. Next, the paths themselves are added as distances, such that the AGVs moving between stations will follow the track. Transfer points are added for each of the stations in the AGVS. These points placed exactly where idle AGVs, AGVs undergoing battery recharge, and AGVs that are stopped at a station awaiting a clear track or being unloaded should be shown.

At the bottom left section of the layout (Figure 15.7) is a status display, showing the relative utilization of the different AGVS track zones. The display bars are created using the levels constructs in Cinema. The numbering scheme corresponds to the TRACK resource numbers in the model. As the simulation is executed, the animation not only reflects the status and location of all transporters by moving their symbols along the defined distance paths, but also shows traffic on the AGVS track by continuously updating the levels in the status display.

15.6.4   Results

Simulation results may be gathered and displayed in many forms, including statistical data, graphical representations (plots, etc.), and the animations themselves. For this FMS, the critical decisions concerning the AGVS are whether the proposed track layout is capable of supporting the predicted production load, and how many AGVs will be required. A sample SIMAN Summary Report, displaying the tabular output from a simulation run, is shown in Figure 15.12. This report includes utilization of each of the sections of TRACK and the intersection, flowtime of the parts that completed production during the run, and counts of the number of parts that

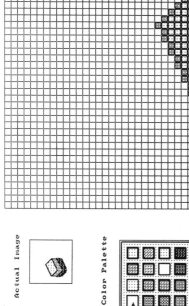

FUNCTION: Line   Row=   Column=

Actual Image

Color Palette

Figure 15.10   AGV transporter idle symbol.

FUNCTION: Point   Row=   Column=

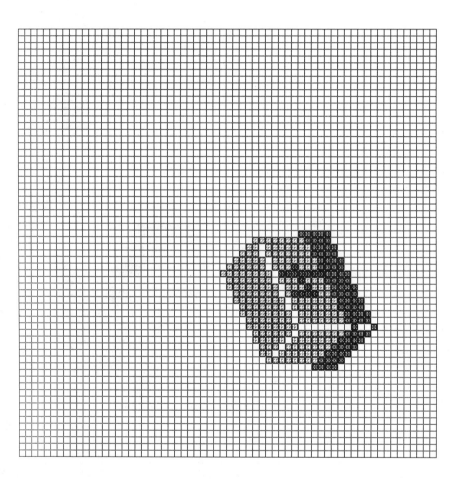

Actual Image

Color Palette

Figure 15.11   Part family 4 entity symbol.

SIMAN Summary Report

Run Number 14 of 20

Project: FMS
Analyst: D.DAVIS
Date   :   5/27/1986

Run ended at time :   .1000E+04

Tally Variables

Number	Identifier	Average	Standard Deviation	Minimum Value	Maximum Value	Number of Obs.
1	FLOWTIME	74.08569	15.68211	41.68545	143.8548	156

Discrete Change Variables

Number	Identifier	Average	Standard Deviation	Minimum Value	Maximum Value	Time Period
1	MACH 1 UTIL	.83015	.48276	.00000	1.00000	1000.0
2	MACH 2 UTIL	.75583	.49687	.00000	1.00000	1000.0
3	MACH 3 UTIL	.81103	.48752	.00000	1.00000	1000.0
4	MACH 4 UTIL	.85483	.47542	.00000	1.00000	1000.0
5	MACH 5 UTIL	.73502	.49877	.00000	1.00000	1000.0
6	TRACK 1 UTIL	.58227	.49319	.00000	1.00000	1000.0
7	TRACK 2 UTIL	.26443	.44103	.00000	1.00000	1000.0
8	TRACK 3 UTIL	.33800	.47303	.00000	1.00000	1000.0
9	TRACK 4 UTIL	.21000	.40731	.00000	1.00000	1000.0
10	TRACK 5 UTIL	.12345	.32895	.00000	1.00000	1000.0
11	TRACK 6 UTIL	.38431	.48643	.00000	1.00000	1000.0
12	TRACK 7 UTIL	.23707	.42528	.00000	1.00000	1000.0
13	TRACK 8 UTIL	.05400	.22602	.00000	1.00000	1000.0
14	TRACK 9 UTIL	.04504	.20738	.00000	1.00000	1000.0
15	TRACK 10 UTIL	.14402	.35111	.00000	1.00000	1000.0
16	INTERXN UTIL	.19877	.39907	.00000	1.00000	1000.0
17	CART UTIL	2.75278	1.12867	.00000	4.00000	1000.0

Counters

Number	Identifier	Count	Limit
1	NUMBER ON TIME	131	Infinite
2	NUMBER LATE	25	Infinite

Figure 15.12   SIMAN summary report.

were completed on time and the number that were late. Utilization statistics are also included for the machine tools and the AGVs themselves.

15.7   CONCLUSIONS

In this chapter, we have discussed the use of SIMAN and Cinema in the design and analysis of complex manufacturing systems. SIMAN and Cinema are examples of the many recent significant advances in simulation tools available to the industrial and manufacturing engineer.

Simulation tools in the area of manufacturing systems will continue to rapidly evolve over the coming years. It is believed that interactive graphics will play an important role in this evolution. This move toward graphical systems will be driven by the inherent advantages of a graphics user interface, as well as the development of low-cost, powerful graphics-based engineering workstations.

REFERENCES

Apple, J. M. (1972). *Material Handling Systems Design*, Wiley-Interscience, New York.

Apple, J. M. (1977). *Plant Layout and Materials Handling*, 3rd ed., Ronald, New York.

Blanding, W. (1978). *Blanding's Practical Physical Distribution*, Traffic Services Corporation, Washington, D.C.

Boltz, H. A. (1958). *Materials Handling Handbook*, Wiley-Interscience, New York.

Conway, R. W., Maxwell, W. L., and Worona, S. L. (1986). *User's Guide to XCELL—Factory Modeling System*, Scientific Press, Palo Alto, CA.

Benjamin F. Fairless Memorial Lecture at Carnegie-Mellon University (1972). Productivity Improvement, Pittsburgh, PA.

Groover, M. P. (1980). *Automation, Production Systems, and Computer-Aided Manufacturing*, Prentice-Hall, Englewood Cliffs, NJ.

Groover, M. P. and Zimmers, E. W., Jr. (1980, Nov.). Factories in the year 2000, *Industr. Engineering*, pp. 34–43.

Muther, R. and Haganas, K. (1969). *Systematic Handling Analysis*, Management and Industrial Research Publications, Kansas City, MO.

Pegden, C. D. (1985). *Introduction to SIMAN*, Systems Modeling Corp., State College, PA.

Rolston, L. J., and Miner, R. J. (1985). MAP/1 Tutorial Proceedings, Winter Simulation Conference (D. T. Gantz, S. L. Solomon, and G. C. Blais, eds.).

General Electric Company, P.O. Box 8106, Charlottesville, VA 22906.

Zeigler, B. P. (1976). *Theory of Modeling and Simulation*, Wiley, New York.

16

Multiprocessor Systems

16.1   INTRODUCTION

Recent advances in digital hardware technology have contributed to various architectural modifications that enhance the speed of the single-instruction, single-data-stream (SISD) architecture introduced in Chapter 10. All these enhancements were essentially transparent to the system user except that the speed of processing and in some cases the precision of the computation increased. We shall trace these enhancements to SISD in Section 16.2.

As the complexity of simulation increased, the speed of the hardware with all the enhancements became a limiting factor in simulation activities, especially when real-time simulations were needed. Multiple-processor systems came into being to accommodate parallel processing, thereby retaining the parallelism of simulation algorithms through the simulation process. To capitalize on the parallel processing capability of modern-day computers it is required that the algorithm parallelism be retained through the coding, compiling, and execution of the simulation program.

Consider the program execution environment of a digital processor. We first select a suitable algorithm for the problem. Let P be the degree of parallelism of the algorithm. Also, let L be the degree of parallelism with which the programming language allows coding of the algorithm, O be the degree of parallelism retained by the compiler, and M be the parallelism offered by the hardware architecture. Then, $P > L > O > M$ to retain the parallelism through the program execution. For example, in FORTRAN programming environment, $L = 1$ and $O = 1$, thereby serializing the processing even though $P > 1$ and $M > 1$. As the parallel architectures (i.e., single instruction, multiple data—SIMD; multiple instruction, single data—MISD; and multiple instruction, multiple data—MIMD, introduced in Chapter 10) evolved, languages that can express parallel algorithms and compilers that can vectorize the code to suit the hardware architecture have also evolved. The choice of a simulator for any application involves a compromise among factors such as cost, flexibility, speed, and accuracy. As the problem character changes, and as new hardware and software technology becomes available, these factors should be reconsidered. The introduction of microprocessors and, more recently, the network of microprocessors has changed the cost effectiveness of super computers that were used in fast, real-time simulations. We shall introduce the microprocessor network structures later in this chapter.

The area of computer architecture has been very dynamic over the last decade. Newer systems and hardware components comprising multiprocessors are continually being introduced. Although the hardware technology has advanced to make the multiprocessor architectures practical, software support in terms of languages that can describe parallelism and compilers that can retain the parallelism are still lacking.

A major application area is the simulation of multiprocessor systems themselves. Although such simulations can be performed on single-processor machines, it is more natural to use multiprocessor machines to gain flexibility and speed. Therefore, the emphasis of this chapter is on application of multiprocessors to simulation (as shown by the example simulation system of Section 16.5.5), rather than a detailed treatment of multiprocessor systems.

16.2   ENHANCEMENTS TO SISD

The SISD architecture assumes a single processor retrieving instructions and data sequentially from the memory and operating on them. One way of increasing the speed of such a system is to replace each component of the system with a faster component that technology provides. Several architectural features that enhance the speed have also been devised over the last 30 years. Some of these features are presented next.

16.2.1   Control Unit

The instruction cycle consists of a "fetch" and an "execute" phase. Although the fetch phase of an instruction must precede its execution phase, it is possible to fetch the next instruction in sequence to be executed, during the execute phase of the current instruction. Such instruction execution overlap has been used in modern-day machines to speed up the processing. The degree of overlap varies between machines. Some prefetch just the next instruction in sequence, whereas others fetch as many as 10–15 instructions into a buffer in the control unit. In fact, the control unit can be designed as a "pipeline" of three stages: fetch, address calculate, and execute. Thus an instruction moves from first through third stage and the subsequent instruction enters the pipeline as soon as the current instruction leaves a stage, increasing the program execution speed.

16.2.2   Arithmetic-Logic Unit

The digital computers of the 1950s and 1960s had arithmetic logic units of very limited capability, with a majority of arithmetic functions implemented through software. As the hardware technology advanced, yielding complex arithmetic-logic units on integrated circuit chips, it became cost effective to implement more functions in hardware, thereby increasing the processing speed. Innovative arithmetic-logic unit architectures were also employed to enhance the throughput of the arithmetic-logic unit. These architectures include multiple processors either of the same capability or each specialized to perform a dedicated set of operations. The processors were arranged either into a pipeline or simply into a pool of processors that could be assigned a computational task as the need arose. For example, Control Data Corporation's STAR-100 (described later in this chapter) is an example

of a pipelined system. Also, Control Data Corporation's 6600 series
machines employed 10 functional units: one add, one double-precision add,
two multiples, one divide, two increments, one shifter, one Boolean, and
one branch unit. The control unit is designed to invoke a functional unit
as needed when the operands are ready in a set of registers. The functional unit would be released when the allocated computation was complete.

16.2.3 Memory Unit

Advances in hardware technology have resulted in the availability of semiconductor memories at very low prices, thereby replacing the magnetic core
as the primary memory device. Semiconductor memories are much faster
than magnetic memories. The memory subsystem of any digital processor
spans from registers in the arithmetic-logic unit to primary memory words
and to secondary memory devices such as magnetic tape and disk. In this
memory hierarchy, the speed of the memory device increases as we move
toward the arithmetic-logic unit and the cost of the memory decreases as
we move away from the arithmetic-logic unit. A "cache" memory unit was
interposed between the primary memory and the arithmetic-logic unit to
enhance the memory speed while keeping the cost low. The cache is a small
memory block usually about 10 times faster than the primary memory. The
cache can be organized to act as a fast data/instruction buffer that can be
simultaneously filled from the primary memory and accessed by the arithmetic-
logic unit, thereby increasing the throughput of the machine.

16.2.4 Input/Output Unit

The simplest and least expensive mode of input/output is the so-called
"programmed input/output" in which the central processor controls the complete input/output operation. Since the input/output devices are usually
much slower than the central processing unit, the central processing unit
idles while the data are being gathered or accepted by the input/output
device, thus reducing the throughput of the system.

The first enhancement of input/output system is to distribute some of
the input/output control to the input/output devices themselves, making
the input/output devices "interrupt" the central processing unit once the
data are ready or accepted. In this mode the central processing unit can
perform some computation between the interrupts. A further enhancement
in the direction of distributing the input/output control is the concept of
"direct memory access" (DMA), in which the input/output devices transfer
data into/from the memory directly and compete with the processor for the
memory access. This mode of input/output makes the system a multiprocessor system although a simple one, since the DMA device and the
processor are working in parallel. An extension of the capabilities of the
DMA device to include error detection/correction, data format conversion,
etc., resulted in the concept of "input/output channels," which are
specialized input/output processors. When some of the processing functions
are also transferred to the channels, in addition to data gathering/distribut-
ing operations, the channels become "peripheral processors" (Bekey and
Karplus, 1977) or "front end processors" (Enslow, 1977). With the current
levels of hardware technology, it is cost effective to use microprocessors as
channels or peripheral processors, thereby gaining a tremendous processing
speed. The CDC 6600, for example, uses 10 peripheral processors that can
be connected to any of the 12 data channels by a switching network.

The enhancements described above make the SISD architecture tend toward the other three classifications described in Chapter 10, at least partially. We shall now provide models for other architecture classes.

16.3 MULTIPLE-PROCESSOR ARCHITECTURES

There are applications, specially in real-time processing, that make the machine throughput inadequate even with all the enhancements to SISD described above. It has thus been necessary to develop newer architectures with higher degrees of parallelism, to circumvent the limits of technology and to accommodate faster processing. A great deal of research and development has been devoted to the development of newer architectures, and several supercomputer systems have evolved as a result. We shall first provide models for the three multiprocessor architecture classes along with a description of representative systems to illustrate the hardware, software, and firmware requirements. The coverage of these topics is very brief and provides only an overview. The references at the end of this chapter provide further details on these topics for the interested reader.

16.3.1 Single Instruction Multiple Data

Figure 16.1 shows the structure of an SIMD. There are N arithmetic processors (P1–Pn) each with its own memory blocks (M1–Mn). The individual memory blocks combined constitute the system memory. The memory bus is used to transfer instructions and data to the control processor (CP). The CP is usually a full-fledged uniprocessor. It retrieves instructions from memory, sends arithmetic/logic instructions to arithmetic processors, and executes control (branch, stop, etc.) instructions itself.

Processors P1–Pn execute the same instruction at any time, each on its own data stream. Based on the arithmetic/logic conditions, some of the processors may be deactivated during certain operations. Such activation and deactivation of processors is handled by the control processor. The processor interconnection network provides for the data exchange between the n processors.

SIMDs are special-purpose machines that can process data in vector or array form, much faster than SISDs. These machines are efficient as long as the processing can be carried out on vector data and all processors are kept busy. If the processing is such that it is necessary to often deactivate and activate a subset of processors owing to data conditions, the processing speed approaches that of an SISD.

Consider the computation of the sum of two $N \times N$ matrices, which requires N^2 additions and the computation time is of the order of N^2 in an SISD. On the other hand, in an SIMD with N arithmetic processors, corresponding columns of the two matrices can be allocated to a processor's memory and the addition can be carried out simultaneously in N processors, thereby reducing the computation time to the order of N. The following programs represent the matrix sum computation:

SISD

```
        DO 30 I = 1,N
        DO 30 J = 1,N
   30   SUM(I,J) = A(I,J)+B(I,J)
```

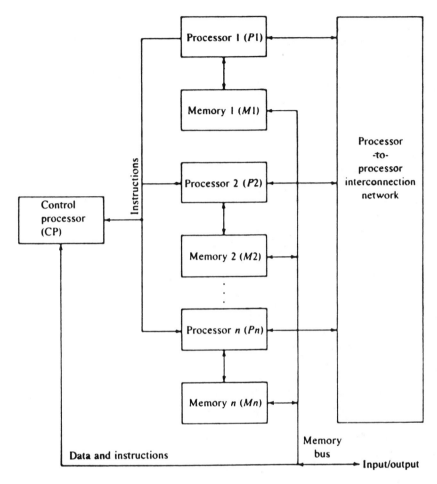

Figure 16.1   SIMD structure.   [From Shiva (1985), by permission of Little, Brown and Company.]

SIMD

```
      DO 30 I = I,N
40    SUM(I,J) = A(I,J)+B(I,J)  [J = 1,N]
```

The statement 40 in the above program indicates the parallel computation of all the N values.

Consider the normalization of the elements of a data matrix with respect to any matrix element. In the data organization of the above example, it is necessary to broadcast the value of the divisor to all processors. This data broadcast is usually accomplished by the processor interconnection network.

As can be seen from these simple examples, programming SIMDs is very similar to that of SISDs as long as the parallelism is recognized. Judicious data allocation between the memory modules is nessary to gain computation efficiency. SIMDs offer a high throughput as long as the processing exhibits a high degree of parallelism at the instruction level.

Illiac-IV (Hord, 1983). The best-known SIMD is Illiac-IV, built between 1966 and 1980. The objective was to build a parallel machine capable
of executing 10^9 instructions/sec, using an SIMD structure with 256 arithmetic processors divided into four quadrants of 64 processors each. Each
quadrant is controlled by one control unit. Only one quadrant was built, and
it achieved a speed of $2 * 10^8$ instructions/sec.

Figure 16.2a shows the Illiac-IV system structure. The control unit
(CU), a Burroughs B-6500, compiles the Illiac-IV programs, schedules array
programs, controls the array configurations, and manages the disk file
transfers and peripherals. The disk file unit acts as the backup memory
for the system.

Figure 16.2b shows the configuration of quadrant. The control unit
provides the control signals for all processing elements (PE_0-PE_{63}), which
work in an instruction-by-instruction lock-step mode. The CU and the
PE array execute in parallel. In addition, the CU generates and broadcasts
the addresses of operands that are common to all PEs, receives status signals from PEs, from the internal I/O operations and from B-6500, and performs the appropriate control function.

Each PE has four 64-bit registers (accumulator, operand register, data-
routing register, and general-storage register), an arithmetic/logic unit, a
16-bit local index register, and an eight-bit mode register that stores the
processor status and also provides the PE enable/disable information as
warranted by the data conditions. Each processing element memory (PEM)
block consists of a 250-nsec cycle-time memory with 2K of 64 bit words.

The PE-to-PE routing network connects each PE to four of its neighbors
(i.e., PE_i to PE_{i+1}, PE_{i-1}, PE_{i+8}, and PE_{i-8}). Thus, the PE array is
arranged as an 8×8 matrix with end-around connections. Interprocessor
data communication of arbitrary distances is accomplished by a sequence of
routings over network.

The processing array is basically 64-bit computation oriented. But it
can be configured to perform as a 128, 32-bit subprocessor array or a 512
eight-bit subprocessor array. The subprocessor arrays are not completely
independent because of the common index register in each PE and the 64-
bit data-routing path.

Two high-level languages have evolved over the years for programming
Illiac-IV: a FORTRAN-like language (CFD) oriented toward computational
fluid dynamics and an ALGOL-like language (GLYPNIR) (Hord, 1983). A
compiler that extracts parallelism from a regular FORTRAN program, and
converts it into a parallel FORTRAN (IVTRAN) has also been attempted.
There is no operating system. Since Illiac-IV is an experimental machine,
there is no software library or set of software tools, thereby making it
difficult to program. The Illiac-IV is located at NASA's Ames center and is
not operational.

16.3.2   Multiple Instruction, Single Data

Figure 16.3 shows the organization of an MISD. There are N processing
stages arranged in a pipeline. The data stream from the memory enters
the pipeline at stage 1 and moves from stage to stage through stage N, and
the resulting data stream enters the memory. The control unit of the
machine is shown to have N subunits, one for each stage.

Assuming that a unit of data stays with each stage for x sec, the total
processing time per unit of data is (x*N) sec. But once a pipeline is full,

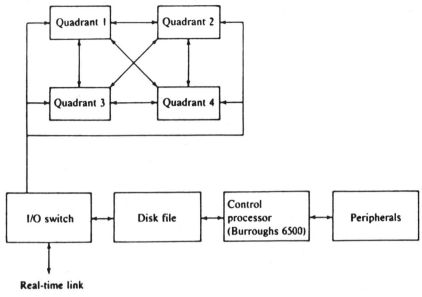

(a)

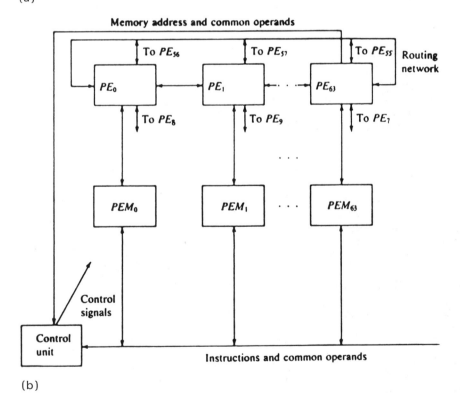

(b)

Figure 16.2   Illiac-IV system.   (a) System structure.   (b) A quadrant.
[From Shiva (1985), by permission of Little, Brown and Company.]

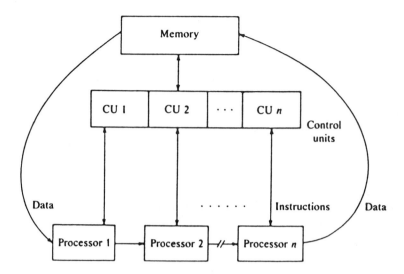

Figure 16.3   MISD structure. [From Shiva (1985), by permission of
Little, Brown and Company.]

there will be an output from the pipeline every x sec, thereby achieving a
very high throughput.

    To retain the processing speed, it is required that the pipeline be kept
full.   Thus, MISDs are efficient when the processing can be streamlined to
follow this assembly line fashion.   Two popular MISDs are: CDC STAR-100
and Texas Instruments Advanced Scientific Computer.   A brief description
of CDC STAR-100 follows.

    CDC Star-100.   Star-100 has two arithmetic pipelines and a string
processing unit (Figure 16.4).   The arithmetic pipelines process 64-bit
data, and the string processor handles 16-bit data.   Pipeline 1 performs
floating-point addition and multiplication as well as address computations.
Pipeline 2 performs floating-point addition and nonpipelined floating-point
division.   It also has a multifunction unit that can perform floating-point
multiplication and division.   Once full, a 64-bit result is output by the
pipeline every 40 nsec.   The 64-bit pipeline can be split into two 32-bit
pipelines, working in parallel and yielding a result every 20 nsec.

16.3.3   Multiple Instruction, Multiple Data

Figure 16.5 shows an MIMD structure consisting of P memory blocks, N
processing elements, and M input/output channels.   The processor-to-
memory interconnection network enables the connection of a processor to
any of the memory blocks.   In addition to establishing the processor-memory
connections, the network should also have a memory-mapping mechanism
that performs logical-to-physical-address mapping.

    The processor to I/O interconnection network is more of an interrupt
network than a data exchange network since most data exchange can be
performed through the memory-to-processor interconnection.

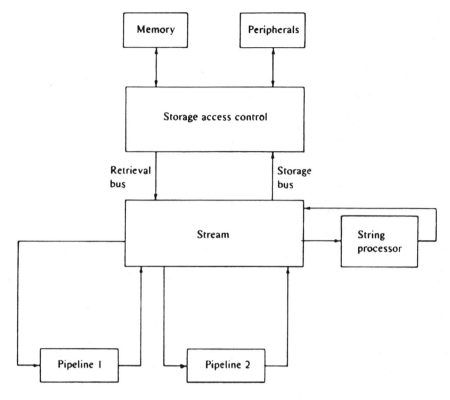

Figure 16.4 Control Data Corporation's STAR-100. (Courtesy of Control Data Corporation.)

MIMDs have the following advantages:

1. A high throughput can be achieved if the processing can be broken into parallel streams, thereby keeping all the processors active concurrently.
2. Since the processors and memory blocks are general-purpose resources, a faulty resource can easily be switched out, thereby achieving better fault tolerance.
3. A dynamic reconfiguration of resources is possible, to accommodate the processing loads.

MIMDs are more general-purpose machines compared to SIMDs. The processors are not in synchronization instruction-by-instruction, as in SIMD. But it is required that the processing algorithms exhibit a high degree of parallelism, so that several processors are active concurrently at any time.

The following program performs the matrix sum on an MIMD:

```
      Do 30 I = 1, N−1
      K = 1
  30  FORK 40
      K = N
```

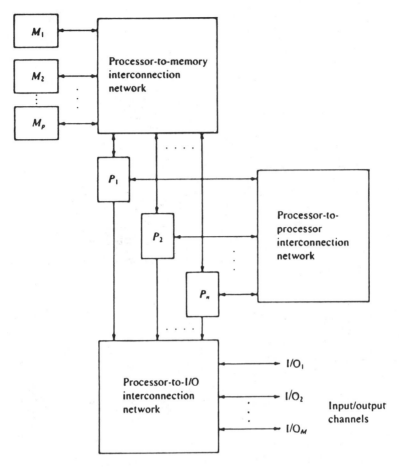

Figure 16.5   MIMD structure.   [From Shiva (1985), by permission of Little, Brown and Company.]

```
40   DO 50 J = 1, N
50   Sum (K,J) = A(K,J) + B(K,J)
     Join N
```

Here, the FORK instruction informs the system to invoke a new process starting at statement 40, each time it is executed. Thus, N − 1 processes are invoked and allocated to processors that are free. If all the processors are busy, the processes are made to wait in a queue until one is free. The processor that is executing the loop 30 continues execution after the loop is satisfied and performs the process N (corresponding to K = N). Each processor executing the loop 50 executes the JOIN instruction. The effect of JOIN is to suspend the process if all the N processes have not executed JOIN. The Nth process executing JOIN will be allowed to continue with the rest of the program, since all the N processes spawned by FORK have converged and the processing can continue.

We have shown the concurrent execution of processes in the above example explicitly. If the system software can detect the parallelism, it is not necessary for the user to explicitly code as above.

Some of the issues of concern in the design of an MIMD system are: (a) efficient allocation of processors to processing needs, in a dynamic fashion as the computation progresses; (b) processor synchronization to prevent two processors trying to change a unit of data simultaneously, while obeying the precedence constraints in data manipulation; (c) interconnection network design; and (d) communication overhead reduction. Partitioning of the problem to identify parallelism in processing algorithms and invoke concurrent processing streams is not a trivial problem.

CM* (Swan et al., 1977). The CM* is an MIMD designed and built by the Department of Computer Science at Carnegie-Mellon University. The goals of Cm* were to investigate new interconnection structures for multiprocessors and to provide a vehicle for exploring software issues in distributed multiprocessing systems. The Cm* uses a packet switched interconnection scheme. The principal connecting elements in Cm* are buses shared by a large number of processing elements. The processing elements communicate with each other by sending and receiving messages. Specialized processors control the routing and buffering of messages. There is a single systemwide physical address space, and every processor can access all the memory in the system.

The basic processing element in Cm* is a computer module (Cm). A Cm consists of a Digital Equipment Corporation's LSI-11 processor and the bus to which up to 128K bytes of memory, and some peripheral devices, are attached (Figure 16.6a). Also present on this bus is a local switch (Slocal), which connects the Cm to the rest of Cm*.

Slocal performs two mapping functions. It decides whether a processor-generated address is to be routed to local memory or a peripheral in this Cm, or whether it should be passed up to the Kmap. On the basis of this decision, it either maps the 16-bit address to an 18-bit LSI-11 bus address or passes on the 16-bit address to the Kmap to which it is attached.

Up to 14 Cms are connected together by a common bus called the map bus, resulting in a cluster (Figure 16.6b). Attached to the map bus is the Kmap, which is a sophisticated switching controller. It controls the map bus and handles the routing and buffering of messages between Cms. The Kmap also acts as the gateway to its cluster and interfaces it to the rest of the system. The Kmap decides on one of two courses of action, depending on whether the mapping yields an address within this cluster or an address in another cluster. If the resulting address is within the cluster, the Kmap constructs a request containing the appropriate Cm number, the 18-bit physical address and a code indicating the requested operation, and puts this request on the map bus. The slocal of the Cm referred to in the request performs the appropriate action at the specified address. The data involved in the reference (read or write) are directly passed between the invoking and responding Cms.

If the address produced after the Kmap mapping lies in another cluster, the Kmap at the requesting cluster sends a request to the Kmap at the destination cluster via one of the intercluster buses. The latter Kmap performs the requested operation on the appropriate Cm in its cluster and returns an indication of this fact to the former Kmap. The data involved in the access form part of requests and responses.

Many clusters are connected together via intercluster buses to form complete Cm* system. Each cluster is connected to two intercluster buses so that a single bus failure will not isolate a cluster. Figure 16.6c illustrate these interconnections.

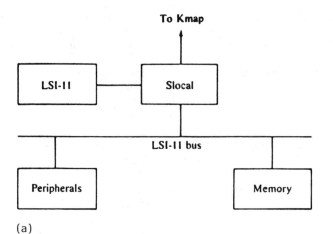

(a)

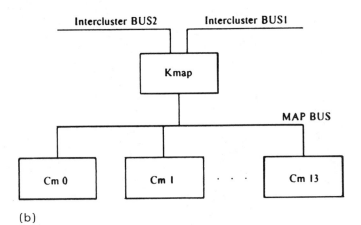

(b)

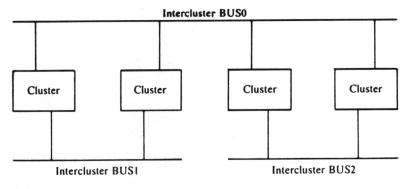

(c)

Figure 16.6   Cm* system.   (a) A computer module (Cm).   (b) A cluster.
(c) The system.   [From Swan et al., (1977).]

Any Cm can initiate the I/O device attached to any other Cm. However, interrupts can still be handled only by the Cm to which the I/O device is attached. From an error recovery point of view, this is undesirable, since if a Cm is lost, all I/O attached to it are lost. This is one respect in which Cm* fails to be a robust system.

16.3.4 Multiprocessor Architecture Implications

We shall now examine the implications of multiple-processor architectures on hardware and software. The MISD structure yields the concept of pipeline. Almost all modern computer systems employ pipelining to increase their throughput, to one degree or another. The major implication of this concept is that the data must be organized to keep the processing pipeline full as much as possible. The majority of the data structuring, alignment, and routing through the pipeline is transparent to the user and is performed by the system software and hardware. Thus, pipelined systems do not demand a severe degree of parallel thinking from the user.

The SIMD approach is the simplest of the multiprocessor configurations. The processor functionalities are well partitioned, and the arithmetic processors are all identical and work in a slave mode to the control processor. Therefore, programming these systems is not much different from that of SISDs. But an algorithm efficient in SISD may not be efficient in SIMD and may require control flow reorganization to effectively use the parallelism of SIMD. Data organization severely impacts the processing throughput. When extensive data sharing between the processors is needed, special interconnection networks are required. Thus, SIMDs are always special-purpose systems tailored to the application at hand.

MIMD structure is the most flexible and usually complex to implement. The functional partitioning at the algorithm and program levels is important to exploit the hardware parallelism. The functions, once partitioned, can be allocated to the processors in two ways: functions dedicated to processors or allowed to migrate between the processors as and when a processor is available. The dedicated approach not only simplifies the control software and is less flexible than the migration approach, but it also makes the dynamic load sharing between processors almost impossible.

The processors can be arranged in a master-slave structure (vertical) or in a logically equivalent structure (horizontal). Horizontal structures are more difficult to build and control, more flexible and reliable, and capable of dynamic load sharing compared to vertical structures.

A variety of processor interconnection networks have been devised. Anderson and Jensen (1975) and Ramamoorthy and Li (1977) describe several topologies and their characteristics. The data transfer speed over the interconnection network influences the throughput of the multiprocessor system. Based on the interprocessor data transfer rates, a multiple-processor system can be classified as one of the following:

Loosely coupled (computer network)
Moderately coupled (multiple processor)
Tightly coupled (multiprocessor)

"Computer networks" are resource-sharing systems with interconnection transfer rates of 50–100 kbits/sec, and each node in the system is a complete computer system. "Multiple processor" systems are processors interconnected with serial or parallel buses and geographically constrained to small distances, whereas in "multiprocessor" systems, the data transfer rates are almost equivalent to those within a processor.

Traditionally, supercomputers, such as Cray-1 and Control Data Corporation's Cyber-205, have been used in real-time simulations. These machines exhibit the characteristics of all four classes of architecture, depending on the computation being performed. With the continuing advances in digital hardware technology, networking of microprocessors has become cost effective and is capable of providing near-real-time speeds. Off-the-shelf interconnection hardware is now available although primitive. Super - computer architectures are described by Shiva (1985), Stone (1980), and Siewiorek et al. (1982).

16.4   SIMULATION-ORIENTED COMMERCIAL SYSTEMS

We shall briefly describe three commercial simulation-oriented systems in this section: Floating-Point Systems' array processor AP-120B, Applied Dynamic International's AD-10, and Intel's iPSC.

16.4.1   Floating-Point Systems' AP-120B

A class of machines known as "array processors" were introduced in the late 1970s. The designation is misleading because these machines are not designed to process arrays of numbers nor do they contain arrays of processors, as in SIMDs. They are a special class of MIMDs containing a host processor and several peripheral processors. The peripheral processors are special-purpose processors designed to perform a limited set of computations fast. They generally use extensive parallelism and pipelining in their structure. The host processor is generally a commercial SISD and loads the programs and the appropriate data into the peripheral processor memory. The peripheral processors execute their programs in parallel with the host processor. Figure 16.7 shows the general structure of a peripheral array processor. Multiple data paths are provided between the peripheral array processor and the host environment to enable maximum parallelism of computations. The peripheral array processor can thus be viewed as a hardware subroutine called by the main program executed by the host (Karplus, 1977).

Figure 16.8 shows the structure of Floating-Point Systems' AP-120B, a popular peripheral array processor. The machine uses a 38-bit floating-point representation, with a 10-bit exponent and 28-bit mantissa. An adder is configured as a pipeline of two stages. A floating-point result is

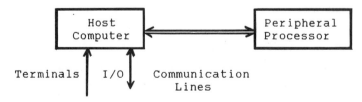

Figure 16.7   Pheripheral array processor model.

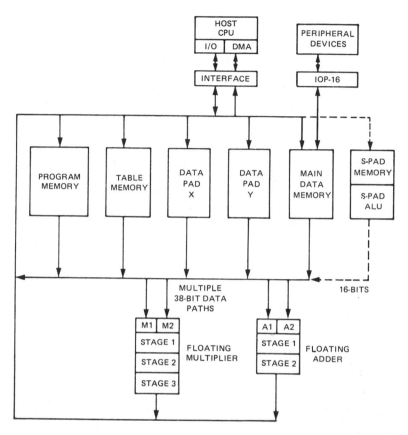

Figure 16.8   Floating-point systems' AP-120B. (Courtesy of Floating-Point Systems, Inc.)

produced by the adder every 167 nsec. There is a three-stage pipelined multiplier that operates concurrently with the adder and produces a product every 167 nsec. There are three memory blocks each with its own controller, thereby eliminating a majority of memory access conflicts. The program memory has 4K, 64-bit words and stores the array processor programs containing 64-bit instructions. Each instruction is divided into six fields, each field controlling a part (adder, multiplier, memory, accumulator, branch, and control) of the processor. The data memory words are 38 bits long. The memory is directly addressable up to 64K and page addressable up to 512K. The table memory block is the fastest of the memory blocks in the peripheral array processor; it has 64K, 38-bit word and is used to store the most often used constants and tables, such as for trigonometric computations. Data pads X and Y each contain 32 registers that are 48 bits long. These 64 registers form the accumulators of the system. The scratch pad (S-PAD) contains 16 index registers and an arithmetic unit that performs address computations and loop counting. There are seven data paths, four of which feed the adder and the multiplier. The remaining three transfer results into accumulators or the memory concurrently. The multiplicity of data paths enables a highly concurrent computation resulting in about 12 million floating-point operations/sec. Up to 256 input/output devices can be connected to the peripheral array processor, either in a programmed or in a direct-memory-access mode. The host interface performs the required data format conversion and also enables either programmed or DMA data transfer between the host and the peripheral array processor. The peripheral array processor can be programmed either in FORTRAN or in assembly language. The peripheral array processor functions are also invoked by calling its math library routines from the host program.

The AP-120B has been a popular system since the 1970s. Several enhancements have been made to the system, and a series of peripheral array processors with varying capabilities are now available from Floating-Point Systems (Charlesworth and Gustafson, 1986) and other manufacturers. Table 16.1 lists the characteristics of a few systems. Peripheral-array processors have been used extensively in signal- and image-processing applications and to a limited extent in simulation. Shibata (1985) describes an application of peripheral array processors in modeling of distributed parameter systems.

16.4.2   Applied Dynamic International's AD-10

Gilbert and Howe (1978) observe that a 16-bit fixed-point computation is adequate for most simulation applications. An exception is the implementation of integration algorithms, where a 48-bit word is desirable to accumulate small increments accurately. To accommodate severe speed requirements, real-time operation and economics imposed by many continuous-system simulations, the machine architecture should include the following features: large memory blocks for data and results storage; high data transfer rates between the memory and other modules; fast addition and multiplication rates for irregularly structured computations; functional pipelining and parallelism of various aspects of computational task; architectural growth capability and very high-speed analog-to-digital, digital-to-analog, and digital-to-digital conversion and data transfers.

The AD-10 is a peripheral processor designed to meet the above requirements. The system shown in Figure 16.9 is built around the Multibus, which contains 16 data lines, 20 address lines, and 12 control lines and is capable of 20 transactions/μsec. Real-time continuous simulation is achieved through the coordination of programs residing in five functionally dedicated processor blocks connected to the Multibus. The host computer loads each of these processor memories with the programs and the data memory with the required simulation data. The data memory contains up to 64 interleaved 4K pages of 16-bit words (plus parity) and can transfer data at the full bus bandwidth of 20 MHz.

The control processor coordinates the activities of the other processors and requests the service from the host as needed. The memory address processor generates the physical data memory addresses from the user logical addresses. The decision processor executes comparison and modification instructions on a dual 128-word register file. The arithmetic processor is pipelined and can produce 20 additions and 10 multiplications each microsecond. The numeric integration processor includes a 48-bit arithmetic unit and is controlled by microprogrammed instructions stored in the 80 $\times$ 1K program memory. The integration routines are all resident in this processor block, so that the integration formulas and step size can be changed at runtime without any host interaction.

Table 16.1 Array Processor Characteristics

Company/model	Throughput (MFLOPS)	Description	Programming	Cost ($000)
Mercury Computer Systems/ZIP 3232+	16	32-bit full Floating-point AP	ZIP/C A C-like language	15
Numerix/NMX-432	30	32-bit full Floating point AP	FORTRAN, Microcode	90-100
Floating Point Systems/3000	54	MIMD system	MAXL (high level), XPAL (assembly), Microcode	57-130
Floating-Point Systems/5000	8-62	AP tailorable to application	FORTRAN, MAXL, XAPL, Microcode	45-165

MELOPS, million floating-point operations/sec; AP, array processor
Source: Adopted from Wilson (1986).

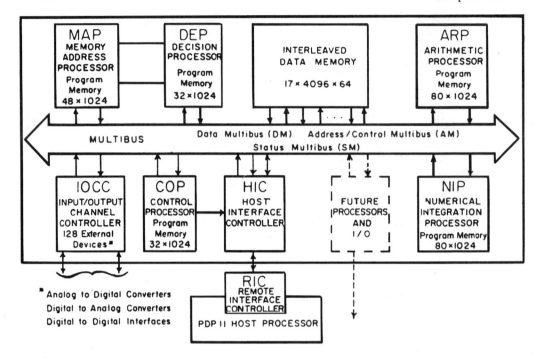

Figure 16.9   Applied Dynamics' AD-10.   (Courtesy of Applied Dynamics International.)

    The functional processors work relatively independent of each other, and the flexible parallel structure allows the system to be tailored to different applications without major impact on other elements in the system. The problems of task assignment, synchronization, and data communication have been minimized because of the essentially sequential nature of programming required by AD-10.

    The 16-bit fixed-point computation of the AD-10 imposes scaling restrictions. The FX is a major expansion of the AD-10 introduced in 1983. The expansion shown in Figure 16.10 offers 64-bit floating-point operation, thus eliminating scaling, extending the dynamic range, and providing a factor of 2–4 increase in speed.

    With the introduction of AD-100, Applied Dynamics took an integrated hardware/software approach to building multiprocessor systems for simulation. The System 100 consists of AD-100, shown in Figure 16.11, and ADSIM software. The AD-100 is built around the 40 MHz plusbus, consisting of 65 data bits and 40 address/control bits. The floating-point representation uses one sign bit, 12 exponent bits, and 53 mantissa bits. The bus bandwidth is 520 megabytes/sec, and the system can produce 20 MFLOPS. All processors have their own program memory, program counter, and instruction decoder. The communication and control block (COM) contains a 32K, 64-bit word dual-port memory. The arithmetic-logic unit (ALU) has two input registers, 4K words of memory, and produces a normalized floating-point sum every 100 nsec. The multiplier (MUL) contains 4K memory and produces a 53-bit rounded floating-point result every 75 nsec. The storage block (STO) has 4K, 65-bit words and 12K, 53-bit words. The

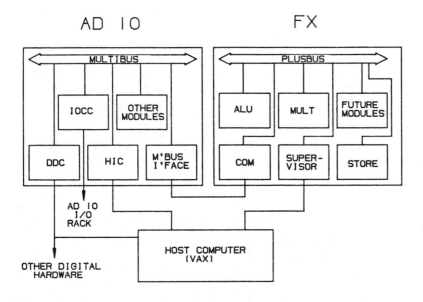

The Hardware Subsystem

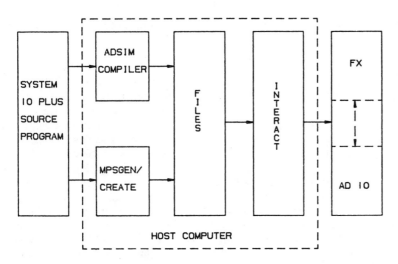

Figure 16.10   AD-10 with FX extension.   (Courtesy of Applied Dynamics International.)

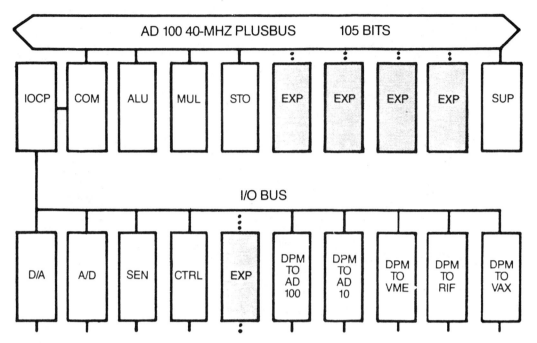

Figure 16.11   AD-100 system.   (Courtesy of Applied Dynamics International.)

supervisor (SUP) is the channel to and from the controller, which is a
Digital Equipment Corporation's VAX system.   The input/output controller
(IOCP) connects the peripherals to COM.   Digital-to-analog converters (12
or 14 bits), analog-to-digital converters (12 or 14 bits), sense line registers
(SEN), control line registers (CTRL), and dual-ported memories (DPM)
allow communications with other AD-100s, AD-10s, and VME bus and other
mainframes.

ADSIM is a continuous-system modeling language for AD-100.   It operates
under VMS operating system on VAX.   ADSIM compiler accepts the source
code, which may call on the standard function library modules and produces the AD-100 runtime code and an interaction database.   It sorts the
modeling equations into proper order and optimizes the execution of statements for the parallel processing capability of AD-100.   The interact module
allows the user to perform prerun system activities such as initialization and
loading; runtime control and monitoring; diagnostics; and postrun save operations.   A librarian that aids the library module development and other
utilities is also available.

16.4.3   Intel's iPSC (Intel, 1985)

The performance of a computer network depends on the performance of the
two major components: the processors at each node of the network and
the interconnection between the nodes.   With the introduction of low-cost,
high-performance microprocessors, it is now possible to configure network
nodes that achieve near-real-time speeds.   Several interconnection topologies
with varying performance have evolved over the last decade.   Although the

majority of them are specialized for particular applications, it is now possible to build networks of microprocessors using off-the-shelf interconnection networks. The Multibus (Intel, 1985) and the Versabus (Motorola, 1985) are the popular interconnection bus schemes available commercially. Multibus II and VME bus schemes (Wilson, 1985) allow the interconnection of 32-bit processors. Various other bus schemes (NUbus, Futurebus) have been proposed, and products that support these schemes are continually being introduced by microprocessor system vendors.

The concept of networking processors to achieve high-speed simulations has been an active research area since early 1970s. Korn (1972) showed that a network of three DEC PDP-11s can achieve a performance better than that of analog computers in most simulation applications. Bekey and Karplus (1977) support the network approach for real-time flight simulations because of cost effectiveness. In addition to the Cm* system described earlier in this chapter, the cosmic cube approach of Caltech has also proved the cost effectiveness of multiprocessor networks for real-time processing and simulation. We shall now provide a brief description of Intel's iPSC, a commercially available system based on the cosmic cube.

Intel's iPSC is a family of expandable concurrent "personal supercomputers." Figure 16.12 shows the system architecture. The two major functional elements of the system are the cube and the cube manager. The cube can be configured to contain one, two, or four computational units. Each unit contains 32 nodes. A node contains an Intel 80286 processor coupled with the 80287 numeric coprocessor and 512 kbytes of memory, as shown in Figure 16.12b. Each node contains eight bidirectional communication channels with dedicated communication processors (82586), seven of which serve to link the nodes together; the eighth channel is connected to the cube manager through the global Ethernet, for program loading, data input/output, and diagnostics.

The nodes are connected by Hypercube. This is an n-dimensional cube, where n is the number of directly connected nodes (i.e., the dimension of the cube). There will be 2^n nodes in an n-dimensional cube. Figure 16.13 shows a four-dimensional cube. The Hypercube topology provides a near-ideal balance between the performance and communication overhead. Internode delays are proportional to the cube dimension, with a maximum delay of n and an average delay of $n/2$, assuming uniformly distributed message distances. Typical application delays are much shorter. Since the memory is local to each node, the memory access conflicts are reduced.

The cube manager consists of an Intel 286/310 microcomputer system which is linked to each node over a global Ethernet channel. The cube manager provides for user interaction through a terminal and also enables networking with remote host environments. It contains a 40-megabyte Winchester disk drive and a 320-kbyte floppy disk.

The system software can be partitioned into cube software and cube manager software. The cube software consists of a ROM-resident monitor at each node and a kernel loaded into the RAM at each node after system initialization. The kernel provides user application processes a cubewide process-to-process message passing service. Cube manager software consists of a UNIX-based programming and development environment with Fortran and C languages. Cube control utilities and system diagnostics are also available at the cube manager.

iPSC systems have been installed at several research laboratories to experiment in simulation, parallel programming techniques, and real-time processing.

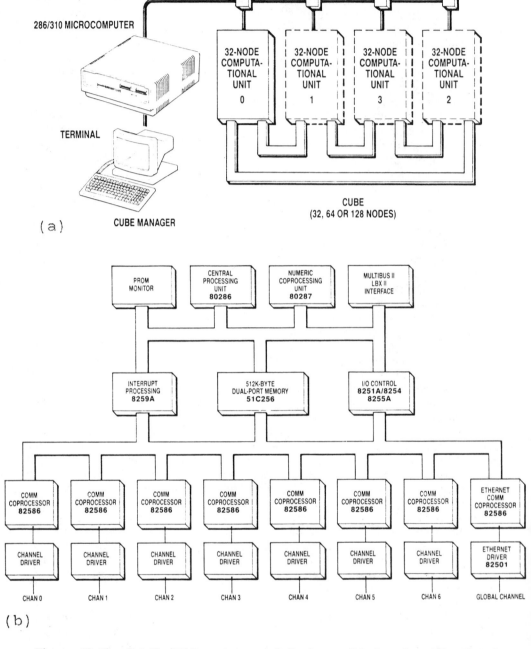

Figure 16.12   Intel's iPSC system.   (a) System.   (b) A node.   (Reprinted
by permission of Intel Corporation, 1985.)

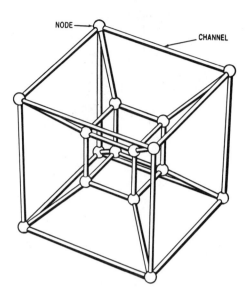

d4(16 NODES)

Figure 16.13   The four-dimensional cube.   (Reprinted by permission of Intel Corporation, 1985.)

16.5   REAL-TIME MULTIPROCESSOR SIMULATOR
(Blech and Arpasi, 1985; Arpasi, 1984; Cole, 1984; Milner and Arpasi, 1986)

The RTMPS project at the NASA Lewis Research Center is aimed at developing multiprocessor hardware and software that provides a portable, low-cost means for real-time simulation applications.   System firmware is used to coordinate data transfers, provide extensive diagnostic capabilities, and interface multiple processors to a simulation-oriented operating system.   This approach allows simulation software development tools to be tested and verified in the multiprocessor environment.   It also provides a valuable facility for investigating various methods of partitioning simulations to run on multiple microcomputers.

The major components of the system, shown in Figure 16.14, are the hardware, the operating system (RTMPOS), and the programming language (RTMPL).

16.5.1   Hardware

The RTMPS uses a dual-bus architecture (Figure 16.15).   The interactive information bus (IBUS) provides a data path primarily for user input and output.   The real-time information bus (RBUS) provides a data path for

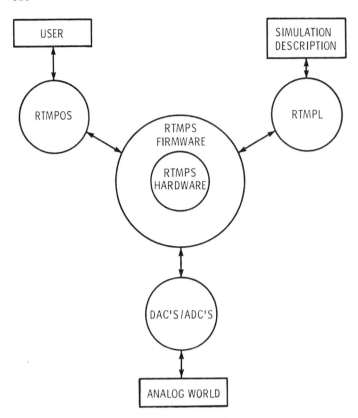

Figure 16.14   The RTMPS structure. [From Blech and Arpasi (1985).]

transferring data between the simulation processors. A special device on
the IBUS, called the front-end processor (FEP), handles all input and output between the operator and the RTMPS. The other processors on the
IBUS (called IBP processors) perform tasks related to data transfer to and
from the FEP. The IBP can also handle part of the simulation processing
An IBP communicates with the processors on the RBUS (called RBPs)
through a dual-port memory. The combination of IBP, and RBP, and interface memory forms a channel. Interprocessor communications can be viewed
as occurring locally within a channel (via dual-port memory) or globally
between channels (over the IBUS or RBUS). The number of channels can
range from zero to the maximum allowed by the physical bus limitations.
The first channel (channel 0) plays a special role. The RBP processor in
channel 0, called the real-time controller, is responsible for maintaining
synchronization between all RTMPS channels. It sets up and begins program
execution and monitors the timing to ensure that all channels complete their
calculations in a specified time. The real-time controller also handles analog
input and output. The IBP processor in channel 0 controls the operating
mode (RUN, HOLD, or STOP) and provides real-time analysis functions
as commanded by the FEP. Real-time analysis functions include data collection, event triggering, rate-of-change monitoring, and peak detection.
    The RTMPS architecture provides a highly user-interactive environment.
However, the architecture can provide higher performance at the cost of

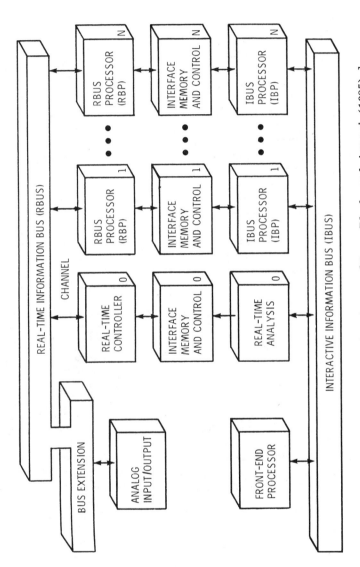

Figure 16.15 RTMPS hardware structure. [From Blech and Arpasi (1985).]

user interaction. In this case, both the IBUS and the RBUS would be
used for interprocessor transfers related to simulation calculations. The
FEP would serve only as a loader and initializer for the simulation code.
This mode of operation would be useful for applications where a proven,
dedicated real-time simulation is required and only minimal operator interaction is needed.

The IBP and RBP processors are implemented by using Motorola VM02
microcomputer boards and consist of an 8-MHz 68000 microprocessor, 128K
of random-access memory, three programmable timers, a system bus interface, and an interrupt controller. The VM02 RAM is dual-ported to the
board's local bus and external system bus. Therefore, all of the 128K RAM
is accessible by the local 68000 microprocessor or by any other microcomputer board on the system bus. Each board also has an input-output
channel to allow communications with devices external to the board and the
system bus. This input-output channel is used to interface the microcomputer board to the interface memory.

The interface memory is the communications link between processors on
the IBUS and the RBUS, through each processor's input-output channel.
The input-output channel is composed of address, data, and control lines
that define a memory-mapped segment of the processor's address space.
The input-output channel is specified to be an asynchronous communications
path. Thus, whenever a processor begins an input-output channel access
cycle, it must wait for an acknowledgment from the channel before complet-
ing the cycle. The control and arbitration logic decodes memory access from
both processors. Either processor can read from or write to the memory.
However, only one processor can access memory at any one time. If simultaneous access (contention) occurs, one processor is delayed while the
other completes its cycle.

There are three 1K blocks of memory. The address spaces occupied by
each of the first two blocks are switch-selectable by software. If the switch
is off, memory block 1 occupies the address space 0–1023, and block 2 resides in locations 1024–2047. When the switch is on, memory block 1 is
located at addresses 1024–2047, and block 2 is at 0–1023. This feature is
especially useful for a simulation problem where past values, such as the
derivatives for an integration algorithm, must be retained. For example,
assume a variable is assigned to location 1024 and its past value to location
0. By toggling the switch and then updating the variable at location 1024,
the past value is automatically retained at location 0.

The firmware on each channel's RBP controls the setting of the memory
switch. It toggles the switch on each channel at every update interval.
The update interval defines what is "real time" for the system. The simulation's dynamic equations, which are functions of time, are recalculated at
every update interval by using a new value for time. Time is advanced by an
increment defined by the update interval, and the memory switch is toggled.

The third block of memory is used to store flags, control parameters,
and other information that synchronizes communication between the two
processors sharing the memory. Two eight-bit latches within the interface
memory are used to generate interrupts. One latch is assigned to each of
the processors that share the memory. One bit from the RBP processor's
latch is also used for the memory switch, as previously described. This
bit is controlled by the RBP processor firmware. Both processors have
four interrupts, each of which is used to begin interprocessor communications within an RTMPS channel.

The RBP processors physically reside in Versabus card cages. Thus, data transfers that occur over the RBUS follow the Versabus specification. The specification allows for multiple bus masters and five levels of bus priority. Because all data transfers are asynchronous, a sending device requires an acknowledgment from a receiving device before the transfer is completed. In this arrangement, the bus controller must reside in the first slot of the card cage. The bus controller monitors requests for bus access and grants access according to priority level.

The controller in channel 0 has the responsibility of synchronizing all the RBP processors during RTMPS operation and also performs the analog input and output. Because board-level analog peripherals are not available for the Versabus, a Versabus-to-Multibus converter is used. The converter allows a multibus card cage to act as an extension of the Versabus system. Thus, the many Multibus-compatible analog peripherals can be used with the Versabus. Another benefit of the Multibus extension is the ability to communicate with other Multibus board-level products, including microcomputer boards.

The IBUS is contained within a Motorola Exormacs development system. The chassis is a 15-slot Versabus, with several slots containing the development system processor board, disk controller, RAM, and communications controller. The development system's Versabus forms the IBUS, with the previously described board set constituting the FEP. The remaining card slots are used for the IBP processors in the RTMPS channels.

Although the Exormacs development system was designed for software development, it has many useful features that make it ideal for an FEP. The system communications controller and a terminal are used as the RTMPS operator's console. The disk controller, hard disk, and floppy disk systems provide storage for programs and data. The resident disk operating system, Versados, contains utilities and system routines that can be called by the RTMPOS operating system. Since the disk operating system is multiuser and multitasking, the RTMPS can be simultaneously used for software development and running real-time simulations.

16.5.2 Firmware

Each IBP and RBP has its own firmware. The firmware interfaces the hardware with the remaining components of the system. The transfer of data between processor during simulation execution is coordinated by firmware. The transfer requirements are established by RTMPL during translation of the simulation source code. For example, the source code for one processor may reference a variable that is calculated on another processor. RTMPL recognizes this situation and generates the necessary code to begin the transfer of variable from the source processor to the destination processor. The firmware provides services to perform physical data movement and to maintain data currency within a simulation update interval. Data currency is maintained through the use of currency flag. All external variables have a currency flag that alternates in value every update interval. A destination processor requiring an external variable must first verify that the variable is current by testing this flag. A source processor must send the currency flag with the variable. The use of the currency flag allows data to be transferred asynchronously (i.e., at any time) during an update interval. The ability to transfer data asynchronously provides the maximum flexibility for partitioning of simulation code to run on multiprocessors.

Three major simulation modes are supported by the firmware: STOP, RUN, and HOLD. In the STOP mode all simulation processors are available for program loading and initialization. In the RUN mode the various segments of the simulation code are executed repetitively on the basis of the cyclical timeouts of a programmable timer. The timeout generates an interrupt in each of the RTMPS processors. The period between interrupts defines the simulation update interval. It is the responsibility of the real-time controller to verify that all channels have completed their calculations before the end of the update interval. During the computation interval each processor performs calculations for its own segment of the simulation. When an RBP has completed its calculations, it waits until the associated IBP processor is finished. The RBP then signals the real-time controller, via interrupt, of that channel's completion. Before this interrupt the real-time controller has been performing simulation calculations, supporting analysis functions, and outputting information to the D/A converters. When all channels have completed their calculations, the real-time controller inputs A/D converter information for the next update interval. If all channels have not completed their calculations before the next update interval, an interrupt is issued to the operating system. RTMPOS then advises the user of a failure. In addition, the firmware maintains a watchdog timer for each processor that, if not reset periodically, interrupts RTMPOS to issue an advisory.

The HOLD mode is similar to the RUN mode in that the simulation is repetitively executed. In HOLD, however, user-specified variables are held constant and computation is recycled on completion and is not timer-driven. The watchdog timer does operate in HOLD. Before each computation is executed, latches (associated with each variable whose value is to be held) are set by the firmware. During computation in this mode the latches are tested when a variable value has been computed and, if set, the old value is not replaced by the new one.

The firmware serves the following RTMPOS selectable analytical functions: data sampling, peak detection, and rate-of-change detection. These services are set up in STOP and may be activated in either RUN or HOLD. The real-time analysis processor in channel 0 executes the analysis firmware (if an analytical task is selected). This allows the other processors to perform simulation calculations uninterrupted. The firmware also provides system and user advisories and performs power-up initialization.

16.5.3 Programming Language

RTMPL allows the user to describe simulation tasks for each processor in a straightforward, structured manner. The RTMPL utility acts as an assembly language programmer, translating the high-level simulation description into time-efficient assembly language code for the processors. The utility sets up all the interfaces between the simulator hardware, firmware, and operating system. All required interprocessor communications (i.e., data transfers) are automatically established by the utility. RTMPL simulations are self-documenting since the utility produces listings, error messages, warnings, and running of a simulation.

The RTMPL utility is written in Pascal. The current implementation runs on a Motorola Exormacs development system and produces 68000 assembly language code. The utility can easily be retargeted for different processors by changing the contents of a target definition file. The

RTMPL language (i.e., command and operation set) is macrobased and can
be modified/expanded to meet particular simulation needs.

RTMPL is structured, high-order language designed to facilitate the
development of error-free, time-efficient simulations. RTMPL source is
written by the user in four segments: control, global data, local data,
and execution. Separate source files are used to store the control global
data segments. For each processor to be used in the simulation, a single
source file (program) is used to specify the local data and execution
segments.

In the control segment, the user identifies all the other source files
in the simulation and any output files. The user can also specify options
to govern the listing functions of utility.

In the global data segment, the user specifies the data that can be
referenced globally by any processor program. The data can consist of
messages and/or constants. Messages are used to advise the user of programmed conditions being met during the simulation execution. The user can
initiate these advisories using the RTMPL ADVISE command. Constants may
be specified by the user to be parametric or fixed. Parameters may be
changed at runtime. Changing a global parameter causes it to be changed
in all programs in which it is referenced.

The local data can consist of three segments: constants, variables, and
argument groups. A constant may be specified to be a simple constant, a
one-dimensional array of constants (data table), or a parameter. Only a
parametric constant may be changed by the user at runtime. Variables may
be specified to have any number of past values. Past value retention is
handled by the utility on the basis of this specification. A variable may be
assigned a transfer path (either the real-time or interactive information bus)
to satisfy external references. If the transfer path is not specified, the
utility selects the real-time information bus. Argument groups are used to
pass arguments to/from target library procedures. They are also used
extensively by the operating system. Argument groups consist of defined
variables and/or constants. The user-specified size of the argument group
determines the minimum number of items to be contained within the group.
The user may "fill" the group with any number of items, up to the maximum, using either RTMPL or RTMPOS at runtime.

As appropriate, the user must specify data type, precision, scale factor,
and values for all constants and variables. Except for values, the RTMPL
utility uses this information to ensure data-type compatibility and to recognize
the need for housekeeping operations when generating the assembly language
code. Values are used by the RTMPOS for its various runtime functions.

The execution segments contain statements that define the computations
to be performed. The execution segment consists of executives and tasks.
Executives contain the main simulation functions. Two types of executives
are allowed: background and foreground. Any number of background executives can exist, but at least one is required in each execution segment
of each processor. The background executive is the main computational
program of the processor. Any of the background executives may be
selected at runtime, through the RTMPOS, for execution. Up to eight
foreground executives are permitted in each execution segment. Foreground executives allow for programmed changes in execution, in a priority
interrupt environment, based on decisions made in the alternate processor
in the local channel. That is, the processor can be programmed to interrupt the background executive in the computation processor, and vice

versa. The execution of foreground executives is accomplished using the RTMPL ACTIVATE command.

Tasks are used to partition the executable segment into parts that have some operational significance. Tasks are reentrant and may be entered from any executive in the execution segment using the RTMPL ENTER command. Tasks may be enabled or disabled in executives or at runtime using RTMPOS. The RTMPL DISPATCH command allows the user to force a data-flow type of task execution. Any task referenced by the DISPATCH command will be executed when its external variable arguments become available from computations on another processor. These tasks are executed on a first-ready, first-serve basis, and dispatching continues until all referenced tasks have been executed.

Tasks and executives are written in terms of statements. There are three types of RTMPL statements:

equivalence:
  <VARIABLE> = <EXPRESSION>

conditional:
  IF <BOOLEAN CONDITION>
       THEN. . . ELSE. . . .

command:
  <RTMPL COMMAND> <COMMAND ARGUMENT>

The equivalence statement requires data-type compatibility across the equals sign. In the conditional statement the Boolean condition may be a Boolean expression or a comparison of arithmetic expressions or a simulator logical state. The "then" and "else" clauses may consist of one or more statements.

Expressions may contain any operations defined in the target file for the required data type. Operands must be defined locally, globally, or externally (in another processor program). The expanded format for RTMPL operand specification allows the user to reference, as an operand, any variable or constant defined in the simulation by specifying the channel name and processor type in which the variable or the constant is defined. External reference of a constant causes the RTMPL utility to create a constant of the same name in the local program. External variable reference causes a transfer map to be formulated for this variable in its source processor program and permits insertion of appropriate data transfer macros in the source and destination programs.

16.5.4 Operating Systems

Refer to Chapter 9 for details of a general operating system. RTMPL database and program source-code/object-code/load-module files are saved on disk files that can be manipulated by the RTMPOS. Once the simulation has been formulated, the RTMPOS provides the following general programming and operational functions: program control, database management, simulator control, runtime monitoring, and simulation results management.

The first step in a simulation session is to load the database from the RTMPL-generated disk files into the FEP memory. The RTMPOS then uses the database to allow the user to interactively modify and execute the simulation. The database management functions provide for: loading, editing, savings, and listing of the database. Editing refers to making changes to the simulation at runtime, such as displaying/changing values of constants and initial condition values of variables. Advisory messages may be changed, and additions, deletions, or changes to items contained in argument groups may be made. The database may be edited before or after the simulation program load modules have been loaded into the simulator. In either case simulation programs are automatically updated by RTMPOS to be consistent with changes in the database. The original database and load modules on the disk files remains unmodified. An edited database may be saved by the user either by overwriting the original disk files or by creating a new set.

At runtime the RTMPOS program control functions are used for loading the program load modules from the disk files into the desired executive in each program and enabling or disabling tasks in the executive. The use of executives and tasks in structuring a simulation program can provide the user with a great deal of flexibility at runtime. The status of program executives and tasks can be reviewed and changed by the user at any time.

The user controls the simulator in much the same manner as an analog computer using the STOP, RUN, and HOLD modes. The runtime monitoring function allows a user to receive advisories via interrupts from the simulator and to record all commands that are executed and the resulting actions that are taken during a simulation session. The advisories can be user or operating system messages or they can be requests to read data from the simulator. The messages are displayed on the message device. The read advisories are built into the simulation by the user. They are transparent to the user at runtime and are automatically serviced by special RTMPOS tasks.

A powerful feature of runtime monitoring is the self-documenting session history—a disk text file that saves all user commands and resulting prompts and messages from the RTMPOS. Any advisory messages that occur are automatically entered into the session history. They may be listed and edited using the manufacturer-supplied utilities that are resident on the FEP. Furthermore, a session history may be executed by the RTMPOS. That is, commands can be input to the RTMPOS from a session history file rather than being entered manually from the keyboard. This allows the user to quickly bring a simulation to any previous state by executing the session history from that session. The user may also create a file with the text editor which can be used to execute the routine parts of a simulation session (e.g., loading a database and the simulator load modules). After the execution of a session history is completed, control is returned to the user at the keyboard.

Simulation results are obtained by means of the read advisory. The user includes requests for sampling of argument groups in the RTMPL source program. When the simulation is running and a read advisory is encountered, an interrupt is sent out to an appropriate RTMPOS task. Each channel has its own task. The RTMPOS task transfers the argument group data from the simulator to an auxiliary memory block in the FEP. Whenever the simulator enters the STOP mode, the RTMPOS converts the data to the proper units using database information, enters the data with identification into a user-defined text file, and clears the auxiliary memory. The data are coordinated with the session history file by means of a run number and an RTMPOS comment in the session history that indicates the number of data records that were saved. The data may be listed using the FEP-resident list utility.

Once the RTMPOS has been invoked by the user, there are two ways
to exit from it. One command results in all RTMPOS tasks being permanent-
ly terminated. A different command allows the user to temporarily suspend
the RTMPOS command task, perform other tasks such as editing and listing
a text file, and then resume the RTMPOS at the point the user left off. In
the latter case, the simulator can continue to run with any advisory messages
output to keep the user informed of the simulation's progress. The current
version of RTMPOS is "menu"-driven (Cole, 1985).

16.5.5 A Simulation Example (Milner and Arpasi, 1986)

A small turboshaft engine in the 20,000-lb thrust class was modeled for
simulation on RTMPS. The model was representative of modern air-breathing
engines, yet simple enough to test the multiprocessing performance aspects
of RTMPS. Included in the model are mathematical representations of the
following: a single-centrifugal stage, five-axial-stage compressor that includes variable inlet guide vanes and variable stator vanes for the first
two stages; an annular combustor; a two-stage axial air-cooled gas generator turbine that drives the compressor rotor; compressor exit bleed air that
cools the turbine; a two-stage turbine that is uncooled and has a coaxial
drive that extends forward through the gas generator turbine and connects
to the output shaft.

The model was previously simulated on an IBM 370 mainframe, and the
results were available. These results were used to establish the validity
of simulations on RTMPS. The simulation on RTMPS was carried out in
three configurations: one-, two-, and four-processor configurations.

The single-processor mode simulation was used to check the hardware/
software functions of RTMPS and to establish a baseline results file for
comparison with the other two modes. Steady-state results from this simulation agreed well with those of the mainframe simulation, to within 0.2%. The
errors are mainly due to two factors: the difference in word sizes of the
two machines and the arithmetic mode. IBM 370 uses 32-bit data representation and floating-point arithmetic. Motorola 68000 is a 16-bit machine, and
the simulation used a scaled fraction arithmetic. To examine the dynamic
characteristics of the RTMPS, a 40-sec transient was simulated starting at
one of the four steady-state conditions. After 2 sec of steady-out period,
the engine fuel flow was ramped from its normal value of 462 to a value of
362 over a 7-sec period and was held at that value for the next 18 sec to
allow steady state. The fuel flow was then ramped to the normal value over
the next 4 sec and held for the remaining 9 sec. The transient results
agreed with those from the mainframe simulation, within 0.5%.

To conduct a two-processor simulation, the simulation equations needed
to be partitioned. To enable the partition, an estimate of the time required
to compute each simulation equation in the single-processor mode was made.
Based on this estimate, the equations were partitioned into two groups to
satisfy the following criteria: (a) Computation times on each processor
should be equal, (b) processors should not idle waiting for data from the
other processor and (c) interprocessor data transfers should be minimal.
For the example at hand, the computation times were within a few clock
cycles of being equal, there was no idle time, and there were nine interprocessor data transfers. The results agreed well with those of the single-
processor simulation, and the simulation was about 1.7 times faster.

The partitioning of the simulation equations resulted in the "critical
path" (i.e., the longest sequential computation string) residing in a single

processor, for the quad-processor case. The simulation results agreed with those of previous cases, but the simulation was about 2.1 times faster than that of the single-processor case.

Note that the simulation times did not decrease linearly as the number of processors increased. This is due to: (a) the communication overhead introduced for the dual-processor case, and (b) the critical path execution time for the quad-processor case. Partitioning the simulation equations to use all the processors concurrently is not a trivial problem, as is minimization of interprocessor data transfers. In fact, for our example, it is not possible to increase the simulation speed any further by increasing the number of processors, since the "critical simulation execution path" is already entirely on one processor in the quad-processor case.

16.6   SUMMARY

This chapter provides a description of various computer architectures. Simulation is viewed as an intensive computation task, and the multiprocessing structures that can accommodate such computations are described. The current trend of building real-time systems using microprocessors to achieve low-cost systems is illustrated with the example of Section 16.5.5. The computer architecture area has been very dynamic in recent years. New systems and hardware components are announced almost daily. Therefore, this chapter concentrated more on concepts and less on architectural details of the systems. Although the advances in hardware technology have made the design of multiprocessing architectures a reality, the lack of tools to express parallel algorithms and to detect parallelism in algorithms and programs automatically is still a problem. The parallel-processing area remains an active area for research.

REFERENCES

AD-100 Systems Manuals (1985). Applied Dynamics, Inc., Ann Arbor, MI.

Anderson, G. A. and Jensen, E. D. (1975). Computer interconnection structures: Taxonomy, characteristics and examples, *Computing Surveys* 7:197–213.

AP-120B Array Processor (1978). Floating-Point Systems, Inc., Portland, OR.

Arpasi, D. J. (1984). RTMPL—A structured programming and documentation. Utility for real-time multiprocessor simulations, in *Proceedings of Summer Computer Simulation Conference*, Boston.

Bekey, G. A. and Karplus, W. J. (1977). Computers for real-time flight simulation, Report for Computer Sciences Corp., Mountainview, CA.

Blech, R. A. and Arpasi, D. J. (1985). Hardware for a real-time multiprocessor simulator, in *Proceedings of the SCS Multiconference*, San Diego, CA, pp. 43–49.

Bowen, B. A. and Buhr, R. J. A. (1980). *The Logical Design of Multiple-Microprocessor Systems*, Prentice-Hall, Englewood Cliffs, NJ.

Charlesworth, A. E. and Gustafson, J. L. (1986). Introducing replicated VLSI to supercomputing: The FPS-164/MAX scientific computer, *IEEE Computer* 19(3), 10–23.

Cole, G. L. (1984). Operating system for a real-time multiprocessor propulsion system simulator, in *Proceedings of Summer Computer Simulation Conference*, Boston,

Cole, G. L. (1985). *Operating System for a Real-Time Multiprocessor Propulsion System Simulator—Users Manual*, NASA Lewis Research Center, Cleveland, TP 2626.

CRAY-1 Computer System Hardware Reference Manual (1977). Cray Research Inc., Minneapolis, MN.

Enslow, P. H. (1977). Multiprocessor organization, *Computing Surveys 9*, 103-129.

Flynn, M. J. (1966). Very high speed computing systems, *IEEE Trans. Computers 54*, 1901-1909.

Gilbert, E. O. and Howe, R. M. (1978). Design considerations in a multiprocessor computer for continuous system simulation in, *Proceedings of AFIPS National Computer Conference*, Montvale, NJ. pp. 385-393.

Hord, R. M. (1983). *The Illiac-IV: The First Supercomputer*, Computer Sciences Press, Rockville, MD.

Intel iPSC Data Sheet (1985). Order Number 280101-001, Beaverton, OR.

Karplus, W. J. (1977). Peripheral processors for high-speed simulation, *Simulation*, November, pp. 143-153.

Kobayashi, K. (1981). Design of real-time, interactive, parallel simulation computer, Ph.D. dissertation, Ohio State University.

Korn, G. A. (1972). Back to parallel computation: Proposal for a completely new on-line simulation system using standard minicomputers for low-cost multiprocessing simulation, *Simulation 19(2)*, 37-45.

Milner, E. J. and Arpasi, D. J. (1986). Simulating a Small Turboshaft Engine in a Real-Time Multiprocessor Simulator Environment, Unpublished Report, NASA Lewis Research Center, Cleveland, OH.

Ramamoorthy, C. V. and Li, H. F. (1977). Pipeline architectures, *Computing Surveys 9*, 61-102.

Siewidrek, D. P., Bell, C. G. and Newell, A. (1982). *Computer Structures: Principles and Examples*, McGraw-Hill, New York, NY.

Shibata, Y. (1985). Pattern recognition approach and array processing for distributed source identification in water pollution systems, Ph.D. dissertation, University of California, Los Angeles.

Shiva, S. G. (1985). *Computer Design and Architecture*, Little, Brown, Boston.

Spriet, J. A. and Vansteenkiste, G. C. (1982). *Computer-Aided Modeling and Simulation*, Academic Press, New York.

STAR-100 Computer Hardware Reference Manual (1972). Control Data Corporation, Minneapolis, MN.

Stone, H. S., ed. (1980). *Introduction to Computer Architecture*, Science Research Associates, Chicago.

Swan, R. J., Fuller, S. H., and Siewiorek, D. P. (1977). The implementation of the Cm* multimicroprocessor, in *Proceedings of AFIPS National Computer Conference*, Montvale, NJ, pp. 637-644.

Weitzman, C. (1980). *Distributed Micro/Minicomputer Systems*, Prentice-Hall, Englewood Cliffs, NJ.

Wilson, A. C. (1985). System Architect's guide to the VME Bus, *Digital Design 15*, 34-40.

Wilson, A. C. (1986). Array processors: The best way to process images, *Digital Design 16*, 47-52.

Wittmeyer, W. R. (1978). Array processor provides high throughput rates, *Computer Design 17*, 93-100.

Appendix A

Computer-Aided Control Systems
Techniques and Tools

A.1   INTRODUCTION

Up to this point, this book has mainly discussed diverse types of simulation
techniques, and indeed, simulation has become extremely important in almost
every aspect of scientific and engineering endeavor. According to Korn
and Wait (1978), *simulation is experimentation with models*. Thus, each
simulation program consists of two parts:

1.   A coded description of the model, which we call the *model representa-
     tion* inside the simulation program (notice the difference as compared
     to Chapters 1 and 2, where the term "model representation" was
     used to denote graphical descriptions such as block diagrams or sig-
     nal flow graphs)
2.   A coded description of the experiment to be performed on the model,
     which we call the *experiment representation* inside the simulation
     program

Analyzing the different types of simulation examples presented so far,
it can be realized that most of these examples independent of whether they
were discrete or continuous in nature, consisted of a fairly elaborate model
on which a rather simple experiment was performed. The standard simulation experiment is as follows: starting with a complete and consistent set
of initial conditions, the change of the various variables of the model (state
variables) over time is recorded. This experiment is often referred to as
determining the *trajectory behavior* of a model. Indeed, when the term
"simulation," as is often done, is used to denote a solution technique rather
than the ensemble of all modeling-related activities (as is done in this book),
simulation can simply be equated to the determination of trajectory behavior.
Most currently available simulation programs offer little beside efficient
means to compute trajectory behavior.
     Unfortunately, few practical problems present themselves as pure simula-
tion problems. For example, it often happens that the set of starting
values is not specified at one point in time. Such problems are commonly

referred to as *boundary value problems* as opposed to the *initial value problems* discussed previously. Boundary value problems are not naturally simulation problems in a puristic sense (although they can be converted to initial value problems by a technique called *invariant embedding* (Tsao, 1986). A more commonly used technique for this type of problem, however, is the so-called *shooting technique,* which works as follows:

1. Assume a set of initial values.
2. Perform a simulation.
3. Compute a *performance index,* e.g., as a weighted sum of the squares of the differences between the expected boundary values and the computed boundary values.
4. If the value of the performance index is sufficiently small, terminate the experiment; otherwise, interpret the unknown initial conditions as parameters, and solve a *nonlinear programming problem,* looping through 2 . . . 4 while modifying the parameter vector in order to minimize the performance index.

As can be seen, this "experiment" contains a multitude of individual simulation runs.

To elaborate on yet another example, assume that an electrical network is to be simulated. The electrical components of the network have tolerance values associated with them. It is to be determined how the behavior of the network changes as a function of these component tolerances. An algorithm for this problem could be the following:

1. Consider those components that have tolerances associated with them to be the parameters of the model. Set all parameters to their minimal values.
2. Perform a simulation.
3. Repeat 2 by allowing all parameters to change between their minimal and maximal values until all "worst case" combinations are exhausted. Store the results from all these simulations in a database.
4. Extract the data from the database and compute an *envelope* of all possible trajectory behaviors for the purpose of a graphical display.

As in the previous example, the experiment to be performed consists of many different individual simulation runs. In this case, there are exactly 2^n runs to be performed, where n denotes the number of parameters.

These examples show that simulation does not live in an isolated world. A scientific or engineering experiment may involve many different simulation runs, and many other things in between. Unfortunately, the need for enhanced experimentation capabilities is not properly reflected by today's simulation software. Although model representation techniques have become constantly more powerful over the past years, very little was done with respect to enhancing the capabilities of simulation experiment descriptions (Cellier, 1986a). Some simulation languages, such as ACSL (Mitchell and Gauthier, 1981), offer facilities for model linearization and steady-state finding. Other simulation languages, such as DSL/VS (IBM, 1984), offer limited facilities for frequency domain analysis, e.g., a means to compute the Fourier spectrum of a simulation trajectory. To our knowledge, there is not a single simulation system currently available that would offer a general-purpose nonlinear programming package for curve fitting, steady-state finding, the solution of boundary value problems, etc., as an integral part of the software, and this is only one among many tools that would be useful to have. Moreover, the few experiment enhancement tools that are currently available are often difficult to use and are very specialized, that is limited in applicability.

Whenever such a situation is faced, we, as software engineers, realize that something must be wrong with the *data structures* offered in the language. Indeed, all refinements in model representation capabilities, such as techniques for proper discontinuity handling and facilities for submodel declarations, led to enhanced *programming structures*, whereas the available data structures are still much the same as they were in 1967, when the CSSL specifications (Augustin et al., 1967) were first formalized.

When we talk about *computer-aided design software* as opposed to simulation software, it is exactly this enhanced experiment description capability that we have in mind. Simulation is no longer the central part of the investigation, but simply one software module (tool) among many others that can be called at will from within the "experiment description software," which from now on will be called "computer-aided design software." Algorithms for particular purposes will be called *computer-aided design techniques,* and the programs implementing these algorithms will be called *computer-aided design tools.* As many of the design tools are application dependent, our discussion will be restricted to one particular application, namely, the design of control systems.

Until very recently, the data structures available in computer-aided control system design (CACSD) software were as poor as those offered in simulation software. However, even the available programming structures in these software tools were pitiable. Users were led through an inflexible question-and-answer protocol. Once an incorrect specification was entered by mistake, there was no chance to recover from this error. The protocol deviated from the designed path and probably led sooner or later to a complete software crash, after which the user had lost all his previously entered data and was forced to start from scratch.

A true breakthrough was achieved with the development of MATLAB, a matrix manipulation tool (Moler, 1980). Its only data structure is a *double-precision complex matrix.* MATLAB offers a consistent and natural set of operators to manipulate these matrices. In MATLAB, a matrix is coded as follows:

$\underline{A} = [1,2,3;4,5,6;7,8,9]$

or alternatively:

$\underline{A} = [\ 1\quad 2\quad 3$
$\qquad 4\quad 5\quad 6$
$\qquad 7\quad 8\quad 9\ ]$

Elements in different columns are separated by either comma or space, whereas elements in different rows are separated by either semicolon or carriage return (CR). With matrices being the only available data structure, scalars are obviously included as a special case. Each element of a matrix can itself again be a matrix. It is, therefore, perfectly legitimate to write

$\underline{a} = [0*\underline{ONES}(3,1),\underline{EYE}(3);[-2\ -3\ -4\ -5]]$

where ONES(3,1) stands for a matrix with three rows and one column full of 1 elements; 0*ONES(3,1) is thus a matrix of same size consisting of 0

elements only. EYE(3) represents a 3 × 3 unity matrix which is concatenated
to the 0 matrix from the right, thus making the total structure now a matrix
with three rows and four columns. Concatenated from below is the matrix
[−2 −3 −4 −5], which has one row and four columns. Thus, the above
expression will create the matrix

$$
A = [\quad 0 \quad 1 \quad 0 \quad 0 \\
\qquad\quad 0 \quad 0 \quad 1 \quad 0 \\
\qquad\quad 0 \quad 0 \quad 0 \quad 1 \\
\qquad -2 -3 -4 -5]
$$

Suppose it is desired to solve the linear system

$$\underline{A} * \underline{x} = \underline{b}$$

For a nonsingular matrix A, it is known that the solution can be obtained
as

$$\underline{x} = \underline{A}^{-1} * \underline{b}$$

which, in MATLAB, can be expressed as

$$\underline{x} = INV(\underline{a}) * \underline{b}$$

or, somewhat more efficiently,

$$\underline{x} = \underline{a} \backslash \underline{b}$$

(b from left divided by a), in which case a Gaussian elimination is performed
in place of the computation of the complete inverse. With MATLAB, we
finally got a tool that allows us to learn what we always wanted to know
about linear algebra [such as: what are the EIG(a+2EYE(a)) where EYE(a)
stands for a unity matrix with the same dimensions as a, and EIG(...)
computes the eigenvalues of the enclosed expression]. In fact, such a tool
existed already for some time. It was called APL and offered much the
same features as MATLAB. However, APL was characterized by a very
cryptic syntax. The APL user was forced to learn to think in a fashion
similar to the computer that executed the APL program, which is probably
why APL never really made it into the world of applications. Cleve Moler,
on the other hand, taught the computer to "think" like the human operator.

MATLAB was not designed to solve CACSD problems. MATLAB is simply
an interactive language for matrix algebra. Nevertheless, this is exactly
the type of tool that the control engineer needs for solving his problems.
As an example, let us solve a simple LQG (linear quadratic gaussian)
regulator design problem. For the linear system

$$\frac{d\underline{x}}{dt} = \underline{A} * \underline{x} + \underline{B} * \underline{u}$$

it is desired to compute a linear state feedback such that the performance
index (PI) is

$$PI = \int_0^\infty (\underline{x}'\underline{Q}\underline{x} + \underline{u}'\underline{R}\underline{u})dt \overset{!}{=} \min$$

where $\underline{u}'$ denotes the transpose of the vector $\underline{u}$. This LQG problem can be solved by means of the following algorithm:

1. Check the controllability of the system. If the system is not controllable, return with an error message.
2. Compute the Hamiltonian of this system:

$$\underline{H} = [\begin{array}{cc} \underline{A} & -\underline{B}\underline{R}^{-1}\underline{B}' \\ -\underline{Q} & -\underline{A}' \end{array}]$$

3. Compute the 2n eigenvalues and right eigenvectors of the Hamiltonian. The eigenvalues will be symmetrical not only with respect to the real axis, but also with respect to the imaginary axis, and since the system is controllable, no eigenvalues will be located on the imaginary axis itself.
4. Consider those eigenvectors associated with negative eigenvalues, concatenate them into a reduced modal matrix of dimension 2n × n, and split this matrix into equally sized upper and lower parts:

$$\underline{V} = [\begin{array}{c} \underline{V}_1 \\ \underline{V}_2 \end{array}]$$

5. Now, the Riccati feedback can be computed as

$$\underline{K} = -\underline{R}^{-1}\underline{B}'\underline{P}$$

where

$$\underline{P} = \mathrm{Re}\{\underline{V}_2 * \underline{V}_1^{-1}\}$$

The following MATLAB "program" (file: RICCATI.MTL) may be used to implement this algorithm:

```
EXEC('contr.mtl')
IF ans <> n, SHOW('System not Controllable'), RETURN, END
[v,d] = EIG([a,-b*(r\b');-q,-a']);
d = DIAG(d); k=0;
FOR j=1,2*n, IF d(j)<0, k = k+1; v(:,k) = v(:,j); END
p = REAL(v(n+1:2*n,1:k)/v(1:n,1:k));
k = -r\b'*p
RETURN
```

which is a reasonably compound way of specifying this fairly complex algorithm. Yet, contrary to an equivalent APL code, we find this code acceptably readable.

After MATLAB had become available, it took amazingly little time until several CACSD experts realized that this was an excellent way to express control problems. Clearly, MATLAB was not designed for CACSD problems, and still a lot had to be done to make it truly convenient, but at least a basis had been created. In the sequel, several CACSD programs have evolved: CTRL-C (Systems Control, 1984; Little et al., 1984), MATRIX$_X$ (Integrated Systems, 1984; Shah et al., 1985), IMPACT (Rimvall, 1983; Rimvall and Bomholt, 1985; Rimvall and Cellier, 1985), PC-MATLAB (Little, 1985), MATLAB-SC (Vanbegin and Van Dooren, 1985). All these programs are "spiritual children" of MATLAB.

We want to demonstrate in this appendix how simulation software designers can learn from recent experiences in CACSD program development, and how the CACSD program developers can learn from the experiences gained in simulation software design.

It would be convenient if a MATLAB-like matrix notation could be used within simulation languages for the description of linear systems or linear subsystems. It would, indeed, be useful if the simulation software could apply to linear (sub)systems an integration algorithm that is more efficient than the regularly used explicit Runge Kutta, Adams-Bashforth, or Gear algorithms, e.g., an implicit Adams-Moulton technique. Linear (sub)systems could automatically be identified by the compiler through the use of a matrix notation.

On the other hand, it is true that most recent CACSD programs offer only limited simulation capabilities (e.g., applicable to linear systems only). It would be useful indeed if all our knowledge about the simulation of dynamic systems/processes could be integrated into the CACSD software. Since a design usually involves more than just simulation, but certainly simulation among other techniques, it would definitely be beneficial if a flexible interface between a CACSD program and a simulation language could be created such that powerful simulation runs could be made efficiently at arbitrary points in a more complex design study. These are some of the topics that this appendix addresses.

Note that this appendix discusses purely digital solutions only. Other simulation techniques (such as analog and/or hybrid simulation techniques) are discussed in Chapter 4 of this book. Although we shall not refer to these techniques explicitly any further, computer-aided control system design algorithms can be implemented on hybrid computers very easily. The dynamic process (that is, the model description) will then be programmed on the analog part of the computer, while the experiment description that triggers indivual simulation runs will be programmed on the digital part of the computer. The digital CACSD program will look just the same as in the purely digital solution, while the simulation program will look just the same as any other analog simulation program. For these reasons, a further elaboration on these concepts can be spared.

In the next section, a systematic classification of CACSD techniques is presented. Different techniques (algorithms) for computer-aided control system design are discussed.

In Section A.3, CACSD tools are classified. This discussion highlights the major differences between several classes of CACSD tools.

Both Sections A.2 and A.3 help to prepare for Section A.4, in which a number of currently available CACSD tools are compared with respect to features (algorithms) offered by these software systems.

After this discussion, the reader may question whether a diversification of tools as can be currently observed is truly justified and desirable. For this reason, the problem of software standardization versus software diversification is discussed in Section A.5.

In Section A.6, we show how simulations can intelligently be used within CACSD software. This section helps to throw a bridge across to other chapters in this book.

Finally, Section A.7 presents our perspective of forthcoming developments in the area of CACSD software design.

A.2 DEVELOPMENT AND CLASSIFICATION OF CACSD TECHNIQUES

Let us look briefly into the history of CACSD problems. CACSD, as we know it today, has its roots in a technology that was boosted by the needs created in World War II, when military leaders started to think about more powerful weaponry, and engineers produced answers in the form of automatically controlled, in place of manually controlled, weapon systems. (Fortunately, automatic control has since found many other nonmilitary applications as well. Nevertheless, even today, a substantial percentage of research grants in the automatic control field stems either directly or indirectly from national defense sources.)

In the beginning, that is, in the 1930s–1950s, engineers were dealing with isolated (small) continuous-time systems with one single input and one single output (so-called SISO systems). The design of these systems was (at least here in the West) predominantly done in the frequency domain, most prominently represented by people such as W. R. Evans and H. Nyquist. Most of the techniques developed were graphical in nature.

With the need to deal with more complex systems with multiple inputs and outputs (so-called MIMO systems), these graphical techniques failed to provide sufficient insight. It was among others R. Kalman who led the scientists and engineers back into the time domain, where systems were now represented in the so-called state space, that is, by sets of first-order ordinary differential equations (ODEs) in place of nth-order ODEs. For further detail, refer to Chapter 2 of this book and to the References of Chapter 2. This modern representation seemed to be better amenable to a systematic (algorithmic) design methodology. This representation was very naturally extensible from SISO system representations to MIMO system representations, and many of the algorithms (such as LQG design) would work as well on MIMO systems as on SISO systems. With the advent of modern digital computers, it was possible to apply these algorithms also to "higher" order systems (say: fifth- to tenth-order systems), whereas the previous hand computations were limited to second- to third-order systems. (This was actually the major reason for choosing a frequency domain representation in previous decades, as frequency domain design can be done manually also in the case of higher-order systems.) What is described above is the technology of the sixties.

What has happened since? What were the major breakthroughs in the seventies and in the early eighties? While research in control theory before was pretty much consolidated, now diversification took place; that is, different types of approaches were made available to tackle different types of problems.

One of the major drawbacks of the previously used technology was ironically found in the high degree of automation characterizing its algorithms. One jotted down some values, called on a subroutine, and the answer came in the form of some other numbers, parameter values, gain factors, etc. The procedure was "sterile." Somehow lacking was the intuitive feel for what was going on. What if the LQG design failed to produce acceptable answers? Where do we go from there? Often, the conclusion had been that the structure of the chosen controller was inappropriate for the task, and thus optimization of the parameters of the inappropriate controller was doomed to failure. Therefore, the control engineer had to take structural decisions in place of purely parametric ones. Unfortunately, such decisions can hardly be taken without profound insight into what is going on. None of the automated algorithms available at that time was able to determine an adequate controller structure.

For these reasons, several researchers went back to the frequency domain and came up with some new design tools [such as a generalization of the Nyquist diagram (Rosenbrock, 1969)], and some new system representations [such as some varieties of polynomial matrix representations (Wolovich, 1974; Wonham, 1974)]. Other groups decided on a different approach to tackle the same problem. Instead of producing individual solutions in the form of sets of parameter values, they tried to develop algorithms that would produce entire "fields" of output parameters as a function of input parameters, to come up with, for example, three-dimensional graphs in the parameter space. For instance, this is often done in the so-called robust controller design (Ackermann, 1980). Unfortunately, these techniques usually involve multiple sweeping, which is number-crunching at its worst. For a somewhat cheaper solution, it may be possible to employ sensitivity analysis instead (Cellier, 1986a). The newest developments in this area try to do away with numerical algorithms altogether. Instead of computing numerically one point in the parameter space at a time, the new algorithms reproduce what the engineer used to do in the paper and pencil age, that is, formula manipulation. The latest developments in nonnumerical data processing are employed to obtain algorithmically and automatically a formula that relates the desired output parameters to the given input parameters. However, these techniques are still in their infancy, and no commercial product of this kind is available as of today. The furthest developed program of this type known to us is an extensive LISP program running on an LMI computer (a special-purpose LISP machine). That program is currently under development at NASA's Johnson Space Center, but it is far from completed, and its user interface is still rather awkward (Norsworthy et al., 1985).

Another development was initiated by the need to deal with even larger systems. How do you control a large system consisting of many subsystems in an "optimal" way? Many of the previously used algorithms fail to work properly when applied to 50th- or 200th-order systems. They either compute forever, or fail to converge, or produce a result that is accurate to exactly zero significant digits ! One way to tackle this problem is to try to cut the system into smaller subsystems and find answers for those subsystems first. This led to decentralized control (Athens, 1978) and hierarchical control (Siljak and Sundareshan, 1976) schemes. More information about these approaches can be found in Chapter 6 of this book. Another answer, of course, might be to design new centralized algorithms that work better on high-dimensional systems (Laub, 1980).

The availability of reliable low-cost microprocessors led to the need to implement subsystem controllers by use of a digital computer. This stimulated research into discrete-time algorithms, as the continuous-time algorithms applied to discrete-time systems tend to exhibit very poor stability behavior.

Finally, the new age of robotic technology led to the need for developing better algorithms for the control of nonlinear systems (Asada and Slotine, 1986). The models describing the dynamics of robot movements are highly nonlinear. Most of the more refined algorithms that were previously developed work poorly, or not at all, when applied to nonlinear systems. Unfortunately, the robustness of an algorithm is often inversely proportional to its refinement; that is, the more specialized an algorithm is, the less likely will it be to handle modified situations. One way to solve this problem is to view the nonlinear time-invariant system as a linear time-variant system, and to design control algorithms for this class of problems, such as *self-tuning regulators* (Åström, 1980), *model-reference adaptive controllers* (Monopoli, 1974; Narendra, 1980), and *robust controllers* (Ackermann, 1980).

So far, we have presented the problems to be solved. Problems can be classified into single-input, single-output (SISO), multiple-input, multiple-output (MIMO), and decentralized problems. For each class of problems, a different suite of algorithms was developed to solve them. Until now, we have totally ignored the problem of numerical aptness of an algorithm, of numerical accuracy and numerical stability. The numerical behavior of algorithms is highly dependent on the system order, that is, the number of state variables describing the system/process. Almost any algorithm can be used to solve a third-order problem. Many algorithms fail when applied to a 10th-order problem, and almost all of them fail to solve a 50th-order problem correctly, and this is true for almost every single algorithm in all three classes of problem types. This fact was detected only recently. Since the late seventies, many researchers, including C. Moler, G. H. Golub, A. Laub, J. H. Wilkinson, and P. van Dooren, have designed a series of new algorithms for SISO- and MIMO-system design that are less sensitive to the system order. A major breakthrough in this area was the development of the singular value decomposition (SVD), described in Golub and Wilkinson (1976).

From now on, algorithms that work only for low-order systems will be referred to as LO algorithms, techniques that work also for high-order systems will be called HO algorithms, and finally, methods that can be used to treat very-high-order systems (mostly discretized distributed parameter systems) will be called VHO algorithms.

Let us introduce next the concepts used in the design of the different classes of algorithms more explicitly. Most of the algorithmic research done so far was concerned with algorithms based on canonical forms (Kailath, 1980). All these canonical forms, in turn, are based on minimum parameter data representations. What is a minimum parameter data representation? A SISO system can be represented in the frequency domain through its transfer function

$$g(s) = (b_0 + b_1 s + \cdots + b_{n-1} s^{n-1})/(a_0 + a_1 s + \cdots + a_{n-1} s^{n-1} + s^n)$$

where the denominator polynomial is of nth order (the system order), and the numerator polynomial is of $(n - 1)$st order. This representation is unique; i.e., the system has exactly 2n degrees of freedom (the degrees of freedom equal the number of linearly independent parameters of any unique data representation). The parameters of this representation are the coefficients of the numerator and denominator polynomials. Any set of parameter values describes one system, and no two different sets of parameters describe the same system. Any data representation sharing this property is a minimum parameter representation. The controller-canonical representation of this system can be written as

$$
\dot{\underline{x}} =
\begin{bmatrix}
0 & 1 & 0 & 0 & \dots & 0 \\
0 & 0 & 1 & 0 & \dots & 0 \\
0 & 0 & 0 & 1 & \dots & 0 \\
\dots & \dots & \dots & \dots & \dots & \dots \\
-a_0 & -a_1 & -a_2 & -a_3 & \dots & -a_{n-1}
\end{bmatrix}
\underline{x} +
\begin{bmatrix}
0 \\
0 \\
0 \\
\dots \\
1
\end{bmatrix} u
$$

$$
y = [\; b_0 \quad b_1 \quad b_2 \quad b_3 \quad \dots \quad b_{n-1}]
$$

Counting the number of parameters of this representation, it can easily be verified that this representation has exactly $2n$ parameters, and they are the same as above. Also, the Jordan-canonical representation

$$
\dot{\underline{x}} =
\begin{bmatrix}
\lambda_1 & 0 & 0 & 0 & \dots & 0 \\
0 & \lambda_2 & 0 & 0 & \dots & 0 \\
0 & 0 & \lambda_3 & 0 & \dots & 0 \\
\dots & \dots & \dots & \dots & \dots & \dots \\
0 & 0 & 0 & 0 & \dots & \lambda_n
\end{bmatrix}
\underline{x} +
\begin{bmatrix}
1 \\
1 \\
1 \\
\dots \\
1
\end{bmatrix} u
$$

$$
\underline{y} = [r_1 \quad r_2 \quad r_3 \quad r_4 \quad \dots \quad r_n] \underline{x}
$$

(assuming all eigenvalues λ_i to be distinct) has exactly the same number of parameters. This is true for all the canonical forms. For LO systems, these representations are perfectly acceptable. However, we require redundancy in order to optimize the numerical behavior of algorithms for HO systems. Thus, all algorithms that are based on canonical forms are clearly LO algorithms.

HO algorithms can be obtained by sacrificing this "simple" system representation through the introduction of redundancy. These new system representations contain more than $2n$ parameters with linear dependencies existing between them. This redundancy can now be used to optimize the numerical behavior of control algorithms by balancing the sensitivities of the parameters (Laub, 1980). Some of the better HO algorithms are based on Hessenberg representations (Patel, 1984).

VHO systems (that is, systems of higher than about 50th order) typically result from discretization of distributed parameter systems. Most of the algorithms developed for this class of systems until now exploit the fact that, in general, VHO systems have sparsely populated system matrices. Thus, algorithms, have been designed that address matrix elements through their indices. Careful bookkeeping ensures that only elements that are different from zero are considered in the evaluations. These so-called *sparse matrix techniques* have a certain overhead associated with them. Thus, they are not cost effective for the treatment of LO systems, and even many HO systems are handled more efficiently by the regular algorithms.

As a rule of thumb, sparse matrix techniques become profitable for systems of higher than about 20th order.

Contrary to the above-described algorithms for HO problems, sparse matrix techniques do not influence the numerical behavior of the involved algorithms, but only their execution time. Therefore, the numerical problems discussed for the case of HO systems remain the same. (In most of the published papers discussing VHO problems, sparse matrix techniques have been applied to one or another of the classical canonical forms.) Unfortunately, the introduction of redundancy also reduces the sparsity of the system matrices and eventually annihilates it altogether. Therefore, these two approaches are in severe competition with each other. More research is needed to find a solution to this serious problem.

Most of the research described so far was concerned with time domain algorithms. It has often been said that frequency domain operations are numerically less stable than time domain operations. It is our opinion that this statement is incorrect. It is not the frequency domain per se that makes the algorithms less suitable, it is the data representation currently used in frequency domain operations that has these undesirable effects. As previously discussed, if one wants to minimize the numerical sensitivity of an algorithm, one has to balance the sensitivities of the system parameters; i.e., each output parameter should be about equally sensitive to changes in the input parameters (Laub, 1980). Also, the sensitivities of algorithmic parameters should be balanced. In the time domain, this has been achieved by the process of orthonormalization, by operating on Hermitian forms (Golub and Wilkinson, 1976). In the frequency domain, it is less evident how the balancing of sensitivities can be achieved. If we represent a polynomial through its coefficients, even the evaluation of the polynomial at any value of the independent variable with a norm much larger or much smaller than 1 will lead to extremely unbalanced parameter sensitivities. Let us consider the polynomial

$$q(s) = a_n s^n + a_{n-1} s^{n-1} + \cdots + a_1 s + a_0$$

If this polynomial is evaluated at $s = 0$, obviously the only parameter that has any influence on the outcome is a_0; that is, a_0 is sensitive to this operation, while all other parameters are not. However, if we evaluate the polynomial at $s = 1000$, obviously a_n exerts the strongest influence, while a_0 can easily be neglected. This problem disappears when we represent the polynomial through its roots:

$$q(s) = k(s - s_1)(s - s_2) \cdots (s - s_n)$$

Here, the parameters are (k) and $(s_1 \ldots s_n)$ instead of $(a_0 \ldots a_n)$. However, if we want to add two polynomials, we won't get around to (at least partially) defactorize the polynomials and refactorize them again after performing the addition. The processes of defactorization and refactorization have badly balanced sensitivities and are thus numerically harmful.

Traditionally, these were the only two data representations considered, and both are obviously unsatisfactory. There is little we can do to improve the numerical algorithms based on these data representations as both are minimum parameter representations. However, we have other choices. For instance, it is possible to represent a polynomial through a set of supporting

values. Let us evaluate q(s) at any (n + 1) points. If we store these
(n + 1) values of s together with those of q(s), we know that there exists
exactly one polynomial of nth order that fits these (n + 1) points. We can
"reconstruct" the polynomial at any time (that is, find its coefficients), and
this, therefore, gives rise to another data representation (Sawyer, 1986).
If we choose more points, we can make use of redundancy and reconstruct
the polynimial by regression analysis, reducing the numerical errors involved
in this computation. The basic operations (addition, subtraction, multiplication, division) all become trivial in this data representation if all involved
polynomials are evaluated over the same base of supporting values (we just
apply them to each data point separately), and as most algorithms are based
solely on repeated application of these basic operations, they can also be
performed easily within this data representation. The redundancy inherent
in this data representation can eventually be used to balance parameter
sensitivities by selecting the supporting values (values of s) carefully. A
preliminary study (Sawyer, 1986) indicates that the best choice might be to
place the supporting values equally spaced along the unit circle. More research is still needed, but we feel that this approach could possibly lead to
a breakthrough in the numerical algorithms for frequency domain operations.

To summarize this discussion, CACSD techniques can be classified in
several ways: techniques for SISO, MIMO, and decentralized systems; techniques for frequency versus time domain operations; techniques for continuous-
time versus discrete-time systems; techniques for linear versus nonlinear
systems; and finally, techniques for low-order, high-order, and very-high-
order systems.

Another classification distinguishes between user-friendly and non-user-
friendly algorithms; user-friendly algorithms allow us to concentrate on
physical design parameters rather than on algorithmic design parameters.
As a typical example of user-friendly algorithms, we may mention the
variable-order, variable-step integration algorithms which enable us to
specify the required accuracy (a physical design parameter) as opposed to
the integration step length and order (which are algorithmic design
parameters).

Finally, one should distinguish between numerical and nonnumerical
algorithms. Nonnumerical algorithms make use of formula manipulations,
and "reasoning" techniques that are usually connoted as artificial intelligence
(AI) techniques. Since none of the CACSD programs discussed in Section
A.4 makes extensive use of such techniques, we shall save a more intimate
discussion of AI techniques for the "outlook" section (A.7).

A.3   DEVELOPMENT AND CLASSIFICATION
OF CACSD TOOLS

Although control theory as we know it today is a child of the early part of
this century, computer-aided control system design (CACSD) really did not
start until 1970. At that time, it would take roughly half a day just to
find the eigenvalues of a matrix, as this involved the following procedure:

1.   Develop a program to calculate eigenvalues by calling a library sub-
     routine with about six call parameters (1/2 hr).
2.   Walk to the computer center to prepare the data input (20 min).
3.   Wait for a card puncher to become available (1/2 hr).
4.   Prepare input data (10 min).

5. Submit card deck to input queue, and wait for output to be returned (turnaround time roughly 1 hr).
6. Correct typos after waiting for another card puncher to become available, and resubmit card deck; wait again for output (90 min).
7. Walk back to office (20 min).

The solution of a true control problem (e.g., the simple LQG design problem described above) took considerably longer time, possibly as much as 1 or 2 weeks. No wonder most colleagues detested the computer at that time and preferred to specialize on other topics that did not require any involvement in this denervating process.

Around 1972, the writers undertook the effort to go around and ask colleagues in the department not to throw away their control programs any longer (after they were done with a particular study), but rather document their subroutines and hand them over for inclusion in a "control library" to be built. By 1976, an impressive (and somewhat formidable) set of (partially debugged) control algorithms had been collected (Cellier et al., 1977). At this time, we decided to ask colleagues from other universities to join in the effort and share their control routines and libraries with us as well. We started the PIC service, a program information center for programs in the control area, and circulated a short newsletter twice a year providing information in the form of a "who has what" in control algorithms and codes. Meanwhile, as first computer terminals became available to us, and using our control library, we were able to reduce the time needed to solve most (simple) control problems to 1 or 2 days of work.

At that time, it was considered important to work toward reducing further the turnaround time by creating an interactive "interface" to our control library. This required conversion of the program to a PDP 11, as the CDC machine of the computer center could be used for batch operation only. Clearly, the interface was meant to be a relatively small add-on to our library, and most of our effort and time were spent on improving the control subroutines themselves. Nevertheless, this activity resulted in INTOPS (Agathoklis et al., 1979), one of several interactive control system design programs made available around the same time. The first generation of true CACSD programs was, however, very limited in scope. To keep the interface simple, the programs were strictly question-and-answer driven, with the effect that they were almost useless for research. True research problems simply do not present themselves in the form of classroom examples that follow a prepaved route as foreseen by the developers of the CACSD software. INTOPS proved very useful for undergraduate control education though. Suddenly, the use of computers to many became real fun. However, as a research tool, INTOPS failed to provide the necessary flexibility, and we became convinced that a true full-fledged programming language was required for that purpose. Unfortunately, such a language could no longer be considered as a small and inexpensive add-on to the control library, and we wondered what could be done about it.

In the fall of 1980, K. J. Åström and G. Golub undertook the commendable effort to bring recognized numerical analysts and control experts together in the first Conference on Numerical Techniques in Control to be ever held. On this occasion, we met with Cleve Moler, who demonstrated his newly released MATLAB software. It took us only minutes to realize the true value of this instrument for our task. When we returned to Zurich, we implemented MATLAB first on a PDP 11/60, and a short while later on the freshly acquired VAX 11/780. Within 1 year, MATLAB became the single most often used program on that machine (which belongs to the department of electrical engineering). Students were able to learn the usage of this tool within half an hour, and suddenly, also the researchers became interested in our "gadgets." MATLAB was fully command-driven.

An often-heard criticism of command-driven languages is that they are too complicated for the occasional user to use. Who can remember all those commands and their parameters except someone who uses the tool on a daily basis? It was simply not true. Our students were enchanted, and they found MATLAB actually much easier to use than the question-and-answer-driven INTOPS program. Extensive interactive HELP information is available to aid in the use of any particular function, and this proved completely satisfactory to our users.

Notice that MATLAB really was not designed to be a CACSD tool at all. There are many shortcomings of MATLAB for our purpose. These were summarized as follows (Cellier and Rimvall, 1983):

1. The programming facility (EXEC-file) of MATLAB is insufficient for more demanding tasks; EXEC-files have no formal arguments; EXEC-files cannot be called as functions but only as subroutines; WHILE, FOR, and IF blocks cannot be properly nested; there is neither a GOTO statement nor an (alternative) loop exit statement available; and the input/output capabilities of EXEC-files are too limited.

2. The SAVE/LOAD concept of MATLAB is insufficient as this immediately results in large numbers of files that are difficult to maintain in an organized fashion. A true database interface would be valuable. Moreover, the user wants the possibility to interface data produced/used by his own programs with MATLAB.

3. Control engineers prefer results in graphical form. The output facilities offered by MATLAB are insufficient in every respect. A database interface would at least soften this request, as it would allow the use of a separate stand-alone program, outside MATLAB, to view data produced by MATLAB graphically.

4. MATLAB does not lend itself easily to operations in the frequency domain.

5. Many control systems call for nonlinear controllers (e.g., windup techniques for treatment of saturations, adaptive controllers for time-varying systems, etc.). MATLAB does not provide for a mechanism to describe nonlinear systems.

6. A library of good and robust control algorithms is needed. (This final request is actually the one that is easiest to satisfy.)

In the sequel, a number of CACSD programs were made available that provide answers to one or several of the above demands. These (and others) are reviewed in the following section of this appendix.

To summarize this section: CACSD tools can be classified into sub-program libraries versus integrated design suites. The first generation of CACSD tools was of the former type; the more recent ones are mostly of the latter type. This new type of CACSD tools can be further classified as either comprehensive design tools or design shells. The former type tries to provide algorithms that handle all imaginable control situations. This may eventually result in very large programs offering many different features; KEDDC (Schmid, 1979, 1985) is an example of this type. The

design shells type provides an open-ended operator set that allows the user
to code his own algorithms within the frame of the CACSD software; MATLAB
(Mohler, 1980) is an example of that type. Of course, a combination of
these two categories is possible (and probably most useful) to the control
engineer.

CACSD programs can, furthermore, be either batch-operated, or fully
interactive, or both. The interactive mode is useful for a quick analysis
and understanding of what is going on in a particular project. However,
there are many control design problems, such as optimal design of nonlinear
systems, that call for an extensive amount of number crunching. Those
problems are best executed in batch mode.

CACSD program can be either code-driven or data-driven or a combination of the two (e.g., by use of incremental compiler techniques). Code-
driven programs are compiled programs that implement their algorithms and
operators in program code. They are faster executing, but they are less
flexible and less easy to augment. On the other hand, data-driven programs
implement algorithms and operators as data statements which are interpreted
during program execution. They are powerful tools for experimentation, but
not necessarily for production. It is usually a good idea to develop a new
CACSD software first as a data-driven program. Later, once the features
and format of the new software are stabilized, it can be reimplemented as a
code-driven program for improved efficiency. Compiler-compilers can
eventually be used to (at least partially) automate the step from the data-
driven to the code-driven implementation.

The user interface of CACSD programs can be either question-and-
answer-driven, command-driven, menu-driven, form-driven, graphics-driven,
or window-driven. The user of the program is asked questions to determine
what needs to be computed next. Thus, the program flow is completely predetermined. This type of user interface is easiest to implement, but it is
inflexible and probably not very useful in a research environment.

Newer CACSD programs are often command-driven. Here, the initiative
stays completely with the user. The CACSD program sends a "prompt" to
the terminal indicating its readiness to receive the next command in just
the same manner as an interactive operating system (e.g., on a PDP 11 or
on a VAX) would. In fact, an interactive operating system is nothing but
a command-driven interactive program. Turning the argument over,
command-driven CACSD programs can also be viewed as special-purpose
operating systems. One disadvantage of this type of user interface is the
need to remember what commands are available at every interface level.
This problem is today mostly remedied by providing an extensive interactive
HELP facility.

Another alternative is to use a menu-driven interface. Here, the
CACSD program displays a menu of the currently available commands on the
screen instead of just sending a prompt. It then waits for the user to
pick one of the items on the list, normally by use of a pointing device such
as a cross-hair cursor or a Swiss mouse. This interface type is quite
easy to implement, and it can be very powerful. One of its major drawbacks
lies in the amount of information that must be exchanged between the program and the user. This interface type is simply too slow; moreover, pointing devices have not yet been standardized, making the software terminal-
dependent.

A form-driven interface is most profitably used during the setup period
of the CACSD program for supplying (or modifying) default values for

large numbers of defaulted parameters of more intricate CACSD commands or operators. Here, the screen is split into separate alphanumeric fields. Each field is used to supply one parameter value. The user can jump from one field to the next to supply (override) parameter values. This interface requires a direct addressing mode to position the alphanumeric curser on the screen. Although there meanwhile exists an ANSI standard for this task, many hardware manufacturers already offered such a feature when the standard became available and refused to modify their hardware and system software to comply with the new standard. A laudable exception is DEC, which adopted the new ANSI standard when switching from the VT52 series terminals to the VT100 series terminals. Owing to these hardware dependency problems, most portable CACSD programs do not exploit this interface type either.

A graphics-driven interface was originally used to display results from a CACSD analysis such as a Bode diagram or a simulation trajectory in a graphical form (and this is about all that can be done on a mainframe computer with a serial asynchronous user interface). However, one obstacle has always been the high degree of terminal dependency of any graphics solution. One way to overcome this problem was to employ a graphics library providing for a large variety of terminal drivers to be placed between the CACSD software and the terminal hardware. In the past, several such libraries have been developed (DISSPLA, DI-3000, etc.). Unfortunately, all these commercially available libraries were expensive, and no true standard existed. Recently, the graphics kernel system (GKS) has been accepted as a standard (ANSI, 1985), and this will certainly drive the prices for GKS implementations down. Fancy graphics, however, call for high-speed communication links. Meanwhile, special-purpose graphics workstations have been developed (e.g., APOLLO Domain and SUN) that provide the necessary speed for enhanced graphics capabilities.

A first generation of simulation programs was recently made available that offer a true animation feature. A mechanism is provided to synchronize the simulation clock with real time, and the user can watch results from a simulation either on-line (that is, while the simulation is going on) or off-line (that is, driven from the simulation database), in a true relation to real time (either slower, equal, or even faster than real time). For an increased feeling of reality, the color graphics screen is divided into a static background picture and an overlaid dynamic foreground picture on which the simulation results are displayed. TESS (Pritsker & Assoc., 1986) and CINEMA (Systems Modelling Corp., 1985) are two such programs. Some flight simulators use a "wallpaper" concept to make the background picture even more realistic. Polygons can now be filled with patterns that represent a blue sky with slight haziness and a few fluffy clouds, or a green meadow with flowers and some trees. Here, the background picture is partly dynamic as well, fed from a three-dimensional database, and a projection program automatically calculates the currently visible display (Evans, 1985).

Graphical input has also become a reality. Control circuits can be drawn on the screen by means of block diagrams, which then are automatically translated by a graphical compiler into a coded model representation. $MATRIX_x$ (Integrated Systems, 1984, Shah et al., 1985) already provides this feature when operated from a SUN workstation (the program module implementing this feature is called SYSTEM-BUILD). The most fancy implementation, however, is provided in HIBLIZ (Elmqvist, 1982; Elmqvist and Mattson, 1986). This program uses a virtual screen concept similar

to the one used in a modern spreadsheet program. The virtual screen is
a portion of memory that maintains the entire graph. The physical window
can be "moved" over the virtual screen such that only part of the total
graph is visible at any one time. A zoom feature is provided to determine
the percentage of the virtual screen to be depicted on the physical screen.
The program is hierarchical. Breakpoints are used to determine the amount
of detail to be displayed. In a typical application, when the entire virtual
screen is made visible, a box may be seen containing a verbal description
of the overall model. Once the user starts to zoom in on the model, a break-
point is passed where the previously visible text suddenly disappears and
is replaced by a diagram showing a couple of smaller boxes with interconnections between them. When the user zooms in on one of these (so far
empty) boxes, a new text may suddenly appear that describes the model
contained in this box, and so on. At the innermost level, boxes are described either through sets of differential equations, a table, a graph, a
transfer function, or a linear system description. The connections (paths
between boxes) are labeled with their corresponding variable names placed
in a small box located at both ends of the path. Pointing to any of these
variables, all connections containing this variable are highlighted. During
compilation, an arrow points to the part of the graph that is currently
being compiled. During simulation, the user can zoom in on any path until
the "small box" containing the names of variables in this path becomes
visible. Pointing to any of these variables now, the user immediately obtains
a display of a graph of that variable over time.

A window-driven interface permits splitting the physical screen into
several logical windows. Each window is now associated with one logical
unit in just the same manner as different physical devices used to be.
Each window by itself can theoretically be alphanumerical or graphical,
question-and-answer-driven, command-driven, menu-driven, or form-driven.
Thus, the window interface is actually at a slightly different level of abstraction than the previously described interface types. Windows can often
be overlaid. In that case, the most recently addressed window will automatically become the top window that is totally visible. On some occasions
(e.g., the AMIGA operating system), windows are attached to a logical
screen. The concept here is to allow multiple screens that can be pulled
down on the physical screen in a fashion similar to a roll shutter. In
practice, window management calls for high-resolution bit-mapped displays
(at least 700 by 1000 pixels).

The different interface modes described above are by no means in-
compatible with each other. We recently experimented with combinations of
several interface types. IMPACT (Rimvall, 1983; Rimvall and Bomholt,
1985; Rimvall and Cellier, 1985) is largely command-driven. However, in
IMPACT, an extensive QUERY facility is available that goes far beyond the
interactive HELP facility offered in previous programs. By use of the
QUERY feature, the user can obtain guidance at either the individual command level or the entire session level; thus, one can decide on an almost
continuous scale at which level of guidance to operate the software (with
the pure question-and-answer-driven mode as the one extreme and the pure
command-driven mode as the other). A form-driven interface is being
provided for particular occasions, e.g., to determine the format of graphs
to be produced by IMPACT. A window-driven interface is provided for the
management of multiple sessions. Multiple sessions are created by use of
a SPAWN facility that works similarly to the one provided in VAX/VMS.

However, our SPAWN facility goes far beyond that of VMS. At any instant, even while entering parameters to a function, the user can SPAWN a new subprocess as a scratchpad for intermediate computations.

These interface discussions do not pertain to CACSD programs alone. In fact, they are crucial considerations in any interactive program. The most modern operating systems experiment with precisely the same elements. For instance, the operating system of the MacIntosh can be classified as a window-driven graphical operating system where the windows themselves are sometimes menu-driven and sometimes form-driven.

A.4   CACSD TOOLS—A SURVEY

As discussed above, a CACSD program can offer a variety of interfaces. There are also different types of design problems to be solved, and a CACSD program may be either general or specialized. Thus, it is important to select the right tool for the problem to be solved. The available CACSD programs are by no means uniform.

Therefore, we gathered information about some of the currently available CACSD programs. We developed a list of features using the categories described above and mailed it to the producers of all CACSD programs we were aware of. There were at that time roughly 40, and half of them responded. Their answers are listed in Table A.1. On the left, the features are listed; thus each feature occupies one row. On the top, the different CACSD programs are shown; thus each program occupies one column. A 0 entry denotes the fact that the feature is not available in the particular CACSD program, a 1 entry denotes availability, and a 2 entry denotes particular strength. In the last row of the table, the accumulated "score" of each CACSD program is presented.

In Table A.1, whenever there existed several versions of one program, we listed the features of the strongest version. For example, MATRIX$_X$, (Integrated Systems, 1984; Shah et al., 1985) has considerably stronger power when executed from a SUN workstation than when operated through a VT100 on a VAX system. The features listed are consequently those of the workstation implementation. The mainframe version has substantially fewer accumulated points and is roughly comparable to CTRL-C or IMPACT (two other mainframe programs). As another example, CTRL-C (Systems Control, 1984; Little et al., 1984) offers only a few features for the treatment of nonlinear systems by itself, but there exists a strong link between CTRL-C and the simulation language ACSL (Mitchell and Gauthier, 1981), making it possible to run ACSL simulations under the supervision of CTRL-C. Thus, CTRL-C is used for the experimental design, while ACSL is utilized for individual simulation runs. The features listed under the heading "CTRL-C" are consequently those of the combined software system CTRL-C/ACSL.

Columns 2–7 list features of MATLAB and its "spiritual children." These programs are actually all very similar in nature and therefore easy to compare. However, the difference in power between the various programs is dramatic. As can be seen from Table A.1, the original MATLAB (Moler, 1980) accumulated 39 points only (and is thus the second "weakest" CACSD program listed), whereas the "strongest" of these programs, MATRIX$_X$ (Integrated Systems, 1984; Shah et al., 1985), accumulated 149 points (more than any other of the listed programs). Also CTRL-C (Systems Control,

1984; Little et al., 1984) and IMPACT (Rimvall, 1983; Rimvall and Bomholt, 1985; Rimvall and Cellier, 1985) are very strong programs, and even the "small" PC-MATLAB (Little, 1985) is amazingly powerful. The developers of MATLAB-SC (Vanbegin and Van Dooren, 1985) concentrated on the algorithmic aspects, implementing some powerful, numerically stable algorithms for particular control problems, but this is only one of the six previously mentioned shortcomings of MATLAB. Therefore, MATLAB-SC accumulated relatively few additional points over MATLAB itself and cannot be considered a full-fledged CACSD tool yet. All the other programs (in Columns 2–7) added strength in face of the six deficiencies in one form or the other.

MATRIX$_x$, CTRL-C, and MATLAB-SC are true descendants of MATLAB, in that they started off from the (FORTRAN-coded) MATLAB program and added enhancements where they saw fit.

PC-MATLAB was completely recoded in C, making its code somewhat easier to read and maintain. PC-MATLAB was developed for operation on the IBM-PC and its lookalikes, which actually was one of the reasons for recoding PC-MATLAB from scratch (the C-compiler available for the PC is far better than the FORTRAN-compiler). Since the time of our questionnaire, two additional implementations of PC-MATLAB were made available, namely, one for the MacIntosh and the other for the SUN workstation, and therefore, the name of the software was changed from PC-MATLAB to PRO-MATLAB (Little et al., 1986). Also since the time of our questionnaire, a tight link between PRO-MATLAB and the simulation language SIMNON (Elmqvist, 1975, 1977) was established. With this new link, PRO-MATLAB passes the 100-point level, making the software comparable in power to CTRL-C and IMPACT.

Also, IMPACT was completely recoded using ADA. IMPACT's particular strength lies in its powerful data structures, which will be discussed later.

One more derivative of MATLAB, called M (Gavl and Herget, 1984), has also been completely recoded using OBJECTIVE-C. This implementation is very powerful. M may be of particular interest to universities, as it is the only one of MATLAB's derivatives that is available as public domain software (as the original MATLAB was). However, its developers do not want to advertise M too highly before its implementation has come to an end, and thus, they did not respond to our questionnaire. Therefore, M's features are regrettably not contained in Table A.1.

Before we continue with the analysis of other programs listed in Table A.1, let us discuss some of the concepts that went into the design of MATLAB's derivatives. Why, for instance, did we decide to reimplement IMPACT from scratch rather than making use of the code already available in MATLAB? It is our experience that new program structures can be easily added on to an existing code, while it is almost hopeless to try to add new data structures to it. Thus, one is stuck with a code once one decides that a major review of the available data structures is needed. As the complex double-precision matrix is the only available data structure in MATLAB, the "true" children of MATLAB had to squeeze any new data types into this data representation. For example, in CTRL-C, polynomials such as p(s)

$$p(s) = 5s^3 + 8s^3 - 2s + 7$$

are represented through their coefficients coded as a vector.

$$\underline{P} = [5 \quad 8 \quad -2 \quad 7]$$

Table A.1 CACSD Program Features

	2 MATLAB	3 CTRL-C	4 IMPACT	5 MATLAB-SC	6 MATRIX$_x$	7 PC-MATLAB	8 CATPAC	9 CC	10 KEDDC	11 L.A.S.	12 LUND	13 TRIP	14 WCDS/DSC	15 ICARE	16 MADPAC	17 PAAS	18 SANCAD	19 SSPACK	20 SUBOPT	21 SUNS
GLOBAL CLASSIFICATION																				
Continuous Systems	0	2	2	1	2	2	1	2	2	1	2	2	2	2	0	1	1	0	2	2
Discrete Systems	0	2	2	1	2	2	1	2	2	1	2	2	2	0	2	0	1	2	0	2
Time Domain	0	2	2	1	2	2	1	2	2	1	2	2	2	2	2	2	0	2	2	0
Frequency Domain	0	2	2	0	2	2	1	2	2	1	2	2	2	2	2	2	1	0	0	2
Single-Input/Single-Output	0	2	2	0	2	2	1	2	2	1	2	2	2	2	0	2	1	2	2	2
Multivariable	0	1	2	0	1	2	1	2	2	1	2	2	2	2	2	2	1	2	2	2
Nonlinear Systems	0	2	1	0	2	1	0	0	1	0	1	0	0	0	0	0	1	0	0	2
Adaptive Control	0	0	0	0	1	1	1	0	2	1	1	1	0	0	0	0	2	1	1	0
Identification	0	1	1	0	1	1	1	0	2	0	2	0	0	0	2	0	0	2	0	0
Real-Time Interface	0	0	0	0	2	0	1	0	2	0	2	0	0	0	0	0	0	0	0	0
USER INTERFACE																				
Interactive Operation																				
Input																				
Question-and-Answer	0	0	1	0	1	0	0	1	2	1	1	0	0	1	1	2	1	2	2	1
Menu-/Form-driven	0	0	2	0	2	0	0	0	1	0	0	0	0	1	2	2	1	2	0	1
Object oriented	0	0	0	0	1	0	0	1	0	0	0	0	0	0	0	0	1	0	0	0
Command-driven	2	2	2	2	2	2	1	2	1	0	2	2	2	2	2	2	0	1	0	0

#	Feature														
29	Full interactive language	1	2	2	1	0	2	2	0	0	0	0	0	0	0
30	Graphical input (Model spec.)	0	0	0	0	1	2	0	1	2	1	1	1	0	0
31	Window-handler	0	0	2	0	0	2	1	0	2	1	2	2	0	0
32	Parallel sessions	0	1	0	1	0	0	2	0	2	1	2	2	0	0
33	Graphical Output														
34	Print-plots	0	2	0	2	1	0	1	1	1	0	0	2	1	0
35	"Pen" plots	0	2	2	2	1	1	2	1	0	2	2	2	2	1
36	Frequency plots (Bode, etc.)	0	2	2	2	2	2	2	1	0	1	2	2	0	1
37	Time transients	0	2	2	2	2	2	2	2	0	1	2	2	0	0
38	General plots	0	2	1	1	2	0	2	2	1	0	2	2	1	1
39	3-dimensional plots	0	1	0	0	0	0	0	0	0	0	0	0	0	0
40	Split-screen graphics	0	2	2	2	0	2	0	0	0	1	2	0	0	0
41	Window handler	0	0	0	0	0	0	2	0	0	1	0	0	0	0
42	Color graphics	0	0	0	2	2	2	0	2	2	2	0	2	0	2
43															
44	Batch operation														
45	Callable from other program	1	0	0	0	0	0	0	0	1	0	0	0	0	0
46	Detached execution	0	1	1	0	1	0	0	0	0	1	0	0	0	0
47															
48	Extendability														
49	Interactive macro definition	1	2	2	2	2	0	2	0	1	0	0	0	0	0
50	Interactive subprograms	0	2	2	2	2	0	1	0	0	0	0	0	0	0
51	Macro/subprogram libraries	0	2	0	2	1	2	1	2	1	0	1	0	1	0
52	Program interface new algor.	1	2	1	0	0	1	1	0	1	0	0	1	0	1
53	Data-exchange over data-base	0	0	0	1	0	0	1	0	0	0	0	0	1	0
54															
55	Documentation														
56	User's Manual	2	2	2	2	2	2	2	1	1	1	0	1	2	1
57	Machine readable	2	0	0	0	0	0	0	0	0	0	0	0	0	0
58	Tutorial	0	1	1	1	2	1	0	1	1	0	1	1	2	0
59	On-line help	1	2	0	2	1	1	1	1	1	1	0	1	1	1

Table A.1 (Continued)

	2 MATLAB	3 CTRL-C	4 IMPACT	5 MATLAB-SC	6 MATRIX$_x$	7 PC-MATLAB	8 CATPAC	9 CC	10 KEDDC	11 L.A.S.	12 LUND	13 TRIP	14 WCDS/DSC	15 ICARE	16 MADPAC	17 PAAS	18 SANCAD	19 SSPACK	20 SUBOPT	21 SUNS
60																				
61																				
62																				
63																				
64																				
65																				
66 DATA/PROGRAM STRUCTURES																				
67																				
68 Numerical Structures																				
69 Matrix environment	2	2	2	2	2	2	1	1	1	1	1	2	1	2	1	0	0	2	0	0
70 Polynomial matrices	0	0	2	0	1	0	0	—	—	—	—	0	2	2	0	0	1	2	0	0
71 Transfer-function structures	0	1	2	2	—	0	1	2	—	—	—	2	2	2	0	2	0	0	0	1
72 State-space representation	1	1	2	2	2	1	1	2	—	—	2	2	—	2	1	0	1	2	2	0
73 System topologies	0	1	—	0	2	—	0	—	—	0	—	0	—	0	0	0	1	0	0	0
74																				
75 Symbolic structures																				
76 General formula descriptions	0	0	0	0	—	0	0	1	0	0	0	0	1	0	0	0	1	1	0	0
77																				
78 Nonnumeric modelling structures																				
79 Nonlinearities	0	1	1	0	2	0	1	0	0	—	—	0	0	0	0	0	2	—	0	—
80 Nonlinear system descriptions	0	2	1	0	2	0	1	0	1	1	—	2	0	0	0	0	2	1	0	2
81 Continuous	0	2	1	0	2	0	0	0	—	—	—	2	0	0	0	0	2	0	0	2
82 Discrete	0	—	1	0	—	—	—	0	—	—	—	2	0	0	0	0	2	1	2	2
83 Sampled data systems	0	2	2	0	2	—	0	0	—	—	—	0	0	0	0	0	2	2	0	2
84 Partial differential equations	0	0	—	0	0	—	—	0	0	0	—	0	0	0	0	0	0	0	0	2
85 Hierarchical model definition	0	1	1	0	2	0	0	1	0	0	1	—	0	0	0	0	1	0	0	0
86																				
87 Command structures	2	2	2	2	2	2	0	0	0	0	1	0	1	0	0	0	0	0	0	0
88 Mathematical notation	2	2	2	2	2	2	0	0	0	0	0	0	0	0	0	0	0	0	0	0

#	Feature	Values
89	Structural elements (FOR, IF ...)	0 0 0 0 0 0 0 − 1 2 2 2 1 2 0 2 2 2 2 1 2 1 2
90	Macro execution	0 0 0 1 0 0 0 0 2 1 2 0 2 1 0 1 2 2 0 0 1 1 2
91	Interpretation	1 0 0 0 0 0 0 0 2 2 2 0 2 1 1 2 2 2 2 1 2 1 2
92	Direct executing code	2 1 0 0 0 1 0 1 0 1 2 0 2 2 1 0 2 2 2 1 2 2 1
93	Automatic compilation	0 0 0 0 0 0 0 0 0 0 0 0 0 0 0 0 0 0 0 0 0 0 0
94		
95	User-defined structures	0 0 0 0 0 0 0 0 2 1 2 2 2 1 0 2 2 2 2 0 1 0 1
96	Clustering of predefined struct.	0 0 0 1 0 1 0 1 2 2 2 1 2 1 0 2 2 2 2 0 1 0 0
97	New numeric structures	0 0 0 0 0 0 0 0 0 0 0 0 0 0 0 0 0 0 0 0 0 0 0
98	New symbolic structures	0 0 0 0 0 0 0 0 0 0 0 0 0 0 0 0 0 0 0 0 0 0 0
99		
100		
101	**BUILT-IN ALGORITHMS**	
102		
103	*Mathematical Routines*	
104	Basic linear algebra routines	2 2 2 2 2 2 2 2 2 2 2 1 2 2 2 2 2 2 2 1 2 2 2
105	Eigenvalue computations	2 2 2 2 2 2 2 2 2 2 2 1 2 2 2 2 2 2 2 0 2 2 2
106	Decompositions, transformat.	2 2 2 1 2 2 2 2 2 1 2 0 2 2 2 2 1 2 2 0 2 2 2
107	Polynomial operations	0 2 1 1 1 1 2 1 1 1 2 1 1 2 1 1 1 2 1 0 1 0 0
108	Transfer-function operations	0 1 1 0 1 1 2 1 1 1 2 1 1 2 1 1 2 1 1 0 0 0 0
109	Nonlinear equation solving	0 0 0 0 0 0 0 0 0 1 0 1 0 1 0 1 1 0 0 0 0 0 0
110		
111	*General data analysis*	
112	Random number generator	0 1 2 1 1 1 1 1 1 1 1 0 1 1 1 1 1 1 0 1 0 1 0
113	Statistical analysis	0 0 0 0 0 0 0 0 0 0 0 0 0 0 0 0 1 0 0 0 0 0 0
114	Fourier-transforms	1 1 1 0 0 0 0 0 1 1 0 1 0 0 1 0 1 0 0 0 1 0 0
115	Filter design	0 1 0 0 1 0 1 0 1 0 0 0 0 0 1 0 1 0 0 0 0 0 0
116		
117	*Symbolic calculations*	
118	Formula manipulation	0 0 0 0 0 0 0 0 0 0 0 0 1 0 0 0 0 0 0 0 0 0 0
119	Parameterized studies	0 1 0 0 0 1 1 0 0 1 0 0 0 0 0 0 0 0 0 1 0 0 1

Table A.1 (Continued)

	1	2 MATLAB	3 CTRL-C	4 IMPACT	5 MATLAB-SC	6 MATRIX$_x$	7 PC-MATLAB	8 CATPAC	9 CC	10 KEDDC	11 L.A.S.	12 LUND	13 TRIP	14 WCDS/DSC	15 ICARE	16 MADPAC	17 PAAS	18 SANCAD	19 SSPACK	20 SUBOPT	21 SUNS
120																					
121																					
122																					
123																					
124																					
125																					
126	Transformation routines																				
127	Similarity transformations	2	2	2	2	2	2	0	2	2	1	1	0	1	1	0	0	0	0	0	0
128	State–space <–> frequency dom.	0	2	2	0	2	2	0	2	1	1	2	1	1	1	1	0	0	0	0	0
129	Continuous <–> discrete	0	2	2	0	2	2	1	2	2	1	2	1	1	0	0	0	0	1	0	0
130	Subsystem interconnections	0	1	1	0	2	1	1	1	0	0	1	0	1	0	0	0	0	0	0	0
131																					
132	Identification																				
133	Scope																				
134	Continuous systems	0	1	1	0	2	1	0	0	2	1	2	2	0	0	0	0	0	0	2	0
135	Sampled–data systems	0	1	0	0	1	1	1	0	2	1	2	2	0	0	1	0	0	2	0	0
136	Linear–in–parameter systems	0	0	1	0	1	1	1	0	2	1	2	2	0	0	1	0	0	2	1	0
137	General parameters	0	0	0	0	1	1	1	0	1	0	1	2	0	0	0	0	0	2	0	0
138	Methods																				
139	Maximum likelihood	0	1	1	0	2	0	1	0	2	0	2	0	0	0	0	0	0	0	0	0
140	Least–squares	0	1	1	0	1	1	1	0	0	1	2	2	0	0	2	0	0	2	0	0
141																					
142	Model reduction																				
143	Modal reduction	0	1	1	0	2	1	1	2	0	1	2	1	1	0	0	0	0	0	0	0
144	Singular perturbation	0	0	0	0	1	0	0	1	1	0	0	0	0	0	0	0	0	2	0	0
145																					

654

146	Analysis routines
147	Controllability, observability
148	Stability
149	Poles, zeroes
150	Frequency responses
151	Time responses
152	Continuous linear systems
153	Discrete systems
154	Sampled-data systems
155	Nonlinear systems
156	Simulation
157	Stability tests
158	Steady-state analysis
159	Sensitivity analysis
160	Linearization
161	
162	Design routines
163	Classical controller design
164	Pole placement/Observer des.
165	LQG design/Kalman filters
166	Opt. control (e.g. time-o.c.)
167	Parameter optimization
168	Multivariable design
169	Time domain
170	Frequency domain
171	Adaptive control design
172	Nonlinear design
173	Distributed systems design

Table A.1   (Continued)

#	Feature	2 MATLAB	3 CTRL-C	4 IMPACT	5 MATLAB-SC	6 MATRIX$_x$	7 PC-MATLAB	8 CATPAC	9 CC	10 KEDDC	11 L.A.S.	12 LUND	13 TRIP	14 WCDS/DSC	15 ICARE	16 MADPAC	17 PAAS	18 SANCAD	19 SSPACK	20 SUBOPT	21 SUNS
180	**REAL TIME CAPABILITIES**																				
182	Real-time synchronization	0	0	0	0	2	0	0	0	1	0	1	0	0	0	0	0	0	0	0	0
183	Data gathering	0	0	0	0	2	0	0	0	1	0	1	0	0	0	0	0	0	0	0	0
184	Real-time control	0	0	0	0	2	0	0	0	2	0	1	0	0	0	0	0	0	0	0	0
185	Code generation	0	0	0	0	2	0	0	0	0	0	1	0	0	0	0	0	0	0	0	0
188	**PORTABILITY, RELIABILITY**																				
190	General portability	2	2	2	2	1	1	2	1	2	1	1	2	2	1	1	1	1	1	2	1
191	Portable programming language	2	1	1	2	1	1	1	1	1	0	0	1	2	0	0	0	1	1	2	1
192	Portable hardware interface	0	1	1	0	1	0	1	0	0	0	0	0	0	0	0	0	0	0	0	0
193	Foreign language interface	2	2	2	2	2	2	2	1	1	2	0	2	2	1	1	0	0	0	0	0
194	Standard numerical libraries	0	2	2	2	2	2	1	1	1	1	0	1	1	1	1	0	0	2	2	2
195	Standard graphical software	0	0	0	0	0	0	0	1	0	0	0	0	1	0	0	0	0	1	2	1
196	Standard data-base interface	0	0	0	0	0	0	0	1	0	0	0	0	0	0	0	0	0	0	0	1
199	**AVAILABILITY**																				
201	Executional version available	2	2	0	2	2	2	1	1	2	2	2	2	2	1	2	1	1	2	2	1
202	Source code available	2	0	0	1	0	0	0	1	2	2	1	2	0	1	2	1	0	2	2	0
203	Public domain software	2	0	0	0	0	0	0	0	1	0	0	0	0	0	1	0	0	0	0	0

This notation is perfectly legitimate, but unfortunately it leads to confusion. For example, if one wants to add two polynomials: p(s) + q(s), this operation cannot be written as P + Q unless both polynomials p(s) and q(s) are of the same order. p(s)*q(s) is entirely different from any matrix multiplication; namely, it is a convolution between P and Q, e.g., expressed in CTRL-C as CONV(p,q). Thus, to be able to overload the conventional operators, we must be able to distinguish polynomials from other vectors through use of a different data type. (The term *operator overloading* has been borrowed from the ADA language. It denotes the ability to reuse the same operator for different purposes depending on the data types of its operands.)

Linear systems are expressed, e.g., in $MATRIX_X$, through an ordinary MATLAB matrix:

$$\underline{S} = [\ A\quad B$$
$$\qquad\quad C\quad D\ ]$$

but certainly, two parallel connected subsystems cannot be expressed as S1 + S2, while two cascaded subsystems cannot be expressed as S2*S1, which would be analogous to frequency domain operations.

In IMPACT, each conceptual data element (matrix, polynomial, transfer function, linear system, trajectory) is expressed by a data structure of its own, thus allowing mathematical operators to be overloaded. This concept also allowed us to implement additional data structures such as polynomial matrices (which are essentially tensors) for the design of MIMO systems in the frequency domain (an area in which most CACSD tools are weak). Polynomials can, in IMPACT, be represented through their coefficients, their roots, or a set of supporting values.

Besides all these differences, the gap between the various MATLAB derivatives seems to close rather than to widen. Recently, a new downscaled version of $MATRIX_X$ was announced for the IBM-PC. This software must obviously be in direct competition with PC-MATLAB. On the other hand, a new upscaled version of CTRL-C is currently under development that will execute on a VAX-Station II, and that will offer features very similar to the combined $MATRIX_X$/SYSTEM_BUILD software.

Columns 8–14 summarize other general-purpose CACSD programs. CATPAC (Buenz, 1986) is a strongly data-driven program. Both the numerical algorithms and the computed results are stored in hierarchically organized databases. Database management (cf. Chapter 7 of this book) is one area in which the MATLAB derivatives are chronically weak. The data organization in CATPAC eases the incorporation of additional algorithms into the program, and it allows results from CACSD studies to be maintained in a better-organized manner. CATPAC does not provide for a full-fledged language though; that is, new algorithms must be hand-compiled before they can be added to the database. Moreover, the execution of data driven programs is slower than the execution of code-driven programs. Thus, the runtime efficiency of CATPAC algorithms must lie somewhere between that of MATLAB derivatives when executing functions programmed in their own interpretive macro languages, which execute even slower, and that of MATLAB derivatives when executing algorithms programmed in their implementation languages (FORTRAN, C, ADA), which execute much faster. Our experience with CTRL-C has shown that an algorithm programmed directly into the CTRL-C system executes roughly 20 times faster than the same algorithm when implemented as a CTRL-C function (actually macro). The optimal solution would probably be to stay code-driven (as in MATLAB), but to provide a function translator that can compile new functions from the application language (MATLAB, CTRL-C, etc.) into the implementation language (FORTRAN, C, ADA), once such a function has been sufficiently debugged. This is the path we intend to take with IMPACT (which is one of the major reasons why language constructs in IMPACT's own procedural language look as similar as possible to equivalent constructs of its implementation language ADA).

Program CC (Thompson, 1986) is a CACSD program that has been implemented in BASIC and runs on IBM-PCs (and PC lookalikes). CC has been designed for multivariable control system analysis and synthesis. It supports the data type "transfer function matrix," which enables frequency domain design also of multivariable systems. Among the MATLAB derivatives, only IMPACT and M offer this feature. Special features of Program CC are graphical displays of transfer functions, partial fraction expansions, and inverse Laplace and Z-transforms, symbolic manipulation of transfer functions, and state-space and frequency domain analysis of multirate sampled-data systems. The symbolic manipulation of transfer functions is a rather unique feature. Only four programs of the 20 surveyed offer any symbolic manipulation capabilities at all. Such a capability is invaluable in parametric studies, and it is certainly one of the features that needs to be strengthened in the future.

KEDDC (Schmid, 1979, 1985) is one of the most comprehensive CACSD programs currently available. It is coded in portable ANSI-FORTRAN, with the exception of the graphics subsystem, which is coded in PASCAL. Interfaces (graphics drivers) exist for many different terminal types and brands. The user interface can be configured. KEDDC can be used in a question-and-answer mode, but is more flexible when used through its menu- and form-driven interface. Depending on the terminal type, KEDDC also supports window management with parallel sessions. Its user's manual alone has 1200 pages, its programmer's manual has 1400 pages! KEDDC thus takes the cureall approach rather than providing for a flexible control shell. Indeed, when the user wants to solve a problem for which no algorithm is currently provided in KEDDC, he may find KEDDC less user-friendly than some of the MATLAB derivatives. The combination of existing algorithms into new ones can less easily be accomplished. However when using existing algorithms, KEDDC offers better guidance, and in some cases offers better-tuned algorithms.

L-A-S (linear algebra and systems; Chow et al., 1983; West et al., 1985) is one of CACSD programs that offer a full-fledged programming language for implementation of new algorithms. L-A-S is thus the pure contrary to KEDDC. It follows the shell approach rather than trying to provide a cure-all. Unfortunately, we find the L-A-S language highly cryptic.

The LUND software system consists of a suite of different CACSD programs for various purposes. SIMNON (Elmqvist, 1975, 1977) is a direct-executing simulation language comparable in power to DESCTOP (Korn, 1985, 1987). It supports the simulation of sampled-data systems, the interconnection of submodels, and multirate integrations of subsystems. SIMNON was the first floating-point direct-executing simulation language. SIMNON executes on VAX/VMS and UNIVAC machines, and only a few months ago, a PC-version of SIMNON was added. SIMNON is coded in FORTRAN-77 (but unfortunately not in a very portable manner).

A preprocessor, DYMOLA (Elmqvist, 1978, 1980), a topology-oriented system description language, generates either SIMNON application programs or directly FORTRAN code that can be linked with the SIMNON system. DYMOLA models (subsystem descriptions) are connected by means of cuts, a new data abstraction mechanism (similar to PASCAL records) that allows grouping variables together in the same way as fibers are grouped into wires, and wires are grouped into cables. There are two versions of DYMOLA. The first version was coded in SIMULA; a newer version is coded in PASCAL. DYMOLA was the first modular modeling system developed.

A graphical prepreprocessor, HIBLIZ (Elmqvist, 1982; Elmqvist and Mattson, 1986), generates a DYMOLA program out of a graphical description of the model. This system was already mentioned in this appendix. To our knowledge, it was the first (experimental) graphical modeling system ever built. It executes on VAX/VMS using a modified CRT with a mouse. In the meantime H. Elmqvist, the designer of all three simulation/modeling systems, has left the Swedish Control Institute at Lund. With him, K. J. Åström lost one of the most talented and innovative simulation experts available.

IDPAC (Wieslander, 1980a) is a highly powerful program for parameter estimation in linear stochastic MIMO models and for model validation purposes. It provides for spectral analysis, correlation analysis, and data analysis in general. POLPAC (Åström, 1985) is a (somewhat more experimental) program for polynomial design. MODPAC (Wieslander, 1980b) is a program for transformations between different system representations; for basic operations on these data types, that is, polynomial operations and matrix operations; for evaluation of basic system properties such as controllability, observability, and stability; and for graphical output.

All programs contained in the Lund software suite use internally the same parser, INTRAC (Wieslander and Elmqvist, 1978), which also provides for a standardized (though somewhat primitive) "language" environment. Even though some of the MATLAB derivatives are more powerful than INTRAC with respect to the features offered through their MACRO languages, the LUND software suite has paved the road for this new software technology. The LUND software suite was the first modern CACSD tool on the CACSD software market.

TRIP (transformation and identification program; Van den Bosch, 1985) is a CACSD program for the design of continuous- and/or discrete-time SISO systems in either time or frequency domain. It has a tight link to the interactive nonlinear simulation program PSI for the design of nonlinear control systems. TRIP is available on VAX and IBM-PC. It is a low-cost CACSD program for the design of systems up to tenth order.

The Waterloo Control System Design Software Packages (WCDS and DSC; Aplevich, 1986) can be used for optimal-control, multivariable frequency domain design and algebraic matrix-fraction design for systems of up to 100th order. These software systems make use of the system representation $(\underline{F} - s\underline{E})\underline{x} + \underline{G}\underline{u} = \underline{0}$, where $\underline{E}$, $\underline{F}$, and $\underline{G}$ are matrices, u denotes the input vector, $\underline{x}$ denotes the state vector, and s is a linear operator. This is the direct linearization of the nonlinear vector equation $\underline{f}(\underline{x}, s\underline{x}, \underline{u}) = \underline{0}$. Any linear model in state-space form, transfer-function representation, or matrix-fraction form can be put into the above representation by inspection, and hence, in principle, any algorithm based on these other system representations can be implemented. However, important simplifications result from this choice of data representation. All binary algebraic operations required for manipulating subsystems are performed using exactly three operations:

combining two systems into one, adjoining linear constraint equations, and reducing a model to an externally equivalent model, with a desired form. The first two operations are numerically harmless; the third requires a generalization of state-space minimality theory and is implemented using numerically stable staircase algorithms.

Columns 15–21 of Table A.1 describe more special-purpose CACSD systems. Each of them [ICARE (Gorez, 1986a), MADPAC (Bartolini et al., 1983), PAAS (Gorez, 1986b), SANCAD (Gray, 1986), SSPACK (Technical Software Systems, 1985), SUBOPT (Fleming, 1979), and SUNS (Atherton and Wadey, 1981; Atherton et al., 1986)] has a strength in one particular area of CACSD, as can be extracted from the feature table. We shall not describe those software systems in any greater detail.

Although we added up the credit units for all these 20 CACSD systems, it is impossible to ensure that we really compared apples to apples in all cases. In particular, the special-purpose CACSD programs are obviously doomed to collect fewer points. Moreover, the question whether to assign 1 or 2 units to a particular feature is somewhat subjective, and the fact that a particular table column (CACSD program) contains many 1 entries may, in fact, speak for the modesty of the software designer, more than for a lack of quality of the particular product. Nevertheless, 5 of these 20 programs (MATRIX$_X$, CTRL-C, IMPACT, LUND, and KEDDC) collected more than 100 points each, and indeed, we believe these five programs to be the most versatile among the CACSD software systems surveyed. With its new link to SIMNON, PRO-MATLAB should be added to this list.

Our CACSD survey is obviously incomplete, as only about 50% of the CACSD software designers known to us by the time we mailed out our circular letter responded to our questionnaire. Another survey of CACSD software was published recently by D. K. Frederick (1985). In that survey, each software system is described in much less detail (as the survey does not contain any feature table), but the survey contains 37 entries in place of our 20, and these are only partially the same as those contained in our survey. Thus, Frederick's survey is highly recommended for an additional source of information (as is the entire volume in which that survey appeared). There has recently also been created an ELCS (extended list of control software) newsletter to provide an information exchange forum for CACSD software developers (Rimvall et al., 1985). The newest issue of this newsletter contains already more than 80 different software entries.

Our software survey is subjective rather than objective. This is due to the fact that the 1 and 2 entries could not be assigned in a completely objective manner. Moreover, the authors of this survey have a strong inclination in favor of MATLAB-like languages and did not hide this fact. We often expressed personal opinions rather than restricting our survey to a simple listing of dry facts and leaving the interpretation of these facts to the reader, but we tried honestly to serve the CACSD community in the best possible way. The authors acknowledge with gratitude the kind cooperation of those responding to our questionnaire.

A.5 STANDARDIZATION VERSUS DIVERSIFICATION

In the previous section, we introduced a number of different CACSD tools. We have seen that they vary with respect to their application areas, as well as with respect to their user interfaces. Is such a diversification really justified and desirable, or might a canalization of the various and diversified efforts into one CACSD software standard be more appropriate? Is there any hope for a CACSD standard comparable, for example, to the continuous system simulation language (CSSL) standard in simulation software (Augustin et al., 1967)? What might such a standard look like?

We indeed believe that with respect to the manner in which control problems are formulated, a standard is both feasible and desirable. The matrix notation of MATLAB-like languages is so natural that we do not see a need for any other notation in this respect. Although the division operators "/" and "\" for right and left division are not "standard" operators in the classical mathematical sense, after MATLAB became available and popular, even a few publications have used this notation for simplicity. Hopefully, a MATLAB-like notation will also be introduced into CSSL's as an additional tool for the description of state-space models.

In IMPACT, we used additional operators for a third dimension, thus operating effectively on complex tensors in place of complex matrices. Multivariable systems can be expressed in terms of polynomial matrices where each matrix element may be a polynomial in the linear operator s (or z in the discrete case). We introduced the "^" operator to separate polynomial coefficients, and (alternatively) the "|" operator to separate polynomial roots. Thus, the polynomial matrix

$$\underline{P} \;=\; [\; (3s^2 + 10s + 3 \qquad (2s - 3) $$
$$\qquad\qquad s^3 \qquad\qquad (-s^2 - 7s - 10) \;]$$

can, in IMPACT, be coded as

$$\underline{P} \;=\; [\; [3\text{^}10\text{^}3], \quad [-3\text{^}2] $$
$$\qquad\quad [\text{^^^}1], \qquad [-10\text{^}-7\text{^}-1] \;]$$

or alternatively as

$$\underline{PF} \;=\; [\; 3*[-3\,|-(1/3)], \quad 2*[\,|1.5] $$
$$\qquad\qquad [0\,|0\,|0], \qquad\quad -1*[-2\,|-5] \;]$$

to denote the factorized form

$$\underline{PF} \;=\; [\; 3(s + 3)(s + 1/3) \qquad\quad 2(s - 1.5) $$
$$\qquad\qquad s*s*s \qquad\qquad -(s + 2)(s + 5) \;]$$

The two operators "^" and "|" naturally extend the previously introduced operators "," (used to separate matrix columns) and ";" (used to separate matrix rows). Once selected, the data representation is maintained until the user decides to convert the polynomial matrices into another data representation, e.g., by writing PF=FACTOR(P) or P=DEFACTOR(PF). Factorized polynomial matrices and defactorized polynomial matrices are two different data structures in IMPACT. Note that FACTOR(PF) will result in an error message. This notation has meanwhile been adopted by the developers of M as well (Gavel et al., 1985).

Of course, it is natural to define once and forever

$$s = [\;\text{^}1\;]$$

(which, in fact, is an IMPACT system variable). Thus, one can also write a polynomial as

p1 = 3*s**2 + 10*s + 3

or alternatively as

p1 = 3*(s + 3)*(s + (1/3))

which will nevertheless, in both cases, result in a polynomial of type defactorized polynomial, as the s-operator was coded in a defactorized form. To prevent this from happening, the user could write

sf = [|0]

and thereafter

p1f = 3*(sf + 3)*(sf + (1/3))

which, however, is not recommended as frequent defactorizations and refactorizations will take place in this case. Notice the consequent overloading of the "+" and "*" operators in these examples. Depending on the types of operands, a different algorithm is employed to perform the operation.

Also, with respect to the embedded procedural language, an informal standard can be achieved. The procedural language of CTRL-C, for instance, is very powerful. It basically extends the PASCAL programming style, operating conveniently on the new matrix data structures. Very useful, for instance, is the extension of the PASCAL-like "FOR" statement:

FOR I = [1,3,7,28], ...

This FOR loop shall be executed precisely four times with I = 1, I = 3, I = 7, and I = 28, respectively. (Many users still deplore the missing GOTO statement in CTRL-C. Although it has been theoretically proven that a GOTO statement is not needed, its lack makes programming sometimes quite difficult.) IMPACT employs an ADA style instead of the previously advocated PASCAL style. It actually does not matter too much which style is adopted in a forthcoming standard, but any standard would be highly welcome to allow smooth exchange of the extensive available soft-coded macro libraries. There is really no good reason to stick to the prevalent variety of only marginally different procedural languages.

Also with respect to the user interface, de facto "pseudostandards" have already been established. Window interfaces look more and more similar to the MacIntosh interface. (Although the MacIntosh was not the first machine to introduce windowing mechanisms, it was this machine that made this new technique popular.) The Swiss mouse is a very convenient, flexible, and fast-input device, and it is expected to make the previously fashionable cross-hair cursors and light pens soon obsolete, as cross-hair cursors are both uncomfortable to use and slow, and as light pens demand very expensive screen sensors. To our displeasure though, there exist mice with one, two, and three buttons. Any standard would be equally acceptable, but a standard must be found. Once the fingers are used to one system, it is hard to adjust to another (like driving a car that has the gas pedal on the left and the clutch on the right).

With respect to the actual functions offered, we shall probably not see a standard quickly. The current diversification into different application areas and design methodologies is most likely to be around for some time, and we actually welcome this, as too early a standard can freeze the lines and hamper the introduction of innovative new concepts.

Another interface, which is rarely even noticed by the casual CACSD software user, is the interface to a database where results of computations, as well as programming modules, notebook files, etc., may be stored. To promote the state of the art of CACSD software further, it is imperative that a database interface standard be defined. Lacking such a standard, most current CACSD software developers don't even offer a database inter-face at all, but fully rely on the file-handling mechanism (directory struc-ture) of the embedding operating environment. This mechanism is computa-tionally efficient (as the record manager, on every computer, is strongly optimized to suit the underlying hardware), but the mechanism is entirely insufficient for our task. The immediate effect of the lack of an appro-priate database concept is a jungle of small and smallest data and program files scattered over different subdirectories, which makes it hard to re-trieve data and programs that have previously been stored for later reuse. As an example, a particular A-matrix of a linear system will probably not be related to the problem under investigation at all, but will be stored as a nonmnemonic file "A.DAT" located somewhere in the directory structure of the underlying operating system. Little has been done to address this pertinent problem. Probably most advanced in this context is the work of Maciejowski (1984).

An IFAC working group discussing "Guidelines for CACSD-Software" was generated recently which consists of three subgroups for the discus-sion of

    CACSD program interfaces (including graphics)
    CACSD program data exchange
    CACSD program algorithm exchange

A similar IEEE working group exists as well. Hopefully, these two bodies will be able to promote a forthcoming CACSD standard.

A.6 SIMULATION AND CACSD

Let us discuss next how simulation has been implemented in some of the current CACSD programs.

In CTRL-C, there is a simulation function that takes the following form:

$$[\underline{y},\underline{x}] = SIMU(\underline{a},\underline{b},\underline{c},\underline{d},\underline{u},\underline{t})$$

where $\underline{a}$, $\underline{b}$, $\underline{c}$, and $\underline{d}$ are the system matrices describing a linear continuous-time MIMO system, $\underline{t}$ is a time base (that is, $\underline{t}$ is a vector of time instants), $\underline{u}$ is the input vector sampled over the time base (that is, $\underline{u}$ is actually a matrix, each row denotes one input variable, and each column denotes one time instant), $\underline{y}$ is the output vector (that is, $\underline{y}$ is a matrix with rows denoting output variables, and columns denoting time instants), and x is the state vector (which is also a matrix with according definitions). Initial conditions can be specified in a previous call of the same function:

    SIMU('IC',x0)

and also the integration method can be declared in a similar manner:

    SIMU('ADAMS',relerr,abserr,maxstp)

An equivalent function DSIMU exists for discrete-time systems. The system matrices can be constructed out of subsystem descriptions by use of a series of interconnection functions (SERIES, PARALLEL, INTERC, and MINREAL).

In IMPACT, we chose a slightly different approach. Since systems and trajectories are identifiable as separate data structures, we can once again overload the meaning of the primitive operators. Time bases ("domains") are created by means of the functions LINDOM and LOGDOM and/or by use of the "&" operator (concatenation operator):

    t = LINDOM(0.,1.,0.1) & LINDOM(2.,20.,1.) & 50. & 100

which generates a domain consisting of 23 points: [0., 0.1, 0.2, . . ., 0.9; 1., 2., 3., . . ., 19., 20, 50., 100.]. Trajectories are functions over domains, thus:

    u = [SIM(t);COS(t)]

which creates a trajectory column vector $\underline{u}$ evaluated over the previously defined domain t. Linear systems are generated by use of the LINCONT and LINDISC functions:

    s1 - LINCONT($\underline{a1}$,$\underline{b1}$,$\underline{c1}$,$\underline{D}$=>d1,$\underline{X}$0=>[0.5;2.;-3.7])

The three matrices $\underline{a1}$, $\underline{b1}$, and $\underline{c1}$ are compulsory positional parameters, whereas the input-output matrix ($\underline{D}$) and the initial condition vector ($\underline{X}$0) are optional (defaulted) named parameters.

Series connection between two subsystems is expressed as s2*s1, that is, multiplication in reverse order (exactly what it would be if the two subsystems were expressed through two transfer function matrices: $\underline{g2}$*$\underline{g1}$); parallel connection is expressed by use of the "+" operator, and feedback is expressed by the " \\" operator:

    g$\underline{tot}$ = $\underline{g}$ \\ (-$\underline{h}$)

($\underline{g}$ fed back with $-\underline{h}$), independently of whether $\underline{g}$ and $\underline{h}$ are expressed as transfer function matrices or as linear system descriptions. Simulations finally are expressed by overloading the "*" operator once more:

    y = s1*$\underline{u}$

which simulates the system s1 (which in our example must have two inputs) from 0. to time 100., interpolating between the specified values of the input trajectory vector u, and sampling the output trajectory vector y over the same domain. Thus:

tout = (s2*s1)*tin

series-connects the two subsystems s1 and s2 and then performs one simulation over the combined system. On the other hand:

tout = s2*(s1*tin)

simulates the subsystem s1 using tin as input, samples the resulting output trajectory over the same domain, and then simulates the subsystem s2 using the previous result as an input by reinterpolating it between its supporting values. Of course, numerically the results of these two operations will be slightly different, but conceptually, the associative law of multiplication holds.

In standard CSSLs, simulation is always viewed as the execution of a special-purpose program (the simulation program) producing simulation results (mostly in the form of a result file). There, the simulation program is viewed as the central part of the undertaking. No wonder such a concept does not lend itself easily to an embedding into a larger whole in which simulation is just one task among many.

In CTRL-C, simulation is viewed as a function mapping an input vector (or matrix) into an output vector (or matrix). Clearly, simulation is here just one function among many others that can be performed on the same data.

In IMPACT finally, simulation is viewed as a binary operator that maps two different data structures, namely, one of type system description (eventually also nonlinear), and the other of type trajectory into another data structure of type trajectory.

Of course, all three descriptions mean ultimately the same thing, yet the accents are drastically different. To prove our case, the reader versed in the use of one or the other of the CSSLs may try to code the IMPACT statement tout = s2*(s1*tin) as a "CSSL simulation program." In most CSSLs, this simply cannot be done. The task would require two separate programs to be executed one after the other. The output from the first simulation run (implementing taux = s1*tin) would have to be manually edited into a "tabular function" and used by the second simulation run (implementing tout = s2*taux).

To give another example: When solving a finite-time Riccati differential equation, one common approach is to integrate the Riccati equation backward in time from the final time t_f to initial time t_0, because the "initial condition" of the Riccati equation is stated as $\underline{K}(t = t_f) = \underline{0}$, and because the Riccati equation is numerically stable in backward direction only. The solution $\underline{K}(t)$ is stored away during this simulation and then reused (in reversed order) during the subsequent forward integration of the state equations with given $\underline{x}(t = t_0)$. Some of the available CSSLs allow solving this problem (mostly in a very indirect manner); other simply cannot be used at all to tackle this problem.

How can one handle this problem in CTRL-C? The first simulation is nonlinear (and autonomous), and the second is linear (and input-dependent) but time-varying; thus, we cannot use the SIMU function in either case. CTRL-C provides for a second means of simulation though. In the newest release of CTRL-C, an interface to the well-known simulation language ACSL was introduced. This interface allows making use of the modeling and simulation power of a full-fledged simulation language, while one is still able to control the experiment from within the more flexible environment of the CACSD program. Several of the discussed CACSD programs follow this path, and it might indeed be a good answer to our problem if the two languages that are combined in such a manner are sufficiently compatible with each other, and if the interface between them is not too slow. Unfortunately, this is currently not yet the case with any of the CACSD programs that use this route.

Let us illustrate the problems. We start by writing an ACSL program that implements the matrix Riccati differential equation

$$\frac{d\underline{K}}{dt} = -\underline{Q} + \underline{K}*\underline{B}*\underline{R}^{-1}*\underline{B}'*\underline{K} - \underline{K}*\underline{A} - \underline{A}'*\underline{K} \qquad \underline{K}(t_f) = 0$$

Since ACSL does not provide for a powerful matrix environment, we have to separate this compact matrix differential equation into its component equations. [ACSL does provide for a vector-integration function, and matrix operations such as multiplication and addition could be (user-)coded by use of ACSL's MACRO language. However, this is a slow, and inconvenient replacement for the matrix manipulation power offered in languages such as CTRL-C.] Furthermore, since ACSL does not handle the case $t_f < t_0$, we must substitute t by

$$t* = t_f - t_0 - t$$

and integrate the substituted Riccati equation

$$\frac{d\underline{K}}{dt*} = \underline{Q} - \underline{K}*\underline{B}*\underline{R}^{-1}*\underline{B}'*\underline{K} + \underline{K}*\underline{A} + \underline{A}'\underline{K} \qquad \underline{K}(0) = 0$$

forward in time from $t* = 0$ to $t* = t_f - t_0$. Through the new interface (A2CLIST), we export the resulting $K_{ij}(t*)$ back into CTRL-C, where they take the form of ordinary CTRL-C vectors. Also in CTRL-C, we have to manipulate the components of $\underline{K}(t)$ individually, as $\underline{K}(t)$ is a trajectory matrix, that is, a three-dimensional structure. However, CTRL-C handles only one-dimensional structures (vectors) and two-dimensional structures (matrices), but not three-dimensional structures (tensors). Back substitution can be achieved conveniently in CTRL-C by simply reversing the order of the components of each of the vectors as follows:

```
[n,m]  =  SIZE(kij)
  nm   =  n*m
  kij  =  kij(nm:-1:1)
```

Now, we can set up the second simulation:

$$\frac{d\underline{x}}{dt} = [\underline{A} - \underline{K}(t)*\underline{B}]*\underline{x} \qquad \underline{x}(t_0) = \underline{x}_0$$

What we would like to do is to ship the reversed $K_{ij}(t)$ back through the interface (C2ALIST) into ACSL, and use them as driving functions for the simulation. Unfortunately, ACSL is not (yet!) powerful enough to allow us to do so. Contrary to the much older CSMP-III system, ACSL does not offer a dynamic table load function (CALL TVLOAD). Thus, once the $K_{ij}(t)$ functions have been sent back through the interface into ACSL, they are no longer trajectories, but simply arrays, and we are forced to write our own interpolation routine to find the appropriate value of K for any given time t. After all, the combined CTRL-C/ACSL software is indeed capable of solving the posed problem, but not in a very convenient manner. This is basically due to the fact that ACSL is not (yet!) sufficiently powerful for our task, and that the interface between the two languages is still awkward. Because of the weak coupling between the two software systems, it might indeed have been easier to program the entire task out in ACSL alone, although this would have meant doing without any of the matrix manipulation power offered in CTRL-C.

What about IMPACT? In IMPACT, it was decided not to rely on any existing simulation language, but rather to build simulation capabilities into the CACSD program itself. This is partly because of the fact that (as the above example shows) the currently available simulation languages are really not very well suited for our task, and partly due to our decision to employ ADA as implementation language. As currently no CSSL has been programmed in ADA, we would have had to rely on the "pragma concept" (which is ADA's way to establish links to software coded in a different language). However, we tried to limit the use of the pragma concept as much as possible as this feature does not belong to the standardized ADA kernel (and, thus, may be implementation-dependent).

Until now, only the use of linear systems in IMPACT was demonstrated. However, nonlinear systems can be coded as special macros (called SYSTEM MACROs). The two linear system types (LINCONT and LINDISC) are, in fact, just special cases of system macros. The Riccati equation can be coded as follows:

```
SYSTEM ricc_eq(a,b,q,rb) RETURN k IS
k = zero(a);
BEGIN
    k` = −q + k*b*rb*k − k*a − a'*k;
END ricc_eq
```

The state equations can be coded as

```
SYSTEM sys_eq(a,b,rb,x0) INPUT k RETURN x IS
x = x0
BEGIN
    x` = (a − rb*k)*x;
END sys_eq
```

The total experiment can be expressed in another macro (of type FUNCTION MACRO):

```
FUNCTION fin_tim_ricc(a,b,q,r,xbeg,time_base) IS
BEGIN
    back_time = REVERSE(time_base);
    rb = r\b´;
    k1 = ric_eq(a,b,q,rb)*back_time;
    k2 - REVERSE(k1);
    x = sys_eq(a,b,rb,xbeg)*k2;
    RETURN <x,k2>;
END fin_tim_ricc;
```

Notice the difference in the call of the two simulations. The first system
(ricc_eq) is autonomous. Therefore, simulation can no longer be expressed
as a multiplication of a system macro with a (nonexistent) input-trajectory
vector. Instead, the system macro here is multiplied directly with the
domain variable, that is, the time base. The second system, on the other
hand, is input-dependent. Therefore, the multiplication is done (as in the
case of the previously discussed linear systems) with the input trajectory.
FIN_TIM_RICC can now be called just like any of the standard IMPACT
functions (even nested). The result of this operation are two variables,
y and k, of the trajectory vector and trajectory matrix type, respectively.

```
x0 = [0;0]; a = [0,1;-2,-3]; b = [0;1];
q = [10,0;0;100]; r = 1;
forw_time = LINDOM(0,10,0.1,METHOD=>'ADAMS',ABSERR=>0.001);
[y,k] = fin_tim_ricc(a,b,q,r,x0,forw_time);
plot(y)
```

As can be seen from the above example, the entire integration information,
in IMPACT, is stored in the domain variables, which makes sense as these
variables anyway contain part of the runtime information (namely, the communication points and the final time). Moreover, this gives us a neat way
to differentiate clearly between the model description on the one hand and
the experiment description on the other.

Obviously, this is a much more powerful tool for our demonstration task
than even the combined ACSL/CTRL-C software. Unfortunately, contrary to
CTRL-C, IMPACT has not yet been released. Roughly the first 75,000
lines of ADA code have meanwhile been coded and debugged, and the
IMPACT kernal will be released soon. This kernel will implement all the
IMPACT language structures (including all the macro types, the complete
query feature, and multiple sessions), but it will not contain all the foreseen control library functions, nor will it contain multiple windows. The
complete software is expected to become available soon.

A.7 OUTLOOK

How is the field of CACSD going to develop further over the next decade
or so? To understand where we are heading, we need to assess where we
currently stand. In the past, and this still holds for the first generation
of CACSD tools, the application programmer was talking about program
development. A program is a tool that calculates something in a sequential
manner when executed on a digital computer. Some programs were parameterized, that is, accepted input data to partly determine what was to be
calculated. The major emphasis was on the program, whereas the data were

of relatively minor importance. There was a clear distinction between the program (a piece of static code in memory), and the data (a portion of memory that changed its content during execution of the program).

With the new generation of CACSD tools, we departed from this viewpoint drastically. New CACSD programs are in themselves true programming languages; that is, the application programmer no longer relies on the computer manufacturer to provide the languages to be used, but creates his own special-purpose languages. The difference is simply that less and less of the computational task is frozen in code, while more and more of it is parameterized, that is, data-driven. The data in itself reached such a degree of complexity that its appropriate organization became essential. The user interface, previously an unimportant detail, turned into a central question that decided whether a particular CACSD tool was good or bad, even more than the algorithmic richness provided within the program. What we gained by this change in accent was a dramatic increase in flexibility offered by the CACSD tools; what had to be paid in return was a certain decrease in runtime efficiency. However, with the advent of more powerful computers (an engineering workstation of today compares in number-crunching power easily with a mainframe computer of not more than a decade ago), this sacrifice could be gladly made. Moreover, it was often true that the compilation and linkage of a simulation program took 10 times longer than the actual execution of the program (at least for sufficiently simple applications). With the advent of the new direct executing (that is, fully data-driven) simulation languages such as SIMNON (Elmqvist, 1975, 1977) and DESCTOP and DESIRE (Korn, 1985, 1986), one can obtain simulation results immediately, and even if the simulation program executes 50% slower than it would if it were properly compiled, the increased flexibility of the tool (ease of model change) pays off easily even with respect to the total time spent at the computer terminal. These types of simulation tools are exactly what is needed within the CACSD program.

However, we are currently at the edge of taking yet another step. We now talk about the development of multiwindow user interfaces, of object-oriented programming style, of language-sensitive editors, of CAD databases, etc. Are these really issues that can (or should) be tackled at the level of a programming language? Are these not rather topics to be discussed at the level of the underlying operating systems? If we say that we need a CAD database to store our models and resulting data files, do we not simply express the fact that the file storage and retrieval system of the operating system in which the tool is being embedded is not powerful enough for our task? Are not interactive languages such as MATLAB and DESCTOP (very primitive) special-purpose operating systems in themselves? We indeed do believe that future programming systems will blur the previously clear-cut distinction between programming languages and the operating systems they are embedded in. This problem was realized by the developers of ADA, who understood that a complex tool such as ADA cannot be designed as a programming language with a clean interface to the outside, implementable independently of the operating system it is to run under. Instead, its developers considered an ADA environment to be offered together with the ADA language. The ADA environment is basically nothing but a (partial) specification of the operating system in which the ADA language is to be embedded. The same is true with respect to CACSD tools. In IMPACT, we were not yet able to address this question in full depth, as the ADA environment itself is not yet completely defined, and as

we would like to borrow as much as possible from ADA concepts. In M
(Gavel et al., 1985), this question was addressed and led to the development of yet another tool, EAGLES (Lawver, 1985), an object-oriented, multitasking, multiwindowing operating system, under which M is to run. The
import/export of M-variables between different sessions (windows) is not
programmed in M itself, but is supported by EAGLES. The entire graphics
system is a facility provided by the EAGLES operating system rather than
being implemented as an M-tool. EAGLES operates on a rather involved
database that serves as a buffer for all data to be shuffled back and forth
between the different tools (such as M) and the operating system EAGLES
itself.

The future will tell how efficient (or rather inefficient) EAGLES is
when implemented as a language running under an existing operating system (as it is currently planned) rather than being implemented as the
operating system itself. Obviously, EAGLES has to rely strongly on the
record manager and system functions of the operating system (VMS). How
will the developers of EAGLES deal with new releases of the underlying
operating system offering enhanced and, at this level, not fully upward-
compatible new features? Are EAGLES users going to have the same problem
as EUNICE users (EUNICE = UNIX under VMS), who are always two or
three versions behind the current version of VMS, because it takes the
EUNICE developers usually 1 year to keep up with the newest (meanwhile
already again outdated) developments in VMS? We don't know the answers
to these questions. We just focus on some of the dangers behind this current development.

One way to overcome the previously mentioned problem may be to
standardize the operating system itself. The UNIX operating system presents one step in this direction. There already exist a large number of
(unfortunately not very uniform) UNIX implementations for various computers. The idea is splendid. Unfortunately, the original UNIX was much
too small to be taken seriously (e.g., with respect to the problem of en-
forcing data security and data integrity). Current UNIX dialects tackle
these problems in various ways, but the experiment is doomed to failure
unless the computer manufacturers can agree on an invariant UNIX kernel
that goes far beyond the original UNIX definition. It must include not
only the procedural command language, but also system calls (lexical functions), the interface to the record manager, and naming conventions for
files, symbols, and logicals.

What about new facilities offered in future CACSD tools? We expect
to see more and more flexibility with respect to the data interface. The
ultimate of data-driven programming is a language in which there is essentially no longer any difference between code and data at all. Each operation
that can be performed in the language is itself expressed as an entry in a
database and can thus be altered at any moment.

One such environment is LISP. Basically, the only primitive operations
in LISP are addition and removal of entries from lists. These operations
are themselves expressed as entries in lists. When interpreted as operation, the first entry in the list is the operator, while all further entries are
its parameters. For these reasons, LISP programs exhibit a serious runtime
inefficiency. A numerical algorithm implemented in LISP will probably
execute two to three orders of magnitude slower than the same algorithm
implemented in a conventionally compiled language. Moreover, LISP is
often rather unwieldy with respect to how a particular numerical algorithm

has to be specified. However, LISP certainly also presents the ultimate in flexibility. Suddenly, self-modifying code has become a feasibility and can be employed to achieve amazing results. Moreover, in LISP, numerical data are entries in lists just like any other data. Thus, nonnumerical data processing is as efficient as numerical data processing, and in this arena, LISP competes a little more favorably with conventional programming techniques. Also, steps have been taken to alleviate some of this inherent inefficiency. Incremental compilers in place of pure interpreters can increase the runtime efficiency by roughly one order of magnitude. Furthermore, a LISP interpreter is an extremely simple program as compared to a conventional compiler. A (basic) LISP interpreter can be coded in roughly 600 lines of (LISP) code. Owing to this simplicity, it may make sense to implement part of this task in hardware rather than in software. The machine instructions of a special-purpose LISP machine can be tuned to optimize efficiency of executing LISP primitives. Such machines are already available and help to overcome at least part of the inefficiency of LISP.

With respect to the user interface, many of LISP's difficulties can be avoided by changing the world view once more. While LISP is basically process-oriented, PROLOG is activity-oriented. That is, in LISP, the programmer takes the standpoint of the operator ("What do I do next with my data?"), whereas in PROLOG, the programmer takes the standpoint of the data ("What needs to happen to me next?"). This helps to concentrate activities to be performed into one piece of code rather than having them spread all over. Unfortunately, digital computers are still sequential machines, whereas activity programming is not procedural in nature. As a consequence, PROLOG is expected to be more inefficient than even LISP. (However, PROLOG can rather easily be implemented in LISP, and thus, there exist PROLOG environments also on LISP machines, and they function amazingly well.) PROLOG primitives are more compound than LISP primitives. The natural consequence of this enhanced degree of specialization are shorter and better readable PROLOG programs on the one hand, but less flexibility on the other. Not every program that can be conceived in LISP can easily be implemented in PROLOG, while the converse is true.

These new languages are expected to shortly lead to yet another generation of CACSD tools. The strength of these new tools will lie in nonnumerical design, that is, in parameterized control system design studies, where several parameters are kept as unknowns in the design process. This will hopefully help to give the designer more insight into what is happening in his system. Nonnumerical controller design algorithms are still in their infancy, and it is not known yet how far these new concepts will lead us. How does one avoid the problem of formula explosion; that is, how can one obtain parametric answers without being confronted with pages and pages of never-ending formulas? Recent developments in programs such as MACSYMA (Symbolics, 1983) and REDUCE (Rand Corp., 1985) may help to answer some of these questions. A clue may be to introduce intermediate variables in the right places, variables that are not further expanded but kept as additional (dependent) parameters. These techniques were recently surveyed by Birdwell et al. (1985).

Another area that will be boosted by concepts such as advertised in PROLOG and LISP is the integration of CACSD software with expert systems. Expert systems are programs that evaluate a set of parameterized rules (conditional statements with mostly nonnumerical operands) by plugging in appropriate parameter values. The set of available parameter values is

called the knowledge of the expert system. Each evaluation may generate
new knowledge, and eventually even new rules. To accommodate this new
knowledge (new rules), the rules of the expert system are evaluated recursively until no further facts (knowledge) can be derived from the current
state of the program.

Why is it that many computer experts smile at the current efforts in
expert system technology? To design an expert system, one needs expert
knowledge. For this reason, most of the early expert systems were written
by experts in the application area rather than by experts in the implementation tool. Such programs did not always exploit the latest in software technology. Expert systems are thus often envisaged as question-and-answer-
driven programs with very limited capabilities. However, our above definition
of the term "expert system" did not mention the user interface at all. In
fact, the user interface (that is, the port through which new knowledge is
entered into the knowledge base of the expert system) is completely decoupled from the mechanisms of rule evaluations (the inference engine) and
can be any of the previously mentioned interface types (question-and-
answer, command, menu, form, graphical, and window interface).

Indeed have not expert systems been further developed than what most
people think? Is not MATLAB in fact an expert system for linear algebra?
Is not every single CACSD tool an expert system for control system design?
They surely exhibit all properties of expert systems. To prove our case,
let us examine the MATLAB statement

$$x = b/a$$

a little more closely. Certainly, the interpreter of this statement performs
symbolic processing. Once it has determined the type of operation to be
performed (division), it has to check the types of the operands. If a is a
scalar, all elements of b must be divided by a. If a is a square matrix, a
Gaussian elimination must take place to determine x. And finally, if a is a
rectangular matrix, x is evaluated as the solution of an over- or under-
determined set of equations in a least-squares sense. Quite obviously,
these are rules to be evaluated.

Of course, most people would not call MATLAB an expert system (and
neither would we). However, there is more expert system technology
readily available than what is commonly exploited. To give an example:
Most expert systems today constantly perform operations on symbolic data.
It is true that the data to be processed are input in a symbolic form. However, that does not mean that they have to be processed within the expert
system in a symbolic form as well. Compiler writers have known this fact
for years. The scanner interprets the input text, maps tokens (symbols)
into more conveniently processable integers, and stores them in fast
addressable data structures. This process is called "hashing." During the
entire operation of compilation (and eventually also symbolic debugging),
the system operates on these numerical quantities in place of the symbolic
ones. Only upon output (e.g., for generating the cross-reference table),
the original symbols are retrieved through the hash table. This mechanism
could easily be used in expert systems to increase their efficiency, but this
is rarely done today. SAPS (Uyttenhove, 1979; Klir, 1985; Yandell,
1987), for instance, can be used for qualitative simulations of discrete
input/output models (that is, models described through sets of input and

output trajectories rather than by means of a symbolic structure). The trajectories can be either discretized continuous variables or variables that are discrete in nature. Often, one would like to characterize a signal as being [≪much_too_small≫, ≪too_small≫, ≪just_right≫, ≪too_large≫, and ≪much_too_large≫]. These symbols are mapped into the set of integers [0, 1, 2, 3, and 4]. The authors of SAPS called this process "recoding." However, in SAPS, the recoding has to be done by hand, and the output will be expressed in terms of the recoded variables instead of the original ones.

How can the emerging expert system technology be exploited by CACSD software? As a first step, the error-reporting facility, the HELP facility, and the TUTORIAL facility of CACSD tools should be made dynamic. Today, such facilities exist in most CACSD programs, but they are static; that is, the amount and detail of information provided by the system are insensitive to the context from which it was triggered. The idea is quite old. IBM interactive operating systems have offered for many years a two-level error-reporting facility. When an error occurs, a short (and often cryptic) message is displayed which may suffice for the expert, but is inadequate for the novice. Thus, after receiving such a message, the user can type a "?" which is honored by a more detailed analysis of the problem. IMPACT's QUERY facility is another step into this direction. Another implementation has been described by Munro et al. (1986).

Also, K. J. Åström and L. Ljung are working on such a facility for IDPAC (private communication). The idea is the following: Rather than letting the students queue in front of Karl Johan's office, his knowledge about the use of the IDPAC algorithms (when to use what module) should be coded into the program itself, providing the students with an adaptive tutorial facility for identification algorithms. Thus, a computer-aided instruction (CAI) facility is being built into the CAD program. A similar approach has been proposed by Taylor and Frederick (1984).

Another related idea was expressed by K. J. Åström (private communication). He wants to add a command spy to his IDPAC software. Here, the idea is as follows: instead of waiting until the student realizes that he is doing something wrong, and therefore seeks the professor's advice, the professor stands, in a figurative sense, behind the student and watches over his shoulder to see what he is doing. As long as the student is doing fine, the professor (that is, the command spy) keeps quiet, but when the student tries to perform an operation that is potentially dangerous to the integrity of his data, or that is likely to lead to illegitimate conclusions, the command spy becomes active and warns the student about the consequences of what he is doing.

A similar feature could be built into a language sensitive editor. This would allow checking a CACSD program early on not only for syntactic correctness, but also for semantic correctness. Some of the semantic tests are, of course, data-dependent, and these can only be performed at execution time.

Other improvements can be expected from screening data for automated selection of the most adequate algorithms. This is similar to the previously mentioned operator overloading facility, but here, the algorithm is selected not on the basis of the types of the operands, but on the basis of the data itself. As a typical example, we could mention the problem of inverting a matrix. Obviously, if the matrix is unitary, its inverse can be obtained by simply computing the conjugate complex transpose of the matrix, which is much faster and gives rise to less error accumulation than computation of the inverse by, e.g., Gaussian elimination. If the matrix is (block-)-diagonal, each diagonal block can be inverted independently. If the matrix presents itself in a staircase form, yet another simplified algorithm can be used, etc. Thus, the matrix should be checked for particular structural properties, and the most appropriate algorithm should be selected on the basis of the outcome of this test. A good amount of knowledge about data classification algorithms exists, a knowledge that is not being exploited by many of today's CACSD programs.

Finally, we expect that even new control algorithms will arise from expert system technology. Today's control algorithms are excellent for local control of subsystems. They are not so good for global assessment of complex systems. A complex system such as the forthcoming space station or a nuclear power plant needs to be monitored, and expert system technology may be used to decide when something odd has happened or is about to happen. Then a global control strategy must take over and decide what to do next. Currently, human operators do a much better job in this respect that automatic controllers. However, they do not solve Riccati equations in their heads. Instead, they decide on the basis of qualitative, that is, highly discretized, information processed by use of a mental model of the process. Cellier (1986b) investigates the possibilities of qualitative simulation and rule-based control system design.

To sum up, CACSD is still a very active research field, and more results are to be expected shortly. We sincerely hope that our survey and discussion may stimulate more research.

REFERENCES

Ackermann, J. (1980). Parameter space design of robust control systems, *IEEE Trans. Automatic Control AC-25*, 1058–1072.

Agathoklis, P., Cellier, F. E., Djordjevic, M., Grepper, P. O., and Kraus, F. J., (1979). INTOPS, educational aspects of using computer-aided design in automatic control, in: *Proceedings of the IFAC Symposium on Computer-Aided Design*. Zürich, Switzerland, August 29–31, 1979 (M. A. Cuénod, ed.), Pergamon Press, Oxford, 441–446.

ANSI (1985). *American National Standard for Information Systems, Computer Graphics, Graphical Kernel System (GKS)*. Functional Description (ANSI X3.124–1985) and FORTRAN Binding (ANSI X3.124.1–1985).

Aplevich, J. D. (1986). Waterloo control system design packages (WCDS and DSC), personal communication, Dept. of Electrical Engineering, University of Waterloo, Waterloo, Ontario, Canada.

Asada, H. and Slotine, J. J. E. (1986). *Robot Analysis and Control*, Wiley, New York, 266 pp.

Åström, K. J. (1980). Self-tuning regulators—design principles and applications, in *Applications of Adaptive Control* (K. S. Narendra and R. V. Monopoli, eds.), Academic Press, New York.

Åström, K. J. (1985). Computer-aided tools for control system design, in: *Computer-Aided Control Systems Engineering* (M. Jamshidi and C. J. Herget, eds.), North-Holland Publishing, Amsterdam, pp. 3–40.

Athens, M., ed. (1978). On large scale systems and decentralized control, *IEEE Trans. Automatic Control, AC-23*, Special Issue.

Atherton, D. P. and Wadey, M. D. (1981). Computer-aided analysis and design of relay systems, in: *IFAC Symposium on CAD of Multivariable Technological Systems*, Pergamon Press, New York, pp. 355–360.

Atherton, D. P. et al. (1986). SUNS: The Sussex University Control Systems Software, in: *Proceedings of the 3rd IFAC Symposium on Computer-Aided Design in Control and Engineering Systems (CADCE'85)*, Copenhagen, July 31–August 2, 1985, Pergamon Press, pp. 173–178.

Augustin, D. C., Fineberg, M. S., Johnson, B. B., Linebarger, R. N., Sanson, F. J. and Strauss, J. C. (1967). The SCi continuous system simulation language (CSSL), *Simulation, 9*, 281–303.

Bartolini, G., et al. (1983). A package for multivariable adaptive control, in: *Proceedings of the 3rd IFAC/IFIP Symposium on Software for Computer Control (SOCOCO'82)*, Madrid, Spain, Pergamon Press, Oxford, pp. 229–235.

Birdwell, J. D. et al. (1985). Expert systems techniques and future trends in a computer-based control system analysis and design environment, in: *Proceedings of the 3rd IFAC Symposium on Computer-Aided Design in Control and Engineering Systems (CADCE'85)*, Copenhagen, July 31–August 2, 1985, Pergamon Press, Oxford, pp. 1–8.

Buenz, D. (1986). CATPAC—Computer-aided techniques for process analysis and control. personal communication, Philips Forschungslaboratorium Hamburg, Hamburg, Federal Republic of Germany.

Cellier, F. E. (1986a). Enhanced run-time experiments in continuous system simulation languages, in: *Proceedings of the 1986 SCSC Multiconference* (F. E. Cellier, ed.), SCS Publishing, San Diego, CA, pp. 78–83.

Cellier, F. E. (1986b). Combined continuous/discrete simulation—Applications, tools, and techniques, Invited Tutorial, in *Proceedings of the Winter Simulation Conference (WSC'86)*, Washington DC.

Cellier, F. E. and Rimvall, M. (1983). Computer-aided control systems design, Invited Survey Paper, in: *Proceedings of the Winter Simulation Conference (ESC'83)*, Aachen, FRG (W. Ameling, ed.), Springer-Verlag, Lecture Notes in Informatics, New York, pp. 1–21.

Cellier, F. E., Grepper, P. O., Rufer, D. F. and Tödtli, J. (1977). AUTLIB, automatic control library, educational aspects of development and application of a subprogram package for control, in: *Proceedings of the IFAC Symposium on Trends in Automatic Control Education*, Barcelona, Spain, March 30–April 1, 1977, Pergamon press, pp. 151–159.

Chow, J. H., Bingulac, J. H., Javid, S. H. and Dowse, H. R. (1983). *User's Manual for L-A-S Language*, System Dynamics and Control Group, General Electric, Schenectady, NY.

Denham, M. J. (1984). Design issues for CACSD systems, *Proc. IEEE 72* (12), 1714–1723.

Elmqvist, H. (1975). *SIMNON—An Interactive Simulation Program for Nonlinear Systems—User's Manual*, Report CODEN: LUTFD2/(TFRT-7502). Dept. of Automatic Control, Lund Institute of Technology, Lund, Sweden.

Elmqvist, H. (1977). SIMNON—An interactive simulation language for nonlinear systems, in: *Proceedings of the International Symposium SIMULATION'77*, Montreux, Switzerland (M. Hamza, ed.), Acta Press, Anaheim, CA, 85–90.

Elmqvist, H. (1978). A structured model language for large continuous systems, Ph.D. Thesis, Report: CODEN: LUTFD2/(TRFT-1015). Dept. of Automatic Control, Lund Institute of Technology, Lund, Sweden, 226 pp.

Elmqvist, H. (1980). A structured model language for large continuous systems. *IMACS TC3 Newsletter 10*.

Elmqvist, H. (1982). A graphical approach to documentation and implementation of control systems, in: *Proceedings of the 3rd IFAC/IFIP Symposium on Software for Computer Control (SOCOCOCO'82)*, Madrid, Spain, Pergamon Press, Oxford.

Elmqvist, H. and Mattson, S. E. (1986). A simulator for dynamic systems using graphics and equations for modelling, in: *Proceedings of the 3rd Symposium on Computer-Aided Control System Design*, Washington, DC.

Evans, D. C. (1985). The art of visual simulation, Keynote Address, Winter Simulation Conference (WSC'85), San Francisco, Evans & Sutherland Computer Corp., IEEE Publishing, Piscataway, NJ.

Fleming, P. J. (1979). A CAD program for suboptimal linear regulators. in: *Proceedings of the IFAC Symposium on Computer-Aided Design*, Zürich, Switzerland, August 29–31, 1979 (M. A. Cuénod, ed.), Pergamon Press, Oxford, 259–266.

Frederick, D. K. (1985). Software Summaries, in: *Computer-Aided Control Systems Engineering* (M. Jamshidi and C. J. Herget, eds.), North-Holland, Amsterdam, pp. 349–384.

Gavel, D. T. and Herget, C. J. (1984). The M language—An interactive tool for manipulating matrices and matrix ordinary differential equations, International Report, Dynamics and Controls Group, Lawrence Livermore National Laboratory, University of California, Livermore, CA.

Golub, G. H. and Wilkinson, J. H. (1976). Ill-conditioned eigensystems and the computation of the Jordan canonical form, *SIAM Rev. 18*(4), 578–619.

Gorez, R. (1986a). The ICARE project—An interactive computing aid for research and engineering, Personal Communication, Laboratoire d'Automatique, de Dynamique et d'Analyse des Systèmes, Université Catholique de Louvain, Bâtiment Maxwell, Louvain-la-Neuve, Belgium.

Gorez, R. (1986b). PAAS—Programme d'aide à l'analyse des systèmes. Personal Communication, Laboratoire d'Automatique, de Dynamique et d'Analyse des Systèmes, Université Catholique de Louvain, Bâtiment Maxwell, Louvain-la-Neuve, Belgium.

Gray, J. O. (1986). SANCAD and SATRES, Personal Communication, Dept. of Electronic and Electrical Engineering, University of Salford, Salford, United Kingdom.

IBM (1984). *Dynamic Simulation Language/VS (DSL/VS). Language Reference Manual*, Program Number 5798-PXJ, Form SH20-6288-0, IBM Corp., Cottle Road, San Jose, CA.

Integrated Systems, Inc. (1984). *Matrix$_x$ User's Guide, Matrix$_x$ Reference Guide, Matrix$_x$ Training Guide, Command Summary, and on-line Help*, Integrated Systems, Inc., Palo Alto, CA.

Kailath, T. (1980). *Linear Systems*, Prentice-Hall, Englewood Cliffs, NJ. 682 pp.

Klir, G. J. (1985). *Architecture of Systems Problem Solving*, Plenum Press, New York, 539 pp.

Korn, G. A. (1985). A new interactive environment for computer-aided experiments, *Simulation 45*(6), 303–305.

Korn, G. A. (1987). Control-System simulation on small personal-computer workstations, Int. J. Modeling and Simulation 8(4).

Korn, G. A. and Wait, J. V. (1978). *Digital Continuous System Simulation*, Prentice-Hall, Englewood Cliffs, NJ. 212 pp.

Laub, A. (1980). Computation of balancing transformations, *Proc. JACC 1*, Paper FA8-E.

Lawver, B. (1985). EAGLES, an interactive environment and program development tool, Personal Communication, Dynamics and Controls Group, Lawrence Livermore National Laboratory, University of California, Livermore, CA.

Little, J. N. (1985). *PC-MATLAB, User's Guide, Reference Guide, and On-line HELP, BROWSE, and Demonstrations*, The MathWorks, Inc., Sherborn, MA.

Little, J. N. et al. (1984). CTRL-C and matrix environments for the computer-aided design of control systems, in: *Proceedings of the 6th International Conference on Analysis and Optimization (INRIA)*, Nice, France, Lecture Notes in Control and Information Sciences, Vol. 63, Springer-Verlag, New York.

Little, J. N., Herskovitz, S., Laub, A. J. and Moler, C. B. (1986). MATLAB and control design on the MacIntosh, in: *Proceedings of the 3rd Symposium on Computer-Aided Control Systems Design*, Washington DC.

Maciejowski, J. M. (1984). Data structures for control system design, in: *Proceedings of the 6th European Conference of Electrotechnics, Computers in Communication and Control (EUROCON'84)*, Brighton, UK.

Mitchell, E. E. L. and Gauthier, J. S. (1986). *ACSL: Advanced Continuous Simulation Language—User/Guide Reference Manual*, Mitchell & Gauthier, Assoc., Concord, MA.

Moler, C. (1980). *MATLAB User's Guide*, Dept. of Computer Science, University of New Mexico, Albuquerque, NM. 40 pp.

Monopoli, R. V. (1974). Model reference adaptive control with an augmented error signal, *IEEE Trans. Automatic Control AC-19*, 474–484.

Munro, N., Palaskas Z. and Frederick, D. K. (1986). An adaptive CACSD dialogue facility, in: *Proceedings of the 3rd Symposium on Computer-Aided Control System Design*, Washington D.C.

Narendra, K. S. (1980). Recent developments in adaptive control, in: *Methods and Applications in Adaptive Control* (H. Unbehauen, ed.), Springer-Verlag, New York.

Norsworthy, R., Kohn, W. and Arellano, J. (1985). A symbolic package for analysis and design of digital controllers, Honeywell, Inc., and NASA Johnson Space Center, Private Communication.

Patel, R. V. and Misra, P. (1984). Numerical algorithms for eigenvalue assignment by state feedback, *Proc. IEEE 72(12)*, 1755–1764.

Pegden, C. D., et al. (1985). *CINEMA User's Manual*, Systems Modeling Corp., State College, PA.

Rand Corp. (1985). *REDUCE User's Manual*, The Rand Corp., Santa Monica, CA.

Rimvall, M. (1983). *IMPACT, Interactive Mathematical Program for Automatic Control Theory, User's Guide*, Dept. of Automatic Control, Swiss Federal Institute of Technology, ETH-Zentrum, Zürich, Switzerland, 208 pp.

Rimvall, M. and Bomholt, L. (1985). A flexible man-machine interface for CACSD applications, in: *Proceedings of the 3rd IFAC Symposium on Computer-Aided Design in Control and Engineering Systems (CADCE'85)*, Copenhagen, July 31–August 2, 1985, Pergamon Press, Oxford, 98–103.

Rimvall, M. and Cellier, F. E. (1985). The matrix environment as enhancement to modeling and simulation, in: *Proceedings of the 11th IMACS World Conference*, Oslo, August 5–9, 1985, North-Holland, Amsterdam.

Rimvall, M., et al. (1985). *ELCS—Extended List of Control Software, Newsletter* (M. Rimvall, D. K. Frederick, C. Herget, and R. Kook, eds.), Dept. of Automatic Control, ETH-Zentrum, Zürich, Switzerland.

Rosenbrock, H. H. (1969). Design of multivariable control systems using the inverse Nyquist array, *Proc. IEE 116*, 1929–1936.

Sawyer, W. (1986). Polynomial operations with a trajectory representation, Term Project (M. Rimvall, adv.), Dept of Automatic Control, ETH-Zentrum, Zürich, Switzerland.

Schmid, C. (1979). *KEDDC, User's Manual and Programmer's Manual*, Dr. -Ing. Chr. Schmid, Lehrstuhl für Elektrische Steuerung und Regelung, Ruhr University Bochum, Federal Republic of Germany.

Schmid, C. (1985). KEDDC—A computer-aided analysis and design package for control systems, in: *Computer-Aided Control Systems Engineering* (M. Jamshidi and C. J. Herget, eds.), North-Holland, Amsterdam, pp. 159–180.

Shah, S., Shah, S. C., Floyd, M. A. and Lehman, L. L. (1985). Matrix$_X$: Control Design and Model Building CAE Capability, in: *Computer-Aided Control Systems Engineering*, (M. Jamshidi, and C. J. Herget, eds.), North-Holland, Amsterdam, pp. 181–207.

Siljak, D. D. and Sundareshan, M. K. (1976). A multilevel optimization of large-scale dynamic systems, *IEEE Trans. Automatic Control AC-21*, 70–84.

Spang, H. A. III (1984). The federated computer-aided control design system, *Proc. IEEE 72*(12), 1724–1731.

Strandridge, C. R., et al. (1986). *TESS with SLAM-II, User's Manual*, Version 2.2, Prisker & Associates, Inc., West Lafayette, IN.

Symbolics, Inc. (1983). *MACSYMA Reference Manual, Version 10*, MIT and Symbolics, Inc., Cambridge, MA.

Systems Control Technology (1984). *CTRL-C, A Language for the Computer-Aided Design of Multivariable Control Systems, User's Guide*, Systems Control Technology, Palo Alto, CA.

Taylor, J. H. and Frederick, D. K. (1984). An expert system architecture for computer-aided control engineering, *Proc. IEEE 72*(12), 1795–1805.

Technical Software Systems (1985). *SSPACK User's Manual Including Sample Problems*, Technical Software Systems, Livermore, CA.

Thompson, P. M. (1986). Program CC, Version 3, Personal Communication, Systems Technology, Inc., Hawthorne, CA.

Uyttenhove, H. J. (1979). *SAPS—System Approach Problem Solver*, Computing and Systems Consultants, Inc., Binghampton, NY.

Vanbegin, M. and Van Dooren, P. (1985). *MATLAB-SC, Appendix B: Numerical Subroutines for Systems and Control Problems*, Technical Note N168, Philips Research Laboratories, Bosvoorde, Belgium, 40 pp.

Van den Bosch, P. P. J. (1985). Interactive computer-aided control system analysis and design, in: *Computer-Aided Control System Engineering*, (M. Jamshidi and C. J. Herget, eds.), North-Holland, Amsterdam, pp. 229–242.

West, P. J., Bingulac, S. P., and Perkins, W. R. (1985). L-A-S: A computer-aided control system design language in: *Computer-Aided Control Systems Engineering* (M. Jamshidi and C. J. Herget, eds.), North-Holland, Amsterdam, pp. 243–261.

Wieslander, J. (1980a). *IDPAC Commands—User's Guide*, Report: CODEN: LUTFD2/(TFRT-3157), Dept. of Automatic Control, Lund Institute of Technology, Lund, Sweden, 108pp.

Wieslander, J. (1980b). *MODPAC Commands—User's Guide*, Report: CODEN: LUTFD2(TFRT-3158), Dept. of Automatic Control, Lund Institute of Technology, Lund, Sweden, 81 pp.

Wieslander, J. and Elmqvist, H. (1978). *INTRAC, A Communication Module for Interactive Programs, Language Manual*, Report: CODEN: LUTFD2/(TFRT-3149), Dept. of Automatic Control, Lund Institute of Technology, Lund, Sweden, 60 pp.

Wolovich, W. A. (1974). *Linear Multivariable Systems*, Springer-Verlag, New York.

Wonham, W. M. (1974). *Linear Multivariable Systems: A Geometric Approach*, Springer-Verlag, New York.

Yandell, D. W. (1985). *SAPS-II: Raw Data Analysis in CTRL-C, User's Manual and Progress Report*, Senior Project (F. E. Cellier, adv.), Dept. of Electrical and Computer Engineering, University of Arizona, Tucson, AZ.

Appendix B

Simulation Software
A Survey

The utilization of computers and simulation languages for commercial, industrial and military systems may depend on the volume of applications and
the cost of developing custom-made software. For low-volume (low
computational and memory requirements) applications, microprocessors offer
low cost, small size, low power consumption, and high reliability. Interactive continuous-system simulation languages such as ISIS and ISIM are
already utilized on a whole range of mini- and microcomputers. For high-
volume (high computational and memory requirements) applications, mainframe
coputers are utilized; and numerous general-purpose programming languages,
such as Ada, PASCAL, FORTRAN, ALGOL, PL1 and BASIC, and simulation
languages, such as GPSS, SIMSCRIPT, GASP, SLAM, and SIMULA are available, as presented in Sections 1.33 and 8.2.

Simulation languages usually differ in their logic, construction, flexibility, accessibility, and ease of usage. These differences include: (a)
basic objective of the language, (b) base code from which the simulation
language is constructed, (c) degree of documentation, (d) event classification, (e) time-advance mechanism, (f) algorithms for generating random
numbers and random deviates, (b) program initialization, (h) data entry,
(i) output reports, and (j) methods for data analysis, as indicated by
Golden (1985) and Pratt (1984). Most simulation languages, however, provide the following standard capabilities: (a) structured data input, (b)
time-advance mechanism, (c) acceptable random-number generators and
random-deviate generators (d) output reports and automatic statistical
analysis routines, and (e) proper documentation and instructions that are
easy to understand and use, as indicated in Section 3.3.

Recent years have witnessed the development of a tremendous number
of discrete, continuous, and combined discrete-continuous simulation
languages. Among the widely used software are the following.

B.1   ADA

Ada, the new Department of Defense standard language, contains many
features, such as tasks and packages, that allow the construction of a

powerful, process-oriented, discrete-system simulation tool. Bryant (1982) develops a design for such a simulation package in Ada. He has shown that the Task construct of Ada allows one to create process-oriented simulation packages; and Generic Packages allow the definition of packages that can be parameterized to implement queries of arbitrary user types. Statistics collection packages are also treated in a similar way. Ada can thus promote the essential features of languages like SIMSCRIPT II.5, the portability of GASP IV, and the advantages of strong-typing and process-orientation for efficient and reliable construction of large discrete simulation programs.

B.2   ACSL (ADVANCED CONTINUOUS SIMULATION LANGUAGE)

ACSL is a continuous-system simulation language that has extended features beyond those of CSMP-III. ACSL is designed for modeling and evaluating the performance of continuous systems described by time-dependent, nonlinear differential equations. Emphasis is placed on the ability to run and evaluate the model on-line. In ACSL, provision has been made to overcome the problem of high data volume; monitoring information can be directed to a a terminal and high volume output to a local line printer. ACSL has a flexible explicit structure, which includes INITIAL, DYNAMIC (with embedded DERIVATIVE and DISCRETE), and TERMINAL sections. Typical application areas of ACSL are missile and aircraft simulation, control systems, chemical processes, and heat transfer analysis. Program construction can be from either block diagram representation, conventional FORTRAN statements, or a mixture of both, as described by Mitchell and Gauthier (1986).

B.3   CSMP-III   (CONTINUOUS-SYSTEM MODELING PROGRAM)

CSMP is one of a wide variety of continuous-system simulation languages. It has been developed to solve systems of first-order differential equations. A CSMP-III program is composed of three types of statements: (a) data statements, which initialize conditions for state variables and assign numerical values to constants, (b) structural statements, which specify the standard functions for solutions, and (c) control statements, which identify input/output formatting and options for program execution. CSMP-II is equation-oriented and allows the simulation of continuous process directly and simply from either block diagram representations or a set of ordinary differential equations. It has high flexibility (degree of freedom in describing a process to be simulated), high adaptability, and utilizes seven numerical integration techniques (rectangular, trapezoidal, Simpson's, Runge-Kutta variable interval, Milne predictor-corrector, and Adams). CSMP-III has the features of nesting, debug-aid, and storage capabilities, as indicated by Speckhart and Green (1976).

B.4 DARE-P (DIFFERENTIAL ANALYSES REPLACEMENT EVALUATION)

DARE-P is a continuous-system simulation language that has extended features beyond those of CSMP-III and ACSL. The processor of DARE-P system is written entirely in ANSI-FORTRAN and is therefore machine-independent, as explained by Korn and Wait (1978).

B.5 DPS (DYNAMIC PROCESS SIMULATOR)

DPS is a continuous-system simulation language that was developed initially at the Computation Centre of the Japanese University of Scientists and Engineers, and now jointly with CAD Centre of Cambridge, England. DPS follows the approach employed in steady-state process flowsheeting programs, in which the user defines the problem by linking previously defined models through material and information flow. Models in DPS are of two types: (a) element: A model defined in terms of algebraic and ordinary differential equations, and (b) unit: A model defined by a collection of elements with its own internal connections.

The DPS program is designed to solve systems of algebraic or ordinary differential equations which may or may not involve time-delay relations. DPS has no restrictions on the occurrence of algebraic loops, the number of unknown variables that are part of a loop, or the number of loops.

DPS provides a choice of Euler, Runge-Kutta, and backward Euler methods for integration of the differential equations, and the direct substitution and Newton-Raphson methods for solving systems of algebraic equations. DPS is available for use with PRIME 50 SERIES (PRIMOS), DEC VAX (VMS), and IBM (MVS/TSO) computers, as indicated by Wood et al. (1984).

B.6 DYNAMO

DYNAMO is a special-purpose compiler for the simulation of information-feedback systems. It was developed to serve a method of systems analysis called industrial dynamics [see Forrester (1983)]. The rules of DYNAMO together with the conventions of industrial dynamics form a computer simulation language that is called simply DYNAMO. It belongs to that wide variety of special-purpose continuous-system simulation languages. DYNAMO treats certain types of dynamic information feedback systems that can be described in terms of a set of finite difference equations. DYNAMO makes use of two different types of instructions, equations and directions, to obtain step-by-step numerical solutions to the set of difference equations describing the system under study. The basic components of the DYNAMO language are almost identical to those found in FORTRAN because they include the following: variables, constants, subscripts, equations, and functions. However, DYNAMO variables are further subdivided into levels, auxiliaries, rates, supplementary variables, and initial values. Among the special functions or subroutines available in DYNAMO are: exponential, logarithmic, third-order delays, step functions, ramp functions, and switch functions. DYNAMO uses a fixed step size and Euler-type integration algorithm. Such step size should not be selected too large; otherwise the accuracy of the model will suffer.

B.7   ESL

ESL is a new continuous-system simulation language (CSSL) that is being developed under contract from the European Space Agency. Its main features include: (a) separation of model and experiment, (b) capability of building models from submodels, (c) advanced discontinuity handling, and (d) parallel segmentation. To implement ESL, an interpreter and a translator version of the language are required. The interpreter translates the user's program into H-Code and the translator converts the H-Code to FORTRAN-77. The entire system is written in FORTRAN-77, as indicated by Crosbie et al. (1985).

ESL was developed on a PRIME 550 computer in the Simulation Laboratory of the University of Salford, England.

B.8   GASP V (GENERAL ACTIVITY SIMULATION PROGRAM)

GASP is a combined discrete-continuous simulation language. It provides an organizational structure that specifies routines for writing differential or difference equations. GASP V has procedures to handle partial differential equations, logic functions (input switches, flipflops, and gates), memory functions (hysteresis and delays), and generator functions (step, ramp, etc.). Other advantages of GASP are its modular characteristics and its machine-independence, which make it possible to alter or expand simulation programs to suit the needs of any given system. Since the entire GASP program is written in FORTRAN, transfer of a model from one machine to another is limited only by the existence of a FORTRAN translator and sufficient computer memory [see Pritsker (1974) and Section 3.3].

B.9   GEMS (GENERAL EQUATION MODELING SYSTEM)

GEMS is a continuous-system simulation language that provides a simple and convenient way to formulate complex mathematical models for simulation using the IMP (implicit modeling package) algorithms. GEMS is well suited for the solution of sets of stiff, nonlinear algebraic and differential equations. GEMS was developed in the Department of Chemical Engineering at the University of Connecticut. It is run on a Control Data 6600 computer and is available under limited conditions through L. F. Stutzman [see Babcock, Stutzman, and Koup (1981)].

B.10   GPSS (GENERAL-PURPOSE SIMULATION SYSTEM)

GPSS is a process-oriented simulation programming language used to build computer models for discrete-event simulations, as described in Section 3.3.

A GPSS model is constructed by arranging a set of standard blocks into a block diagram that defines the logical structure of the system. The GPSS language consists of 48 standard statements, each of which has a corresponding representation called a block. Some of these blocks and their functional descriptions are as follows:

Block type	Functional description
GENERATE	Create transactions
SEIZE	Cause a transaction to capture a facility
RELEASE	Cause a transaction to release a facility
ENTER	Cause a transaction to capture a set of parallel servers
LEAVE	Cause a transaction to release a set of parallel servers
ADVANCE	Actual service time of a transaction
QUEUE	Increase the number in a queue
DEPART	Decrease the number in a queue
TERMINATE	Stop a run
TABULATE	Record the observations of a system variable

In GPSS customers or temporary entities that require service are called transactions, and permanent entities that provide the required service are called facilities or storages. The principal appeal of GPSS is its modeling simplicity. It is easy to learn, and it takes less time to code and debug than other general-purpose simulation languages. Disadvantages of GPSS include the following (a) It is limited in computing power, (b) it lacks a capability of floating-point or real arithmetic, and (c) it has a limited capability for generating random variates. However, in the most recently developed versions of GPSS, i.e., GPSS/H, programs are reported to run faster than corresponding GPSS 360/V programs, and also a wider variety of variate random generators has been provided.

GPSS/H was developed by James Henriksen (1979) of Wolverine Software Corporation. It is implemented for IBM 370, 43XX, and 30XX machines, and for DEC's VAX machines.

B.11 ISIS AND ISIM (INTERACTIVE SIMULATION)

ISIS is an interactive simulation language developed in the mid-1970s as a portable, highly interactive, continuous-system simulation language (CSSL) for minicomputers. It includes a number of advanced features, such as discontinuity processing and stiff integration, and has been installed on a wide range of minicomputers, including PDP 11/20, VAX, HP 1000/2000, Harris, Cyber, Perkin-Elmer, and Norsk Data. ISIM is a continuous-system simulation language derived from ISIS and provides many interactive features. It allows models to be defined using a natural differential equation notation. Outputs may be given in either tabulated or graphical form; flexible run-control features are provided. The ISIM language has much in common with ISIS. The model description is made up of three regions: (a) the initial region, which contains code and initial condition setting, (b) the dynamic region, which is composed of the differential and algebraic equations defining the model, and (c) the terminal region.

ISIM can run of a small eight-bit microcomputer with the CP/M operating system, 48K bytes of memory, and floppy disks. It also runs on 16-bit computers operating under CP/M 86 or MS-DOS and on IBM-PC under PC-DOS, as described by Hay and Crosbie (1984).

B.12   MACSYMA (MAC'S SYMBOLIC MANIPULATION SYSTEM)

MACSYMA is an interactive symbol manipulation language used for performing symbolic as well as numerical mathematical manipulations. It was developed specifically for interactive use and has capabilities for manipulating algebraic expressions involving numbers, variables, and functions. It can differentiate, integrate, take limits, solve systems of equations, factor polynomials, expand functions as Laurent or Taylor series, plot curves, manipulate matrices, etc. MACSYMA is applied to the problems of formulating models of aeronautical systems for simulation studies, as explained by Howard (1976).

B.13   MACTRAN

MACTRAN is a simulation language designed for the purpose of editing data on an observation-by-observation basis. It may be viewed as having standard analog function capability, such as integration, differentiation, a variety of filters, delay, plus all the elementary operations and functions. MACTRAN language is related to FORTRAN and has FORTRAN-like statements. It differs from FORTRAN in that a number of FORTRAN capabilities are not needed and hence are not included. On the other hand, several operations not available in FORTRAN, but handy in editing, have been added to MACTRAN [see Otnes and McNamme (1974)].

B.14   MARSYAS (MARSHALL SYSTEM FOR AEROSPACE SIMULATION)

MARSYAS is a block- or equation-oriented simulation language that can be used to simulate a system of differential equations or block-oriented systems. MARSYAS has the features of nesting, high flexibility, high adaptability, integration, storage capacity, and multiple simulation. It utilizes five numerical integration techniques (Euler, Runge-Kutta fixed interval, Adams-Bashforth predictor-corrector, Sarafyan fifth-order variable step, and Butcher's fifth order). MARSYAS does not have a debug-aid, and this presents a negative aspect.

B.15 MODEL S (MULTIOPTIONAL DIFFERENTIAL-EQUATION LANGUAGE)

MODEL S is a continuous-system simulation model that is based on the inter-active simulation language MODEL 1. It was developed to study and simu-late large, flexible spacecraft with digital controllers; and thus MODEL S can accept mathematical models involving both differential equations and sampled-data equations. MODEL S does not include any restrictions that re-quire reordering the model's equations when writing the simulation program. Consequently, the language is kept as simple and unrestricted as possible while the translator performs routine tasks such as checking, reordering, and sequencing. MODEL S generates FORTRAN code, which minimizes mathematical operations within the integration loop. Simulations of digital control systems show the language's power, flexibility, and ease of use and understanding, as described by Zimmerman (1983).

B.16 PASSIM (PASCAL-BASED SIMULATION)

PASSIM is a set of procedures that can be used to construct discrete-event simulation models in PASCAL. The procedures assume an event-oriented approach that combines schedules and entity concepts from GPSS with pointer-based data structures and control structures from PASCAL. PASSIM includes a dynamic entity type and three system entity types. The dy-namic entity is the transaction; the system entities are facility, storage, and chain. PASSIM also provides block procedures that represent events. PASSIM has more capabilities than the FORTRAN-based SPURT package, but less than GASP. To use PASSIM the modeler should be a competent PASCAL programmer [see Uyeno and Vaessen (1980)].

B.17 PLUG

PLUG is a simulation language designed to handle time series on a time-slice by time-slice basis. PLUG operates in two modes. In the first, data are input, processed, and output; in the second, no data are input, rather data are generated internally, manipulated, and then output. PLUG manipu-lations include the following: (a) arithmetic operations, (b) elementary real functions, (c) complex arithmetic, (d) complex functions, (e) digital filtering, (f) data generation, and (g) testing and transfer functions. PLUG has a total of 46 different operations that may be performed, and up to four data functions may be involved in a single operation, as explained by Otnes and McNamme (1974).

B.18 Q-GERT (QUEUEING SYSTEMS: GRAPHICAL EVALUATION AND REVIEW TECHNIQUE)

Q-GERT is a process-oriented discrete-simulation programming language that employs an activity-on-branch network concept, in which a branch represents an activity that models a processing time or delay. Nodes are used to separate branches and to model decision points and queues. Flowing through the network are entities, called transactions. There are 10 node types in Q-GERT, as follows:

Node type	Description
SOURCE	Create transactions
REGULAR	Accumulate transactions
QUEUE	Determine disposition of transaction: hold or route to server
SELECT	Determine disposition of transaction and/or server
MATCH	Match transactions in queue nodes with same attribute values and route transactions when a model occurs
ALLOCATE	Allocate resources to transactions waiting for resources in queue nodes
FREE	Make resources available to be allocated
SINK	Accumulate transactions, collect statistics, terminate a run
ALTER	Change capacity of resources
STATISTICS	Accumulate transactions and collect statistics

To construct a model in Q-GERT, the modeler combines the Q-GERT network elements into a network model that pictorially represents the system of interest. Q-GERT is easy to learn, has a real-valued clock, and provides functions to generate all commonly used random variables [see Pritsker (1979) and Section 3.3].

B.19   SDL (SIMULATION DATA LANGUAGE)

SDL is a database management system with data organization capabilities and data manipulation functions. SDL provides data structures in which models' inputs and outputs can be stored, including time series of observations, statistics, and histogram. These structures allow storage, in a single database, inputs and outputs from multiple runs of the same model and runs of different simulation experiments. Besides, SDL provides FORTRAN subroutine calls designed specifically to be used in retrieving inputs and storing outputs. SDL has commands that meet the specialized needs of simulation analysts. For example, the commands that perform statistical computations on time series of model outputs let the user select the data by (a) model replication, (b) batches within a single replication, or (c) regeneration cycle. SDL provides OIL, a high-level programming language for data manipulation. The OIL processor translates OIL statements into calls to the appropriate SDL subroutines. The operational characteristics of SDL are as follows:

SDL is written in 1966 ANSI FORTRAN IV. It is independent of any simulation language and may be interfaced with any simulation language or other program capable of calling FORTRAN subroutines. It has been implemented on several computers, including an IBM 370/168, a CDC CYPER 175, and a DEC VAX 11/780. SDL has been interfaced successfully with the SLAM, Q-GERT, and GASP IV simulation languages. SDL stores the database in a single FORTRAN file accessed by relative record numbers. The accessing of this file varies from machine to machine, but the machine-dependent part of the SDL consists of only about 30 lines of FORTRAN. Versions have been coded for most large computers [see Standridge and Phillips (1983)].

B.20 SIMAN (SIMULATION ANALYSIS)

SIMAN is a general-purpose simulation programming language that can utilize event scheduling, process interaction, or continuous approach, or a combination of these approaches (see Section 3.3). SIMAN uses more than 140 blocks or statements, many of which are similar to GPSS blocks. SIMAN has five individual processors: (a) a model processor for constructing the block diagram model, (b) an experimental processor for defining the experiment, (c) a link processor for combining the model and experimental processors, (d) a run processor, and (e) an output processor. These processors give the user flexibility to perform sensitivity analysis. SIMAN was developed by Pegden (1984) and runs on any computer with a FORTRAN 66 compiler. It is currently running on IBM, Univac, DEC, Honeywell, and CDC computers. It also runs on microcomputers using MS-DOS, such as IBM-PC, Wang PC, Victor 9000, HP 150, Zenith 100, and TI professional. Both mainframe and micro versions are completely compatible, as indicated by Banks and Carson (1985).

B.21 SIMPAC

SIMPAC is a fixed-time-increment simulation language that uses standard flowchart symbols. Models formulated in SIMPAC consist of four basic components: activities, transactions, queues, and operation resources. Although SIMPAC is characterized by a fairly flexible range of output reports, it is a somewhat more difficult language to learn than GPSS, GASP, or DYNAMO [see Young (1963)].

B.22 SIMPL/1

SIMPL/1 is a simulation programming language that uses the process interaction approach, is implemented as a superset of PL/1, and follows the structure and design philosophy of PL/1. It is an extension of the PL/1 instruction set that includes macrostatements for performing discrete-event digital simulation. Here a preprocessor translates SIMPL/1 simulation program into corresponding PL/1 code. Compilation occurs via a PL/1 compiler, but error messages that are detected during compilation and execution refer to the original SIMPL/1 code. SIMPL/1 combines the special-purpose features of a simulation system with the flexibility and power of the PL/1 high-level language. The user has access to the standard mathematical and statistical routines of PL/1 libraries, a list-processing capability, and specialized facilities necessary for modeling many types of systems, as shown in IBM (1972a,b).

B.23   SIMSCRIPT II.5

SIMSCRIPT is primarily event-oriented discrete-simulation programming language (see Section 3.3). In SIMSCRIPT, the static structure of the system is described by entities; their associated attributes, and logical groupings of entities are referred to as sets. The dynamic structure is modeled by events which are changes of state taking place instantaneously at discrete points in simulated time initiated by the execution of an event routine. To construct a discrete-event simulation model in SIMSCRIPT, one must write a preamble, a main program, and the usual event routines. The preamble gives a static description of the entire model and the required events, entities, and sets. It also defines variables for which statistics are to be collected. The preamble does not contain any executable statements. The main program initializes the event list and the state variables, reads input parameters, and starts the simulation. Event routines define the logic associated with processing each event in the model.

Programs written in SIMSCRIPT are easy to read and understand and tend to be almost self-documenting; this is due to its English-like and free-form syntax.

SIMSCRIPT was initially developed by the RAND Corporation in the 1960s. The latest version of the language, SIMSCRIPT II.5, is owned by CACI, Inc.

SIMSCRIPT II.5 is available for IBM, Univac, CDC, Honeywell, Digital Prime, and NCR mainframe computers and for the IBM-PC, as indicated by Kiviat (1978a) and Russell (1981).

B.24   SIMULA

SIMULA is a superset of ALGOL; so it is really a general-purpose programming language, despite its name. The instructions have the form of ALGOL statements, and the concept of system classes defines a set of characteristics of special interest in certain application areas. SIMULA extends the block concept which is the fundamental mechanism for decomposition in ALGOL, but unlike ALGOL, SIMULA provides input/output statements as a standard part of the language. To allow flexible string handling, character and text variables are available with different handling procedures [see Dahl and Nygaard (1966)]. SIMULA advantages include efficient compilation and programming ease due to the adequacy of its formal syntactic structure, high computational power, and ease of extensibility. However, SIMULA is complicated by the lack of adequate documentation, confusing semantics, and difficulties of program segmentation.

SIMULA was developed at the Norwegian Computing Center from 1963 onward. The latest version, SIMULA 67, contains Algol 60 as a subset (Algol's own variables are not included). It is implemented for IBM, CDC, Univac, and VAX mainframes and some microcomputers.

B.25   SLAM   (SIMULATION LANGUAGE FOR
ALTERNATIVE MODELING)

SLAM is a combined discrete-continuous simulation language (see Section 3.3). It incorporates the process-oriented features of Q-GERT and the combined discrete-continuous features of GASP. As such, the modeler can build discrete-event models using either the event or the process orientation or both, continuous models employing differential equations, and combined models using all these orientations. Pritsker and Pegden (1979) and Wilson and Pritsker (1984) cite six specific interactions that can take place if all above orientations are combined within the same simulation model. These are as follows:

1. Entities in the network model (process orientation) can indicate the occurrence of discrete events.
2. Events can alter the flow of entities in the network model.
3. Entities in the network model can cause instantaneous changes to values of the state variables.
4. State variables reaching prescribed values can initiate entities in the network model.
5. Events can cause instantaneous changes to the values of state variables.
6. State variables reaching preset threshold values can initiate events.

The real appeal of SLAM is due to the ability it gives the modeler to construct combined network-event-continuous models with interactions between each orientation. This enhances the modeling power and flexibility.

SLAM II, the latest release of SLAM, permits process-interaction, event-scheduling, and continuous-modeling approaches, or any combination of the three, to be used in constructing a single simulation model.

SLAM runs on mainframe computers of IBM, CDC, DEC, Honeywell, Harris, Univac, CRAY, and Xerox. SLAM II requires a FORTRAN 66 compiler. In addition, an IBM-PC version of SLAM II is available and is compatible with the mainframe version.

B.26 CONCLUSION

Most real-world systems are too complex to allow realistic models to be evaluated analytically; so these models must be studied by simulation. In a simulation, we usually describe the problem entity (system) in a form of mathematical relationships (model), which is in turn transformed into a computerized model by utilization of a general- or special-purpose simulation language. General-purpose simulation languages have the following advantages: (a) They are available on virtually every computer, (b) they may require less execution time, (c) they allow greater programming flexibility, and (d) most modelers already know the language. The advantages of special-purpose simulation lnaguages include the following: (a) They decrease programming time, (b) they provide better error detection, (c) their basic building blocks are more relevant to simulation, (d) they provide dynamic storage allocation during execution, and (e) models are easier to change.

Criteria for the selection of a simulation language include the following: (a) flexibility and power of the language, (b) computer time efficiency of the language, (c) computer storage requirements, (d) documentation, ease of learning and usage, (e) portability, (f) types of systems to be simulated, (g) familiarity with language, and (h) cost of installing and maintaining the language.

BIBLIOGRAPHY

Abed, S. Y., Barta, T. A., and McRoberts, K. L. (1985). A qualitative comparison of three simulation languages: GPSS/H, SLAM, SIMSCRIPT, *Computers Industr. Engineering 9*(1), 34–43.

ACSL/User Guide (1975). Mitchell and Gauthier Associates, Concord, MA.

Babcock, P. D., Stutzman, L. F., and Koup, T. G. (1981). General equation modeling systems (GEMS)—A mathematical simulation language, *Simulation 36*(2), 41–49.

Baldonado, O. and Wespierson, J. (1973). FORTRAN vs. GPSS in reliability simulation, in: *Proceedings Annual Reliability and Maintainability Symposium*, IEEE, N.Y., New York, pp. 550–553.

Banks, J. and Carson, J. S., II (1985). Process-interaction simulation languages, *Simulation 44*(5), 225–235.

Bratley, P., Fox, B. L., and Schrage, L. E. (1983). *A Guide to Simulation*, Springer-Verlag, New York.

Brender, R. F. and Nassi, I. R. (1981). What is Ada? *IEEE Computer 14* (6), 17–24.

Brennan, R. D. (1968). Continuous system modeling programs: State-of-the-art and prospectus for development, *Simulation Programming Languages*, 3(2), pp. 371–394.

Bryant, R. M. (1982). Discrete system simulation in Ada, *Simulation 39*(4), 111–121.

Cellier, F. and Blitz, A. E. (1976). GASP V: A universal simulation package, in *Proceedings IFAC Conference*, IEEE, New York, NY, pp. 230–236.

Cooper, D. and Clancy, M. (1982). *Oh. Pascal!* Norton, New York.

Crosbie, R. E., Javey, S., Hay, J. L., and Pearce, J. G. (1985). ESL—A new continuous system simulation language, *Simulation 44*(45), 242–246.

CSSL-IV Continuous Simulation Language (1976). User Guide/Reference Manual, Nielson Associates, Chatsworth, CA.

Dhal, O. G., and Nygaard, K. (1966). SIMULA and ALGOL-based simulation language, *Commun Assoc. Computing Machinery*, 9(9), 671–678.

Dhal, O. G., Myhrhaug, B., and Nygaard, K. (1968). SIMULA 67: Common base language, Publication No. S-2, Norwegian Computing Center, Oslo, Norway.

Dynamic Process Simulator Casebook (1983). CAD Centre Ltd., Cambridge, England.

Feuer, A. R., and Gehani, N. H. (1982). A comparison of the programming languages C and PASCAL, *Computing Surveys 14*(1), 73–92.

Fishman, G. S. (1978). *Principles of Discrete Event Simulation*, Wiley, New York.

Forrester, J. W. (1983). *Industrial Dynamics*, The MIT Press, Cambridge, MA.

Golden, D. G. (1985). Software engineering considerations for the design of simulation languages, *Simulation 45*(4), 169–178.

Golden, D. G., and Schoeffler, J. D. (1973). GSL—A combined continuous and discrete simulation language, *Simulation 20*(1), 1–8.

Gordon, G. (1975). *The Application of GPSS to Discrete System Simulation*, Prentice-Hall, Englewood Cliffs, NJ.

Gordon, G. (1978). *System Simulation*, 2nd ed., Prentice-Hall, Englewood Cliffs, NJ.

Graybeal, W. and Pooch, U. W. (1980). *Simulation: Principles and Methods*, Winthrop, Cambridge, MA.

Hammond, W. M. and McNabb, J. W. (1974). Modeling and simulation with CSMP, in: *Proceedings of the 4th Annual Conference on Modeling and Simulation*, Simulation Councils, Inc., San Diego, CA, pp. 403–407.

Hay, J. L. (1978). Interactive simulation on minicomputers: Part 1-ISIS, a CSSL language, *Simulation 31(1)*, 1–7.

Hay, J. L. and Crosbie, R. E. (1984). ISIM; a simulation language for microprocessors, *Simulation 43(3)*, 133–136.

Henriksen, J. O. (1979). *The GPSS/H User's Manual*, Wolverine Software, Falls Church, VA.

Henriksen, J. O. (1985). The development of GPSS/85 in: *Proceedings of the 18th Annual Simulation Symposium*, Simulation Councils, Inc., San Diego, CA. pp. 66–77.

Howard, J. C. (1979). The formulation of simulation models of aeronautical systems, in: *Proceedings of the Seventh Annual Conference on Modeling and Simulation*, Simulation Councils, Inc., San Diego, CA, pp. 510–515.

IBM, SIMPL/1 Operations Guide, SH-19, 5038-0 (1972a).

IBM, SIMPL/1 Program Reference Manual, SH19-5060-0 (1972b).

IBM-G20-0367-Systems/360 Continuous System Modeling Program—User's Manual (1980).

IBM-G-20-0282-1130 Continuous System Modeling Program—Program Description and Operations Manual (1982a).

IBM-SH19-7001-Continuous System Modeling Program III (CSMPIII), Program Reference Manual (1982b).

Kernighan, B. W. and Ritchie, D. M. (1978). *The C Programming Language*, Prentice-Hall, Englewood Cliffs, NJ.

Kiviat, P. J., Villanueva, R. V., and Markowitz, H. (1978a). *SIMSCRIPT II.5 Programming Language* (E. C. Russell, ed.), CACI, Inc., Los Angeles.

Kiviat, P. J., Villanueva, R., and Markowitz, H. (1978b). *The SIMSCRIPT II Programming Language*, Prentice-Hall, Englewood Cliffs, NJ.

Korn, G. A. and Wait, J. V. (1978). *Digital Continuous-System Simulation*, Prentice-Hall, Englewood Cliffs, NJ.

Law, A. M. and Kelton, W. D. (1982). *Simulation Modeling and Analysis*, McGraw-Hill, New York.

Leventhal, L. A. (1978). Microprocessors in aerospace applications, *Simulation 30(4)*, 111–115.

Levi, A., Cardillo, D., and Unger, B. W. (1978). Programming languages for computer system simulation, *Simulation 30(4)*, 101–110.

MARSYAS (1974). Computer Sciences Corporation, Huntsville, AL.

Mitchell, E. L. and Gauthier, J. S. (1986). *Advanced Continuous Simulation Language (ACSL)*, Reference Manual, Mitchell & Gauthier Assoc., Concord, MA.

Otnes, R. K. and McNamme, L. P. (1974). The PLUG and MACTRAN simulation languages, in: *Proceedings of the 4th Annual Conference on Modeling and Simulation*, Simulation Councils, Inc., San Diego, CA, pp. 398–402.

Pegden, C. (1984). *Introduction to SIMAN*. Systems Modeling Corporation, State College, PA.

Pratt, C. A. (1984). Catalog of simulation software, *Simulation 43(4)*, 180–192.

Pritsker, A. A. B. (1974). *The GASP IV Simulation Language*, Wiley, New York.

Pritsker, A. A. B. (1979). *Modeling and Analysis Using Q-GERT Networks,* 2nd ed., Halstead, New York.

Pritsker, A. A. B. and Pegden, C. D. (1979). *Introduction to Simulation and SLAM,* Halstead, New York.

Pugh, A. L. (1973). *DYNAMO II User's Manual,* MIT Press, Cambridge, MA.

Robert, E. B. (1964). New directions in industrial dynamics, *Industr. Management Rev.* 6(1), 5–14.

Russell, E. C. (1976). *Simulation and SIMSCRIPT II. 5,* CACI, Inc., Los Angeles.

Russell, E. C. (1981). *Building Simulation Models with SIMSCRIPT II. 5,* CACI, Inc., Los Angeles.

Schriber, T. J. (1976). *Simulation Using GPSS,* Wiley, New York.

Schroer, B. J., Black, J. T., and Zhang, S. X. (1985). Just-In-Time (JIT), with Kanban, manufacturing system simulation on a microcomputer, *Simulation* 45(2), 62–80.

Speckhart, H. and Green, W. H. (1976). *A Guide to Using CSMP,* Prentice-Hall, Englewood Cliffs, NJ.

SPURT: A FORTRAN Simulation Package for University Research and Training, (1969). University of British Columbia Computing Centre, Vancouver, Canada.

Standridge, C. R. (1981a). The simulation data language (SDL): Applications and examples, *Simulation* 37(10), 119–130.

Standridge, C. R. (1981b). Using the simulation data language (SDL), *Simulation* 37(9), 73–81.

Standridge, C. R. and Phillips, J. R. (1983). Using SLAM and SDL to assess space shuttle experiments, *Simulation* 41(1), 25–35.

Standridge, C. R. and Wortman, D. B. (1981). A database management system for modelers, *Simulation* 37(8), 55–57.

Ung, M. (1984). *Aerospace Simulation,* Simulation Series, Vol. 13, The Society for Computer Simulation, La Jolla, CA, p. 1.

Unger, B. W. (1978). Programming languages for computer system simulation, *Simulation* 30(4), 101–110.

Uyeno, D. H. and Vaessen, W. (1980). PASSIM: A discrete event simulation package for PASCAL, *Simulation* 35(6), 183–190.

Wegner, P. (1980). *Programming with Ada,* Prentice-Hall, Englewood Cliffs, NJ.

Wilson, J. R. and Pritsker, A. A. B. (1984). Computer simulation, in: *Handbook of Industrial Engineering* (G. Salvendy, ed.), Wiley, New York.

Wirth, N. and Jensen, K. (1974). *PASCAL User Manual and Report,* Springer-Verlag, Berlin, Germany.

Wood, R. K., Thambynayagam, R. K. M., Noble, R. G., and Sebastian, D. J. (1984). DPS: A digital simulation language for the process industries, *Simulation* 42(5), 221–231.

Young, K. A. (1963). *User's Experience with Three Simulation Languages: GPSS, SIMSCRIPT AND SIMPAC,* System Development Corporation, Report No. TM-1755/000/00.

Zimmerman, B. G. (1983). Model S, a sampled-data simulation language, *Simulation* 40(5), 183–193.

Index

ACSL, 20,632,666,682
A/D conversion, 18,172
A/D converter, 371,403
AD-10, 612
AD-100, 614
ADSIM, 616
AHC assembler, 364
ALGOL, 20
AND, 171
AND gate, 172
AP-120B, 610,612
APL, 634
ARISTOTLE, 508,527
Accelerometer, 40,142
Acceptance-rejection method, 129
Access time, 339
Accuracy, 18, 214
Activity, 99
Activity oriented, 671
Actuators, 452, 478–485
Ada, 20, 100, 681
Adaptive control, 490
Additive congruential method, 127
Aggregability, 218
Aggregation, 214, 215–231
Aggregation matrix, 218
Agriculture, 504, 516
Agriculture models, 521
Algorithms, 267
Algorithm robustness, 639
Algorithmic design parameters, 642
Aliase, 405

Alternative models, comparison of, 193
Analog computer-general purpose, 18
Analog computers, 355
Analog/digital simulation, 390
Analog simulation, 155–171, 376
Analogous systems, 38
Analysis, 3
Analysis of variance, 196
Animation, 646
  layout, 582
  system, 581
Antithetic variates, 199
Applications
  humanities/social sciences, 3
  science-oriented, 3
  technical/engineering, 3
Arithmetic-logic unit, 356
Armature-controlled motor, 41
Armature-controlled motor, state diagram, 58
Array, 268
Array processor, 369, 610
Array processor characteristics, 613
Arrays, 271, 273
Arrival process, 114
Artificial intelligence, 21
Assembler, 358
Assembly language, 358
Assessment of languages, 307–313
Asymptotes, 84
Attributes, 98

Autocorrelation function, 548
Autocorrelation test, 128
Automatic guided vehicles, 568
Auxiliary equation, 81

BASIC, 20
Backward equations, 485–489
Balanced aggregation, 227–231
Balanced approach for unstable
    systems, 231
Balancing transformation, 228
Bariloche model, 501, 519
Batch-driven program, 645
Bernoulli process, 103
Bias errors, 491
Bilinear transformation, 390, 411
Binary-coded decimal, 372
Binary number representation, 439,
    265
Binary semaphore, 334
Binary tree, 280
Binomial distribution, 104
Black box, 4
Block diagram, 9
Block diagram algebra, 49, 50-54
Block function name, 572
Blocked state, 328, 343
Blocking method, 183
Boolean type, 267
Boot-strapping technique, 99
Boundary layer, 215, 237
Boundary value problems, 632
Bounded buffer problem, 336
Breakaway point, 85
Burrough B-6500, 602
Business cycle, 541

C, 100
CACSD, 633
CAD, 21, 287
CAD tools, 413
CAE, 21
CAM, 21
CATPAC, 657
CDC 6600, 599
CDC Cyber-205, 610
CDC Star-100, 604
CHAINAGG, 226–227
CM, 607
COBOL, 20
CREATE block, 573
CSMP, 20, 151

CSMP-III, 682
CSSL, 10, 151, 633
CTRL-C, 663–667, 649
C-SIMSCRIPT, 20
Cache, 599
Calendar of events, 99
Calling population, 114
Cascade compensation, 415
Cascade controller, 403
Cascade programming, 378
Causal model, 507, 508
Center of mass, 484
Centrality, 247
Centralized systems, 249
Centroid, 84
Chained aggregation, 222–227
Character type, 267
Characteristic polynomial, 79
Characteristic roots, 551
Charged coupled device camera, 451
Cinema system, 568, 581
Circuit level, 282
Closed loop system, 403
Club of Rome, 499
Code-driven program, 645
Command-driven interface, 645
Command spy, 673
Common random numbers, 198
Compiler, 358, 636, 645
Complementary strips, 406–409
Completely aggregable, 222
Complex systems, 213
Components, 4
Composite system method, 256
Computational delays, 443
Computed torque technique, 489
Computer-aided design software, 633
Computer-aided instruction, 673
Computer, analog, 17
Computer, digital, 17
Computer, hybrid, 17
Computer-memory, 18
Computerized model, 5
Computer network, 609
Computing time, 18
Concatenated, 267
Concurrency, 327, 330
Concurrency control, 287
Confidence interval, 190
Constant multiplier technique, 126
Constraint minimization, 417
Consumer price index, 557
Conting semaphore, 336
Continuous allocation, 342

Continuous
   random variable, 102
   simulation languages, 291–317
   system, 11
Control law, 422, 550
Control matrix, 546
Control of robots, 486–491
Control ratio, 405
Control unit, 356
Control variates, 204
Controllability, 88, 635, 257
Controllability matrix, 89
Controllability of singularly
      perturbed system, 257
Controller-canonical representation,
      640
Controller structure, 440–442,
      638
Convergence, 561
Conveyor path, 589
Conveyors, 580
Convolution method, 128
Coordination of motion, 454
Coordinator problem, 242
Coprocessor, 369
Correlation analysis, 13
Cost function, 422
Counter, 100
Cray-1, 19, 610
Credible models, 4
Critical region, 332
Cross-correlation function, 548
Cumulative distribution function,
      102, 105
Cycle, 126

D/A conversion, 172
D/A converter, 371, 403
DARE-P, 20, 683
DC motor, 40
DDL, 286
DML, 286
DPS, 683
DSL, 20
DSL/VS, 632
DYMOLA, 659
DYNAMO, 20, 683
DYSAC, 20
Damped natural frequency, 77
Damping factor, 76–77, 409
Data abstraction
   conceptual level, 283
   physical level, 283

[Data abstraction]
   view level, 283
Data
   constraints, 284
   handling, 265
   independence, 284
   isolation, 282
   models, 284
   processing, 72
   redundancy, 282
   semantics, 284
   structures, 265, 633
Data-driven program, 645, 657
Data, nonnumerical, 265
Data, numerical, 266
Database, 13, 281, 265
   interface, 663
   logical structure of, 286
   management system, 281
   manager, 287
   schemes, 283
Decentralized control, 247–256, 638
Decision assessment, 538
Decision making, 3, 581
Declarations, 271
Decomposed system, 248
Decomposition, 213, 239
Deduction, 8
Deductive models, 13
Degenerate circuits, 59
Degrees of freedom, 450
Delay, 99
Demand, 502
Demand segmentation, 348
Denavit-Hartenberg, 462–464
Derivative controller, 78
Describing function, 16
Design methods, 413, 636
Design shells, 644
Device driver, 342
Device manager, 322, 342–343
Diagram model, 572
Difference equation
   backward, 409
   forward, 409
   order of, 63
   time invariant, 405
Difference equations, 11, 63, 139
Digital-analog simulator, 20
Digital/analog converter, 18, 371,
      403
Digital
   computers, 355
   control structures, 402, 403
   controller, 403

[Digital]
  controller software, 443
  differential analyzer, 18
  simulation, 384
Digitization approach, 432
Direct approach, 438
Direct-executing simulation language,
  658
Direct memory access (DMA), 599
Direct programming, 377
Directory, 340
Disaggregated model, 519, 523
Discontinuities, 14
Discrete-event model, 12
Discrete-event simulation languages,
  291–317
Discrete random variable, 102
Discrete state vector model, 412
Discrete-time systems, 11, 62, 555
Discrete-time systems, simulation
  of, 146–147
Discrete-time, state space model,
  71
Discrete transfer function, 69,
  385–388
Dispatcher, 322
Distributed-parameter model, 11
Distributed simulation, 319, 352
Distribution matrix, 46
Disturbance bounds, 426
Disturbance rejection, 418
Disturbances, 424
Dynamic compensation, 249,
  252–256
  data structure, 274
  exactness, 218
  models, 503
  models of robots, 478
  programming, 423
  simulation, 190

EAGLES operating system, 670
ESL, 684
Econometrics, 500
Economic
  forecasting, 555
  model, 510
  stability, 551
  variables, 541
Economics, 500, 504, 512
Eigenvalues, 634
Eigenvectors, 635
Electromagnetic device, 8

End effector, 450
Energy, 514
Energy storage elements, 59
Entity, 98
Entry, 271
Enumerated type, 267
Envelope, 632
Environmental models, 523
Equilibrium, 502
Erlang distribution, 108
Error covariance, 556–557
Error function, 417
Estimation of parameters, 180
Euler angles, 458, 468
Euler's method, 138
Event, 97
Exception handler, 444
Expectation, 102
Experiment representation, 631
Experimental design, 98
  frame, 569
  modeling, 8
Expert systems, 21, 311, 314, 671–
  674
Exponential distribution, 106
External fragmentation, 341

F-test, 196
FLIP-FLOPS, 171
FORESIGHT models, 527
FORESIGHT, OIL, 533
FORESIGHT/PTV models, 534
FORESIGHT/WIM, 533
FORK, 606
FORTRAN, 20, 100
FRONT, 279
Factorial design, 186
Fast subsystem, 257
Federal Reserve Board, 545
Feedback
  compensation, 418
  controller, 403
  gains, 553
  system, 51, 74, 82
Field, 272
File manager, 322, 325, 340, 342
File systems, 282, 339
Final-value theorem, 35, 68
Finite word length, 439
Firmware, 623
First-level problem, 242
Fixed modes, definition of, 250
Fixed-point, 440

Fixed polynomial, 250, 253
Flexible manufacturing system, 568
Floating-point, 18, 440
Food models, 521
Force sensors, 451
Forecasting, 3
Forecasting of economic behavior, 542
Form-driven interface, 645
Format, 282
Forward equations, 485–489
Fourier analysis, 13
Fourier transform, 409
Fractional factorial design, 186
Frames, 349
Frequency
  domain, 45, 637–638, 426
  loci, 409
  test, 127
Full-order model, 61
Function translator, 658

GASP, 20
GASP IV, 20, 100, 101
GASP V, 684
GEMS, 100, 101, 684
GERTS, 100
GLYPNIR, 602
GOL, 501
GPS, 684
GPSS, 20, 100, 101
Gain, 82
Gain matrix, 547
Gamma distribution, 107
Gamma function, 107
Gap test, 128
Gaussian elimination, 634
General purpose languages, 291–317, 574
Generalized Hessenberg representation (GHR), 222–225
Generalized forces calculations, 481
Generalized trees, 281
Geometric distribution, 104
Global economic models, 500
Global interdependence, 499
Global problems, 240, 519
Goal coordination, 239–242
Gradient vector, 417
Graphical compiler, 646
  input, 646
  modeling system, 659

[Graphical]
  systems, 596
Graphics-driven interface, 646
Graphics kernel system, 646
Graphics workstation, 646
Graphs, 278
Gross domestic product, 521
Gross National Product, 525, 544
Gross regional product, 513
Guillemin-Truxal, 414

HELP facility, 673
HIBLIZ, 659
HO algorithm, 639–664
HOLD block, 569
Hamiltonian, 242–243, 635
Hardware-in-the-loop (HWIL), 9
Hermitian forms, 641
Hessenberg representations, 640
Heterogeneous data type, 271
Hierarchical
  control, 237–247, 239, 638
  database, 282
  structure, 237
  systems, 500
  systems, properties of, 238
Hierarchy of programs, 323
Hierarchy of storage, 339
High-dimensional systems, 638
High-frequency rolloff, 420
High-level programming language, 358
Homogeneous difference equation, 64
Homogeneous transformation, 455, 460, 461, 464
Hybrid computers, 636
Hybrid simulation, 171–173, 443
Hybrid systems, 11
Hydraulic jack, 452
Hypercube, 617

IEA/OARU, 501
IFAC, 21
ILLIAC-IV, 19, 602
IMPACT, 647, 649, 657, 661–668
INTRPRD, 245
INVERSION, 171
I/O system, 342
I/O throughput, 341
ISIM, 685
ISIS, 685
IVTRAN, 602

Identification, 8, 13
  off-line, 13
  on-line, 13
Imbedding parameter, 240
Importance sampling, 206
Impulse sampler, 403
Incremental compilers, 671
Induction, 8
Initial conditions, 145, 159
Initial-value problems, 632
Initial-value theorem, 35, 68
Input data analysis, 130
Input, controlled, 7
Input, disturbance, 7
Input/output methods, 256
Input/output systems, 356, 500
Input/output models, 672
Input parameters, 180
Instrumentation, 16
Integer type, 267
Integral controller, 267
Integral controller, 78
Integrated design suites, 644
Integrating amplifier, 157
Integration algorithms, 636
Integration interval, 141
Integrator, 18
Intel iPSC, 616
Intel 286/310, 617
Interaction
  balance, 239-242
  of factors, 181
  prediction, 242-247
  variables, 240
Interactive program, 645
Interconnected network, 282
Interface, 19
Interface engine, 672
International Monetary Fund, 500
Interrupt, 599
Interval estimate, 132
Invariant embedding, 632
Inventory systems, 120
Inverse
  Jacobian matrix, 477
  Laplace transform, 35
  mapping, 406
  rotation matrix, 456
  transform technique, 128
  Z-transform, 69

JOIN, 606
Jacobian model, 471, 474-476

Joint variable, 450
Jury's test, 415

Kalman filter, 561
Kinematic energy, 479
Kinematic equations solution, 468-
  470
Knowledge, 672
Knowledge base, 288

LIFO, 278
LISP, 670-671
LISP machine, 638, 671
LO algorithm, 639-664
LQG design, 637, 634-635
LSSPAK/PC, 226-245
LUND software system, 658-659
Lagrange-Euler formulation, 478
Lagrange multipliers, 242
Lagrangian multiplier, 422
Language selection criteria, 307
Language trends, 311-314
Languages-general purpose, 20
Laplace transform
  pairs, 36
  definition, 33
  partial-fraction expansion, 36
  theorems, 34
Large scale center, 638
Large scale, definition, 213-214
Large-scale system, 12, 213-262
Large-scale systems
  modeling and model reduction,
    214-237
  structural properties, 256-257
Laser cutting, 493
Lead compensator, 415
Lead/lag, 414
Least-squares algorithm, 491
Least-squares method, 13, 543
Limit cycle, 442
Linear
  congruential method, 127
  quadratic gaussian, (LQG),
    634
  perturbation equations, 490
  state feedback, 634
  systems, 11, 636
  time-variant system, 639
Linearization, 16
Link, 450, 462, 484
Link traversing, 276

Linked
  cells, 580
  list, 274
  lists, doubly, 278
  lists, singly, 278
Logic sets, 171
Loop constructs, 268
Loop transfer function, 426
Loop transmission, shaping of, 427
Low-order models, 543, 560
Lower strata, 508
Lumped-parameter model, 11
Lyapunov methods, 227, 256

M, 670
M programming, 378
MACSYMA, 686
MACTRAQN, 686
MARSYAS, 686
MATLAB, 633–635, 643–644
MATRIX$_x$, 648, 657
MIDAS, 20
MIMD, 369, 604
MIMIC, 20
MIMO systems, 637
MISD, 369
MOIRA, 501, 524
Machine language, 358
Macroeconomic models, 541, 543
Magnitude scaling, 165
Main effects, 181
Mainframe computers, 3, 19
Man-in-the-loop, 3
Manufacturing systems, 567
Mason's gain formula 56
Master problem, 241
Master-slave technique 453
Material-handling, 574, 579
Materials, 516
Mathematical modeling, 8
Matrices, 266
Matrix, rank, 89, 90
Maximum-likelihood method, 13
Maximum principle, 423
Memory-mapped, 344
Memory unit, 356
Menu-driven interface, 645
Microcomputer, 19
Microprocessor, 19
Midproduct technique, 126
Midsquare technique, 126

Minimum parameter representation, 639, 641
Missile
  launching system, 11
  guidance, 9
  autopilot, 9
Minimum time control, 14
Mnemonic, 363
Mobile robots, 493
Modal aggregation, 220
Model, 3, 97, 98
  acceptance, 7
  building, 10, 16, 97
  fitting, 13
  overdescription, 29, 61
  representation, 631
  specification, 5
  validation, 5, 551
  variables, 502
  verification, 5
  degree of complexity, 10
  deterministic, 12
  dimension of, 11
  discrete-event, 12
  pilot, 8
  prototype, 8
  refinement, 16
  scale, 8
  stochastic, 12
Model-reference adaptive controller, 639
Modeling, 4
Modeling errors, 491
Modeling workstations, 575
Models, strongly coupled, 235–237
Models, weakly coupled, 232–235
Modular modeling systems, 659
Monetary policy, 541
Money supply, 544
Multidimensional arrays, 269
Multilevel systems, 238, 500
Multiple
  comparisons, 196
  poles, 37
  processor, 609
  rankings, 196
  sampling, 444
Multiprocessing, 321, 369, 609
Multiprogramming, degree of, 321, 345
Multivariable systems, 661
Mutual exclusion, 332

NASA Lewis Research Center, 619
National economy, 541

National income, model, 65
Newton-Euler formulation, 479-485
Next event, 99
Nichols chart, 427
Node, 280
Noise, 72
Noise covariance matrix, 561
Noisy observations, 555
Nonlinear
    programming problem, 632
    system simulation, 168
    systems, 11, 639
Nonlinearities, typical, 15
Nonlocal variable referencing, 298
Nonnumerical data processing, 638
Nonnumerical design, 671
Nonparametric identification, 13
Nonprocedural language, 285
Nonrenewable resources, 504
Nonrepeatability, 324
Nonsingular matrix, 634
Normal distribution, 109
Normal orientation approach, 457
Numerical
    accuracy, 639
    integration, 138
    integration, error in, 140-145
    stability, 639
Nyquist diagram, 638

OECD, 506
OPERATION block, 569
OR, 171
OR gate, 172
Object code, 358
Object-oriented database, 289
Object-oriented design, 289
Object-oriented programming, 289
Observability, 88, 257
Observability matrix, 89, 546
Observability and controllability
    Grammians, 227
Observers, 61
Open-loop systems, 403
Operating point, 61, 319-353,
    360, 626
Operating systems, special
    purpose, 645
Operational amplifier, 155
Operator overloading, 657
Operators, arithmetic, 267
Operators, relational, 267

Optimal controllers, 422
Optimization, 500
Order of system, 59
Orientation error, 487-489
Output variable, 7
Overshoot, 77

P, 334
PASAMS, 100, 101
PASSIM, 687
PC-MATLAB, 649
PID controller, 419
PLUG, 687
POP, 278
PROGRAM CC, 658
PROLOG, 671
PUSH, 278
Padé approximation, 413, 418
Page-table, 349
Parallel computation, 18
Parallel programming, 377
Parameter estimation, 13, 547, 659
Parameter identification, 491, 638
Parameters, 4, 632
Parametric identification, 13
Partial fraction expansion, 406
Pascal, 100
Peak-overshoot, 414
Per-unit method, 165
Perfect aggregation, 218
Performance
    criterion, 413
    evaluation, 3
    index, 422, 632
    index, one-step, 491
Peripheral array processor, 610
Personal computers, 3
Personal income, 546
Perturbation, 214, 232-237
Phase-plane method, 16
Physical address space, 345
Physical design parameters, 642
Physical system, 3
Physically unrealizable controller,
    416
Pick-and-place robot, 449
Pilot model, 9
Pipeline, 598
Pixel, 585
Plant, nominal, 427
Plant, reference, 427
Plant template, 427-432

Point estimates, 132
Point-to-point technique, 452
Poisson distribution, 105
Poisson processes, 112
Poker test, 128
Pole-zero pairing, 442
Poles, 33
Pollution, 505
Polynomial matrices, 661
Polynomial matrix representation, 638
Polynomials, 249
Population, 504, 512
Position error, 487-489
Positional control, 486
Potential energy, 480
Potentiometer, 160
Power spectral density, 551
Prediction, 561
Predictor-corrector, 140
Primary strip, 406-409
Prismatic joints, 450, 464, 475
Probability density function, 105
Probability mass function, 113
Procedural language, 284
Process, 322, 327
Process-control actuator, 14
Process interaction, 100
Process-oriented, 671
Process queues, 338
Process states, 328
Processor
  emulation, 436
  manager, 322, 324
  memory switch, 357
Product assembly automation, 492
Product or rotation matrices, 457
Production growth, 547
Production, types of, 567
Productivity, 567
Programmed input/output, 599
Programming environments, 306-307
Programming language control structures, 300-305
Programming language data structure, 296-298
Programming evolution, 292-295
Programming languages, 16, 20, 267, 291-317
Programming languages—higher level, 20
Programming structures, 633

Project Link, 500
Proportional controller, 78
Prototype, 4, 9
Prototyping, 311
Proximity sensors, 451
Pseudo inertia matrix, 480
Pseudorandom number, 125
Pull down menus, 583

Q-GERT, 687
QUERY facility, 647
QUEUE block, 573
Quadratic cost function, 422
Qualitative simulation, 672
Quantitative feedback method, 423
Quantization, 439
  effects, 374
  errors, 18, 137, 375
  noise, 375
Question-and-answer-driven interface, 645
Queueing models, 114

REAR, 279
RICRKUT, 246
RTMPL, 619, 624
RTMPOS, 619, 626
RTMPS, 619
Radar tracking system, 71
Random
  access, 340
  digits, 208
  disturbance, 550
  numbers, 125
  shocks, 550
  variable, 101
  variate generation, 128
Range measurements, 72
Rate feedback, 78
Real systems, 4
Real-time clock, 444
Real type, 267
Real-time solution, 17
Record, 266, 271
Rectangular rule, 138
Reduced modal matrix, 635
Reduced-order models, 60
Reduced-order observers, 61
Reduction of system's order method
  classical, 61
  modal, 61

[Reduction of system's order
      method]
   topological, 61
Redundancy, 640, 642
Redundant robot, 450
Regression analysis, 642
Regular aggregation, 216–220
Reliability, 4
Replacement policies, 348
Replication, 182
Resolved acceleration control, 487
Resolved motion rate control, 487
Resource manager, 321
Resource schedules, 578
Response surface methodology, 185
Revolving joints, 450, 464, 475
Riccati equation, 243, 635, 665–668
Right-shifting property, 68
Rise-time, 414
River pollution problem, 244
Robot arm model, 388
Robot manipulator, 449
Robot, structure of, 450
Robot vision, 492
Robotics, 639
Robots, dynamic models of, 478–
      485
Robots, kinematic modeling of, 462
Robots' control, 486–491
Robots' programming, 453
Robust controller design, 638
Root-locus method, 82, 414, 552
Rotation matrices, 455
Rounding, 436
Route path, 589
Routh approximation, 230
Routh-Hurwitz criterion, 79, 415
Runs test, 128

S-plane, 32
SCARA structure, 451, 466
SDL, 20, 688
SIGNAL, 333
SIMAN, 20, 100, 567, 689
SIMD, 369, 600
SIMD structure, 601
SIMNON, 20
SIMPAC, 689
SIMPL/1, 689
SIMSCRIPT, 100
SIMSCRIPT II.5, 20, 690
SIMSTAR, 19

SIMULA, 20, 690
SISD, 369
SISO systems, 637
SLAM, 20, 690
SLAM-II, 100
SPAWN facility, 647
STAR-100, 19, 598
Sample-and-hold device, 403
Sample rate, 374
Sample statistics, 100
Sampled data system, 405
Sampling period, 66
Scale model, 9
Scale parameter, 107
Scaling, 17, 442
Scene analysis, 453, 492
Scheduling
   long-term, 327
   medium-term, 327
   short-term, 327
Second-level problem, 242
Segment, 345, 580
Segment fault interrupt, 348
Segment table, 346
Segmentation, 345
Self-modifying code, 671
Semaphores, 333
Sensitivity, 3
Sensitivity balancing, 640, 641
Sensors, 451
Separation ratio, 234
Sequential access, 340
Sequential sampling, 197
Service mechanism, 114
Set, 98, 272
Set-points, 419
Set, size, 272
Settling time, 76, 414–422
Shannon's sampling theorem, 405
Shape parameter, 107
Sharing data, 330
Shifting data, 273
Shooting technique, 632
Signal-flow-graph, 9, 51, 55
Simplicity, 214
Simulated concurrency, 327
Simulation, 3, 16, 97
   database, 646
   experiments, analysis of, 189
   experiments, design of, 180
   interactive mode, 21
   language strategies, 302–305
   languages, 20, 151–155, 291–317
   menu-driven, 21

[Simulation]
  packages, 20
  report, 98
  run, 182
  software, 19, 681–691
  support environments, 306–307
Simulator-piloted flight, 10
Simulator-radio frequency, 9
Simultaneous equations, 168
Single-data-stream (SISD), 597
Single-instruction, 597
Single-period model, 124
Singular perturbation, 235
Singular points, 32
Singular-value-decomposition, 639
Slow subsystems, 257
Software engineering, 309–311,
    444
Source program, 358
Space heating system, 43
Sparse matrix techniques, 640–641
Special purpose operating
    system, 669
Spectral analysis, 13, 197
Stability, 79, 415, 638
Stability matrix, 546
Stabilization problem, 248
Stacks, 278
State
  diagram, 58
  equations, 411
  error correlation matrix, 561
  estimation, 561
  space models, 45, 545
  space simulation, 141–145, 170
  transition matrix, 413
  variable, 12, 45
  vector, 46
State-space, 411
State-space controllers, 422
Statement model, 572
Static
  models, 503
  simulation, 189
  structure, 273
Statistics, 100
Steady-state, 13
Steady-state simulation, 190
Steady-state error, 74
Step response, 77
Step-disturbance response, 430
Stepping motor, 452
Stochastic disturbance, 543

Strategy formulation, 529
Stratified sampling, 205
Stratum, 507
String, 266
Strongly coupled, 232
Structural controllability, 257
Subsystem-problem, 242
Summing amplifier, 156
Superposition, 11
Supervisory control, 248
Supply, 502
Supply and demand, 513
Swapping, 328, 351
Swiss mouse, 662
Symbolic manipulation, 658
System, 97
  boundary, 7
  capacity, 114
  dynamic, 7
  linear-time-invariant (LTI), 412
  model, 569
  multilevel, 237–238
  multivariable, 8, 412
  nonlinear, 11
  order, 639
  order-reduction, 29
  reality, 5
  response, 11
  single-input-single-output, 8, 637
  state, 98
  static, 7
  utility software, 326

TRANSFER block, 569
Tables, 269, 271
Taylor's series, 138
Teaching by doing, 452–453, 486
Thermal system, 42
Time control in simulation languages,
    302–305
Time-dependent statistic, 100
Time scaling, 167, 382
Time series, 557
Trace of a matrix, 557
Tracking bounds, 426
Tracking problems, 418
Trajectory behavior, 6, 631
Trajectory generation, 452
Transfer functions, 29, 38, 639
Transfer functions, simulation of,
    147–151
Transfer rate, 339

Transient state, 13
Transition matrix, 144
Transport aircraft longitudinal model, 435
Trapezoidal rule, 140
Transporter path, 589
Transporters, 579
Trees, 278
Trees restructuring, 280
Trees traversal, 280
Truncation, 436
Truncation error, 141
Tuning regulators, 639
Turboshaft engine, 628
Tustin transformation, 409
2's complement, 439
Two-input mechanical systems, 43

UNIDO, 501
UNIX operating system, 670
Uncertain parameters, 424
Undamped natural frequency, 76, 550
Unemployment, 542
Uniform distribution, 105
Unit circle, 407, 409
Unit impulses, 403
United Nations, 500
Universal frequency bound, 431
Unstable systems, 81, 229–231
Upper strata, 508
User interface, 581

V, 334
VHO algorithm, 639–641
VLSI, 282
VMS operating system, 670
Valid model
  structurally, 6
  predictively, 6
  replicatively, 6
Validation, 5, 98, 131
Variance, 103

Variance reduction techniques, 197
Vector processors, 369
Vehicle-buffer system, 16
Verification, 5, 98, 131
Virtual addressing, 345
Virtual memory, 345
Virtual screen, 646
Visitation sequences, 577

W-transformation, 411
WAIT, 333
WAVE, 454
WCDS, 659–660
Warping, 409
Watch-dog timer, 444
Water cutting, 493
Weakly coupled conditions, 233–234
Weakly coupled systems, 232
Weibull distribution, 110
Weighting matrix, 491
Welding, 492
White box, 4
Window-driven interface, 647
Window interfaces, 662
Wordlength, 436
Workcenter submodel, 576
Workcenters, 574
Working set size, 351
World 2 model, 500
World 3 model, 500
World Bank, 500
World Integrated Model, 500
World modeling, 501
World views, 302–305

XCELL, 100, 101

Z-plane, stability, 87
Z-transform, 65, 146, 405
Z-transform pairs, 67
Zero-order hold, 148, 371, 405
Zeros, 33